Lecture Notes in Electrical Engineering

Volume 321

About this Series

"Lecture Notes in Electrical Engineering (LNEE)" is a book series which reports the latest research and developments in Electrical Engineering, namely:

- Communication, Networks, and Information Theory
- Computer Engineering
- Signal, Image, Speech and Information Processing
- Circuits and Systems
- Bioengineering

LNEE publishes authored monographs and contributed volumes which present cutting edge research information as well as new perspectives on classical fields, while maintaining Springer's high standards of academic excellence. Also considered for publication are lecture materials, proceedings, and other related materials of exceptionally high quality and interest. The subject matter should be original and timely, reporting the latest research and developments in all areas of electrical engineering.

The audience for the books in LNEE consists of advanced level students, researchers, and industry professionals working at the forefront of their fields. Much like Springer's other Lecture Notes series, LNEE will be distributed through Springer's print and electronic publishing channels.

More information about this series at http://www.springer.com/series/7818

António Paulo Moreira · Aníbal Matos
Germano Veiga
Editors

CONTROLO'2014 - Proceedings of the 11th Portuguese Conference on Automatic Control

 Springer

Editors
António Paulo Moreira
Faculty of Engineering
University of Porto
Porto
Portugal

Germano Veiga
INESC Porto Campus da FEUP
Porto
Portugal

Aníbal Matos
Faculty of Engineering
University of Porto
Porto
Portugal

ISSN 1876-1100   ISSN 1876-1119   (electronic)
ISBN 978-3-319-10379-2   ISBN 978-3-319-10380-8   (eBook)
DOI 10.1007/978-3-319-10380-8

Library of Congress Control Number: 2014947128

Springer Cham Heidelberg New York Dordrecht London

Printed on acid-free paper

Springer is part of Springer Science+Business Media (www.springer.com)

Preface

CONTROLO 2014 – 11th Conference on Automatic Control in Porto, Portugal, July 21th to 23th, 2014, was organized by INESC TEC, in partnership with FEUP (Faculdade de Engenharia da Universidade de Porto), ISEP (Instituto Superior de Engenharia do Porto) and APCA the Portuguese Association of Automatic Control which is a national member organization of the International Federation of Automatic Control (IFAC).

The conference is Sponsored by APCA the Portuguese Association of Automatic Control with three partner institutions SBA - Sociedade Brasileira de Automática, CEA - Comité Español de Automática and SPR - Sociedade Portuguesa de Robótica.

We would like to thank the invaluable contribution of the International Program Committee Members, External Reviewers, Keynote Speakers, Sessions Chairs and Authors. We would also like to acknowledge Easychair for their Conference Management System, which was freely used for managing the paper submission and evaluation processes of CONTROLO 2014.

These proceedings present peer-reviewed papers from a wide range of topics including different areas of control compiled in the first 6 chapters, Decision, Estimation and Modelling in chapters 7, 8 and 9, Robotics and Sensing in Chapters 10 and 11 and finally Education in chapter 12.

António Paulo Moreira
Aníbal Matos
Germano Veiga

Organization

General Chair

António Paulo Moreira — FEUP / Dep. Eng. Electrotécnica e de Computadores – INESC TEC

Vice-Chair

Anibal Matos — FEUP / Dep. Eng. Electrotécnica e de Computadores – INESC TEC

Organizing Committee

Anibal Matos — FEUP / Dep. Eng. Electrotécnica e de Computadores – INESC TEC

António Paulo Moreira — FEUP / Dep. Eng. Electrotécnica e de Computadores – INESC TEC

Fernando Gomes de Almeida — FEUP / Dep. Mecânica – IDMEC-LAETA

Fernando Martins — FEUP / Dep. Eng. Química - LEPAE

Germano Veiga — INESC TEC / Unidade de Robótica e Sistemas Inteligentes

José Boaventura Da Cunha — UTAD / Dep. Engenharias (Eng. Electrotécnica) – INESC TEC

Manuel Fernando Silva — ISEP-IPP / Departamento de Engenharia Electrotécnica – INESC TEC

Maria do Rosário de Pinho — FEUP / Dep. Eng. Electrotécnica e de Computadores – ISR-Porto

Pedro Costa — FEUP / Dep. Eng. Electrotécnica e de Computadores – INESC TEC

Scientific Committee

A. Pedro Aguiar — Universidade do Porto Portugal

Alberto Isidori — University di Roma "La Sapienza" Italy

José Tenreiro Machado	Instituto Superior de Engenharia do Porto Portugal
José Vieira	Instituto Politecnico de Castelo Branco Portugal
Jurek Sasiadek	Carleton University Canada
Karolina Baras	Universidade da Madeira Portugal
Krzysztof Kozlowski	Poznan University of Technology Poland
Leonardo Honorio	Universidade Federal de Juiz de Fora Brazil
Luis Gomes	Universidade Nova de Lisboa Portugal
Luiz de Siqueira Martins Filho	Universidade Federal do ABC Brazil
Manuel Dominguez Gonzalez	Universidad de Leon Spain
Manuel Fernando Silva	Instituto Politécnico do Porto Portugal
Maria do Rosário Pinho	Universidade do Porto Portugal
Maria Graça Ruano	Universidade do Algarve Portugal
Miguel Ayala Botto	Instituto Superior Tecnico Portugal
Miguel Bernal	Sonora Institute of Technology Mexico
Morgado Dias	University of Madeira Portugal
Oscar Barambones	Universidad del País Vasco Spain
Paulo Costa	Universidade do Porto Portugal
Paulo Garrido	Universidade do Minho Portugal
Paulo Lopes dos Santos	Universidade do Porto Portugal
Paulo Oliveira	Instituto Superior Tecnico Portugal
Paulo Oliveira	Universidade Tras-os-Montes e Alto Douro Portugal
Paulo Salgado	Universidade Tras-os-Montes e Alto Douro Portugal
Pedro Albertos	Universidad Politecnica de Valencia Spain
Pedro Costa	Universidade do Porto Portugal
Pedro Peres	Universidade Estadual de Campinas Brazil
Ramiro Barbosa	Instituto Superior de Engenharia do Porto Portugal
Ramon Vilanova	Universitat Autònoma de Barcelona Spain
Robert Harrison	University of Sheffield UK
Rolf Johansson	Lund University Sweden
Rui Neves-Silva	Universidade Nova de Lisboa Portugal
Silvia Botelho	Universidade Federal do Rio Grande Brazil
Sonia Pinto	Instituto Superior Tecnico Portugal
Teresa Perdicoulis	Universidade Tras-os-Montes e Alto Douro Portugal
Teresa Mendonça	Universidade do Porto Portugal

Contents

Part III: Fuzzy and Neural Control

Part IV: Optimal Control

Part V: Controller Design

Part VI: Control Applications

Part VII: Decision

Part VIII: Estimation

Part IX: Modelling

Part XI: Sensing

Part XII: Education

Part I
Control Theory

Controllability for the Constrained Rolling Motion of Symplectic Groups

André Marques[1] and Fátima Silva Leite[2]

[1] Escola Superior de Tecnologia e Gestão, Instituto Politécnico de Viseu,
3504 - 510 Viseu, Portugal
`codecom@mat.estv.ipv.pt`
[2] Instituto de Sistemas e Robótica, Universidade de Coimbra,
3030-290 Coimbra, Portugal
and
Departamento de Matemática, Universidade de Coimbra,
3001-501 Coimbra, Portugal
`fleite@mat.uc.pt`

Abstract. Rolling motions describe how two manifolds both embedded in the same manifold roll on each other without slip and without twist. With these constraints on the velocities of motion, an interesting issue is that of knowing whether such rolling motions are controllable or not. In this paper, after embedding the symplectic group and its affine tangent space at a point in an appropriate pseudo-Riemannian space, we derive the kinematic equations for rolling and prove that the corresponding rolling motions are controllable.

1 Introduction

The classical definition of a rolling map, as appeared in Sharpe [12], describes how two manifolds of the same dimension, both isometrically embedded in the same Euclidean space, roll on each other without slip and without twist. Such a rolling motion is described by the action of the group of isometries in Euclidean space that preserve orientations, the special Euclidean group. This definition has been extended and used in the situation when the two manifolds are isometrically embedded in a general Riemannian manifold [5], and adapted to describe intrinsic rolling, for instance in [2]. The extension of Sharpe's definition to include pseudo-Riemannian manifolds, manifolds equipped with a scalar product that is not positive definite, is somewhat natural and has already appeared in [8] for Lorentzian spheres, in [9] for hyperbolic spheres, and in [1] for pseudo-orthogonal groups. The idea is to adapt the definition of Sharpe by replacing the Euclidean group by the group of orientation preserving isometries of the ambient space, and taking orthogonality with respect to the pseudo-metric.

In this article we adopt the extrinsic approach, but take a step further with respect to the existing literature and concentrate on rolling quadratic Lie groups on the affine tangent space at a point. This may be used successfully to generate interpolating curves on these groups, as explained in [1]. But our final objective

© Springer International Publishing Switzerland 2015

A.P. Moreira et al. (eds.), *CONTROLO'2014 - Proc. of the 11th Port. Conf. on Autom. Control*,
Lecture Notes in Electrical Engineering 321, DOI: 10.1007/978-3-319-10380-8_1

here is to particularize the rolling motions for the case of symplectic Lie groups and prove controllability of the corresponding kinematic equations of motion. The Lie group of symplectic matrices plays an important role in many applications, in particular for finding the numerical solution of discrete time, robust and optimal control problems [10] and in mechanical systems and robotics [13].

The organization of this paper is as follows. We introduce the basic notions of quadratic Lie groups in Section 2, followed by the definition of rolling map in Section 3. This is used in Section 4 to derive the kinematic equations of rolling quadratic Lie groups. Finally, in Section 5, we prove that the kinematic equations for the rolling motion of a symplectic Lie group over the affine tangent space at a point are completely controllable.

2   Quadratic Lie Groups

For $\alpha = \pm 1$, let $J \in GL(n)$ be any real matrix satisfying $J^2 = \alpha I_n$ and $J^\top = \alpha J$, where I_n is the identity matrix of order n e $J^\top$ the transpose of J. One can define a matrix Lie group G_J, associated to each such matrix J, as

$$G_J := \{X \in GL(n) : X^\top J X = J\}. \tag{1}$$

These groups, called *quadratic Lie groups*, have Lie algebra given by

$$\mathcal{L}_J = \{A \in \mathfrak{gl}(n) : A^\top J = -JA\}, \tag{2}$$

where the Lie product $[\cdot, \cdot]$ is the commutator, i.e., $[A, B] = AB - BA$. The following vector sace, associated to J, will also be used later.

$$\mathcal{E}_J := \{A \in \mathfrak{gl}(n) : A^\top J = JA\}.$$

One can also define a scalar product in $\mathfrak{gl}(n)$ associated to the matrix J and denoted by $\langle A, B \rangle_J$, that generalizes the Frobenius inner product hereafter denoted by

$$\langle A, B \rangle = tr(A^\top B), \quad \text{for } A, B \in \mathfrak{gl}(n), \tag{3}$$

where $tr(\cdot)$ denotes the trace. Such scalar product is defined by

$$\langle A, B \rangle_J := tr(J^\top A^\top J B). \tag{4}$$

It is an easy exercise to show that the bilinear form (4) is indeed a scalar product, that is, it is symmetric an non-degenerated, but not positive definite. Also, the following properties involving G_J, $\mathcal{L}_J$, $\mathcal{E}_J$ and $\langle \cdot, \cdot \rangle_J$ can be easily checked.

Proposition 1.   *1. If $X \in G_J$ and $A \in \mathcal{L}_J$, then $X^{-1}AX \in \mathcal{L}_J$.*
2. If $t \rightsquigarrow X(t)$ is a smooth curve in G_J, then

$$\dot{X}(t)X^{-1}(t) \in \mathcal{L}_J \text{ and } X^{-1}(t)\dot{X}(t) \in \mathcal{L}_J.$$

3. $X \in G_J$ if and only if $\langle XA, XB \rangle_J = \langle AX, BX \rangle_J = \langle A, B \rangle_J, \forall A, B \in \mathfrak{gl}(n)$.
4. $\langle \mathcal{L}_J, \mathcal{E}_J \rangle_J = 0$.

Equipped with the scalar product $\langle \cdot, \cdot \rangle_J$, $\mathfrak{gl}(n)$ becomes a pseudo-Riemannian manifold hereafter denoted by $(\mathfrak{gl}(n), \langle \cdot, \cdot \rangle_J)$. It follows from *3.*, in Proposition 1, that for every U, W in G_J and every Z in $\mathfrak{gl}(n)$, the mapping

$$
\begin{array}{ccc}
\mathfrak{gl}(n) & \longrightarrow & \mathfrak{gl}(n) \\
X & \rightsquigarrow & UXW^{-1} + Z
\end{array} \tag{5}
$$

is an isometry of $(\mathfrak{gl}(n), \langle \cdot, \cdot \rangle_J)$. Due to this, the semi-direct product

$$
\mathrm{G}_J \times \mathrm{G}_J \rtimes \mathfrak{gl}(n) = \{(U, W, Z) : U, W \in \mathrm{G}_J, \ Z \in \mathfrak{gl}(n)\}, \tag{6}
$$

equipped with the multiplication $\circ$ defined by

$$
(U_2, W_2, Z_2) \circ (U_1, W_1, Z_1) = (U_2 U_1, W_2 W_1, U_2 Z_1 W_2^{-1} + Z_2),
$$

is a group of isometries of $(\mathfrak{gl}(n), \langle \cdot, \cdot \rangle_J)$. For the purpose of this paper, only isometries that preserve orientation are considered. So, instead of working with the whole semi-direct product we use its connected subgroup. From now on, and for the simplicity of notation, $\overline{\mathrm{G}}_J$ *will denote the connected component, containing the identity* $(I_n, I_n, 0_n)$, *of the group defined in* (6).

Remark 1. It is known that, after fixing n, there are only two quadratic Lie groups up to isomorphisms. One is the *pseudo-orthogonal group*, denoted by $\mathrm{O}_\kappa(n)$ and associated with the matrix

$$
J = \left[\begin{array}{c|c} -I_\kappa & 0 \\ \hline 0 & I_{n-\kappa} \end{array} \right], \text{ with } \kappa \geq 0,
$$

(the cases $\kappa = 0, n$ correspond to the Riemannian situation), the other is the *symplectic group*, denoted by $\mathrm{Sp}(2m)$ and associated to the matrix

$$
J = \left[\begin{array}{c|c} 0 & I_m \\ \hline -I_m & 0 \end{array} \right], \text{ with } n = 2m \geq 2.
$$

3 Rolling Maps in $(\mathfrak{gl}(n), \langle \cdot, \cdot \rangle_J)$

The classical definition of a rolling map, as appeared in Sharpe [12], describes how two manifolds of the same dimension, M_1 and M_2, both isometrically embedded in the same Euclidean space $\mathbb{R}^n$, roll on each other without slip and without twist. Such a rolling motion is described by the action of the group of isometries in Euclidean space $\mathbb{R}^n$ which preserve orientations, $SE(n) = SO(n) \ltimes \mathbb{R}^n$. As explained in the Introduction, this definition has been extended in several different directions. Here, we consider our manifolds to be isometrically embedded in the pseudo-manifold $(\mathfrak{gl}(n), \langle \cdot, \cdot \rangle_J)$ and adapt the classical definition by replacing $SE(n)$ by the group $\overline{\mathrm{G}}_J$ of orientation preserving isometries of this ambient space and taking orthogonality with respect to the pseudo-metric $\langle \cdot, \cdot \rangle_J$.

The problem of rolling two manifolds on each other, without the assumption that they are embedded in another manifold, was studied in [2]. Although this

last paper is about intrinsic rolling, at the beginning the authors also present
a definition of rolling concerning embedded manifolds. This definition is proved
to be equivalent to the classical definition of Sharpe in [12], but has an easier physical interpretation and is geometrically more intuitive. The next definition, although in a pseudo-Riemannian context, also follows the ideas in [2] and
privileges the geometric interpretation of the nonholonomic constraints of non-
slip and non-twist. Due to lack os space we can't give here more details about
pseudo-Riemannian geometry (also called semi-Riemannian geometry), but refer
the interested reader to [11], in particular to the notion of parallel transport of
vector fields that appear in the next definition.

Definition 1. *Let M_1 and M_2 be two pseudo-Riemannian submanifolds of
$(\mathfrak{gl}(n), \langle \cdot, \cdot \rangle_J)$, having the same dimension and index. A smooth rolling map of
M_1 over M_2 is a smooth curve*

$$g : [0, \tau] \longrightarrow \overline{G}_J \\ t \rightsquigarrow (U(t), W(t), Z(t)) \tag{7}$$

satisfying the following conditions, 1., 2., and 3., for all $t \in [0, \tau]$.

1. *Rolling conditions: There exists a smooth curve $\alpha : [0, \tau] \longrightarrow M_1$, such that:*
   (a) $g(t)(\alpha(t)) = U(t)\alpha(t)W^{-1}(t) + Z(t) \in M_2$,
   (b) $T_{g(t)(\alpha(t))}(g(t)(M_1)) = T_{g(t)(\alpha(t))}M_2$.

   *The curve α in M_1 is called the rolling curve, while the corresponding curve
   $\alpha_{\text{dev}} : [0, \tau] \longrightarrow M_2$ on M_2, defined by $\alpha_{\text{dev}}(t) := g(t)(\alpha(t))$, is called the
   development of α on M_2.*

2. *Non-slip condition:*

$$\dot{\alpha}_{\text{dev}}(t) = \mathrm{d}_{\alpha(t)}\, g(t)(\dot{\alpha}(t)), \tag{8}$$

   where $\mathrm{d}_{\alpha(t)}\, g(t)$ denotes the differential of $g(t) : \mathfrak{gl}(n) \to \mathfrak{gl}(n)$ at $\alpha(t)$.

3. *Non-twist conditions:*
   (a) (tangential condition)
       *A vector field $Y(t)$ is tangent parallel along the curve $\alpha(t)$ if, and only
       if, $V(t) = \mathrm{d}_{\alpha(t)}\, g(t)(Y(t))$ is a tangent parallel vector field along $\alpha_{\text{dev}}(t)$.*
   (b) (normal condition)
       *A vector field $Y(t)$ is normal parallel along the curve $\alpha(t)$ if, and only
       if, $V(t) = \mathrm{d}_{\alpha(t)}\, g(t)(Y(t))$ is a normal parallel vector field along $\alpha_{\text{dev}}(t)$.*

Remark 2. This definition still makes sense if the word "smooth" is replaced
by "piecewise smooth", as long as the non-slip and non-twist conditions are
constrained to each subinterval where the rolling map, the rolling curve and its
development are smooth.

Remark 3. Taking into consideration (5), and since $g(t) = (U(t), W(t), Z(t))$,
when the rolling condition happens the non-slip and non-twist conditions can be
rewritten in a more practical form, as follows.

2. The <u>non-slip condition</u> is equivalent to:

$$\dot{U}(t)U^{-1}(t)(\alpha_{\text{dev}}(t) - Z(t)) + (Z(t) - \alpha_{\text{dev}}(t))\dot{W}(t)W^{-1}(t) + \dot{Z}(t) = 0. \quad (9)$$

3. The <u>non-twist conditions</u> are equivalent to:

(a) (tangential condition)

$$\forall v \in T_{\alpha_{\text{dev}}(t)}M_2, \quad \dot{U}(t)U^{-1}(t)v - v\dot{W}(t)W^{-1}(t) \in (T_{\alpha_{\text{dev}}(t)}M_2)^{\perp}; \quad (10)$$

(b) (normal condition)

$$\forall v \in \left(T_{\alpha_{\text{dev}}(t)}M_2\right)^{\perp}, \quad \dot{U}(t)U^{-1}(t)v - v\dot{W}(t)W^{-1}(t) \in T_{\alpha_{\text{dev}}(t)}M_2. \quad (11)$$

4 Rolling Quadratic Lie Groups

The main objetive in this section is to derive the kinematic equations for the rolling motion of a quadratic Lie group over its affine tangent space at a point. First, we must embedded both manifolds in the same pseudo-Riemannian manifold and ensure that they have the same dimension (this is obvious) and the same index. The following proposition guarantees that we are in the conditions of Definition 1.

Proposition 2. *The quadratic Lie group G_J and its affine tangent space at a point $X \in G_J$ are pseudo-Riemannian submanifolds of $(\mathfrak{gl}(n), \langle \cdot, \cdot \rangle_J)$.*

Proof. It is enough to show that, for each of the manifolds G_J and $T_X^{\text{aff}} G_J$, the restriction of the scalar product $\langle \cdot, \cdot \rangle_J$ to each tangent space is a non-degenerated form and that all the tangent spaces have the same index.
Before proving this, notice that the tangent space to G_J at point X is given by

$$T_X G_J = \{X\Psi : \Psi \in \mathcal{L}_J\} = \{\Psi X : \Psi \in \mathcal{L}_J\}, \quad (12)$$

and, consequently,

$$T_X^{\text{aff}} G_J := \{X + V : V \in T_X G_J\} = \{X + X\Psi : \Psi \in \mathcal{L}_J\} = \{X + \Psi X : \Psi \in \mathcal{L}_J\}.$$

Also, due to the fact that $T_Y\left(T_X^{\text{aff}} G_J\right) = T_X G_J, \quad \forall Y \in T_X^{\text{aff}} G_J$, it is enough to prove the statement above for the manifold G_J. Taking into consideration that $T_X G_J$ is the translation to the point X of $T_I G_J = \mathcal{L}_J$, and that, acording to Proposition 1, $\langle ., . \rangle_J$ is G_J-invariant, the first condition will be satisfied if we prove that, for $A \in \mathcal{L}_J$,

$$\langle A, B \rangle_J = 0, \forall B \in \mathcal{L}_J \implies A = 0. \quad (13)$$

But since $\langle A, B \rangle_J = \langle J^\top A J, B \rangle$ and $\langle J^\top A J, B \rangle = 0, \forall B \in \mathcal{L}_J$, the statement (13) follows from the fact that the restriction of the Euclidean inner product $\langle \cdot, \cdot \rangle$ to any subspace (in this case, $\mathcal{L}_J$) is also an inner product and, consequently, non-degenerated.

Concerning the index, it is clear that all tangent spaces $T_X G_J$ have the same index since they are translations to the point X of the tangent space at the identity. $\qquad \square$

Also note that

$$(T_X \, \mathrm{G_J})^\perp = \{X\Upsilon : \Upsilon \in \mathcal{E}_J\} = \{\Upsilon X : \Upsilon \in \mathcal{E}_J\} . \tag{14}$$

We are ready to present the kinematic equations for the constrained rolling motion of the quadratic Lie group $\mathrm{G_J}$ over its affine tangent space at a point. The natural way to derive these equations is omitted here to avoid some technicalities. Instead, in the next theorem we present the kinematics and show that indeed they describe the rolling motion.

Theorem 1. *Let X_0 be an arbitrary point in $\mathrm{G_J}$, $\Omega : t \in [0, \tau] \rightsquigarrow \Omega(t) \in \mathcal{L}_J$ a smooth curve in $\mathcal{L}_J$ and $g : t \in [0, \tau] \rightsquigarrow g(t) = (U(t), W(t), Z(t)) \in \overline{\mathrm{G}}_J$ a smooth curve in $\overline{\mathrm{G}}_J$ satisfying*

$$\begin{cases} \dot{U}(t) = -\Omega(t)U(t) \\ \dot{W}(t) = X_0^{-1}\Omega(t)X_0 W(t) \\ \dot{Z}(t) = 2\Omega(t)X_0 \end{cases} \tag{15}$$

and the inicial condition $g(0) = (U_0, W_0, Z_0)$, with $Z(0) \in T_{X_0} \, \mathrm{G_J}$. Then, g is a rolling map of $\mathrm{G_J}$ over $T_{X_0}^{\mathrm{aff}} \, \mathrm{G_J}$, with rolling curve $\alpha(t) = U^{-1}(t)X_0 W(t)$ and development curve $\alpha_{\mathrm{dev}}(t) = X_0 + Z(t)$.

The equations (15) are called the *kinematic equations of rolling*.

Proof. We are going to show that g satisfies all the conditions in Definition 1, with $M_1 = \mathrm{G_J}$ and $M_2 = T_{X_0}^{\mathrm{aff}} \, \mathrm{G_J}$. To simplify notations, whenever convenient we omit the explicit dependence on t.

1. *Verifying the rolling conditions*

    (a) It is clear from the last equation in (15) that $\dot{Z} \in T_{X_0} \, \mathrm{G_J}$. Also, by assumption, $Z(0)$ belongs to $T_{X_0} \, \mathrm{G_J}$. So, since $\alpha_{\mathrm{dev}}(t) = g(t)(\alpha(t)) = X_0 + Z(t)$, we have the guarantee that $\alpha_{\mathrm{dev}}(t) \in T_{X_0}^{\mathrm{aff}} \, \mathrm{G_J}$, as required.

    (b) Since $\mathrm{G_J} \times \mathrm{G_J}$ acts transitively on $\mathrm{G_J}$, we have that $g(t)(\mathrm{G_J}) = \mathrm{G_J} + Z(t)$. So, $T_{\alpha_{\mathrm{dev}}(t)} \, (g(t)(\mathrm{G_J})) = T_{X_0} \, \mathrm{G_J}$ and, consequently,

$$T_{\alpha_{\mathrm{dev}}(t)} \, (g(t)(\mathrm{G_J})) = T_{\alpha_{\mathrm{dev}}(t)} \left(T_{X_0}^{\mathrm{aff}} \, \mathrm{G_J} \right) .$$

2. *Verifying the non-slip condition*

    Since $\alpha_{\mathrm{dev}}(t) = X_0 + Z(t)$, proving the equality (9) amounts to prove that

$$\dot{U}U^{-1}X_0 - X_0\dot{W}W^{-1} + \dot{Z} = 0.$$

    But this can be easily checked, using the constraints in equations (15).

3. *Verifying the non-twist conditions*

    We prove the tangential part only (the normal condition can be proved in a similar way). The fact, already mentioned, that $T_{\alpha_{\mathrm{dev}}(t)} \left(T_{X_0}^{\mathrm{aff}} \, \mathrm{G} \right) = T_{X_0} \, \mathrm{G}$, shows that the equality in (10) holds if

$$\forall V \in T_{X_0} \, \mathrm{G}, \quad \dot{U}U^{-1}V - V\dot{W}W^{-1} \in (T_{X_0} \, \mathrm{G})^\perp .$$

Now, let $V = \Psi X_0$, with $\Psi \in \mathcal{L}_J$. From the first two cinematic equations, we obtain

$$\dot{U}U^{-1}V - V\dot{W}W^{-1} = -(\Omega\Psi + \Psi\Omega)X_0.$$

So, to complete the proof it is enough to show that the matrix $\Omega\Psi + \Psi\Omega$ belongs to the vector space $\mathcal{E}_J$. And indeed, since the matrices Ω e Ψ belong to $\mathcal{L}_J$, we can write

$$(\Omega\Psi + \Psi\Omega)^\top J = \Psi^\top \Omega^\top J + \Omega^\top \Psi^\top J = -\Psi^\top J\Omega - \Omega^\top J\Psi$$
$$= J\Psi\Omega + J\Omega\Psi = J(\Omega\Psi + \Psi\Omega),$$

which proves the required condition. □

When $\Omega(t)$ is constant, the kinematic equations (15) can be solved explicitly.

Corollary 1. *In the particular case when $\Omega(t) = \Omega$ is constant and $g(0) = (I_n, I_n, 0_n)$, the rolling map reduces $g(t) = \left(e^{-t\Omega}, X_0^{-1}e^{t\Omega}X_0, 2t\Omega X_0\right)$ and the rolling curve and its development are, respectively, geodesics in G_J and $T_{X_0}^{\mathrm{aff}}\,\mathrm{G}_J$, given by $\alpha(t) = e^{2t\Omega}X_0$ and $\alpha_{\mathrm{dev}}(t) = X_0 + 2t\Omega X_0$.*

5   Controllability of the Kinematic Equations

It is clear from the kinematic equations (15) that $t \rightsquigarrow \Omega(t) \in \mathcal{L}_J$ completely determines the solutions. So, the choice of Ω and of an initial condition uniquely determines how G_J rolls over the affine tangent space at a point, under the action of $\overline{\mathrm{G}}_J$. As a consequence, the kinematic equations may be seen as a control system, where (U, W, Z) are the states and $t \rightsquigarrow \Omega(t)$ is the control function.

In this context, it makes sense to study the controllability properties of the kinematic equations. Due to lack of space, we only consider here the rolling of the symplectic group $\mathrm{G}_J = \mathrm{Sp}(2m)$. Without loss of generality, we may also consider the case when $X_0 = I_n$. In this situation the kinematic equations (15) reduce to

$$\begin{cases} \dot{U}(t) = -\Omega(t)U(t) \\ \dot{W}(t) = \Omega(t)W(t) \quad , \\ \dot{Z}(t) = 2\Omega(t) \end{cases} \tag{16}$$

where $(U, W, Z) \in \mathrm{Sp}(2m) \times \mathrm{Sp}(2m) \times \mathfrak{sp}(2m)$, $\mathfrak{sp}(2m)$ being the Lie algebra of the symplectic group. Since $\mathrm{Sp}(2m)$ is connected, the kinematic equations also evolve on a connected Lie group. To prove the controllability, we first rewrite the system above as a homogeneous right invariant control system, in order to apply a theorem in [6], which reduces the proof of controllability to checking an algebraic condition, the rank controllability condition.

In order to rewrite the system, first notice that, since $\mathfrak{sp}(2m) = \{\Omega \in \mathfrak{gl}(2m) : \Omega^\top J = -J\Omega\}$, matrices in this Lie algebra may be partitioned as

$$\Omega = \left[\begin{array}{c|c} \Omega_1 & \Omega_2 \\ \hline \Omega_3 & -\Omega_1^\top \end{array}\right],$$

with $\Omega_1 = [\overline{\omega}_{ij}] \in \mathfrak{gl}(m)$, $\Omega_2 = [\widetilde{\omega}_{ij}] \in \text{sym}(m)$, and $\Omega_3 = [\widehat{\omega}_{ij}] \in \text{sym}(m)$, where $\text{sym}(m)$ denotes the set of $m \times m$ symmetric matrices. Clearly there is an isomorphism between $\mathfrak{sp}(2m)$ and $\mathfrak{gl}(m) \times \text{sym}(m)^2$, assigning to each Ω the triplet $(\Omega_1, \Omega_2, \Omega_3)$.

Since the matriz Z can also be partitioned as $Z = \left[\begin{array}{c|c} Z_1 & Z_2 \\ \hline Z_3 & -Z_1^\top \end{array}\right]$, the last equation in (16) is equivalent to

$$\dot{Z}_1 = 2\Omega_1, \quad \dot{Z}_2 = 2\Omega_2, \quad \dot{Z}_3 = 2\Omega_3, \tag{17}$$

and the states (U, W, Z) may be represented by the $5n \times 5n$ block diagonal matrix

$$X = \text{diag}\left(U, W, \left[\begin{array}{c|c} I_m & Z_1 \\ \hline 0 & I_m \end{array}\right], \left[\begin{array}{c|c} I_m & Z_2 \\ \hline 0 & I_m \end{array}\right], \left[\begin{array}{c|c} I_m & Z_3 \\ \hline 0 & I_m \end{array}\right]\right), \tag{18}$$

which belongs to the connected Lie group $G = \text{Sp}(2m)^2 \times \mathfrak{gl}(m) \times \text{sym}(m)^2$. The Lie algebra of G consists of the matrices of the form

$$\text{diag}\left(\left[\begin{array}{c|c} \Omega_1 & \Omega_2 \\ \hline \Omega_3 & -\Omega_1^\top \end{array}\right], \left[\begin{array}{c|c} \Omega_4 & \Omega_5 \\ \hline \Omega_6 & -\Omega_4^\top \end{array}\right], \left[\begin{array}{c|c} 0_m & \Omega_7 \\ \hline 0 & 0_m \end{array}\right], \left[\begin{array}{c|c} 0_m & \Omega_8 \\ \hline 0 & 0_m \end{array}\right], \left[\begin{array}{c|c} 0_m & \Omega_9 \\ \hline 0 & 0_m \end{array}\right]\right),$$

where, $\Omega_1, \Omega_4, \Omega_7 \in \mathfrak{gl}(m)$, and $\Omega_2, \Omega_3, \Omega_5, \Omega_6, \Omega_8, \Omega_9 \in \text{sym}(m)$.

It is immediate to check that the kinematic equations (16) are equivalent to:

$$\dot{X} = \text{diag}\left(-\Omega, \Omega, \left[\begin{array}{c|c} 0_m & 2\Omega_1 \\ \hline 0 & 0_m \end{array}\right], \left[\begin{array}{c|c} 0_m & 2\Omega_2 \\ \hline 0 & 0_m \end{array}\right], \left[\begin{array}{c|c} 0_m & 2\Omega_3 \\ \hline 0 & 0_m \end{array}\right]\right) X. \tag{19}$$

To achieve our goal of writing the kinematic equations as a right invariant system, we still need to introduce a few more notations. Let E_{ij} denote the square matrix with all entries equal to zero except the entry (i, j) which is 1. Define:

$$C_{ij} = E_{ij} - E_{m+i,m+j}, \quad B_{ij}^\rightarrow = E_{i,m+j} + E_{j,m+i}, \quad B_{ij}^\downarrow = E_{m+i,j} + E_{m+j,i},$$

$$\overline{Y}_{ij} = -C_{ij} + C_{n+i,n+j} + 2E_{2n+i,5m+j},$$

$$\widetilde{Y}_{ij} = \begin{cases} -\frac{1}{2}B_{ii}^\rightarrow + \frac{1}{2}B_{n+i,n+i}^\rightarrow + B_{3n+i,3n+i}^\rightarrow, & \text{se } i = j \\ -B_{ij}^\rightarrow + B_{n+i,n+j}^\rightarrow + 2B_{3n+i,3n+j}^\rightarrow, & \text{se } i \neq j \end{cases}, \tag{20}$$

$$\widehat{Y}_{ij} = \begin{cases} -\frac{1}{2}B_{ii}^\downarrow + \frac{1}{2}B_{n+i,n+i}^\downarrow + B_{4n+i,4n+i}^\rightarrow, & \text{se } i = j \\ -B_{ij}^\downarrow + B_{n+i,n+j}^\downarrow + 2B_{4n+i,4n+j}^\rightarrow, & \text{se } i \neq j \end{cases}.$$

We are now in conditions to state the following.

Proposition 3. *The kinematic equations (16) are equivalent to the following homogeneous right-invariant affine control system evolving on the connected Lie group $G = \text{Sp}(2m)^2 \times \mathfrak{gl}(m) \times \text{sym}(m)^2$, where $(\overline{\omega}_{11}, \cdots, \widehat{\omega}_{m,m}) \in \mathbb{R}^{\frac{n^2+n}{2}}$ are the control functions.*

$$\dot{X} = \left(\sum_{1 \leq i,j \leq m} \overline{\omega}_{ij}\overline{Y}_{ij} + \sum_{1 \leq i \leq j \leq m} \widetilde{\omega}_{ij}\widetilde{Y}_{ij} + \sum_{1 \leq i \leq j \leq m} \widehat{\omega}_{ij}\widehat{Y}_{ij}\right) X. \tag{21}$$

According to a theorem in [6], this system is controllable if and only if the control vector fields generate $T_X G$, for every $X \in G$, which amounts to showing that the following controllability rank condition holds.

$$\left\{ \{\overline{Y}_{ij} : 1 \leq i, j \leq m\} \cup \{\widetilde{Y}_{ij} : 1 \leq i \leq j \leq m\} \cup \{\widehat{Y}_{ij} : 1 \leq i \leq j \leq m\} \right\}_{L.A.} = \mathcal{L}(G).$$

Finally, we can state and prove the main result concerning controllability.

Theorem 2. *The kinematic equations (16), for the rolling motion of* $\mathrm{Sp}(2m)$ *over its affine tangent space at the identity* I_n, *are controllable.*

Proof. It is enough to show that every element of a basis of $\mathcal{L}(G)$ can be written as a linear combination of the matrices $\overline{Y}_{ij}$, $\widetilde{Y}_{ij}$ e $\widehat{Y}_{ij}$ and their Lie brackets. We prove this for the following basis of $\mathcal{L}(G)$:

$$\left\{ C_{ij} : 1 \leq i, j \leq m \right\} \cup \left\{ B_{ij}^{\rightarrow} : 1 \leq i \leq j \leq m \right\} \cup \left\{ B_{ij}^{\downarrow} : 1 \leq i \leq j \leq m \right\}$$
$$\cup \left\{ C_{n+i,n+j} : 1 \leq i, j \leq m \right\} \cup \left\{ B_{n+i,n+j}^{\rightarrow} : 1 \leq i \leq j \leq m \right\} \cup \left\{ B_{n+i,n+j}^{\downarrow} : 1 \leq i \leq j \leq m \right\}$$
$$\cup \left\{ E_{2n+i,5m+j} : 1 \leq i, j \leq m \right\} \cup \left\{ B_{3n+i,3n+j}^{\rightarrow} : 1 \leq i \leq j \leq m \right\} \cup \left\{ B_{4n+i,4n+j}^{\rightarrow} : 1 \leq i \leq j \leq m \right\},$$

Since for the elementary matrices we have $[E_{ij}, E_{fl}] = \delta_{jf} E_{il} - \delta_{il} E_{fj}$, we can also write the commuting structure of the basis elements as:

$$\begin{aligned}
\left[C_{ij}, C_{fl} \right] &= \delta_{jf} C_{il} - \delta_{il} C_{fj} \\
\left[C_{ij}, B_{fl}^{\rightarrow} \right] &= \delta_{jf} B_{il}^{\rightarrow} + \delta_{jl} B_{if}^{\rightarrow} \\
\left[C_{ij}, B_{fl}^{\downarrow} \right] &= -\delta_{if} B_{lj}^{\downarrow} - \delta_{il} B_{fj}^{\downarrow} \\
\left[B_{ij}^{\rightarrow}, B_{fl}^{\downarrow} \right] &= \delta_{jf} C_{il} + \delta_{il} C_{jf} + \delta_{jl} C_{if} + \delta_{if} C_{jl},
\end{aligned}$$

and the following, which proves the required rank corrollability condition.

1. For $1 \leq i, j \leq m$,

$$C_{ij} = \begin{cases} -\frac{1}{4} \left[\left[\overline{Y}_{ii}, \widetilde{Y}_{ii} \right] - 2\widetilde{Y}_{ii}, \widehat{Y}_{ii} \right], & \text{if } i = j \\ -\frac{1}{2} \left[\left[\widetilde{Y}_{ii}, \widehat{Y}_{ii} \right] - \overline{Y}_{ii}, \overline{Y}_{ij} \right], & \text{if } i \neq j \end{cases}$$

$$C_{n+i,n+j} = \begin{cases} \frac{1}{4} \left[\left[\overline{Y}_{ii}, \widetilde{Y}_{ii} \right] + 2\widetilde{Y}_{ii}, \widehat{Y}_{ii} \right], & \text{if } i = j \\ \frac{1}{2} \left[\left[\widetilde{Y}_{ii}, \widehat{Y}_{ii} \right] + \overline{Y}_{ii}, \overline{Y}_{ij} \right], & \text{if } i \neq j \end{cases}$$

2. For $1 \leq i \leq j \leq m$,

$$B_{ij}^{\rightarrow} = -\frac{1}{2} \left[\left[\widetilde{Y}_{ii}, \widehat{Y}_{ii} \right] - \overline{Y}_{ii}, \widetilde{Y}_{ij} \right]$$
$$B_{ij}^{\downarrow} = \frac{1}{2} \left[\left[\widetilde{Y}_{jj}, \widehat{Y}_{jj} \right] - \overline{Y}_{jj}, \widehat{Y}_{ij} \right]$$

$$B_{n+i,n+j}^{\rightarrow} = \frac{1}{2}\left[\left[\widetilde{Y}_{ii}, \widehat{Y}_{ii}\right] + \overline{Y}_{ii}, \widetilde{Y}_{ij}\right]$$

$$B_{n+i,n+j}^{\downarrow} = -\frac{1}{2}\left[\left[\widetilde{Y}_{jj}, \widehat{Y}_{jj}\right] + \overline{Y}_{jj}, \widehat{Y}_{ij}\right]$$

$$E_{2n+i,5m+j} = \frac{1}{2}\left(\overline{Y}_{ij} + C_{ij} - C_{n+i,n+j}\right)$$

$$B_{3n+i,3n+i}^{\rightarrow} = \begin{cases} \widetilde{Y}_{ii} + \frac{1}{2}B_{ii}^{\rightarrow} - \frac{1}{2}B_{n+i,n+i}^{\rightarrow}, & \text{if } i = j \\ \frac{1}{2}\left(\widetilde{Y}_{ij} + B_{ij}^{\rightarrow} - B_{n+i,n+j}^{\rightarrow}\right), & \text{if } i \neq j \end{cases}$$

$$B_{4n+i,4n+j}^{\rightarrow} = \begin{cases} \widehat{Y}_{ii} + \frac{1}{2}B_{ii}^{\downarrow} - \frac{1}{2}B_{n+i,n+i}^{\downarrow}, & \text{if } i = j \\ \frac{1}{2}\left(\widehat{Y}_{ij} + B_{ij}^{\downarrow} - B_{n+i,n+j}^{\downarrow}\right), & \text{if } i \neq j \end{cases} \qquad \square$$

Acknowledgements: The work of the second author was supported by FCT, under project PTDC/EEA-CRO/122812/2010.

References

1. Crouch, P., Silva Leite, F.: Rolling maps for pseudo-Riemannian manifolds. In: Proc. IEEE-CDC 2012, Hawaii, USA, December 10-13 (2012)
2. Godoy, M., Grong, E., Markina, I., Silva Leite, F.: An intrinsic formulation of the rolling manifolds problem. Journal of Dynamical and Control Systems 18(2), 181–214 (2012)
3. Grong, E.: Controllability of rolling without twisting or slipping in higher dimensions. SIAM J. Control Optim. 50(4), 2462–2485 (2012)
4. Hüper, K., Silva Leite, F.: On the geometry of rolling and interpolation curves on S^n, $SO(n)$ and Grassmann manifolds. Journal of Dynamical and Control Systems 13(4), 467–502 (2007)
5. Hüper, K., Krakowski, K., Silva Leite, F.: Rolling maps in a Riemannian framework. Textos de Matemática 43, 15–30 (2011), (J. Cardoso, K. Hüper, P. Saraiva, Eds.) Department of Mathematics, Universisity of Coimbra
6. Jurdjevic, V., Sussmann, H.: Control systems on Lie groups. Journal of Differential Equations 12, 313–329 (1972)
7. Jurdjevic, V., Zimmerman, J.A.: Rolling sphere problems on spaces of constant curvature. Math. Proc. Camb. Phil. Soc. 144, 19–729 (2008)
8. Korolko, A., Silva Leite, F.: Kinematics for rolling a Lorentzian sphere. In: Proc. CDC-ECC, Orlando, USA, December 2-15, pp. 6522–6528 (2011)
9. Marques, A., Silva Leite, F.: Rolling a pseudohyperbolic space over the affine tangent space at a point. In: Proc. CONTROLO 2012, Paper # 36, Funchal, Portugal, pp. 16–18 (July 2012)
10. Mehl, C., Mehrmann, V., Ran, A., Rodman, L.: Perturbation analysis of Lagrangian invariant subspaces of symplectic matrices. Linear and Multilinear Algebra 57(2), 141–184 (2009)
11. O'Neill, B.: Semi-Riemannian geometry with applications to relativity. Academic Press, Inc., N.Y (1983)
12. Sharpe, R.W.: Differential geometry. Springer, New York (1996)
13. Tar, J.K., Kozlowski, K., Rudas, I.J., Ilkei, T.: Application of the symplectic group in a novel branch of soft computing for controlling of electro-mechanical devices. In: Proc. IEEE International Workshop on Robot Motion and Control, Dworek, Poland, October 18-20, pp. 71–77 (2001)

Minimal State-Space Realizations
of Convolutional Codes[*]

Telma Pinho[1], Raquel Pinto[1], and Paula Rocha[2]

[1] CIDMA, Department of Mathematics, University of Aveiro, Portugal
[2] CIDMA and Faculty of Engineering, University of Porto, Portugal

Abstract. In this work the minimality of state-space realizations of an input/output operator (encoder) and of the corresponding output behavior (code) are analyzed. Moreover, a procedure to obtain a minimal realization of a convolutional code starting from a minimal realization of an encoder of the code is provided.

1 Introduction

The problem of obtaining minimal state-space realizations for convolutional codes is a question of crucial importance not only due to implementation issues, but also because such realizations allow to construct codes with suitable properties, like, for instance, good error correcting capacity, [3].

State-space realizations for a convolutional code can be obtained via the realization of a corresponding encoder. However, since the same code admits encoders with different McMillan degrees (i.e., with different minimal state-space realization dimensions), an arbitrary choice of the encoder to be realized may lead to a non-minimal code realization.

This issue has been solved in [2,5,6], where a procedure to obtain all the minimal (McMillan degree) encoders of a code starting from an arbitrary encoder has been proposed. Such procedure allows obtaining a minimal state-space realization of a code starting from an arbitrary encoder $G(d)$ by first performing suitable transformations on $G(d)$ so as to obtain a minimal encoder $G^*(d)$, and then realizing $G^*(d)$ by a minimal state-space model. Although conceptually very elegant, this method implies dealing with polynomial matrices, which may constitute a drawback from the computational point of view.

Here we propose an alternative approach and provide a method to obtain a minimal realization of a convolutional code starting from a minimal realization of an arbitrary encoder of the code, and then, if necessary, reducing the dimension of this realization so as to obtain a minimal realization of the code. This only implies dealing with constant matrices.

[*] This work was supported by Portuguese funds through the CIDMA - Center for Research and Development in Mathematics and Applications, and the Portuguese Foundation for Science and Technology ("FCT – Fundação para a Ciência e a Tecnologia"), within project PEst-OE/MAT/UI4106/2014.

This paper is organized as follows: in the next section we present some preliminary results on convolutional codes and their encoders. In section 3, the realization problem is presented. Concretely, realizations of encoders and of codes are introduced and the minimality of such realizations is investigated. Our method is presented in section 4. Section 5 contains the concluding remarks.

2   Convolutional Codes and Their Encoders

We consider convolutional codes constituted by sequences indexed by $\mathbb{Z}$ and taking values in $\mathbb{F}^n$, where $\mathbb{F}$ is a field. Such sequences $\{\mathbf{w}(i)\}_{i \in \mathbb{Z}}$ can be represented as elements of the set of bilateral formal power series over $\mathbb{F}^n$, denoted by $\mathcal{F}^n$, i.e.

$$\hat{\mathbf{w}}(d) = \sum_{i \in \mathbb{Z}} \mathbf{w}(i) d^i.$$

Note that $\mathcal{F}^n$ constitutes a module over the ring $\mathbb{F}[d]$ of polynomials in d over $\mathbb{F}$.

Given a subset $\mathcal{C}$ of the sequences indexed by $\mathbb{Z}$, taking values on $\mathbb{F}^n$, we denote by $\hat{\mathcal{C}}$ the subset of $\mathcal{F}^n$ defined by $\hat{\mathcal{C}} = \{\hat{w} : w \in \mathcal{C}\}$.

Definition 1. *A convolutional code $\mathcal{C}$ is a subset of sequences indexed by $\mathbb{Z}$ such that $\hat{\mathcal{C}}$ is a submodule of $\mathcal{F}^n$ which coincides with the image of $\mathcal{F}^k$ (for some $k \in \mathbb{N}$) by a polynomial matrix $G(d)$, i.e.,*

$$\hat{\mathcal{C}} = \mathrm{Im}\ G(d) = \{\hat{\mathbf{w}}(d) = G(d)\hat{\mathbf{u}}(d),\ \hat{\mathbf{u}}(d) \in \mathcal{F}^k\};$$

with some abuse of language we also write $\mathcal{C} = \mathrm{Im}\ G(d)$.

It can be shown that given a convolutional code $\mathcal{C}$ there always exist full column rank matrices $G(d) \in \mathbb{F}[d]^{n \times k}$ such that $\mathcal{C} = \mathrm{Im}\ G(d)$. The *encoders* of $\mathcal{C}$ are here defined to be such matrices. This definition of encoder is slightly different from the one in [2] where non full column rank polynomial matrices are allowed as encoders. However, our definition is motivated by the fact that only full column rank encoders are relevant for the purpose of obtaining minimal realizations of a code.

3   Realization Problem

In this section we consider discrete time state-space models. A discrete-time state-space model is a description of a linear, discrete and time-invariant system through equations of the form

$$\begin{cases} \sigma x(t) = Ax(t) + Bu(t) \\ w(t) = Cx(t) + Du(t) \end{cases}, \tag{1}$$

where A, B, C and D are matrices over $\mathbb{F}$ of size $m \times m$, $m \times k$, $n \times m$ and $n \times k$, respectively; $\sigma x(t) = x(t+1)$, for all $t \in \mathbb{Z}$, u is the input-variable, w is the output-variable and x is the state-variable. The system described by (1) will be denoted by $\Sigma(A, B, C, D)$, and its dimension is defined to be the dimension of the state space, i.e., m.

Depending on what type of situation we are interested in, these models can be viewed from different perspectives, namely as realizations of input/output relations (corresponding to encoders) or as realizations of output behaviors (corresponding to codes).

3.1   Realizations of Encoders

Definition 2. $\Sigma(A, B, C, D)$ *is said to be a realization of the encoder* $G(d) \in \mathbb{F}[d]^{n \times k}$ *if*

$$\begin{aligned} \mathcal{B}_{(u,w)} &:= \{(u, w) : \; \hat{w}(d) = G(d)\hat{u}(d)\} \\ &= \{(u, w) : \; \exists \, x \; s.t. \; (u, x, w) \; satisfies \; (1)\}. \end{aligned}$$

In this case we write $\Sigma(A, B, C, D) = \Sigma(G)$.

Note that the set $\mathcal{B}_{(u,w)}$ is what is known in the behavioral approach to systems and control [1] as the (external) input/output behavior associated with (1).

Note further that, since for bilateral sequences, $\widehat{\sigma x} = d^{-1}\hat{x}$, equations (1) are equivalent to

$$\begin{cases} \hat{x}(d) = Ad\hat{x}(d) + Bd\hat{u}(d) \\ \hat{w}(d) = C\hat{x}(d) + D\hat{u}(d) \end{cases} , \tag{2}$$

which, by eliminating the variable $\hat{x}$, yields:

$$\hat{w}(d) = \left(C(I_m - Ad)^{-1}Bd + D\right)\hat{u}(d).$$

Therefore $\Sigma(A, B, C, D)$ is a realization of the encoder $G(d)$ if and only if

$$G(d) = C(I_m - Ad)^{-1}Bd + D.$$

A polynomial encoder $G(d) \in \mathbb{F}[d]^{n \times k}$ admits many realizations with possibly different dimensions. Efficiency leads to focusing on obtaining realizations of minimal dimension.

Definition 3. *Let* $G(d) \in \mathbb{F}[d]^{n \times k}$ *be a polynomial encoder.* $\Sigma(A, B, C, D)$ *is said to be a minimal realization of* $G(d)$ *if no other realization of* $G(d)$ *has smaller dimension, i.e., if the size of the state x is minimal among all the realizations of* $G(d)$. *The minimal dimension of a realization of* $G(d)$ *is called the McMillan degree of* $G(d)$ *and is represented by* $\mu(G)$.

It is well known that the minimal realizations of an encoder $G(d) \in \mathbb{F}[d]^{n \times k}$ are characterized by being simultaneously observable and controllable[1] [4].

3.2 Realizations of Convolutional Codes

Definition 4. $\Sigma(A, B, C, D)$ *is said to be a realization of the convolutional code* $\mathcal{C}$ *if*

$$\mathcal{B}_w := \{w : \mathbb{Z} \to \mathbb{F}^n \mid \exists x, u \text{ s. t. } (u, x, w) \text{ satisfies } (1)\} = \mathcal{C}.$$

This is denoted by $\Sigma(A, B, C, D) = \Sigma(\mathcal{C})$.

It is not difficult to see that a realization of an encoder of a convolutional code is also a realization of the corresponding code, however the converse is not true.

It turns out that a code $\mathcal{C}$ can be regarded as a behavior, the main object of study of the already mentioned behavioral approach developed by J.C. Willems [1]. The behaviors corresponding to codes constitute a particular class of behaviors, known as controllable behaviors, that are precisely sets of sequences that constitute the image of a polynomial shift-operator (in coding language, the encoder). Within the behavioral approach, a particular type of state-space representations for a behavior $\mathcal{B}$ have been introduced, called state/driving-variable (s/dv) representations, whose input is an auxiliary variable (the driving-variable); the behavior $\mathcal{B}$ corresponds to the output behavior of the s/dv model. Thus, the realizations of a code $\mathcal{C}$ are nothing else than s/dv realizations of the controllable behavior $\mathcal{B} = \mathcal{C}$.

Definition 5. $\Sigma(\mathcal{C})$ *is said to be a minimal realization of the code* $\mathcal{C}$ *if the size of* (x, u) *is minimal among all the realizations of* $\mathcal{C}$. *The minimal size of* (x, u) *is denoted by* $\eta(\mathcal{C})$.

A complete characterization for the minimality of code realizations is given by the conditions for the of minimality of s/dv realizations for controllable behaviors that can be derived from [Theorem 4.2, [1]], and are stated as follows using the terminology of codes.

Theorem 1. *[Theorem 4.2, [1]] A realization* $\Sigma(A, B, C, D)$ *of a convolutional code* $\mathcal{C}$ *is minimal if and only if the following conditions are satisfied.*

(i) $\begin{bmatrix} B \\ D \end{bmatrix}$ *has full column rank;*

---

[1] Recall that $\Sigma(A, B, C, D)$ of dimension m is controllable if and only if rank $[B \mid AB \mid \cdots \mid A^{m-1}B] = m$, or, equivalently, if and only if rank $[\lambda I_m - A \mid B] = m$, $\forall \lambda \in \bar{\mathbb{F}}$. $\Sigma(A, B, C, D)$ is observable if and only if

rank $\begin{bmatrix} C \\ CA \\ \vdots \\ CA^{m-1} \end{bmatrix} = m$, or, equivalently, if and only if rank $\begin{bmatrix} \lambda I_m - A \\ C \end{bmatrix} = m$, $\forall \lambda \in$

$\bar{\mathbb{F}}$. Here $\bar{\mathbb{F}}$ denotes the algebraic closure of $\mathbb{F}$.

(ii) (A, B) *is a controllable pair;*

(iii) $\ker D \subseteq \ker B$, *i.e., there exists a matrix L such that $B = LD$;*

(iv) *Let L be as in (iii), and let Λ be a minimal left-annihilator (mla)*[2] *of D. Then the pair $(A - LC, \Lambda C)$ is observable.*

Remark 1. Note that (i) and (iii) are equivalent to (i') - D has full column rank - and (iii).

The next example shows that a minimal realization of an encoder $G(d)$ of a code $\mathcal{C}$ is not necessarily a minimal realization of the code $\mathcal{C}$.

Example 1. Consider the following polynomial encoder of a code $\mathcal{C}$

$$G(d) = \begin{bmatrix} 1 + d - d^3 & -1 + d^3 \\ d + d^2 - d^3 & -1 - d^2 + d^3 \\ d + d^2 & -1 - d - d^2 \end{bmatrix}.$$

It can be easily checked that

$$\Sigma\left(\begin{bmatrix} 0 & 0 & 0 \\ 1 & 0 & 0 \\ 0 & 1 & 0 \end{bmatrix}, \begin{bmatrix} 1 & -1 \\ 0 & 0 \\ 0 & -1 \end{bmatrix}, \begin{bmatrix} 1 & 0 & -1 \\ 1 & 1 & -1 \\ 1 & 1 & 0 \end{bmatrix}, \begin{bmatrix} 1 & -1 \\ 0 & -1 \\ 0 & -1 \end{bmatrix}\right)$$

is a realization of $G(d)$ which is controllable and observable and therefore is minimal. However $\Sigma(A, B, C, D)$ is not a minimal realization of $\mathcal{C}$, as not all the conditions of Theorem 1 are satisfied. Indeed, condition (iii) is fulfilled for

$$L = \begin{bmatrix} 1 & 0 & 0 \\ 0 & 0 & 0 \\ 0 & 0 & 1 \end{bmatrix};$$

however, considering the minimal left annihilator $\Lambda = \begin{bmatrix} 0 & 1 & -1 \end{bmatrix}$ of D, we have that

$$A - LC = \begin{bmatrix} -1 & 0 & 1 \\ 1 & 0 & 0 \\ -1 & 0 & 0 \end{bmatrix} \quad \text{and} \quad \Lambda C = \begin{bmatrix} 0 & 0 & -1 \end{bmatrix},$$

are such that the pair $(A - LC, \Lambda C)$ is not observable.

$\Diamond$

Minimal encoders are defined as the ones for which a minimal realization is also minimal as a code realization; this is formalized in the following definition.

Definition 6. *Let $\mathcal{C} \subset \mathcal{F}^n$ be a convolutional code and $G(d) \in \mathbb{F}[d]^{n \times k}$ and encoder of $\mathcal{C}$. $G(d)$ is said to be a minimal encoder of $\mathcal{C}$ if*

$$\mu(G) + k = \eta(\mathcal{C}).$$

---

[2] Λ is a mla of D if $\Lambda D = 0$ and for all Λ^* such that $\Lambda^* D = 0$ there exists $\tilde{\Lambda}$ satisfying $\Lambda^* = \tilde{\Lambda}\Lambda$.

The situation illustrated in the previous example is due to the fact that when realizing an input/output operator (encoder) $G(d)$ one has no freedom in performing transformations in the input. This restriction is not present in the realization of the corresponding output behavior (code), where the input-variables may be transformed. Therefore, given a minimal realization of a non-minimal encoder $G(d)$, it is still possible to reduce its dimension in order to obtain a minimal realization of the corresponding code. This reduction procedure is carried out in the next section.

4   Minimal Code Realization Procedure

The following procedure shows precisely how to obtain a minimal realization $\tilde{\Sigma}(\mathcal{C}) = \tilde{\Sigma}(\tilde{A}, \tilde{B}, \tilde{C}, \tilde{D})$ of a code $\mathcal{C}$ by performing operations and reducing the number of variables in a minimal realization $\Sigma(G) = \Sigma(A, B, C, D)$ of a corresponding encoder $G(d)$.

Let us consider a minimal realization $\Sigma(A, B, C, D)$ of $G(d)$. Then

$$G(d) = C(I_m - Ad)^{-1}Bd + D$$

$$= \begin{bmatrix} C(I_m - Ad)^{-1}d \mid I_k \end{bmatrix} \begin{bmatrix} B \\ D \end{bmatrix}.$$

Since encoders have full column rank, clearly $\begin{bmatrix} B \\ D \end{bmatrix}$ must have full column rank and hence condition (i) of Theorem 1 is satisfied. Moreover, the minimality of $\Sigma(A, B, C, D)$ as realization of the encoder $G(d)$ implies the controllability of the pair (A, B). Thus, a minimal realization of the encoder $G(d)$ satisfies condition (ii) of Theorem 1.

Suppose now that condition (iii) of the Theorem 1 is not satisfied i.e., $Ker\, D \nsubseteq Ker B$. Then we can suppose, without loss of generality, that

$$D = \begin{bmatrix} I_r & 0 \\ 0 & 0 \end{bmatrix} \text{ and } B = \begin{bmatrix} B_1 & B_2 \end{bmatrix}, \tag{3}$$

with $B_2 = \begin{bmatrix} 0 \\ S \end{bmatrix}$ full column rank of size $m \times (k-r)$, where S is a square invertible matrix of size $k - r$, and $B_1 = \begin{bmatrix} B_{11} \\ B_{21} \end{bmatrix}$ of size $m \times r$[3].

Therefore, (1) is of the form

$$\begin{cases} \sigma x_1 = A_{11}x_1 + A_{12}x_2 + B_{11}u_1 & \text{(4a)} \\ \sigma x_2 = A_{21}x_1 + A_{22}x_2 + B_{21}u_1 + Su_2 & \text{(4b)} \\ w_1 = C_{11}x_1 + C_{12}x_2 + Iu_1 & \text{(4c)} \\ w_2 = C_{21}x_1 + C_{22}x_2 & \text{(4d)} \end{cases}$$

---

[3] If this is not the case, changes of coordinates in the u, x, w spaces allow bringing D and B to the desired form. The coordinate change in the w space modifies the code under consideration, but can be reversed at the end of the reasoning that will be presented.

where the variables x, u and w have been partitioned according to the given matrix partitions. Equations (4a-4d) show that x_2 is a free variable. Indeed, given x_2 and u_1, it is possible to find x_1, w_1 and w_2 such that equations (4a), (4c) and (4d) are satisfied. Moreover, since S is invertible, there exists u_2 such that (4b) holds. Therefore, this latter equation can be eliminated from the description of the code $\mathcal{C}$, and x_2 can assume the role of a driving variable. This means that

$$\begin{cases} \sigma x_1 = A_{11}x_1 + \bar{B}\bar{u} \\ w = C_{11}x_1 + \bar{D}\bar{u} \end{cases}, \tag{5}$$

with $\bar{B} = \begin{bmatrix} A_{12} & B_{11} \end{bmatrix}$, $\bar{u} = \begin{bmatrix} x_2 \\ u_1 \end{bmatrix}$ and $\bar{D} = \begin{bmatrix} C_{12} & I \\ C_{22} & 0 \end{bmatrix}$ is still a realization of the code with smaller dimension than the initial one (recall that the dimension of a code realization is defined as the size of the joint state and driving-variable vector).

Note that the new system obtained in (5) still satisfies the condition (ii) of Theorem 1 since if the pair

$$(A, B) = \left(\begin{bmatrix} A_{11} & A_{12} \\ A_{21} & A_{22} \end{bmatrix}, \begin{bmatrix} B_{11} & 0 \\ B_{21} & S \end{bmatrix} \right)$$

is controllable, then the pair

$$(A_{11}, \bar{B}) = \left(A_{11}, \begin{bmatrix} A_{12} & B_{11} \end{bmatrix} \right)$$

is also controllable. Indeed the controllability condition

$$\text{rank} \begin{bmatrix} \lambda I_m - A & | & B \end{bmatrix} = m, \quad \forall\, \lambda \in \bar{\mathbb{F}},$$

becomes

$$\text{rank} \begin{bmatrix} \lambda I_{m_1} - A_{11} & -A_{12} & B_{11} & 0 \\ -A_{21} & \lambda I_{m_2} - A_{22} & B_{21} & S \end{bmatrix} = m_1 + m_2 = m, \quad \forall \lambda \in \bar{\mathbb{F}},$$

which implies that

$$\text{rank} \begin{bmatrix} \lambda I_{m_1} - A_{11} & | & A_{12} & | & B_{11} \end{bmatrix} = \text{rank} \begin{bmatrix} \lambda I_{m_1} - A_{11} & | & -A_{12} & | & B_{11} \end{bmatrix}$$
$$= m_1,$$

meaning that $(A_{11}, \bar{B})$ is a controllable pair.

Moreover, in case $\begin{bmatrix} \bar{B} \\ \bar{D} \end{bmatrix}$ is not full column rank, there exists an invertible matrix T such that

$$\begin{bmatrix} \bar{B} \\ \bar{D} \end{bmatrix} T = \begin{bmatrix} \bar{\bar{B}} & 0 \\ \bar{\bar{D}} & 0 \end{bmatrix},$$

with $\begin{bmatrix} \bar{\bar{B}} \\ \bar{\bar{D}} \end{bmatrix}$ full column rank. Partitioning $T^{-1}\bar{u}$ accordingly as $T^{-1}\bar{u} = \begin{bmatrix} \bar{\bar{u}} \\ \tilde{u} \end{bmatrix}$, equations (5) become

$$\begin{cases} \sigma x_1 = A_{11}x_1 + \bar{\bar{B}}\bar{\bar{u}} \\ w = C_{11}x_1 + \bar{\bar{D}}\bar{\bar{u}}, \end{cases}, \tag{6}$$

which again yields a realization of the code $\mathcal{C}$ with smaller dimension as the previous one, that now satisfies condition (i) of Theorem 1.

Since

$$\left[\lambda I_{m_1} - A_{11} \mid \bar{\bar{B}} \mid 0\right] = \left[\lambda I_{m_1} - A_{11} \mid \bar{B}T\right] = \left[\lambda I_{m_1} - A_{11} \mid \bar{B}\right] \begin{bmatrix} I & 0 \\ 0 & T \end{bmatrix},$$

where I denotes the identity matrix of suitable size, and

$$\text{rank } \left[\lambda I_{m_1} - A_{11} \mid \bar{B}\right] = \text{rank } \left[\lambda I_{m_1} - A_{11} \mid \bar{\bar{B}} \mid 0\right]$$
$$= \text{rank } \left[\lambda I_{m_1} - A_{11} \mid \bar{\bar{B}}\right],$$

the controllability of the pair $(A_{11}, \bar{B})$ implies that the pair $(A_{11}, \bar{\bar{B}})$ is controllable, and the realization (6) also satisfies condition (ii) of Theorem 1.

In case this realization does not satisfy condition (iii) of Theorem 1, the procedure can be restarted and repeated, yielding successive realizations of the code with smaller dimension, till a realization of the code is obtained that simultaneously satisfies conditions (i), (ii) and (iii). To avoid introducing too much notation, this realization will be again denoted by $\Sigma(A, B, C, D)$ (as the original one).

Suppose now that $\Sigma(A, B, C, D)$ does not satisfy condition (iv) of Theorem 1. From (1), and because condition (iii) is satisfied, we have that

$$\begin{cases} \sigma x = Ax + LDu \\ w = Cx + Du \end{cases} . \tag{7}$$

Since $Du = w - Cx$ implies $LDu = Lw - LCx$, (7) is equivalent to

$$\begin{cases} \sigma x = (A - LC)x + Lw \\ w = Cx + Du \end{cases} . \tag{8}$$

Let Λ be a *mla* of D with full row rank. Then, there exists a matrix X such that $V = \begin{bmatrix} X \\ \Lambda \end{bmatrix}$ is invertible and $VD = \begin{bmatrix} X \\ \Lambda \end{bmatrix} D = \begin{bmatrix} I_k \\ 0 \end{bmatrix}$. Let $\bar{w} := Vw = \begin{bmatrix} X \\ \Lambda \end{bmatrix} w$ be partitioned in the obvious way as $\bar{w} = \begin{bmatrix} \bar{w}_1 \\ \bar{w}_2 \end{bmatrix} = \begin{bmatrix} Xw \\ \Lambda w \end{bmatrix}$. It follows from (8) that

$$\begin{cases} \sigma x = (A - LC)x + LV^{-1}\bar{w} \\ \bar{w}_1 = XCx + u \\ \bar{w}_2 = \Lambda Cx \end{cases} . \tag{9}$$

The second equation of (9) shows that $\bar{w}_1$ is a free variable, which may be taken as a new driving-variable, replacing u. Letting V^{-1} be suitably partitioned as $\begin{bmatrix} R & Q \end{bmatrix}$, this yields

$$\begin{cases} \sigma x = (A - LC)x + LR\bar{w}_1 + LQ\bar{w}_2 \\ \bar{w}_2 = \Lambda Cx \end{cases} . \tag{10}$$

Since $\Sigma(A, B, C, D)$ does not satisfy condition (iv) of Theorem 1, the pair $(A - LC, AC)$ is not observable; thus by reducing equations (10) to the Kalman observability decomposition form through a coordinate change in the state-space, and eliminating the nonobservable states we obtain a description

$$\begin{cases} \sigma\bar{x} = \bar{A}\bar{x} + \bar{B}_1\bar{w}_1 + \bar{B}_2\bar{w}_2 \\ \bar{w}_2 = \bar{C}\bar{x} \end{cases}, \tag{11}$$

for the same set of $(\bar{w}_1, \bar{w}_2)$ trajectories as (10), where the size of the state $\bar{x}$ is smaller than the one of x.

Equations (11) can still be written as

$$\begin{cases} \sigma\bar{x} = (\bar{A} + \bar{B}_2\bar{C})\bar{x} + \bar{B}_1\bar{u}_1 \\ \bar{w}_1 = \bar{u}_1 \\ \bar{w}_2 = \bar{C}\bar{x} \end{cases}, \tag{12}$$

which, by noting that

$$w = V^{-1}\bar{w} = \begin{bmatrix} R\,Q \end{bmatrix} \begin{bmatrix} \bar{w}_1 \\ \bar{w}_2 \end{bmatrix} = R\bar{w}_1 + Q\bar{w}_2 = R\bar{u}_1 + Q\bar{C}\bar{x}$$

finally yields:

$$\begin{cases} \sigma\bar{x} = \bar{\bar{A}}\bar{x} + \bar{B}_1\bar{u}_1 \\ w = \bar{\bar{C}}\bar{x} + \bar{D}\bar{u}_1 \end{cases}, \tag{13}$$

with $\bar{\bar{A}} = \bar{A} + \bar{B}_2\bar{C}$, $\bar{\bar{C}} = Q\bar{C}$ and $\bar{D} = R$.

This is a state-space realization for the same code as $\Sigma(A,B,C,D)$, but with smaller dimension.

If one of the conditions of Theorem 1 is not satisfied by the realization $\Sigma(\bar{\bar{A}}, \bar{B}, \bar{\bar{C}}, \bar{D})$, then one can perform the relevant steps described above, reducing each time the dimension of the code realization. In this way a minimal state/driving-variable realization of the initial code is obtained in a finite number of steps.

It is however worth mentioning the following. As we have just seen, the state-space system that satisfies all conditions of Theorem 1 obtained by this procedure (and that we once more denote by $\Sigma(A, B, C, D)$, with dimension m, by resetting the notation) is a minimal realization of $\mathcal{C}$. Nevertheless it can happen that $C(I_m - Ad)^{-1}Bd + D$ is no longer polynomial and hence is not an encoder of $\mathcal{C}$. In that case, due to the controllability of the pair (A, B), there exists a matrix K of suitable size such that $A - BK$ has only zero eigenvalues, and is therefore nilpotent. This implies that the square $(m \times m)$ polynomial matrix $M(d) := I_m - Ad$ is such that

$$\text{rank } M(\lambda) = m \quad \forall\lambda \in \bar{\mathbb{F}},$$

meaning that $\det M(d)$ must be a nonzero constant, or equivalently, that $M(d)$ is unimodular. Therefore, applying the feedback $u = \bar{u} - Kx$ to the system

$$\begin{cases} \sigma x = Ax + Bu \\ w = Cx + Du \end{cases} \tag{14}$$

yields the system

$$\begin{cases} \sigma x = (A - BK)x + B\bar{u} \\ w = (C - DK)x + D\bar{u} \end{cases} . \tag{15}$$

It can be shown that $\Sigma(A - BK, B, C - DK, D)$ is still a minimal realization of the code. Moreover, the polynomial matrix $G(d) = (C - DK)(I - d(A - BK))^{-1}Bd + D$ is a minimal encoder of the code.

5   Conclusion

In this paper we have analyzed the minimality of realizations of convolutional codes. It turns out that, when realizing an encoder (input/output operator) $G(d)$, one has no freedom in performing transformations in the input; however, this restriction is not present in the realization of the corresponding code (output behavior) $\mathcal{C} = \operatorname{Im} G(d)$, where the input variables may be transformed. This can be exploited in order to reduce the dimension of the obtained state-space realizations. In this way, code realizations can have lower dimension than encoder realizations. Here we proposed a procedure that overcomes this problem and allows obtaining a minimal realization of a convolutional code starting from a minimal realization of an arbitrary encoder of the code.

Acknowledgment.   We would like to thank Professor Ettore Fornasini for helpful discussions about this topic.

References

1. Willems, J.C.: Models for Dynamics. Dynamics Reported 2, 171–269 (1989)
2. Fornasini, E., Pinto, R.: Matrix Fraction Descriptions in Convolutional Conding. Linear Algebra and its Applications 392, 119–158 (2004)
3. Rosenthal, J., Schumacher, J.M.: A state space approach for constructing MDS rate $\frac{1}{n}$ convolutional codes. In: 1998 Information Theory Workshop, pp. 116–117 (1998)
4. Kailath, T.: Linear Systems. Prentice Hall, Englewood Cliffs (1980)
5. Forney, G.: Convolutional Codes I: Algebraic Structure. IEEE Trans. Inform. Theory 16(6), 720–738 (1970)
6. Forney, G.: Minimal Bases of Rational Vector Spaces, with Applications to Multivariable Systems. SIAM J. Control 493–520 (1975)

Solvability of Implicit Difference Equations

Aram Arutyunov[1], Fernando Pereira[2], and Sergey Zhukovskiy[1]

[1] Peoples' Friendship University of Russia, M. Maklaya str., 6, 117198,
Moscow, Russia
[2] University of Porto, Rua Dr. Roberto Frias, 4200-465 Porto, Portugal
arutun@orc.ru, flp@fe.up.pt, s-e-zhuk@yandex.ru

Abstract. This article concerns sufficient conditions for the solvability
of implicit difference equations. These are cast in a very general framework that relies on the α-covering and Lipschitz properties of the implicit
recursive map with respect relatively to the first and second arguments.
Difference equations have a prominent role in many scientific and engineering fields and may arise due to a wide variety of reasons. These may
range from the discretization of time and other variables in dynamic systems given by either ordinary or partial differential equations required
by computational procedures, or from the intrinsically discrete nature
of many systems. While in some of these, discrete state transitions are
caused by discrete events, in others, their decentralized nature requires
iterative procedures built in the control synthesis in order to achieve
some kind of global consensus.

Keywords: Implicit difference equations, initial value problems, covering mappings.

1 Introduction

The presented article is devoted to the following issue. Given nonempty sets X,
Y, a mapping $f : X \times X \to Y$, a point $a \in X$, a sequence $\{y_n\} \subset Y$, consider an
implicit difference equation

$$f(x_{n+1}, x_n) = y_n, \quad x_0 = u. \tag{1}$$

A solution to this problem is a sequence $\{x_n\}$ satisfying equality (1) for any $n =
0, 1, 2, \ldots$. whose solution is a sequence $\{x_n\} \subset X$, $n = \overline{0, \infty}$. Unlike the explicit
difference equations (i.e. equations of the form $x_{n+1} = g(x_n)$), the solvability of
the problem (1) is not obvious. Moreover, simple examples show that problem
(1) can have no solutions even if $X = Y = \mathbb{R}$ and function f is smooth. So, the
question of implicit difference equations solvability causes a natural interest.

Difference equations have a prominent role in many scientific and engineering
fields may arise in the modeling of both natural and artificial systems.

Many important phenomena in nature have an intrinsically discrete behavior. Generation of biological populations (cells and other more complex living
organisms) generations (see [9], the various states of matter in physics, many

© Springer International Publishing Switzerland 2015   23
A.P. Moreira et al. (eds.), *CONTROLO'2014 - Proc. of the 11th Port. Conf. on Autom. Control,*
Lecture Notes in Electrical Engineering 321, DOI: 10.1007/978-3-319-10380-8_3

mechanic impact systems, to name just a few, are inherently discrete and thus their evolution is adequately described by difference equations. Clearly, even for systems evolving in continuum time whose dynamics are given by either ordinary or partial differential equations, difference equations emerge as a result of the discretization of time and other variables required by computational procedures. Finally,it is important to note that some artificial dynamic systems are essentially discrete in nature, [8]. While for some of these, discrete state transitions are caused by discrete events, in others, their decentralized nature requires the specification of iterative procedures built in the control synthesis in order to achieve some kind of global consensus, [10].

Traditionally authors consider this type of problems assuming that $X = Y = \mathbb{R}^n$ and mapping f is sufficiently smooth (see, for instance, [1]). In this paper consider the non-smooth case. Namely, we assume that X and Y are metric spaces and f is a closed mapping. In order to investigate this problem we apply covering mappings theory (see, for instance, [2] – [4]) and a technique established in [5]. Note that the covering mappings theory has multiple applications in investigation of implicit ordinary differential equations (see [5]), control problems (see [6]), abstract and integral Volterra equations unsolved to the unknown function (see [7]).

2   Main Results

Let (X, ρ_X), (Y, ρ_Y) be metric spaces with metrics ρ_X and ρ_Y, respectively, number $\alpha > 0$, $\beta \geq 0$ be given. Denote by $B_X(x, R)$ the closed ball in X centered at $x \in X$ with a radius $R \geq 0$, i.e.

$$B_X(x, R) = \{u \in X : \rho_X(x, u) \leq R\}.$$

Moreover, set

$$B_X(M, R) = \bigcup_{x \in M} B_X(x, R).$$

Recall the definition of covering mapping from [2].

Definition 1. *A mapping* $\Psi : X \to Y$ *is called* α*-covering if*

$$B_Y\big(\Psi(x), \alpha r\big) \subset \Psi\big(B_X(x, r)\big).$$

for any $x \in X$, $r \geq 0$.

Note that the defined property is equivalent the following

$$\forall x_0 \in X, \quad \forall y \in Y \quad \exists x \in X : \quad \Psi(x) = y \quad \text{and} \quad \rho_X(x_0, x) \leq \frac{\rho_Y(y, \Psi(x_0))}{\alpha}.$$

Given a mapping $f : X \times X \to Y$, a point $u \in X$ and a sequence $\{y_n\} \subset Y$, consider the initial value problem (1). Our first goal is to obtain solvability conditions for this problem and solution estimates.

Recall that a mapping $g : X \to Y$ is called β-Lipschitz, if $\rho_X(g(x), g(u)) \leq \beta \rho_X(x, u)$ for all $x, u \in X$.

Lemma 1. *Assume that*

(A1) $f(\cdot, x)$ *is α-covering for any $x \in X$, $f(x, \cdot)$ is β-Lipschitz for any $x \in X$.*

Then for any initial condition $u \in X$ and for any right-hand side $\{y_n\} \subset Y$ there exists a solution $\{x_n\} \subset X$ to problem (1) such that

$$\rho_X(x_0, x_1) \le \frac{\rho_Y(y_0, f(u, u))}{\alpha},$$

$$\rho_X(x_{n+1}, x_n) \le \frac{\beta}{\alpha}\rho_X(x_n, x_{n-1}) + \frac{\rho_Y(y_n, y_{n-1})}{\alpha} \tag{2}$$

for all $n \ge 0$.

The proofs of this and the subsequent statement will be presented in the following section. Let us now show that under additional assumptions the solutions to the problem (1) converges exponentially.

Theorem 1. *Assume that* **(A1)** *holds and, moreover,*

(A2) X *is complete, $f : X \times X \to Y$ is closed;*
(A3) $\beta < \alpha$;

Let there be given a point $y^* \in Y$ and a sequence $\{y_n\}$ such that $y_n \to y^*$ as $n \to \infty$ and, moreover,

$$\rho_Y(y_n, y^*) \le K\mu^n \quad \forall n = 0, 1, 2, ... \tag{3}$$

for some $K > 0$, $\mu \in [0, 1)$, $\mu \ne \alpha^{-1}\beta$.

Then for any initial condition $u \in X$ there exist a point $x^* \in X$ and a solution $\{x_n\} \subset X$ to problem (1) such that

$$f(x^*, x^*) = y^*, \quad x_n \to x^* \quad as \quad n \to \infty$$

and, moreover,

$$\rho_X(x_n, x^*) \le \left(\frac{\beta}{\alpha}\right)^n \frac{\rho_Y(y_0, f(u, u))}{\alpha - \beta} + \left(\frac{\beta}{\alpha}\right)^n \frac{\alpha c}{\alpha - \beta} - \mu^n \frac{c}{1 - \mu} \quad \forall n \ge 0, \tag{4}$$

where

$$c = \frac{\alpha K(\mu + 1)}{\beta - \alpha\mu}.$$

The presented theorem has the following simple corollary. Fix an arbitrary point $y^* \in Y$ and consider difference equation

$$f(x_{n+1}, x_n) = y^*, \quad x_0 = u. \tag{5}$$

Recall that an equilibrium point in this equation is a point $x^* \in X$ such that $f(x^*, x^*) = y^*$. Obviously, $x_n \equiv x^*$ is a constant solution to this problem. If the assumptions **(A1)**–**(A3)** hold then for any initial condition $u \in X$ there exists an equilibrium point $x^* \in X$ and a solution to the initial value problem $\{x_n\} \subset X$ such that x_n converges to x^* exponentially.

3   Proofs of the Main Results

First let us prove Lemma 1.

Proof. Set $x_0 = u$. Let us prove the existence of the desired sequence by induction. Since $f(\cdot, x_0)$ is α-covering, there exist $x_1 \in X$ such that

$$f(x_1, x_0) = y_0 \quad \text{and} \quad \rho_X(x_0, x_1) \leq \frac{\rho_Y(y_0, f(u, u))}{\alpha}.$$

Since $f(\cdot, x_1)$ is α-covering, there exist $x_2 \in X$ such that $f(x_2, x_1) = y_1$ and

$$\rho_X(x_1, x_2) \leq \frac{\rho_Y(y_1, f(x_1, x_1))}{\alpha}.$$

Applying triangle inequality, since $f(x_1, \cdot)$ is β-Lipschitz and $f(x_1, x_0) = y_0$, we obtain

$$\rho_X(x_1, x_2) \leq \frac{\rho_Y(y_1, y_0) + \rho_Y(f(x_1, x_0), f(x_1, x_1))}{\alpha} \leq \frac{\rho_Y(y_1, y_0) + \beta \rho_X(x_0, x_1)}{\alpha}$$

The points x_0, x_1, x_2 are constructed.

Assume now that there exist points $x_0, x_1, ..., x_k$ such that (2) holds for $n = \overline{0, k-1}$. Since $f(\cdot, x_k)$ is α-covering, there exist $x_{k+1} \in X$ such that $f(x_{k+1}, x_k) = y_k$ and

$$\rho_X(x_k, x_{k+1}) \leq \frac{\rho_Y(y_k, f(x_k, x_k))}{\alpha}.$$

Applying triangle inequality, since $f(x_k, \cdot)$ is β-Lipschitz and $f(x_k, x_{k-1}) = y_{k-1}$, we obtain

$$\rho_X(x_k, x_{k+1}) \leq \frac{\rho_Y(y_k, y_{k-1}) + \rho_Y(f(x_k, x_{k-1}), f(x_k, x_k))}{\alpha} \leq$$

$$\leq \frac{\rho_Y(y_k, y_{k-1}) + \beta \rho_X(x_k, x_{k-1})}{\alpha}.$$

The point x_{k+1} is constructed. Thus, the desired sequence exists. □

Let us now prove Theorem 1.

Proof. Assumption (2) implies that there exists a solution $\{x_n\} \subset X$ to problem (1) satisfying (2). It follows from assumption (3) that

$$\rho_Y(y_n, y_{n-1}) \leq \rho_Y(y_n, y^*) + \rho_Y(y^*, y_{n-1}) \leq K(\mu + 1)\mu^{n-1}$$

for any $n \geq 1$. So, inequality (2) implies

$$\rho_X(x_{n+1}, x_n) \leq \frac{\beta}{\alpha}\rho_X(x_n, x_{n-1}) + \frac{K(\mu + 1)}{\alpha}\mu^{n-1} \leq$$

$$\leq \frac{\beta}{\alpha}\left(\frac{\beta}{\alpha}\rho_X(x_{n-1},x_{n-2}) + \frac{K(\mu+1)}{\alpha}\mu^{n-2}\right) + \frac{K(\mu+1)}{\alpha}\mu^{n-1} \leq \dots$$

$$\dots \leq \left(\frac{\beta}{\alpha}\right)^n \rho_X(x_0,x_1) + \frac{K(\mu+1)}{\alpha}\sum_{j=0}^{n-1}\left(\frac{\beta}{\alpha}\right)^j \mu^{n-1-j} =$$

$$= \left(\rho_X(x_0,x_1) + \frac{K(\mu+1)}{\alpha^{-1}\beta - \mu}\right)\left(\frac{\beta}{\alpha}\right)^n - \left(\frac{K(\mu+1)}{\alpha^{-1}\beta - \mu}\right)\mu^n.$$

Therefore,

$$\rho_X(x_n,x_{n+k}) \leq \sum_{j=0}^{k-1}\rho_X(x_{n+j},x_{n+j+1}) =$$

$$= \left(\frac{\beta}{\alpha}\right)^n\left(\rho_X(x_0,x_1) + \frac{K(\mu+1)}{\alpha^{-1}\beta - \mu}\right)\frac{1-(\alpha^{-1}\beta)^k}{1-\alpha^{-1}\beta} - \mu^n\left(\frac{K(\mu+1)}{\alpha^{-1}\beta - \mu}\right)\frac{1-\mu^k}{1-\mu} =$$

$$= \left(\left(\frac{\beta}{\alpha}\right)^n\frac{\rho_Y(y_0,f(u,u))}{\alpha - \beta} + \left(\frac{\beta}{\alpha}\right)^n\frac{\alpha c}{\alpha - \beta}\right)\left(1 - \left(\frac{\beta}{\alpha}\right)^k\right) - \mu^n\frac{c}{1-\mu}(1-\mu^k).$$

Since $\beta < \alpha$ and $\mu \in [0,1)$, the sequence $\{x_n\}$ is the Cauchy sequence. Therefore, x_n converges to some point $x^* \in X$. Further, passing to the limit as $k \to \infty$ for $n = 0$ in the above inequality we obtain (5). Moreover, $y_n = f(x_{n+1},x_n)$ converges to y^* and f is closed. Thus, $y^* = f(x^*,x^*)$. □

Acknowlededegement. This research was supported by the FP7 Marie Curie project COVMAPS with ref. FP7-PEOPLE-2011-IIF 911177, FCT funding from the project PEst-OE/EEI/UI0147/2014, RFBR grant (project No. 12-01-00427).

References

1. Anh, P.K., Loi, L.C.: On the solvability of initial-value problems for nonlinear implicit difference equations. Adv. Diff. Eq. 2004(3), 195–200 (2004)
2. Arutyunov, A.V.: Covering mappings in metric spaces and fixed points. Dokl. Math. 76(2), 665–668 (2007)
3. Arutyunov, A.V.: Stability of coincidence points and properties of covering mappings. Math. Notes 86(1-2), 153–158 (2009)
4. Mordukhovich, B.S.: Variational Anaysis and Generalized Differentiation, vol. 1. Springer (2005)
5. Avakov, E.R., Arutyunov, A.V., Zhukovskii, E.S.: Covering mappings and their applications to differential equations unsolved for the derivative. Diff. Equations 45(5), 627–649 (2009)
6. Arutyunov, A.V., Zhukovskiy, S.E.: Local solvability of control systems with mixed constraints. Diff. Equations 46(11), 1561–1570 (2010)
7. Arutyunov, A.V., Zhukovskiy, E.S., Zhukovskiy, S.E.: Covering mappings and well-posedness of nonlinear Volterra equations. Nonl. Anal. 75(3), 1026–1044 (2011)

8. Cassandras, C., Lafortune, S.: Introduction to Discrete Event Systems. Springer (2008)
9. Eisen, M.: Mathematical Models in Cell Biology and Cancer Chemotherapy. Lecture Notes in Biomathematics, vol. 30. Springer (1979)
10. Sarlette, A., Sepulchre, R.: Consensus optimization on manifolds. SIAM Journal on Control and Optimization 48(1), 56–76 (2009)

Control of a Hysteresis Model Subject to Random Failures

Elmer Lévano[1] and Alessandro N. Vargas[1,2]

[1] Universidade Tecnológica Federal do Paraná, UTFPR, Av. Alberto Carazzai 1640,
86300-000 Cornelio Procópio-PR, Brazil
`avargas@utfpr.edu.br`
[2] Basque Center for Applied Mathematics, BCAM, Alameda de Mazarredo 14,
E-48009 Bilbao, Vizcaya, Spain

Abstract. The note presents conditions to assure weak stability in the mean for a hysteresis Bouc-Wen model controlled by a proportional-integral controller subject to random failures. When a failure happens, the controller turns off and remains off for a while. After that the controller turns on and keeps on until the occurrence of the next failure. The failures occur according to a Poisson distributed process. A numerical example illustrates the result.

Keywords: Hysteresis, Bouc-Wen model, stochastic system, stochastic control.

1 Introduction

Hysteresis is a nonlinear phenomenon encountered in a wide variety of processes in which the input-output dynamic relation between variables involve memory effects. An important model able to account hysteresis is called the Bouc-Wen model, a topic of intensive investigation in the recent years [1], [2], [3], [4], [5], [6], [7]. The survey paper [8] and the monograph [9] present in a unified and detailed way the most important results dedicated for such hysteresis model.

The normalized version of the Bouc-Wen model introduced in [1] (see also [4] and [9]) relates the single-output $\phi(x)(t)$ to the single-input $x(t)$ in the following way:

$$\Phi_{BW}(x)(t) = k_x x(t) + k_w w(t), \quad \forall t \geq 0, \tag{1}$$

and

$$\dot{w}(t) = \rho(\dot{x}(t) - \sigma|\dot{x}(t)||w(t)|^{n-1}w(t) - (\sigma - 1)\dot{x}(t)|w(t)|^n), \tag{2}$$

where the parameters are $k_x > 0$, $k_w > 0$, $\rho > 0$, $\sigma \geq 1/2$, and $n \geq 1$ In this single-input single-output relation, it is assumed that both $x(t)$ and $\phi(x)(t)$ are accessible to measurements but the internal nonlinear state, $w(t)$, is hidden and can not be measured.

Controlling Bouc-Wen models is a topic of interest [10], [11], [12], and the Proportional Integrative (PI) controller shows to be appropriate to handle such models as shown in [6]. A limitation of the results derived in [6] is that they are

© Springer International Publishing Switzerland 2015
A.P. Moreira et al. (eds.), *CONTROLO'2014 - Proc. of the 11th Port. Conf. on Autom. Control*,
Lecture Notes in Electrical Engineering 321, DOI: 10.1007/978-3-319-10380-8_4

not appropriate to deal with the Bouc-Wen model subject to random elements. In contrary, our approach considers the PI controller subject to random failures. This approach is useful, for instance, to represent the case in which the source of energy supplying the controller fails. In this random context, our contribution is to derive conditions to assure stability of the Bouc-Wen model with PI controller subject to random failures.

The main contribution of this paper is to present conditions for a weak stability in the mean concept for the Bouc-Wen model. We assume that the PI controller is subject to random failures. The failures follow a stochastic process with Poisson distribution. At the instant of occurrence of a failure, the PI controller turns off and keeps in this situation for a while. After that, it is allowed to turn on. Under this random on-off behavior, we show that the resulting Bouc-Wen model system is stable in a weak sense. This sets the main contribution of this paper.

2 Basic Definitions and Main Result

The Frobenius norm is denoted by $\| \cdot \|$, and the absolute value is denoted by $|\cdot|$. The symbol $1\!\!1_{\{\cdot\}}$ stands for the Dirac measure. We use $\mathrm{Re}(\cdot)$ to represent the real part of a complex number. When A is a square matrix, we let

$$\mathrm{Re}(A) := \{\max(\mathrm{Re}(\lambda_i)), i = 1, \ldots, n : \lambda_i \text{ is an eigenvalue of } A\}.$$

The scheme of the PI controller associated with the Bouc-Wen model is depicted in Fig. 1. The instant times

$$0 < t_1 < t_2 < \cdots < t_k < \cdots$$

denote the time of occurrence of a random failure. When a failure happens, a command to the controller to turn off is emitted and the two constants k_P and k_I vanish to zero. The system remains with the PI controller off during a certain period of time, say $\mu > 0$, in order to guarantee a certain degree of stability for the system. In applications, μ can be a variable chosen for safety requirements and its role in this investigation will be exploited in the sequence. After that off waiting time, the PI controller turns on and this happens precisely at $s_k = t_k + \mu$, see Fig. 2 for an illustration. Notice that s_k and t_k mark the instants for which the PI controller turns its status to 'on' and 'off', respectively. The time interval for which the PI controller keeps on is random, i.e., $t_{k+1} - s_k$ is a random variable. The next assumption sets this property.

(A.1) Assumption: The process $\{t_k\}$ governing the failures of the PI controller follows the arrival times of a homogeneous Poisson process with rate $\lambda > 0$.

Remark 1. After the occurrence of a failure at t_k, the PI controller turns off and keeps in this situation in the time interval $[t_k, s_k)$. No other failure is allowed to happen when the controller is off. Since μ is a deterministic value chosen a

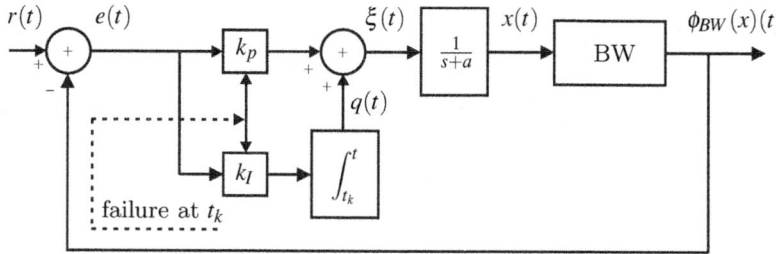

Fig. 1. Scheme diagram representing the Bouc-Wen model with Proportional Integrative controller subject to random failures

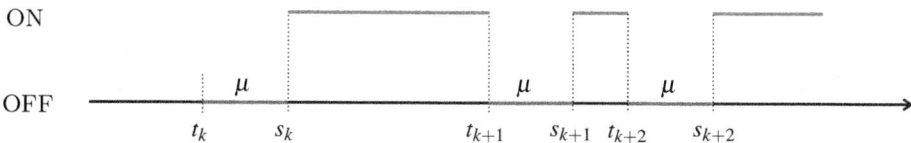

Fig. 2. Status of the PI controller. The controller remains 'off' from t_k to s_k and 'on' from s_k to t_{k+1}.

priori and the controller turns on at s_k, with $s_k = t_k + \mu$, the next failure t_{k+1} may happen at any random instant after s_k. Assumption (A.1) then implies that the inter-arrival times $\delta_k := t_{k+1} - s_k$, $k \geq 0$, are i.i.d. with exponential probability distribution [13, p. 202]

$$\Pr[\delta_k = t] = \lambda e^{-\lambda t}, \quad \forall t \geq 0, \forall k \geq 0.$$

With the Dirac function $\mathbb{1}_{t \in [s_k, t_{k+1})}$ indicating that the controller is on (one) when t lies within the interval $[s_k, t_{k+1})$ and off (zero) otherwise, we can define the PI parameters as

$$k_P(t) = \mathbb{1}_{t \in [s_k, t_{k+1})} k_P \quad \text{and} \quad k_I(t) = \mathbb{1}_{t \in [s_k, t_{k+1})} k_I, \quad \forall t \geq 0,$$

where k_P and k_I are fixed constants.

According to the scheme shown in Fig. 1, we can write the PI controller equations as

$$\xi(t) = k_P(t) e(t) + q(t), \tag{3}$$

$$q(t) = k_I(t) \int_{t_k}^t e(\tau) d\tau, \tag{4}$$

$$e(t) = r(t) - \phi_{BW}(x)(t), \quad \forall t \geq 0. \tag{5}$$

Notice that $x(t)$ satisfies the relation

$$\dot{x}(t) + a x(t) = \xi(t), \quad \forall t \geq 0, \tag{6}$$

where $a > 0$ is a given constant.

Remark 2. Even though the signal $\xi(t)$ presents infinitely many discontinuities when $\lambda > 0$, the solution $x(t)$ from the differential equation (6) is continuous.

At this point, after characterizing the control setup, we present the stability concept investigated in this paper.

Definition 1. *We say the stochastic nonlinear Bouc-Wen model in (1)–(6) is weakly stable in the mean if there exists a sequence of time instants $\{t_k\}$ and a constant $c > 0$ (which does not depend on $\{t_k\}$) such that*

$$|E[\Phi_{BW}(x)(t_k)]| \leq c, \quad \forall k > 0.$$

The next matrix is useful to characterize the main result of this paper:

$$A = \begin{bmatrix} -(a + k_P k_x) & 1 \\ -k_I k_x & 0 \end{bmatrix}. \tag{7}$$

Now we are in position to present the main result of this paper.

Theorem 1. *Let $\{t_k\}$ be a stochastic process representing the occurrence of failures. Then the stochastic nonlinear Bouc-Wen model in (1)–(6) is weakly stable in the mean if and only if $\mathrm{Re}(A) - \lambda < 0$.*

The proof of Theorem 1 is shown in the sequence.

Remark 3. The stability condition $\mathrm{Re}(A) - \lambda < 0$ in Theorem 1 can be determined through the analysis of the characteristic equation

$$p(s) = s^2 + (a + k_I k_x + 2\lambda)s + \lambda(a + k_p k_x) + k_I k_x = 0.$$

According to the Routh-Hurwitz condition [14], the roots of $p(s)$ have negative real parts if and only if

$$a + k_I k_x + 2\lambda > 0 \quad \text{and} \quad \lambda(a + k_p k_x) + k_I k_x > 0 \tag{8}$$

are satisfied. Since the roots of $p(s)$ are also roots of $A - \lambda I$, the condition in (8) is necessary and sufficient for $\mathrm{Re}(A) - \lambda < 0$.

2.1   Proof of Theorem 1

To prove Theorem 1, some preliminary results are necessary.

Proposition 1. *([15, p. 84]). Let A be a matrix of dimension $n \times n$. Then there exists a similarity transformation matrix Z such that*

$$A = ZJ_A Z^{-1},$$

where J_A is the corresponding Jordan form. Moreover, for any scalar c, there holds

$$\exp(cA) = Z\exp(cJ_A)Z^{-1}.$$

Lemma 1. *Let A be a square matrix of dimension two.*

(i) The square matrix $M_1 = \mathrm{E}[\exp(A\delta_k)]$, $\forall k \geq 0$, exists if and only if $\mathrm{Re}(A) - \lambda < 0$.

(ii) Let h be a two dimensional bounded continuous-time signal. The square matrix

$$M_2 = \mathrm{E}\left[\int_0^{\delta_k} \exp(A(\delta_k - \tau))h(\tau)d\tau\right] \tag{9}$$

exists only if $\mathrm{Re}(A) - \lambda < 0$.

Proof. The proof of Lemma 1 is in Appendix.

Proposition 2. *([1]). If the input signal $x(t)$ is bounded and continuous, then*

$$\sup_{t \geq 0} |w(t)| \leq \max\{w(0), 1\}. \tag{10}$$

Proof of Theorem 1 (continued)

The proof follows by an induction argument on k and is divided into two parts.
   Case 1. Controller is on:
In this case t belongs to the interval $[s_k, t_{k+1})$, which results in $k_P(t) = k_P$ and $k_I(t) = k_I$. Substituting (3) into (6), we have

$$\dot{x}(t) = -ax(t) + q(t) + k_P e(t), \quad \forall t \in [s_k, t_{k+1}). \tag{11}$$

It follows from (4) that $\dot{q}(t) = k_I e(t)$ when $t \in [s_k, t_{k+1})$, which together with (1) and (5) allows us to write

$$\dot{q}(t) = k_I(r(t) - k_x x(t) - k_w w(t)), \quad \forall t \in [s_k, t_{k+1}). \tag{12}$$

On the other hand, by applying (1) and (5) in (11), we get

$$\dot{x}(t) = -ax(t) + q(t) + k_P(r(t) - k_x x(t) - k_w w(t)), \quad \forall t \in [s_k, t_{k+1}). \tag{13}$$

By stacking the differential equations (12) and (13), we have

$$\dot{Y}(t) = AY(t) + h(t), \quad \forall t \in [s_k, t_{k+1}) \tag{14}$$

where

$$Y(t) = \begin{bmatrix} x(t) \\ q(t) \end{bmatrix}, \quad A = \begin{bmatrix} -(a + k_P k_x) & 1 \\ -k_I k_x & 0 \end{bmatrix},$$

and

$$h(t) = \begin{bmatrix} k_P \\ k_I \end{bmatrix} r(t) + \begin{bmatrix} -k_P k_w \\ -k_I k_w \end{bmatrix} w(t).$$

The solution of (14) is given by

$$Y(t) = \exp(A(t - s_k))Y(s_k) + \int_{s_k}^{t} \exp(A(t - \tau))h(\tau)d\tau. \tag{15}$$

Taking $t \uparrow t_{k+1}$ in (15), and recalling that $\delta_k = t_{k+1} - s_k$, we have

$$Y(t_{k+1}) = \exp(A\delta_k)Y(s_k) + \int_{0}^{\delta_k} \exp(A(\delta_k - \tau))h(\tau - s_k)d\tau. \tag{16}$$

Passing the expected value operator on both sides of (16) yields

$$\mathrm{E}[Y(t_{k+1})] = \mathrm{E}[\exp(A\delta_k)Y(s_k)] + \mathrm{E}\left[\int_{0}^{\delta_k} \exp(A(\delta_k - \tau))h(\tau - s_k)d\tau\right]$$

$$= \mathrm{E}[\exp(A\delta_k)]\mathrm{E}[Y(s_k)] + \mathrm{E}\left[\int_{0}^{\delta_k} \exp(A(\delta_k - \tau))h(\tau - s_k)d\tau\right] \tag{17}$$

where the last equality follows from the i.i.d property of the Poisson process. Lemma 1 allows us to get from (17) that

$$|\mathrm{E}[Y(t_{k+1})]| \leq \|M_1\| |\mathrm{E}[Y(s_k)]| + \|M_2\|, \quad \forall k \geq 0, \tag{18}$$

where M_1 and M_2 are matrices satisfying Lemma 1.

Notice from the definition of the vector $Y(t)$ that

$$\mathrm{E}[Y(s_k)] = \lim_{t \uparrow s_k} \mathrm{E}\begin{bmatrix} x(t) \\ q(t) \end{bmatrix} = \begin{bmatrix} \mathrm{E}[x(s_k)] \\ 0 \end{bmatrix}. \tag{19}$$

We now introduce the induction argument. Let $k = n = 0$ in (18) and (19) to get that

$$|\mathrm{E}[Y(t_1)]| \leq \|M_1\| + \|M_2\|,$$

since $|x(s_0)| = |x(0)| = 0$.

Case 2. Controller is off:

We now show that

$$|\mathrm{E}[x(s_k)]| \leq 1, \quad \forall k > 0. \tag{20}$$

When the controller is off, t belongs to the interval $[t_k, s_k)$ and this results in $k_P(t) = k_I(t) = 0$. The equations (3)-(6) guarantee that $q(t) = \xi(t) = 0$ and so $\dot{x}(t) + ax(t) = 0$ whenever $t \in [t_k, s_k)$. The solution of this autonomous system is given by

$$x(t) = x(t_k) \exp(-a(t - t_k)), \quad \forall t \in [t_k, s_k). \tag{21}$$

Let us assume for the moment that $|\mathrm{E}[x(t_k)]|$ is bounded above by a constant $c_1 := \max(1, \|M_1\| + \|M_2\|)$ which does not depend on k, where M_1 and M_2 are the matrices as defined in Lemma 1. This assumption applied in (21) yields

$$|\mathrm{E}[x(t)]| = |\mathrm{E}[x(t_k)]| \exp(-a(t - t_k)) \leq c_1 \exp(-a(t - t_k)), \tag{22}$$

when $t \in [t_k, s_k)$.

Recalling that $s_k = t_k + \mu$ for all $k > 0$, we now choose the value of $\mu > 0$ to guarantee the result. Indeed, if we let $\mu > 0$ be large enough such that

$$\mu > \frac{\log(c_1)}{a}, \tag{23}$$

which is equivalent to

$$c_1 \exp(-a\mu) < 1,$$

then one can take $t \uparrow s_k$ in (22) to produce

$$\lim_{t \uparrow s_k} |E[x(t)]| \leq c_1 \exp(-a\mu) < 1. \tag{24}$$

A direct consequence of (24) is that $|E[x(s_1)]| \leq 1$, since the assumption $|E[x(t_k)]| \leq c_1$ is considered valid for $k = n = 1$ according to (19).

Repeating the arguments of Case 1, now with $k = n = 1$, one can show that

$$|E[Y(t_2)]| \leq \|M_1\| + \|M_2\|.$$

And taking this inequality in Case 2 with $k = n = 2$ one gets that $|E[x(s_2)]| \leq 1$. Proceeding similarly, one can conclude that

$$|E[Y(t_k)]| \leq \|M_1\| + \|M_2\|, \quad \forall k > 0. \tag{25}$$

Taking the expected value operator on both sides of (1), we get

$$E[\Phi_{BW}(x)(t)] = k_x E[x(t)] + k_w E[w(t)], \quad \forall t \geq 0. \tag{26}$$

Combining Proposition 2, which assures that $|w(t)| \leq 1$ for all $t > 0$, (25), and (26), we obtain

$$|E[\Phi_{BW}(x)(t_k)]| \leq k_x(\|M_1\| + \|M_2\|) + k_w, \quad \forall k \geq 0,$$

which shows the result.   □

3   Experimental Evaluation

We simulated the stochastic Bouc-Wen model in (1)–(6) with $r(t) = \sin(t)$, $t \geq 0$,

$$k_x = 2, \quad k_w = 2, \quad \rho = 2, \quad n = 1.5, \quad \sigma = 1$$

and

$$k_P = 0.9, \quad k_I = 0.9, \quad a = 1, \quad \mu = 1.$$

The exponential distribution to generate the failures was taken with $\lambda = 2$. Figure 3 shows a sample path for the output $\Phi_{BW}(t)$. It can be seen that after the occurrence of a failure, the system returns to track the desired reference signal.

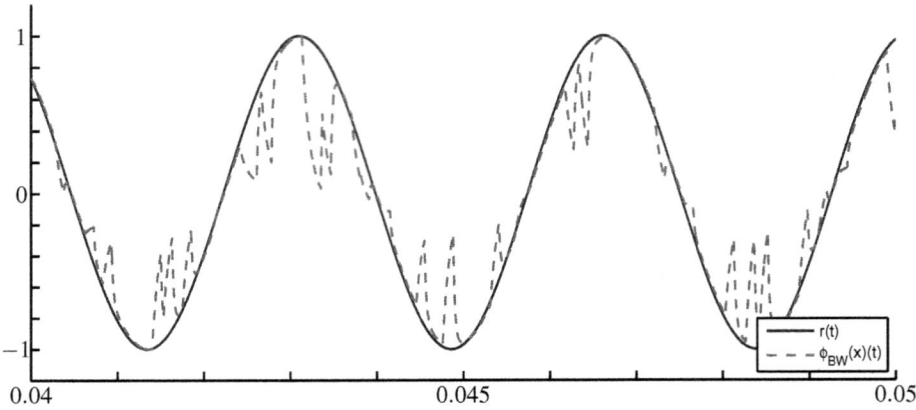

Fig. 3. Response of the Bouc-Wen model with PI controller for a sample path. The continuous line represents the reference $r(t)$ and the dotted one represents the output of the Bouc-Wen model $\Phi_{BW}(x)(t)$.

4 Concluding Remarks

The paper presents necessary and sufficient conditions to guarantee weak stability in mean for the Bouc-Wen model. The conditions basically rely on the analysis of eigenvalues of a two dimensional matrix. Further investigation is under progress to convert the weak stability, valid for a time subsequence $\{t_k\}$ on the real line, into a strong one valid for all times $t > 0$.

Acknowledgements. Research supported in part by the Brazilian agencies CAPES and CNPq and by the Basque Center for Applied Mathematics – "BCAM Visiting fellow program".

Appendix: Proof of Lemma 1

[Proof of (i)]: Recalling that the inter-arrival process $\delta_k = t_{k+1} - s_k$, $k > 0$, obeys a Poisson process, we can write (see Remark 1)

$$M_1 = \mathrm{E}[\exp(A\delta_k)] = \int_0^{+\infty} \exp(At)\Pr[\delta_k = t]dt$$
$$= \lambda \int_0^{+\infty} \exp(At)\exp(-\lambda t)dt. \qquad (27)$$

Using the result of Proposition 1 in (27), we have

$$M_1 = \lambda \int_0^{+\infty} Z\exp(J_A t)Z^{-1}\exp(-\lambda t)dt$$
$$= \lambda \int_0^{+\infty} Z\exp((J_A - \lambda I)t)Z^{-1}dt. \qquad (28)$$

On the other hand, if we denote by σ_1 and σ_2 the two eigenvalues of A, we get that the matrix $J_A - \lambda I$ assumes either the form

$$\begin{bmatrix} \sigma_1 - \lambda & 0 \\ 0 & \sigma_2 - \lambda \end{bmatrix} \quad \text{or} \quad \begin{bmatrix} \sigma_1 - \lambda & 1 \\ 0 & \sigma_1 - \lambda \end{bmatrix}. \tag{29}$$

It follows from (29) that the exponential matrix $\exp((J_A - \lambda I)t)$ is identical to either [14, Ch. 3.6]

$$\begin{bmatrix} \exp((\sigma_1 - \lambda)t) & 0 \\ 0 & \exp((\sigma_2 - \lambda)t) \end{bmatrix}$$

$$\text{or} \quad \begin{bmatrix} \exp((\sigma_1 - \lambda)t) & t\exp((\sigma_1 - \lambda)t) \\ 0 & \exp((\sigma_1 - \lambda)t) \end{bmatrix}. \tag{30}$$

Notice from (30) that the integral $\int_0^{+\infty} \exp((J_A - \lambda I)t)\, dt$ exists if and only if both $\mathrm{Re}(\sigma_1) - \lambda < 0$ and $\mathrm{Re}(\sigma_2) - \lambda < 0$ hold true, which is equivalent to the condition $\mathrm{Re}(A) - \lambda < 0$. This argument completes the proof of (i).

[Proof of (ii)]: A direct evaluation of (9) yields

$$\begin{aligned} M_2 &= \mathrm{E}\left[\int_0^{\delta_k} \exp(A(\delta_k - \tau)) h(\tau) d\tau \right] \\ &= \int_0^{+\infty} \left(\int_0^t \exp(A(t - \tau)) h(\tau) d\tau \right) \lambda \exp(-\lambda t) dt \\ &= \lambda \int_0^{+\infty} \exp(At) \exp(-\lambda t) \left(\int_0^t \exp(-A\tau) h(\tau) d\tau \right) dt. \end{aligned}$$

Applying Proposition 1 in the above identity, we have

$$\begin{aligned} M_2 &= \lambda \int_0^{+\infty} Z\exp((J_A - \lambda I)t)Z^{-1} \left(\int_0^t Z\exp(-J_A\tau)Z^{-1} h(\tau) d\tau \right) dt \\ &= \lambda \int_0^{+\infty} Z\exp((J_A - \lambda I)t) \left(\int_0^t \exp(-J_A\tau)Z^{-1} h(\tau) d\tau \right) dt. \end{aligned} \tag{31}$$

After some algebraic manipulation, we get that (31) is identical to

$$M_2 = \int_0^{+\infty} \Gamma(t) \left(\int_0^t h(\tau) d\tau \right) dt, \tag{32}$$

where $\Gamma(t)$ is two-dimensional square matrix where all of its entries are bounded from above and below by a term in the form $t\exp(-\rho t)$, $\rho > 0$. Combining this and the assumption that $h(\cdot)$ is bounded, we can conclude from (32) that

$$\begin{aligned} \|M_2\| &\leq \int_0^{+\infty} \|\Gamma(t)\| \left(t \sup_{s \geq 0} \|h(s)\| \right) dt \\ &\leq \sup_{s \geq 0} \|h(s)\| \int_0^{+\infty} t^2 \exp(-\rho t) dt = \frac{2\sup_{s \geq 0} \|h(s)\|}{\rho^3}, \end{aligned}$$

which shows the result. $\qquad\square$

References

1. Ikhouane, F., Rodellar, J.: On the hysteretic Bouc-Wen model. Nonlinear Dynamics 42(1), 63–78 (2005)
2. Ikhouane, F.: Characterization of hysteresis processes. Math. Control Signals Syst. (2013), doi:10.1007/s00498-012-0099-6
3. Rochdi, Y., Giri, F., Ikhouane, F., Chaoui, F.Z., Rodellar, J.: Parametric identification of nonlinear hysteretic systems. Nonlinear Dynamics 58, 393–404 (2009)
4. Ikhouane, F., Gomis, O.: A limit cycle approach for the parametric identification of hysteretic systems. Systems Control Lett. 57, 663–669 (2008)
5. Ikhouane, F., Mañosa, V., Rodellar, J.: Dynamic properties of the hysteretic Bouc-Wen model. Systems Control Lett. 56, 197–205 (2007)
6. Ikhouane, F., Rodellar, J.: A linear controller for hysteretic systems. IEEE Trans. Automat. Control 51(2), 340–344 (2006)
7. Gomis-Bellmunt, O., Ikhouane, F., Castell-Vilanova, P., Bergas-Jané, J.: Modeling and validation of a piezoelectric actuator. Electrical Engineering 89(8), 629–638 (2007)
8. Ismail, M., Ikhouane, F., Rodellar, J.: The Bouc-Wen model: A survey. Arch. Comput. Methods Eng. 16, 161–188 (2009)
9. Ikhouane, F., Rodellar, J.: Systems with hysteresis: Analysis, identification and control using the Bouc-Wenmodel. John Wiley & Sons, Chichester (2007)
10. Oriol Gomis-Bellmunt, F.I., Montesinos-Miracle, D.: Control of Bouc-Wen hysteretic systems: Application to a piezoelectric actuator. In: 13th Power Electronics and Motion Control Conference, pp. 1670–1675 (2008)
11. Yao, G.Z., Yap, F.F., Chen, G., Li, W.H., Yeo, S.H.: MR damper and its application for semi-active control of vehicle suspension system. Mechatronics 12(7), 963–973 (2002)
12. Ikhouane, F., Nosa, V.M., Rodellar, J.: Adaptive control of a hysteretic structural system. Automatica 41(2), 225–231 (2005)
13. Ross, S.M.: Introduction to probability models. Acadmic Press (1985)
14. Chen, C.-T.: Linear System Theory and Design, 3rd edn. Oxford University Press, Inc., New York (1998)
15. Hirsch, M., Smale, S.: Differential Equations, Dynamical Systems, and Linear Algebra. Academic Press, Inc. (1974)

On the Study of Complex Impulsive Systems

Dmitry Karamzin[1], Valeriano de Oliveira[1], Geraldo Silva[1],
and Fernando L. Pereira[2]

[1] IBILCE-UNESP, São José do Rio Preto, Brazil
[2] Institute for Systems and Robotics Porto, FEUP, University of Porto, Portugal
dmitry_karamzin@mail.ru, {antunes,gsilva}@ibilce.unesp.br, flp@fe.up.pt

Abstract. The article investigates complex impulsive systems in which the so-called controlling systems jumps effect emerges. In particular, this research includes the correctness of the solution to the impulsive control system and approximation lemmas. A 3D model example is provided which illustrates the relevance of the considered approach to the study of complex impulsive systems.

Keywords: optimal control, impulsive control, correctness of solution.

1 Introduction

Impulsive controls are widely used in various engineering applications, in particular space navigation where one of the goals has generally been to minimize fuel consumption [1]. Impulsive controls, then, are helpful in the analysis since they provide mathematically precise solutions to the extended problem in which the fuel consumption rate is allowed to take unbounded values. The latter assumption is quite realistic because the process of fuel consumption in jet propulsion systems has an explosive character. (That is, when a considerable quantity of fuel is consumed within short time intervals and, therefore, the value of the consumption rate can be very high.) Consideration of impulsive controls simplifies the mathematical analysis of the problem and, in particular, by facilitating the application of the optimality conditions easier.

In this work we investigate impulsive control processes of a specific nature. Their main feature is that the state trajectory can be controlled while the impulse develops. This means that there are controls of the conventional type that enable to steer the trajectory "inside" of the system's jump. Examples of such processes can be found in problems related to aerospace flights where the mass of a flying object may change while the impulse due to the fuel consumption. Since the mass changes, the center of gravity also changes. However, this quick alteration in the parameters of the system requires correction in control during the time of the impulse. In this way, conventional controls attached to the system jumps appear. The changing of the center of gravity or of the mass distribution can also disturb the rotation motion of the object. For example, unwanted rotations may arise. Then, it is important to know the right control mode for the new mass parameters in order to save fuel and, at the same time, to keep the flying object with the non-perturbed rotation. A real-life example of such a system is depicted

Fig. 1. RCT blocks on the Apollo Lunar Module
(Source:http://www.hq.nasa.gov/office/pao/History/alsj/a14/AS14-66-9254HR)

in Fig. 1. In other words, it is important to provide a correct joint control of the motion in the two modes: translation and orientation. In [10–12], the concept of impulsive control used in this work was introduced and developed, particularly in [11] where the issue of controlling system jumps was firstly discussed.

Impulsive control problems have been investigated in several different frameworks by many researchers, among which, we single out, A. Bressan and F. Rampazzo, [2], B. Miller and E. Rubinovich, [3], V. Dykhta and O. Samsonyuk, [4], A. Kurzhanskii and A. Daryin, [5], P. Wolenski and S. Zabic, [6], W. Code and P. Loewen, [7], R. Vinter and G. Silva, [8], R. Vinter and F. Pereira, [9]. This is by no means an exhaustive list but includes some of the most relevant works to this article.

In section 2 we formulate the impulsive control problem in detail and in section 3 we present results on the solution correctness. A model example is discussed in section 4 before the conclusions.

2   Impulsive Control Problem

In this section, we state the impulsive control problem which can be regarded as an extension of the conventional optimal control problem, [13].

The functional to be minimized is given by $c(p)$, where $c : \mathbb{R}^{2n} \to \mathbb{R}^1$ and $p = (x_0, x_1)$, with $x_0 = x(t_0)$, $x_1 = x(t_1)$, is the vector of both endpoints of the trajectory, and the constraints are given by

$$\begin{cases} dx = f(x, u, t)dt + g(x, u, t)d\vartheta, \\ u(t) \in U, \ \text{range}(\vartheta) \subset K, \\ p \in S, \ t \in T, \end{cases} \tag{1}$$

where $f : \mathbb{R}^n \times \mathbb{R}^m \times \mathbb{R}^1 \to \mathbb{R}^n$, and $g : \mathbb{R}^n \times \mathbb{R}^m \times \mathbb{R}^1 \to \mathbb{R}^{n \times k}$ are given continuous functions, the closed sets $T = [t_0, t_1] \subset \mathbb{R}$, $S \subset \mathbb{R}^{2n}$, $U \subset \mathbb{R}^m$, and $K \subset \mathbb{R}^k$ are, respectively, the given fixed time interval, endpoint state constraint set, control pointwise constraint set, and a nonempty convex cone, and ϑ is the impulsive control, [11], whose total variation measure is designated by $|\vartheta|$. Now, we recall from [11] the concept of impulsive control that we adopt for further developments of this article.

Consider the Borel vector measure μ such that range$(\mu) \subset K$, where K be a nonempty closed convex cone in $\mathbb{R}^k$, and let $V(\mu)$ be the set of scalar-valued nonnegative Borel measures ν such that $\exists \mu_i$ satisfying range$(\mu_i) \subset K$ and $(\mu_i, |\mu_i|) \overset{*}{\rightarrow} (\mu, \nu)^1$. Note that if K is contained in one of the orthants, then $V(\mu) = \{|\mu|\}$ (single-point set).

Given an arbitrary scalar measure $\nu \in V(\mu)$ and a number $\tau \in T$, we say that a family of vector functions $\{v_\tau\}$, depending on the real parameter $\tau \in T$, is said to be attached to the vector measure (μ, ν), if, for every τ, there is some measurable vector function $v_\tau : [0, 1] \to K$, such that,

$$\sum_{j=1}^{k} |v_\tau^j(s)| = \nu(\tau), \text{ a.a. } s \in [0, 1], \text{ and } \int_0^1 v_\tau^j(s) ds = \mu^j(\tau), \ j = 1, .., k.$$

Definition 1. *The impulsive control is a triple $\vartheta = (\mu, \{u_\tau, v_\tau\})$, where $\{u_\tau\}$ is a family of measurable and uniformly bounded in τ vector functions defined on the segment $[0, 1]$ with values in $\mathbb{R}^m$, if there exists $\nu \in V(\mu)$ such that $\{v_\tau\}$ is attached to (μ, ν). The measure ν is called variation of the impulsive control ϑ and is denoted by $|\vartheta|$.*

The ordinary control function $u(\cdot)$ is assumed to be measurable and essentially bounded w.r.t. both Lebesgue ℓ and Lebesgue-Stieltjes $|\vartheta|$ measures.

Take impulsive control $\vartheta = (\mu, \{u_\tau, v_\tau\})$, a number $\tau \in T$ and a vector $x \in \mathbb{R}^n$. Denote by $x_\tau(\cdot) = x_\tau(\cdot, a)$ the solution to the following system $\dot{x}_\tau(s) = g(x_\tau(s), u_\tau(s), \tau) v_\tau(s)$, $s \in [0, 1]$, $x_\tau(0) = a$.

The function of bounded variation $x(t)$ on the interval T is called solution to the differential equation in (1), corresponding to the control (u, ϑ) and the starting value x_0, if $x(t_0) = x_0$ and, for every $t \in (t_0, t_1]$,

$$x(t) = x_0 + \int_{t_0}^t f(x, u, \tau) d\tau + \int_{[t_0, t]} g(x, u, \tau) d\mu_c + \sum_{\tau \le t} \left[x_\tau(1, x(\tau^-)) - x(\tau^-) \right] \quad (2)$$

where μ_c designates the continuous component of μ. Note, that the sum in (2) is well defined since there is only a countable set of points τ, where v_τ is nonzero.

The collection (x, u, ϑ) is called control process, if (2) holds. A control process is said to be admissible, if all the constraints of problem (1) are satisfied. An admissible process $(\hat{x}, \hat{u}, \hat{\vartheta})$ is said to be optimal if, for any admissible process (x, u, ϑ), the inequality $c(\hat{p}) \le c(p)$ holds, where $\hat{p} = (\hat{x}(t_0), \hat{x}(t_1))$.

3 Solution Correctness

In this section, we present key results demonstrating that the solution concept proposed in Section 2 for the dynamic impulsive control system

$$dx = f(x, u, t) dt + g(x, u, t) d\vartheta. \quad (3)$$

---

[1] Here, the symbol $\overset{*}{\rightarrow}$ means weak*-convergence, i.e. all the coordinates of μ_i converge in $C^*(T)$ endowed with its weak star topology, and also the variation measure $|\mu_i|$ converges weakly* to ν, where $C^*(T)$ is the dual space to $C(T)$.

is correct both in the classic sense and with respect to some metric ρ defined on the space $\mathcal{P}$ of all feasible control pairs $\wp = (u, \vartheta)$ as follows:

$$\rho(\wp_1, \wp_2) := |\nu_1(T) - \nu_2(T)| + \int_{t_0}^{t_1} |F(t, \nu_1) - F(t, \nu_2)| dt + \max_{s \in [0,1]} \Big\{ |\operatorname{Ext}[\zeta_1(\cdot), \vartheta_1](s)$$
$$- \operatorname{Ext}[\zeta_2(\cdot), \vartheta_2](s)| \Big\} + \int_0^1 |\operatorname{Ext}[u_1(\cdot), \vartheta_1](s) - \operatorname{Ext}[u_2(\cdot), \vartheta_2]| ds.$$

Here, $\nu = |\vartheta|$, $F(t, \nu) = \nu([t_0, t])$, $d\zeta_j = d\vartheta_j$, $\zeta_j(t_0) = 0$, $j = 1, 2$, $\operatorname{Ext}[\phi, \vartheta](s) :=$ $\phi(x_\tau(\xi_\tau(s)), u_\tau(\xi_\tau(s)), \tau)$ for $\tau \in \operatorname{Ds}(|\vartheta|)$, and $s \in \Gamma_\tau$, and $\phi(x(\theta(s)), u(\theta(s)), \theta(s))$ otherwise, and $\xi_\tau(s) = \frac{s - \pi(\tau^-)}{\ell(\Gamma_\tau)} : \Gamma_\tau \to [0, 1]$ where $\Gamma_\tau = \pi^{-1}(\{\tau\})$, being π an appropriate time reparametrization (for details see [11]).

This correctness is critical to ensure a consistent passing to the limits and to construct adequate approximations to impulsive controls.

The following lemma concerns the correctness in the Cauchy sense and is a simple generalization of classical results.

Lemma 1. *Let $x_0 \in \mathbb{R}^n$ and $(u, \vartheta) \in \mathcal{P}$. Suppose $m(t) : T \to \mathbb{R}^1$ be an integrable function such that:*

$$|f(x, u, t)| + |g(x, u, t)| \leq m(t)(1 + |x|), \quad \forall x, u, t \in \mathbb{R}^n \times U \times T. \tag{4}$$

Then, on the interval T, there exists a unique solution $x(\cdot)$ to (3), corresponding to the control $(u, \vartheta) \in \mathcal{P}$ and to the starting value $x(t_0) = x_0$.

If (4) does not hold, but f, g are C_1 in x, and $t_0 \notin \operatorname{Ds}(|\vartheta|)$, $\exists \delta > 0$ such that the solution $x(\cdot)$ exists only locally on the interval $[t_0, t_0 + \delta]$.

Moreover, small perturbations of x_0 produce small deviations of the solution

$$|x(t) - x'(t)| \leq \operatorname{const} |x_0 - x_0'| \; \forall t \in T, \tag{5}$$

where $x'(t)$ is the solution to (3) corresponding to the starting value x_0' and the same control $(u, \vartheta) \in \mathcal{P}$.

Now, let us see the possibility of correct passing to limits in the nonlinear impulsive control systems of the type (3).

Lemma 2. *Let $(u_i, \vartheta_i) \xrightarrow{\rho} (u, \vartheta)$, where (u_i, ϑ_i), (u, ϑ) are from $\mathcal{P}$. Assume that $x_{0,i} \to x_0 \in \mathbb{R}^n$ and $\operatorname{Ext}[u_i, \vartheta_i](s)$ are uniformly bounded. Assume (4) be in force. Denote by $x_i(\cdot)$ the solution to (3), corresponding to the control (u_i, ϑ_i) and the starting value $x_{0,i}$, which by means of Lemma 1 exists on $T \; \forall i$.*

Then, $\operatorname{Ext}[x_i, \vartheta_i](s) \rightrightarrows \operatorname{Ext}[x, \vartheta](s)$ uniformly in $s \in [0, 1]$, where $x(\cdot)$ is the solution to (3), corresponding to the control $(u, \vartheta) \in \mathcal{P}$ and the starting value $x(t_0) = x_0$. The latter implies $x_i(t) \to x(t) \; \forall t \in (T \setminus \operatorname{Ds}(|\vartheta|)) \cup \{t_0, t_1\}$.

The following lemma shows how, with the help of an integral functional, it becomes possible to approximate impulsive controls by conventional controls. Such lemma would be useful at proofs of various results in impulsive control theory and also for applications at theoretical justification of the convergence of approximating sequences (constructed via solving auxiliary linear-quadratic problems).

Lemma 3. *Let* $(\bar{u}_i, \bar{v}_i) \xrightarrow{\rho} (\bar{u}, \bar{\vartheta})$, *and*

$$\int_{t_0}^{t_1} \left(|y_i(t) - \bar{y}_i(t)|^2 + |u_i(t) - \bar{u}_i(t)|^2 \right) m_i(t) dt \to 0, \tag{6}$$

where $y_i(t) = \int_{t_0}^{t} (v_i, |v_i|) d\tau$, $y_i(t) = \int_{t_0}^{t} (\bar{v}_i, |\bar{v}_i|) d\tau$, $m_i(t) = 1 + \frac{|v_i(t)| + |\bar{v}_i(t)|}{2}$, *functions* u_i, $\bar{u}_i$, v_i, *and* $\bar{v}_i$ *are of class* $\mathbb{L}_\infty(T)$, *and functions* u_i *are uniformly bounded. Then,* $(u_i, v_i) \xrightarrow{\rho} (\bar{u}, \bar{\vartheta})$.

4   Model Example

Consider a flying ball in space $\mathbb{R}^3$. The ball has four thrusters located at points A, B, C, D as shown in the right of Fig. 2. Geometrically, $ABCD$ is a regular tetrahedron inscribed in the sphere. Assign numbers 1, 2, 3 and 4 to the thrusters at points A, B, C and D, respectively. Denote by O the center of the ball. For the center of gravity G we assume that $G \in \text{int } ABCD$.

Consider a coordinate system with the origin in the center O and axes as in the left of Fig. 2. Axes are chosen such that point A belongs to Oz and the projection of point B onto plane Oxy belongs to Ox. It is assumed that the ball has a fuel container, see in the left of Fig. 2. The center of gravity location G is a function of the fuel margin: $G := G(m)$. When the container is empty, the center of gravity location coincides with the center of the ball $G = O$. But, when the fuel is present, point G slides down by the $0z$ axis. The law of the center of gravity alteration is not specified precisely, but, in general, it can be any smooth function depending on m with the values in the interior of $ABCD$ (not necessary concentrated in axis Oz).

For each thruster, the controls are the direction l_i of the propulsion force that is given by angles θ_i, φ_i, $i = 1, .., 4$ (thruster inclination, see in the right of Fig. 3) and also the power coefficient p_i, $i = 1, .., 4$. To be more precise, for the

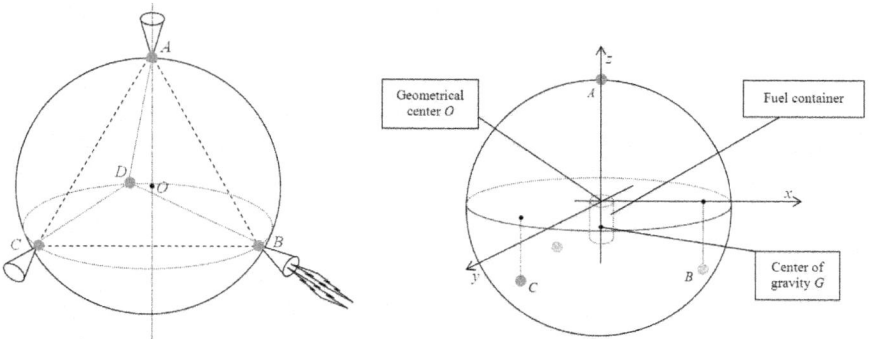

Fig. 2. Left: Four thrusters. Right: Coordinate system $Oxyz$ and the fuel container.

thruster number 1, angles θ_1 and φ_1 are exactly as in the right of Fig. 3, that is $\theta_1 = \angle OAl_1$, $\varphi_1 = \angle xOl_1'$, where the sign prime $'$ means the projection onto plane Oxy. For the thruster number 2, we have that $\theta_2 = \angle GBl_2$, and angle φ_2 is specified similarly by the projection of line l_2 onto plane $\{BG\}^\perp$ in which axes are chosen in some fixed way. (For example, in such a way that axis x is collinear to the projection of vector GO onto $\{BG\}^\perp$ and the orientation of axes is counter clockwise with respect to vector BG). Angles θ_3, φ_3 and θ_4, φ_4 are specified in the same way as for the thruster at point B.

The vector of the propulsion force generated by each thruster has to be directed inside of or, at least, to be tangent to the sphere. Therefore, for example, for the thruster at point A it should hold that $0 \leq \theta_1 \leq \pi/2$, $0 \leq \varphi_1 \leq 2\pi$. However, it is easy to see that already for the thruster at point B angle θ_2 cannot be $\pi/2$ for every φ_2. Indeed, otherwise for some φ_2 the propulsion line l_2 is directed out of the sphere. Then, obviously, angle θ_2 has to be restricted by the angle φ_2 and this restriction should also depend on mass m since the location of the center of gravity, point G depends on m. Thus, for some smooth function f_2, we have the following constraints

$$0 \leq \theta_2 \leq f_2(m, \varphi_2), \ \ 0 \leq \varphi_2 \leq 2\pi.$$

It is not specified, but this is a straightforward stereometric task to find f_2, and, in general, by setting $f_1 = \pi/2$, the constraints for the control angles become

$$0 \leq \theta_i \leq f_i(m, \varphi_i), \ \ 0 \leq \varphi_i \leq 2\pi, \ \ i = 1, .., 4.$$

Constraints $\theta_i \leq f_i(m, \varphi_i)$ are called mixed because they relate state coordinate (mass) and controls. Note that the thrusters work in the orientation mode only (with the zero translation) always corresponds to the angle value $\theta_i = \pi/2$. However, this "pure orientation" mode is not always achievable for any direction of the angular velocity due to the technological restrictions induced by the mixed constraints.

In what concerns the power coefficients p_i, they should be nonnegative and should satisfy the norm equation $p_1 + p_2 + p_3 + p_4 = 1$.

As it was mentioned above the translation thrusters must satisfy the restriction of the force line intersecting the center of gravity, see in the left of Fig. 3. This guarantees zero rotation (no fuel spent for rotation). Such driving mode has a low control authority, and, thus, it does not allow us to reorient the ball. In the case when it is necessary to reorient and to translate the ball at the same time, the force line does not necessarily intersect the center of gravity. Then, it can be decomposed in two orthogonal components one of which passes through the center of gravity and responds for translation, and the other one responds for rotation.

The first aim now is to find equations of translation and rotation motions. Let $e_i(m)$ be unit vectors directed from the thrusters to the center of gravity, i.e. for instance $e_2(m) = BG/|BG|$. These vectors are defined in coordinate system $Oxyz$ related with the ball (that is in the rotating system). However, in order to write the equations of translation motion we also need to use a coordinate

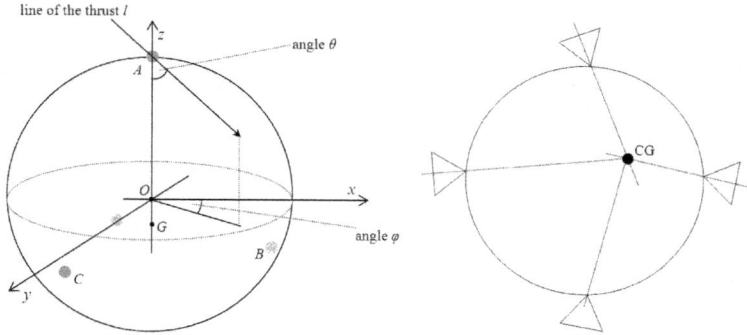

Fig. 3. Left: Angles θ and φ. Right: Force lines passing through the CG.

system that is not related with the ball and in which the ball is a particle. Consider such a fixed system $0x_1x_2x_3$. Let α denote the vector angle specified by the orientation of the ball in system $0x_1x_2x_3$. More precisely, vector α has three components: α_{23}, α_{13}, α_{12}, the rotation angles about the axes $O + 0x_1$, $O + 0x_2$, $O + 0x_3$ in planes $O + 0x_2x_3$, $O + 0x_1x_3$, $O + 0x_1x_2$ correspondingly; $\alpha = 0$ if $Oxyz = O + 0x_1x_2x_3$. The positive rotations are counter-clockwise.

Denote by $T(\alpha)$ the following matrix of rotation by α

$$
\begin{bmatrix} \cos\alpha_{12} & -\sin\alpha_{12} & 0 \\ \sin\alpha_{12} & \cos\alpha_{12} & 0 \\ 0 & 0 & 1 \end{bmatrix} \times \begin{bmatrix} \cos\alpha_{13} & 0 & -\sin\alpha_{13} \\ 0 & 1 & 0 \\ \sin\alpha_{13} & 0 & \cos\alpha_{13} \end{bmatrix} \times \begin{bmatrix} 1 & 0 & 0 \\ 0 & \cos\alpha_{23} & -\sin\alpha_{23} \\ 0 & \sin\alpha_{23} & \cos\alpha_{23} \end{bmatrix}
$$

Note that $Oxyz = O + T(\alpha)0x_1x_2x_3$.

Now the equations of translation motion are:

$$
\dot{x} = w, \quad \dot{w} = g(x) + T(\alpha) \sum_{i=1}^{4} p_i \cos\theta_i e_i(m) \cdot \frac{v}{m}, \quad \dot{m} = -v.
$$

Here, $x = (x_1, x_2, x_3)$, $w = (w_1, w_2, w_3)$, m, v, and $g(x)$, are, respectively, the particle position, the particle velocity, the particle mass, the mass of the ball which is not constant and changes abruptly in time while the fuel is consumed, the nonnegative fuel consumption rate, and the vector function specifying the gravitational field providing the gravity acceleration at point x of the space. Above, the absolute value of the propulsion force by the thruster is proportional to its fuel consumption rate.

For thruster i, in the coordinate system $Oxyz$, consider the rotation matrix $U_i(m)$ that maps plane Oxy into the corresponding thruster tangent plane in such a way that vector $U_i(m) \times (\cos\varphi_i, \sin\varphi_i, 0)^\top$ is collinear with the orientation component of the thrust. In other words, $U_1(m) = E$, where E is identity matrix, $U_2(m)$ maps Oxy onto $\{BG\}^\perp$ with the required axes orientation and so on. Let us put $d_i(m, \varphi_i) = U_i(m) \times (\cos\varphi_i, \sin\varphi_i, 0)^\top$. Denote the distance between thruster i and point G by $r_i(m)$; the moment of inertia matrix about the axes

$G + 0x_1$, $G + 0x_2$, $G + 0x_3$ by $I(m)$; the angular velocity with respect to the center of gravity by ω. The sum of the torques by the forces is given by $\tau = T(\alpha) \sum_{i=1}^{4} p_i r_i(m) \sin \theta_i \left[d_i(m, \varphi_i) \times e_i(m) \right] \cdot v$, where denotation $[a \times b]$ is used for the vector product of a and b.

For the differential of the angular momentum, we have

$$dL = I(m + dm)(\omega + d\omega) - \sum_{i=1}^{4} p_i r_i^2(m)\omega dm - I(m)\omega,$$

where, in the computitation of the angular momentum of the consumed fuel, it was also used the fact that, for two orthogonal vectors a and b, it holds that $a = [b \times [a \times b]]$. By conservation of the angular momentum, it follows that $dL = \tau dt$. Then, by using that $\dot{m} = -v$, and writing $I'(m) = dI/dm$ as the derivative of the moment of inertia matrix with respect to mass, the following equation for the orientation angle α is obtained:

$$\dot{\alpha} = \omega,$$

$$\dot{\omega} = I^{-1}(m)\left(I'(m)\omega - \sum_{i=1}^{4} p_i r_i^2(m)\omega \right.$$

$$\left. + T(\alpha) \sum_{i=1}^{4} p_i v r_i(m) \sin \theta_i \left[d_i(m, \varphi_i) \times e_i(m) \right] \right) \cdot v,$$

Now, by means of the equations of motion derived in the previous section, the optimal fuel consumption problem associated with the motion of the ball from point A to point B takes the following form:

$$\left\{ \begin{array}{l} \text{Minimize} \quad m(0) \\ \text{subject to } \dot{x} = w, \ x(0) = A, \ x(1) = B, \\ \dot{w} = g(x) + T(\alpha) \sum_{i=1}^{4} p_i \cos \theta_i e_i(m) \cdot \dfrac{v}{m}, \ w(0) = w_A, \ w(1) = w_B, \\ \dot{m} = -v, \ m(1) = M \\ \dot{\alpha} = \omega, \ \alpha(0) = \alpha_A, \ \alpha(1) = \alpha_B, \\ \dot{\omega} = I^{-1}(m)\left(I'(m)\omega - \sum_{i=1}^{4} p_i r_i^2(m)\omega \right. \\ \qquad \left. + T(\alpha) \sum_{i=1}^{4} p_i r_i(m) \sin \theta_i \left[d_i(m, \varphi_i) \times e_i(m) \right] \right) \cdot v, \\ w(0) = w_A, \ w(1) = w_B, b, \ 0 \le \theta_i \le f_i(m, \varphi_i), \ 0 \le \varphi_i \le 2\pi, \\ \sum p_i = 1, \ p_i \ge 0, \ v \ge 0, \ i = 1, .., 4, \end{array} \right. \tag{7}$$

where M is the mass of the ball without fuel.

Note, the control v is assumed unbounded. Therefore, we expect that problem (7) has no solution in the class of ordinary bounded controls. However, the solution can be found in the class of impulsive controls.

By setting the problem (7) in the impulsive framework as stated in Section 2, we come to the following impulsive control problem:

$$
\begin{cases}
\text{Minimize} \quad m(0) \\
\text{subject to} \quad \dot{x} = w, \quad x(0) = A, \quad x(1) = B, \\
\qquad dw = g(x) + T(\alpha)\dfrac{\sum_{i=1}^{4} p_i \cos\theta_i e_i(m)}{m} d\vartheta, \quad w(0) = w_A, \ w(1) = w_B, \\
\qquad dm = -d\vartheta, \quad m(1) = M \\
\qquad \dot{\alpha} = \omega, \quad \alpha(0) = \alpha_A, \ \alpha(1) = \alpha_B, \\
\qquad d\omega = I^{-1}(m)\left(I'(m)\omega - \sum_{i=1}^{4} p_i r_i^2(m)\omega \right. \\
\qquad\qquad \left. + T(\alpha)\sum_{i=1}^{4} p_i r_i(m)\sin\theta_i [d_i(m,\varphi_i) \times e_i(m)] \right) d\vartheta, \\
\qquad \omega(0) = \omega_A, \ \omega(1) = \omega_B, b, \ 0 \le \theta_i \le f_i(m,\varphi_i), \ 0 \le \varphi_i \le 2\pi, \\
\qquad 0 \le \theta_i \le f_i(m,\varphi_i), \ 0 \le \varphi_i \le 2\pi, \ \sum p_i = 1, \ p_i \ge 0, \ i = 1,..,4, \\
\qquad \vartheta = (\{(\varphi_{i,\tau}, \theta_{i,\tau}, p_{i,\tau})\}|_{\tau\in[0,1], i=1,..,4}, \mu), \ \mu \ge 0.
\end{cases}
\tag{8}
$$

Here, $\vartheta = (\{(\varphi_{i,\tau}, \theta_{i,\tau}, p_{i,\tau})\}|_{\tau\in[0,1], i=1,..,4}, \mu)$ is impulsive control. Velocity, angular velocity and mass trajectories are discontinuous functions now. For example, velocity jump at point τ such that $\mu(\{\tau\}) > 0$ is computed by means of adjoint controls and adjoint system on segment $[0,1]$ as follows

$$
\dot{w}_\tau(s) = g(x(\tau)) + T(\alpha(\tau))\sum_{i=1}^{4} p_{i,\tau}(s)\cos\theta_{i,\tau}(s)e_i(m_\tau(s))m_\tau(s)^{-1}\mu(\{\tau\}),
$$
$$
\dot{m}_\tau(s) = -\mu(\{\tau\}),
$$

where $w_\tau(0) = w(\tau^-)$, $m_\tau(0) = m(\tau^-)$, $w_\tau(1) = w(\tau^+)$, $m_\tau(1) = m(\tau^+)$ and $f(\tau^-)$, $f(\tau^+)$ denotes left- and right-hand limits of f at point τ.

5   Conclusions

In this article we discussed an abstract framework for the optimal impulsive control of dynamic systems in which rapid variations might occur in the trajectory. This discussion follows previous work of some of the authors [11], and was centered in the correctness of the adopted solution concepts. A simple example but yet compelling example in space navigation is included in order to illustrate the usefulness of the proposed framework in both problem modeling, amenability to the application of optimality conditions (see [11]) and, in the numerical perspective, for the approximation by conventional optimal control problems.

Acknowledgments. The first, second and third authors were supported by CNPq (Brazil) "Sem Fronteiras" grant no 401689/2012-3. The last author was supported by FCT (Portugal) grants Incentivo/EEI/UI0147/2014, and FCT funded project PEst-OE/EEI/UI0147/2014.

References

1. Lawden, D.F.: Optimal Trajectories for Space Navigation. Butterworth, London (1963)
2. Bressan, A., Rampazzo, F.: On differential systems with vector-valued impulsive controls. Boll. Un. Matematica Italiana 2-B, 641–656 (1988)
3. Miller, B., Rubinovich, E.: Impulsive Control in Continuous and Discrete-Continuous Systems. Springer (2002)
4. Dykhta, V.A., Samsonyuk, O.N.: Optimal Impulse Control with Applications, Moscow, Fizmatlit (2000) (in Russian)
5. Kurzhanski, A.B., Daryin, A.N.: Dynamic Programming for Impulse Controls. Annual Reviews in Control 32(2), 213–227 (2008)
6. Wolenski, P.R., Zabic, S.: A Sampling Method and Approximation Results for Impulsive Systems. SIAM J. Control Optim. 46(3), 983–998 (2007)
7. Code, W.J., Loewen, P.D.: Optimal Control of Non-convex Measure-driven Differential Inclusions. Set-Valued Anal. 19, 203–235 (2011)
8. Silva, G.N., Vinter, R.B.: Measure Driven Differential Inclusions. J. Math. Analysis and Applic. 202, 727–746 (1996)
9. Vinter, R.B., Pereira, F.L.: A Maximum Principle for Optimal Processes with Discontinuous Trajectories. SIAM J. Control Optim. 26, 205–229 (1988)
10. Arutyunov, A.V., Karamzin, D.Y., Pereira, F.L.: On constrained impulsive control problems: controlling system jumps. Journal of Mathematical Sciences 165(6), 654–687 (2010)
11. Arutyunov, A.V., Karamzin, D.Y., Pereira, F.L.: Pontryagin's maximum principle for constrained impulsive control problems. Nonlinear Analysis, Theory, Methods and Applications 75(3), 1045–1057 (2012)
12. Arutyunov, A.V., Karamzin, D.Y., Pereira, F.L.: On the extension of classical calculus of variations and optimal control to problems with discontinuous trajectories. In: 51st IEEE Conf. on Decision and Control, Maui, Hawaii, December 10-13, pp. 6406–6411 (2012)
13. Pontryagin, L.S., Boltyanskii, V.G., Gamkrelidze, R.V., Mishchenko, E.F.: Mathematical Theory of Optimal Processes, Moscow, Nauka (1983)

Part II
Model Predictive Control

Promoting Collaborative Relations at Intermodal Hubs Using an Iterative MPC Approach

João Lemos Nabais[1,3,*], Luís F. Mendonça[2,3], and Miguel Ayala Botto[3]

[1] Department of Informatics and Systems Engineering,
Polytechnical Institute of Setúbal, Setúbal, Portugal
`joao.nabais@estsetubal.ips.pt`
[2] Escola Superior Náutica Infante D. Henrique, Department of Marine Engineering,
Paço d'Arcos, Portugal
`luismendonca@enautica.pt`
[3] LAETA, IDMEC, Instituto Superior Técnico, Universidade de Lisboa, Lisboa,
Portugal
`ayalabotto@tecnico.ulisboa.pt`

Abstract. Intermodal hubs are part of transportation networks in a world scale. All partners in a transportation network should contribute to move cargo from a source to a final destination at the agreed time. With the increase of globalization, source and final destinations are becoming far distant from each other. This geographical separation leads to an increase of transportation cargo demand despite the current economic crisis. The increase in freight commerce creates a challenge to transportation networks: the same infrastructure (intermodal hubs and transport providers) has to be able to move a higher amount of cargo without compromising the client satisfaction. In this paper it is shown that collaborative relations can be promoted at intermodal hubs using an iterative approach. Cargo is categorized in terms of type, final destination and due time to be delivered. Using a mathematical model to update the existent cargo per destination and due time at a terminal it is possible to make predictions about the future. These predictions can be used on the Model Predictive Control (MPC) problem formulated for cargo assignment at each terminal within an intermodal hub. Collaborative relations among terminals are implemented through information exchange from the solution of the MPC problem of each terminal, without compromising private information. A central coordinator collects the information from all terminals and updates the transport capacity allocated for each terminal. The iterative procedure is based in terms of achieving a better solution from both a hub, transport operator and client perspective. The iterative approach is illustrated with numerical experiments considering an intermodal hub composed of three container terminals.

Keywords: Transport Hubs, Terminals, Collaborative Relations, Intermodal Transport, Transport Modal Split.

* This work was supported by FCT, through IDMEC, under LAETA Pest-OE/EME/LA0022 and supported by the project PTDC/EMS-CRO/2042/2012.

1   Introduction

Intermodal hubs (seaports, terminals, consolidation centers, airports...) create an added value to the region where they are integrated by enabling the capacity of receiving cargo through different transport modalities, offering freight and logistics services [1]. Intermodal hubs act generally as logistic centers and are a case of particularly competitive market where many partners are present (merchants, forwarders, terminal managers, shippers, infrastructure owners...) [2]. The sector is dominated by a lack of confidence between partners where few information is shared believing that doing so will prevent losing their clients towards the rivals. In spite of the challenges (competitive market and lack of information available) also authority policies [3] (e.g., recently a sustainable environment is a political priority) have to be added. The rapid increase in the international freight commerce is pushing the existing infrastructure to its limits.

Improving the performance of intermodal hubs should be based on the development of optimization methods that enable to respond to client demands. Currently, cargo transport on some transport modalities is far from optimal; it is common to have empty trucks arriving/departing at/from the intermodal hub. A solution can be based on an increase in coordination between the transport provider and the terminals within the intermodal hub. The performance of an intermodal hub and the terminals within is seen in terms of client satisfaction, that is to say, the capacity to assign all cargo to the transport network such that the cargo is delivered at the agreed time and at the agreed location to the final client [4].

Cooperation at the hub is beneficial for all partners; 1) the hub will assume the role of a reliable logistic center 2) the terminal will be integrated in a reliable network partner, and finally 3) for the transport operator it is possible to make a better use of the transport capacity owned. The problem addressed can be stated as: given a known and finite transport capacity at the intermodal hub how can the different terminals, without sharing private information, such that the cargo is assigned to the transport capacity in order to reach the final destination on time?

In this paper it is shown that an iterative approach can address the problem stated. The cargo assignment problem for each terminal is done by a control agent which formulates an MPC problem [6]. At each negotiation step control agents share with a coordinator agent the marginal costs which are related to the amount of cargo that will not reach the final destination on time [5]. The coordinator agent will redistribute the transport capacity among the control agents according to the setup of collaborations among terminals. Iterations will proceed until all cargo is assigned such that the final destination is reached on time or no improvement is achieved. This paper is organized as follows. In Section 2 the intermodal hub is modeled as a collection of terminals. The iterative approach to promote collaborative relations at the intermodal hub is addressed in Section 3. The affect of collaborative relations on hub performance is tested through simulation experiments in Section 4. In Section 5 conclusions are drawn and future research topics are indicated.

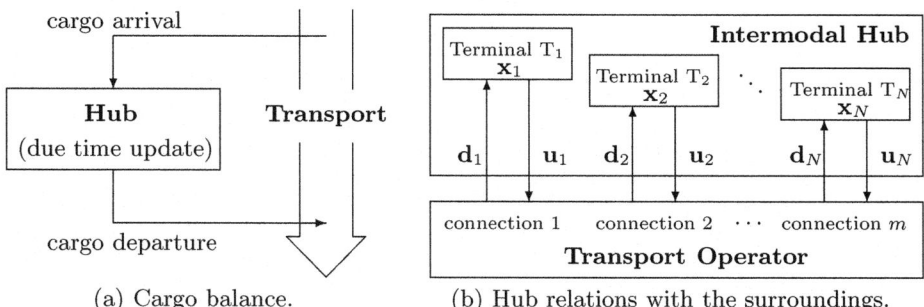

(a) Cargo balance.   (b) Hub relations with the surroundings.

Fig. 1. Intermodal hub schematics

2   Modeling Intermodal Hubs

In terminal haulage, the terminals co-determines the land transport. The cargo assignment to the existent transportation network depends on some cargo properties. Cargo is categorized in respect to three categories: *cargo type* n_{ct} is the number of different cargo types at the hub, for example, dry bulk, liquid bulk, containers and general cargo; *cargo destination* n_{d} is the number of available destinations in the transportation network where the hub is integrated; *cargo due time* n_{dt} is the number of different due times considered to the final destination, typically measured in days.

The cargo evolution over time at a terminal depends on the cargo arrival pattern, the assignment of cargo towards the final destination and the update of due times (see Fig. 1(a)). The cargo arrival pattern is considered out of the terminal scope and is seen as an exogenous input, the terminal by default accepts all cargo. The cargo assignment (per type, destination, due time and connection) is considered a decision of the terminal. The due time update is a phenomenon that happens at each time step k, decreasing the amount of time available to deliver a given cargo to the final destination. If no cargo assignment is done by the terminal, all cargo arrived at the terminal will fail the due time. If cargo fails the due time then it is considered a lost cargo and affects negatively the terminal performance and reputation.

The hub can be seen as a collection of the existing N terminals. Each terminal has its own state-space vector $\mathbf{x}_i$, cargo arrival pattern $\mathbf{d}_i$, and cargo assignment vector $\mathbf{u}_i$ (see Fig. 1(b)). The hub state $\mathbf{x}$, cargo arrival $\mathbf{d}$, and cargo assignment $\mathbf{u}$ can be separable in terms of the terminals within. The coupling is present in the form of available resources $\mathbf{\Theta}$ transport capacity at the intermodal hub, so the following relation should hold at each time step k,

$$\mathbf{u}(k) = \sum_{i=1}^{N} \mathbf{P}_{\mathrm{pu},i}\mathbf{u}_i(k) \le \mathbf{\Theta}(k) \tag{1}$$

where $\mathbf{P}_{\mathrm{pu},i}$ is the projection from the terminal i control action set $\mathcal{U}_i$ into the hub control action set $\mathcal{U}$. Each terminal inside the intermodal hub is modeled based on cargo amount conservation and due time updates [4],

$$\begin{aligned}
\mathbf{x}_i(k+1) &= \mathbf{A}_i\mathbf{x}_i(k) + \mathbf{B}_{k,i}\mathbf{u}_i(k) + \mathbf{B}_{\mathrm{d},i}\mathbf{d}_i(k) \\
\mathbf{y}_i(k) &= \mathbf{C}_i\mathbf{x}_i(k) \\
\mathbf{x}_i(k) &\geq \mathbf{0} \\
\mathbf{B}_{k,i} &\in \mathcal{B}_{\mathrm{m}}
\end{aligned} \tag{2}$$

where $\mathbf{y}_i$ is the terminal cargo quantity per destination, matrices $\mathbf{A}_i$, $\mathbf{B}_{k,i}$, $\mathbf{B}_{\mathrm{d},i}$, and $\mathbf{C}_i$ are the state-space matrices, and $\mathcal{B}_{\mathrm{m}}$ is the set of all possible connection schedules at the hub. Matrix $\mathbf{B}_{k,i}$ is related to the outflow of cargo and is assumed time-varying due to the possibility of different transport schedules over time.

3   Collaborative Relations at Intermodal Hubs

Each terminal within the intermodal hub is interested in solving a cargo assignment problem which can be stated as follows:

Problem: *At each time step k, given a known and finite transport capacity per transport modality, destination, and cargo type, how should the existing cargo at the terminal be assigned to the transport capacity at its disposal such that cargo is delivered at the agreed location and at the agreed time?*

3.1   Control Agent for Each Terminal

The cargo assignment at each terminal is done by a control agent that formulates and solves an MPC problem [6]. The cost function used by control agent i assigned to terminal i is composed of three components (due time to destination $f_{\mathrm{x},i}(\mathbf{x}_i, \mathbf{u}_i)$, cargo type per transport modality $f_{\mathrm{u},i}(\mathbf{u}_i)$, and connection schedule $f_{\alpha,i}(\alpha)$). For the prediction horizon N_{p} the cost function for control agent i is defined as, $f_i(\mathbf{x}_i, \mathbf{u}_i, \alpha) = \sum_{j=0}^{N_{\mathrm{p}}-1} f_{\mathrm{x},i}(\mathbf{x}_i, \mathbf{u}_i) + f_{\mathrm{u},i}(\mathbf{u}_i) + f_{\alpha,i}(\alpha)$.

The cargo assignment problem respecting the cargo due time and destination can be written using an MPC strategy [4] for each control agent at the hub,

$$\min_{\bar{\mathbf{u}}_{k,i}} \ f_i(\mathbf{x}_i, \mathbf{u}_i, \alpha) \tag{3}$$

$$\begin{aligned}
\text{subject to } \ \mathbf{x}_i(k+1+j) &= \mathbf{A}\mathbf{x}_i(k+j) \\
&\quad + \mathbf{B}_{k+j,i}\mathbf{u}_i(k+j) + \mathbf{B}_{\mathrm{d},i}\mathbf{d}_i(k+j) \tag{4} \\
\mathbf{y}_i(k+j) &= \mathbf{C}_i\mathbf{x}_i(k+j), j = 0, \dots, N_{\mathrm{p}} - 1 \tag{5} \\
\mathbf{x}_i(k+j) &\geq \mathbf{0} \tag{6} \\
\mathbf{u}_i(k+j) &\geq \mathbf{0} \tag{7} \\
\mathbf{P}_{\mathrm{xu},i}(k+j)\mathbf{u}_i(k+j) &\leq \mathbf{x}_i(k+j) \tag{8} \\
\mathbf{u}_i(k+j) &\leq \mathbf{u}_{\mathrm{adm},i}^{\alpha(k+j)} \tag{9} \\
\mathbf{P}_{\mathrm{mu},i}(k+j)\mathbf{u}_i(k+j) &\leq \mathbf{\Theta}(k+j) \tag{10} \\
\mathbf{B}_{k+j,i} &\in \mathcal{B}_{\mathrm{m}}. \tag{11}
\end{aligned}$$

where $\tilde{\mathbf{u}}_{k,i}$ is the vector composed of control action vectors for each discrete time over the prediction horizon, $[\mathbf{u}_i(k)^{\mathrm{T}}, \ldots, \mathbf{u}_i(k + N_{\mathrm{p}} - 1)^{\mathrm{T}}]^{\mathrm{T}}$, $\mathbf{P}_{\mathrm{xu},i}(k)$ is the projection from the control action space into the state-space and is time-varying depending on the schedule α at time step k, $\mathbf{P}_{\mathrm{mu},i}(k)$ is the time varying projection matrix from the control action space into he current connection schedule space with dimension $n_{\mathrm{m}} \times n_{\mathrm{u}}(k)$, $\boldsymbol{\Theta}(k)$ is the available transport capacity with dimension $n_{\mathrm{m}}(k)$ at the hub using schedule $\alpha(k)$, $\mathbf{u}_{\mathrm{adm},i}^{\alpha(k)}$ contains the maximum admissible cargo capacity for each destination for all connections. Constraints (7)–(10) are introduced to guarantee the assumptions made on the terminal behavior: i) constraint (7) imposes that only loading operation is possible; ii) control actions can only assign cargo that is available at the terminal which is imposed through (8); iii) proper assignment of cargo in respect to destination is imposed using constraint (9); iv) the transport capacity per connection and schedule is bounded through constraint (10).

3.2   Coordinator Agent

The MPC problem that a control agent solves is coupled with the remaining MPC problems due to constraint (10). In order to overcome this coupling, a primal decomposition of the original optimization problem is used which guarantees a feasible solution at each negotiation step l. A coordinator agent, will update the resource allocation among terminals such that the following relation holds

$$\boldsymbol{\Theta}(k) = \sum_{i=1}^{N} \boldsymbol{\Theta}_i(k). \tag{12}$$

Using (12) it is possible to write constraint (10) as

$$\mathbf{P}_{\mathrm{mu},i}(k + j)\mathbf{u}_i(k + j) \le \boldsymbol{\Theta}_i(k + j) \tag{13}$$

leading to N decoupled cargo assignment problems which are solved by control agents 1 to N using only local information available: the terminal state $\mathbf{x}_i$ and the cargo $\mathbf{d}_i$. The cooperation problem at the hub has been transformed into a resource allocation problem. Control agents share with the coordinator agent the marginal costs $\mathbf{g}_i$ associated to the resource allocated, no private information regarding the terminal activity is shared. This way privacy issues are assured. The coordinator agent will update the resource allocation between iterations, based on a switch of resources from the control agent with the lower marginal cost to the one with a higher marginal cost for all transport connections available over the prediction horizon [5]

$$\theta_{\tilde{m}}^{l+1}(k) = \mathcal{P}_\theta \left[\theta_{\tilde{m}}^{l}(k) + \beta^l \mathbf{W} \mathbf{g}_{\tilde{m}} \right], \quad \tilde{m} = 1, \ldots, \sum_{j=1}^{N_{\mathrm{p}}} n_{\mathrm{m}}(j) \tag{14}$$

where $\mathbf{W}$ is a square weighting matrix, see [7] for details on how to determine the elements of matrix $\mathbf{W}$ such that the approach converges. The operator $\mathcal{P}_\theta(\mathbf{v})$

denotes the Euclidean projection of $\mathbf{v}$ into the set θ and β is an adequate stepsize. A particular feature of matrix $\mathbf{W}$ is that the sum per column is zero, which means that there is a switch of transport capacity between terminals. Matrix $\mathbf{W}$ can capture collaborative relations at hubs. In case all terminals share freely information, matrix $\mathbf{W}$ is a full matrix. In case alliances are established at the hub (for example whenever several terminals are owned by the same company), matrix $\mathbf{W}$ captures the cooperation strategy at the hub. If the alliances do not share information with the remaining terminals or other alliances, then matrix $\mathbf{W}$ can be written as a diagonal block matrix.

After the initial resource allocation, iterations between control agents will only start in case of at least one control agent not being capable, over the negotiation horizon N_g ($N_g \leq N_p$), to assign all cargo such that it is delivered at the final destination at the agreed time. Iterations will continue until there is no lost cargo over the negotiation horizon for all control agents or the resource allocation update is bellow a threshold δ.

4 Simulation Experiments

Consider an intermodal hub composed of 3 terminals [8] (T_1, T_2, and T_3) where all outgoing transport capacity (barge, train and truck modality) is provided by one single transport operator and is finite. The following assumptions are made: 1) all terminals are of the same type – container terminals; 2) all connections departing from the hub are available to all terminals; 3) the daily schedule of connections at the intermodal hub is fixed over time. The focus of this work is on the cargo assignment at hub A towards the final destination. According to the network $n_d = 4$, hub A is also an available final destination. There are three due times (from 1 to 3 days at maximum) for all cargo at the hub, so $n_{dt} = 3$. Since all terminals at the hub are of the same type $n_{ct} = 1$, the state-space vector for all terminals and hub has length $n_x = 12$. A full day (24 hours) is considered as the time step for model (2). Cargo quantities are measured in TEU (twenty-foot equivalent unit). A total of 16 daily connections are available $n_m = 16$ [9]. There are $n_u = 16 \times 4 \times 3 = 192$ cargo assignment decisions per terminal at each time step. The intermodal hub is able to export a maximum of 1430 TEU daily. However, the maximum capacity considering a one day due time is only 890 TEU. Destination C is the destination with the lowest capacity to deliver cargo to with a one day due time, only 350 TEU.

4.1 Simulation Setup

Terms in the cost function are set equal for all terminals: for the pair destination/due time the state-space variable related to one day due time is the only penalized as all destinations are reachable in one day; for the transport penalty a distinction is made concerning different transport modalities present at the terminal (the barges are less penalized and trucks are the most penalized). The number of resources to share among terminals is $n_m N_p$. As threshold for stopping negotiations was used $\delta = 0.1$ TEU. The optimization problem is solved

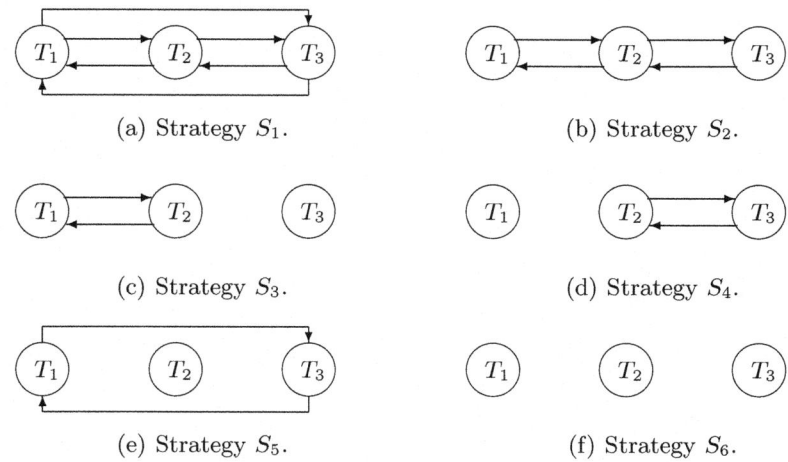

(a) Strategy S_1.

(b) Strategy S_2.

(c) Strategy S_3.

(d) Strategy S_4.

(e) Strategy S_5.

(f) Strategy S_6.

Fig. 2. Communication graph for collaborative strategies

at each time step of the simulation using the MPT v2.6.3 toolbox [10] with the CDD Criss-Cross solver for linear programming problems.

4.2 Numerical Experiments

In order to test the impact of collaborative relations on hub performance six co-operation strategies are implemented. The difference between cooperation strategies are related on how information is shared between terminals at the hub and are the following (see Fig. 2): strategy S_1 (all terminals share information freely), strategy S_2 (information is gathered at terminal T_2, the only communicating with all terminals), strategy S_3 (an alliance is established between T_1 and T_2 sharing information freely), strategy S_4 (an alliance is established between T_2 and T_3 sharing information freely), strategy S_5 (an alliance is established between T_1 and T_3 sharing information freely), and strategy S_6 where no information exchange is allowed. Communication strategies can be divided into three subgroups: i) strategies S_1 and S_2 are common in the sense that all terminals participate in the negotiation process, the communication graph is fully connected, ii) strategies S_3, S_4 and S_5 correspond to situations where an alliance/cooperation between two terminals is established and one terminal is left outside the negotiation process, the communication graph is not fully connected, iii) strategy S_6 corresponds to the inexistence of cooperation between terminals, each terminal tries to solve its problem alone and also without any external help, the graph is composed of isolated nodes.

It is assumed that at the beginning of each time step the transport capacity is distributed evenly among the terminals. Although having access to the same transport capacity at each time step, each terminal has its own cargo arrival pattern. Terminal T_1 receives close to 35% of the hub cargo, terminals T_2 and T_3 receive 25% and 40% of the hub cargo, respectively.

Table 1. Effective cargo lost (in TEU)

	S_1	S_2	S_3	S_4	S_5	S_6
$N_p = 2$	309.22	311.01	779.41	966.18	1010.53	1468.25
$N_p = 3$	0	0	302.87	381.96	352.02	684.83
$N_p = 4$	0	0	211.74	381.96	0	593.70

Table 2. Sum of iterations over the whole simulation

Strategies	Total			Mean			Maximum		
	$N_p = 2$	$N_p = 3$	$N_p = 4$	$N_p = 2$	$N_p = 3$	$N_p = 4$	$N_p = 2$	$N_p = 3$	$N_p = 4$
S_1	177	64	62	5.90	2.13	2.07	30	4	4
S_2	194	72	73	6.47	2.40	2.43	44	4	4
S_3	119	77	69	3.97	2.57	2.30	16	7	4
S_4	99	110	93	3.30	3.67	3.10	11	25	17
S_5	111	86	192	3.70	2.87	6.40	15	8	25
S_6	46	46	50	1.53	1.53	1.67	2	2	2

The first performance criterion is the client satisfaction in terms of receiving the cargo at the agreed time. With the increase of the prediction horizon it is possible to reduce the amount of cargo that is not assigned to a transport such that the final destination is reached on time. For cooperation strategies S_1 and S_2, using a prediction horizon of 3 time steps is enough to eliminate the existence of lost cargo (see Table 1). For other cooperation strategies, there is a reduction of the amount of cargo lost. In particular, for alliances of two terminals worse results are achieved. For S_5 with a prediction horizon of 4 time steps no lost cargo occurs. This means that terminal T_1 and T_3 are able with a transport capacity of 66% to assign 75% of the seaport cargo on time. This shows that assigning cargo respecting the due time is dependent on the cargo arrival pattern (quantity per destination) and the features of the transport capacity available (capacity and routes). It is important a proper design of connections which should be made in cooperation between the hub (cargo pattern) and the transport operator (transport capacity and network).

With the increase of the prediction horizon, negotiations tend to be shorter but happen for more time steps, see Table 2 and Fig. 3). Using a bigger prediction horizon introduces more freedom in the negotiation since more transport connections are now taken into consideration (e.g. a train connection in a three day horizon).

For terminal T_2 there is no need to establish cooperation with other terminals, in the sense that having access to a transport capacity of 33% for an amount of 25% of the hub cargo the cargo assignment problem is less demanding. However, due to lost cargo at terminals T_1 and T_3, the hub performance as a group is affected. This also affects terminal T_2. In an ultimate situation, terminal T_2 can be integrated in a group of poor performing partners. This is the motivation for terminal T_2 start cooperating with other terminals. For a prediction horizon

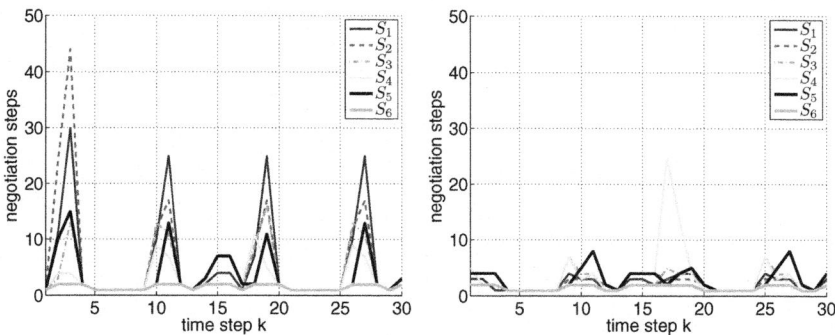

Fig. 3. Iterations over the simulation for $N_p = 2$ (left) and $N_p = 3$ (right)

Table 3. Negotiation details over a prediction horizon $N_p = 2$ (in TEU)

Strategies	Terminal Improvement			Cargo lost			Total Lost
	T_1	T_2	T_3	T_1	T_2	T_3	
S_1	1159.66	−118.56	625.44	115.16	118.56	58.97	332.69
S_2	1048.21	−26.56	646.87	235.21	26.56	53.18	314.96
S_3	1012.54	−32.51	0	386.66	32.51	553.40	972.57
S_4	0	0	716.63	1707.61	0	0	1707.61
S_5	399.87	0	284.19	1218.10	0	318.60	1536.69
S_6	0	0	0	1707.61	0	553.40	2261.01

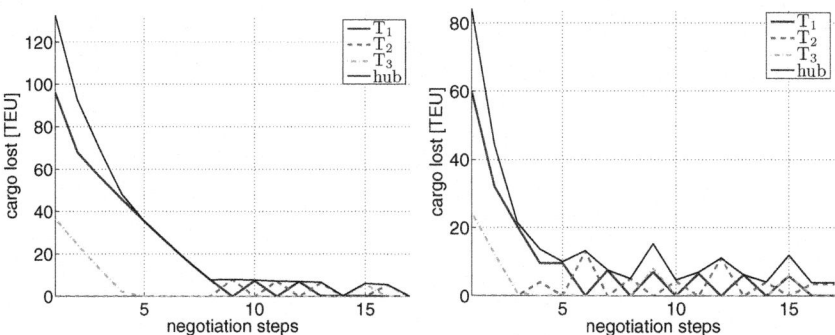

Fig. 4. Iterations for strategy S_2 at time step $k = 11$ (left) and $k = 19$ (right) for $N_p = 2$

of two steps, in strategies S_1 to S_3 terminal T_2 has a worst position after the negotiation phase, loosing more cargo at the beginning of the negotiations no cargo was lost (see Table 3). Although locally for terminal T_2 this looks a bad situation, for the hub the amount of cargo lost reduced significantly, terminals T_1 and T_3 could improve significantly their initial performance, see Fig. 4.

5 Conclusions and Future Research

In paper it is shown that an iterative MPC framework can support collaborative relations amongst terminals at intermodal hubs. Collaborative relations are used to promote an intelligent use of the limited existing infrastructure, in this case the finite transport capacity at the hub. Terminals can exchange information freely or according to alliances. The framework allows an altruist behavior per terminal. The iterations among terminals are based on information exchange related to the MPC problem only. The private information associated to a terminal, such as the amount and type of cargo, the cargo arrival pattern, and final destination is protected.

Future research will focus on convergence properties and the introduction of measures to compensate a terminal for a lost in performance while acting in the benefit of the group.

References

1. Casaca, A., Marlhow, P.: Fourth generation ports - a question of agility? International Journal of Physical Distribution Logistics Management 33(4), 22 (2003)
2. Rodrigue, J.-P., Comtois, C., Slack, B.: The Geography of Transport Systems. Taylor and Francis Group (2006)
3. Transcontinental Infrastructure Needs to 2030/2050 - North-West Europe Gateway Area, Rotterdam Workshop, Final Report, OECD (November 2010)
4. Nabais, J.L., Negenborn, R.R., Botto, M.: Model Predictive Control for a Sustainable Transport Modal Split at Intermodal Container Hubs. In: Proc. 10th IEEE Int. Conf. on Networking, Sensing and Control, Paris, pp. 591–596 (April 2013)
5. Nabais, J.L., Negenborn, R.R., Carmona-Benítez, R.B., Botto, M.A.: Setting Cooperative Relations Among Terminals at Seaports Using a Multi-Agent System. In: Proc. of the 16th IEEE Int. Conf. on Intelligent Transportation Systems, pp. 1731–1736. The Hague (October 2013)
6. Maciejowski, J.M.: Predictive control with constraints. Prentice Hall, Harlow (2002)
7. Xiao, L., Boyd, S.: Optimal Scalling of a Gradient Method for Distributed Resource Allocation. Journal of Optimization Theory and Applications 129(3), 469–488 (2006)
8. Stahlbock, R., Voss, S.: Operations research at container terminals: a literature update. OR Spectrum 30(1), 1–52 (2008)
9. Nabais, J.L., Negenborn, R.R., Botto, M.A.: A novel predictive control based framework for optimizing intermodal container terminal operations. In: Proceedings of the 3rd Int. Conference on Computational Logistics, Shanghai, China, pp. 53–71 (September 2012)
10. Kvasnica, M., Grieder, P., Baotić, M., Morari, M.: Multi-parametric toolbox (MPT). In: Alur, R., Pappas, G.J. (eds.) HSCC 2004. LNCS, vol. 2993, pp. 448–462. Springer, Heidelberg (2004)

Structures´ Influence on Model-Plant Mismatch Detection Methods in MPC Using Partial Correlation

Marcos V. Loeff[1], André S.R. Kuramoto[2], and Claudio Garcia[1]

[1] Polytechnic School, University of São Paulo, São Paulo, Brazil
marcos.loeff@usp.br, clgarcia@lac.usp.br
[2] Petrobras, Henrique Laje Refinery, São Paulo, Brazil
andrekuramoto@yahoo.com.br

Abstract. One of the challenges that still needs to be overcome in order to improve the performance of the model predictive control (MPC) is its maintenance. Re-identification of the process is one of the best options available to update the internal model of the MPC, in order to increase the production and improve efficiency. However, re-identification is costly Researchers have proposed two different methods able to detect plant mismatch through partial correlation analysis. Using these techniques, instead of re-identifying all the sub-models in the process, only a few inputs with significant mismatch would have to be perturbed and only the degraded portion of the model would be updated. Nevertheless, there isn't enough information and analysis about the influence of the choice of the structures for identification (as FIR, ARX, ARMAX and OE) on partial correlation results. This paper demonstrates that the Carlsson method is a particular solution of the Badwe et al. method, when the models used on the identification process are FIR structures. Moreover, some other types of structures will were analyzed in order to check if they are suitable for the partial correlation procedure to detect plant mismatches.

Keywords: Model Predictive Control, Bias Analysis, Model-Plant Mismatch, Partial Correlations Analysis.

1 Introduction

Model Predictive Control (MPC) is an important advanced control technique for complex multivariable plants, which has been widely and successfully applied to various processes in many industries. As MPC uses a model based control strategy, its performance is related to the quality of the plant model.

According to a recent survey (Kano and Ogawa, 2009), there are still problems that need to be solved in order to improve the performance of this type of advanced control, that is, increase production and reduce costs. One of them is the MPC maintenance. Changes in process characteristics are one important factor accounting for the plant model deterioration (Kano and Ogawa, 2009). Hence, re-identification of the

© Springer International Publishing Switzerland 2015   61
A.P. Moreira et al. (eds.), *CONTROLO'2014 - Proc. of the 11th Port. Conf. on Autom. Control,*
Lecture Notes in Electrical Engineering 321, DOI: 10.1007/978-3-319-10380-8_7

plant is the key for maintenance. Nevertheless, re-identifying the process model with a large number of inputs and outputs is costly. Detecting model-plant mismatch can reduce the cost of re-identification by selecting a set of sub-models with more significant mismatches. In this case, re-identification of all the sub-models is not necessary, but only the ones detected by the algorithm.

In 2009, Badwe *et al.* proposed a new method using partial correlation analysis between the model residuals and the manipulated variables to detect plant mismatches in multivariable systems. In one of its steps, the regression step, it is necessary to use dynamic models (Equations (1) and (3)), which requires performing several system identifications as shown in Section 2. In order to make this method more appealing to industry, it was simplified by Carlsson (2010), applying the conventional definition of partial correlation, which, as is demonstrated in Section 2, applies the least squares method. It is considered simpler to implement than the method of Badwe *et al.* (2009), as it requires less computational effort but it is still robust.

One question that still needs to be addressed is which type of structure could be used in the partial correlation procedure, explaining the reasons. This paper demonstrates that the method of Carlsson is a particular solution of the method of Badwe *et al.*, when the models used in the identification process are FIR structures. However, what would be the effects of the other types of structures on the results of plant-mismatch detection? Could the order of a model influence the final results as well? Is there a way to measure the efficiency of the method using a given structure? The purpose of this paper is to answer those questions and also analyze briefly if the semi-partial correlation algorithm could be used instead of partial correlation.

2   Relationship between the Methods of Badwe *et al.* and Carlsson

According to Badwe *et al.* (2009), before applying the partial correlation analysis with dynamic models, data from MVs (manipulated variables) and residuals are obtained with sufficient excitation. Then, the disturbance free components of the MVs are found. The next steps of the correlation analysis using dynamic models are:

1. Estimate $\epsilon_{u_i}(k)$ that is the component of the i-th input of the model, $u_i(k)$, that is uncorrelated with the rest of MVs, $\tilde{u}_i(k)$. That is:

$$u_i(k) = G_{u_i}\tilde{u}_i(k) + \epsilon_{u_i}(k) \tag{1}$$

$$\hat{\epsilon}_{u_i}(k) = u_i(k) - \hat{G}_{u_i}\tilde{u}_i(k) \tag{2}$$

2. An analog procedure is used to estimate the component of the j-th model residual, $e_j(k)$, that is uncorrelated with the rest of MVs, $\tilde{u}_i(k)$. That is:

$$e_j(k) = G_{e_j}\tilde{u}_i(k) + \epsilon_{e_j}(k) \tag{3}$$

$$\hat{\epsilon}_{e_j}(k) = e_j(k) - \hat{G}_{e_j}\tilde{u}_i(k) \tag{4}$$

G_{u_i} and G_{e_j} are the true models and $\hat{G}_{u_i}$ and $\hat{G}_{e_j}$ are the estimated linear time-invariant (LTI) models from data of MVs and residuals.

3. Evaluate the correlation between $\hat{\epsilon}_{e_j}(k)$ and $\hat{\epsilon}_{u_i}(k)$.

If G_{u_i} is identified using a finite impulse response (FIR) structure of order n through the least squares (LS) method using N samples, then:

$$
\begin{bmatrix} u_i(0) \\ u_i(1) \\ \vdots \\ u_i(N-1) \end{bmatrix} = \begin{bmatrix} \boldsymbol{\varphi}^T(0) \\ \boldsymbol{\varphi}^T(1) \\ \vdots \\ \boldsymbol{\varphi}^T(N-1) \end{bmatrix} \boldsymbol{\theta}_{u_i} + \begin{bmatrix} \epsilon_{u_i}(0) \\ \epsilon_{u_i}(1) \\ \vdots \\ \epsilon_{u_i}(N-1) \end{bmatrix} \tag{5}
$$

$$
\mathbf{u}_i = \Psi \boldsymbol{\theta}_{u_i} + \boldsymbol{\epsilon}_{u_i} \tag{6}
$$

where $\boldsymbol{\theta}_{u_i}$ is the parameter vector of G_{u_i} and $\boldsymbol{\varphi}(k)$ is the regressor vector related to $\tilde{u}_i(k)$:

$$
\boldsymbol{\varphi}^T(k) = [\tilde{\mathbf{u}}^T{}_i(k) \ \tilde{\mathbf{u}}^T{}_i(k-1) \ \cdots \ \tilde{\mathbf{u}}^T{}_i(k-n+1)] \tag{7}
$$

The estimate of $\boldsymbol{\theta}_{u_i}$ is:

$$
\hat{\boldsymbol{\theta}}_{u_i} = (\Psi^T \Psi)^{-1} \Psi^T \mathbf{u}_i \tag{8}
$$

The vector of residuals is calculated by expression (10):

$$
\hat{\boldsymbol{\epsilon}}_{u_i} = \mathbf{u}_i - \Psi(\Psi^T \Psi)^{-1} \Psi^T \mathbf{u}_i \tag{9}
$$

$$
\hat{\epsilon}_{u_i}(k) = u_i(k) - \boldsymbol{\varphi}^T(k)(\Psi^T \Psi)^{-1} \Psi^T \mathbf{u}_i \tag{10}
$$

The procedure to calculate $\hat{\epsilon}_{e_j}(k)$ is analogous:

$$
\hat{\epsilon}_{e_j}(k) = e_j(k) - \boldsymbol{\varphi}^T(k)(\Psi^T \Psi)^{-1} \Psi^T \mathbf{e}_j \tag{11}
$$

Then the regular correlation between $\hat{\boldsymbol{\epsilon}}_{u_i}(k)$ and $\hat{\boldsymbol{\epsilon}}_{e_j}(k)$, when using a FIR structure and the LS method, Equations (10) and (11), is the same analysis of the partial correlation between $u_i(k)$ and $e_j(k)$ shown by Carlsson (2010):

$$
\rho(\hat{\epsilon}_{u_i}(k-d), \hat{\epsilon}_{e_j}(k)) = \rho\big(u_i(k-d), e_j(k) \big| \tilde{\mathbf{u}}_i(k-d), \tilde{\mathbf{u}}_i(k-d-1)\ldots\big) \tag{12}
$$

Hence, the method of Carlsson is a particular solution of the method of Badwe *et al.* (2009), when the models used to estimate $\hat{\epsilon}_{u_i}(k)$ and $\hat{\epsilon}_{e_j}(k)$ are FIR structures with its parameters estimated by the LS method.

3   Model Bias in Correlation Analysis

The bias is defined based on the model parameters (Aguirre, 2007):

$$\mathbf{b} = \mathrm{E}\{\widehat{\boldsymbol{\theta}}_{u_i}\} - \boldsymbol{\theta}_{u_i} \tag{13}$$

However, considering a LTI model, $\widehat{\boldsymbol{\theta}}_{u_i}$ is a linear combination of $\mathbf{u}_i$:

$$\widehat{\boldsymbol{\theta}}_{u_i} = A\mathbf{u}_i \tag{14}$$

where A is called linear estimator of the model parameters. Particularly, $A = (\Psi^T\Psi)^{-1}\Psi^T$ is a linear parameter estimator of a FIR structure as (8).

Besides that,

$$\mathbf{b} = \mathrm{E}\{A\mathbf{u}_i\} - \boldsymbol{\theta}_{u_i} = \mathrm{E}\{A(\Psi\boldsymbol{\theta}_{u_i} + \boldsymbol{\epsilon}_{u_i})\} - \boldsymbol{\theta}_{u_i} = \mathrm{E}\{A\Psi - I\}\boldsymbol{\theta}_{u_i} + \mathrm{E}\{A\boldsymbol{\epsilon}_{u_i}\} \tag{15}$$

Therefore, the bias will not be approximately zero if $\boldsymbol{\epsilon}_{u_i}$ and A are correlated. The same conclusion is valid for $\boldsymbol{\epsilon}_{e_j}$.

We conclude that, in order to this technique to give non biased results, it needs to guarantee that $\boldsymbol{\epsilon}_{e_j}$ and the estimator A are uncorrelated. The same is valid for $\boldsymbol{\epsilon}_{u_i}$. If these conditions are not respected, $\widehat{\boldsymbol{\theta}}_{u_i}$ or $\widehat{\boldsymbol{\theta}}_{e_j}$ will be biased and, therefore, so will $\widehat{\boldsymbol{\epsilon}}_{u_i}$ and $\widehat{\boldsymbol{\epsilon}}_{e_j}$.

The geometrical interpretation of this phenomenon is shown in Figure 1.

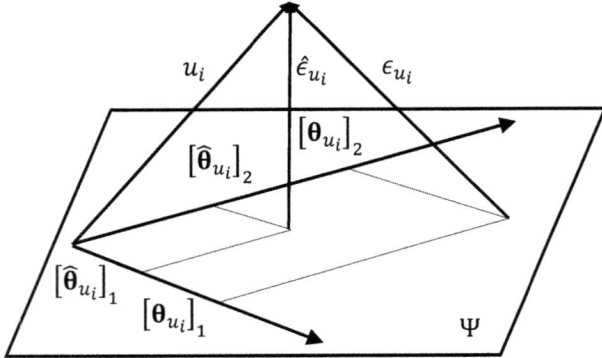

Fig. 1. Geometrical interpretation of partial correlation using the LS method

4   Suitable Dynamic Models

In the last section, it was shown that for a linear estimator, the bias will be approximately zero if the residual and estimator are uncorrelated. For a FIR structure, $A = (\Psi^T\Psi)^{-1}\Psi^T$, where Ψ contains only inputs of the models G_{u_i} or G_{e_j}, What would happen if we use the ARX or ARMAX structures instead?

Let's suppose that G_{u_i} is identified using an ARX structure of order n through the least squares (LS) method using N samples. As the method of identification is the same compared to the FIR structure, the main difference of using the ARX structure is

$\varphi^T(k)$ in equation (5), in which vectors related to the output are added, as it is shown below.

$$\varphi^T(k) = [\tilde{\mathbf{u}}^T_i(k) \ ... \ \tilde{\mathbf{u}}^T_i(k-n) \ \mathbf{u}^T_i(k) \ ... \ \mathbf{u}^T_i(k-m)] \tag{16}$$

Finally, the condition for the presence of bias is also the same for this structure. Nevertheless, as $\varphi^T(k)$ in this case includes vectors not only related to the input but also to the output, the results of a model mismatch analysis for structures can be different.

In order to understand the condition of bias using an ARX structure, consider the first order model (n=1 and m=1) for the system with two MVs:

$$u(k) = b\tilde{u}(k-1) + au(k-1) + v(k) \tag{17}$$

$$u(k-1) = b\tilde{u}(k-2) + au(k-2) + v(k-1) \tag{18}$$

$$u(k) = \{b\tilde{u}(k-1) + a[au(k-2) + b\tilde{u}(k-2) + v(k-1)]\} + v(k) \tag{19}$$

where:

$$\varphi^T(k) = \begin{bmatrix} \tilde{u}(k-1) & [au(k-2) + b\tilde{u}(k-2) + v(k-1)] \end{bmatrix} \tag{20}$$

and:

$$\theta_{u_i} = \begin{bmatrix} b \\ a \end{bmatrix} \tag{20}$$

in equation (5).

Assuming that the model is ARX then the error $v(k)$ in equation (17) is assumed a white noise random process $w(k)$.

As demonstrated in Section 3, the condition for the LS estimator not to be biased is that the regressors and the residuals are not correlated. Supposing that an ARX model fits well the true model (G_{u_i} or G_{e_j}), this condition would be satisfied if the noise is white, which means that $v(k)$ and $v(k-1)$ would be uncorrelated.

If an OE structure fits well the true model, then we would have:

$$F(q)u(k) = B(q)\tilde{u}(k) + F(q)w(k) \tag{21}$$

Unless $F(q) = 1$, $v(k) = F(q)w(k)$ will not be white noise. Therefore, the vector $v(k)$ will be correlated with some regressors, resulting in bias.

Suppose $v(k)$ from expression (17) is defined as:

$$v(k) = w(k) + cw(k-1) \tag{22}$$

Therefore, expression (19) is equal to:

$$u(k) = \{a[a\tilde{u}(k-2) + b\tilde{u}(k-2) + w(k-1) + cw(k-2)]\} + b\tilde{u}(k-1) \\ + w(k) + cw(k-1) \tag{23}$$

Note that $w(k - 1)$ influences not only the regressors but also the residuals, which means that they are correlated. So if the LS estimator were used in this example, the bias would not be zero.

The extended least square (ELS) addresses this problem by extending the matrix of the regressors as follows:

$$\begin{bmatrix} u(k) \\ u(k + 1) \\ \vdots \\ u(k + N - 1) \end{bmatrix} = \begin{bmatrix} u(k - 1) & \tilde{u}(k - 1) & v(k - 1) \\ u(k) & \tilde{u}(k) & v(k) \\ \vdots & \vdots & \vdots \\ u(k + N - 2) & \tilde{u}(k + N - 2) & v(k + N - 2) \end{bmatrix} \begin{bmatrix} a \\ b \\ c \end{bmatrix} + \begin{bmatrix} v(k) \\ v(k + 1) \\ \vdots \\ v(k + N - 1) \end{bmatrix} \quad (24)$$

$$\mathbf{u} = \mathbf{\Psi \theta} + \mathbf{\epsilon} \quad (25)$$

It shows that if expression (23) fits well statistically the error, the ELS estimator would be able to solve the problem of bias. The parameters of an ARMAX model, which has the structure defined in expression (24), can be estimated by the ELS estimator.

$$A(q)u(k) = B(q)\tilde{u}(k) + C(q)v(k) \quad (26)$$

If $A(q) = C(q) = F(q)$, then this structure is the same as OE. Therefore, it would be possible to use the ELS algorithm to estimate the parameters of this model without bias.

If the error can be statistically approximated by a colored noise and if the error is not correlated with the input signal, then the problem of bias is solved.

5   Mismatch Detection Example

In this case study, the following process $G_p(s)$ and model $\hat{G}_p(s)$ are simulated:

$$\hat{G}_p(s) = \begin{bmatrix} \dfrac{4,05e^{-6s}}{50s + 1} & \dfrac{1,77e^{-7s}}{60s + 1} \\ \dfrac{5,39e^{-4s}}{60s + 1} & \dfrac{5,725e^{-5s}}{60s + 1} \end{bmatrix} (27) \qquad G_p(s) = \begin{bmatrix} \dfrac{5,67e^{-6s}}{50s + 1} & \dfrac{1,77e^{-7s}}{60s + 1} \\ \dfrac{5,39e^{-4s}}{50s + 1} & \dfrac{5,725e^{-3s}}{60s + 1} \end{bmatrix} (28)$$

A pseudorandom binary signal (PRBS) signal is used as reference for the system. Noise is introduced into the system for both channels as white noise $w(k)$ with $\sigma = 0.01$, filtered according to   expression (30) to create colored noise $v(k)$.

$$v(k) = \frac{z}{z - 0.95} w(k) \quad (30)$$

Note that there is a gain mismatch in MV1-CV1, a time constant mismatch in MV1-CV2 and a time delay mismatch in MV2-CV2. Therefore, MV2-CV1 is the only sub-model with no mismatch.

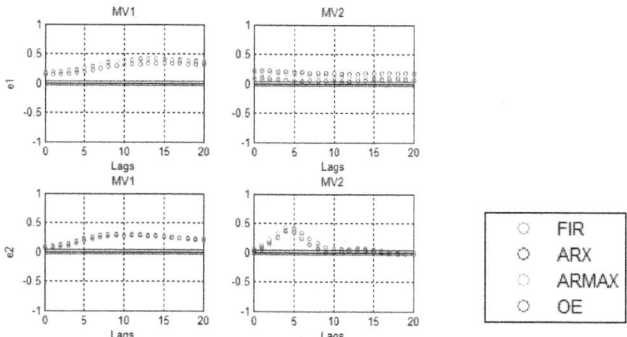

Fig. 2. Partial correlation for FIR, ARX, ARMAX and OE structures

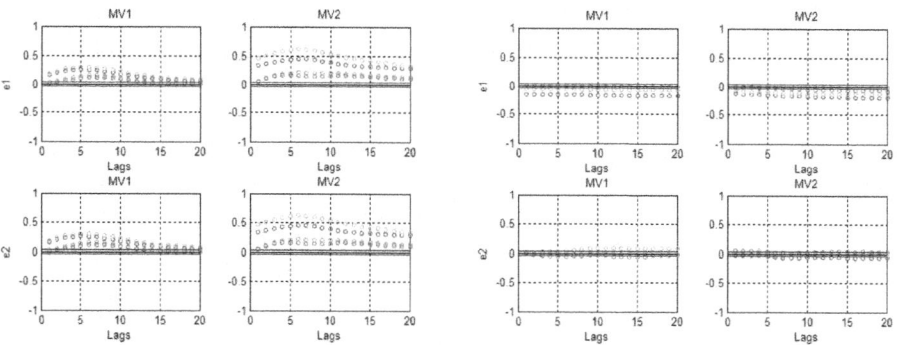

Fig. 3. Cross-correlation between the residuals and the estimated u

Fig. 4. Cross-correlation between the residuals and the estimated e

Figures 2 to 4 were plotted, each one showing the results of the following structures:

- FIR with $n_b=60$ and $n_k=0$
- ARX with $n_a=10$, $n_b=60$ and $n_k=0$
- ARMAX with $n_a= n_c =10$, $n_b=60$ and $n_k=0$
- OE with $n_b= 4$, $n_f =7$ and $n_k=0$

For each structure, it was plotted the cross-correlation between the residual $\hat{\epsilon}_{u_i}$ or $\hat{\epsilon}_{e_i}$ and the estimated vectors $\hat{u}_i = \Psi\hat{\theta}_{u_i}$ or $\hat{e}_i = \Psi\hat{\theta}_{e_j}$. If the residual and the estimated vectors were not correlated, estimated vectors wouldn't be correlated to the Ψ plain either, as it contains the residuals (Figure 1).

Figures 3 and 4 show that the structures FIR and OE are the ones which have the lowest correlation values from all the analyzed structures. ARX is the one with the highest correlation values in module followed by ARMAX.

This same ranking of structures reflects the efficiency of the partial correlation. This could be concluded by analyzing the correlations of MV2-CV1, whose correlation values are higher for the ARX and ARMAX structures.

Concerning the FIR structure, the fact that the cross-correlation between the residuals and the estimated vector of u and e are low for every lag, this is enough to guarantee that the parameters identified are biased in a small proportion. However, this is not true for the ARX structure, since the residuals are auto-correlated, which generates a biased estimation of the parameters and, therefore, misleading results.

The ARMAX structure improves the results of the ARX a little, as the method ELS is used instead of the LS. The advantage of this method is that it avoids the bias, even if the residuals are autocorrelated. Nevertheless, there is still bias when the residual is correlated with one of the inputs $\tilde{u}_i(k)$ and the issue is that it is not possible to make sure that this situation will not happen.

6   Influence of Model Order on Mismatch Detection

In this section, a different process is going to be used to study the influence of the order of a model using an OE structure on the partial correlation results. The process $G_p(s)$ and its model $\hat{G}_p(s)$ are shown on expressions (30) and (31). The noise model to generate the colored noise $v(k)$ is the same from the previous section.

$$G_p(s) = \begin{bmatrix} \dfrac{4.05 * 1.5e^{-6s}}{50s + 1} & \dfrac{1.77e^{-7s}}{60s + 1} & \dfrac{5.88e^{-6s}}{50s + 1} \\ \dfrac{5.39e^{-4s}}{50s + 1} & \dfrac{5.72e^{-3s}}{60s + 1} & \dfrac{6.9e^{-3s}}{40s + 1} \\ \dfrac{4.38e^{-5s}}{33s + 1} & \dfrac{4.42e^{-5s}}{44s + 1} & \dfrac{7,2}{19s + 1} \end{bmatrix} \quad (29)$$

$$\hat{G}_p(s) = \begin{bmatrix} \dfrac{6,07e^{-6s}}{50s + 1} & \dfrac{1.77e^{-7s}}{66s + 1} & \dfrac{3,5e^{-6s}}{50s + 1} \\ \dfrac{5.39e^{-4s}}{50s + 1} & \dfrac{5.72e^{-3s}}{42s + 1} & \dfrac{6.9e^{-5s}}{40s + 1} \\ \dfrac{4.38e^{-8s}}{33s + 1} & \dfrac{4.42e^{-5s}}{57s + 1} & \dfrac{11,52}{19s + 1} \end{bmatrix} \quad (30)$$

Figure 5 shows the partial correlations using dynamic models for different lags when identifying a 3x3 model using the OE structure with a fixed size.

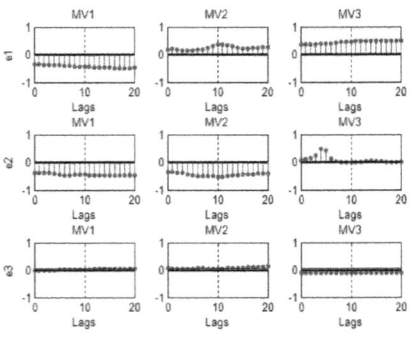
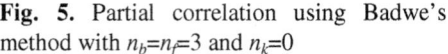
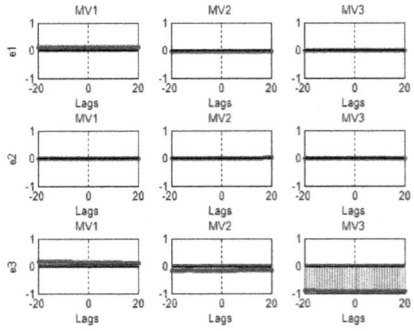

Fig. 5. Partial correlation using Badwe's method with $n_b = n_f = 3$ and $n_k = 0$

Fig. 6. Cross-correlation between the residuals and the estimated u with $n_b = n_f = 3$ and $n_k = 0$

The cross-correlation between the residuals and the estimated u ($\Psi\widehat{\theta}_{u_i}$) are approximately zero, excluding the one related to the third input and the third output, which is almost -1 for all the lags, as it is shown in Figure 6.

Hence, Figure 5 shows incorrectly that there is no mismatch in the pair MV3-CV3 as $\widehat{G}_{p_{3,3}}(s) = 1.6 * G_{p_{3,3}}(s)$.

One way to overcome this problem is to test different model orders, until a model with small cross-correlation values is found. The new model found has $n_b=n_f=2$ and $n_k=0$. Hence, Figure 7 is plotted detecting correctly the mismatch for every inputs and outputs. The related cross-correlation is shown is Figure 8 and in this case, the cross-correlations values are approximately zero for the lags of all nine sub models. Thus, a model in which $\widehat{\epsilon}_{e_j}(k)$ or $\widehat{\epsilon}_{u_i}(k)$ are not orthogonal to the plane $G_{u_i}\widetilde{u}_i(k)$ can lead to an erroneous analysis of model mismatch.

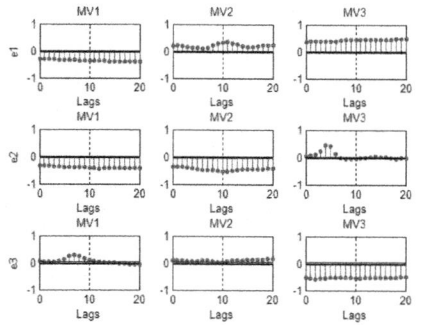 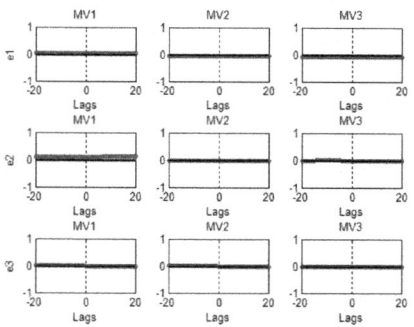

Fig. 7. Partial correlation using Badwe's method with $n_b=n_f=2$

Fig. 8. Cross-correlation between the residuals and the estimated u with $n_b=n_f=2$ and $n_k=0$

7   Conclusions

The LS method gives the estimated components $\widehat{\epsilon}_{e_j}(k)$ and $\widehat{\epsilon}_{u_i}(k)$ that are orthogonal to the plane of regressors Ψ. In an output error (OE) structure, as FIR, regressors are related only to $\widetilde{u}_i(k)$. Otherwise, in an equation error structure, as ARX, regressors consist of $\widetilde{u}_i(k)$ and $u_i(k)$ or $e_j(k)$. In this case, ϵ_{u_i} and ϵ_{e_j} can be correlated to estimator A resulting in bias. For example, even when the noise summed at the plant output is white, parameters of the ARX model $\widehat{\theta}_{e_j}$ are biased because ϵ_{e_j} are correlated to the regressors. Then, it is convenient to use OE structures applying this correlation analysis with dynamic models.

Concerning the order of the model, it is not true that the size of the model is inversely proportional to the magnitude of the cross-correlation values between the residuals and the estimated values for u and e. In the example of the previous section, the model with order $n_b=n_f=2$ had better results when compared to the model with order 3.

It is recommended to check those cross-correlations to reduce the chance of bias in the final results.

Elevated cross-correlations between the estimated values between $\hat{u}_i$ or $\hat{e}_i$ and residuals $\hat{\epsilon}_{e_j}(k)$ or $\hat{\epsilon}_{u_i}(k)$ invalidate results of an analysis of a plant model using the Badwe *et al.* (2009) method. Then, for a confident analysis, it is necessary adjust the model order of the OE structure to get low cross-correlation values between $\hat{u}_i$ or $\hat{e}_i$ and residuals $\hat{\epsilon}_{e_j}(k)$ or $\hat{\epsilon}_{u_i}(k)$.

References

1. Aguirre, L.A.: Introduction to system identification – linear and non-linear techniques for real systems, 3rd edn., Belo Horizonte, Brazil (2007) (in Portuguese)
2. Badwe, A.S., Gudi, R.D., Patwardhan, R.S., Shah, S.L., Patwardhan, S.C.: Detection of model-plant mismatch in MPC applications. Journal of Process Control 19(8), 1305–1313 (2009)
3. Carlsson, R.: A practical approach to detection of plant model mismatch for MPC. Master thesis. Linköping University (2010)
4. Kano, M., Ogawa, M.: The state of the art in advanced chemical process control in Japan. In: IFAC Symposium on Advanced Control of Chemical Processes (ADCHEM), CD–ROM. Istanbul, Turkey (2009)
5. Ljung, L.: System Identication. Theory for the User, 2nd edn. Prentice Hall, Englewood Cliffs (1999)

On Performance of Distributed Model Predictive Control in Power System Frequency Regulation

Luís M. Monteiro[1] and José Manuel Igreja[1,2]

[1] Instituto Superior de Engenharia de Lisboa (ISEL),
R. Conselheiro Emídio Navarro, 1,
1959-007 Lisboa, Portugal
luismgmonteiro@gmail.com
[2] INESC-ID, Rua Alves Redol, 9,
1000-029 Lisboa, Portugal
jmigreja@deea.isel.ipl.pt

Abstract. This paper describes the implementation of a distributed model predictive approach for automatic generation control. Performance results are discussed by comparing classical techniques (based on integral control) with model predictive control solutions (centralized and distributed) for different operational scenarios with two interconnected networks. These scenarios include variable load levels (ranging from a small to a large unbalance generated power to power consumption ratio) and simultaneously variable distance between the interconnected networks systems. For the two networks the paper also examines the impact of load variation in an island context (a network isolated from each other).

Keywords: AGC, Load-Frequency Control, Interconnected Control Areas, Distributed MPC.

1 Introduction

Guaranteeing the voltage waveform quality, in order to be characterized by a sinusoidal voltage with nominal frequency with a RMS value not less than the contractually established and not unbalanced (in the case of three-phase systems) as well as the continuity of power supply, is a crucial task for all operators in electric system. This task is becoming increasingly demanding, because modern societies are more and more dependent on electricity and so the electric consumption system must satisfy high standards of quality of service, never violating the system security conditions.

Voltage regulation requirements are not as severe as for the frequency which normally accepts a band of variation of ± 5% of the nominal value. Local voltage regulation depends heavily of the reactive power transits in the network (voltage regulation-reactive power also called *control-QV*). On the other hand, the frequency amplitude must be maintained with in a very narrow range, typically ± 0.1% of nominal value. The frequency regulation is closely related with the balance maintenance

© Springer International Publishing Switzerland 2015

A.P. Moreira et al. (eds.), *CONTROLO'2014 - Proc. of the 11th Port. Conf. on Autom. Control*,
Lecture Notes in Electrical Engineering 321, DOI: 10.1007/978-3-319-10380-8_8

between generator power and active consumption power (frequency control-active power also called *control-Pf*).

As the transits of active and reactive power in a network are largely independent and influenced by different control variables, they can be studied separately for a wide range of situations. In general, the *QV* loop is much faster than the *Pf* loop, due to the mechanical inertia constant of the *Pf* loop. If it is assumed that the transient of *QV* loop ends practically when the *Pf* loop reacts, then the coupling between them can be neglected [2], [4].

The classical techniques of AGC, based on method of *integral control* [1], do not provide adequate performance in all operational contingencies, or take into account the nonlinearities (*e.g.*, parameters variation of the generation system, dead-bands speed regulator or operational constraints). In this context, the AGC can progress to a level with the adoption of new control techniques. A control based on *predictive models* (MPC), with special emphasis in distributed predictive control, is in principle able to tackle the complexity of the distributed nature of production and power consumption, anticipating undesired behaviors, incorporating and handling nonlinearities and constraints of the *control-Pf* project [8],[3]. This paper describes the implementation of a distributed model predictive control for automatic generation control focusing in the frequency regulation performance for two interconnected control areas, assuming typical linear models and no operational constraints. The rest of paper is organized as follows. In section 2 the problem formulation is presented. The section 3 describes the distributed control implementation scheme. In section 4 the different scenarios are described and results are depicted. Finally section 5 draws conclusion.

2   Problem Formulation

The problem discussed in this paper consists in frequency regulation for two interconnected control areas hereafter referred as AC_1 and AC_2 respectively. A positive variation of ΔP_{L12} represents an increase in power transferred from AC_1 to AC_2, which is equivalent to an increase in AC_2 load regarding to AC_1 load. Note that the base power must be common to the interconnected control areas.

Considering an isolated system, the variation of mechanical power ΔP_m and the variation of electrical power ΔP_e (equal at each instant to the power variation in load ΔP_c) is translated by the following mathematical model of the energy balance, where W_{cin} is the kinetic energy, ΔP_D is the sensitivity of following mathematical model of the energy balance, W_{cin} is the kinetic energy and ΔP_D is the sensitivity of the active power load due the effect in frequency variation.

$$\Delta P_m - \Delta P_c = \frac{dW_{cin}}{dt} + \Delta P_D. \tag{1}$$

In (1) when a load increase occur, the generation should increase as well (secondary control), which does not happen instantaneously. While the balance is not restored, the additional charge will be satisfied at the cost of the decrease of the kinetic energy stored in the rotating masses of the generators (primary control), [2].

Based on the system of two interconnected control areas, the area control error of each area of control ACE$_1$ and ACE$_2$ is, respectively, ACE$_1 \equiv y_1 = \Delta P_{L12} + B_1 \Delta f_1$ and ACE$_2 \equiv y_2 = \Delta P_{L21} + B_2 \Delta f_2$. The state space model is given by

$$
\begin{bmatrix} \Delta \dot{f}_1 \\ \Delta \dot{f}_2 \\ \Delta \dot{P}_{m1} \\ \Delta \dot{P}_{m2} \\ \Delta \dot{X}_{Z1} \\ \Delta \dot{X}_{Z2} \\ \Delta \dot{P}_{L12} \end{bmatrix} =
\begin{bmatrix}
-\dfrac{1}{\tau_{p1}} & 0 & \dfrac{K_{p1}}{\tau_{p1}} & 0 & 0 & 0 & -\dfrac{K_{p1}}{\tau_{p1}} \\
0 & -\dfrac{1}{\tau_{p2}} & 0 & \dfrac{K_{p2}}{\tau_{p2}} & 0 & 0 & \dfrac{K_{p2}}{\tau_{p2}} \\
0 & 0 & -\dfrac{1}{\tau_{t1}} & 0 & \dfrac{1}{\tau_{t1}} & 0 & 0 \\
0 & 0 & 0 & -\dfrac{1}{\tau_{t2}} & 0 & \dfrac{1}{\tau_{t2}} & 0 \\
-\dfrac{1}{R_1 \tau_{r1}} & 0 & 0 & 0 & -\dfrac{1}{\tau_{r1}} & 0 & 0 \\
0 & -\dfrac{1}{R_2 \tau_{r2}} & 0 & 0 & 0 & -\dfrac{1}{\tau_{r2}} & 0 \\
2\pi P_S & -2\pi P_S & 0 & 0 & 0 & 0 & 0
\end{bmatrix}
\cdot
\begin{bmatrix} \Delta f_1 \\ \Delta f_2 \\ \Delta P_{m1} \\ \Delta P_{m2} \\ \Delta X_{Z1} \\ \Delta X_{Z2} \\ \Delta P_{L12} \end{bmatrix}.
\quad (2)
$$

More details about the model and model parameters can be found in [1] and [3].

3   MPC Distributed Algorithm

In MPC with a central control, all of the information of the control area(s), *i.e.*, all measurements of the sensors converge at a single processing location and hence the optimal control actions are obtained, after the minimization of a centralized quadratic cost function. In this subsection the same control problem will be treated, but based on a distributed approach, where agents are responsible for local controls, concerning to each control area, only exchanging information with neighboring interconnected agents. This approach emerges from the necessity to use distributed control due to the fact that is not practical to control large scale systems in a central mode, such as power systems, which they are inherently scattered geographically [5], [6]. The Fig. 1 is a schematic representation of M subsystems (or control areas) sequentially serially distributed:

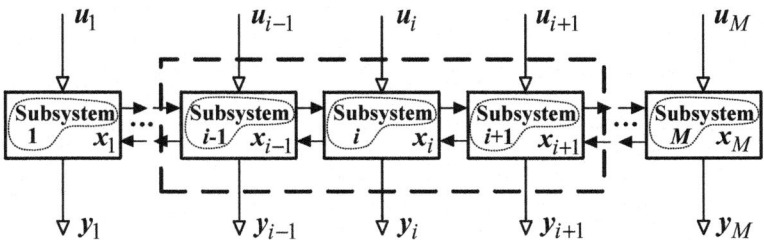

Fig. 1. Diagram of network MPC distributed for serially connected

As shown in the figure, the overall system of AGC is divided in M interconnected subsystems (*i.e.* control areas: AC_i, $i =1,...,M$, with $M \in \mathbb{N}$: connected in series). For each control area $i =1,...,M$, the triplet (u_i, x_i, y_i) represent the input vectors, state and output, respectively.

In this arrangement of control areas in series, the control area AC_i interacts only with its neighbors: AC_i and AC_{i+1}. This interaction is accomplished only through manipulation of the input vector in each area. The state model of each control area is represented by [7]:

$$x_i(k+1) = A_{ii}x_i(k) + \sum_{h=i-1}^{i+1} B_{ih}u_i(k); \quad \text{with } h = \{i-1, i, i+1\}; \; i = \{1,...,M\} \in \mathbb{N}. \quad (3)$$
$$y_i(k) = C_{ii}x_i(k) + D_{ii}u_i(k);$$

The matrices $A_{ii[n \times n]}$, $C_{ii[r \times n]}$ and $D_{ii[r \times m]}$ that comprise the SSM of distributed control, are constituted by the submatrices corresponding to each subsystem, whose dimensions depend on the area of control which they are related. For the matrix $A_{ih[n \times m]}$, with $h \in \{i\text{-}1, i, i\text{+}1\}$: when $h = \{i\}$ the matrix $B_{ii[n \times m]}$ reflects the variation of the state variables subject to inputs (u_i) of the control area i; when $h = \{i\text{-}1, i\text{+}1\}$ the matrix B_{ih} reflects the variation of the state variables subject to the inputs of neighboring control area $(u_{i-1}$ and $u_{i+1})$. From state equation (2), the estimated state vectors $\hat{x}_i$ for the $k+1$ instant $(k+1|k)$ for each control area, with $i = \{1,...,M\}$, are given by

$$\hat{x}_i(k+1|k) = A_{ii}\hat{x}_i(k) + \sum_{h=i-1}^{i+1} B_{ih}u_i(k|k). \quad (4)$$

Multiplying both sides of the equation matrix of the estimated state (4) by the matrix output C_{ii}, result the output vector predict $\hat{y}_i$ for control area i (assuming decoupling $A_{ih} = 0$, for $i \neq h$), results

$$\hat{y}_i = \hat{y}_{i0} + \sum_{h=i-1}^{i+1} G_{ih}u_i, \quad (5)$$

where G_{ih} is the dynamic matrix of the subsystem i, considering the interactions with neighboring subsystems interconnected; $\hat{y}_{i0}$ represents the vector of predicted outputs for subsystem i without control movements (*i.e.*, the system in *free response*).

In distributed control, the solution of the optimization problem is decomposed by control agents, where each of these agents is responsible for a local optimization subproblem $J_i(k)$ (with $i=\{1,..,M\} \in \mathbb{N}$, corresponding to a single subsystems) [8]. Control agents exchange information locally, therefore they are neighboring interconnected, synchronizing their calculations and iterating until the convergence of overall suboptimal solution. The obtained solution for the sequence of future inputs of each control area is:

$$u_i^{*(r)} = \left[u_1^{*(r)} \left(k + j \mid k \right)^{\mathrm{T}}, ..., u_i^{*(r)} \left(k + j \mid k \right)^{\mathrm{T}} \right]^{\mathrm{T}} \tag{6}$$

with $j=\{1,...,N_p\}$ indicating the instant of time, within the prediction horizon, for which the solution of u_i^* are obtained, $i=\{1,...,M\}$ corresponds to the totality of interconnected subsystems and (r) corresponds to the number of iterations to get the optimal inputs sequence within each sampling period.

Removing the time dependence, the objective function is

$$J = e_{i-1}{}^{\mathrm{T}} Q_{i-1} e_{i-1} + e_i{}^{\mathrm{T}} Q_i e_i + e_{i+1}{}^{\mathrm{T}} Q_{i+1} e_{i+1} + u_i{}^{\mathrm{T}} R_i u_i, \tag{7}$$

The solution to determine u_i optimum that minimizes $J_i(k)$ will be

$$\min_{u_i} J(k) = \min_{u_i} \left\{ e_{i-1}{}^{\mathrm{T}} Q_{i-1} e_{i-1} + e_i{}^{\mathrm{T}} Q_i e_i + e_{i+1}{}^{\mathrm{T}} Q_{i+1} e_{i+1} + u_i{}^{\mathrm{T}} R_i u_i \right\} \tag{8}$$

The error between reference and the predicted output is defined as $e = w - \hat{y}$ and the predicted output $\hat{y}$ is characterized by the expression (5), then the error signal of the subsystem i and its neighbors will be

$$\left[e_{i-1}^{\mathrm{T}} = ..; e_i^{\mathrm{T}} = w_i - \left(\hat{y}_{i0} + \sum_{h=i-1}^{i+1} G_{ih} u_i \right); e_{i+1}^{\mathrm{T}} = .. \right]^{\mathrm{T}} \tag{9}$$

Substituting (5) in (8) and considering the quadratic penalization matrix unitary, i.e., $Q_{i-1} = Q_i = Q_{i+1} = [I] = \mathrm{diag}[1_{[11]}; 1_{[22]}, ..., 1_{[NpNp]}]$. Then deriving $J_i(k)$ in order to u_i and finally equaling to zero ($\partial J_i(\cdot)/\partial u_i = 0$), result, simplifying, the optimum input vector that minimizes the error between the reference and the predicted output:

$$G_{i-1,i}^{\mathrm{T}} G_{i-1,i-2} u_{i-2}^* + \left(G_{i-1,i}^{\mathrm{T}} G_{i-1,i-1} + G_{ii}^{\mathrm{T}} G_{i,i-1} \right) u_{i-1}^* + \left(G_{i-1,i}^{\mathrm{T}} G_{i-1,i} + G_{ii}^{\mathrm{T}} G_{ii} + \right.$$

$$\left. + G_{i+1,i}^{\mathrm{T}} G_{i+1,i+1} + R_i \right) u_i^* + \left(G_{i,i}^{\mathrm{T}} G_{i,i+1} + G_{i+1,i}^{\mathrm{T}} G_{i+1,i+1} \right) u_{i+1}^* + G_{i+1,i}^{\mathrm{T}} G_{i+1,i+2} u_{i+2}^* =$$

$$= G_{i-1,i}^{\mathrm{T}} \left(w_{i-1} - \hat{y}_{i-1,0} \right) + G_{ii}^{\mathrm{T}} \left(w_i - \hat{y}_{i0} \right) + G_{i+1,i}^{\mathrm{T}} \left(w_{i+1} - \hat{y}_{i+1,0} \right). \tag{10}$$

Building blocks of the matrices from (10), gets

$$\Omega_{i,i-2} u_{i-2}^* + \Omega_{i,i-1} u_{i-1}^* + \Omega_{ii} u_i^* + \Omega_{i,i+1} u_{i+1}^* + \Omega_{i,i+2} u_{i+2}^* = \psi_{i-1} + \psi_i + \psi_{i+1}. \tag{11}$$

From (11), result the condensed matrix notation: $\Omega_i u_i = \psi_i$. With Ω_i is the dependent coefficient matrix and ψ_i is the vector of independent terms.

In the case of existence of two interconnected control areas, the optimum vector sequences inputs u_i^*, with $i = \{1,\ 2\}$, is defined as:

$$\begin{bmatrix} \Omega_{11} & \Omega_{12} \\ \Omega_{21} & \Omega_{22} \end{bmatrix} \cdot \begin{bmatrix} u_1^* \\ u_2^* \end{bmatrix} = \begin{bmatrix} \psi_1 \\ \psi_2 \end{bmatrix}. \tag{12}$$

Where the matrix elements Ω_i are given by:

$$\begin{cases} \Omega_{11} = G_{11}^{\mathrm{T}} G_{11} + G_{21}^{\mathrm{T}} G_{22} + R_1; \\ \Omega_{12} = G_{11}^{\mathrm{T}} G_{12} + G_{21}^{\mathrm{T}} G_{22}; \\ \Omega_{21} = G_{12}^{\mathrm{T}} G_{11} + G_{22}^{\mathrm{T}} G_{21}; \\ \Omega_{22} = G_{12}^{\mathrm{T}} G_{12} + G_{22}^{\mathrm{T}} G_{22} + R_2. \end{cases} \tag{13}$$

The matrix elements ψ_i are given by:

$$\begin{cases} \psi_1 = G_{11}^{\mathrm{T}} \left(w_1 - \hat{y}_{10} \right) + G_{21}^{\mathrm{T}} \left(w_2 - \hat{y}_{20} \right); \\ \psi_2 = G_{12}^{\mathrm{T}} \left(w_1 - \hat{y}_{10} \right) + G_{22}^{\mathrm{T}} \left(w_2 - \hat{y}_{20} \right). \end{cases} \tag{14}$$

4   Numerical Simulation

A system composed by interconnection of two control areas (with different characteristics) is used to illustrate the dynamic behavior using models presented in Section 2.

Some simulations including several changes in the interconnected control areas operational conditions were developed including: deviation of active power consumed by the load (ΔP_c) for different values in each control area and, at the same time, the distance between control areas (l_L - length of the tie line). The combination of these changes is assigned by *scenarios*. The purpose of these scenarios is to analyze the behavior of deviation frequency (Δf), power generated (ΔP_G), power through the interconnection line (ΔP_{L12}) and the area control error (ΔACE).

4.1   Control Areas Configuration

Table 1 shows the simulation parameters of the areas of control.

Table 1. Simulation parameters for the control areas

Governors and turbines	
Área of control 1 (AC₁)	**Área of control 2 (AC₂)**
$\tau_{r1}=0{,}1$ s	$\tau_{r2}=0{,}075$ s
$R_1=0{,}04$ p.u.	$R_2=0{,}05$ p.u.
$\tau_{t1}=0{,}3$ s	$\tau_{t2}=0{,}25$ s

Table 1. (*continued*)

Power system (generator + load)	
P_{base}=1000 MW	
f_{base}=50 Hz (1 p.u.)	
P_{n1}=1000 MW (1 p.u.)	P_{n2}=500 MW (0,5 p.u.)
P_{c1}=600 MW (0,6 p.u.)	P_{c2}=300 MW (0,3 p.u.)
H_1=4,0 s	H_2=3,0 s
$D_1 \Rightarrow$ 2% in $f \Leftrightarrow$ 2% in P_c	$D_2 \Rightarrow$ 1% in $f \Leftrightarrow$ 1% in P_c
(strongly inductive load)	(weakly inductive load)
Tie-line	
Reactance $\Leftrightarrow X_L$=0,5 ΩKm^{-1}	
Resistance $\Leftrightarrow R_L$=0,05 ΩKm^{-1}	

These values that were taking into account are typical values of exploitation of energy systems. All the parameters defined in the table are in nominal power of the respective control area. The base value for the power is: P_{base}=1000 MW. For the frequency will be chosen the nominal frequency of the network: f_{base}=50 Hz.

The simplified model of two interconnected areas is depicted in Fig. 2.

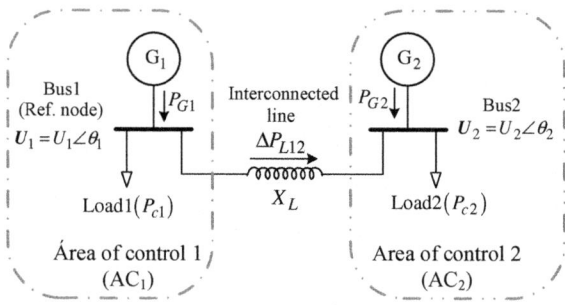

Fig. 2. Interconnection of the two control areas

From this figure, it can be noted that the Bus1 is the emitter node or reference, whereas the Bus2 is the receptor node. The voltages on the buses (Bus1 and Bus2) are in modulus U_1 and U_2 and phase angle θ_1 and θ_2, respectively, for AC$_1$ and AC$_2$. Table 2 shows the voltage on buses (Bus1 and Bus2) depending on the distance between the control areas.

Table 2. Voltage in each node depending the distance between areas

l_L=120 km	l_L=350 km
θ_1=0°=0 rad	θ_1=0°=0 rad
θ_2=10°=0.175 rad	θ_2=20°=0.349 rad
U_{base}=150 kV	U_{base}=400 kV
U_1=150 kV=1 p.u.	U_1=400 kV=1 p.u.
U_2=154.5 kV=1.03 p.u.	U_2 = 404.5 kV=1.011 p.u.

The step change of the power consumed by the load has several levels: 0%, 5%, 20%. Actually, a load variation of 20% never happens in reality (which would cause a blackout). Typically the load has a slow variation of a few tenths per cent of the power consumed. It was also observed the influence of the distance between control areas. Two distances between control areas where considered, one closest (l_L=120 km) and the other further apart (l_L=350 km).

Table 3. Exploration scenarios for two interconnected areas

ΔP_c	Scenarios	l_L
	Scenario1	
ΔP_{c1}	20%·P_{c1}=0.12 p.u.	
ΔP_{c2}	0%· P_{c2}=0.0 p.u.	
	Scenario2	
ΔP_{c1}	0%· P_{c1}=0.0 p.u.	**120 km**
ΔP_{c2}	5%· P_{c2}=0.03 p.u.	
	Scenario3	
ΔP_{c1}	20%· P_{c1}=0.12 p.u.	
ΔP_{c2}	5%· P_{c2}=0.03 p.u.	
	Scenario4	
ΔP_{c1}	20%· P_{c1}=0.12 p.u.	**350 km**
ΔP_{c2}	5%· P_{c2}=0.03 p.u.	

4.2   Simulation Results. Interconnected Control Areas

For interconnected systems, it is required, in addition to typical requirements in terms of frequency response, that the deviation of the power line interconnection must be zero. This basically requires that each area search fulfill the agreed values of power exchanges. Therefore, in each *ACE*, the frequency and tie line power deviations must vanish. Tuning parameters for MPC controller with secondary control are given in Table 4.

Table 4. Tuning parameters in block "*MPC Controller*" of MATLAB/*Simulink* [7]

Prediction horizon	N_p =10 time instants
Control horizon	N_c =2 time instants
Sampling time	T_S = 0.001
Penalty square error matrix	Q =1
Input weight matrix	R = 0.1
Coefficient of robust-ness/speed response	0.8

For the proposed scenarios, Table 3, figures 3, 4, and 5 displays some of the simulations performed for the decentralized case. Centralized MPC and integral control simulations are not shown due to lack of space.

Frequency Variation (Δf)

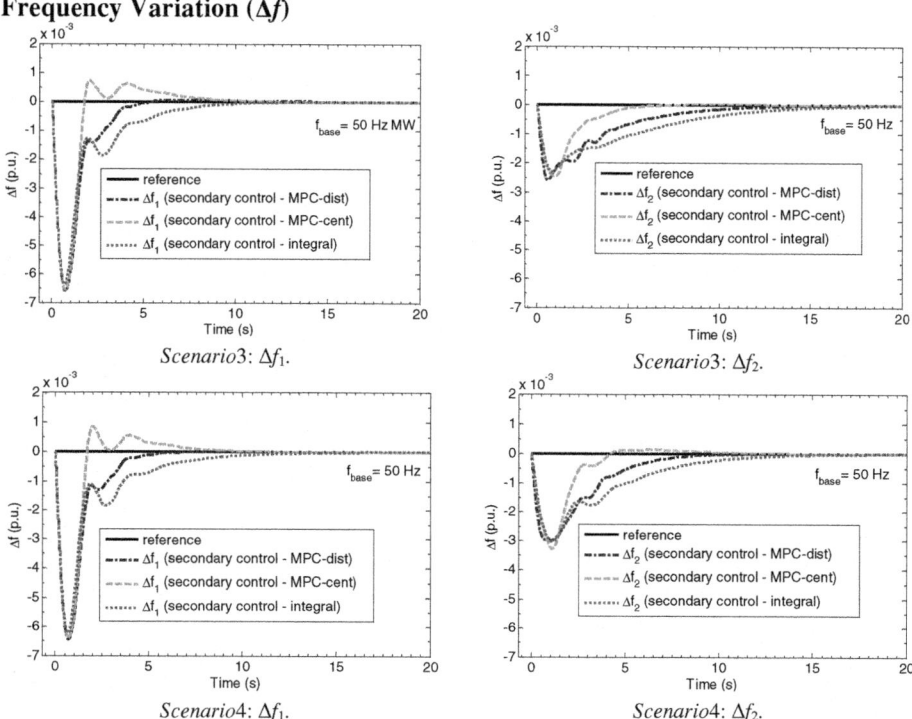

Fig. 3. Frequency deviation (Δf) with AC_1 and AC_2 interconnected with secondary control

Tie Line Deviation (ΔP_{L12})

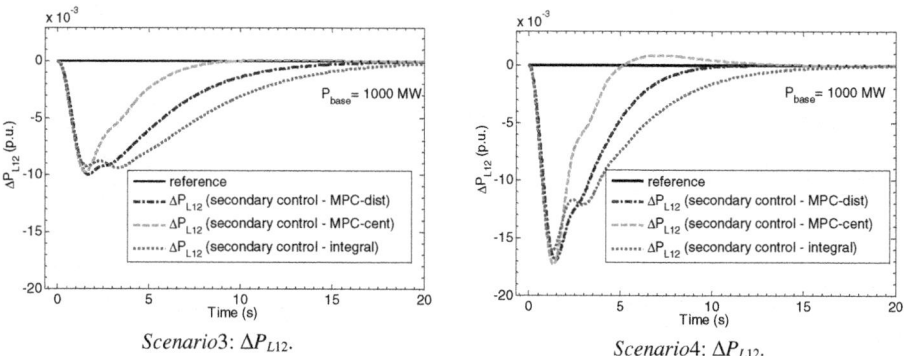

Fig. 4. Tie-line deviation (ΔP_{L12}) with AC_1 and AC_2 interconnected with secondary control

Variation of Area Control Error (ΔACE)

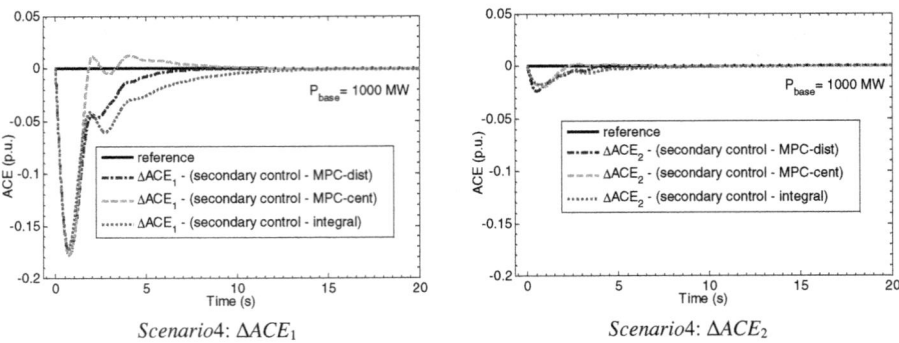

Scenario4: ΔACE_1 Scenario4: ΔACE_2

Fig. 5. Area control error deviation (ΔACE) with AC_1 and AC_2 interconnected

5   Conclusions

In conclusion, in the case of isolated systems, the *control-Pf* performs better when the controller used is the integral control (or classical control). It is a controller that has a lower frequency transient error, with a settling time similar to centralized MPC and with a response without overshoot. For interconnected control areas, the best option is the centralized MPC. The reason for this choice has to do with the best response performance of the frequency variation and tie line power deviation compared to other controllers: low oscillation; shorter settling time; rising time faster. Moreover, this controller that provides the best response for *ACE* regulation. The distributed MPC is the most suitable for distributed systems and it shows similar performance in comparison with the centralized MPC. Another factor worth mentioning is that with the increasing of distance between areas, the oscillation increases until it reaches a steady state; the settling time decreases; the overshoot is increased and the rise time is larger. However, it is evident that the MPC controllers are more insensitive to the increase of tie-line reactance, consequence of this increase the length.

Acknowledgment. This paper was supported by PEst-OE/EEI/LA0021/2013.

References

1. Anderson, G.: Dynamics and Control of Electric Power Systems, Zurich, Switzerland (2012)
2. Elgerd, O.L.: Electric Energy Systems Theory. McGraw-Hill, USA (1970)
3. Igreja, J.M., Costa, S.J., Lemos, J.M., Cadete, F.M.: Multi-agent Predictive Control with Application in Intelligent Infrastructures. In: Intelligent Systems, Control and Automation, vol. 61 (2013)
4. Kundur, P.: Power System Stability and Control. McGraw-Hill, New York (1994)
5. Negenborg, R.R.: Multi-Agent Predictive Control with Applications to Power Network. PhD Thesis, Delft University of Technology, Netherland (2007)
6. Venkat, A.N., et al.: Distribute MPC Strategies with Applications to Power Systems Automatic Generation Control. IEEE Transactions on Control Systems Technologies 16(6) (2008)
7. Wang, L.: Model Predictive Control System Design and Implementation Using MATLAB. Springer, London (2009)
8. Zhang, Y., Li, S.: Networked model predictive control based on neighborhood optimization for serially connected large-scale process. Journal Process Control (17) (2006)

Sliding Mode Generalized Predictive Control Based on Dual Optimization

Josenalde Oliveira[1], José Boaventura-Cunha[2],
Paulo Moura Oliveira[2], and Hélio F. Freire[2]

[1] Federal University of Rio Grande do Norte, Agricultural School of Jundiaí,
Macaíba RN 59280-000, Brazil
josenalde@eaj.ufrn.br
[2] INESC TEC - INESC Technology and Science (formerly INESC Porto,
UTAD pole) Department of Engineering, School of Sciences and Technology 5000-801
Vila Real, Portugal
{jboavent,oliveira}@utad.pt, freireh@gmail.com

Abstract. This work presents a new approach to tune the parameters of the discontinuous component of the Sliding Mode Generalized Predictive Controller (SMGPC) subject to constraints. The strategy employs Particle Swarm Optimization (PSO) to minimize a second aggregated cost function. The continuous component is obtained by the standard procedure, by Sequential Quadratic Programming (SQP), thus yielding a dual optimization scheme. Simulations and performance indexes for a non minimum linear model result in a better performance, improving robustness and tracking accuracy.

Keywords: Model Predictive Control, Sliding Mode, Particle Swarm Optimization, Evolutionary Computing, Robustness.

1 Introduction

Since the considerations by Young et. al. [1], the research on Sliding Mode Control theory (SMC) and its applications have been of increasing interest, since it provided an engineering look at SMC, mainly clarifying aspects such as chattering phenomena, both in continuous and discrete time. SMC is a nonlinear control scheme known to be robust to model uncertainties, disturbances and unmodeled dynamics, being quite suitable for industrial environments. The key idea consists in choosing a function of the state variables (sliding surface) in which all trajectories must reach in finite time (reaching phase) and, once reached, can not escape, *sliding* to the desired final value (sliding phase). A control law is then designed to force the trajectories towards this surface (corrective action) and, moreover, to keep them thereafter (equivalent control [2]). This control law must be discontinuous or, at least, it must contain a discontinuous component.

Model Predictive Controllers (MPC) are interesting for linear, nonlinear, time-delayed and non minimum phase systems. It offers a straightforward design method to anticipate future control actions within some time horizon (control

A.P. Moreira et al. (eds.), *CONTROLO'2014 - Proc. of the 11th Port. Conf. on Autom. Control,*
Lecture Notes in Electrical Engineering 321, DOI: 10.1007/978-3-319-10380-8_9

horizon), in order to track a future behavior (in some prediction horizon), predicted by the explicit model. For parametric models, Generalized Predictive Control (GPC) is the common choice [3] and it is very used in industry, since it allows the direct treatment of practical constraints such as actuator and output limits, associated with the minimization or maximization of some objectives, expressed in its simpler form as terms of an aggregated quadratic cost function. For this optimization scenario, Sequential Quadratic Programming (SQP) is usual.

In [4], an attempt to aggregate robustness to GPC by combining it with SMC was firstly reported. This well succeeded *melting* motivated other works and applications [5, 6, 7, 8, 9]. Most of the work referenced is concerned with common process control problems, such as delayed and non minimum phase systems, often represented by a First Order Plus Dead Time (FOPDT) transfer function. For this type of models, [5] proposed for the discontinuous component of the control law a set of tuning equations for the initial values, as a function of the characteristic parameters of the FOPDT model. When other model structures are considered, these equations are no longer valid and intelligent approaches are interesting in order to help online tuning of the controller.

Particle Swarm Optimization (PSO) is a natural inspired computation technique introduced by [10] and since then has been deployed for its simplicity and high efficiency in searching global optimal solution in problem spaces. This feature attracted the attention of control engineers in the sense of a simple way of searching and optimal tuning of controller's parameters [11]. In classical SMC, a possibility is for instance the tuning of sliding surface parameters, as obtained in [12].

Following some design steps of the SMPC presented in [9], here named Sliding Mode Generalized Predictive Controller (SMGPC), this article keeps SQP for the optimization of the continuous component of the control law responsible for the sliding phase, but proposes PSO as a optimization tool for selecting optimal parameters for the discontinuous component of the control law, thus yielding a dual optimization scheme (henceforth named Dual-SMGPC), applicable to a wider class of systems. Besides, now its parameters are adaptive, providing robustness during reaching phase. In traditional SMGPC, these parameters are kept constant and calculated offline, normally through simulations. Simulations on a non minimum phase linear system [5] will be presented and the results compared with classical SMGPC. This paper is organized as follows: section 2 states the GPC and SQP optimization problem; section 3 details the PSO algorithm used and discuss adjustment criteria; section 4 presents and comment simulations results; section 5 gives some conclusions and encourage future works.

2   Sliding Mode Generalized Predictive Control

The SMGPC presented here is based on a Controller Auto-Regressive Integrated Moving-Average model (CARIMA), considered linear around each operating point and described as:

$$\Delta A(q^{-1})y(k) = B(q^{-1})\Delta u(k - d - 1) + \xi(k), \tag{1}$$

where d is the delay from input to output (here considered as a multiple of the sampling time), u is the input signal, q^{-1} is the backward-shift operator, $\Delta : 1 - q^{-1}$ and ξ is the zero mean white noise. A and B are polynomials in q^{-1} defined as:

$$A(q^{-1}) = 1 + a_1 q^{-1} + a_2 q^{-2} + \ldots + a_{na} q^{-na}, \qquad (2)$$
$$B(q^{-1}) = b_0 + b_1 q^{-1} + b_2 q^{-2} + \ldots + a_{nb} q^{-nb}. \qquad (3)$$

2.1 Controller Design

According to the SMC theory [2], the first step to design the controller is to define a sliding surface, $S(t) = 0$, along which the process can slide to find its desired final value. Very often, it is chosen in such a way that represents a desired system dynamics and/or control objective. For instance, $S(t) = 0$ could be the tracking error $e_o = y - w$, with w being some reference signal. From ([7], [9]), the j-step ahead prediction of $S(k) = 0$ with information until the actual instant $t = k$ is given by:

$$\hat{S}(k + j) = P_s(q^{-1})(\hat{y}(k + j) - w(k + j)) + Q_s(q^{-1})\Delta u\,(k + j - 1 - d). \quad (4)$$

Polynomials $P_s(q^{-1})$, $Q_s(q^{-1})$ have degree np and nq respectively, and allow to design the desired dynamics in the sliding condition. A common adjustment is choosing $P_s(q^{-1})$ and $Q_s(q^{-1})$ as a first order system:

$$\frac{Q_s(q^{-1})}{P_s(q^{-1})} = \frac{(1 - \alpha)q^{-1}}{1 - \alpha q^{-1}}, \qquad (5)$$

with $0 < \alpha \leq 1$, since all roots of $P_s(q^{-1})$ must be inside the unit circle [9]. As $\alpha \to 0$ the dynamics is faster.

The cost function aggregates, through a weighted combination, two simultaneous objectives:

$$J_C = \sum_{j=N_1}^{N_y} \lambda_1(j) \left[\hat{S}(k + j)\right]^2 + \sum_{j=1}^{N_u} \lambda_2(j) \left[\Delta u(k + j - 1)\right]^2 \qquad (6)$$

where $N_1 - N_y$ is the period of time in which one desires the output tracks the reference signal and N_u is the control horizon. Here, the proposed cost function employs Dynamic Weighted Aggregation (DWA), since even for only a bi-objective problem, it presented better results than Conventional Weighted Aggregation (CWA) with fixed weights [13] defined as follows:

$$\lambda_1(k) = |sin\,(2\pi k/a)\,|, \qquad\qquad \lambda_2(k) = 1 - \lambda_1(k), \qquad (7)$$

where a is a user-defined adaptation frequency.

Likewise GPC, to minimize (6) the output prediction within the interval $j = [N_1, N_y]$ is formed by two parts:

$$\hat{y}(k+j) = y_f + y_l, \tag{8}$$

where y_f is the forced response (considering the initial conditions null but subject to future control actions) and y_l is the free response (natural system response from the initial conditions with no future control actions).

Forced Response, y_f. The forced response may be calculated from the step response of the parametric model (1):

$$y_f = H\Delta u, \tag{9}$$

where,

$$H_{(N_y - N_1 + 1) \; x \; N_u} = \begin{bmatrix} h_0 & 0 & \cdots & 0 \\ h_1 & h_0 & \cdots & 0 \\ \vdots & \vdots & \ddots & \vdots \\ h_{N_y-1} & h_{N_y-2} & \cdots & h_{N_y-N_u} \end{bmatrix}, \tag{10}$$

$$\Delta u_{(N_u \; x \; 1)} = \begin{bmatrix} \Delta u(k) \; \Delta u(k+1) \ldots \Delta u(k+N_u-1) \end{bmatrix}^T. \tag{11}$$

The parameters that compound H may be get from:

$$h_j = -\sum_{i=1}^{j} a_j h_{j-i} + \sum_{i=0}^{j-1} b_i, \qquad h_j = 0, \forall (j-i) < 0. \tag{12}$$

Free Response, y_l. Although y_f and y_l might be calculated by solving a Diophantine equation [9], as usual, here the available parametric model is used, thus simplifying the design. The concept of free response says that the future control signal increments after $t = k - 1$ are zero, then $\Delta u(k+j) = 0$, $j \geq 0$. Another assumption is that $y_l(k+j) = \hat{y}(k+j)$, $j \leq 0$, since y_l takes into account the initial conditions. Considering these conditions and after some manipulation in (1), one has for the general case:

$$y_l(k+j+1) = \tilde{A}_l^T Y_l + B_l^T \Delta u_l, \tag{13}$$

with

$$\tilde{A}_l = \begin{bmatrix} -\tilde{a}_1 & -\tilde{a}_2 \ldots -\tilde{a}_{na+1}, \end{bmatrix}^T, \tag{14}$$

$$Y_l = \begin{bmatrix} y_l(k+j) \; y_l(k+j-1) \ldots y_l(k+j-na) \end{bmatrix}^T, \tag{15}$$

$$B_l = \begin{bmatrix} b_0 \; b_1 \ldots b_{nb} \end{bmatrix}^T, \tag{16}$$

$$\Delta u_l = \begin{bmatrix} \Delta u(k-d+j) \; \Delta u(k-d+j-1) \ldots \Delta u(k-d+j-nb) \end{bmatrix}^T, \tag{17}$$

and $\tilde{a}_1 = a_1 - 1, \tilde{a}_2 = a_2 - a_1, \ldots, \tilde{a}_{na} = a_{na} - a_{na-1}, \tilde{a}_{na+1} = -a_{na}$.

Prediction of the Sliding Surface. SMGPC control law, $\Delta u_{SMGPC}(k)$, is the combination of two additive parts: a continuous part $\Delta u_c(k) = \Delta u(k)$ (11) developed like an GPC by the minimization of (6) using SQP, which is responsible for keeping the process variable on the reference value, and a discontinuous part $\Delta u_d(k)$ to be detailed further, responsible for guiding the system to the sliding surface. To calculate $\Delta u_c(k)$, (8) is substituted into (4) and, after putting it into matrix form, one has:

$$\hat{S}(k) = (P_s H + Q_s)\,\Delta u_c(k) + P_s\,(y_l(k) - w(k)), \qquad (18)$$

with the matrices P_s and Q_s composed from (5), $p_{s0} = 1, p_{s1} = -\alpha, q_{s0} = 0, q_{s1} = 1 - \alpha, np = 1, nq = 1$,

$$P_s = \begin{bmatrix} p_{s0} & 0 & \dots & 0 \\ p_{s1} & p_{s0} & \dots & 0 \\ \vdots & \vdots & \ddots & \vdots \\ 0 & p_{snp} & \dots & p_{s0} \end{bmatrix}, \quad Q_s = \begin{bmatrix} q_{s0} & 0 & \dots & 0 \\ q_{s1} & q_{s0} & \dots & 0 \\ \vdots & \vdots & \ddots & \vdots \\ 0 & q_{snq} & \dots & q_{s0} \end{bmatrix}, \qquad (19)$$

and being of orders $(N_y - N_1 + 1)\ x\ (N_y - N_1 + 1)$ and $(N_y - N_1 + 1)\ x\ N_u$, respectively. Substituting (18) into (6), the cost function becomes:

$$J_C = \frac{1}{2}\Delta u_c^T G \Delta u_c + \Delta u_c^T b + f_0, \qquad (20)$$

with $Z = P_s H + Q_s$, $f_0 = \lambda_1 (P_s (y_l - w))^T (P_s (y_l - w))$, $b = 2\lambda_1 Z^T P_s (y_l - w)$ and, $G = 2(\lambda_1 Z^T Z + \lambda_2 I)$. A necessary and sufficient condition for discrete time reaching motion is $|\hat{S}(k + 1)| \le \hat{S}(k)$ [9] and a control law satisfying this condition does guarantee that all trajectories will enter and remain within at least a non-increasing domain. In order to guarantee the reaching condition of (18), a constraint is added into the optimization problem of (20), namely:

$$- S(k)\mathbf{1} + P_s(y_l - w) \le (P_s H + Q_s)\Delta u_c \le S(k)\mathbf{1} - P_s(y_l - w), \qquad (21)$$

where $\mathbf{1}_{N_u x 1}$ is a vector whose entries are ones.

Discontinuous Control Signal Component. Δu_d is given by:

$$\Delta u_d(k + j) = K_d \frac{\hat{S}(k + j)}{|\hat{S}(k + j)| + \delta}, \quad \delta \ge 0, \qquad (22)$$

where K_d is a gain responsible for the velocity of the reaching mode but also increases chattering, reduced by an appropriate δ. A pair (K_d, δ) directly affects system performance and they are commonly selected through simulations. For

FOPDT models, [5] proposed some initial tuning equations, but here it is presented a new methodology based on PSO for a wider class of systems and model independent, providing an adaptive online scheme for tuning these parameters.

PSO is a popular population based metaheuristic search algorithm which can search for optima [10]. In this case, besides (20), a second quadratic bi-objective cost function must be minimized by PSO, as follows:

$$J_D = \lambda_3 \left[(y_l - w)^T (y_l - w) \right] + \lambda_4 \left[\Delta u_d^T \Delta u_d \right], \tag{23}$$

where $\lambda_3(k) = sign(sin(2\pi k/a))$ and $\lambda_4(k) = 1 - \lambda_3(k)$, a defined as in (7), calculated using Bang-Bang Weighted Aggregation (BWA) [13] which is based on sign function and force the algorithm to keep moving towards the optimal solutions and, in this case, presented better results than DWA. One entity of the N_p individuals is named *particle* and its position in the swarm represents a possible solution for the problem. Therefore, a pair (K_d, δ) represents a particle in a 2-dimensional search space and modifies its movement according to its own experience and its neighboring particle experience. In classical PSO, only two equations are used and updated in each iteration: position update (X) and velocity update (V). For the ith particle of the swarm (of size N_p), one has for this case the vectors $X_i = (x_{i1}, x_{i2})$ and $V_i = (v_{i1}, v_{i2})$. The previously best visited position of the ith particle is denoted by $P_i = (p_{i1}, p_{i2})$, and g is the index of the best particle of the swarm (global best). The *movement* equations are:

$$v_{id}(k) = \omega v_{id}(k-1) + c_1 r_1(p_{id}(k) - x_{id}(k)) + c_2 r_2(p_{gd}(k) - x_{id}(k)), \tag{24}$$
$$x_{id}(k) = x_{id}(k-1) + v_{id}(k), \tag{25}$$

where $d = 1, 2$, $i = 1, \ldots, N_p$. c_1, c_2 are constants, called respectively cognitive and social parameters, r_1, r_2 are random numbers drawn from a uniform distribution, for instance, in the set $[0, 1]$, and ω is the inertia weight, balancing the effect of initial velocity in the exploration and exploitation process. In this article, the fixed approach is used. The limits for the search space (therefore, for the controller parameters) are expressed by X_{max} and the maximum velocity V_{max}, normally associated with X_{max}, controls the global exploration of the particles in the swarm.

Therefore, the complete control signal increment for the SMGPC at $t = k$ is:

$$\Delta u_{SMGPC}(k) = \Delta u_c(k) + \Delta u_d(k). \tag{26}$$

It is noteworthy that at each instant k, although N_u components be calculated for Δu_c and Δu_d, only the first component of each ($u_c(k)$, $u_d(k)$) is considered and the others neglected, obeying the receding horizon principle. Then, the actual control signal sent to the process is given by:

$$u_{SMGPC}(k) = u_{SMGPC}(k-1) + u_c(k) + u_d(k). \tag{27}$$

3   Simulation Results

All simulations were carried out on an Intel Core i7 2.3 GHz, 4GB DDR3 RAM, MacOS X 10.7.3, Matlab R-2011 64-bit and represent a time of 150 seconds. Non disturbed and disturbed scenarios are tested. The disturbances are a load step change of 0.3 acting between 60s and 90s and a parameter variation of -20% equally applied to all discrete model parameters after $t = 100$s. The initial conditions are set to zero. The following parameters are set for PSO: $\omega = 0.7$, $c_1 = c_2 = 1.4962$, $N_p = 100$, Iterations= 80. For each scenario, four curves are plotted, representing the following cases: **SQP** (traditional SMGPC with fixed (K_d, δ) defined offline and tuned to avoid overshoot), **Dual1** (every 20 epochs, the population is restarted keeping the average of the present population as a new individual in the next random individuals), **Dual2** (restart, but totally random) and **Dual3** (without restart). For the initial population, all cases used the same seed for the random number generator (Mersenne Twister, seed 0). For the restarts, different seeds based on the computer clock are used to force new members. Plant model is discretized with a sampling time of 1s and Zero Order Hold (ZOH). Besides, the control signal amplitude is restricted to ± 10 and its slew rate to ± 1.

3.1   System Description and Design Parameters

The simulated system is given by

$$G_1(s) = \frac{1-s}{(s+1)^2}, \qquad G_1(z) = \frac{-0.1036z + 0.5032}{z^2 - 0.7358 + 0.1353}, \qquad (28)$$

with the following parameters for SQP: $K_d = 0.1, \delta = 1$ and PSO: $0.05 \leq K_d \leq 1$, $0.5 \leq \delta \leq 5$, $N_1 = 1, N_y = N_u = 40, \alpha = 0.3$. Fig. 1 shows that all PSO-based solutions are faster than SQP, besides their higher robustness (Fig. 3). If some overshoot is acceptable, there is not need for restarting, since Dual3 presented good response in both cases. For a faster, but non oscillatory tracking, some degree of elitism used in Dual1 seems to be interesting. This elitism may be seen as a way of preserving some aspect of the present population in the next population. Although in some cases Dual3 can offer a better robustness, this is not guaranteed, since the parameters are kept constant after initial particles movement (Fig. 2). δ has the same behavior as K_d and it is omitted here. For the hardware and software described above, the average computation time for SQP is 0.1s and for the Dual approach is 0.4s, being reasonable for the chosen sampling time. Tables 1 and 2 present some performance indexes for comparison, namely, energy consumption (sum of control actions, E_u), control effort standard deviation (σ_u) and Integral of Time Multiplied Absolute Error (ITAE), confirming a better tracking while keeping equivalent control efforts.

Table 1. Performance Indexes - non disturbed case

	SQP-SMGPC	Dual1	Dual2	Dual3
ITAE	1.4×10^4	4.9×10^3	4.9×10^3	2.7×10^3
E_u	4.54×10^2	4.64×10^2	4.70×10^2	4.70×10^2
σ_u	0.9433	0.9852	0.9995	1.0045

Table 2. Performance Indexes - disturbed case

	SQP-SMGPC	Dual1	Dual2	Dual3
ITAE	3.1×10^4	1.5×10^4	1.3×10^4	1.4×10^4
E_u	8.11×10^2	7.5×10^2	7.5×10^2	7.5×10^2
σ_u	1.7061	1.7847	1.8559	1.8053

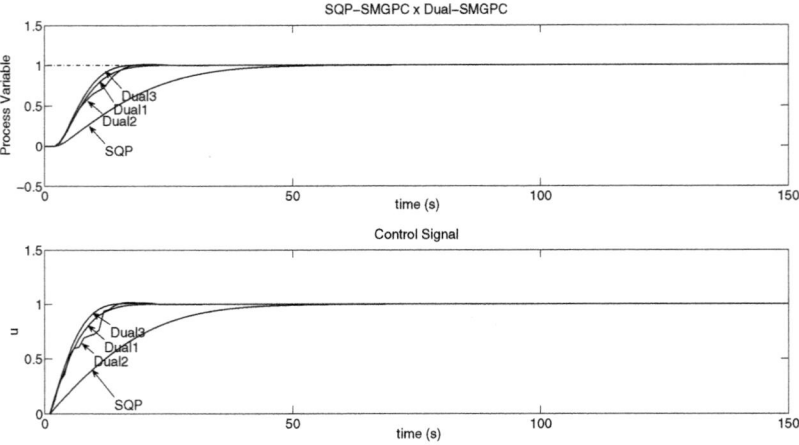

Fig. 1. Non disturbed case - top plot: system output - bottom plot: control signal

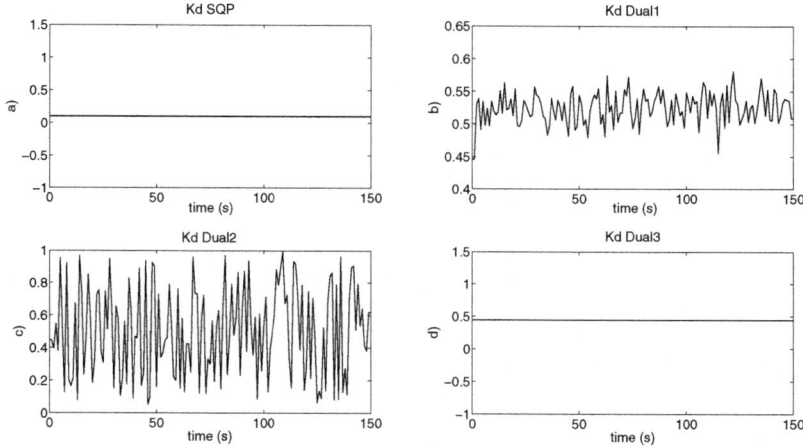

Fig. 2. K_d parameter for a) SQP - b) Dual1 - c) Dual2 - d) Dual3

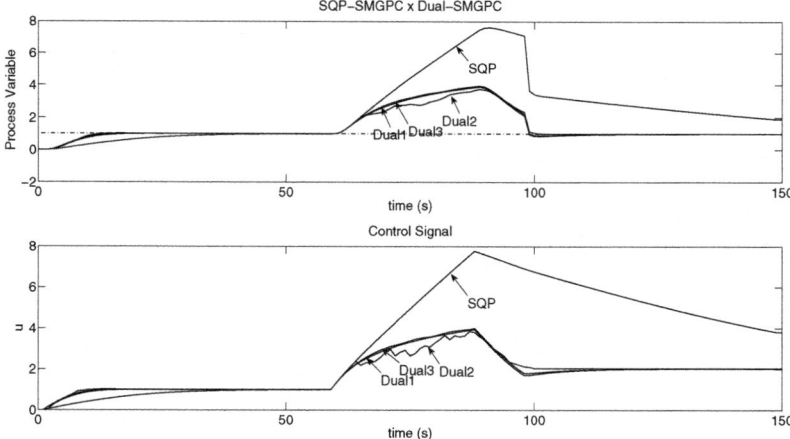

Fig. 3. Disturbed case - top plot: system output - bottom plot: control signal

4   Conclusions and Further Works

A new model independent scheme based on Particle Swarm Optimization has been developed to tune the parameters of the discontinuous control law component of a Sliding Mode Generalized Predictive Controller, providing robustness and turning the design process easier. Simulations results and performance indexes for a non minimum phase system, commonly found in process control industry, were presented, showing a better tracking without significant control effort and computational time increase. Therefore, future works will explore theoretical stability and convergence aspects, including long time-delayed systems. Studies on practical issues and real implementation will also be considered.

References

1. Young, K.D., Utkin, V.I., Özgüner, Ü.: A Control Engineer's Guide to Sliding Mode Control. IEEE Transactions on Control Systems Technology 7, 328–342 (1999)
2. Zinober, A.S.: Variable Structure and Lyapunov Control. Springer, London (1994)
3. Clarke, D.W., Mohtadi, C., Tuffs, P.S.: Generalized Predictive Controller - Part i The Basic Algorithm. Automatica 23, 137–148 (1987)
4. Corradini, M.L., Orlando, G.: A VSC Algorithm based on Generalized Predictive Control. Automatica 33, 927–932 (1997)
5. Camacho, O., Rojas, R., García, W.: Variable Structure Control Applied to Chemical Processes with Inverse Response. ISA Transactions 38, 55–72 (1999)
6. de la Parte, M.P., Camacho, O., Camacho, E.F.: Development of a GPC-based Sliding Mode Controller. ISA Transactions 41, 19–30 (2002)
7. García-Gabín, W., Camacho, E.F.: Sliding Mode Model Based Predictive Control for Non Minimum Phase Systems. In: European Control Conference, Cambridge, UK (2003)
8. García-Gabín, W., Zambrano, D., Camacho, E.F.: Sliding Mode Predictive Control for Chemical Process with Time Delay. In: 16th IFAC World Congress, Prague, Czech Republic, pp. 1677–1682 (2005)
9. García-Gabín, W., Zambrano, D., Camacho, E.F.: Sliding Mode Predictive Control of a Solar Air Conditioning Plant. Control Engineering Practice 17, 652–663 (2009)
10. Kennedy, J., Eberhart, R.C.: Particle Swarm Optimization. In: IEEE International Conference on Neural Networks, Perth, Australia, pp. 1942–1948 (1995)
11. Oliveira, P.B.M., Boaventura Cunha, J., Coelho, J.P.: Design of PID Controllers using Particle Swarm Algorithm. In: IASTED International Conference Modelling, Identification and Control, Innsbruck, Austria, pp. 263–268 (2002)
12. Serbencu, A.E., Serbencu, A., Cernega, D.C., Minzu, V.: Particle Swarm Optimization for the Sliding Mode Controller Parameters. In: 29th Chinese Control Conference, Beijing, China, pp. 1859–1864 (2010)
13. Jin, Y., Olhofer, M., Sendhoff, B.: Evolutionary Dynamic Weighted Aggregation for Multi-Objective Optimization: Why does it Work and How?. In: GECCO Conference, San Francisco, CA, pp. 1042–1049 (2001)

On Wind Turbine Model Predictive Pitch Control: An Event-Based Simulation Approach

Carla Viveiros[1,2,3], Rui Melício[1,3], José Manuel Igreja[2], and Victor M.F. Mendes[1,2]

[1] Department of Physics, Universidade de Évora, 7000 Évora, Portugal
ruimelicio@uevora.pt
[2] Department of Electrical Engineering and Automation, Instituto Superior de Engenharia de Lisboa, 1959-007 Lisbon, Portugal
[3] IDMEC/LAETA, Instituto Superior Técnico, Universidade de Lisboa, Lisbon, Portugal
{cviveiros,jigreja,vfmendes}@deea.isel.ipl.pt

Abstract. This paper is about a hierarchical structure with an event-based supervisor in a higher level and a model predictive control (MPC) in a lower level applied to a wind turbine. The event-based supervisor analyzes the operation conditions to determine the state of the wind turbine. The objectives and constraints of the model predictive controller can also be adjusted by this supervisory controller. This hierarchical structure is able to ensure the performance and reliability required for a wind turbine to be integrated into an electric grid. Comparisons between model predictive pitch control and a default proportional integral pitch controller applied to a wind turbine benchmark are given and simulation results by Matlab/Simulink are shown in order to prove the effectiveness of the proposed approach.

Keywords: Wind turbine, model predictive control, supervisory control, event-based supervisor.

1   Introduction

Currently, the wind energy conversion system (WECS) deployment is in expansion, contributing to an increase share of converting renewable energy into electric energy. In the first half of 2013, wind power exploitation has a growth of 16.6% [1]. The increasing integration of wind power has led to the modern megawatt WECS which is an expensive one and therefore reliability is an important issue. WECS are classified into two types: constant speed and variable speed. Although, constant-speed designs dominated the market, new requirements led to the emergence of variable speed ones [2,3]. A WECS running at variable speed has add advantages: mechanical stress is reduced, torque oscillations are not transmitted to the electric grid, between the cut-in and the rated wind speed the rotor speed is controlled to achieve the best aerodynamic efficiency. A variable speed WECS connected to the electric grid has either a doubly fed induction generators (DFIGs) or a full-power converter. A variable-speed WECS having a DFIG [4] is implemented with the converter feeding the rotor winding and the stator winding is connected to the electric grid. The suitable use of control systems

© Springer International Publishing Switzerland 2015
A.P. Moreira et al. (eds.), *CONTROLO'2014 - Proc. of the 11th Port. Conf. on Autom. Control*,
Lecture Notes in Electrical Engineering 321, DOI: 10.1007/978-3-319-10380-8_10

on WECS can provide for better adequacy in what regard the diminishing of losses of profit. Control systems ability to collect, analyze and process data from the wind turbine is an important issue for modern megawatt WECS. Also, the integration of a WECS into electric energy systems compels the use of control systems in order to avoid performance degradation on the quality of energy injected into the electric grid [4]. Power capturing is of extreme importance for modern megawatt WECS and a suitable control system in indispensable to lessen the losses of profit. A pitch control system is the most suitable for regulating the power capturing by the rotor due to the different positions of the blades given by the pitch angle, influencing the level of power captured.

The control system for a WECS has to consider the fact that the wind turbine is driven by the wind power which is an uncontrolled input and also acts as a disturbance one. Thus, the design of a control strategy for a wind turbine must consider a series of important aspects such as wind speed, wind turbine components, the influence of the wind speed on these components and the performances that the closed loop system must have. In order to accomplish this goal, in the past years, researchers have been studying different control strategies from classical gain scheduling technique [5], fuzzy logic control [6] through adaptive LQG control [7]. Model predictive control (MPC) application in renewable energy sources has been increasing due to optimality, stability and robustness properties of this control [8,9]. MPC introduces the advantage of taking into account system constraints [10] and optimizes the control to meet the objectives, which has a reasonably interest not only for WECS, but also for systems converting solar energy into electric energy [11].

Event-based supervisor is based on supervisory control theory [12] and has been proven as a handy insert on systems for control application such as set-point, monitoring, fault detection, diagnosis, scheduling, planning and production optimization. The supervisory control system offers capabilities for conveniently scheduling of the different phases on the operation, i.e., startup, production and shutdown, based upon the information observed for the current state and the processing in order to achieve the specification required on the operation.

This paper is about a hierarchical structure with an event-based supervisor in a higher level and a MPC in a lower level applied to a wind turbine. The event-based supervisor analyzes the operation conditions to determine the state of the wind turbine and passes that information into the lower level where model predictive pitch controller acts on the value of the pitch angle in order to maintain the output power around the nominal value, i.e., the rated power. The simulations are carried out in Matlab/Simulink and makes use of the toolbox Stateflow chart for the supervisory control. Also, the wind turbine benchmark model developed by [13] is used. Comparisons between model predictive pitch control and a default controller consisting of a proportional integral pitch controller in the lower level are given.

The rest of the paper is organized as follows: Section 2 describes the wind turbine benchmark; Section 3 presents the control strategies; Section 4 presents a case study of a wind turbine benchmark and simulation results for the comparison purpose. Finally, concluding remarks are given in Section 5.

2   Modeling

The variable speed WECS considered is a conventional horizontal axis turbine with a three-bladed rotor design and the rotor is positioned upwind of the tower. The turbine has a rated power of 4.8 MW. A more detailed description for the wind turbine benchmark model can be seen in [13].

2.1   WECS Model

The WECS can be analyzed on a benchmark block diagram with functional systems namely: the blade and pitch system, drive train system, generator and power converter system and the controller.

The block diagram of the benchmark model presented in [13] is shown in Fig. 1.

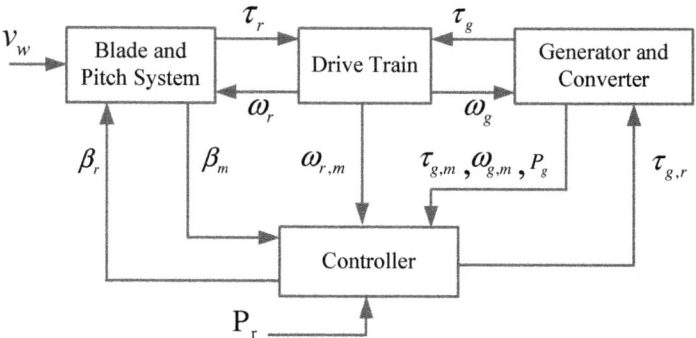

Fig. 1. Block diagram of the wind turbine benchmark [13]

In Fig.1, the following notation is used:

$v_w \, [m/s]$	wind speed	$\tau_r \, [Nm]$	rotor torque
$\tau_g \, [Nm]$	generator torque	$\omega_r \, [rad/s]$	rotor speed
$\omega_g \, [rad/s]$	generator speed	$\beta \, [°]$	pitch angle
$P_g \, [W]$	generator power	$P_r \, [W]$	rated power

The r, m subscripts stand respectively for the identification of references or rotor and measurements values.

Blade and Pitch System Model. This model is a combination of the aerodynamic and pitch system model.

The aerodynamics of the wind turbine is modeled in order to determine the torque acting on the blades. The aerodynamic torque is given by:

$$\tau_r(t) = \frac{\rho \pi R^3 C_p\left(\lambda(t), \beta(t)\right) v_w(t)^2}{2} \tag{1}$$

where ρ is the air density, $C_p\left(\lambda(t), \beta(t)\right)$ is the power coefficient, which is a function of the tip speed ratio and the pitch angle. The tip speed ratio is given by:

$$\lambda(t) = \frac{\omega_r(t)R}{v_w(t)} \tag{2}$$

where R is the radius of the blades, $v_w(t)$ is the wind speed and $\omega_r(t)$ the rotor speed. From (2), when a variation in the wind speed occurs two consequences can happen: if the rotor speed is kept constant, then the tip speed ratio will change, leading to a consequent change of the power coefficient C_p and hence in the power capturing; if the rotor speed is conveniently adjusted according to the wind speed variation, then the tip speed ratio can be held at a reference point and as a consequence the power coefficient C_p can be kept at a desired value.

This second consequence can be exploited to implement a strategy for generate maximum power output from the WECS. The pitch actuator is modeled as a second order system given by:

$$\ddot{\beta}(t) = -2\xi\omega_n(t)\dot{\beta}(t) - \omega_n^2\beta(t) + \omega_n^2\beta_r(t) \tag{3}$$

Drive Train Model. The drive train consists in a low-speed shaft and a high-speed shaft interconnected having respectively inertias J_r and J_g and viscous friction coefficient B_r and B_g. The shafts are interconnected by a transmission having gear box with ratio N_g, combined with torsion stiffness K_{dt}, and torsion damping B_{dt}. The torsion angle is $\theta_\Delta(t)$, and the torque applied to the generator is $\tau_g(t)$, at a speed $\omega_g(t)$. The linear model for the drive train model is given by:

$$J_r\dot{\omega}_r(t) = \tau_r(t) + \frac{B_{dt}}{N_g}\omega_g(t) - K_{dt}\theta_\Delta(t) - \left(B_{dt} + B_r\right)\omega_r(t) \tag{4}$$

$$J_g\dot{\omega}_g(t) = \frac{K_{dt}}{N_g}\theta_\Delta(t) + \frac{B_{dt}}{N_g}\omega_r(t) - (\frac{B_{dt}}{N_g^2} + B_g)\omega_g(t) - \tau_g(t) \tag{5}$$

$$\dot{\theta}_\Delta(t) = \omega_r(t) - \frac{1}{N_g}\omega_g(t) \tag{6}$$

Generator and Power Converter Model. The generator and the power converter dynamics is modeled by a first order system where α_{gc} is the inverse of the first order time constant. The generator and the power converter model are given by:

$$\dot{\tau}_g(t) = -\alpha_{gc}\tau_g(t) + \alpha_{gc}\tau_{g,r}(t) \tag{7}$$

the generator power is given by:

$$P_g(t) = \eta_g \omega_g(t)\tau_g(t) \tag{8}$$

where η_g denotes the efficiency of the generator.

3   Control Strategy

The control strategy of such a complex system creates a significant challenge. One main objective is to achieve the maximization of the power associated with the energy conversion. This maximization occurs evidently when the wind speed is in the range between the cut-in, v_{min} and the cut-out, v_{max}, and for that range, literature on wind turbine control is consistently with the fact that four regions of operation can be distinguished and should be considered for the WECS operation. These regions are shown in Fig. 2.

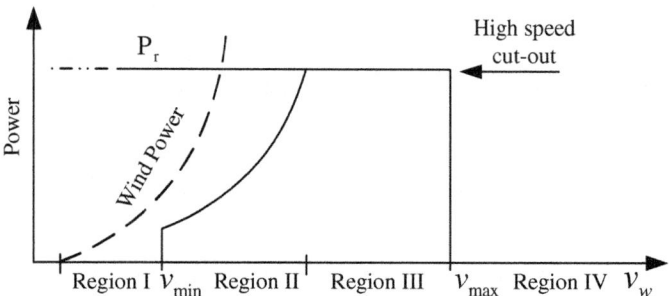

Fig. 2. Regions of power by wind speed [5]

Region I correspond to the start up of the turbine. Region II corresponds to power optimization conditions, in a wind speed range that enables the conversion at global optimum rating within safety conditions. The control objective in this region is to capture all possible wind power with a pitch angle equal to 0 degrees and control the generator torque to keep the turbine speed closest to the one that will assure optimal tip speed ratio, attaining global maximum power. Region III is characterized by a conversion at constant power due to the fact that the rated power is not greater than the input power, i.e., wind has a power that is not less than the one that is possible to convert, operating in this region is to ensure that the conversion is within the feasible limits. The control objective in this region is to operate the wind turbine at the nominal power holding the generator torque at its nominal value and keeping rotational speed of the generator constant through pitch control. Finally, region IV is characterized by too high wind speed where the wind turbine must be shut down for economic and safety reasons.

3.1   Model Predictive Control

In this paper, only regions II and III are considered. Model predictive controller takes the states of pitch system, drive train system and nonlinear aerodynamics as input. By optimizing the best output according to the operation mode and control objective the model predictive controller generates reference values for pitch angle reference, $\beta_r(k)$ and generator torque reference, $\tau_{g,r}(k)$. In region II, power optimization, it is considered $\beta_r(k)=0°$ for the pitch reference and for the generator torque reference the following equations:

$$\tau_{g,r}(k) = K_{opt}\left(\frac{\omega_g(k)}{N_g}\right)^2 \tag{9}$$

and

$$K_{opt} = \frac{1}{2}\rho AR^3 \frac{Cp_{max}}{\lambda_{opt}^3} \tag{10}$$

where A is the area swept by the blades and λ_{opt} is found as the optimum point in the power coefficient. In region III the pitch reference and generator torque reference should be adjusted simultaneously and the latter is given by:

$$\tau_{g,r}(k) = \frac{P_r(k)}{\eta_g\omega_g(k)} \tag{11}$$

From (1) to (8) a linear state space model representing the WECS dynamics at certain operating wind speed is given by:

$$\begin{aligned}x(k+1) &= Ax(k) + Bu(k) \\ y(k) &= Cx(k)\end{aligned} \tag{12}$$

where $x = \begin{bmatrix}\omega_r & \omega_g & \theta_\Delta & \tau_g & \beta\end{bmatrix}^T \in \mathbb{R}^5$ is the state vector, $u = \begin{bmatrix}\tau_{g,r} & \beta_r\end{bmatrix}^T \in \mathbb{R}^2$ is the control input and $y = \begin{bmatrix}\omega_g & P_g\end{bmatrix}^T \in \mathbb{R}^2$ is the measured output. One important objective in this region is to keep the generator speed closest to the nominal speed with the least stress on the generator and pitch control system. To accomplish this objective one must follow the mathematical formulation for the objective function, i.e., the cost function:

$$\begin{aligned}\min_{u(k)\cdots \hat{u}(k+Np-1)} J(k) &= \sum_{j=1}^{Np}\left[r(k+j)-\hat{y}(k+j|k)\right]^T Q(j)\left[r(k+j)-\hat{y}(k+j|k)\right] \\ &+ \sum_{j=0}^{Np-1}\left[u(k+j|k)\right]^T R(j)\left[u(k+j|k)\right]\end{aligned} \tag{13}$$

The cost function in (13) is associated with the prediction horizon, given by the periods j with $j = 1,...,Np$. The first term is a quadratic functional, weighted by matrix Q, of the difference between reference $r(k + j)$ and predicted output $\hat{y}(k + j|k)$ and is used for penalization of the error. Also, the second term is a quadratic functional, weighted by matrix R, measuring the control effort $u(k + j|k)$.

The optimization problem consists in the minimization of the cost function subjected to WECS dynamic model (12) and to the following constraints: $\beta_{min} \leq \beta_r \leq \beta_{max}$, $(\omega_{nom} - \omega_\Delta) \leq \omega_g \leq (\omega_{nom} + \omega_\Delta)$ and $V_{min} \leq V_w \leq V_{max}$. Where ω_{nom} is the nominal generator speed and ω_Δ is a small offset that introduce some hysteresis in the switching scheme avoiding frequent switching from control modes.

3.2   Supervisory Control

The supervisory control system analyzes the operation conditions in order to determine the operational state of the WECS. It is considered four operational states such as park, startup, generating and brake. In the park state, the wind turbine should not be rotating and the generator is not connected to the grid. The startup state has the wind above a certain speed and therefore the wind turbine should be rotating in order to produce power, region II, also the generator is still not connected to the grid. The generating state is the power production state, region III, where the turbine speed is in a certain range and generator is connected to the grid. The brake state is based on a number of conditions being possible to exit the generating state, enter the startup state or enter into the park state. The operational states used to model the event-based controller are shown in Fig. 3.

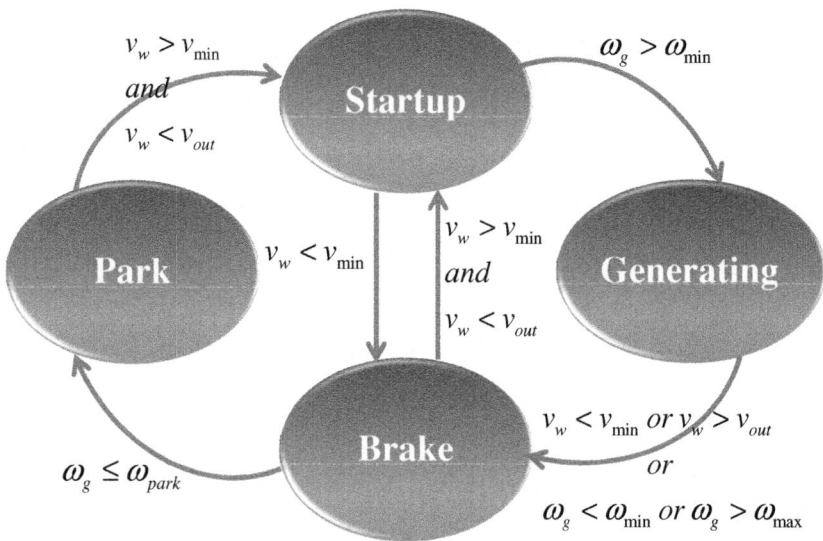

Fig. 3. Operational states and conditions

4   Simulation and Results

In order to validate the MPC controller results, its behavior is compared to conventional linear proportional integral (PI) controller [13]. The simulation was performed using Matlab/Simulink environment. The wind turbine benchmark is linearized for a power set-point, Pr, of 4.8 MW and a wind speed, v_w, of 13 m/s. The wind turbine parameters are given by: $R = 57.5$, $\rho = 1.225$, $\xi = 0.6$, $\omega_n = 11.11$, $\alpha_{gc} = 50$, $\eta_g = 0.98$ and sampling time $Ts = 0.01s$. White noise is added to the wind speed to simulate a wind disturbance. This noise is shown in Fig. 4.

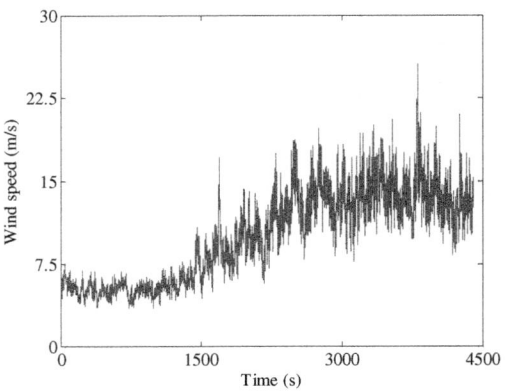

Fig. 4. Wind speed

Comparisons between generated and reference power of the default controller and the proposed controller are shown in Fig. 5 a) and b). From closed loop system response point of view, the default controller confers higher level of oscillations than the one presented by model predictive controller.

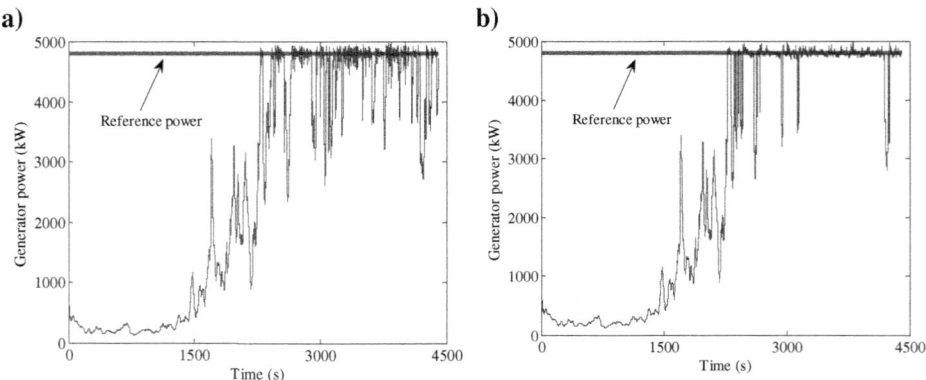

Fig. 5. Generator power and reference power **a)** PI; **b)** MPC

Comparisons between pitch angles are presented in Fig.6 a) and b). The default controller achieves angles varies around 22° with some peaks above 30° while with the model predictive controller, pitch angle stays under 20°.

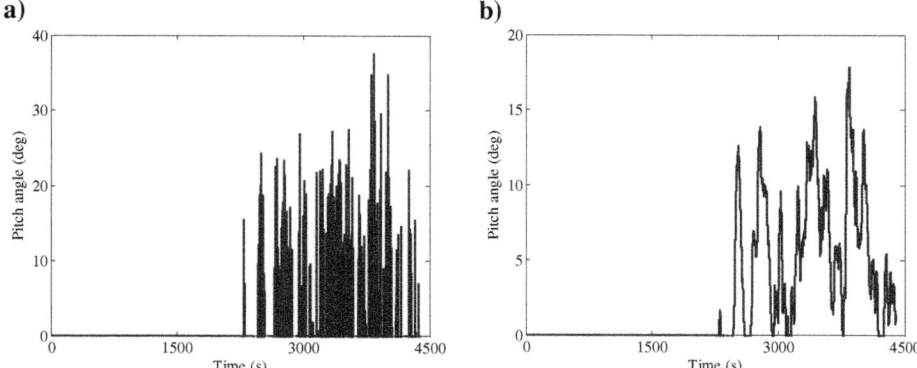

Fig. 6. Pitch angle **a)** PI; **b)** MPC

The switching between regions is presented in Fig.7 a) and b). MPC controller when compared to PI presented less switches from region II (0) to region III (1) leading to less variations on the pitch angle of the blades. This approach can minimize the mechanical stresses of the wind turbine.

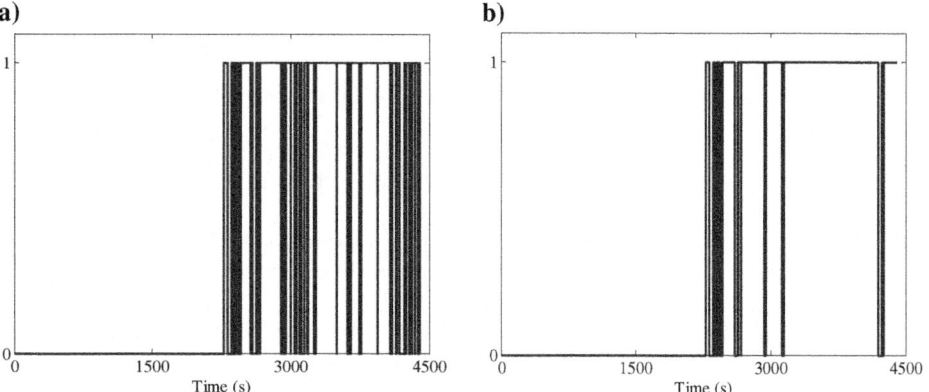

Fig. 7. Control modes **a)** PI; **b)** MPC

5   Conclusions

This paper presented a hierarchical structure with an event-based supervisor in a higher level and a model predictive control in a lower level applied to a wind turbine. An event-based supervisor was implemented in order to analyze which operations state should be turned on or turned off. Comparisons between model predictive pitch control and a default proportional integral pitch controller applied to a wind turbine benchmark were presented and simulation results by Matlab/Simulink proved the effectiveness of the design. The proposed event-based approach demonstrates that MPC controller outperformed proportional integral controller.

Acknowledgment. This work was partially supported by Fundação para a Ciência e a Tecnologia, through IDMEC under LAETA, Instituto Superior Técnico, Universidade de Lisboa, Portugal.

References

1. Gsanger, S.: World Wind Energy Half-Year Report 2013. In: 13th World Wind Energy Conference & Renewable Energy Exhibition, China (2014)
2. Garcia-Sanz, M., Houpis, C.H.: Wind Energy Systems: Control Engineering Design. CRC Press FL, Taylor & Francis, Boca Raton (2012)
3. Burton, T., Sharpe, D., Jenkins, N., Bossanyi, E.: Wind Energy Handbook. Wiley, London (2001)
4. Melício, R., Mendes, V.M.F.: Doubly fed induction generator systems for variable speed wind turbine. In: 9th Spanish-Portuguese Congress on Electrical Engineering, 9CHLIE, Marbella, Spain, pp. 161–164 (2005)
5. Bianchi, F., Battista, H., Mantz, R.: Wind turbine control systems. Springer, London (2007)
6. Scherillo, F., Izzo, L., Coiro, D., Lauria, D.: Fuzzy Logic Control for a Small Pitch Controlled Wind Turbine. In: International Symposium on Power Electronics, Electrical Drives, Automation and Motion, Italy, pp. 588–593 (2012)
7. Mateescu, R., Pintea, A., Stefanoiu, D.: Discrete-Time LQG Control with Disturbance Rejection for Variable Speed Wind Turbines. In: 1st International Conference on Systems and Computer Science, Romania, pp. 1–6 (2012)
8. Soliman, M., Malik, O.P., Westwick, D.T.: Multiple Model Predicitve Control for Wind Turbines with Doubly Fed Induction Generators. IEEE Transactions on Sustainable Energy 2(3), 215–226 (2011)
9. Wang, F., Stelson, K.A.: Model Predictive Control for Power Optimization in a Hydrostatic Wind Turbine. In: 13th Scandinavian International Conference on Fluid Power, Sweden, pp. 1–6 (2013)
10. Maciejowski, J.M., Goulart, P.J., Kerrigan, E.C.: Constrained Control Using Model Predictive Control. In: Tarbouriech, S., Garcia, G., Glattfelder, A.H. (eds.) Advanced Strategies in Control Systems with Input and Output Constraints, vol. 346, pp. 273–291. Springer, Heidelberg (2007)
11. Igreja, J.M., Lemos, J.M., Silva, R.: Nonlinear Predictive Control of a Solar Plant Based on Reduced Complexity Models. In: 16th IFAC World Congress, Czech Republic (2005)
12. Ramadge, P., Wonham, W.M.: Supervisory control of a class of discrete event processes. SIAM Journal on Control and Optimization 25, 206–230 (1987)
13. Odgaard, P.F., Stroustrup, J., Kinnaert, M.: Fault tolerant control of wind turbines: a benchmark model. IEEE Transactions on Control Systems Technology 21, 1168–1182 (2013)

Part III
Fuzzy and Neural Control

Neural PCA Controller Based on Multi-Models

Luis Brito Palma[1,2], Fernando Vieira Coito[1,2], and Paulo Sousa Gil[1,3]

[1] Universidade Nova de Lisboa - FCT - DEE, Caparica 2829-516, Portugal
[2] Uninova CTS Research Institute, Caparica 2829-516, Portugal
LBP,FJVC@fct.unl.pt
[3] CISUC Research Institute, Coimbra 3030-290, Portugal
PSG@fct.unl.pt

Abstract. In this paper, a new approach to design nonlinear adaptive PI multi-controllers, for SISO systems, based on neural local linear principal components analysis (PCA) models is proposed. The PCA neural networks only implements the integral term of the PI multi-controller, a proportional term is added to obtain a PI structure. A modified normalized Harris performance index is used for evaluating the controller performance. Some experimental results obtained with a nonlinear three tank benchmark model are presented, showing the adaptive PI-PCA multi-controller performance compared to neural linear PI controllers.

Keywords: nonlinear adaptive PI control, principal component analysis, multi-models.

1 Introduction

Industrial plants and processes are typically nonlinear systems, most of them controlled by linear or nonlinear proportional-integral (PID) controllers, [1, 2, 30]. Most loops use in fact only PI control because the derivative action is not used very often. The PID controller is used for a wide range of control problems: process control, motor drives, flight control, robotic systems in automotive industries, etc. The PID controllers appears in different forms: as standard single-loop hardware controllers, as software modules in programmable logic controllers (PLC's) and also as distributed multi-loop control systems in industrial plants.

Principal component analysis (PCA) and other multivariate statistical methods (factor analysis, discriminant analysis, etc) can be used to develop nonparametric models for process monitoring (including fault detection and diagnosis), [4, 20, 26, 27]. PCA can also be used for controller loop monitoring, [9], as well as for feedback control in the scores space, [8, 26, 27].

The PCA control formulation, in a reduced control scores space, is analogous to eigenvalue-assignment control. The great advantage of the control structure proposed in this work is the fact that only input-output process data is needed for controller tuning, resulting in a data-based controller tuning approach using neural networks.

© Springer International Publishing Switzerland 2015   103
A.P. Moreira et al. (eds.), *CONTROLO'2014 - Proc. of the 11th Port. Conf. on Autom. Control*,
Lecture Notes in Electrical Engineering 321, DOI: 10.1007/978-3-319-10380-8_11

In this paper, a nonlinear adaptive PI multi-controller based on linear local neural PCA models, to deal with nonlinear SISO systems in real-time applications, is proposed. This work is based on the previous work focused on a linear neural PCA controller, [6].

2   Control Loop Performance Analysis (CLPA) Approach Based on a Modified Normalized Harris Index

In a typical industrial plant there are dozens or even hundreds of controllers, most of them based on the PID structure. Even if these controllers initially perform well, various factors can contribute to their performance deterioration, including [16, 18]: actuator/sensor or process faults, sticking valves producing oscillations, external disturbances, to name just a few. Around 50% of all industrial controllers, or more, have some kind of performance problem, [14], [28].

Traditional performance measures such as overshoot, rise time, settling time, control error, integral error criteria, etc, have been used by some researchers for CLPA. The most widespread criterion considered for CLPA is the variance. Most of the reported industrial applications of control loop performance monitoring are based on the pioneer method suggested by Harris, [15], or some algorithms inspired on Harris approach, such as the approaches found in the references [3, 19], to name just a few. The great popularity of the methods based on the Harris performance index is due to both the conceptual and computational simplicity, as also the small amount of information required. Performance indices can be used for control loop performance monitoring or for control structure selection, [11].

The first indices for control loop performance assessment were proposed by Harris [15], Desborough and Harris [12] and Stanfelj et al [29]. A review of the status in control loop performance assessment (CLPA) technology and industrial applications was published in 2006, by Jelali [21]. Merits and drawbacks of each CLPA method are highlighted. The application of fuzzy logic and neural networks were also investigated in the CLPA field, [10].

The modified normalized Harris index used in this work is described next, summarily, [9]. In this work, the algorithms were implemented in discrete-time: at time instant t_k, the sample is k. The key variable for CLPA is the control error, $e(k) = r(k) - y(k)$. The control error should have no predictable component. Let's compute a residual signal $\delta(k)$ between the measured control error $e(k)$ and a forward prediction of the control error $\hat{e}(k)$, described by

$$\delta(k) = e(k) - \hat{e}(k) \tag{1}$$

In a control loop that is performing well it is expected that the control error contains only random noise. The normalized CLPA index used in this work $\vartheta(k)$, computed on-line for a sliding window (time horizon), is described by (2), where $var(.)$ is the variance and $mse(.)$ is the mean-squared error. For typical data from process control loops an autoregression $AR(n,b)$ time series model that makes predictions b steps ahead is suitable for modeling the forward prediction $\hat{e}(k)$, [9]. A good performing loop has a value of the CLPA index $\vartheta(k)$ close to 0, while for a poorly performing loop is close to 1. This CLPA index is used here for evaluating the controllers performance in closed-loop.

$$\vartheta(k) = 1 - \frac{var(\delta(k))}{mse(e(k))} \tag{2}$$

3 Linear Integral Controller Based on Classical PCA Model

In the past some approaches have been proposed for design classical integral controllers based on principal components analysis (PCA), [5, 7, 8, 26, 27].

The architecture of the classical integral PCA controller is depicted in Fig. 1. Detailed information about the design of this kind of controllers can be found in [5, 7, 8]. The block "**Pa**" represents the plant, "**C**" is the PCA controller, and $\boldsymbol{Q}$ is given by (3). Let's assume that k is the k sample in the algorithms.

$$\boldsymbol{Q} = (\boldsymbol{U}^T)^+ \tag{3}$$

In (3), + represents the pseudo-inverse of the matrix and T is the transposed. The matrix $\boldsymbol{U}$ is obtained from (4), where $x_d(k) = y(k)$ is the process output.

$$\boldsymbol{U} = (\boldsymbol{\Upsilon}^T \, x_d(k))^T \tag{4}$$

Assuming two principal components, $a = 2$, $\boldsymbol{\Upsilon}$ is the projection of the regression matrix $\boldsymbol{X}$ (input/output data) on the 2D PCA scores space, accordingly (5).

$$\boldsymbol{\Upsilon} = \boldsymbol{X} \, \boldsymbol{P} \tag{5}$$

Assuming that the training data matrix is $\boldsymbol{X} \in \Re^{n \times m}$, where n is the number of rows and m is the number of columns, the matrix $\boldsymbol{P}$ corresponds to the first a columns of the loadings (singular vectors) matrix $\boldsymbol{V}$, i.e. $\boldsymbol{V} = (\boldsymbol{P})_{:,1\ldots a}$, for a SVD decomposition as follows (6), assuming that $\boldsymbol{\Lambda}$ is the singular values matrix.

$$cov(\boldsymbol{X}) = \frac{1}{n-1} \, \boldsymbol{X}^T \, \boldsymbol{X} = \boldsymbol{V} \, \boldsymbol{\Lambda} \, \boldsymbol{V}^T \tag{6}$$

Assuming an ARX(2,2,1) model for the plant/process under study, the projection of the on-line regressor data vector $\boldsymbol{x}(k) = [y(k) \, y(k-1) \, y(k-2) \, u(k-1) \, u(k-2)]$ on the 2D scores space, is given by (7).

$$t(k) = \boldsymbol{x}(k) \, \boldsymbol{P} \tag{7}$$

The classical integral PCA controller was initially formulated by Piovoso, [26, 27]. Some developments using ARX models were proposed later, for static PCA models and also for adaptive PCA models, [5, 7, 8]. In the incremental form, assuming $u(k) = x_{mp}(k)$, the control action of the classical integral controller based on static PCA is given by (8).

$$x_{mp}(k) = x_{mp}(k-1) + K_i \, \Delta x_m(k) \tag{8}$$

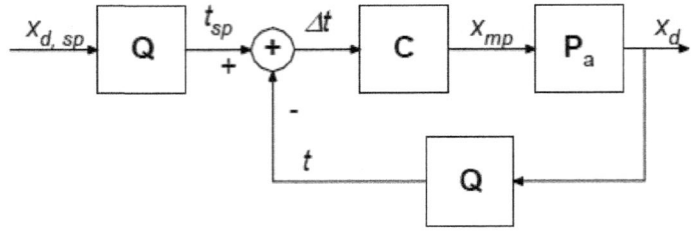

Fig. 1. Architecture of the classical integral PCA controller

The incremental action depends on the manipulated matrix $\boldsymbol{P}_{mp}$ as described in (9).

$$\Delta x_m(k) = \Delta t(k)\, \boldsymbol{P}_{mp} \tag{9}$$

The $\boldsymbol{P}$ matrix is decomposed into two matrices, $\boldsymbol{P} = [\boldsymbol{P}_{ex}|\boldsymbol{P}_{mp}]$: exogenous matrix ($\boldsymbol{P}_{ex}$, first r columns) and manipulated matrix ($\boldsymbol{P}_{mp}$, the last $m-r$ columns). The variables m, r and $m-r$ are, respectively, the length of $\boldsymbol{x}(k)$, the number of process output variables and process input variables in the regression vector $\boldsymbol{x}(k) = [y(k)\; y(k-1)\; y(k-2)\; u(k-1)\; u(k-2)]$; for an autoregressive ARX(2,2,1) model the variables assume the values $m = 5$, $r = 3$ and $m-r = 2$.

4   Linear Controller Based on Neural PCA Model

The main ideas related to the design of integral controllers using neural PCA models can be found in [6]. The integral controller presented here is based on neural linear PCA that can be implemented accordingly to the architecture depicted in Fig. 2, [13, 22]. The auto-associative neural network implements data compression at the bottleneck layer (BL) using function G and data decompression at the output layer (OL) using function H.

The main idea is the implementation of the equation $t(k) = \boldsymbol{x}(k)\, \boldsymbol{P}$ using a neural structure. In fact, since to implement PCA on neural structures an auto-associative architecture is needed, the projection on the 2D scores space is obtained at the output of the bottleneck layer (BL), assuming two principal components, $a = 2$. In this work, only linear activation functions without bias were used, in order to guarantee a linear neural network, so both σ and ϕ are linear activation functions.

First of all, the algorithm needs input / output training data to build a neural linear PCA model. Let's assume that the training data is archived in the matrix $\boldsymbol{X} \in \Re^{n \times m}$. In this work, to build the data matrix, the regressor vector was inspired on the regressor of an ARX(2,2,1) model, as described in (10), without loss of generality. The simulations were done assuming $n = 800, m = 5$, for a sampling period of $T_s = 1[s]$.

$$\boldsymbol{x}(k, :) = [y(k)\; y(k-1)\; y(k-2)\; u(k-1)\; u(k-2)] \tag{10}$$

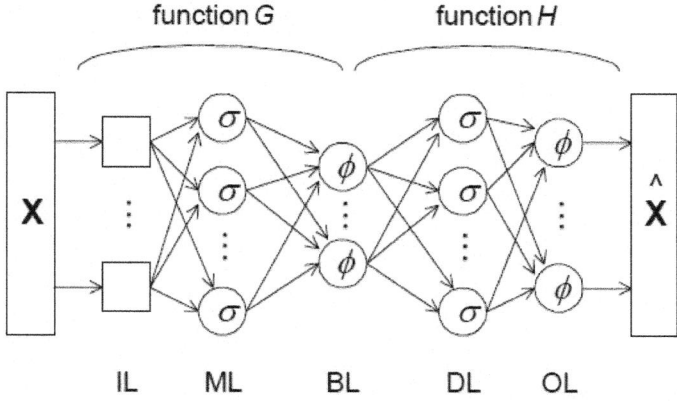

Fig. 2. Architecture of auto-associative neural network for PCA

The training data should not contain very high frequency data since this can cause problems in the Levenberg-Marquardt optimization algorithm used in the training of the neural networks, [25].

The integral controller based on linear neural PCA will be presented using a matrix formulation, inspired on the linear integral PCA controller described in section 3. In the algorithm M is the neural mapping matrix (first layer matrix ML, with weight matrix W_1) and B is the neural bottleneck matrix (second layer matrix BL, with weight matrix W_2).

The P_n matrix, equivalent to the main singular vectors matrix, is obtained from the neural network weight matrices as described in (11).

$$P_n = B\,M \tag{11}$$

The P_n matrix is decomposed into two matrices, $P_n = [P_{ex}|P_{mp}]$: exogenous matrix (P_{ex}, first r columns) and manipulated matrix (P_{mp}, the last columns).

$$P_{ex} = (P_n)_{:,1...r} \tag{12}$$

$$P_{mp} = (P_n)_{:,r+1...m} \tag{13}$$

The matrix M_λ is obtained from the P_n matrix, where $+$ is the pseudo-inverse of the matrix.

$$M_\lambda = (P_{ex})^+ P_{mp} \tag{14}$$

The error vector is given by

$$\Delta x(k,:) = [e_c(k)\ e_c(k-1)\ e_c(k-2)] \tag{15}$$

The increment in the manipulated variable is expressed by

$$\Delta x_{mp}(k) = \Delta x(k)\,(M_\lambda)_{:,1} \tag{16}$$

The non-saturated control action is formulated as

$$u_{c0}(k) = u_c(k-1) + K_i \, \Delta x_{mp}(k) + a_w(k-1) \tag{17}$$

The saturated control action is given by

$$u_c(k) = satur(u_{c0}(k), ...) \tag{18}$$

The anti-windup term is expressed by

$$a_w(k) = K_{aw} \, [u_c(k) - u_{c0}(k)] \tag{19}$$

A proportional term was added to the integral controller in order to improve the overall controller performance. The new non-saturated control action is given by (20).

$$u_{c0}(k) = K_p \, e_c(k) + u_c(k-1) + K_i \, \Delta x_{mp}(k) + a_w(k-1) \tag{20}$$

5 Nonlinear Adaptive PI Multi-Controller Based on Neural PCA Models

Inspired on the previous works [5, 6, 7, 8], a nonlinear adaptive PI multi-controller based on PCA multi-models, one PCA model for each setpoint range, is proposed here. The general architecture is depicted in Fig. 3. The supervisor, with a control-loop performance analysis module based on the modified Harris index as described in section 2, implements the adaptive mechanism used to compute the adaptive integral term. The block "Sat" is a saturation mechanism. As described in section 4 the integral term of the controller depends on the P_n matrix. This matrix is computed accordingly $P_n = B \, M$. For each local linear neural PCA model a set of matrices $\{B_i; M_i\}$ is available after off-line training of the linear local PCA neural models. Assuming ζ local linear neural models for ζ setpoint ranges, each adaptive matrix $\{B(k); M(k)\}$ is computed on-line from a linear combination of the matrices of local models accordingly (21). The weights are computed accordingly (22), where $|e_i(k)|$ is the absolute value of the control error at each time instant k. This methodology weighs more nearby models.

The supervisor should guarantee a minimum dwell-time associated with switching between matrices in order to guarantee stability, [23, 24]. In this work a rounding approach was applied to the weights α_i in order to guarantee a minimum dwell-time.

$$M(k) = \sum_{i=1}^{\zeta} \alpha_i(k) M_i(k) \quad B(k) = \sum_{i=1}^{\zeta} \alpha_i(k) B_i(k) \quad \sum_{i=1}^{\zeta} \alpha_i(k) = 1 \tag{21}$$

$$\alpha_i(k) = \frac{\Gamma(k)}{\sum_{i=1}^{\zeta} \Gamma(k)} \quad \Gamma(k) = 1 - \frac{|e_i(k)|}{\sum_{i=1}^{\zeta} |e_i(k)|} \quad |e_i(k)| = |r_i(k) - y(k)| \tag{22}$$

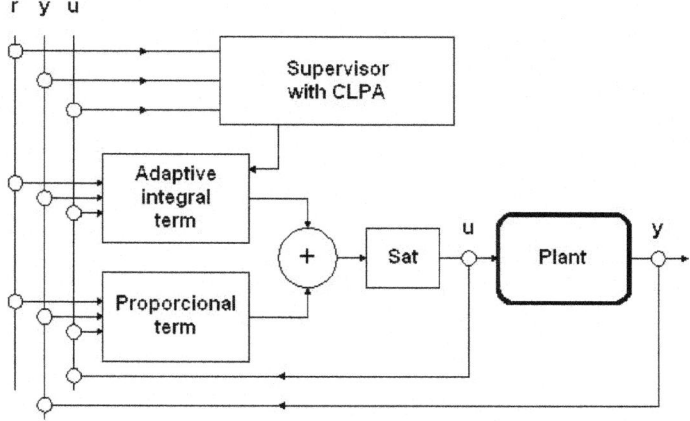

Fig. 3. Architecture for nonlinear neural PI-PCA multi-controller

6 Experimental Results

In order to evaluate the performance of the proposed PI multi-controller structure, experimental results obtained with a simulation model of a three-tank system benchmark are presented, [17], as depicted in Fig. 4. The water levels (h1, h2
and h3) in each tank are the process state variables. In this benchmark, usually,
the main goal is to maintain the water level, in the central tank, around a certain
predefined set-point. The experiments show the water level evolution in the left
tank T1. It is assumed that valve V13 and the outflow valve in the central tank
T3 are open. The algorithms were implemented in Matlab in discrete-time k.
The sampling time is 1 s. The input/output variables were normalized to the
range [0;1].

In Fig. 5 can be observed an experiment done with the neural PCA multi-
controller proposed, for two set-points $r1 = 0.15/0.6$ and $r2 = 0.3/0.6$. From top
to bottom can be observed the following signals: reference (r) and process output
(y), control error (e), control action (u), proportional action (up) and integral
action (ui), high-pass filtering of reference signal (hpfr), weights α_i (wx), and
finally the modified Harris index (ha). The controller gains are the following:
$K_p = 1, K_i = 0.2 \ \eta$, with $\eta = mean(abs(t1))$. The value $t1$ is the projection of
data along the first singularvector, i.e., the first principal component.

The adaptive PCA multi-controller (AMC) was compared with two linear
neural controllers (C1 & C2) tuned for each set-point $(r1, r2)$, for the same
profile of set-points depicted in Fig. 5. The mean squared error of the modified
Harris index obtained for each controller are the following, respectively, for the
controllers AMC, C1 and C2: $0.368, 0.390$ and 0.738. The multi-controller (AMC)
reveals the best performance.

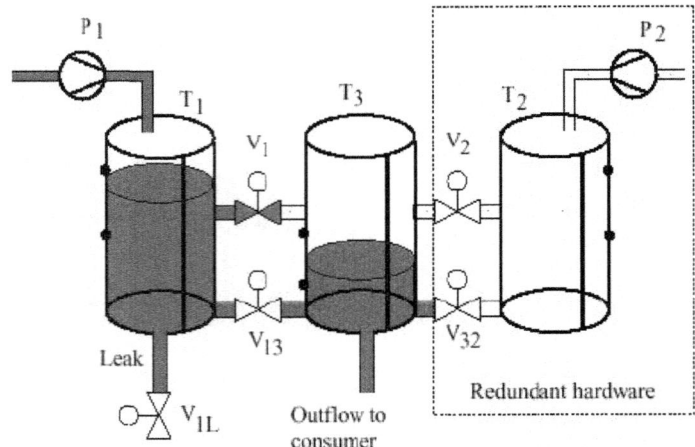

Fig. 4. Three-tank benchmark system

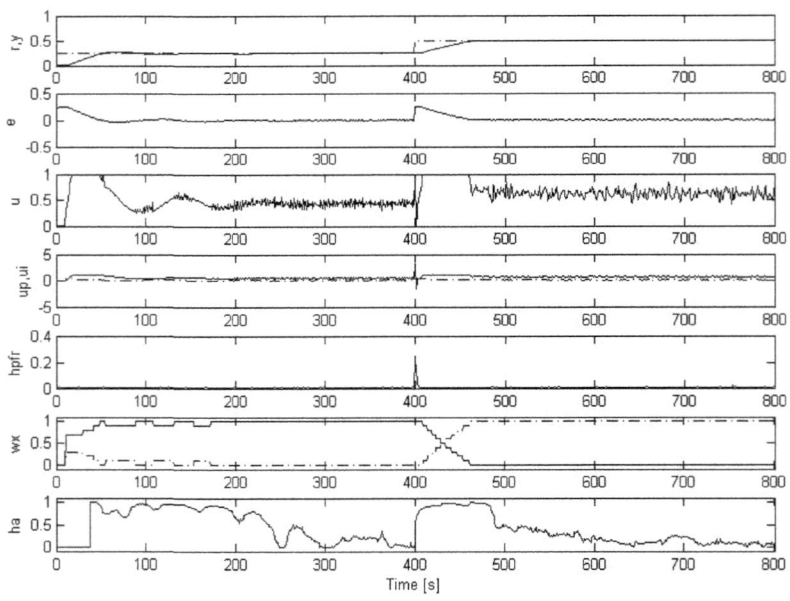

Fig. 5. Experiment with the multi-controller for two set-points, $r1$ and $r2$

7   Conclusions

In this paper, a new approach for design nonlinear adaptive PI multi-controllers based on PCA multi-models, for SISO systems, was proposed.

Since neural networks are very sensitive to data, high frequency data should be avoided in the training phase of each PCA neural network. Only data around each set-point should be considered for the training of each local linear PCA model used to build the integral term of the PI multi-controller. The behaviour of the integral neural PCA controller, in terms of overdamped or underdamped responses, depends on the spectral content of the data used in the training phase of the neural networks. The obtained results are very promising. A good performance was obtained for different simulations as observed in the behaviour of the modified normalized Harris index. More tests should be done to evaluate this approach with other classes of systems.

Some future research pointers are: a) improvement of the approaches for better tuning the controller gains; b) discover the classes of systems that are apropriate for this kind of controllers; c) the comparison with classical adaptive PI controllers based on gain scheduling; d) the robustness of the multi-controller against faults and failures should also be analysed; e) the generalization for MIMO systems should also be investigated.

Acknowledgments. This work has been supported by Faculdade de Ciências e Tecnologia da Universidade Nova de Lisboa, by Uninova-CTS, by CISUC-Coimbra and by national funds through FCT - Fundação para a Ciência e a Tecnologia (project PEst-OE/EEI/UI0066/2011). The authors would like to thank all the institutions.

References

[1] Ang, K., Chong, G., Li, Y.: Pid control system analysis, design, and technology. IEEE Transactions on Control Systems Technology 13(4), 559–576 (2005)

[2] Astrom, K., Hagglund, T.: The future of pid control. Control Engineering Practice 9, 1163–1175 (2001)

[3] Bezergianni, S., Georgakis, C.: Controller performance assessment based on minimum and open-loop output variance. Control Eng. Practice 8, 791–797 (2000)

[4] Brito Palma, L.: Fault Detection, Diagnosis and Fault Tolerance Approaches in Dynamic Systems based on Black-Box Models. Ph.D. thesis, Universidade Nova de Lisboa, Portugal (2007)

[5] Brito Palma, L., Coito, F., Gil, P.: Design of adaptive sliding window pi-pca controller. In: IEEE 20th Mediterranean Conference on Control and Automation (MED), Barcelona - Spain, July 3-6 (2012)

[6] Brito Palma, L., Coito, F., Gil, P.: Pi controller for siso linear systems based on neural linear pca. In: European Control Conference (paper accepted), Strasbourg, France, June 24-27 (2014)

[7] Brito Palma, L., Coito, F., Gil, P., Neves-Silva, R.: Process control based on pca models. In: 15th IEEE Int. Conf. on Emerging Technologies and Factory Automation (ETFA), Bilbao, Spain, September 13-16 (2010)

[8] Brito Palma, L., Coito, F., Gil, P., Neves-Silva, R.: Design of adaptive pca controllers for siso systems. In: 18th World Congress of the International Federation of Automatic Control (IFAC WC), Milano, Italy, August 28-September 2 (2011)

[9] Brito Palma, L., Moreira, J., Gil, P., Coito, F.: Hybrid approach for control loop performance assessment. In: 5th International Conference on Intelligent Decision Technologies (KES-IDT), Sesimbra, Portugal, June 26-28 (2013)

[10] Cano-Izquierdo, J., Ibarrola, J., Kroeger, M.: Control loop performance assessment with a dynamic neuro-fuzzy model (dfasart). IEEE Trans. on Automation Science and Engineering 9, 377–389 (2012)

[11] Choobkar, S., Sedigh, A., Fatehi, A.: Input-output pairing based on the control performance assessment index. In: IEEE Int. Conf. on Advanced Computer Control (ICACC), China, Shenyang (2010)

[12] Desborough, L., Harris, T.: Performance assessment measures for univariate feedback control. Canadian Journal of Chemical Engineering 70, 1186–1197 (1992)

[13] Diamantaras, K.: Principal Component Neural Networks: theory and applications. Wiley (1996)

[14] Ender, D.: Process control performance: not as good as you think. Control Engineering 180 (1993)

[15] Harris, T.: Assessment of control loop performance. Canadian Journal of Chemical Engineering 67, 856–861 (1989)

[16] Harris, T., Seppala, C., Desborough, L.: A review of performance assessment and process monitoring techniques for univariate and multivariate control systems. In: IFAC ADCHEM Conference, Canada (1997)

[17] Heiming, B., Lunze, J.: Definition of the three-tank benchmark problem for controller reconfiguration. In: European Control Conference, Karlsruhe, Germany (1999)

[18] Horch, A.: Condition Monitoring of Control Loops. Ph.D. thesis, Royal Institute of Technology, Sweden (2000)

[19] Horch, A., Isaksson, A.: A modified index for control loop performance assessment. In: ACC American Control Conference, USA, Philadelphia (1998)

[20] Jackson, J.: A User's Guide to Principal Components. Wiley (2003)

[21] Jelali, M.: An overview of control performance assessment technology and industrial applications. Control Engineering Practice 14, 441–466 (2006)

[22] Kramer, M.: Nonlinear principal component analysis using auto-associative neural networks. AICHE Journal 37(2), 233–243 (1991)

[23] Lemos, J., Rato, L., Mosca, E.: Integrating predictive and switching control: Basic concepts and an experimental case study. In: Progress in Systems and Control Theory. Birkhäuser Verlag (2000)

[24] Liberzon, D.: Switching in Systems and Control. Birkhauser Boston-MA (2003)

[25] Norgaard, M., Ravn, O., Poulsen, N., Hansen, L.: Neural Networks for Modelling and Control of Dynamic Systems. Springer (2003)

[26] Piovoso, M.: The Use of Multivariate Statistics in Process Control. In: The Control Handbook, pp. 561–573. CRC Press (1996)

[27] Piovoso, M., Kosanovich, K.: Applications of multivariate statistical methods to process monitoring and controller design. Int. Journal of Control 59(3), 743–765 (1994)

[28] Siemens, A.: How to improve the performance of your plant using the appropriate tools of simatic pcs7 apc portfolio - white paper. Siemens AG (2008)

[29] Stanfelj, N., Marlin, T., MacGregor, J.: Monitoring and diagnosing process control performance: the single loop case. Ind. Eng. Chem. 32, 301–314 (1993)

[30] Vilanova, R., Visioli, A. (eds.): PID Control in the Third Millennium - Lessons Learned and New Approaches. Springer (2012)

Optimization of Control Systems
by Cuckoo Search

Ramiro S. Barbosa and Isabel S. Jesus

GECAD – Knowledge Engineering and Decision Support Research Center
Institute of Engineering - Polytechnic of Porto (ISEP/IPP)
Dept. of Electrical Engineering, Porto, Portugal
{rsb,isj}@isep.ipp.pt

Abstract. This paper investigates the use of a new metaheuristic search algorithm, called Cuckoo Search (CS), in the optimization of control systems with a generalized PID controller. The CS algorithm is based on cuckoo bird's behavior in combination with the Lévy flight behavior of some birds and fruit flies. Recent studies show that it is very promising and could outperform existing algorithms such as genetic algorithms (GA). Comparative results with GA prove the effectiveness of CS in the optimization of feedback control systems.

Keywords: cuckoo search, optimization, PID controller, fractional calculus, fractional-order control, genetic algorithm.

1 Introduction

The PID algorithm is the most used type of controller in industry [1]. This type of algorithm has been proved to give satisfactory performance in the control of many linear and nonlinear processes. Nevertheless, the demanding of better ways to enhance the PID is still nowadays a hot field of research. More recently, the fractional order PID (FO-PID) controller has been receiving an increasing attention and its usefulness and applicability has been demonstrated by the numerous studies and applications reported in current literature [2, 3, 4].

We assist nowadays to an increasing use of metaheuristic algorithms inspired by nature with an aim to carry out global search. The efficiency of metaheuristic algorithms can be attributed to the fact that they imitate the best features in nature, especially the selection of the fittest in biological systems which have evolved by natural selection over millions of years. Two important characteristics of metaheuristics are: intensification and diversification. Intensification intends to search around the best solutions and select the best candidates or solutions, while diversification makes sure that the algorithm can explore the search space more efficiently, often by randomization [5].

Recently, a new metaheuristic search algorithm, called Cuckoo Search (CS), has been developed by Yang and Deb (2009) [6, 5]. Preliminary studies show that it is very promising and could outperform existing stochastic algorithms [7]. In this paper, we apply CS to tune optimal fractional-order PID controller systems.

© Springer International Publishing Switzerland 2015
A.P. Moreira et al. (eds.), *CONTROLO'2014 - Proc. of the 11th Port. Conf. on Autom. Control,*
Lecture Notes in Electrical Engineering 321, DOI: 10.1007/978-3-319-10380-8_12

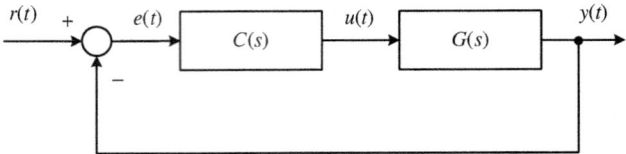

Fig. 1. Simple feedback control system

Two types of plants are addressed: an unstable plant with a time delay and a zero numerator dynamics, and a fractional-order plant. These plants represent many real dynamical processes [2, 3]. In fact, it is well-known that many systems and materials incorporate memory and hereditary effects and they are better modeled or characterized by fractional-order transfer functions [3, 2]. Also, we compare the applicability of CS and genetic algorithm (GA) in the design of proposed optimal control feedback systems.

The paper is organized as follows. In section 2, we present the fundamental of fractional-order control. Here we establish several definitions for the fractional derivatives, present the fractional PID controller, and outline the numerical algorithm for digital implementation of fractional-order control. After, in section 3, the basics of the CS algorithm are summarized. Section 4 shows the simulation results and the comparative analysis between CS and GA for two illustrative examples. Finally, section 5 draws the main conclusions and addresses perspectives to future research.

2   Fractional-Order Control Systems

2.1   Fractional PID Controller

In Figure 1 we show the block diagram of a simple feedback control system, where $G(s)$ denotes the process transfer function and $C(s)$ the controller. In this work, we adopt a fractional PID controller since the conventional PID is a special case of this more general controller.

The fractional-order controller of PID-type, usually named $\text{PI}^\lambda \text{D}^\mu$ controller, may be given as [3, 4]:

$$C(s) = \frac{U(s)}{E(s)} = K_p + \frac{K_i}{s^\lambda} + K_d s^\mu \qquad (1)$$

where K_p, K_i and K_d are the proportional, integral and derivative gains, and (λ, μ) are the fractional orders. In order to assure a good steady-state error, the term $1/s^\lambda$ in (1) should be implemented by means of an integer integrator, i.e. $s^{1-\lambda}/s$ [8]. Taking $(\lambda, \mu) \equiv \{(1,1), (1,0), (0,1), (0,0)\}$ we get the classical {PID, PI, PD, P}-controllers, respectively. The $\text{PI}^\lambda \text{D}^\mu$-controller is more flexible and gives the possibility of adjusting more carefully the dynamical proprieties of a control system [4].

The time domain equation of the $PI^\lambda D^\mu$ controller is:

$$u(t) = K_p e(t) + K_i D^{-\lambda} e(t) + K_d D^\mu e(t) \tag{2}$$

where $D^{(*)}$ ($\equiv {}_0D_t^\alpha$) denotes the differential operator to a fractional-order $\alpha = \{-\lambda, \mu\} \in \Re$.

A commonly used approach of a fractional derivative is given by the Riemann-Liouville definition ($\alpha > 0$):

$$D^\alpha f(t) = \frac{1}{\Gamma(n-\alpha)} \frac{d^n}{dt^n} \int_0^t \frac{f(\tau)}{(t-\tau)^{\alpha-n+1}} d\tau \tag{3}$$

where $n - 1 < \alpha < n$, n is an integer, $f(t)$ is the applied function, and $\Gamma(x)$ represents the Gamma function of x.

For our purpose we use the Grünwald-Letnikov definition, which can be written as ($\alpha \in \Re$) [3, 9]:

$$D^\alpha f(t) = \lim_{h\to 0} \frac{1}{h^\alpha} \sum_{j=0}^{[t/h]} (-1)^j \binom{\alpha}{j} f(t-jh) \tag{4a}$$

$$\binom{\alpha}{j} = \frac{\Gamma(\alpha+1)}{\Gamma(j+1)\Gamma(\alpha-j+1)} \tag{4b}$$

where h is the time increment and $[v]$ means the integer part of v.

The fractional integrals and derivatives can also be defined in the s-domain. Considering null initial conditions, they are given by the simple form ($\alpha \in \Re$):

$$D^\alpha f(t) = L^{-1}\{s^\alpha F(s)\} \tag{5}$$

where L represents the Laplace operator and $F(s) = L\{f(t)\}$.

Using definition (4), a discrete fractional $PI^\lambda D^\mu$ control equation can be obtained from (2) as ($h \approx T$, T is the sampling period):

$$u(k) = K_p e(k) + K_i D^{-\lambda} e(k) + K_d D^\mu e(k) \tag{6}$$

with

$$D^\alpha e(k) \approx \frac{1}{T^\alpha} \sum_{j=0}^{k} (-1)^j \binom{\alpha}{j} e(k-j) \tag{7}$$

The difference control equation (6) is then given by:

$$u(k) = K_p e(k) + \frac{K_i}{T^{-\lambda}} \sum_{j=0}^{k} (-1)^j \binom{-\lambda}{j} e(k-j) + \frac{K_d}{T^\mu} \sum_{j=0}^{k} (-1)^j \binom{\mu}{j} e(k-j) \tag{8}$$

The unlimited memory of (8) makes the computation too heavy as time increases and so unsuitable for a real implementation of the algorithm. In this way, for practical realization of fractional integral and derivative (7) we often apply the short memory principle [3], resulting in expression:

$$u(k) = K_p e(k) + \frac{K_i}{T^{-\lambda}} \sum_{j=v}^{k} c_j^{(-\lambda)} e(k-j) + \frac{K_d}{T^\mu} \sum_{j=v}^{k} c_j^{(\mu)} e(k-j) \qquad (9)$$

where $v = 0$ for $k < L/T$ or $v = k - L/T$ for $k > L/T$; L is the memory length and $c_j^{(\alpha)} = (-1)^j \binom{\alpha}{j}$ are the binomial coefficients which may be calculated recursively as:

$$c_0^{(\alpha)} = 1; \quad c_j^{(\alpha)} = \left(1 - \frac{1+\alpha}{j}\right) c_{j-1}^{(\alpha)}, \quad j = 1, \, 2, \, \cdots \qquad (10)$$

Note that (9) is given in the form of a FIR filter. Other discrete-time approximations in the form of IIR filters are also possible [10, 11].

2.2   Objective Function for Optimal Controller Tuning

In this study, we consider the minimization of an integral performance index (J) given as the weighted sum of Integral of Time multiplied Absolute Error (ITAE) and the Integral of Squared Control Output Signal (ISCOS), represented as:

$$J = \int_0^\infty \left[\omega_1 t \, |e(t)| + \omega_2 u^2(t)\right] dt = (\omega_1 \times \text{ITAE}) + (\omega_2 \times \text{ISCOS}) \qquad (11)$$

where $e(t)$ is the error signal between the actual output $y(t)$ and the desired response $r(t)$, and $u(t)$ is the control output signal from the fractional PID controller. The weights ω_1 and ω_2 give the relative importance of the two terms. In this work, we set $\omega_1 = \omega_2 = 1$, which implies that the setpoint tracking is as important as the control signal. The ITAE criterion penalizes errors that persist for long periods of time and, in general, it is the preferred error performance criterion since it results in the most conservative controller settings.

3   Cuckoo Search Algorithm

Cuckoo search is a metaheuristic algorithm inspired by the brood parasitism of cuckoo species by laying their eggs in the nests of other host birds. The host takes care of the eggs presuming that the eggs are its own. If a host bird discovers the eggs are not their own, it may either throw out the discovered alien eggs or simply abandon its nest and build a new nest at a different location. Each egg in a nest represents a solution, and a cuckoo egg represents a new solution. The better new solution (cuckoo) is replaced with a solution which is not so good in the nest. In the simplest form, each nest has one egg. When generating a new solution Lévy flight is performed [6, 5, 12].

The rules for CS as proposed by Yang and Deb [6] are described as follows:

- Each cuckoo lays one egg at a time, and dumps it in a randomly chosen nest;
- The best nests with high quality of eggs (solutions) will carry over to the next generations;
- The number of available host nests is fixed, and a host can discover an alien egg with a probability $p_a \in [0,\ 1]$. In this case, the host bird can either throw the egg away or abandon the nest so as to build a completely new nest in a new location.

For simplicity, this last assumption can be approximated by a fraction p_a of the n nests being replaced by new nests (with random solutions at new locations). For a maximization problem, the quality or fitness of a solution can simply be proportional to the objective function. Other forms of fitness can be defined in a similar way to the fitness function in genetic algorithms.

Based on these three rules, the basic steps of the Cuckoo Search (CS) can be summarized as the pseudo code shown in Figure 2.

When generating new solutions $\mathbf{x}_i^{(t+1)}$ for, say cuckoo i, a Lévy flight is performed:

$$\mathbf{x}_i^{(t+1)} = \mathbf{x}_i^{(t)} + \alpha \oplus \mathrm{L\acute{e}vy}\,(\beta) \tag{12}$$

where $\alpha > 0$ is the step size which should be related to the scales of the problem of interest. In most cases, we can use $\alpha = 1$. In order to accommodate the difference between solution quality, we can also use:

$$\alpha = \alpha_0 \left(\mathbf{x}_j^{(t)} - \mathbf{x}_i^{(t)} \right) \tag{13}$$

where α_0 is a constant, while the term in the bracket corresponds to the difference of two randomly solutions. This mimics that fact that similar eggs are less likely to be discovered and thus new solutions are generated by the proportionality of their difference. The product $\oplus$ means entry-wise multiplications. Lévy flights essentially provide a random walk while their random steps are drawn from a Lévy distribution for large steps:

$$\mathrm{L\acute{e}vy} \sim u = t^{-1-\beta}, \quad (0 < \beta \le 2) \tag{14}$$

which has an infinite variance with an infinite mean. Here the consecutive jumps (steps) of a cuckoo essentially form a random walk process which obeys a power-law step-length distribution with a heavy tail.

In addition, a fraction p_a of the worst nests can be abandoned so that new nests can be built at new locations by random walks and mixing. The mixing of the eggs/solutions can be performed by random permutation according to the similarity/difference to the host eggs.

Objective function $f(\mathbf{x})$, $\mathbf{x} = (x_1, \ldots, x_d)^T$;
Initial a population of n host nests $\mathbf{x}_i$ $(i = 1, 2, \ldots, n)$;
while $(t < \text{MaxGeneration})$ or (stop criterion);
Get a cuckoo (say i) randomly by Lévy flights;
Evaluate its quality/fitness F_i;
Choose a nest among n (say j) randomly;
if $(F_i > F_j)$,
Replace j by the new solution;
end
Abandon a fraction (p_a) of worst nests
[and build new ones at new locations via Lévy flights];
Keep the best solutions (or nests with quality solutions);
Rank the solutions and find the current best;
end while
Postprocess results and visualization;

Fig. 2. Cuckoo Search (CS) [6]

The generation of the step size s samples using Lévy flights is more efficiently calculated through the so-called Mantegna algorithm. A scheme based in Mantegna's algorithm is summarized in [6, 5] as:

$$s = \alpha_0 \left(\mathbf{x}_j^{(t)} - \mathbf{x}_i^{(t)} \right) \oplus \text{Lévy}(\beta) \approx 0.01 \frac{u}{|v|^{\frac{1}{\beta}}} \left(\mathbf{x}_j^{(t)} - \mathbf{x}_i^{(t)} \right) \tag{15}$$

where β is a parameter between $[1, 2]$ interval and considered to be 1.5; u and v are drawn from normal distributions as [7]:

$$u \sim N\left(0, \sigma_u^2\right), \quad v \sim N\left(0, \sigma_v^2\right) \tag{16}$$

$$\sigma_u = \left\{ \frac{\Gamma(1+\beta)\sin\left(\frac{\pi\beta}{2}\right)}{\Gamma\left(\frac{(1+\beta)}{2}\right)\beta 2^{\frac{(\beta-1)}{2}}} \right\}^{\frac{1}{\beta}}, \quad \sigma_v = 1 \tag{17}$$

where Γ is the standard Gamma function.

The unique features of CS, such as exploration by Lévy flight, mutation by a combination of Lévy flights and vectorized solution difference, crossover by selective random permutation, and elitism, makes CS potentially more powerful than algorithms using one or some of these components [7].

4   Illustrative Examples

In this section, the proposed Cuckoo Search (CS) algorithm is used to obtain the optimum parameters of the generalized PID controller in the control of two plants: an unstable plant with a time delay and zero numerator dynamics and a fractional-order plant. The performance of the results are assessed and compared with GA [13].

In order to take into account the stochastic effects and to obtain a meaningful statistical analysis, the algorithms (CS and GA) have been run 20 times and the best result (i.e. the simulation with the lower value of fitness function J) is chosen. For the CS we have used a number of host nests (or the population size) of $n = 25$, step size $\alpha = 1$ and probability $p_a = 0.25$. As stated in [5] the convergence rate is, somehow, insensitive to algorithm-dependent parameters. In this work, we use the GA of the Global Optimization Toolbox (version 3.2) of MATLAB [14]. We establish the following values for the GA parameters: population size P=25 (the same as CS), crossover probability C=0.8 and mutation probability M=0.05. Other values of GA parameters are used as default (see [14]). The choice of the CS and GA parameters revealed adequate for the tuning of the fractional-order control systems under study.

4.1   Example 1: Unstable System with Time Delay and a Zero

The control system performance may be severely affected by the numerator dynamics (i.e. the existence of a zero) of the process. Several real processes are represented by transfer functions which incorporate a time delay with a zero transfer function model [15]. Here we consider such a model given by transfer function [15]:

$$G_1(s) = \frac{3.87\,(1 + 0.5283s)\,e^{-0.1s}}{0.4769s^2 - 0.348s + 1} \tag{18}$$

The optimal tuning of the fractional-order PID controller is carried out by using (9) and minimizing the objective function (11). Table 1 presents the optimum PID parameters, namely the gains (K_p, K_i, K_d) and orders (λ, μ), and minimum error J_{min}. Table 2 shows the corresponding time-domain specifications of percent overshoot (OS%), rise time (t_r) and settling time (t_s). The adopted search space for the controller parameters were: (K_p, K_i, K_d) $\in [0, 10]$ and (λ, μ) $\in [0, 2]$. As stop criterion for CS and GA, we use the number of iterations $N_{iter} = 200$, which proved to be enough for the convergence in all experiments. The memory length is $L = 10$ s (absolute memory of the fractional derivative). The step responses and control signals with the optimal controllers are shown in Figure 3, for the CS and GA algorithms. It can be observed that the results are very similar with a slightly better response of CS in terms of

Table 1. Optimal controller parameters with $G_1(s)$

| Algorithm | J_{min} | Controller parameters | | | | |
		K_p	K_i	K_d	λ	μ
CS	1.4202	0.0357	1.8087	0.8238	0.9938	0.2519
GA	1.4210	0.2934	1.7216	0.6114	0.9996	0.2817

Table 2. Time-domain specifications with $G_1(s)$

Algorithm	Time specifications		
	OS(%)	t_r (s)	t_s (s)
CS	46.2090	0.0982	2.1291
GA	48.4504	0.1023	2.0803

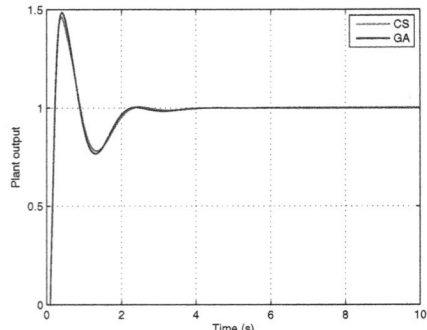 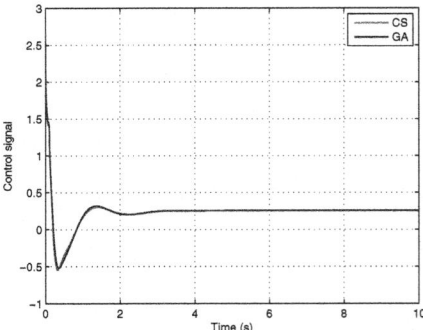

Fig. 3. Unit-step responses (a) and control signals (b) with $G_1(s)$

Table 3. Optimal controller parameters with $G_2(s)$

Algorithm	J_{min}	Controller parameters				
		K_p	K_i	K_d	λ	μ
CS	138.3579	0.0000	0.00800	5.5864	1.0910	0.1516
GA	146.5396	4.2465	0.0987	1.2726	1.1179	0.2607

overshoot and rise time of step response. The error J in CS is lower than in GA. Also, we note that in both cases (CS and GA) the optimal tuning leads to a fractional-order controller. In this case a fractional $C(s) \approx PID^\mu$ controller in both metaheuristics algorithms. This reveals that the best controller is, in most cases, of fractional-order, and thus giving better performance than the conventional PID controller.

4.2   Example 2: Fractional-Order Plant

Many real dynamical processes are modeled by fractional-order transfer functions [2, 3]. Here we consider the fractional-order plant model given in [16]:

$$G_2(s) = \frac{1}{39.69s^{1.26} + 0.598} \tag{19}$$

Tables 3 and 4 present the controller gains and the time-domain specifications of system with the fractional PID controller optimized by the CS and

Table 4. Time-domain specifications with $G_2(s)$

Algorithm	Time specifications		
	OS(%)	t_r (s)	t_s (s)
CS	1.4989	10.5148	15.5808
GA	7.9595	7.6262	30.4326

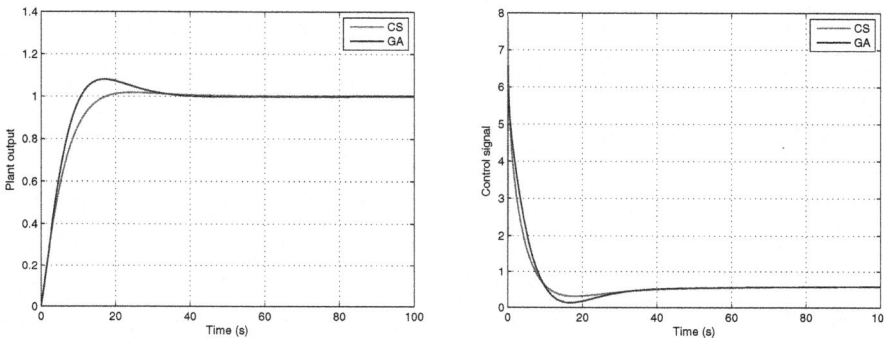

Fig. 4. Unit-step responses (a) and control signals (b) with $G_2(s)$

GA algorithms using (9) and objective function (11), respectively. The adopted search space for the controller parameters were: $(K_p,\ K_i,\ K_d) \in [0,\ 100]$ and $(\lambda,\ \mu) \in [0,\ 2]$. As stop criterion for CS and GA, we use the number of iterations $N_{iter} = 100$, which proved to be enough for the convergence in all experiments. The memory length is $L = 100$ s (absolute memory of the fractional derivative). Figure 4 shows the step responses and control signals with the optimal controllers for both methods. It is observed that the CS leads to a better controller performance than GA in terms of the overshoot and settling time specifications. The error J is also lower than in GA. Once again, we see that the best tuning leads to a fractional-order controller in both algorithms. In this case we obtain a fractional $C(s) \approx I^\lambda D^\mu$ controller for CS and a fractional $C(s) \approx PI^\lambda D^\mu$ controller for GA.

5 Conclusions

This paper proposed the use of a recently developed metaheuristic algorithm, called Cuckoo Search (CS), in the optimization of the controller parameters of integer and/or fractional-order control systems.

The results revealed that the CS can outperform some state of the art stochastic algorithms, such as the widely used GA. The CS gives comparable or even better results than the powerful Genetic Algorithm Toolbox of MATLAB. Moreover, the number of parameters to be tuned is less than GA. Also, the convergence rate is somehow insensitive to the algorithm-dependent parameters, and thus CS is more generic and robust for many optimization problems.

This study shows some preliminaries results and the potential application of CS in control systems optimization. However, further research is needed in applying CS to other systems, controller structures and compare it with other metaheuristic algorithms.

Acknowledgement. This work is supported by FEDER Funds through the "Programa Operacional Factores de Competitividade - COMPETE" program and by National Funds through FCT "Fundação para a Ciência e a Tecnologia" under the project: FCOMP-01-0124-FEDER- PEst-OE/EEI/UI0760/2014.

References

[1] Astrom, K.J., Hagglund, T.: PID Controllers: Theory, Design, and Tuning. Instrument Society of America, USA (1995)

[2] Oldham, K.B., Spanier, J.: The Fractional Calculus. Academic Press, New York (1974)

[3] Podlubny, I.: Fractional Differential Equations. Academic Press, San Diego (1999)

[4] Podlubny, I.: Fractional-order systems and $PI^\lambda D^\mu$-controllers. IEEE Transactions on Automatic Control 44(1), 208–214 (1999)

[5] Yang, X.S., Deb, S.: Engineering optimization by cuckoo search. Int. J. Mathematical Modelling and Numerical Optimization 1(4), 330–343 (2010)

[6] Yang, X.S., Deb, S.: Cuckoo search via Lévy flights. In: Proc. of World Congress on Nature and Biologically Inspired Computing (NaBIC 2009), pp. 210–214 (2009)

[7] Yang, X.S., Deb, S.: Multiobjective cuckoo search for design optimization. Computers and Operations Research (40), 1616–1624 (2013)

[8] Axtell, M., Bise, M.: Fractional calculus applications in control systems. In: Proc. of the IEEE 1990 National Aerospace and Electronics Conference, New York, pp. 563–566 (1990)

[9] Machado, J.A.T.: Analysis and design of fractional-order digital control systems. SAMS Journal of Systems Analysis, Modelling and Simulation 27, 107–122 (1997)

[10] Barbosa, R.S., Machado, J.A.T., Silva, M.F.: Time domain design of fractional differintegrators using least-squares. Signal Processing 86, 2567–2581 (2006)

[11] Chen, Y.Q., Vinagre, B.M., Podlubny, I.: Continued fraction expansion approaches to discretizing fractional order derivatives-an expository review. Nonlinear Dynamics 38, 155–170 (2004)

[12] Buaklee, W., Hongesombut, K.: Optimal DG allocation in a smart distribution grid using cuckoo search algorithm. ECTI Transactions on Electrical Eng., Electronics, and Communications 11(2), 16–22 (2013)

[13] Goldberg, E.: Genetic Algorithms in Search Optimization and Machine Learning. Addison-Wesley (1989)

[14] MathWorks: Global Optimization Toolbox (version 3.2). The MathWorks, Inc. (2011)

[15] Krishna, D., Aaron, A., Satyakala, K., Swetha, B.V.: Controllability studies on unstable SOPTD systems with a zero. Chemical Engineering 52, 11478–11482 (2012)

[16] Monje, C.A., Chen, Y., Vinagre, B.M., Xue, D., Feliu-Battle, V.: Fractional-Order Systems and Controls. Springer (2010)

Use of a Genetic Algorithm to Tune a Mandani Fuzzy Controller Applied to a Robot Manipulator

José B. de Menezes Filho[1], Nuno Miguel Fonseca Ferreira[2],
and José Boaventura-Cunha[3]

[1] Instituto Federal de Educação, Ciência e Tecnologia da Paraíba, Av. Primeiro de Maio, 720,
Jaguaribe - João Pessoa-PB   Bolsista da CAPES – Proc. nº 2477-13-0
jbmenezesf@hotmail.com
[2] Instituto Superior de Engenharia de Coimbra, Portugal
nunomig@isec.pt
[3] INESC TEC - INESC Technology and Science (formerly INESC Porto) and UTAD,
Universidade de Trás-os-Montes e Alto Douro,
School of Sciences and Technology – Vila Real, Portugal
jose.boaventura@inesctec.pt

Abstract. This work presents the use of a genetic algorithm to design a Mandani Fuzzy Controller with two inputs and one output, written in Matlab® environment, applied to a two axis positioning system using a robot. The robot has 6 degrees of freedom and is controlled with the objective of capturing an object on a workspace using a fuzzy controller. A genetic algorithm is used in order to determine the main characteristics of the membership functions of the fuzzy controller. The complete system employed to simulate the two axes positioning system uses the Transfer Function of two axes of the robot and the Fuzzy controller. In this work was implemented and simulated an operating scenario, being the results and the performance of the controller presented regarding the controller energy effort and the evolution of the (x,y) trajectories over time.

Keywords: Fuzzy Controller, Genetic Algorithm, Position Controller, Robot Manipulator.

1   Introduction

The positioning systems such as robotic manipulators are extensively used in a wide variety of industrial equipments. Particularly in industry, systems that have motion control provides a continuous improvement of product quality associated with the minimization of manufacturing costs, which is possible through the use of high- precision equipment that combines flexibility with security and high-speed time response, such as robotic systems. Therefore, the study of robot control systems plays an important role in the industrial development.   In this paper it is shown a strategy to control the claw position of a robot with 6 degrees of freedom.   The claw (x,y) position is computed by processing the image acquired by a webcam with a resolution of 1054×750 pixels, being the spatial resolution of 0.95mm×1.33mm/pixel in the

© Springer International Publishing Switzerland 2015,
A.P. Moreira et al. (eds.), *CONTROLO'2014 - Proc. of the 11th Port. Conf. on Autom. Control*,
Lecture Notes in Electrical Engineering 321, DOI: 10.1007/978-3-319-10380-8_13

work space. A program designed on fuzzy logic, based on the error and on the derivative of the error position signals, compute the control signals to drive the x and y axes motors.

According [1, 2], the objectives of the fuzzy controller are: to control and operate several variables without take in consideration if the controlled system is a linear or a nonlinear process. The automatic control must be performed in a similar way as performed by working human operators, i.e., respecting the specifications and the operational constraints. The control action should be simple, robust and operate in real time. Moreover, the great advantages of using fuzzy logic based controllers are the shrinking of the design time and it is implemented as a user-friendly technology in practical applications. As stated in [3] "The major objective of the model-based fuzzy control is to use the full available range of existing linear and nonlinear design of such fuzzy controllers which have better stability, performance, and robustness properties than the corresponding non-fuzzy controllers designed by the use of these same techniques".

According to the related literature, the pioneering work in application of fuzzy logic in process control is due to [4], whose theoretical supports are described in articles of [5]. Although different methods are presented in the literature, according [4, 6, 7, 8, 9] they can be grouped into two major groups: The Mandani and the Sugeno Fuzzy systems.

In this work it is shown the results obtained by use of Mandani Controller. The work is validated with the presentation of simulated results. This work is organized in the following way: In this first section was presented the overall view of the subject treated. In section two are presented the concepts of vector control. Section three shows the implementation of the Fuzzy Controller. Section four is dedicated to describe the genetic algorithm used to optimize the fuzzy controller membership functions. In section five are presented the results obtained concerning the trajectories of the robot claw and the control signals efforts. Finally the comments and conclusions of this work are presented.

2   Vector Control of Robot

The strategy to control the robot consists in transforming the information of its position in the vector format, as it was used by [6]. This mode of operation is called vector, in which the position signal of the robotic hand is treated as vector and is decomposed in its Cartesians components Vx and Vy.

The position of the targeting object, reference (R), is obtained by processing the object and workspace images acquired with the webcam in order to compute the Cartesian positions, here called Rx and Ry. In this way the web camera provides the information to compute the X and Y position of the object over the workspace.

The difference between the position of the object and the reference position define the vector error. This vector error is then decomposed in its Cartesian components, given in accordance to Eq. (1) e (2).

$$ex = Rx - Vx \tag{1}$$

$$ey = Ry - Vy \tag{2}$$

The two Cartesian components are combined to form the error signal in the vector format, Eq. (3), (4), (5). The module of this signal is used as input to the fuzzy system and the angle will be used to determine the command signals to drive the robot movement.

$$\hat{e}_{x,y} = (\hat{R}_{x,y} - \hat{V}_{x,y}) = \parallel e_{x,y} \parallel < e_{x,y} \tag{3}$$

$$\parallel e \parallel = \left(\sqrt{ex^2 + ey^2} \right) \tag{4}$$

$$ang(e) = atan\left({ey}/{ex} \right) \tag{5}$$

Figure (1) shows the Reference Vector, the Position Vector of robotic hand and the Error Vector and the angles for each vector.

In this work the following notation is used:

$\hat{V}_{x,y}$ = Position vector of robotic hand

V_x = X axis component of position vector of robot hand

V_y = Y axis component of position vector of robot hand

$\hat{R}$ = Position vector of the object.

$\parallel R \parallel$ = Position vector module of the object

ang(R)= Position vector angle of the object

Rx = X axis component of the object position vector

Ry = Y axis component of the object position vector

$\hat{e}$ = Position vector error

||e||= Position vector error module.

ang(e)= position vector error angle

In Figure (2) is showed the block diagram of the controller. The position error vector module is used as an input signal in the fuzzy system. Another input is the error obtained in one sample time delayed. The Position error vector module is the output of the block "Determination of the module and angle". Another output of this block is the position error vector angle, which is utilized as an input in the block "Converter polar to rectangular". This last block receives the result of the fuzzy system and sends control signals to the X and Y axis of the robotic manipulator.

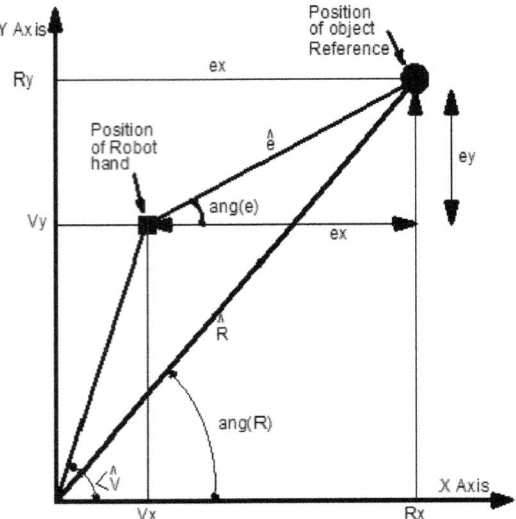

Fig. 1. Reference, Position and Error vectors of robotic hand

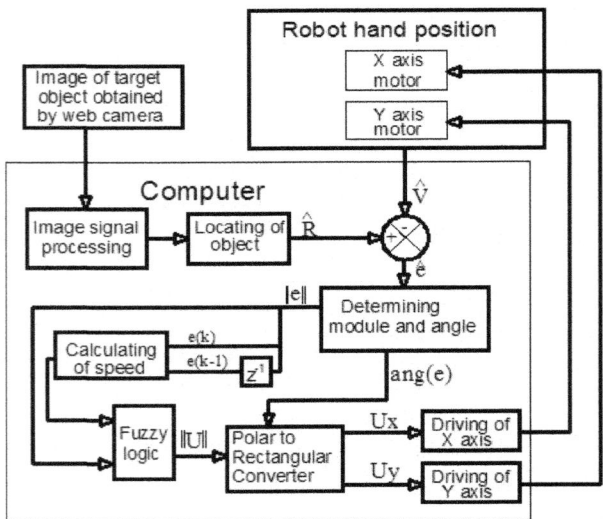

Fig. 2. Fuzzy Robot Controller

The result signal of the fuzzy logic controller is decomposed in two parts: Ux and Uy. The first of these signals is applied in the motor of X axis, according to Eq. (6). The second one is applied in the motor of Y axis in accordance to Eq. (7).

$$Ux = \| U \| \cos (ang(e)) \qquad (6)$$

$$Uy = \| U \| sen(ang(e)) \qquad (7)$$

3   Implementation of Mandani Fuzzy Controller

The Mandani Fuzzy Controller uses two input variables: The error and the error derivative, which is called dError. The four membership functions were initially specified for the variables error and dError as shown in Fig. (3).

In this Figure are shown the points xSp and xMp. The determination of these points plays an important role in the formation of the membership functions for the error and dError signals. By means of the displacing of these points it is possible to modify the shape of the triangle of the membership functions Sp, Mp, dSp and dMp. The values of xSp and xMp are computed with a genetic algorithm that will be detailed in section 4.

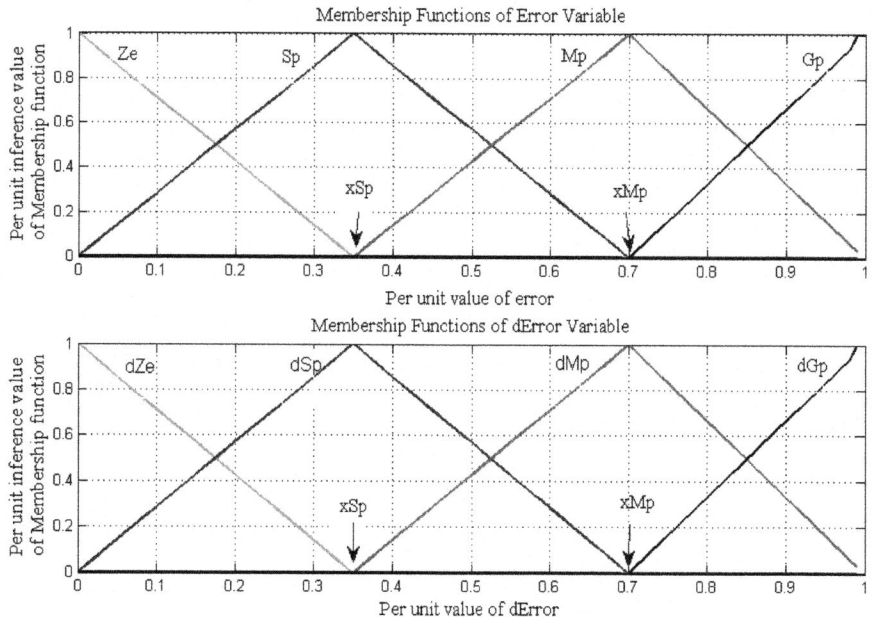

Fig. 3. Memberships Functions of error and dError variables

The membership functions of the output variables have the same shape as the previous ones, being all normalized per unit inference value. It must be noted that the implemented methods use only the positive side of the universe of discourse for the error and the derivative error signals, since the information of the error angle is provided to compute the Ux and Uy control signals. This is a great advantage of this type of controller.

In order to perform a simulation that can be used by several robotics systems, all values are expressed in per unit value. In practical experimentation the true measures may be in any units. In per unit values the results are not affected by unit of measure. In this way the universe of discourse ranges between zero to one.

3.1 Mandani Fuzzy Controller – Inference

In the inference step of Mandani Fuzzy Controller, the composition of each rule and the relationship between them are accomplished in accordance to Table (1). In this step when the rules "if then" are activated, the consequences are obtained with the minimum value between the Error and Derror by using the inference technique MAX-MIN. In this step the control variable which drives the robot is determined. This variable has four membership functions. In the defuzzification process it was employed the Media-of-Maximum method.

Table 1. Fuzzy rules of the Fuzzy Mandani controller

Derror / Error	DZe	DSp	DMp	DGp
Ze	Ze	Sp	Sp	Mp
Sp	Sp	Mp	Mp	Gp
Mp	Mp	Gp	Gp	Gp
Gp	Gp	Gp	Gp	Gp

4 Optimization of the Fuzzy Controller Using Genetic Algorithm

In order to optimizing the fuzzy controller it was employed a genetic algorithm to compute the membership functions. The main purpose of the genetic algorithm is determining the xSp and xMp points that belong to the universe of discourse of the membership functions with the aim of minimizing the control effort, i.e, the energy (En) supplied to the X and Y motors. This is done using a cost function given by the sum of the quadratic values of the control signals multiplied by the time variable, for both robot axes, Eq. (8).

$$En=\sum(u_x^2.t+u_y^2.t) \tag{8}$$

Where $t=k.T_s$, being k the sample number and T_s the sampling time.

The genetic algorithm works in accordance with the following steps [10, 11, 12, 13, 14]: a population with 8 chromosomes is considered, each of them with 8 genes, to compute the values dxSp and dxMp as showed in Figure (4). These values are used to correct the xSp and xMp values. The crossover procedure involves combining 50% of two selected chromosomes about a crossover point, generating two new chromosomes. The minimization of the cost function of Eq. 8 is performed with a genetic algorithm that stops whenever the cost function is minimum or the number of generations reaches 200. A mutation genetic operator is used to ensure genetic diversity within the population. The mutation operator randomly modifies the genes of the chromosome, introducing additional randomness into the population and yielding to a new candidate solution. The probability of mutation is generally set to a low value.

Afterwards, it is executed the fuzzy algorithm to obtain the control output for each axis of the robot. Then the energy used in the robot movement is computed according to Eq. (8). If the energy is the minimal value, the optimization process is finished and the final optimum values xSp and xMp were found. If the energy is not the minimal value, it are selected the first four chromosomes in the crescent order of energy consumption. These four chromosomes are the parents of a new population while the last four chromosomes are discarded. With the four chromosomes selected to be the parents it was performed the crossover to generate a new population of eight chromosomes. At last is employed the mutation with a probability set to 0.01.

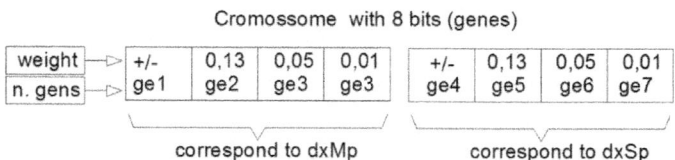

Fig. 4. Bits of the chromosome used in the genetic algorithm

5   Simulation Results

In order to evaluate the performance of the controller it was performed a simulation considering the robot hand starting position at (0.1, 0.5) and the object position, or reference position, in (0.7, 0.4), being the sampling time of 10ms. Two identical second order Transfer Functions, expressed in the Z domain, were used for the X and Y axis. The Function has one sample time delay and its parameters were obtained with the use of identification techniques using the data of a real robot. Eq. (9) shows the Transfer Function utilized. The sample time in which the Transfer Function were obtained was also 10 ms, with $b_1=0.0516383$; $b_2=-0.1055211$; $b_3=0.0741196$; $a_1=1.8676718$ and $a_2= -0.8676718$.

It was performed an optimization of the memberships function using genetic algorithm as explained early. Starting from the initial values of xSp= 0.3 and xMp= 0.75, it was founded the optimal values xSp= 0.49 and xMp= 0.69. The results of the simulation performed by using this method and the discrete transfer function, for the X and Y axes, of Eq. (9) are shown in Fig. (5) and   (6).

$$\frac{Y_{x,y}}{U_{x,y}} = Z^{-1}\frac{b_1+b_2Z^{-1}+b_3Z^{-2}}{1-a_1Z^{-1}-a_2Z^{-2}} \tag{9}$$

This system was simulated for several values for sXp and xMp, that were chosen randomly. The results for these simulations are shown in the Table 2. This table shows the following results: enxf: per unit total energy consumed by axis x;. enyf: per unit total energy consumed by axis y; enf: per unit total energy consumed in both axes x and y. It is possible to observe that the xSp and xMp determined by optimization shows the best performance.

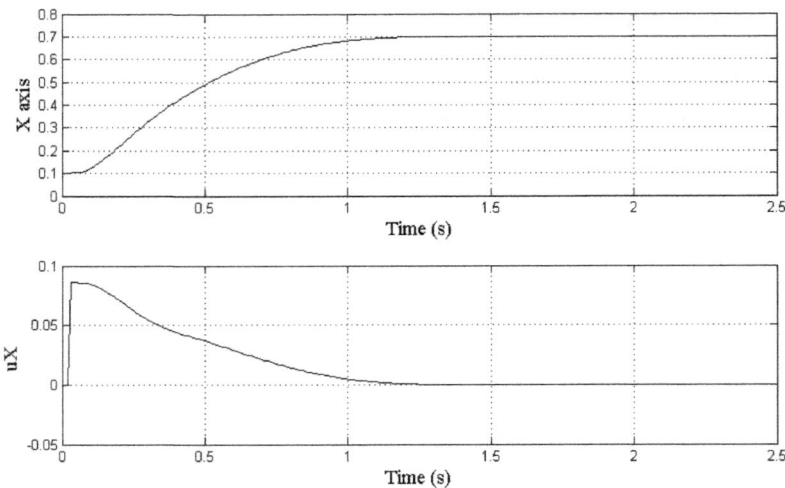

Fig. 5. Trajectory and control variable for the X axis with xSp=0.49 and xMp=0.69

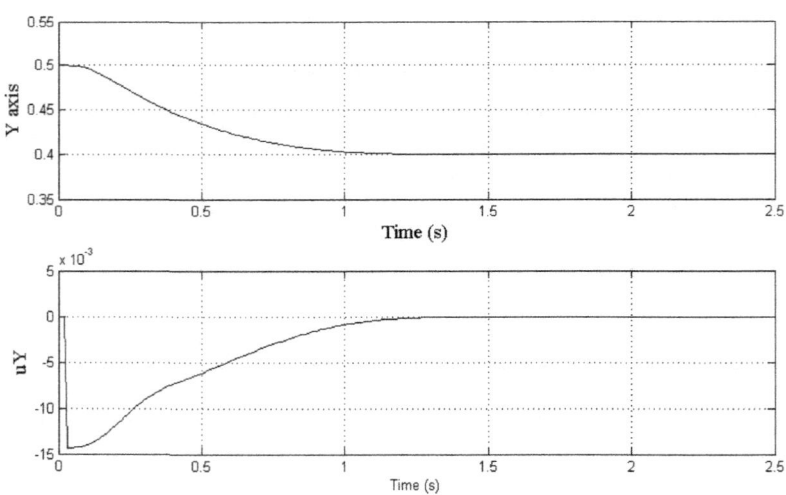

Fig. 6. Trajectory and control variable for the Y axis with xSp=0,49 and xMp=0,69

Table 2. Controller efforts for the considered xSp and xMp values

xSp	xMp	enxf	enfy	enf
0,30	0,75	0.7178	0.0199	0.7377
0,40	0,60	0.6132	0.0170	0.6302
0,49*	0,69*	0.5427	0.0151	0.5578
0,20	0,6	0.7693	0.0214	0.7907

6   Comments and Conclusions

In this work it was presented a fuzzy controller applied to a robot with six Degrees of Freedom. The type of controller system used, called in this work as vector control, has the advantage of using only one controller structure to control 2 axes at the same time. Moreover, the fuzzy controller does not need to be applied in the traditional universe of discourse with negative and positive sides. Here, it is only necessary the use of the positive side of the universe of discourse. This is due to the characteristic of the control vector used in the fuzzy controller, which avoids the need of using negative values for the variables.

The membership functions of the fuzzy controller were optimized by a genetic algorithm especially designed for this purpose. The results achieved for the x and y trajectories of the robotic hand and the energy cost functions were presented. The simulated values for the total energy consumption showed the advantage of using the genetic algorithm in the optimization design of the membership functions. The system was also simulated for other values of xSp and xMp that showed a performance below the ones achieved with the optimized parameters.

Acknowledgments. The authors thanks CAPES Foundation, Ministry of Education of Brazil, the financial supporting to this research.

References

1. Camargos, F.L.: Lógica Nebulosa: uma abordagem filosófica e aplicada. Universidade Federal de Santa Catarina (UFSC), Brasil (2002)
2. Campos, M.M., Saito, K.: Sistemas Inteligentes em Controle e Automação de Processos, 1st edn. Rio de Janeiro, Editora Ciência Moderna (2004)
3. Driankov, D., Rainer, P.: Advances in Fuzzy Control. Physica-Verlag (2013)
4. Mandani, E.H., Assilan, S.: An experiment in linguistic synthesis with a fuzzy logic controller. International Journal of Man-Machine Studies 7(1), 1–13 (1975)
5. Zadeh, L.A.: Fuzzy Sets. Information and Control 8, 338–353 (1965)
6. Menezes Filho, J.B.M., Silva, S.A., Araujo, C.S., Filho, A.F.X.: Controlador Vetorial Neural para Mesa de Coordenadas XY. Revista Controle & Automação 21(4), 406–426 (2010)
7. Paraskevopoulos, P.N.: Digital Control Systems, 1st edn. Prentice Hall, USA (1996)
8. Simões, M.G.: Controle e Modelagem Fuzzy. Editora Edgard Blucher Ltda, São Paulo, SP - Brasil (2001)
9. Fernandez, A., Garcia, A., Jesus, M.J., Herrera, F.: A study of the behavior of linguistic fuzzy rule based classification system in the framework of imbalanced data sets. Fuzzy Sets and Systems 159, 2378–2398 (2008)
10. Holland, J.: Adaptation in natural and artificial systems. The University of Michigan Press, Ann Arbor (1975)
11. Koza, J.R.: Genetic Programming. MIT Press (1992)

12. Sivanandam, S.N., Deepa, S.N.: Introduction to Genetic Algorithms. Springer, Berlin (2008)
13. Gen, M., Kumar, A., Kim, J.: Recent network design techniques using evolutionary algorithms. International Journal of Production Economics 98(2), 251 (2005)
14. Gen, M., Syarif, A.: Multi-stage Supply Chain Network by Hybrid Genetic Algorithms with Fuzzy Logic Controller. In: Verdegay, J.L. (ed.) Fuzzy Sets Based Heuristics for Optimization. STUDFUZZ, vol. 126, pp. 181–196. Springer, New York (2003)

Part IV
Optimal Control

Optimal Control of a SEIR Model with Mixed Constraints and L^1 Cost

Maria do Rosário de Pinho[1], Igor Kornienko[1], and Helmut Maurer[2]

[1] University of Porto, Faculdade de Engenharia, ISR, Porto, Portugal
mrpinho,igor@fe.up.pt
[2] Institut für Numerische und Angewandte Mathematik, Universität Münster,
Münster, Germany
maurer@math.uni-muenster.de

Abstract. Optimal control can help to determine vaccination policies for infectious diseases. For diseases transmitted horizontally, SEIR compartment models have been used. Most of the literature on SEIR models deals with cost functions that are quadratic with respect to the control variable, the rate of vaccination. Here, we propose the introduction of a cost of L^1 type which is linear with respect to the control variable. Our starting point is the recent work [1], where the number of vaccines at each time is assumed to be limited. This yields an optimal control problem with a mixed control-state constraint. We discuss the necessary optimality conditions of the Maximum Principle and present numerical solutions that precisely satisfy the necessary conditions.

Keywords: optimal control, epidemiology, mixed constraints, numerical solutions, bang-bang control.

1 Introduction

SEIR models are widely used to model the spreading of infectious diseases in a population. These models divide the individuals in four compartments relevant to the epidemic. An individual can be in the S compartment if vulnerable (or susceptible) to the disease. Those infected but not able to transmit it are in the E compartment, the exposed population. Infected individuals capable of spreading the disease are in the I compartment and those who are immune are in the R compartment. In SEIR models everyone is assumed to be susceptible to the disease by birth and the disease is transmitted to the individual by horizontal incidence, i.e., a susceptible individual becomes infected when in contact with infectious individuals.

Let $S(t)$, $E(t)$, $I(t)$, and $R(t)$ denote the number of individuals in the susceptible, exposed, infectious and recovered compartments at time t respectively. The total population is $N(t) = S(t) + E(t) + I(t) + R(t)$. The disease transmission in a certain population is described by the following parameters: e, the rate at which the exposed individuals become infectious, g, the rate at which infectious individuals recover, a, the death rate due to the disease, b, the natural birth rate, and d, the natural death rate. Let c be the incidence coefficient of horizontal transmission. Then the rate of transmission of the disease is $cS(t)I(t)$. For simplicity the parameters are assumed constants (although they may vary in reality if the time horizon is big). For more information about such model we refer the reader to [2], [9], [13] and references within.

Based on such models, different vaccines policies have been studied based on optimal control. Such scientific exercise are of interest to compare different vaccination scenarios. In [13] and [1] different policies are confronted. However, in both such works, the cost of the optimal control problems considered is defined by an integrand quadratic with respect to control. A special feature of [1] is that a somewhat realist scenario of limited supplies of vaccine at each instant is considered. From the point of view of optimal control, this is translated by the introduction of a mixed constraint (also known as state dependent control constraint). Here, we assume that the vaccine is effective so that all vaccinated susceptible individuals become immune. Let $u(t)$ represent the fraction of susceptible individuals being vaccinated per unit of time. Taking all the above considerations into account we are led to the following dynamical system:

$$\dot{S}(t) = bN(t) - dS(t) - cS(t)I(t) - u(t)S(t) \tag{1}$$

$$\dot{E}(t) = cS(t)I(t) - (e + d)E(t) \tag{2}$$

$$\dot{I}(t) = eE(t) - (g + a + d)I(t) \tag{3}$$

$$\dot{R}(t) = gI(t) - dR(t) + u(t)S(t) \tag{4}$$

$$\dot{N}(t) = (b - d)N(t) - aI(t) \tag{5}$$

with the initial conditions

$$S(0) = S_0, \; E(0) = E_0, \; I(0) = I_0, \; R(0) = R_0, \; N(0) = N_0. \tag{6}$$

To keep track of the number of vaccinated individuals we introduce an extra variable W that satisfies the equation

$$\dot{W}(t) = u(t)S(t), \quad W(0) = 0. \tag{7}$$

Like other models in epidemiology, SEIR models represent only a rough approximation of reality. However, they can be of use to provide new insights into the spreading of diseases and, when optimal control is applied, to propose better vaccination policies. In this paper, we are not treating a particular disease in a specific population. Rather, we want to illustrate how different optimal control formulations can be used to propose new vaccination policies, when different scenarios and cost functionals are considered.

Here, our stepping stone is the paper by Biswas, Paiva, de Pinho [1] who study situations, where the number of vaccines is limited or the capability of vaccinating is limited at each time. In [1], the overall limit of vaccines $W(T) \leq W_M$ considered in [13] is replaced by a mixed control-state constraint of the form:

$$u(t)S(t) \leq V_0 \quad \text{a.e.} \; t \in [0, T], \tag{8}$$

where V_0 is an upper bound on vaccines available at each instant t. The inequality (8) is also known in the literature as state dependent control constraint. The constraint $W(T) \leq W_M$ in [13] should be satisfied only at the end terminal time T, while the mixed constraint (8) is to be satisfied at almost every instant of time during the whole vaccination program.

Both papers [1] and [13] consider a control quadratic cost functional of L^2–type. It has been argued that a control quadratic cost is not appropriate for problems with a biological or biomedical background. Therefore, in our work we consider a L^1 cost functional that is linear with respect to the control variable u. Since the control variable appears linearly in the dynamics and constraints as well, the (augmented) Hamiltonian function is linear with respect to the control variable. The evaluation of the necessary optimality condition of the Maximum Principle shows that any optimal control must be a concatenation of bang-bang and singular arcs. Here, the term "bang-bang" or "singular" refers to the mixed constraint itself which will be made clear in section 3. In this paper, we do not further discuss singular controls, since our numerical solutions furnish only bang-bang controls. We obtain a numerical solution of our problem by applying direct optimization method, i.e., we discretize the control problem and use nonlinear programming methods. Then we employ the Imperial College London Optimal Control Software – ICLOCS – version 0.1b ([6]). This is an optimal control interface, implemented in Matlab, that calls the IPOPT – Interior Point OPTimizer – an open-source software package for large-scale nonlinear optimization. For a study of different optimal control solvers see [14]. As for information on IPOPT we refer the reader to [16]. Alternatively, we use the Applied Modeling Programming Language AMPL [7] which can be interfaced to IPOPT and a number of other optimization solvers.

Although we do not show that the numerical solution is indeed a (local) optimum, we do however validate our findings. Using the Lagrange multipliers provided by the optimization solver, we can validate our numerical solution by showing that it satisfies precisely the necessary condition of optimality.

2   The Optimal Control Problem with Mixed Constraints

We consider the following optimal control problem with a mixed control-state constraint:

$$(P_1) \begin{cases} \text{Minimize} \quad J_1(x, u) = \int_0^T (AI(t) + Bu(t)) \, dt \\ \text{subject to} \\ \dot{S}(t) \quad = bN(t) - dS(t) - cS(t)I(t) - u(t)S(t), \\ \dot{E}(t) \quad = cS(t)I(t) - (e + d)E(t), \\ \dot{I}(t) \quad = eE(t) - (g + a + d)I(t), \\ \dot{N}(t) \quad = (b - d)N(t) - aI(t), \\ u(t)S(t) \leq V_0, \\ u(t) \quad \in [0, 1] \quad \text{for a.e. } t \in [0, T], \\ S(0) \quad = S_0, \; E(0) = E_0, \; I(0) = I_0, \; N(0) = N_0. \end{cases}$$

We have removed the equation (4) for R, since the number of recovered individuals is given by $R(t) = N(t) - S(t) + E(t) + I(t)$, and the equation (7) for W which is redundant for the control problem. Thus the state variable is

$$x = (S, E, I, N) \in \mathbb{R}^4.$$

Observe that our cost is now the L^1 cost

$$J_1(x, u) = \int_0^T (AI(t) + Bu(t)) \, dt$$

in contrast to [13] and [1], where the control quadratic cost

$$J_2(x, u) = \int_0^T \left(A_2 I(t) + B_2 u^2(t) \right) dt$$

is considered (here A, B, A_1 and B_1 are given constants). The convexity of $J_2(u)$ with respect to u is advantageous for the numerical approach, since it allows to express the control variable by the state and adjoint variable. In both cases, the cost functional is a weighted sum of the overall cost of caring for the infected individuals and the cost of vaccination. Observe, however, that in the case of $J_2(x, u)$ the cost of vaccination will depend on u^2, a small quantity compared to u which takes values less than 1. In this respect, $J_1(x, u)$ is a more realistic cost functional.

To simplify the analysis of the necessary optimality conditions of the control problem (P_1), it is convenient to rewrite it in the form of a general optimal control problem with a mixed control-state constraint:

$$(P_{\text{mixed}}) \begin{cases} \text{Minimize } \int_0^T L(x(t), u(t)) \, dt \\ \text{subject to} \\ \dot{x}(t) = f(x(t)) + g(x(t))u(t) \quad \text{a.e. } t, \\ m(x(t), u(t)) \leq 0 \quad \text{a.e. } t, \\ u(t) \in [0, 1] \quad \text{a.e. } t, \\ x(0) = x_0, \\ x(T) \in \mathbb{R}^n, \end{cases}$$

where

$$x = (S, E, I, N), \qquad L(x, u) = AI + Bu = L_1(x) + L_2(u),$$
$$f(x) = f_1(x) + A_1 x, \quad f_1(x) = c(-SI, SI, 0, 0)^T,$$
$$g(x) = (-S, 0, 0, 0)^T, \quad m(x, u) = uS - V_0,$$

and

$$A_1 = \begin{bmatrix} -d & 0 & 0 & b \\ 0 & -(e+d) & 0 & 0 \\ 0 & e & -(g+a+d) & 0 \\ 0 & 0 & -a & b-d \end{bmatrix}.$$

The initial condition x_0 and parameters will be specified in Table 1. The differential equation $\dot{x}(t) = f(x(t)) + g(x(t))u(t)$ is affine in the control and is nonlinear in the state x due to the term $f_1(x)$. Note that the mixed control-state constraint satisfies the standard *regularity condition*

$$m_x(x(t), u(t)) = S(t) \neq 0 \quad \forall\, t \in [0, T] \text{ with } u(t)S(t) = V_0. \tag{9}$$

Here and in the following, partial derivatives are denoted by subscripts. Moreover, appealing to [4], for example, it is an easy task to prove the existence of a solution (x_*, u_*) to problem (P_{mixed}).

3 Necessary Optimality Conditions for (P_1)

Let (x_*, u_*) be a minimizer for our problem (P_1) (or (P_{mixed})). In the following, we shall evaluate the necessary optimality condition of the *Maximum Principle*. Since we are maximizing $-J_1(x, u)$, the standard Hamiltonian function is given by

$$H(x, p, u) = -\lambda L(x, u) + \langle p, f(x) + g(x)u \rangle, \quad \lambda \in \mathbb{R},$$

where $p = (p_S, p_E, p_I, p_N) \in \mathbb{R}^4$ denotes the adjoint variable. In the *augmented* Hamiltonian, the mixed constraint $m(x, u) \geq 0$ is adjoined by a multiplier $q \in \mathbb{R}$ to the Hamiltonian:

$$\mathcal{H}(x, p, q, u) = H(x, p, u) - q\, m(x, u).$$

Here, the minus sign is due to the fact that the *Maximum Principle* assumes that the control-state constraint is written in the form $-m(x, u) \geq 0$. In view of the regularity condition (9), Theorem 7.1 in [5] (cf. also [8,11]) asserts the existence of a scalar $\lambda \geq 0$, an absolutely continuous function $p : [0, T] \to \mathbb{R}^4$ and an integrable function $q : [0, T] \to \mathbb{R}$ such that the following conditions are satisfied almost everywhere:

(i) $\max\{|p(t)| : t \in [0, T]\} + \lambda > 0,$

(ii) (adjoint equation and transversality condition)

$$-\dot{p}(t) = \mathcal{H}_x[t] = -\lambda L_x[t] + \langle p(t), f_x[t] + g_x[t]u_*(t) \rangle - \langle q(t), m_x[t] \rangle,$$
$$-p(T) = (0, 0, 0, 0),$$

(iii) (maximum condition for Hamiltonian H)

$$H(x_*(t), p(t), u_*(t)) = \max_u \{ H(x_*(t), p(t), u) \mid 0 \leq u \leq 1,\, m(x_*(t), u) \leq 0 \},$$

(iv) (local maximum condition for augmented Hamiltonian $\mathcal{H}$)

$$\mu(t) = \mathcal{H}_u[t] = -L_u[t] + \langle p(t), g[t] \rangle - q(t)\, m_u[t] \in N_{[0,1]}(u_*(t)),$$

(v) (complementarity condition)

$$q(t)\, m(x_*(t), u_*(t)) = q(t)\, (u_*(t)S_*(t) - V_0) = 0 \quad \text{and} \quad q(t) \geq 0.$$

In (iv), $N_{[0,1]}(u_*(t))$ stands for the normal cone from convex analysis to $[0,1]$ at the optimal control $u_*(t)$ (see e.g. [3]) and it reduces to $\{0\}$ when $u^*(t) \in]0,1[$. Since the terminal state $x(T)$ is free, it is easy to prove that the above necessary conditions hold with $\lambda = 1$; for a complete discussion see [1]. Hence, our problem is *normal*. We can further prove the existence of a constant K_q^1 such that

$$|q(t)| \leq K_q^1 |p(t)| \tag{10}$$

for almost every $t \in [0, T]$ (see [5]).

Now we want to extract information from the conclusions (i)–(v) with $\lambda = 1$ that later will be used to validate our numerical solution. The adjoint equations in (ii) for the adjoint variable $p = (p_S, p_E, p_I, p_N)$ are explicitly given by

$$-\dot{p}_S(t) = -(d + cI_*(t) + u_*(t))p_S(t) + cI_*(t)p_E(t) - u_*(t)q(t), \tag{11}$$

$$-\dot{p}_E(t) = -(e + d)p_E(t) + ep_I(t), \tag{12}$$

$$-\dot{p}_I(t) = -cS_*(t)p_S(t) + cS_*(t)p_E(t) - (g + a + d)p_I(t) - ap_N(t) - A, \tag{13}$$

$$-\dot{p}_N(t) = bp_S(t) + (b - d)p_N(t). \tag{14}$$

Next, we evaluate the maximum condition (iii) for the Hamiltonian H. We define the *switching function* ϕ by

$$\phi(x, p) = H_u(x, u, p) = -B - p_S\, S, \quad \phi(t) = \phi(x(t), p(t)) \tag{15}$$

and see that the condition (iii) is equivalent to the maximum condition

$$\phi(t)u_*(t) = \max_u \{\, \phi(t)u \mid 0 \leq u \leq 1,\ u\, S_*(t) \leq V_0 \,\}. \tag{16}$$

This yields the control law

$$u_*(t) = \begin{cases} \min\left\{1, \dfrac{V_0}{S_*(t)}\right\}, & \text{if } \phi(t) > 0 \\[2mm] 0 & , \quad \text{if } \phi(t) < 0. \end{cases} \tag{17}$$

Any isolated zero of the switching function $\phi(t)$ yields a switch of the control from $\min\{1, V_0/S_*(t)\}$ to 0 or vice versa. If, however, $\phi(t) = 0$ holds on an interval $[t_1, t_2] \subset [0, T]$, then we have a *singular control*. We do not enter here into a detailed discussion of singular controls, since they never appeared in our computations. Moreover, our computations show that $0 < u_*(t) < 1$ holds along a boundary arc of the mixed constraint $uS \leq V_0$, i.e., whenever $u_*(t) = V_0/S_*(t)$. Hence, the control is determined by

$$u_*(t) = \begin{cases} V_0/S_*(t), & \text{if } \phi(t) > 0 \\[2mm] 0 & , \quad \text{if } \phi(t) < 0. \end{cases} \tag{18}$$

Due to $0 < u_*(t) < 1$ the multiplier $\mu(t)$ in (iv) vanishes which yields the relation

$$0 = \mu(t) = \mathcal{H}_u[t] = -B - p_S(t)S_*(t) - q(t)S_*(t).$$

This allows us to compute the multiplier $q(t)$ for which we get in view of the complementarity condition (v)

$$q(t) = \begin{cases} -\dfrac{B}{S_*(t)} - p_s(t) = \phi(t)/S_*(t), & \text{if } u_*(t) = V_0/S_*(t), \\ \qquad\qquad 0, & \text{if } u_*(t) < V_0/S_*(t) . \end{cases} \tag{19}$$

4   Numerical Results

In Table 1 we present the values of the initial conditions, parameters and constants which have been used in our computations. Apart from the weights A and B they coincide with those in [13]. As stated before in the Introduction,

Table 1. Parameters with their clinically approved values and constants as in [13]

Parameter	Description	Value
b	natural birth rate	0.525
d	natural death rate	0.5
c	incidence coefficient	0.001
e	exposed to infectious rate	0.5
g	recovery rate	0.1
a	disease induced death rate	0.2
A	weight parameter	5
B	weight parameter	10
T	number of years	20
S_0	initial susceptible population	1000
E_0	initial exposed population	100
I_0	initial infected population	50
R_0	initial recovered population	15
N_0	initial population	1165
W_0	initial vaccinated population	0

we use the software package ICLOCS [6], resp., the programming environment AMPL and IPOPT [7,16] to solve problem (P_{mixed}). As in [1] we consider the final time $T = 20$ (days). We choose a rather fine grid with $N = 10000$ nodes with stepsize $\Delta t = T/N = 0.002$ and use the implicit Euler scheme as integration method. The acceptable convergence tolerance is set to $\varepsilon_{rel} = 10^{-9}$. For the mixed constraint $u(t)S(t) \leq V_0 = 125$ we find the optimal control

$$u_*(t) = \begin{cases} 125/S_*(t) & \text{for } 0 \leq t \leq t_1, \\ \qquad 0 & \text{for } t_1 < t \leq T = 20. \end{cases} \tag{20}$$

This shows that the constraint itself when expressed as the *new control variable* $v = uS$ is a bang-bang control with only one switch at t_1; cf. [10]. We obtain the numerical results

$$J_1(x, u) = 1692.2, \quad t_1 = 17.89,$$
$$S(T) = 1723.8, \quad E(T) = 7.7030, \quad I(T) = 4.7038, \quad N(T) = 1824.2.$$

The total amount of vaccines is $W(T) = 2235.8$ The optimal trajectories and optimal control are presented in Figure 1.

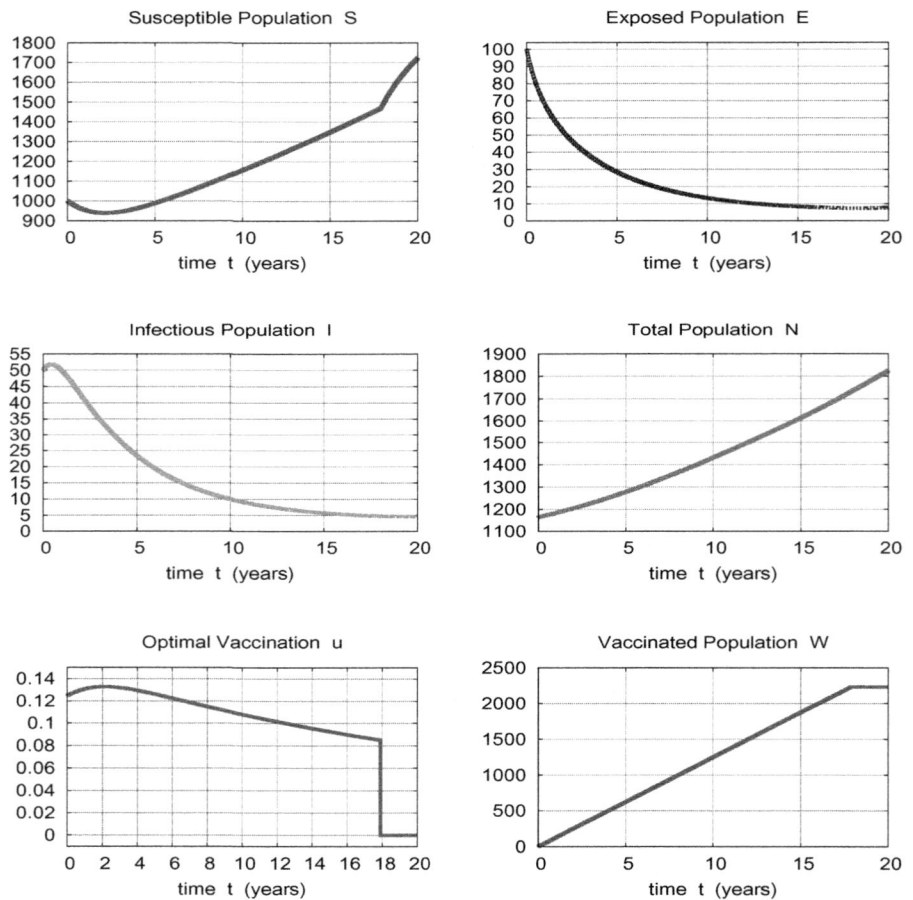

Fig. 1. Optimal trajectories and control (vaccination) for mixed constraint $uS \le 125$. *Top row*: (left) susceptible population S, (right) exposed population E. *Middle row*: (left) infectious population I, (right) total population N. *Bottom row*: (left) vaccination (control) u, (right) vaccinated population W.

The adjoint variables, the switching function $\phi(t)$ and multiplier $q(t)$ are displayed in Figure 2. It can be seen in Figure 2, bottom row, that the switching function $\phi(t)$ satisfy exactly the control law (17) while the multiplier $q(t)$ obeys the multiplier rule (19).

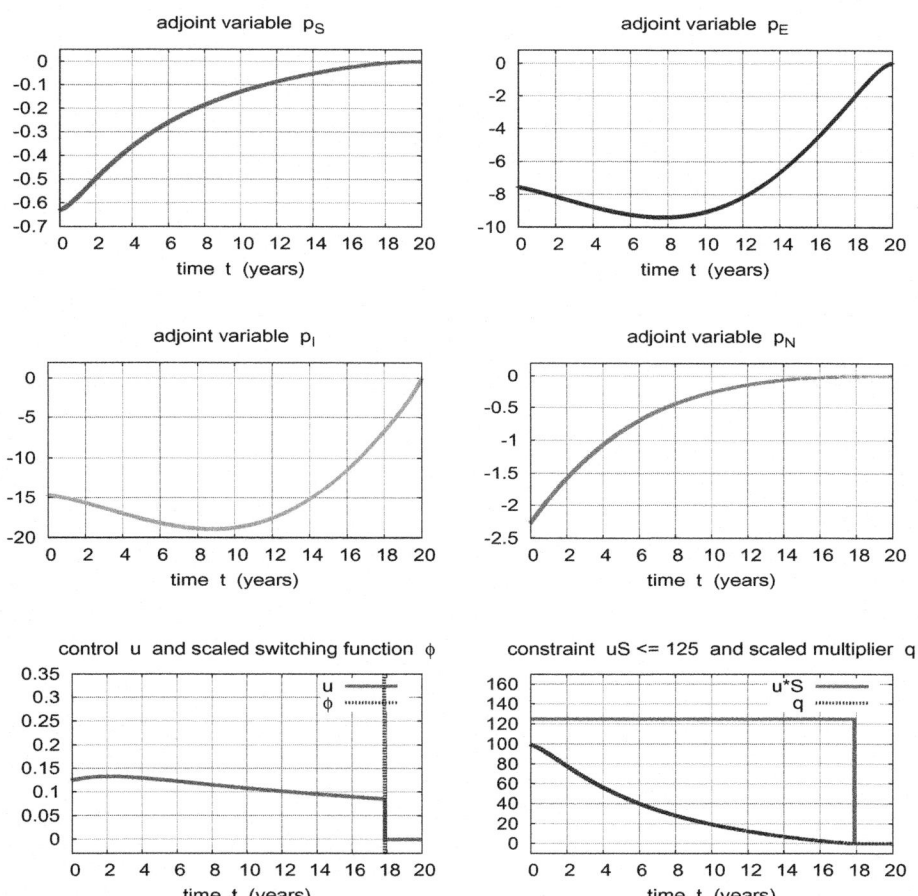

Fig. 2. Adjoint variables, multiplier and switching function for mixed constraint $u(t)S(t) \leq 125$. *Top row:* (left) adjoint variable p_S, (right) adjoint variable p_E. *Middle row:* (left) adjoint variable p_I, (right) adjoint variable p_N. *Bottom row:* (left) control u and (scaled) switching function $\phi(t)$ satisfying (17), (right) constrained function uS and multiplier q satisfying (19).

5   Conclusion

We considered an optimal control problem with mixed constraints and L^1 cost for a SEIR epidemic model of human infectious diseases. In this optimal control problem the control appears linearly. We discussed the necessary conditions of

the Maximum Principle and obtained explicit formulas for the switching function
and the multiplier associated with the mixed constraint in terms of state and
adjoint variables. The optimal control problem is solved by discretization and
nonlinear programming methods. The numerical solution shows that the constraint itself, when regarded as the new control variable $v = uS$, is a bang-bang
control. We never encountered singular controls.

Since the numerical approach furnishes as well the adjoint variables, we could
verify that the computed solution satisfies the necessary optimality conditions
precisely. The study of sufficient conditions is beyond the scope of this paper.
In a future work, we shall investigate the numerical verification of second-order
sufficient conditions using the methods in [10,12]. This also allows us to study
sensitivity analysis and compute parametric sensitivity derivatives.

Acknowledgement. The financial support of FCT projects PTDC/EEA-ELC/
122203/2010, PTDC/EEI-AUT/1450/2012—FCOMP-01-0124-FEDER-028894,
and European Union FP7 (FP7-PEOPLE-2010-ITN, Grant Agreement no.
264735-SADCO) are gratefully acknowledged.

References

1. Biswas, M.H.A., Paiva, L.T., de Pinho, M.D.R.: A SEIR model for control of
infectious diseases with constraints. Mathematical Biosciences and Engineering (to
appear, 2014)
2. Brauer, F., Castillo-Chavez, C.: Mathematical Models in Population Biology and
Epidemiology. Springer, New York (2001)
3. Clarke, F.: Optimization and Nonsmooth Analysis. John Wiley, New York (1983)
4. Clarke, F.: Functional Analysis, Calculus of Variations and Optimal Control.
Springer, London (2013)
5. Clarke, F., de Pinho, M.D.R.: Optimal control problems with mixed constraints.
SIAM J. Control Optim. 48, 4500–4524 (2010)
6. Falugi, P., Kerrigan, E., van Wyk, E.: Imperial College London Optimal Control
Software User Guide (ICLOCS). Department of Electrical and Electronic Engineering. Imperial College London, London (2010)
7. Fourer, R., Gay, D.M., Kernighan, B.W.: AMPL: A Modeling Language for Mathematical Programming. Duxbury Press, Brooks–Cole Publishing Company (1993)
8. Hestenes, M.R.: Calculus of Variations and Optimal Control Theory, 2nd edn., 405
pages. John Wiley, New York (1980)
9. Hethcote, H.W.: The basic epidemiology models: models, expressions for R_0, parameter estimation, and applications. In: Ma, S., Xia, Y. (eds.) Mathematical Understanding of Infectious Disease Dynamics, vol. 16, ch. 1, pp. 1–61. World Scientific
Publishing Co. Pte. Ltd., Singapore (2008)
10. Maurer, H., Osmolovskii, N.P.: Second-order conditions for optimal control problems with mixed control-state constraints and control appearing linearly. In: Proceedings of the 52nd IEEE Conference on Control and Design (CDC 2013), Firenze,
pp. 514–519 (2013)
11. Maurer, H., Pickenhain, S.: Second order sufficient conditions for optimal control
problems with mixed control-state constraints. J. Optimization Theory and Applications 86, 649–667 (1995)

12. Osmolovskii, N.P., Maurer, H.: Applications to Regular and Bang-Bang Control: Second-Order Necessary and Sufficient Optimality Conditions in Calculus of Variations and Optimal Control. SIAM Advances in Design and Control 24 (2013)
13. Neilan, R.M., Lenhart, S.: An Introduction to Optimal Control with an Application in Disease Modeling. DIMACS Series in Discrete Mathematics 75, 67–81 (2010)
14. Paiva, L.T.: Optimal Control in Constrained and Hybrid Nonlinear Systems. Project Report (2013),
    http://paginas.fe.up.pt/~faf/ProjectFCT2009/report.pdf
15. Schaettler, H., Ledzewicz, U.: Geometric Optimal Control. Theory, Methods and Examples. Springer, New York (2012)
16. Wächter, A., Biegler, L.T.: On the implementation of an interior-point filter line-search algorithm for large-scale nonlinear programming. Mathematical Programming 106, 25–57 (2006)

Optimal Control for a Cascade of Hydroelectric Power Stations: Case Study

Ana Filipa Ribeiro

Faculty of Engineering of University of Porto, FEUP, Porto, Portugal
ana.fca.ribeiro@gmail.com

Abstract. In this work we consider a simple model for a hydro-electric system with two power stations with reversible turbines. The problem is considered in the framework of optimal control theory. The objective is to find water flows to be turbined/pumped on each station and the corresponding volumes in the reservoirs that maximize the profit of selling energy. Restrictions on the water level in the reservoirs must be satisfied. The presence of state constraints and the nonconvexity of the cost function contribute to an increase of complexity of the optimal control problem. Using available software we obtain numerical results. Taking into account the profile of the solution we apply the maximum principle of Pontryagin and some new sufficient conditions of optimality to validate such solution.

Keywords: optimal control, sufficient conditions, maximum principle, pure state constraints.

1 Introduction

The use of water to produce energy must be effective and efficient to maximize its benefit. Being the water a renewable resource that is becoming scarce, the concern on a good use of it is more evident. The management of interconnected reservoir systems in a river is of particular importance if there is also the possibility of reusing the downstream water in a situation of drought. This may be implemented in modern reversible hydro-electric power stations associated with reservoirs along a river basin with a cascade structure, where it is possible both to turbine water from upstream to produce electric energy and to pump from downstream to refill an upstream reservoir. Optimization in reservoirs system management has attracted the attention of many researchers in different contexts [2]. In this work we consider a simplified model for a cascade of two hydro-electric power stations where one of the stations has reversible turbines. The problem is considered in the framework of optimal control theory. The control variables are the turbined/pumped water flows for each reservoir and the state variables are the volumes of water present at each reservoir. The objective is to find water flows and corresponding volumes in the reservoirs that maximize the profit of selling energy in the management of water in the system. A high energy price corresponds to a high need of energy and inversely a low price corresponds to a low need. It is a challenging problem since besides constraints in the control it also involves pure state constraints. Furthermore, the cost function is nonconvex which contributes to increase the complexity of the problem. Numerical results are obtained using available software. Taking into account

© Springer International Publishing Switzerland 2015

A.P. Moreira et al. (eds.), *CONTROLO'2014 - Proc. of the 11th Port. Conf. on Autom. Control*,
Lecture Notes in Electrical Engineering 321, DOI: 10.1007/978-3-319-10380-8_15

the profile of the solution given by the numerical approach we apply some mathematical tools to validate such solution. More specifically we use the maximum principle of Pontryagin and some new sufficient conditions of optimality for a particular class of optimal control problems which includes our problem. This work illustrates the importance of combining numerical and theoretical tools to analyse an optimization problem. In many cases, the numerical methods alone does not give assurance that the obtained candidate is a solution to the problem. However, those methods supported by analytical results, necessary and sufficient optimality conditions, and the analysis of particular properties of the problem contribute in general to a rigorous treatment of the problem.

We shall use the notation $BV([0,T];\mathbb{R})$ and $AC([0,T],\mathbb{R})$ for the space of functions $f:[0,T] \to \mathbb{R}$ which are, respectively, of bounded variation and absolutely continuous.

2 Problem Statement

This work is based on a simplification of the model presented in [1]. Here, we consider a system with two hydro-electric power stations, referred to as station 1 and station 2, in a cascade structure. The power station 1 is reversible, it has the possibility to turbine water from upstream to produce electric power and to pump from downstream to refill an upstream reservoir. Naturally this station pumps when the price and demand are low, allowing the reusing of water on a later more convenient time.

For a cascade of 2 hydro-electric power stations, we consider the dynamic of water volumes $V_k(t)$, in the reservoirs $k = \overline{1,2}$, described by the following control system

$$\dot{V}_1(t) = A - u_1(t), \qquad \dot{V}_2(t) = u_1(t) - u_2(t). \qquad (1)$$

The control variables $u_1(t), u_2(t)$ are the turbined/pumped water flows of reservoirs 1 and 2 at time t and A is the incoming flow. Equation (1) is called water balance equation and is present in many references (see, e.g. [1]). The control variables and the state variables $V_1(\cdot), V_2(\cdot)$ satisfy the following technical constraints :

$$V_k(0) = V_k(T), \qquad V_k(t) \in [V_k^m, V_k^M], \qquad u_k(t) \in [u_k^m, u_k^M], \qquad k = 1,2. \qquad (2)$$

Here V_k^m and V_k^M, $k = \overline{1,2}$, stand for the imposed minimum and maximum water volumes, respectively; u_k^m and u_k^M, $k = \overline{1,2}$, are the imposed minimum and maximum turbined/pumped water flows. The equality $V_k(0) = V_k(T)$ is called periodic constraint and it ensures in particular that the reservoir k does not spend all the water on the period $[0,T]$. Also and under similar conditions on a period of time that would follow, the optimal solution would repeat itself.

The objective is to find water flows $(\hat{u}_1(\cdot), \hat{u}_2(\cdot)) = \hat{u}(\cdot)$ and respective volumes $(\hat{V}_1(\cdot), \hat{V}_2(\cdot)) = \hat{V}(\cdot)$, that maximize the profit of selling energy.

The cost function $J(u(\cdot), V(\cdot))$ to be maximized is represented by:

$$\int_0^T c(t) \left[u_1(t) \left(\frac{V_1(t)}{S_1} + H_1 - \frac{V_2(t)}{S_2} - H_2 \right) + u_2(t) \left(\frac{V_2(t)}{S_2} + H_2 \right) \right] dt, \qquad (3)$$

where H_k, $k = \overline{1,2}$, are the liquid surface elevations, S_k, $k = \overline{1,2}$, are the areas of the reservoirs (for simplicity we are assuming the reservoirs with cylindric form) and

$$c(t) = \begin{cases} c_1, & \text{if } t \in [0,a[\cup\{T\} \\ c_2, & \text{if } t \in [a,T[\end{cases} \tag{4}$$

with c_1, c_2 positive constants, is the price of energy and $a = T/2$.
Replacing u_1, u_2 by equivalent expressions obtained from (1) and after some calculus we can write (3) as

$$J(u(\cdot),V(\cdot)) = -\frac{Ac_1}{S_1}\int_0^a V_1(t)dt - \frac{Ac_2}{S_1}\int_a^T V_1(t)dt + H_1(c_2 - c_1)V_1(0)$$

$$+ \frac{c_2 - c_1}{2S_1}V_1^2(0) + H_2(c_2 - c_1)V_2(0) + \frac{c_2 - c_1}{2S_2}V_2^2(0) - H_1(c_2 - c_1)V_1(a) \tag{5}$$

$$- \frac{c_2 - c_1}{2S_1}V_1^2(a) - H_2(c_2 - c_1)V_2(a) - \frac{c_2 - c_1}{2S_2}V_2^2(a) - AH_1a(c_1 + c_2)$$

It is easy to check that the optimal control problem of maximizing (3) (or equivalently the minimization of (5)) subject to dynamics/constraints (1) and (2) has an optimal solution. For that we may apply existence results of *Lee & Markus* [8]. The data of our problem satisfy the compactness and regularity hypothesis imposed on Theorem 4 and Corollary 1 of section 4.2 on the cited work.

3   Numerical Results

The simple model of the previous section was treated numerically with available software. Such software allow us to find (local) solutions of mathematical optimization problems.

We consider the following data:

$$\begin{array}{llllll} u_1^m = -0.351 & u_2^m = 0 & u_1^M = 0.44496 & u_2^M = 0.8316 & S_1 = 81.7 & c_1 = 2 \\ V_1^m = 86.7 & V_2^m = 48.3 & V_1^M = 147 & V_2^M = 66 & S_2 = 44.5 & c_2 = 20 \\ A = 0.158 & T = 24 & H_1 = 3 & H_2 = 1 \end{array}$$

Figure 1 represents the numerical solution for $\hat{V}_1$, $\hat{V}_2$ (states), $\hat{u}_1$, $\hat{u}_2$ (controls) when it is applied an optimization package from book [3]. This software works in a regime of dialogue with the computer. The user has the possibility to select some optimization method (e.g. simplex, conjugated gradient, Newton's method) and its numerical precision. There is also the possibility to control the penalty coefficient (to obtain a feasible solution this coefficient must be large enough).

In the next section, necessary conditions for optimal control problems with state constraints are applied. We make use of the control and trajectory profiles observed in the above numerical results. The exact values for $\hat{V}_i(0)$, $i = 1,2$ and the instant $\theta_1 \in]0, a[$ on which there's a change of $\hat{u}_1(t)$ value are deduced easily after some analysis of the problem. We prove that the necessary conditions of optimality are satisfied in normal form by the observed pair control/trajectory.

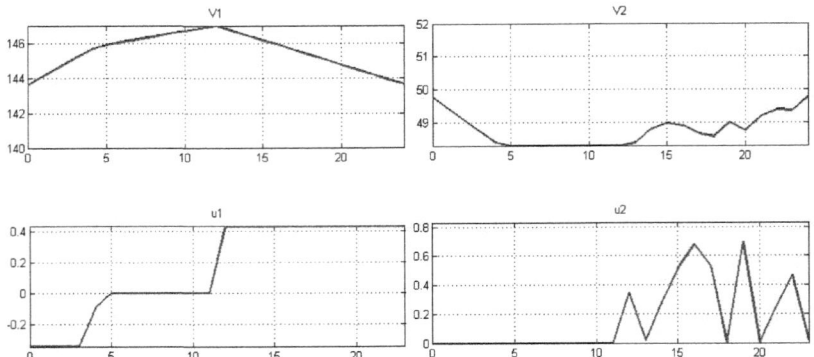

Fig. 1. Numerical Results (state variables: V_1, V_2; control variables: u_1, u_2)

4   Necessary Conditions

Recall that the pair $(\hat{u}(\cdot), \hat{V}(\cdot))$, with $\hat{u}(\cdot) = (\hat{u}_1(\cdot), \hat{u}_2(\cdot))$ and $\hat{V}(\cdot) = (\hat{V}_1(\cdot), \hat{V}_2(\cdot))$, denotes an optimal process for the problem of maximizing $J(u(\cdot), V(\cdot))$ given by (3), subject to dynamic and restrictions (1) and (2).

In [7], Theorem 9.5.1 introduces a smooth maximum principle for state constrained optimal control problems. When applied to our problem, it establishes the existence of absolutely continuous functions p_1, p_2, nonnegative Borel measures μ_1, μ_2, μ_3, μ_4 and a real number λ such that

(i) $(p, \mu, \lambda) \neq (0, 0, 0)$ $\qquad\qquad\qquad\qquad$ $p = (p_1, p_2), \quad \mu = (\mu_1, \mu_2, \mu_3, \mu_4)$

(ii) $p_1(t) = p_1(0) - \dfrac{1}{S_1} \displaystyle\int_0^t \lambda c(t) u_1(\tau) d\tau$

$p_2(t) = p_2(0) + \dfrac{1}{S_2} \displaystyle\int_0^t \lambda c(t)(u_1(\tau) - u_2(\tau)) d\tau$

(iii) $p_1(0) = p_1(T) - \mu_1\{[0,T]\} + \mu_2\{[0,T]\}$

$p_2(0) = p_2(T) - \mu_3\{[0,T]\} + \mu_4\{[0,T]\}$

(iv) The following maximum is attained for $u_1 = \hat{u}_1(t)$ and $u_2 = \hat{u}_2(t)$, $a.e.\ t \in [0,T]$

$$\max_{(u_1,u_2)} u_1 \left(-p_1(t) + \mu_1\{[0,t[\} - \mu_2\{[0,t[\} + p_2(t) - \mu_3\{[0,t[\} + \mu_4\{[0,t[\} \right.$$

$$\left. + \lambda c(t) \left(\frac{\hat{V}_1(t)}{S_1} + H_1 - H_2 - \frac{\hat{V}_2(t)}{S_2} \right) \right)$$

$$+ u_2 \left(-p_2(t) + \mu_3\{[0,t[\} - \mu_4\{[0,t[\} + \lambda c(t) \left(\frac{\hat{V}_2(t)}{S_2} + H_2 \right) \right)$$

(v) $supp\{\mu_i\} \subset I_i$, $i = \overline{1,4}$, where

$$I_1 = \{t \in [0,T] : V_1(t) = V_1^m\} \qquad I_2 = \{t \in [0,T] : V_1(t) = V_1^M\}$$

$$I_3 = \{t \in [0,T] : V_2(t) = V_2^m\} \qquad I_4 = \{t \in [0,T] : V_2(t) = V_2^M\}$$

and $supp\{\mu\}$ denotes the support of the measure μ.

We will prove that the conditions $(i) - (v)$ are satisfied for a determined set of multipliers p_1, p_2, μ_i, $i = \overline{1,4}$ and $\lambda = 1$. Acording to numerical solution we have

$$\hat{u}_1(t) = \begin{cases} u_1^m, \text{ for } t \in [0,\theta_1[\\ 0, \text{ for } t \in [\theta_1,a[\\ u_1^M, \text{ for } t \in [a,T] \end{cases} \qquad \hat{u}_2(t) = \begin{cases} 0, \text{ for } t \in [0,a[\\ w_2(t), \text{ for } t \in [a,T] \end{cases} \qquad (6)$$

where $\theta_1 \in]0,a[$ is the instant of change of control behaviour.

It is clear from (5) that changes on the trajectory $V_2(t)$ maintaining the values of $V_2(0)(=V_2(T))$ and $V_2(a)$ doesn't affect the cost. In fact, it is easy to see that the optimal trajectory $V_2(t)$ on the interval $[a,T]$ is not unique. This explains the irregular behaviour of $\hat{u}_2$ and $\hat{V}_2$ on Figure 1. Thus, we don't fix the value of the control on that interval. Using the simulation results we can also write

$$V_1(a) = V_1^M \quad \text{and} \quad V_2(\theta_1) = V_2^m. \qquad (7)$$

$$V_1(t) = \begin{cases} V_1(0) + (A - u_1^m)t, \ t \in [0,\theta_1[\\ V_1(\theta_1) + A(t - \theta_1), \ t \in [\theta_1,a[\\ V_1^M + (A - u_1^M)(t - a), \ t \in [a,T] \end{cases} \quad V_2(t) = \begin{cases} V_2(0) + u_1^m t, \ t \in [0,\theta_1[\\ V_2^m, \ t \in [\theta_1,a[\\ V_2^m + u_1^M(t - a) - w(t), \ t \in [a,T] \end{cases}$$

where $w(t) = \int_a^t w_2(\tau)d\tau$ and $w_2(\tau) \in [u_2^m, u_2^M]$. From (7) and definitions above we get

$$\hat{V}_1(0) = V_1^M + u_1^m\theta_1 - Aa \quad \text{and} \quad \hat{V}_2(0) = V_2^m - u_1^m\theta_1. \qquad (8)$$

Also from the periodic constraints $V_i(0) = V_i(T)$, $i = 1,2$ we can deduce that

$$\theta_1 = \frac{a(2A - u_1^M)}{u_1^m} \quad \text{and} \quad \int_a^T w_2(\tau)d\tau = AT. \qquad (9)$$

We analyse now the necessary conditions of optimality. Observe that since $\hat{V}_1(t) \neq V_1^m$ and $\hat{V}_2(t) \neq V_2^M$, $\forall t$, we have $\mu_1 \equiv \mu_4 \equiv 0$. As $\hat{V}_1(t) = V_1^M$ only at $t = a$, it comes $\mu_2\{[0,T]\} = \mu_2\{a\}$. From $(ii),(iii)$ and (v) we get

$$\mu_2\{[0,T]\} = \frac{c_1 u_1^m\theta_1}{S_1} + \frac{c_2 u_1^M a}{S_1}; \ \mu_3\{[0,T]\} = \frac{c_1 u_1^m\theta_1}{S_2} + \frac{c_2 u_1^M a}{S_2} - \int_a^T \frac{c_2 w_2(\tau)}{S_2}d\tau \qquad (10)$$

Information expressed by condition (iv) must be analysed separately on $[0,\theta_1[$, $[\theta_1, a[$ and $[a,T]$. The signals of coefficients of u_1, u_2 in (iv) determines the values of $\hat{u}_1$ and $\hat{u}_2$. Since we are departing from assumed $\hat{u}_1$ and $\hat{u}_2$, we obtain in this case information about those signals. On what follows we use (ii) and write the expressions in terms of $p_1(0)$ and $p_2(0)$.

To make the reading easier we define $\alpha = \left(\frac{\hat{V}_1(0)}{S_1} + H_1 - \frac{\hat{V}_2(0)}{S_2} - H_2 \right)$.

- On the interval $[0, \theta_1[: \mu_2\{[0, \theta_1[\} = \mu_3\{[0, \theta_1[\} = 0$. As $\hat{u}_1(t) = u_1^m$ and $\hat{u}_2(t) = u_2^m = 0$, we have

$$\frac{Ac_1\theta_1}{S_1} - p_1(0) + p_2(0) + c_1\alpha \leq 0 \qquad\qquad - p_2(0) + \frac{c_1\hat{V}_2(0)}{S_2} + c_1 H_2 \leq 0 \qquad (11)$$

- On the interval $[\theta_1, a[: \mu_2\{[0, a[\} = 0$. As $\hat{u}_1(t) = 0 \in]u_1^m, u_1^M[$ and $\hat{u}_2(t) = u_2^m$ we get

$$\frac{Ac_1 t}{S_1} - p_1(0) + p_2(0) + c_1\alpha = \mu_3\{[\theta_1, t[\} \geq 0 \qquad (12)$$

Taking the limit when $t \downarrow \theta_1$ $(t \to \theta_1, t > \theta_1)$ on the above expression we obtain

$$\frac{Ac_1\theta_1}{S_1} - p_1(0) + p_2(0) + c_1\alpha \geq 0 \qquad (13)$$

From (11) and (13) we conclude that

$$\frac{Ac_1\theta_1}{S_1} - p_1(0) + p_2(0) + c_1\alpha = 0. \qquad (14)$$

Consider now $\hat{u}_2(t) = 0 = u_2^m$. From (iv), (12) and (14), we get

$$\frac{Ac_1 t}{S_1} - p_1(0) + c_1 \left(\frac{\hat{V}_1(0)}{S_1} + H_1 \right) \leq 0.$$

The function on the first member of the last equation is increasing and continuous. The inequality is satisfied $\forall t \in [\theta_1, a[$ if and only if it is satisfied for $t = a$. We can write

$$p_1(0) \geq c_1 \left(\frac{Aa}{S_1} + \frac{\hat{V}_1(0)}{S_1} + H_1 \right).$$

- On the interval $[a, T[: \mu_2\{[0, T]\} = \mu_2\{a\}; \hat{u}_1(t) = u_1^M$. After some calculus we conclude from (iv):

$$- p_1(0) + \frac{c_2}{S_1} \left(\hat{V}_1(0) - u_1^M a - u_1^m\theta_1 + At \right) + p_2(0) + \frac{c_2}{S_2} \left(AT - u_1^M a - \hat{V}_2(0) - u_1^m\theta_1 \right)$$
$$+ c_2(H_1 - H_2) \geq 0$$

There are infinite different ways to define $\hat{V}_2(t)$ on $[a, T]$. Choosing $\hat{V}_2(t)$ interior to the admissible constraint set, we obtain $\mu_3\{[0, T]\} = \mu_3\{[\theta_1, a]\}$. Figure 1 shows one such trajectory. Also $\hat{u}_2(t)$ can be choosen to be interior to the set of admissible controls. We obtain

$$p_2(0) = \frac{c_2}{S_2} \left(u_1^M a - AT + \hat{V}_2(0) + u_1^m\theta_1 \right) + c_2 H_2 \qquad (15)$$

Now, looking at the conclusions above, we can easily get:

- $\hat{V}_1(0)$, $\hat{V}_2(0)$ and θ_1 written in terms of data of the problem (see (8) and (9))

- $p_2(0)$ written in terms of the data of the problem *(see (15))*
- $p_1(0)$ *(see (14))*
- equations defining $\mu_3\{[0,t[\} = \mu_3\{[\theta_1,t[\}$ and $\mu_3\{[0,T]\}$ *(see (12) and (10))*
- equation defining $\mu_2\{[0,T]\} = \mu_2\{a\}$ *(see (10))*

Using the data of the problem and above information we can write

$$\hat{V}_1(t) = \begin{cases} 143.56 + 0.51t, \ t \in [0, \theta_1[\\ 145.1 + 0.16t, \ t \in [\theta_1, a[\\ 147 - 0.287(t - 12), \ t \in [a, T] \end{cases} \qquad \hat{V}_2(t) = \begin{cases} 49.85 - 0.351t, \ t \in [0, \theta_1[\\ 48.3, \ t \in [\theta_1, a[\\ 48.3 + 0.44496(t - 12) - w(t), \ t \in [a, T] \end{cases}$$

where $w(t) = \int_a^t w_2(\tau)d\tau$ and $\theta_1 = 4.4$.

The multipliers $p_1(t)$ and $p_2(t)$ are now completely determined. The inequalities involving these functions must also be satisfied. For the data of our problem p_1 and p_2 can be defined as

$$p_1(t) = \begin{cases} 47.69 + 0.0086t, \ t \in [0, \theta_1[\\ 47.73, \ t \in [\theta_1, a[\\ 49.04 - 0.11t, \ t \in [a, T] \end{cases} \qquad p_2(t) = \begin{cases} 42.4 - 0.016t, \ t \in [0, \theta_1[\\ 42.33, \ t \in [\theta_1, a[\\ 39.93 + 0.19t - 0.45w(t), \ t \in [a, T[\\ 43.03, \ t = T \end{cases}$$

Using the above data we obtain

$$\mu_2\{[0, t[\} = \begin{cases} 0, \ t \in [0, a[\cup]a, T] \\ 1.27, \ t = a \end{cases} \qquad \mu_3\{[0, t[\} = \begin{cases} 0, \ t \in [0, \theta_1[\cup]a, T] \\ 0.004t - 0.02, \ t \in [\theta_1, a[\\ 0.63, \ t = a \end{cases}$$

where $w(T) = \int_a^T w_2(\tau)d\tau = 3.79$, $\quad V_1(0) = 143.56 \quad$ and $\quad V_2(0) = 49.85$.

Figure 2 overlaps the numerical solution with the analytical solution. These solutions are the same with exception of $V_2(t)$ on $[a, T]$. As outlined before, the solution is not unique on that set.

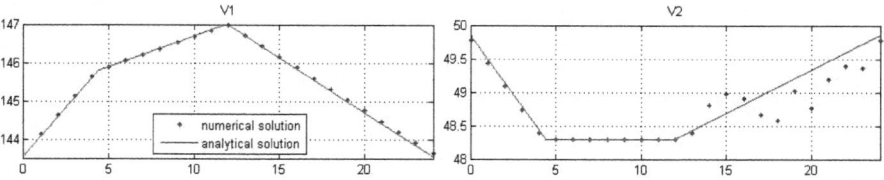

Fig. 2. Numerical Results vs. Analytical Results

5   Sufficient Conditions

In [4,5] sufficient conditions of optimality are developed for the control problem of hydro-electric power stations. They are formulated for a general case of N hydro-electric power stations in cascade. Here we specify such conditions for our problem with 2 reservoirs and the price given by (4). We verify that these conditions are satisfied for the reference pair $(\hat{u}(\cdot), \hat{V}(\cdot))$. Such sufficient conditions are formulated in terms of bounded variation functions. These functions are intrinsically connected with the multipliers of the previous section. The absolutely continuous function $p(\cdot)$ absorbs the measure terms giving rise to $q(\cdot)$ (compare (iv) of the previous section with (ii) of Theorem 1). The functions η_1 and η_2 of the theorem are associated to μ_1, μ_2 and μ_3, μ_4 respectively of previous section. The fact that μ_1, μ_2 have supports on sets of empty intersection allows the reduction of the number of functions. Conditions (i), (ii) and (iii) of Theorem 1 are necessary conditions for the problem, satisfied by more regular multipliers η_1 and η_2.

Theorem 1. *Let $(\hat{u}_k(\cdot), \hat{V}_k(\cdot))$, $k = \overline{1,2}$, be an admissible control process. Assume that the following conditions are satisfied:*

(i) there exist right continuous functions $q_1(\cdot), q_2(\cdot) \in BV([0,T],R)$ and piecewise absolutely continuous functions $\eta_k(\cdot)$, $k = \overline{1,2}$, satisfying

$$dq_1(t) = -\frac{A}{S_1}c(t)\,dt - H_1\,dc(t) - \frac{\hat{V}_1(t)}{S_1}dc(t) + d\eta_1,$$

$$dq_2(t) = -H_2\,dc(t) - \frac{\hat{V}_2(t)}{S_2}dc(t) + d\eta_2,$$

$$q_1(0) = q_1(T), \qquad q_2(0) = q_2(T),$$

$$\eta_1(t) = v_1(t) + \Delta\eta_1(a)H(t-a) + \Delta\eta_1(T)H(t-T),$$

$$\eta_2(t) = v_2(t) + \Delta\eta_2(a)H(t-a) - \Delta\eta_2(T)H(t-T),$$

where $v_1(\cdot), v_2(\cdot) \in AC([0,T],R)$, $\Delta\eta_1(a)$, $\Delta\eta_2(a)$ $\Delta\eta_1(T)$, $\Delta\eta_2(T)$ are constants, and $H(\cdot)$ stands for the Heaviside step function;

(ii) the following equality holds a.e. $t \in [0,T]$;

$$\max_{u_k \in [u_k^m, u_k^M],\, k=\overline{1,2}} u_1(-q_1(t)+q_2(t)) + u_2(-q_2(t)) = \hat{u}_1(-q_1(t)+q_2(t)) + \hat{u}_2(-q_2(t)),$$

(iii) the functions $v_1(\cdot)$, $v_2(\cdot)$, satisfy the inequalities

$$dv_k(t) \le 0, \quad if \quad \hat{V}_k(t) = V_k^m \qquad dv_k(t) \ge 0, \quad if \quad \hat{V}_k(t) = V_k^M$$

$$dv_k(t) = 0, \quad if \quad \hat{V}_k(t) \in]V_k^m, V_k^M[\qquad\qquad for\ k = 1,2;$$

(iv) if $c_1 < c_2$, then for all $k = \overline{1,2}$, the inequalities hold

$$\Delta\eta_k(a) < 0, \quad if \quad \hat{V}_k(a) = V_k^m \quad and \quad \Delta\eta_k(a) > 0, \quad if \quad \hat{V}_k(a) = V_k^M;$$

(v) if $\hat{V}_k(t) \in]V_k^m, V_k^M[$ for some k, then $dc(t) \le 0$, for $k = 1,2$.

Then $J(\hat{u}(\cdot)+\bar{u}(\cdot), \hat{V}(\cdot)+\bar{V}(\cdot)) \ge J(\hat{u}(\cdot), \hat{V}(\cdot))$ wherever $(\hat{u}_k(\cdot)+\bar{u}_k(\cdot), \hat{V}_k(\cdot)+\bar{V}_k(\cdot))$, $k = \overline{1,2}$, is an admissible process and $\max_{k=\overline{1,2}} \max\{\bar{V}_k(a), \bar{V}_k(T)\}$ is sufficiently small.

Taking into account the position of $\hat{V}_1(t)$ and $\hat{V}_2(t)$ relative to the boundary of the admissible volume sets, we can write

$$\int_{[0,T]} d\eta_1(t) = \Delta\eta_1(a) \qquad \text{and} \qquad \int_{[0,T]} d\eta_2(t) = \int_{[\theta_1,a]} d\eta_2(t)$$

From (8) and (9), $\theta_1 = 4.4$, $V_1(0) = 143.56$ and $V_2(0) = 49.85$. Now, from (i) we have

$$q_1(t) = q_1(0) - \frac{A}{S_1}\int_{]0,t]} c(t)dt + \int_{]0,t]} d\eta_1(t) - \int_{]0,t]}\left(H_1 + \frac{\hat{V}_1(t)}{S_1}\right)dc(t),$$

$$q_2(t) = q_2(0) - \int_{]0,t]}\left(H_2 + \frac{\hat{V}_2(t)}{S_2}\right)dc(t) + \int_{]0,t]} d\eta_2(t)$$

and from the periodicity condition $q_1(0) = q_1(T)$ we deduce

$$\int_{]0,T]} d\eta_1(t) = \frac{Aa(c_1+c_2)}{S_1} + \frac{(c_2-c_1)}{S_1}\left(V_1^M - \hat{V}_1(T)\right),$$

$$\int_{]0,T]} d\eta_2(t) = \frac{(c_2-c_1)}{S_2}\left(V_2^m - \hat{V}_2(T)\right).$$

Analysis of the condition (ii) and (iii) leads to:

$$q_1(0) \geq \frac{Ac_1\theta_1}{S_1} + q_2(0) \geq 0 \qquad a.e.\ t \in [0,\theta_1[\qquad \text{and} \qquad q_2(0) \geq 0 \qquad (16)$$

$$\int_{[\theta_1,t]} d\eta_2(\tau) = q_1(0) - q_2(0) - \frac{Ac_1t}{S_1} \leq 0 \qquad a.e.\ t \in [\theta_1,a[\qquad (17)$$

$$\int_{]0,t]} d\eta_1(\tau) = 0 \quad q_2(0) \geq -\int_{[\theta_1,t]} d\eta_2(\tau) \qquad a.e.\ t \in [\theta_1,a[$$

From (16) and (17) we conclude that $\quad q_1(0) = q_2(0) + \dfrac{Ac_1\theta_1}{S_1}$. From ($iii$) and ($iv$) we have $\int_{]0,t]} d\eta_1(\tau) = \Delta\eta_1(a)$ and $\int_{[\theta_1,t]} d\eta_2(\tau) \leq 0$, $\forall t \in [a,T[$.

Due to the infinity of solutions for $V_2(t)$ on $[a,T]$, we can write, for $t \in [a,T[$

$$-q_2(t) = 0 \Rightarrow q_2(0) = (c_2 - c_1)\left(H_2 + \frac{V_2^m - u_1^m\theta_1}{S_2}\right)$$

$$-q_1(0) + q_2(0) + (c_2 - c_1)\left(H_1 + \frac{\hat{V}_1(T)}{S_1} - H_2 - \frac{\hat{V}_2(T)}{S_2}\right) + \frac{Ac_2(t-T)}{S_1} \geq 0.$$

Working the above information and the data of the problem we can write: $\lambda = 1$,

$$q_1(t) = \begin{cases} 38.18 + 0.0039t, & t \in [0,a[\\ -0.04t - 46.52, & t \in [a,T[\\ 38.18, & t = T \end{cases} \qquad q_2(t) = \begin{cases} 38.16, & t \in [0,\theta_1[\\ 38.18 - 0.004t, & t \in [\theta_1,a[\\ 0, & t \in [a,T[\\ 38.16, & t = T \end{cases}$$

$$\int_{]0,t]} d\eta_1(\tau) = \begin{cases} 0, \ t < a \\ 1.27, \ t \geq a \end{cases} \qquad \int_{]0,t]} d\eta_2(\tau) = \begin{cases} 0, \ t < \theta_1 \\ -0.004t + 0.02, \ t \in [\theta_1, a[\\ -0.63, \ t \geq a \end{cases}$$

These multipliers were calculated taking into consideration (i) and (ii) essentially. Let us check that conditions (iii), (iv) and (v) are fully accomplished.

In $[0, a[\cup]a, T]$, $\hat{V}_1(t) \in]V_1^m, V_1^M[$ and we have $dv_1(t) = 0$. For $t = a$, $\hat{V}_1(a) = V_1^M$ and $dv_1(t) \geq 0$. Also $\hat{V}_2(t) \in]V_2^m, V_2^M[$ on $[0, \theta_1[\cup]a, T]$, and on that set $dv_2(t) = 0$. When $t \in [\theta_1, a]$, $\hat{V}_2(t) = V_2^m$ and $dv_2(t) \leq 0$. So we can claim that condition (iii) is verified. Furthermore, $\hat{V}_1(a) = V_1^M$ and $\Delta\eta_1(a) = 1.27 > 0$. Also $\hat{V}_2(a) = V_2^m$ and $\Delta\eta_2(a) = -0.63 - (0.02 - 0.004 * 12) < 0$, so condition (iv) is also verified. Condition (v) is obviously satisfied. Observe that $\hat{V}_1(t) \in]V_1^m, V_1^M[$ in $[0, a[\cup]a, T]$ and $\hat{V}_2(t) \in]V_2^m, V_2^M[$ in $[0, \theta_1[\cup]a, T]$ and in these intervals $dc(t) = 0$.

We conclude that $(\hat{u}, \hat{V})$ is in fact a local minimizer for the problem.

6   Conclusion

In this work we analyse an optimal control problem for a model with two hydro-electric power stations in cascade where one of the stations has reversible turbine. Our objective is to optimize the profit of power production. For that we start with a numerical simulation result, obtained with some available software. To validate the numerical solution we undertook an analysis of the problem. Using the profile of the numerical solution we prove that necessary and sufficient condition of optimality are satisfied. We also observe irregular behaviour of the numerical solution on an subinterval and that is explained by the existence of more that one solution on such subinterval.

Acknowledgment. The author is grateful to Prof. Margarida Ferreira and Prof. Gueorgui Smirnov for the constructive suggestions which improved the paper. This work was supported by the EU 7th Framework Prog.[FP7-PEOPLE-2010-ITN] grant agr.64735-SADCO and project FCOMP-01-0124-FEDER-028894, Ref. FCT PTDC/EEI-AUT/1450/2012.

References

1. Ribeiro, A.F., Guedes, M.C.M., Smirnov, G.V., Vilela, G.V.: On the optimal control of a cascade of hydro-electric power stations. Electric Power Systems Research 88, 121–129 (2012)
2. Labadie, J.W.: Optimal Operation of multireservoir systems: state-of-the-art review. J. Water Resour. Plng. and Mgmt. 130(2), 93–111 (2004)
3. Smirnov, G., Bushenkov, V.: Curso de Optimização: Programação Matemática,Cálculo das Variações, Controlo Óptimo. Escolar Editora (2005) (in Portuguese)
4. Ferreira, M.M.A., Ribeiro, A.F., Smirnov, G.V.: Local Minima of Quadratic Functionals and Control of Hydro-Electric Power Stations (submitted)
5. Ferreira, M.M.A., Ribeiro, A.F., Smirnov, G.V.: Sufficient Conditions of Optimality for a Cascade of Hydro-Electric Power Stations. In: AIP Conference Proceedings, vol. 1558, pp. 1–2 (2013)
6. Bushenkov, V.A., Ferreira, M.M.A., Ribeiro, A.F., Smirnov, G.V.: Numerical approach to a problem of hydroelectric resources management. In: AIP Conference Proceedings, vol. 1558, pp. 630–633 (2013)
7. Vinter, R.B.: Optimal Control. Birkhauser, Boston (2000)
8. Lee, E.B., Markus, L.: Foundations of Optimal Control Theory. Wiley, New York (1968)

Optimal Control for an Irrigation Planning Problem: Characterisation of Solution and Validation of the Numerical Results

Sofia O. Lopes[1,3], Fernando A.C.C. Fontes[2], Rui M.S. Pereira[1,4],
Maria do Rosário de Pinho[2], and C. Ribeiro[1]

[1] CMAT and Departamento de Matemática e Aplicações,
Universidade do Minho, Guimarães, Portugal
[2] ISR-Porto, Faculdade de Engenharia,
Universidade do Porto, Porto, Portugal
[3] Collaborator with ISR-Porto, Universidade do Porto
[4] Collaborator with Centre of Physics,
Universidade do Minho

Abstract. In a previous study, the authors developed the planning of the water used in the irrigation systems of a given farmland in order to ensure that the field cultivation is in a good state of preservation. This planning was modelled and tackled as an optimal control problem: minimize the water flow (control) so that the extent water amount in the soil (trajectory) fulfils the cultivation water requirements. In this paper, we characterize the solution of our problem guaranteeing the existence of the solution and applying the necessary and sufficient conditions of optimality. We validate the numerical results obtained previously, comparing the analytical and numerical solutions.

1   Introduction

The climate system and the ecosystems are under accelerated change while human cultures, economic activities and national interactions are undergoing dramatic and, sometimes, exponential changes. The rapid increase of the world population leads to a soaring demand of water. Agriculture exerts pressure on the environment, especially on water. Thus, appropriate water management throughout the irrigation processes is needed [6].

A model to optimize the water use in the irrigation of a farm field via optimal control (water flow) that takes into account the evapotranspiration, rainfall, losses by infiltration and runoff was developed in [10]. There the problem is introduced, and a solution obtained for the "Yearly Planning" problem considering different weather scenarios with the help of the so called "precipitation factor" that is multiplied by the rainfall monthly average. In [11], the authors present the "Initial Planning Problem" for rainfall: this includes an extra term taking into account the rainfall in the previous time period (this rainfall model

A.P. Moreira et al. (eds.), *CONTROLO'2014 - Proc. of the 11th Port. Conf. on Autom. Control*,
Lecture Notes in Electrical Engineering 321, DOI: 10.1007/978-3-319-10380-8_16

was statistically proven to be significant) where a comparison between this new model and the solution knowing the rainfall *a priori* was shown.

In this paper, we show that the solution of Initial Planning Problem exists and we characterize it using necessary and sufficient conditions of optimality. We also compare the results obtained analytically and numerically.

This paper is organized as follows. In section 2 we present a model for the planning of irrigation based on the hydrologic balance equation. In section 3 we show the analytical study of the solution. In section 4, we present a numerically implementation of the problem and we show that the numerical results agree with the results obtained analytically.

2   A Model for Planning the Irrigation

We consider a simple model for the planning of the water used in the irrigation of farm fields, based on the hydrlogic balance. This means that, the variation of water in the soil is given by

$$\dot{x} = u + \mathbf{rfall} - \mathbf{evtp} - \mathbf{rnoff} \tag{1}$$

where x is the water in the soil, u is the amount of water flow introduced in the soil via its irrigation system, the **rfall** is rainfall, the **evtp** is the evaporation of the soil and the transpiration of the crop and **rnoff** are the losses of water due to the runoff and deep infiltration.

Having in mind the postulate of Horton's equation, infiltration decreases exponentially with time [4]. The dynamical equation can be written as

$$\dot{x}(t) = l(t) - w(t)x(t) \tag{2}$$

where $l(t) = u(t) + \mathbf{rfall}(t) - \mathbf{evtp}(t)$.

From (1) and (2), one may say $\mathbf{rnoff} = \beta x$, where β is a parameter that depends on the type of soil.

From now on, we consider the following dynamic equation

$$\dot{x} = u + g(t) - \beta x, \tag{3}$$

where $g(t) = \mathbf{rfall}(t) - \mathbf{evtp}(t)$ and β is the percentage of losses of water due to the runoff and deep infiltration.

The dynamic equation (3) is solved over the time interval $[0, T]$ subject to the initial condition $x(0) = x_0$. The variable of state x is subject to the inequality constraint $x \geq x_{\min}$, where $x_{\min}$ is the hydrological need of the crop (according to [9]). The control variable u belongs to the interval $[0, M]$, where M is the maximum flow of water that comes from tap.

The optimal control problem (OCP) formulation is then:

$$\min \quad \int_0^T u(t)dt$$

subject to:
$$\dot{x}(t) = u + g(t) - \beta x \text{ a.e. } t \in [0, T]$$
$$x_{\min} - x \leq 0 \qquad \forall t \in [0, T] \qquad (4)$$
$$u(t) \in [0, M] \qquad \text{a.e.}$$
$$x(0) = x_0.$$

A detailed description of this models is given in [11].

3   Analysis of the Solution

In order to characterize the solution of our problem, we prove the existence of solution and we apply the necessary and sufficient conditions of optimality.

Throught, the function H represents the pseudo-Hamilton function:

$$H(t, x, p, u, \lambda) = p(u + g(t) - \beta x) - \lambda u \qquad (5)$$

where p and λ are Lagrangian multipliers.

3.1   Existence of Solution

Existence of solution was introduced by Tonelli (1915) when he proposed the first theorem of existence of solution for calculus of variations problems. Even today, Tonelli's theorem remains the central existence theorem for dynamic problems, although the hypotheses of the theorem can be relaxed, see, for example, [13]. In this section, we apply the theorem 5.4.4. in [2] to guarantee the existence of solution for our OCP. Let us verify the conditions of this theorem, see theorem (in Appendix)):

* The dynamic function and cost function are differentiable in the state and control variable and the control belongs to the interval $[0, M]$.
* The condition i) of theorem 5.4.4. is satisfied, since the set

$$\{[u + g(t) - \beta x, u + \delta] : u \in [0, M] \text{ and } \delta \geq 0\}$$

  is convex. Indeed, this set is the epigraph of a convex function.
* For the condition ii): take $\sigma(t) = \beta$, $\rho(t) = 0$ and $\phi(t, p) = |p-1|M + |p||g(t)|$. Then we have

$$p(u + g(t) - \beta x) - u \leq |p - 1|M + |p||g(t)| + \beta|p||x|.$$

* The set $C_0 = \{x_0\}$ is compact, so the condition iii) is satisfied.

The assumptions of the theorem 5.4.4. in [2] are verified and we conclude that there exists an admissible solution to our OCP.

3.2 Necessary and Sufficient Conditions of Optimality

In this section, we start by verifying that our problem satisfies the constraint qualification that allows to write the Maximum Principle (MP) in the normal form: the multiplier associated to the objective function λ is not zero (see [3] and [8] for discussion of normal forms of the MP for optimal control problems with state constraints). We finish this section applying the MP in the normal form and we verify that MP conditions are also sufficient conditions.

Step 1: Verify the Normality

In Rampazzo and Vinter [8], the MP can be written with $\lambda = 1$, if there exists a continuous feedback $u = \eta(t, \xi)$ such that

$$\frac{dh(\xi(t))}{dt} = h_t(t, \xi) + h_x(t, \xi) \cdot f(t, \xi, \eta(t, \xi)) < -\gamma' \tag{6}$$

for some positive γ', whenever (t, ξ) is close to the graph of the optimal trajectory, $\bar{x}(\cdot)$, and ξ is near to the state constraint boundary. There should exist a control (flow of water provided by the irrigation systems) pulling the state variable away from the state constraint boundary (this guarantees that the crop survives).

In our problem $h(x) = x_{min} - x$ and, from (6), we write

$$\frac{dh(\xi(t))}{dt} = h_x(\xi(t)) \cdot f(t, \xi, , \eta(t, \xi)) = -(\eta(t, \xi) + \triangle(t, \xi)) \leq -\gamma', \tag{7}$$

where $\triangle(t, \xi) = g(t) - \beta\xi$. For a ξ on a neighbourhood of $\bar{x}$, we can always choose η sufficiently large so that the equation (7) is satisfied, as long as $M \geq \beta\bar{x}(t) - g(t)$, a condition we can impose with loss of generality. Thus the inward pointing condition (7) is satisfied and normality follows.

Step 2: Application of the Maximum Principle

A known form of the normal MP for smooth problems with state constraint is:

Let $(\bar{x}, \bar{u})$ be a minimizer for (OCP), then there exists an absolutely continuous function p and $\mu \in C^*(0, 1)$ such that,

$$-\dot{p}(t) = H_x(t, \bar{x}(t), q(t), \bar{u}(t), 1)$$
$$H(t, \bar{x}(t), q(t), \bar{u}(t), 1) = \max_{v \in [0,M]} H(t, \bar{x}(t), q(t), v, 1) \text{ a.e.;}$$
$$supp\{\mu\} \subset \{t \in [0, T] : h(\bar{x}(t)) = 0\} \tag{8}$$
$$q(T) = 0,$$

where $q(t)$ is defined as follows,

$$q(t) = \begin{cases} p(t) - \int_{[0,t)} \mu(ds), & t \in [0, T) \\ p(T) - \int_{[0,T]} \mu(ds), & t = T. \end{cases}$$

Applying theses conditions to our problem, we have:

$$\dot{p}(t) = \beta q(t)$$
$$q(t)(\bar{u}(t) - u(t)) - (\bar{u}(t) - u(t)) \geq 0$$
$$supp\{\mu\} \subset \{t \in [0, T] : \bar{x}(t) = x_{\min}\}$$
$$q(T) = 0.$$

Step 3: Application of the First Order Sufficient Conditions

We can apply sufficient conditions to our problem. Sufficiente conditions given by corollary of theorem 5.4.2 from [2] are validated under assumptions verified by the data of our problem, i.e., (we refer the reader to see corollary 1 in the appendix):

* The dynamic function and cost function are differentiable in the state and control variable and the control belongs to the interval $[0, M]$.
* The conditions (8) of the MP are satisfied.
* For each t the function $(x, u) \rightarrow p(t)(u + g(t) - \beta x) - u - \psi_{[0,M]}(u)$ is concave.

Thus, the necessary conditions of optimality of our problem are also sufficient conditions.

3.3   Characterization of Solution

Now, we characterize the optimal solution for (OCP) studying the Weierstrass condition of the MP for $\bar{u} = 0$, $\bar{u} = M$ and $\bar{u} \in]0, M[$.

If $\bar{u} = 0$, we have that for all $u(t) \in [0, M]$,

$$q(t)u(t) - u(t) \leq 0 \Leftrightarrow q(t) \leq 1.$$

If $\bar{u} = M$, we have that for all $u(t) \in [0, M]$,

$$(q(t) - 1)(M - u) \geq 0 \Leftrightarrow q(t) \geq 1.$$

In the remaining case (i.e. $\bar{u} \in]0, M[$), we have:

$$q(t)(\bar{u}(t) - u(t)) - (\bar{u}(t) - u(t)) \geq 0 \Leftrightarrow$$
$$(q(t) - 1)(\bar{u}(t) - u(t)) \geq 0 \Leftrightarrow q(t) = 1.$$

Next we shall use all this information to validate the numerical solution (already presented in [12]) of our problem.

4 Numerical Implementation and Results

The numerical simulations of our problem requires its discretization. Its discrete formulation is as follows:

$$\min \delta \sum_{i=1}^{N} u_i$$

such that:
$$x_{i+1} = x_i + \delta f(t_i, x_i, u_i), \ a.e. \ i = 1, \ldots, \ N,$$

$$x_1 = x_0$$
$$x_i \geq x_{\min}, \qquad\qquad i = 1, \ldots, \ N,$$
$$u_i \in [0, M], \qquad\qquad a.e. \ i = 1, \ldots, \ N,$$

where $x = (x_1, \ldots, x_N)$ is the trajectory, $u = (u_1, \ldots, u_{N-1})$ is the control, f is the hydrological balance function, $x_{\min}$ is the hydrological need of the crop, x_0 is an initial state, δ is the time step discretization, and $N = T/\delta$. In the dynamic equation, f is defined by

$$f(t_i, x_i, u_i) = u_i + \mathrm{rainfall}(t_i) - \mathrm{evapotranspiration}(t_i) - \mathrm{losses}(x_i), \qquad (9)$$

where the evapotranspiration takes in account the evaporation of the soil and the transpiration of the crop and the losses are the *losses* of water due to the runoff and deep infiltration. The Rainfall model [11] is based on a linear combination of average monthly rainfall from the last 10 years and the amount of rainfall in the previous month. The evapotranspiration model is described as the crop coefficient (in our case potatoes) multiplied by the reference value of evapotranspiration in Lisbon, given by Pennman-Monteith methodology, (see [5]). The losses are model as 15% of the water in the soil, based on Horton's equations. A more detailed description of these models can be seen in [11].

The state constraint ($x_i \geq x_{\min}$) translates the fact that the plants needs a minimum amount flow of water to survive.

We consider a field of potatoes in the region of Lisbon, Portugal. Based on data from [7] we consider:

$$x_{\min} = 0.56/12 \ \mathrm{m}^3/\mathrm{month} \qquad T = 12$$
$$x_0 = 4x_{\min} \ \mathrm{m}^3/\mathrm{month} \qquad \beta = 15\%$$
$$M = 1 \ \mathrm{m}^3/\mathrm{month}.$$

To obtain the numerical solution for the optimal control problem we have approximate the problem by a sequence of finite dimensional nonlinear programming problems, (see [1]). To implement this optimization problem, we use *fmincon* function of MatLab with the algorithm "active set", by default.

The code produces results that are according to what is expected for this region [9].

Next, we plot the numerical solution and the expected multipliers.

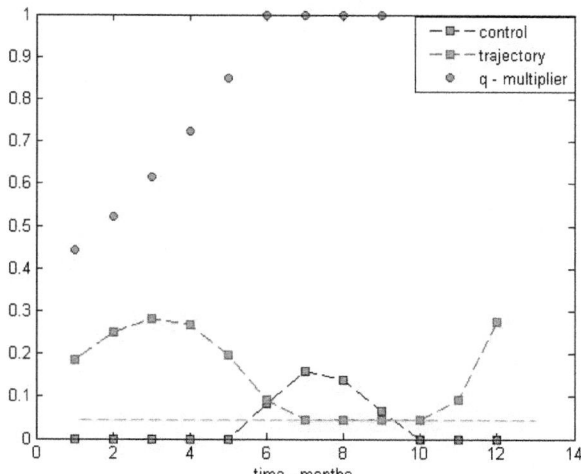

Note that the green line represents the hydrological need of the crop. We can observe that:

$$q(t) \leq 1 \text{ if } \bar{u} = 0, \qquad q(t) = 1 \text{ if } \bar{u} \in]0, M[$$

and since $\bar{u}$ is never equal to 1, q is never great than 1, as expected from section 3.3. From here, we can say that although the analytical explicit solution was not obtained, the numerical solution fulfils the necessary optimality conditions.

Our numerical findings suggest that the trajectory has a " boundary interval" $[t_{in}, t_{out}]$, with $t_{in} > 0$ and $t_{out} < 12$ (i.e. $\bar{x}(t) = x_{min}$ for all $t \in [t_{in}, t_{out}]$ and $\bar{x}(t) \neq x_{min}$ for $t \notin [t_{in}, t_{out}]$) and that q is absolutely continuous function excepted at t_{out} where it exhibits a jump. Taking these information into account we now get a analytical characterization of the solution and q multiplier.

Step 1: $h(\bar{x}(t)) < 0$ for $t \in]t_{out}, 12]$.

Since the inequality constraint is not active, then $p(t) = q(t)$. Thus we most have $p(12) = 0$ and, since $\dot{p}(t) = \beta p(t)$, by the adjoint equation of the MP, we can conclude that $p(t) = q(t) = 0$.

Applying the Weierstrass condition of MP, we get $\bar{u} \leq u, \forall u \in [0, M]$. Thus $\bar{u} = 0$.

Replacing $\bar{u}$ by zero in the dynamics, we have:

$$\dot{\bar{x}}(t) = g(t) - \beta \bar{x}(t).$$

As $\bar{x}(t_{out}) = x_{min}$, then $\bar{x}(t) = e^{-\beta(t - t_{out})} \left(\int_{t_{out}}^{t} e^{\beta(s - t_{out})} g(s) ds + x_{min} \right)$,

for $t \in]t_{out}, 12]$.

Therefore $(\bar{x}(t), \bar{u}(t)) = \left(e^{-\beta(t - t_{out})} \left(\int_{t_{out}}^{t} e^{\beta(s - t_{out})} g(s) ds + x_{min} \right), 0 \right)$ and $q(t) = 0$, for $t \in]t_{out}, 12]$.

Step 2: $h(\bar{x}(t)) = 0$ for $t \in [t_{in}, t_{out}]$.

As for $t \in [t_{in}, t_{out}]$: $h(\bar{x}(t)) = 0$, we have $\bar{x}(t) = x_{min}$. Therefore:

$$\dot{\bar{x}}(t) = 0 \Leftrightarrow \bar{u}(t) = -g(t) + \beta x_{min}.$$

And we may conclude that $(\bar{x}(t), \bar{u}(t)) = (x_{min}, -g(t) + \beta x_{min})$ for $t \in]t_{in}, t_{out}[$.

Since $\bar{u}(t) > 0$ for $t \in]t_{in}, t_{out}[$, then by the MP we also conclude that $q(t) = 1$, $t \in]t_{in}, t_{out}[$.

Step 3: $h(\bar{x}(t)) < 0$ for $t \in [0, t_{in}[$.

Since $h(\bar{x}(t)) < 0$, then $p(t) = q(t)$. Again, as $p(t_{in}) = 1$ and $\dot{p}(t) = \beta p(t)$, by the adjoint equation of the MP, we can conclude that $p(t) = q(t) = e^{\beta(t - t_{in})}$. On the other hand $p(t) = q(t) < 1$, then by Weierstrass condition we get $\bar{u} \leq u, \forall u \in [0, M]$. Therefore $\bar{u} = 0$.

Consequently, our dynamics is written as:

$$\dot{\bar{x}}(t) = g(t) - \beta \bar{x}(t).$$

Since $\bar{x}(0) = x_0$, then $\bar{x}(t) = e^{-\beta t} \left(\int_0^t g(s)ds + x_0 \right)$, for $t \in [0, t_{in}]$.

Therefore $(\bar{x}(t), \bar{u}(t)) = (e^{-\beta t} \int_0^t g(s)ds + x_0, 0)$ and $q(t) = e^{\beta(t - t_{in})}$, for $t \in [0, t_{in}]$.

Briefly,

$$\bar{x}(t) = \begin{cases} e^{-\beta t} \left(\int_0^t g(s)ds \right) + x_0 & t \in [0, t_{in}] \\ x_{min} & t \in]t_{in}, t_{out}[\\ e^{-\beta t} \left(\int_{t_{out}}^t g(s)ds + e^{\beta t_{out}} x_{min} \right) & t \in [t_{out}, 12], \end{cases}$$

$$\bar{u}(t) = \begin{cases} 0 & t \in [0, t_{in}] \\ -g(t) + \beta x_{min} & t \in]t_{in}, t_{out}[\\ 0 & t \in [t_{out}, 12] \end{cases}$$

$$q(t) = \begin{cases} e^{\beta(t - t_{in})} & t \in [0, t_{in}] \\ 1 & t \in]t_{in}, t_{out}[\\ 0 & t \in [t_{out}, 12] \end{cases}$$

Next, we plot the numerical and analytical solution obtained from our model for the year 2010.

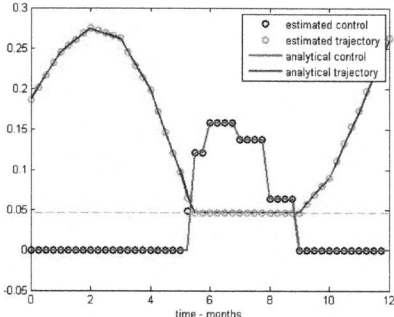 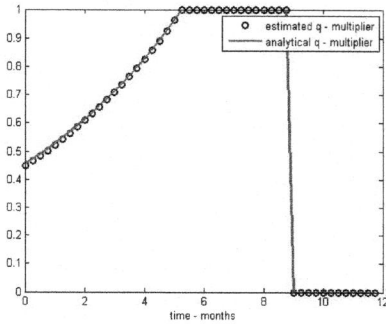

From this figure we confirm that the numerical solution agrees with the results shown in section 3. We notice the analytical and estimated results of trajectory, control and multipliers coincide.

5 Conclusion

We characterized the optimal solution to the irrigation problem described as the minimization of the water flow (control) so that the amount flow of water in the soil (trajectory) fulfils the cultivation water requirements of the crop. We proved the existence of solution and we verify that the Maximum Principle conditions are also sufficient conditions. Furthermore, we characterized the solution applying the necessary conditions of optimality in the form of the Maximum Principle and we conclude that the multiplier $q(t) \leq 1$ when $\bar{u} = 0$ and when $\bar{u} \in]0, 1[$ then $q(t) = 1$. We used the information obtained in numerical solution with respect to the time interval where the state constraint is active to get the optimal solution analytically. Finally, we compare the results obtained analytically and numerically are compared showing conclude that the numerical solution fulfils the MP conditions.

Appendix

Auxiliary Results

Here, we present an adaptation of corollary of theorem 5.5.4 and theorem 5.4.2 from [2] to our problem, that can be written as:

$$(OP) \quad \min \quad \int_0^T L(x(t), u(t))dt$$

subject to:
$$\dot{x}(t) = f(t, x(t), u(t)) \qquad \text{a.e.}\, t \in [0, T]$$
$$h(x(t)) \leq 0 \qquad\qquad \forall t \in [0, T]$$
$$u(t) \in \Omega \qquad\qquad \text{a.e.}$$
$$x(0) \in C_0$$

(10)

Theorem 1. *(Existence) (theorem 5.5.4 in [2]) Consider the (OP) problem and assume that f and L are differentiable in the state and control variables, Ω is convex bounded set, h is lower continuous function in x and C_0 is closed and convex and suppose in addition that:*

i) For each x with $h(x) \leq 0$, the following set is convex:

$$\{[f(t, x, u), L(x, u) + \delta] : u \in [0, M] \text{ and } \delta \geq 0\}.$$

ii) There exist functions $\sigma(t)$, $\rho(t)$ and $\phi(t, p)$ finite and summable in t (with σ and ρ nonnegative) such that for all x satisfying $h(x) \leq 0$ for all w in Ω and for all p one has

$$H(t, x, p, w, 1) \leq \phi(t, p) + |x|(\rho(t) + \sigma(t))|p|.$$

iii) The set C_0 is compact;

Then, if there is at least one admissible (x, u) for (OP) giving a finite value to the cost functional, there is a solution to (OP).

Corollary 1. *(Sufficient condition)(corollary of theorem 5.5.4 in [2]) Consider the (OP) problem and assume that f and L are differentiable in the state and control variables, Ω is convex bounded set, h is lower continuous function in x and C_0 is closed and convex. Let (x, u) be admissible for (OP), and suppose that there is an arc p satisfying:*

i) For almost all t, the function $w \to H(t, x(t), p(t), w, 1)$ attains a maximum over Ω at $w = u(t)$.
ii) $-\dot{p}(t) = H_x(t, x(t), p(t), u(t), 1)$ a.e..
iii) $q(T) = 0$.
iv) For each t, the function $(y, w) \to H(t, y, p(t), w, 1) - \psi_\Omega(w)$ is concave. ($\psi_\Omega(w)$ is the indicator function.)

Then (x, u) solves (OP).

Acknowledgments. The financial support of PEst-OE/MAT/UI0013/2014 and PEST-C/FIS/UI607/2013, European Union FP7 (FP7-PEOPLE-2010-ITN, Grant Agreement no. 264735-SADCO), FCT projects PTDC/EEA-CRO/ 116014/2009 and PTDC/EEI-AUT/1450/2012 are gratefully acknowledged.

References

1. Betts, J.B.: Pratical methods for optimal control using nonlinear programming. SIAM, Philadelphia (1943)
2. Clarke, F.H.: Optimization and Nonsmooth Analysis. Wiley-Interscience, New York (1983)

3. Lopes, S., Fontes, F.A.C.C.: Normal forms of necessary conditions for dynamic optimization problems with pathwise inequality constraints. Journal of Mathematical Analysis and Applications 399, 27–37 (2013)
4. Horton, R.E.: An approach toward a physical interpretation of infiltration capacity. Soil Sci. Soc. Am. Proc. 5, 300–417 (1940)
5. Elliott, R., Itenfisu, D., Brown, P., Jensen, M.E., Mecham, B., Howell, T.A., Snyder, R.L., Eching, S., Spofford, T., Hattendorf, M., Martin, D., Cuenca, R.H., Walter, I.A., Allen, R.G., Wright, J.L.: The ASCE standardized reference evapotranspiration equation. Rep. Task Com. on Standardized Reference Evapotranspiration (2002)
6. Gaspar, J., Machado, A.A., Haie, R.M.: Pereira. Analysis of effective effciency in decision making for irrigation interventions. Water Resources 39, 700–707 (2012)
7. Pereira, L.S.: Necessidades de água e métodos de rega. Publicações Europa - América (2004)
8. Rampazzo, F., Vinter, R.B.: A theorem on the existence of neighbouring feasible trajectories with aplication to optimal control 16, 335–351 (1999)
9. Raposo, J.R.: A REGA — dos primitivos regadios às modernas técnicas de rega. Fundação Calouste Gulbenkian (1996)
10. Pereira, R., Lopes, S., Fontes, F., Machado, G.J.: Irrigation planning in the context of climate change. In: Mathematical Models for Engineering Science, MMES 2011, pp. 239–244 (2011)
11. Pereira, R., Gonçalves, M., Lopes, S., Fontes, F., Machado, G.J.: Irrigation planning: an optimal control approach. In: International Conference of Numerical Analysis and Applied Mathematics, AIP Conference Proceedings, vol. 1558, pp. 622–626 (2013)
12. Pereira, R., Gonçalves, M., Lopes, S., Fontes, F., Machado, G.J.: An optimal control approach to the irrigation planning problem. Submitted to Conferência Brasileira de Dinâmica, Controle e Aplicações (2013)
13. Vinter, R.: Optimal control. Birkhauser, Boston (2000)

Part V
Controller Design

Extended Stability Conditions
for CDM Controller Design

João Paulo Coelho[1], José Boaventura-Cunha[2],
and Paulo B. de Moura Oliveira[2]

[1] Department of Electrotechnics, School of Technology and Management, Polytechnic
Institute of Bragança, 5301-857 Bragança, Portugal
jpcoelho@ipb.pt
[2] INESC TEC - INESC Technology and Science (formerly INESC Porto) and
Universidade de Trás-os-Montes e Alto Douro, UTAD, Escola de Ciências e
Tecnologia, Quinta de Prados, 5000-801 Vila Real, Portugal

Abstract. The coefficient diagram method (CDM) is one of the easiest methods for model based control system design. Its core is based on an algebraic method but it also encompasses a graphical analysis diagram that helps the user to evaluate the three main closed-loop system requirements: dynamic behaviour, robustness and stability. This later characteristic is analysed by a set of stability conditions derived from the previous work of Lipatov and Sokolov on sufficient conditions for stability. However, in CDM, only a fraction of the total conditions are considered. This work will show that this fact increases the inconclusive area within the stability space. Moreover an extended set of CDM stability conditions, in conjunction with its graphical interpretation, will be presented.

1 Introduction

The coefficient diagram method (CDM) is a mixed algebraic/graphical analysis and synthesis control system design procedure. It was devised by Shunji Manabe during the nineties of the twentieth century [6,7]. Since then, many articles have been published in both CDM theoretical extensions [9,1] and practical applications [2,10].

The algebraic component of this method is similar to pole-placement. However, unlike pole-placement, the characteristic polynomial is easily obtained by just defining the predominant time-constant. With the characteristic polynomial, and after setting the controller order, the design process comes down to solve a Diophantine type equation.

After the algebraic design process is finished, CDM provides a graphical inference tool that can be used to analyse the closed-loop system. This analysis is done regarding three major system properties: bandwidth, robustness and stability. This last attribute is added to the diagram by computing a set of parameters derived from the work of Lipatov and Sokolov in the late sixties [5] and later adapted by Shunji Manabe [8].

In the limit, the Lipatov-Sokolov conditions come down to a set of inequalities involving the coefficients of the characteristic polynomial. Verification, or not, of this inequations provides information concerning the system stability. This information appears, not as necessary, but as sufficient conditions for system stability or instability.

Of the three main theorems and corollaries associated with the Lipatov-Sokolov stability criteria, only two were incorporated into the CDM diagram. The reason to leave one of them out can be assigned to its complexity in an algebraic point-of-view.

In this article one shows that the inconclusive region in the stability space can be reduced by adding the neglected condition. Moreover a simplified version of this condition is developed leading to an easy integration on the coefficient diagram. But first, in section 2, the original Lipatov-Sokolov conditions are presented. Later, in section 3 , the adapted CDM stability conditions are described. In section 4, and using the stability problem proposed in [5], one will show that, using only a fraction of the Lipatov-Sokolov conditions, will lead to a larger inconclusive region within the system stability region. Section 5 shows how the inconclusive region can be further reduced by adding the overlooked condition. A new form of stability representation within the CDM diagram will be presented in section 6. Finally the work results are summarized in the conclusion section.

2   The Lipatov-Sokolov Stability Conditions

When designing practical control systems it's not enough to know if the closed-loop will be stable or instable. We also need to know that, if the system is stable, how far is it from instability. In physical plants the system poles drifts from its nominal values due to many factors like aging, environmental work conditions and so on. For this reason the system stability only makes sense when defined within a range of operating intervals. Gain and phase margins, easily inferred in Bode plots, are probably the best known example of stability constraints. However Lipatov and Sokolov also point out some sufficient stability conditions by analysing, not the open loop frequency response, but the closed loop characteristic polynomial [5]. Another algebraic method that uses this information is the Routh-Hurwitz method [3]. However the excessive information provided by this later method, for example the exact number of roots in the left-hand plane, leads to excessive computation. On the other hand the Lipatov and Sokolov conditions are easier to derive (from a mathematical point-of-view). But this advantage is achieved by limiting the information to sufficient conditions for stability and instability.

The Lipatov-Sokolov stability conditions are expressed in the form of inequalities which must be satisfied by certain combinations of characteristic polynomial adjacent coefficients. As already mentioned, those conditions only define sufficient conditions for stability or only necessary conditions for instability. As will be seen ahead, those conditions are very easy to compute. However, for some systems, they can lead to inconclusive results.

We begin by considering a linear, time-invariant, system with the following n-order generic characteristic polynomial:

$$P(s) = p_n s^n + \cdots + p_i s^i + \cdots + p_0 \tag{1}$$

where $p_i > 0$, for $i = 0, \cdots, n$, are the polynomial coefficients and s is the Laplace complex variable.

In [5], Lipatov and Sokolov define a variable, λ, whose value is computed, using the characteristic polynomial coefficients, as follows:

$$\lambda_{n-i} = \frac{p_{i+1} \cdot p_{i-2}}{p_i \cdot p_{i-1}} \tag{2}$$

for $i = 2, \cdots, n-1$ and $n \geq 2$.

Taking the array of possible values of λ the following three theorems describe the Lipatov-Sokolov stability conditions. The first two define necessary conditions and the last a sufficient condition.

Theorem 1. *If there is at least one λ_{n-i}, for $i = 2, \cdots, n-1$, greater than one, then the system is unstable.*

Theorem 2. *Let $C_{n,n-i}$, for $i = 2, \cdots, n-2$, be equal to:*

$$C_{n,n-i} = \frac{\left(i + \frac{(-1)^i - 1}{2}\right) \cdot \left(n - i + \frac{(-1)^{n-i} - 1}{2}\right)}{\left(i + \frac{(-1)^i + 3}{2}\right) \cdot \left(n - i + \frac{(-1)^{n-i} + 3}{2}\right)} \tag{3}$$

If, at least, one situation exists where,

$$\lambda_{n-i-1} \cdot \lambda_{n-i} > C_{n,n-i} \tag{4}$$

then the system is unstable. Please note that this second Lipatov-Sokolov theorem is the one that is dropped in the CDM stability condition.

Theorem 3. *Let λ^* be equal to the real root of equation $\lambda(\lambda + 1)^2 - 1 = 0$ (approximately 0.466) and $\lambda^{**} = \frac{3}{\sqrt[3]{4}} - 1$ (around 0.89).*

Corollary 1. *If $n \geq 5$ and if, for all $i = 2, \cdots, n-1$, $\lambda_{n-i} < \lambda^*$ then the system is stable.*

Corollary 2. *If the first theorem cannot be satisfied and there are one, or more, λ whose value is greater than λ^* then, for the system to be stable, the inequality $\lambda_{n-i} + \lambda_{n-i+1} < \lambda^{**}$, for all $i = 2, \cdots, n-1$ and $n \geq 5$, must hold.*

3   The CDM Stability Conditions

It's wonderful how past theory can be recovered many years after and brought again to daylight. The work of Lipatov and Sokolov, in the late seventies, is a perfect example of this phenomenon. Almost twenty years after the publication of "Some Sufficient Conditions for Stability and Instability of Continuous Linear Stationary Systems", Shunji Manabe incorporated the published results into his coefficient diagram as a way to infer about closed-loop system stability.

In [8] a variable denoted by γ and defined as the stability index, is introduced and related to the characteristic polinomial coefficients by means of the following equality:

$$\gamma_i = \frac{p_i^2}{p_{i+1} \cdot p_{i-1}} \tag{5}$$

for any integer i in the range $i = 1, \cdots, n-1$ where, once again, n refers to the characteristic polynomial order. For $i = n$, $p_{n+1} = 0$ hence $\gamma_n = \infty$. The same can be said for $i = 0$.

Comparing equations (2) and (5) a close relationship between λ and γ is highlighted. As a matter of fact, the entanglement between both variables can be algebraically expressed by the subsequent equality:

$$\lambda_{n-i} = \frac{1}{\gamma_i \cdot \gamma_{i-1}} \tag{6}$$

where, in this case, the variable i can take any integer value between 2 and $n-1$.

Due to this straight connection between both the stability limit variable and λ, Shunji Manabe was able to translate the first and third Lipatov-Sokolov theorems as a function of γ. Hence the stability conditions enumerated in section 2 can be translated as:

Theorem 1. *The instability condition, which occurs if any $\lambda_{n-i} > 1$, is replaced by $\gamma_i \cdot \gamma_{i-1} < 1$ for $i = 2, \cdots, n-1$.*

Theorem 2. *Each of the two corollaries associated to theorem 3 of section 2 are rephrased as:*

Corollary 1. *For the system to be stable the product $\gamma_i \cdot \gamma_{i-1}$ must be grater than $(\lambda^*)^{-1}$. That is,*

$$\gamma_i \cdot \gamma_{i-1} > 2.1479 \tag{7}$$

for $i = 2, \cdots, n-1$.

This relation tell us that if all the γ values are larger than, approximately, 1.466 the system is stable [4].

Corollary 2. *Translated to γ, the second corollary of the third Lipatov-Sokolov theorem becomes:*

$$\gamma_i > \frac{1}{\lambda^{**}} \left(\frac{1}{\gamma_{i-1}} + \frac{1}{\gamma_{i+1}} \right) \tag{8}$$

or,

$$\gamma_i > 1.12375 \cdot \gamma_i^* \tag{9}$$

for $i = 2, \cdots, n-2$.

The variable γ_i^* is designated by stability limit [7] and is computed regarding the following expression:

$$\gamma_i^* = \begin{cases} \frac{1}{\gamma_2}, & \text{if } i = 1 \\ \frac{1}{\gamma_{i-1}} + \frac{1}{\gamma_{i+1}}, & \text{if } i = 2, \cdots, n-2 \\ \frac{1}{\gamma_{n-2}}, & \text{if } i = n-1 \end{cases} \tag{10}$$

Please note that the second theorem of Lypatov-Sokolov is not incorporated within the CDM stability conditions. The reason pointed out in [7] for this fact lies in this theorem exaggerated algebraic complexity.

The section that follows will show that the application of the current CDM stability conditions can lead, with somewhat large probability, to inconclusive results. This statement will be demonstrated by running an example taken from the original Lipatov-Sokolov paper.

4   The CDM Stability Conditions Limitations

In his celebrated paper, Lypatov and Sokolov apply their stability conditions to a 6th order characteristic polynomial in the Laplace domain. This polynomial is reproduced in the following expression.

$$P(s) = s^6 + 12s^5 + 47s^4 + 108s^3 + 122s^2 + p_1 s + p_0 \tag{11}$$

As can be seen, it has two degrees-of-freedom. The two lower order coefficients p_0 and p_1 can take any real value greater than zero. Within this range the closed-loop system can be stable or unstable. Figure 1 (a) shows the system stability distribution within the parameters space. For p_0 and p_1 between $]0, 320]$ the system has a parabolic shape area in which the system is stable. Outside this area, and for $p_0, p_1 > 320$, the system is unstable.

This section begins by testing the ability of the CDM stability conditions to provide conclusive stability results as a function of this two variables. In order to do this, both theorems described in section 3 will be employed.

The first CDM stability theorem requires the evaluation of all the γ_i for $i = 1, \cdots, 5$ followed by the product computation of adjacent γ_i. The former is presented in (12) and the latter in (13).

$$\begin{cases} \gamma_1 = \frac{p_1^2}{p_2 \cdot p_0} = \frac{p_1^2}{122 p_0} \\ \gamma_2 = \frac{p_2^2}{p_3 \cdot p_1} = \frac{122^2}{108 p_1} \\ \gamma_3 = \frac{p_3^2}{p_4 \cdot p_2} = \frac{108^2}{47 \cdot 122} = 2.034 \\ \gamma_4 = \frac{p_4^2}{p_5 \cdot p_3} = \frac{47^2}{12 \cdot 108} = 1.704 \\ \gamma_5 = \frac{p_5^2}{p_6 \cdot p_4} = \frac{12^2}{1 \cdot 47} = 3.06 \end{cases} \tag{12}$$

According to the first CDM stability theorem, for the system to be unstable it is required that one (or both) of the two first adjacent products in (13) to be smaller than the unity.

$$\begin{cases} \gamma_2 \cdot \gamma_1 = \frac{122p_1}{108p_0} \\ \gamma_3 \cdot \gamma_2 = \frac{108 \cdot 122}{47 \cdot p_1} \\ \gamma_4 \cdot \gamma_3 = \frac{47 \cdot 108}{122 \cdot 12} = 3.47 \\ \gamma_5 \cdot \gamma_4 = \frac{12 \cdot 47}{108} = 5.22 \end{cases} \tag{13}$$

That is $\frac{122p_1}{108p_0} < 1$ or $\frac{108 \cdot 122}{47 \cdot p_1} < 1$. In the space defined by the pair $\{p_1, p_0\}$, both inequalities define half-planes limited by straight lines. For the first case,

$$p_1 < 0.885 \cdot p_0 \tag{14}$$

and in the later it is just a horizontal line whose equation is given by:

$$p_1 > 280.3 \tag{15}$$

Together, these two equations define a region where it's guaranteed the system to be unstable. Figure 1 (b) represents the area defined by CDM's first stability theorem.

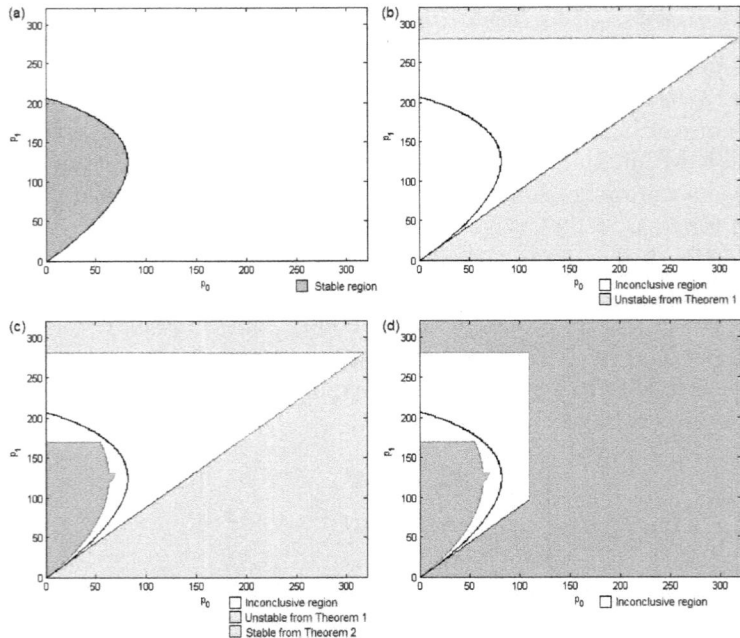

Fig. 1. (a) Effective system stability boundary (dark line) and stability regions. (b) Unstable areas defined by theorem 1. (c) Stable area as defined by theorem 2. (d) Total inconclusive region after applying the three Lipatov-Sokolov stability conditions.

By observing this figure it can be concluded that, when the above stability constraints are true, the system is unstable. On the other hand, there is a large area in which those equations become useless. This region is marked in 1 (b) as "inconclusive region". However this space can be narrowed down by adding further information. This new information is obtained by applying the CDM second stability theorem.

Taking into consideration the set of constraints defined by (8), and since the product between successive stability indexes was already presented in (13), then the set of values for p_0 and p_1 that result in a stable system are found to be governed by the two inequations below:

$$\begin{cases} p_1 > 1.9 \cdot p_0 \\ p_1 < 130.52 \end{cases} \tag{16}$$

Geometrically, this set of inequations defines a triangle shape area inside the stability region.

The second stability corollary, expressed in (9), enables further reduction of the inconclusive region. First the stability limits are computed and their values can be found in Tab. 1.

Table 1. Stability limit values for the characteristic polynomial expressed in (11)

γ_2^*	γ_3^*	γ_4^*	γ_5^*
$\dfrac{122 \cdot p_0}{p_1^2} + \dfrac{1}{2.034}$	$\dfrac{108 \cdot p_1}{122^2} + \dfrac{1}{1.704}$	$\dfrac{1}{2.034} + \dfrac{1}{3.06} \approx 0.8184$	$\dfrac{1}{1.704} = 0.5869$

Now, in order to cope with the constraints defined in (9), the set of additional conditions described by inequalities (17) is established.

$$\begin{cases} \dfrac{122^2}{108 p_1} > 1.12375 \left(\dfrac{122 p_0}{p_1^2} + \dfrac{1}{2.034} \right) \\ \dfrac{2.034}{1.12375} > \dfrac{108 p_1}{122^2} + \dfrac{1}{1.7045} \\ 1.704 > 0.91968 \\ 3.06 > 0.659523 \end{cases} \tag{17}$$

The last two constraints are true. So it is only necessary to force the remain conditions to be valid. In other words, coefficients p_0 and p_1 must verify the following inequalities:

$$\begin{cases} p_1^2 - 249.45 \cdot p_1 + 248.15 \cdot p_0 < 0 \\ p_1 < 168.6 \end{cases} \tag{18}$$

The geometric shape, resultant from the intersection area defined by (16) and (18), is presented in Fig. 1 (c). By direct observation one can perceive that a substancial area within the stable region is not covered by the above stability conditions. In concrete, the ratio between the inconclusive area inside the stable region and the total stable dimension, is around 30%. Additionally there is also a large inconclusive zone inside the unstable region. However, for this one, the original Lipatov-Sokolov conditions are able to further reduce it. Figure 1 (d) shows the total inconclusive area after applying the second Lipatov-Sokolov stability condition.

In the next section an approximation of Lipatov-Sokolov second theorem is presented. His integration on the CDM is able to further reduce the inconclusive region. Moreover, in section 6 a new stability graphical format is introduced making easier to infer the system stability from the CDM's diagram.

5   Extension of the CDM Stability Conditions

This section begins by defining a new variable. It will be designated by ζ_i, and is related to Manabe's γ by:

$$\zeta_i = \gamma_i \cdot \gamma_{i-1} \tag{19}$$

As will be shown latter, in the next section, the introduction of this variable will lead to an easier way to interpret system stability within the CDM diagram.

The original Manabe stability conditions, presented in section 3 are now expressed in terms of ζ as follows:

Theorem 1. *The system is unstable if $\zeta_i < 1$ for any $i = 2, \cdots, n-1$.*

Theorem 2. *Each of the two corollaries associated to theorem 3 of section 2 are rephrased as:*

Corollary 1. *The system is stable if $\zeta_i > 2.1479$ for all $i = 2, \cdots, n-1$.*

Corollary 2. *If all $\zeta_i \geq 1$, for $i = 2, \cdots, n-2$, with the exception of a subset whose value are lower than 2.1478 then, if all $\zeta_i > \zeta_i^*$, where*

$$\zeta_i^* = \begin{cases} 1.12375 \cdot \left(1 + \frac{\gamma_{i-1}}{\gamma_{i+1}}\right), & \text{if } i = 2, \cdots, n-2 \\ 1.12375, & \text{if } i = n-1 \end{cases} \tag{20}$$

the system is stable.

As seen in section 4 the inconclusive region can be further reduced if one is willing to apply the second Lipatov-Sokolov theorem. However, due to its complex algebraic aspect, and in order to maintain the CDM stability conditions as simple as possible, a more conservative version of the the second Lipatov-Sokolov theorem will be derived.

We begin by applying the relation (6) into (4). This leads to the inequality:

$$\frac{1}{\gamma_i \cdot \gamma_{i-1}} \cdot \frac{1}{\gamma_{i+1} \cdot \gamma_i} > C_{n,n-i} \tag{21}$$

Inverting both terms on the inequality and then multiplying by γ_{i-1} results in the following expression:

$$(\gamma_i \cdot \gamma_{i-1})^2 < \frac{\gamma_{i-1}}{\gamma_{i+1}} \cdot \frac{1}{C_{n,n-i}} \tag{22}$$

Taking into consideration the equality in (19), the above expression results into:

$$\zeta_i < \sqrt{\frac{\gamma_{i-1}}{\gamma_{i+1}} \cdot \frac{1}{C_{n,n-i}}} \tag{23}$$

In order to simplify the computation of $C_{n,n-i}$ an approximation will be used. This approximation was obtained by developing a non-linear function that is able to upper-bound approximate the maximum of $C_{n,n-i}$ for a given n. The obtained function has the structure presented in (24) and his approximation capability is illustrated in Fig. 2.

$$C_{n,n-i} = \frac{3n - 4}{3n + 20} \tag{24}$$

Taking this into consideration the CDM stability conditions will be extended by introducing the following new theorem:

Theorem 3. *If there is, at least, a ζ_i, for $i = 2, \cdots, n-1$ such that,*

$$\zeta_i < \zeta_i^{**} \tag{25}$$

for

$$\zeta_i^{**} = \sqrt{\frac{\gamma_{i-1}}{\gamma_{i+1}} \cdot \frac{3n + 20}{3n - 4}} \tag{26}$$

than the system is unstable.

Figure 2 presents the amount of inconclusive area after applying this third CDM theorem. A substancial decrease can be observed since now the entire area to the left of $p_0 \approx 120$ is now defined as unstable. However this decrease is lower than the one observed if one is willing to fully apply the second Lipatov-Sokolov stability theorem. That is the application range of this theorem was sacrificed for the sake of algebraic simplicity.

The following section illustrates how the designer can interpret this adapted stability criteria.

6   Graphical Interpretation of Stability Conditions

Figure 3 presents the adapted coefficient diagram for four different examples taken from the characteristic polynomial (11). Besides the characteristic polynomial coefficients, it is also represented the curves related to ζ_i, ζ_i^* and ζ_i^{**}. Additionally two horizontal lines are added. One that defines a upper instability bound at 1 and the other defines the lower stability bound at 2.1479.

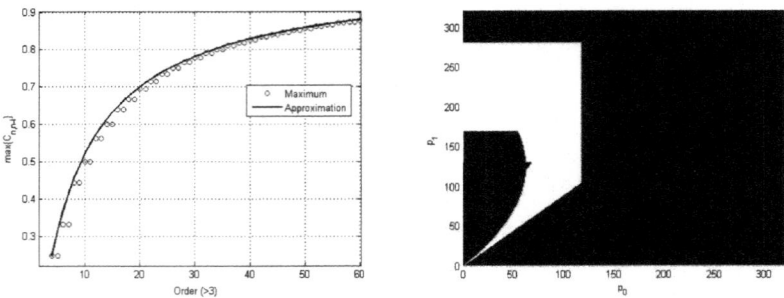

Fig. 2. In the left, the upper bound approximation accuracy of $C_{n,n-i}$ maxima by using the non-linear function (24). The right figure illustrates the total area covered by the extended stability conditions.

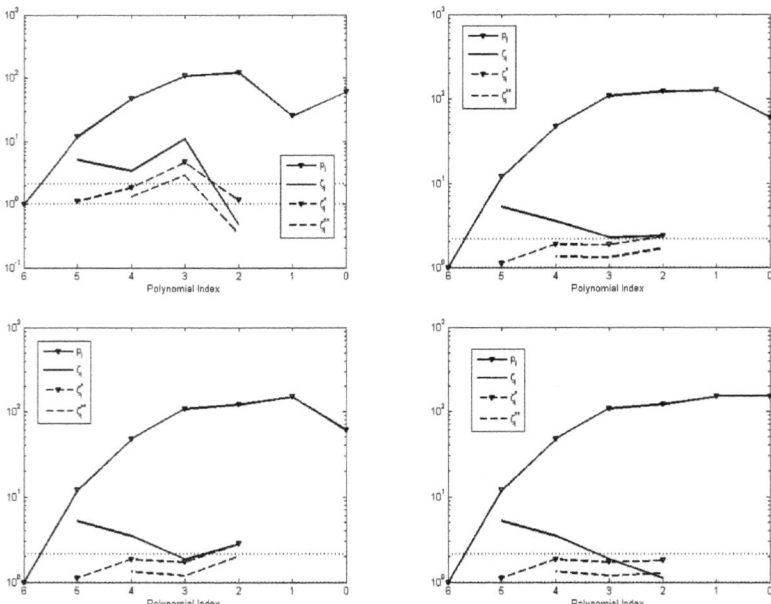

Fig. 3. Coefficient diagram for four different cases for (p_0, p_1). At the upper left $(60, 25)$, upper right $(60, 125)$, lower left $(60, 150)$ and lower right $(150, 150)$.

The upper left figure regards the referred characteristic polynomial with $p_0 = 60$ and $p_1 = 25$. Since the ζ_i curve falls below the upper instability limit the system is unstable. For the upper right diagram the p_0 value is kept constant and the p_1 coefficient is increased to 125. It is straightforward to conclude that the system is stable since the ζ_i curve is always above the lower stability limit.

The lower left figure also refers to a stable system. This one was obtained by further increase the value of p_1 from 125 to 150. In this case the stability conclusion is derived from the fact that the ζ_i curve is consistently above the ζ_i^*.

The final plot in Fig. 3 was obtained by deliberately pushing the (p_0, p_1) point to an area where the ordinary CDM stability condition is inconclusive. In concrete the new characteristic polynomial was forced to be in $(150, 150)$. However, due to the new adapted theorem, and since the ζ_i curve falls below the ζ_i^{**} curve the system is unstable.

7   Conclusion

The CDM control design method includes a graphical plot that the designer can use to infer about several closed-loop characteristics. One of them is system stability. The CDM stability conditions are derived by using a set of theorems arising from the work of Lipatov and Sokolov. However only two of the three theorems were included in the original CDM stability conditions formulation. The lack of one theorem leads to an increased inconclusive region, i.e. an area within the characteristic equation parameters space were the stability result is inconclusive.

In order to circumvent this limitation, this work proposes a simplified version of the second Lipatov-Sokolov theorem. The original algebraic difficulty is simplified but the theorem becomes more conservative. Nevertheless the CDM inconclusive stability area was able to decrease when compared to the original formulation. Moreover a new form of CDM diagram stability curves was introduced. Instead of plotting the stability values γ_i and their limits γ_i^* a new variable ζ_i and two curves ζ_i^* and ζ_i^{**} are presented. This lead to an improved and more accurate inference about system stability.

References

1. Bir, A., Öcal, Ö., Tibken, B.: Digital design of coefficent diagram method. In: 2009 American Control Conference, pp. 2849–2854 (2009)
2. Benjanarasuth, T., Ngamwiwit, J., Cahyadi, A., Isarakorn, D., Komine, N.: Application of coefficient diagram method for rotational inverted pendulum control. In: Control, Automation, Robotics and Vision Conference, vol. 3, pp. 1769–1773 (2004)
3. Hurwitz, A.: On the conditions under which an equation has only roots with negative real parts. Mathematische Annelen 46, 273–284 (1895)
4. Koksal, M., Hamamci, S.E.: A program for the design of linear time invariant control systems: Cdmcad. Computer Applications in Engineering Education 12(3), 165–174 (2004)
5. Lipatov, A.V., Sokolov, N.: Some sufficient conditions for stability and instability of continous linear stationary systems. Automat. Remote Contr. 39, 1285–1291 (1979)
6. Manabe, S.: Coefficient diagram method as applied to the attitude control of controlled-bias-momentum satellite. In: 13th IFAC Symposium on Automatic Control in Aerospace, pp. 322–327 (September 1994)
7. Manabe, S.: The coefficent diagram method. In: 14th IFAC Symposium on Automatic Control in Aerospace, pp. 199–210 (August 1998)

8. Manabe, S.: Sufficient condition for stability and instability by lipatov and its application to coefficent diagram method. In: 9th Workshop on Astrodynamics and Flight Mechanics. ISAS, pp. 440–449 (July 1999)
9. Manabe, S.: Coefficient diagram method in mimo application: An aerospace case study. In: Proceedings of the 16th IFAC World Congress, pp. 1961–1966 (2005)
10. Roengruen, P., Suksri, T., Kongratana, V., Numsomran, A., Thuengsripan, S.: Smith predictor design by cdm for temperature control system. In: Proceeedings ICCAS 2007, pp. 1472–1477 (2007)

Many-Objective PSO PID Controller Tuning

Hélio F. Freire, Paulo B. de Moura Oliveira,
Eduardo J. Solteiro Pires, and Maximino Bessa

INESC TEC - INESC Technology and Science (formerly INESC Porto)
Universidade de Trás-os-Montes e Alto Douro, UTAD
Department of Engineering, School of Sciences and Technology,
Quinta dos Prados,
5001-801 Vila Real, Portugal
`freireh@gmail.com`, `{oliveira,epires,maxbessa}@utad.pt`

Abstract. Proportional, integral and derivative controller tuning can be a complex problem. There are a significant number of tuning methods for this type of controllers. However, most of these methods are based on a single performance criterion, providing a unique solution representing a certain controller parameters combination. Thus, a broader perspective considering other possible optimal or near optimal solutions regarding alternative or complementary design criteria is not obtained. Tuning PID controllers is addressed in this paper as a many-objective optimization problem. A Multi-Objective Particle Swarm Optimization algorithm is deployed to tune PID controllers considering five design criteria optimized at the same time. Simulation results are presented for a set of four well known plants.

1 Introduction

Proportional, integral and derivative (PID) controllers are among the most used in industrial process control [1, 2]. They are very popular mainly due to their simple operating algorithm and broad range of pratical control applications. Improper PID parameters tuning can collapse the system operation [3] and about 80% of the PID-type controllers in industries are poorly or less optimally tuned [4]. Since the pioneering work developed by Ziegler-Nichols, a huge amount of PID controllers design and tuning rules have been proposed [5]. Many of these techniques were developed for specific types of system dynamics, and the design criteria considered is usually set-point tracking or disturbance rejection.

Taking advantage of the increasingly available computational power, natural inspired search and optimization techniques, such as evolutionary computation and swarm search intelligence, were applied to design PID controllers considering multiple design criteria and control restrictions. Examples of the former techniques are: genetic algorithms (GA) [6], ant colony optimization (ACO) [7], genetic programming (GP) [8], particle swarm optimization (PSO) [9, 10] and artificial bee colony (ABC) [11]. Many of these algorithms treat the PID parameter optimization as a single objective problem. Thus, the difficulty of finding 'good' PID parameters considering multiple design criteria (objectives) continues

© Springer International Publishing Switzerland 2015
A.P. Moreira et al. (eds.), *CONTROLO'2014 - Proc. of the 11th Port. Conf. on Autom. Control*,
Lecture Notes in Electrical Engineering 321, DOI: 10.1007/978-3-319-10380-8_18

to be a challenging problem, particularly in cases with more than 3 objectives: many-objective optimization. Relevant design criteria within the PID controller design are: rise-time, first overshoot, disturbance rejection and control effort. These design criteria can be adequately formulated using optimization objective functions, directly or indirectly by minimizing error based criteria. If the former design criteria are consider as individual objectives, a multi-objective optimization should be adopted. Indeed, if an aggregated function is used the search for optimal solutions can be compromised. Thus, finding PID parameters is a multi-objective problem in which different objectives are conflicting, *i.e.* the improvement of one criterion may lead to the deterioration of other criteria. The global aim of a multi-objective optimization is to provide a set of optimal solutions from which the decision maker can select the most appropriate one for a particular problem. This work addresses the PID controller tuning as a many-objective problem using a Multi-Objective Particle Swarm Optimization (MOPSO) algorithm.

The remainder of this paper is organized as follows. Section 2 presents some notes about PID controllers, while section 3 explains the concepts of many-objective optimization and the PSO algorithm. Section 4 presents the problem statement. Section 5 presents simulation results and its discussion and finally section 6 outlines the conclusions and future research.

2   PID Controller

PID controller dynamics can be governed by (1), combining three control actions in a parallel form in series with a first order filter. Each controller component considers three variable parameters: proportional, denoted by K_p, integrative, denoted by K_i and derivative, denoted by K_d. The filter time constant is represented by T_f. Appropriate tuning of these parameters will improve the system dynamic response, reduce the overshoot, eliminate the steady state error and increase the stability [1]. Figure 1 represents a block diagram of a feedback system with a PID controller, where (Gc) is the PID controller defined by (1) and the process to be controlled, is represented by (Gp). Signals u and y are the controller process input and output, respectively. Signal r is the reference input. Two types of disturbances are considered, d_1, the load disturbance and d_2, the output disturbance.

$$G_c(s) = \left(\frac{K_p s + K_i + K_d s^2}{s} \right) \left(\frac{1}{1 + sT_f} \right) \tag{1}$$

Different indices are commonly used to evaluate the system performance [12, 13], such as: integral of absolute error (IAE), integral of squared error (ISE), integral of time squared error (ITSE) and integral of time absolute error (ITAE). All of these criteria aim to minimize the error signal, $e(t) = r(t) - y(t)$. Among the former, the ITAE index (2) is used in this work.

$$ITAE = \int_0^\infty t|e(t)|\mathrm{d}t \tag{2}$$

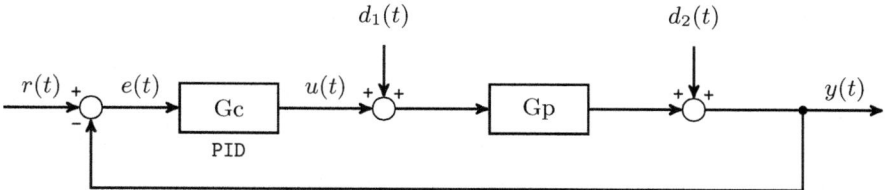

Fig. 1. PID controller structure

For the transient response analysis, maximum overshoot (OS), settling time (t_s) and rise time (t_r) are normally considered significant criteria. Faster tracking systems, require the minimum possible values for the former. Another control design objective considered relevant is the control effort. In this work the control effort is evaluated using (3).

$$Control\ Effort = \int_{t=0}^{\infty} |u(t)|\mathrm{d}t \tag{3}$$

The system robustness can be evaluated using the well known gain and phase margins or the sensitivity function S. A convenient way to estimate both, gain and phase margins, is by using the Vector Margin (VM) index, equation (4), which corresponds to the shortest distance from the Nyquist curve to the critical point $-1 + \mathrm{j}0$ and the inverse of the maximum sensitivity value, M_S,

$$VM = \frac{1}{M_S} = \frac{1}{\max\limits_{\omega} |S(\mathrm{j}\omega)|} = \min\limits_{\omega} \frac{1}{|S(\mathrm{j}\omega)|} \tag{4}$$

where S represents the sensitivity function evaluated with:

$$S(\mathrm{j}\omega) = \frac{1}{1 + Gc(\mathrm{j}\omega)Gp(\mathrm{j}\omega)} \tag{5}$$

3 Many-Objective and Particle Swarm Optimization

In the last years, Multi-Objective Evolutionary Algorithms (MOEA) have been proposed to solve real-world problems [14–16], involving the optimization of several objectives simultaneously. The objectives should be in conflict with each other, meaning that the improvement of one objective value will necessarily deteriorate some other objective values.

Multi-objective optimization main concerns can be stated as to [17]:

- Preserve non-dominated points in the objective space.
- Continue to make algorithmic progress towards the Pareto Front in objective function space.
- Maintain diversity of points on Pareto front.
- Provide the decision maker 'enough' but limited number of Pareto solutions for corresponding decision variable values for a given problem.

The PID controller design problem can be treated as a multi-objective problem. When the number of objectives is greater than 3 it is considered a many-objective problem [18].

Particle swarm optimization (PSO) is a stochastic population-based algorithm inspired in the social behaviour of bird flocking or fish schooling, proposed by Eberhart and Kennedy [19]. Each swarm particle is initialized randomly and represents a possible solution of a given problem. At each iteration, the particle is moved in the search space, according to the fundamental equations (6) and (7), incorporating solutions individual memory, p_{best}, that represents the fitness value of the best particle and the g_{best} represents the best global swarm fitness value.

$$v_k^i = \omega_k^i v_{k-1}^i + c_1 r_1 (p_{best}^i - s_{k-1}^i) + c_2 r_2 (g_{best}^i - s_{k-1}^i) \tag{6}$$

$$s_k^i = s_{k-1}^i + v_k^i, \tag{7}$$

where v and s represents the particle velocity and position, respectively, k is the epoch, i is a swarm particle, c_1 and c_2 are constants that represent the cognitive and social parameters. r_1 and r_2 are uniform distribution random numbers in [0,1] interval, and ω is the inertia weight that is used to control the velocity and plays a key role in the process of providing balance between the exploration and exploitation search phases [20].

The multi-objective PSO algorithm used here is based on the MaxiMin MOPSO [21]. This algorithm has been adapted for multi-criteria control design problems. The MaxiMin MOPSO algorithm is used with a swarm size of 100 and an epochs number of 50. The archive has size of 20. The inertia weight was fixed, $\omega = 0.7298$, $c_1 = c_2 = 1.496$ and r_1 and r_2 are random values in the [0,1] interval. The former parameter settings [22] are generally accepted as good values to promote the swarm convergence.

Random particle initialization can lead to the insertion of unstable PID controller settings in the initial population. In order to circumvent this problem, the PID parameter search interval was defined using an adaptive initial procedure until 10% of the swarm elements are stable and satisfy the robustness constraints described in the following section.

4   Problem Statement

Consider the control structure represented in Fig. 1, assuming the PID controller governed by (1) using a filter time constant of $T_f = 0.1$. The problem under study here can be classified as many-objective optimization, due to the number of objectives considered. The overall design goal is to obtain a set of non-dominated Pareto front solutions, considering the following PID controller design criteria:

obj1: Set-point tracking by minimizing the ITAE (2): a unit step is applied solely to the input reference ($r = 1$, $d_1 = 0$, $d_2 = 0$).

obj2: Load disturbance rejection by minimizing the ITAE (2): a step is applied solely to the load disturbance input ($r = 0$, $d_1 = 0.5$, $d_2 = 0$).

obj3: Output disturbance rejection by minimizing the ITAE (2): a step is applied solely to the output disturbance input ($r = 0$, $d_1 = 0$, $d_2 = 0.5$).

obj4: Minimization of Control effort, by minimizing (3) when $r = 1$, $d_1 = 0.5$ and $d_2 = 0.5$.

obj5: Gain and phase margins maximization, accomplish by maximizing the Vector Margin (VM) equation (4) A constraint threshold value of $VM = 0.3$ is predefined, in order to guarantee that a minimum value for the gain margin is $GM = 1.42$ and phase margins $PM = 17°$ for all population members. A vector margin value of $VM = 0.5$ [23] results in $GM = 2$ and $PM = 29°$, which are commonly accepted as good values within industrial process control. However, a vector margin restriction was set to a lower minimum value 0.3 in order not to make the search procedure too restrictive regarding this criterion, which would compromise values obtained for other conflicting criteria.

Four classic plant models were selected from a set identified by Åström and Hagglünd [24] as representative to be used as typical industrial processes in which PID controllers should be tested: a time delay and double lag (Gp_1), a 4-lag plant (Gp_2), a fourth order system (Gp_3) and a non-minimum phase plant (Gp_4).

$$Gp_1(s) = \frac{e^{-s}}{(1+s)^2} \tag{8}$$

$$Gp_2(s) = \frac{1}{(1+s)^4} \tag{9}$$

$$Gp_3(s) = \frac{1}{(1+s)(1+0.5s)(1+0.25s)(1+0.125s)} \tag{10}$$

$$Gp_4(s) = \frac{1-2s}{(1+s)^3} \tag{11}$$

5   Results and Discussion

The optimization process was carried to determine the optimal PID controller gains K_d, K_i and K_p in the four process plants. From the simulation response, five objectives were evaluated, as described in the previous section. Unlike any conventional/manual methods tuning the PID controller, this method provides a large number of solutions (Pareto frontier) to the control designers, giving them more flexibility and options by considering the various design trade-offs.

The final 20 Pareto front solutions obtained in the MOPSO archive can provide the reader with a perspective of the diversity of PID controller settings obtained. The pairs of figures 2-3, 4-5, 8-9 and 12–13, present the set-point tracking and disturbance rejection response and control signals, for the 20 solutions and

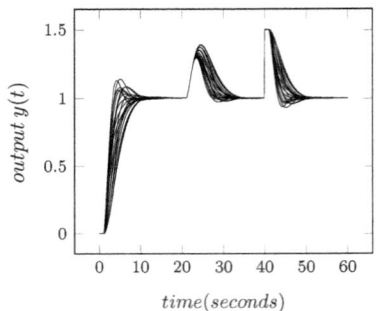

Fig. 2. Gp1 - System output

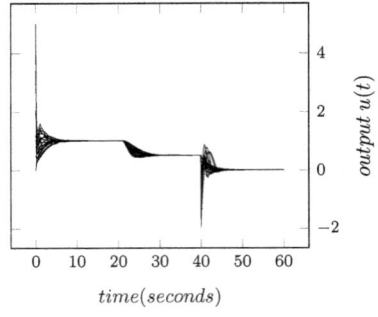

Fig. 3. Gp1 - Control signal

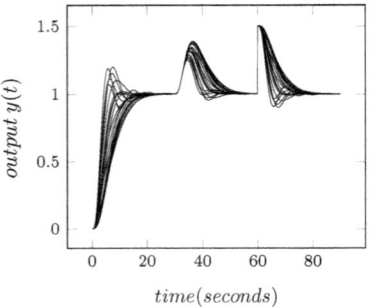

Fig. 4. Gp2 - System output

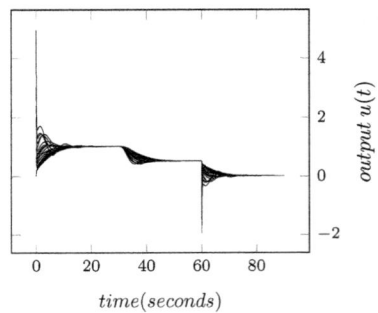

Fig. 5. Gp2 - Control signal

for plant models $Gp_1 - Gp_4$, respectively. The former systems responses were obtained by applying a input reference unit step response at $t = 0$s $(r = 1)$, a load disturbance step of amplitude of $(d_1 = 0.5)$ at $t = 20$s and an output disturbance of $(d_2 = 0.5)$ at $t = 40$s for the Gp_1 and Gp_3 plants. For the plants Gp_2 and Gp_4 the disturbance times are at $t = 30$s and $t = 60$s for the load and output disturbance respectively. As it can be observed from figures 2-3, 4-5, 8-9 and 12–13, the solutions obtained allow the decision maker (in this case the control engineer) to select PID gains providing tracking responses presenting different compromises between first overshoot and rise time as well as disturbance capabilities. The minimum and maximum values obtained for each objective and the entire population non-dominated solutions, are presented in Table 2. Values presented for obj1, obj2 and obj3 concern ITAE values, for obj4 concern the control effort values and obj5 vector margin values. As it can be observed in table 2 approximate minimum value for the vector margin were obtained near the minimum constraint imposed for this objective of 0.3. This result can be clearly observed from the Nyquist plots presented in Fig. 14, for the respective 20 loop Nyquist plots, for plants $Gp_1 - Gp_4$.

Due to space limitations not all plots regarding the 4 plants can be presented here. Taking system Gp_3 as an example, Fig. 6 presents the range of values obtained for each PID parameter gain and the relation between them, for the

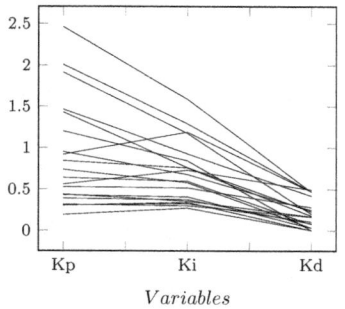

Fig. 6. Gp3 - PID decision space variables

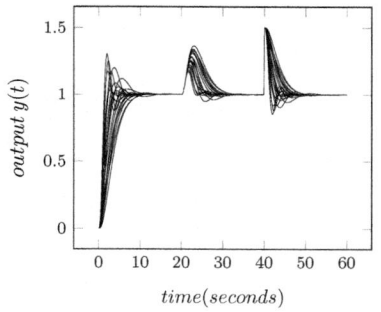

Fig. 7. Gp3 - Objective space normalized plot

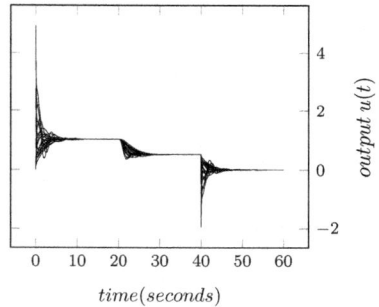

Fig. 8. Gp3 - System output

Fig. 9. Gp3 - Control signal

Table 1. Gp1 - Two PID solutions values

	K_p	K_i	K_d	obj1	obj2	obj3	obj4	obj5	OS	GM	PM
Solution 17	0.261	0.217	0.000	13.53	60.18	101.37	31.52	0.674	0.69	3.07	39.3°
Solution 18	0.864	0.436	0.341	4.27	27.92	54.48	32.83	0.475	6.25	1.91	27.5°

set the 20 Pareto solutions. Fig. 7 presents the relation between the 20 solution objective space values for the 5 objectives considered. It is important to state at this point, that in Fig. 7 all the objective values were normalized between 0 a 1. From Fig. 6 it can be observed that the controller gain which varies the most is K_p, followed by K_i. The parameter which varies the less is K_d. From Fig. 7, it can be observed that apparently objectives 1, 2 and 3 are less in conflict that the former 3 with objective 4 and 5. The results also indicate less conflict between the PID gains obtained for set-point tracking and output disturbance than between set-point tracking and load disturbance. Indeed, while not clear from Fig. 8, the results indicate that objective 1 and 3 are in conflict with results obtained for objective 2. The same was observed for the other plants.

An illustrative example is provided in Figures 10 and 11, presenting respectively, the set-point tracking and disturbance rejection obtained for two Pareto

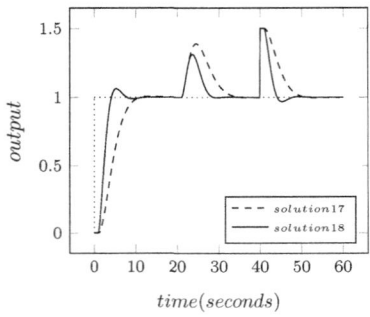

Fig. 10. Gp1 - Two PID selected solutions

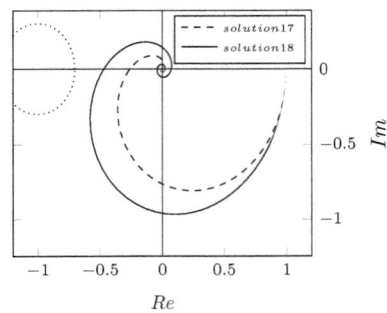

Fig. 11. Gp1 - Nyquist plot for two PID solutions

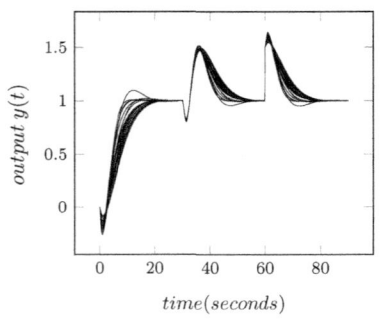

Fig. 12. Gp4 - System output

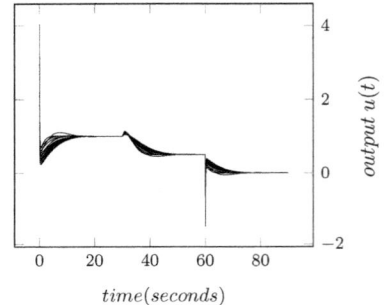

Fig. 13. Gp4 - Control signal

Table 2. Objectives maximum and minimum values for the 4 systems

	obj1		obj2		obj3		obj4		obj5	
	min	max	min	max	min	max	min	max	min	max
Gp1	3.286	14.413	23.155	62.051	48.432	104.427	31.519	34.104	0.304	0.743
Gp2	9.327	40.921	45.254	152.900	107.072	255.721	46.999	49.876	0.310	0.906
Gp3	0.927	9.354	7.277	45.499	24.387	80.451	30.948	33.297	0.301	0.941
Gp4	14.353	49.582	118.827	200.479	169.055	297.890	47.529	48.694	0.300	0.719

solutions and respective Nyquist plots. In this case, the decision maker, with the information in the these figures and in table 1 can decide which solution is preferable.

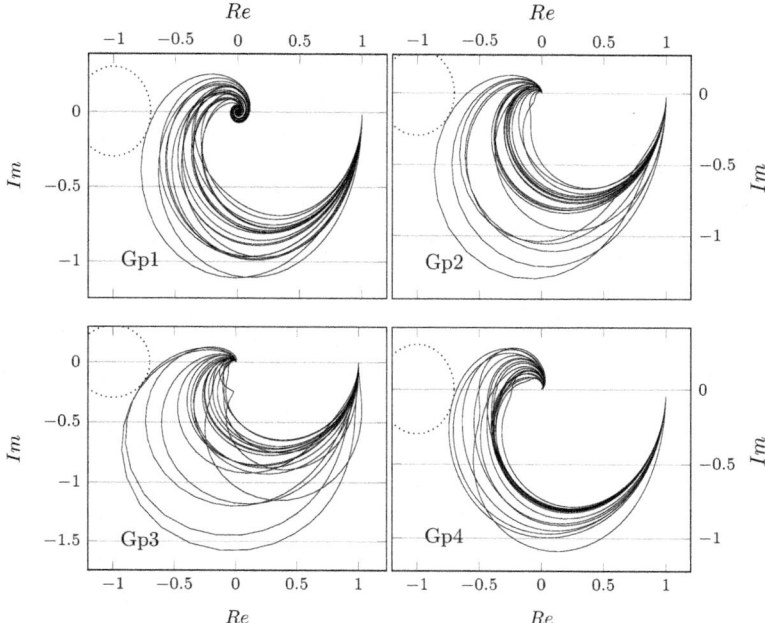

Fig. 14. Nyquist plots for the 4 processes $Gp_1 - Gp_4$

6   Conclusion

A MOPSO was applied to design PID controllers considering a many-objective problem formulation. Four well known plant models were used to test the proposed technique considering five optimization criteria. For all the problems a non-dominated set of solutions representing PID gains was obtained. From these solution sets the control engineer has a wider perspective of alternative optimal PID parameters and corresponding system response, regarding all criteria. While, five objectives were used, the proposed algorithm enables the incorporation of alternative design criteria and to consider more than five objectives. Further work will be direct in developing refined many-objective optimization techniques.

Acknowledgements. This work was supported by the Fundação para a Ciência e a Tecnologia (FCT) under PhD studentship No. SFRH/BD/79463/2011.

References

1. Åström, K.J., Hagglünd, T.: PID Controllers: Theory, Design, and Tuning, 2nd edn. The Instrument, Systems, and Automation Society, Research Triangle Park (1995)
2. Kano, M., Ogawa, M.: The state of the art in chemical process control in japan: Good practice and questionnaire survey. J. of Process Control 20(9) (2010)

3. Åström, K.J., Hagglünd, T.: Automatic Tuning of PID Controllers. Instrument Society of America, Research Triangle Park (1988)
4. Åström, K.J., Hägglund, T.: Advanced PID Control. ISA - The Instrumentation, Systems, and Automation Society Research Triangle Park (2006)
5. O'Dwyer, A.: Handbook of PI and PID Controller Tuning Rules, 2nd edn. Imperial College Press (February 2006)
6. Herreros, A., Baeyens, E., Péran, J.R.: Design of PID-type controllers using multiobjective genetic algorithms. ISA Trans 41(4), 457–472 (2002)
7. Chiha, I., Liouane, N., Borne, P.: Tuning PID controller using multiobjective ant colony optimization. Appl. Comp. Intell. Soft. Comput. (January 2012)
8. Streeter, M.J., Keane, M.A., Koza, J.R.: Automatic synthesis using genetic programming of improved PID tuning rules. In: Ruano, A.E. (ed.) Preprints of the 2003 Intelligent Control Systems and Signal Processing Conf., Portugal, pp. 494–499.
9. Easter Selvan, S., Subramanian, S., Theban Solomon, S.: Novel Technique for PID Tuning by Particle Swarm Optimization. In: Seventh Annual Swarm Users/Researchers Conference, SwarmFest 2003 (2003)
10. Rajinikanth, V., Latha, K.: Tuning and retuning of PID controller unstable systems using evolutionary algorithm. Appl. Comp. Intell. Soft. Comput. (January 2012)
11. Abachizadeh, M., Yazdi, M.R.H., Yousefi-Koma, A.: Optimal tuning of PID controllers using Artificial Bee Colony algorithm. In: IEEE/ASME International Conference on Advanced Intelligent Mechatronics (2010)
12. Alcántara, S., Vilanova, R., Pedret, C., Skogestad, S.: A look into robustness/performance and servo/regulation issues in PI tuning. IFAC 2, 181–186 (2012)
13. Garpinger, O., Hägglund, T., Åström, K.J.: Criteria and trade-offs in PID design, Brescia, Italy (2012)
14. Reyes-Sierra, M., Coello, C.A.C.: Multi-objective particle swarm optimizers: A survey of the state-of-the-art. J. of Comp. Intelligence Research 2(3) (2006)
15. Fleming, P.J., Purshouse, R.C., Lygoe, R.J.: Many-Objective Optimization: An Engineering Design Perspective. Design 3410(6), 14–32 (2005)
16. Ishibuchi, H., Tsukamoto, N., Nojima, Y.: Evolutionary many-objective optimization: A short review. In: 2008 IEEE C. on Evolutionary Computation, pp. 2419–2426 (2008)
17. Coello, C.A.C., Lamont, G.B., Veldhuizen, D.A.V.: Evolutionary Algorithms for Solving Multi-Objective Problems (2002)
18. Adra, S., Fleming, P.: Diversity management in evolutionary many-objective optimization. IEEE Transactions on Evolutionary Computation 15(2), 183–195 (2011)
19. Kennedy, J., Eberhart, R.: Particle swarm optimization. In: Proceedings IEEE International Conference on Neural Networks, vol. 4, pp. 1942–1948 (1995)
20. Bansal, J.C.: et al.: Inertia weight strategies in particle swarm optimization. In: NaBIC, pp. 633–640. IEEE (2011)
21. Freire, H., de Moura Oliveira, P.B., Pires, E.J.S., Lopes, A.M.: MaxiMin MOPSO design of parallel robotic manipulators. In: Corchado, E., Snášel, V., Sedano, J., Hassanien, A.E., Calvo, J.L., Ślęzak, D. (eds.) SOCO 2011. Advances in Intelligent Systems and Computing, vol. 87, pp. 339–347. Springer, Heidelberg (2011)
22. Clerc, M., Kennedy, J.: The particle swarm - explosion, stability, and convergence in a multidimensional complex space. IEEE Trans. Evol. Computation 6(1) (2002)
23. Skogestad, S., Postlethwaite, I.: Multivariable Feedback Control: Analysis and Design. John Wiley & Sons (2005)
24. Åström, K.J., Hägglund, T.: Benchmark systems for PID control. In: Digital Control – Past, present, and future of PID Control, Elsevier (2000)

Gain Equalization and LQG/LTR Controller Design for a 3I3O Multivariable Dynamic System with One Pole at the Origin

Danilo Azevedo Figueiredo and Jorge Roberto Brito-de-Souza

Federal University of Pará (UFPA), Belém, Brazil
{daniloaf,jrgbrito}@ufpa.br

Abstract. This paper addresses the problem of achieving gain equalization on all input-output channels of multivariable square dynamic systems which have a pole at the origin. A mathematical procedure based on linear transformations is proposed to extract the integrator associated with this pole to the outside of the system in a specific output channel. The order of the system model is then reduced, and the removed integrator can be joined with other new integrators that are added to the others output channels of the system in order to accomplish the desired gain equalization. A 5th order, 3I3O system (three inputs and three outputs) is considered as an example, for which a reduced order LQG/LTR controller is also designed.

Keywords: gain equalization, loopshaping, LQG/LTR controller, MIMO systems, pole at the origin, target feedback loop.

1 Introduction

The LQG/LTR method for controllers design was introduced in the control systems literature by Doyle and Stein [1] as a successor to the LQG method, which aims to recover, at least partially, the excellents gain and phase margins that the LQR regulators enjoy [2], but that the regulators with observers cannot guarantee, accordingly proved by Doyle [3]. The method can be applied to design controllers for both SISO and MIMO systems, and it has two alternative versions that are chosen depending upon where the multiplicative unstructured uncertainties of the system model are represented—either on its input or its output. In order to provide a good controller, the method requires that the system to be controlled be a minimum phase system, but in principle there is no requirement that it must be stable.

The design of a LQG/LTR controller for a given system $G(s)$ is made in two steps [4]. First, a *Target Feedback Loop* $T_{ROL}(s)$ is designed so that it has suitable frequency response characteristics, and then it is replaced by a robust LQR regulator $T_{LQ}(s)$ with similar *loopshaping* as of $T_{ROL}(s)$. In the second step a controller $K(s)$ is designed in such a way that the open-loop matrix transfer function $K(s)G(s)$ have the same frequency response characteristics of $T_{LQ}(s)$.

The *Target Feedback Loop* must be designed in order to guarantee the following performance characteristics to the closed-loop controlled system: static precision of

© Springer International Publishing Switzerland 2015
A.P. Moreira et al. (eds.), *CONTROLO'2014 - Proc. of the 11th Port. Conf. on Autom. Control*,
Lecture Notes in Electrical Engineering 321, DOI: 10.1007/978-3-319-10380-8_19

its outputs with respect to their reference signals, capability to reject external disturbances, ability to filter measurement noises, and good robustness to accommodate the
uncertainties associated to the nominal model of the system.

For the achievement of all of these goals, the addition of integrators to the system
is necessary, along with the imposition of some restrictions to the principal gains of
the *Target Feedback Loop*. These gains are defined by the maximum and minimum
singular values of $T_{LQ}(s)$ and basically they must be high at low frequencies and low at
high frequencies [5]. Of course that the equalization of these principal gains amounts to
the equalization of all system gains which are related with each one of its input-output
channels.

For the case of multivariable systems, the above mentioned gain equalization of all
its diverse input-output channels, if accomplished for all frequencies, is a good prescrip-
tions for achieving another important goal, namely, the decoupling of each output signal
of the system with regard to the reference signals associated with the others' outputs.

There exist several procedures for the achievement of gain equalization of multivariable systems. Reference [6] presents an approach that equalizes these gains at low and
high frequencies, but does not guarantee the equalization at medium frequencies. Reference [7] introduces a formula which allows the gain equalization in all frequencies, but
the use of this formula requires the inversion of the state matrix of the system. Therefore, the use of this formula requires that $det(A) \neq 0$, which does not happen when the
system has at least one pole at the origin. This work presents a procedure to overcome
this difficulty.

The main contribution of this paper is a procedure that provides gain equalization, in
all frequencies, for all input-output channels of multivariable dynamic systems which
have a single pole at the origin. The procedure requires the addition of an integrator in
every output channel—considering the uncertainties on the input—through which the
pole at the origin is not observable.

Besides achieving the gain equalization in all frequencies, a byproduct of this procedure is that the number of integrators that are added to the system is reduced, and
ultimately it makes possible to design a reduced order LQG/LTR controller.

2   Gain Equalization via Addition of Integrators: The Basic Approach

Consider a multivariable linear system whose state space representation is defined by

$$\begin{cases} \dot{x} = Ax + Bu \\ y = Cx \end{cases} \tag{1}$$

where $x \in \mathbb{R}^n$ is the state vector of the system, and $u \in \mathbb{R}^m$ and $y \in \mathbb{R}^p$ are the input
and output vectors, respectively. The matrices of the triplet $\{A, B, C\}$ have compatible
dimensions but in this paper it is assumed that the system is square, so that $m = p$ and
its (square!) matrix transfer function is given by

$$G(s) = C(sI_n - A)^{-1}B. \tag{2}$$

Fig. 1 shows a block diagram in which the uncertainties of nominal model $G(s)$ are represented on its input and a block of integrators are added to its output. The diagram also shows the controller $K(s)$ to be designed.

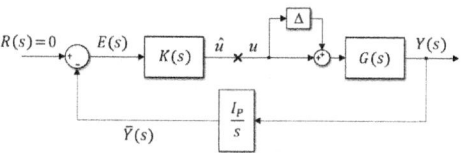

Fig. 1. Block diagram of the closed-loop system with integrators added to the output of the system $G(s)$

With the addition of the integrators, the representation of the augmented system becomes

$$
\begin{bmatrix} \dot{x} \\ \dot{\bar{y}} \end{bmatrix} = \begin{bmatrix} A & 0_{p \times p} \\ C & 0_{p \times p} \end{bmatrix} \begin{bmatrix} x \\ \bar{y} \end{bmatrix} + \begin{bmatrix} B \\ 0_{p \times m} \end{bmatrix} u
$$

$$
\bar{y} = \begin{bmatrix} 0_{p \times n} & I_p \end{bmatrix} \begin{bmatrix} x \\ \bar{y} \end{bmatrix},
$$

(3)

or yet

$$
\begin{bmatrix} \dot{x} \\ \dot{\bar{y}} \end{bmatrix} = \dot{\bar{x}} = \bar{A}\bar{x} + \bar{B}u
$$

$$
\bar{y} = \bar{C}\bar{x},
$$

(4)

whose corresponding matrix transfer function is given by

$$
\bar{G}(s) = \bar{C}(sI - \bar{A})^{-1}\bar{B} = \frac{I_p}{s}G(s).
$$

(5)

The matrix transfer function of the *Target Feedback Loop* to be designed is defined by [4]

$$
T_{ROL}(s) = \bar{H}(sI - \bar{A})^{-1}\bar{B},
$$

(6)

and its complete description depends upon a very suitable choice of $\bar{H}$.

By replacing the matrices $\bar{A}$ and $\bar{B}$ given in (3)–(4) into (6), the outcome is

$$
T_{ROL}(s) = \begin{bmatrix} \bar{H}_1 & \bar{H}_2 \end{bmatrix} \begin{bmatrix} (sI_n - A)^{-1} & 0_{n \times p} \\ C\dfrac{(sI - A)^{-1}}{s} & \dfrac{I_p}{s} \end{bmatrix} \begin{bmatrix} B \\ 0_{p \times m} \end{bmatrix},
$$

(7)

or yet

$$
T_{ROL}(s) = \bar{H}_1(sI_n - A)^{-1}B + \bar{H}_2 C\frac{(sI_n - A)^{-1}}{s}B.
$$

(8)

It is easy to verify that, by choosing the matrix $\overline{H}$ according to the following equation

$$\overline{H} = \left[\overline{H}_1 \vdots \overline{H}_2 \right] = \left[(CA^{-1}B)CA^{-1} \vdots -(CA^{-1}B)^{-1} \right], \tag{9}$$

the *Target Feedback Loop* becomes

$$T_{ROL}(s) = \frac{I_p}{s}. \tag{10}$$

Thanks to its diagonal type, this resulting *Target Feedback Loop* is perfect to equalize in all frequencies the gains of all input-output channel and, at the same time, these gains are high (low) at low (high) frequencies. Therefore it fulfills the basic requirements that it is expected to meet, as metioned before.

Equation (9) is a dual version of another similar equation that appears in [7], the difference being that this referred equation is used when the unstructured uncertaints of the system model are represented on its output.

3   Gain Equalization for Systems with One Pole at the Origin

In the previous section the usual method for gain equalization, in all frequencies, of the *Target Feedback Loop* $T_{ROL}(s)$ was briefly reviewed. Basically it is based upon the addition of one integrator in every output channel of the system—which increases the order of the system from n to $n+p$; and it also requires the determination of the matrix $\overline{H}$ according to the equation given in (9). A simple analysis of this equation indicates that its application requires the inversion of the state matrix of the system, and due to this reason it can not be used in those cases in which the system has at least one pole at the origin. Therefore, the gain equalization problem for the specific case of systems with poles at the origin requires a new approach.

If a system has a pole at the origin it means that the system has an integrator within its internal structure. In so being, it is eventually feasible to use a particular linear transformation to extract this integrator to the outside of the internal structure of the system so that it may be placed in a specific output channel of the system. Therefore, when it comes to the gain equalization process, the addition of an integrator to this specific output channel will be dispensed.

On the other hand, this process looks like a "fictional" order reduction of the system, since its state matrix is reduced to $n-1$, after the integrator is extracted.

More important, this reduced order state matrix is nonsingular and invertible. In what follows a constructive description of this proposed approach is presented.

Let the following system that appears in [8] be considered:

$$\dot{x} = \begin{bmatrix} 0 & 0 & 1.1320 & 0 & -1 \\ 0 & -0.0538 & -0.1712 & 0 & 0.0705 \\ 0 & 0 & 0 & 1 & 0 \\ 0 & 0.0445 & 0 & -0.856 & -1.0130 \\ 0 & -0.2909 & 0 & 1.053 & -0.6859 \end{bmatrix} x + \begin{bmatrix} 0 & 0 & 0 \\ -0.120 & 1 & 0 \\ 0 & 0 & 0 \\ 4.4190 & 0 & -1.6650 \\ 1.5750 & 0 & -0.0732 \end{bmatrix} u$$

$$y = \begin{bmatrix} 1 & 0 & 0 & 0 & 0 \\ 0 & 1 & 0 & 0 & 0 \\ 0 & 0 & 1 & 0 & 0 \end{bmatrix} x \tag{11}$$

This system has no transmission zeros, and its modes are the following: $\lambda_{1,2} = -0.7801 \pm 1.296i$, $\lambda_{3,4} = -0.0176 \pm 0.1826i$ and $\lambda_5 = 0$. The value of λ_5 indicates that $det(A) = 0$, and that the system has an integrator within its internal structure.

3.1   Diagonalization of the System Representation

The original system model in (11) can be replaced by a new quasi-diagonal state space representation which is attained through the linear transformation defined by $z = T_1 x$, where transformation matrix is given by

$$T_1 = \left[Real(t_1) \vdots Imag(t_1) \vdots Real(t_3) \vdots Imag(t_3) \vdots t_5 \right], \tag{12}$$

where $At_i = \lambda_i t_i$, $i = 1, 2, 3, 4, 5$.

The new realization $\{\widehat{A}, \widehat{B}, \widehat{C}\}$, where $\widehat{A} = T_1^{-1} A T_1$, $\widehat{B} = T_1^{-1} B$ and $\widehat{C} = C T_1$, may be partitioned as

$$\dot{z} = \left[\begin{array}{c} \dot{z}_{1:4} \\ \hdashline \dot{z}_5 \end{array} \right] = \underbrace{\left[\begin{array}{c:c} \widehat{a} & 0_{4 \times 1} \\ \hdashline 0_{1 \times 4} & 0 \end{array} \right]}_{\widehat{A}} \left[\begin{array}{c} z_{1:4} \\ \hdashline z_5 \end{array} \right] + \underbrace{\left[\begin{array}{c} {}_{1:4}\widehat{b} \\ \hdashline {}_5\widehat{b} \end{array} \right]}_{\widehat{B}} u$$

$$y = \left[\begin{array}{c} y_1 \\ \hdashline y_{2:3} \end{array} \right] = \underbrace{\left[\begin{array}{c:c} {}_1\widehat{c} & 1 \\ \hdashline {}_{2:3}\widehat{c} & 0_{2 \times 1} \end{array} \right]}_{\widehat{C}} \left[\begin{array}{c} z_{1:4} \\ \hdashline z_5 \end{array} \right] \tag{13}$$

Note in this equation that the variable z_5 represents the signal at the output of the integrator that exists within the system, and it is decoupled with regard to the variables in $z_{1:4}$. Moreover, the outputs y_2 and y_3 are independent of z_5, but the output y_1 does depend on z_5.

3.2   Extracting the System Integrator Toward One of Its Outputs

A direct transformation of (13) leads to the block diagram shown in Fig. 2(a)—where the dashed blocks ought to be neglected—which includes, in a explicit (dotted) block, the internal integrator of the system. The output y_1 is given by

$$y_1 = {}_1\widehat{c} z_{1:4} + z_5 \tag{14}$$

Now, in order to extract the internal integrator of the system to its outside, the (dotted) block corresponding to the referred integrator has to be advanced toward the output channel corresponding to y_1. However, upon advancing this integrator, and in order to keep the value of y_1 given in (14), it also requires the inclusion of a derivative block in the overall diagram. These two steps lead to the block diagram that is shown in Fig. 2(b)—where now the dotted blocks are the ones that ought to be neglected.

Comparing the block diagrams in Figs. 2(a)–(b), it is clear that the variables in $z_{1:4}$ are the same on both of them. The output signals y_2, y_3 are also the same. However,

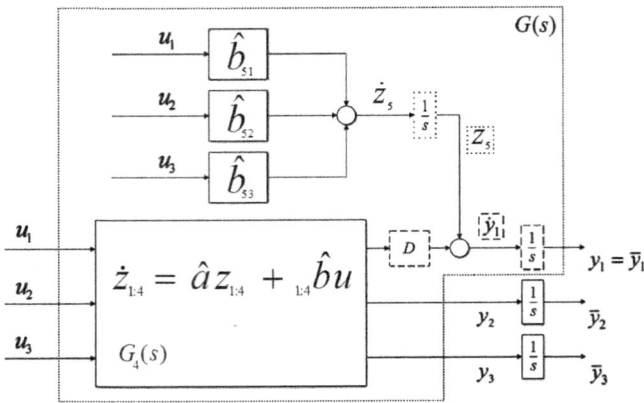

Fig. 2. Block diagram of the system $G(s)$ after its quasi-diagonal representation: (a) ignoring the blocks with dashed outline; (b) ignoring the blocks with dotted outline

the variable at the output of the integrator in the block diagram in Fig. 2(a) is z_s, but in Fig. 2(b) it is y_1. All this, combined with (14), leads to a change of variables defined by $z = T_2 \tilde{z}$ which is specified by

$$
\overbrace{\begin{bmatrix} z_{1:4} \\ \hline z_5 \end{bmatrix}}^{z} = \underbrace{\begin{bmatrix} I_4 & 0 \\ \hline -{}_1\hat{c} & 1 \end{bmatrix}}_{T_2} \overbrace{\begin{bmatrix} z_{1:4} \\ \hline y_1 \end{bmatrix}}^{\tilde{z}}. \tag{15}
$$

The vector $\tilde{z}$ is also related with the original state vector x according to the relation $x = T_3 \tilde{z}$, where $T_3 = T_1 T_2$.

Before closing this subsection it is important to observe that the derivative of y_1 in (14) is given by

$$
\dot{y}_1 = {}_1\hat{c}\,\dot{z}_{1:4} + \dot{z}_5 \tag{16}
$$

and ultimately

$$
\dot{y}_1 = {}_1\hat{c}\,\hat{a}\,z_{1:4} + \left[{}_1\hat{c}\, {}_{1:4}\hat{b} + {}_5\hat{b} \right] u. \tag{17}
$$

3.3 Adding New Integrators

After advancing the original integrator of the system toward its first output channel, and before going for the gain equalization processes, it is still necessary to add a new integrator in each output channel of the system that still does not have one. Fig. 2 shows these new integrators at the output channels two and three.

The state equation for the new variables associated with these new integrators is

$$
\begin{bmatrix} \dot{\bar{y}}_2 \\ \dot{\bar{y}}_3 \end{bmatrix} = \begin{bmatrix} 1 & 0 \\ 0 & 1 \end{bmatrix} \begin{bmatrix} y_2 \\ y_3 \end{bmatrix}.
$$

(18)

For the sake of establishing a standard notation, let the output y_1 be renamed as $\bar{y}_1$. Then the state space representation of the augmented system becomes

$$
\dot{\bar{z}} = \begin{bmatrix} \dot{z}_{1:4} \\ \dot{\bar{y}}_1 \\ \dot{\bar{y}}_2 \\ \dot{\bar{y}}_3 \end{bmatrix} = \overbrace{\begin{bmatrix} \hat{a} & 0_{4\times1} & 0_{4\times1} & 0_{4\times1} \\ 1\hat{c}\,\hat{a} & 0 & 0 & 0 \\ 2\hat{c} & 0 & 0 & 0 \\ 3\hat{c} & 0 & 0 & 0 \end{bmatrix}}^{\bar{A}} \begin{bmatrix} z_{1:4} \\ \bar{y}_1 \\ \bar{y}_2 \\ \bar{y}_3 \end{bmatrix} + \underbrace{\begin{bmatrix} 1{:}4\hat{b} \\ 1\hat{c}\ 1{:}4\hat{b} + 5\hat{b} \\ 0 \\ 0 \end{bmatrix}}_{\bar{B}} u
$$

(19)

$$
\bar{y} = \underbrace{\begin{bmatrix} 0_{3\times4} & I_3 \end{bmatrix}}_{\bar{C}} \overbrace{\begin{bmatrix} z_{1:4} \\ \bar{y} \end{bmatrix}}^{\bar{z}}, \qquad \bar{y} = \begin{bmatrix} \bar{y}_1 \\ \bar{y}_2 \\ \bar{y}_3 \end{bmatrix},
$$

and the matrix transfer function of the augmented system in (19) is given by

$$
\bar{G}(s) = \frac{\bar{I}_2}{s} G(s).
$$

(20)

where $\bar{I}_2$ is a nonstandard augmented identity matrix defined as

$$
\bar{I}_2 = \begin{bmatrix} s & 0 & 0 \\ 0 & 1 & 0 \\ 0 & 0 & 1 \end{bmatrix} \rightarrow \frac{\bar{I}_2}{s} = \begin{bmatrix} 1 & 0 & 0 \\ 0 & 1/s & 0 \\ 0 & 0 & 1/s \end{bmatrix}.
$$

(21)

The frequency response characteristics of the principal gains of the original system $G(s)$ and the augmented system $\bar{G}(s)$ are shown on Fig. 3. It shows that, upon the addition of the new integrators, the magnitude of the minimum principal gain is increased at low frequencies. However, the maximum and minimum principal gains still are unmatched.

3.4   Gain Equalization

A direct comparison of (19)–(20) and (3)–(5) indicates that they have the very same partitioning and overall form. Therefore, by using the accordingly matrices in (9), it leads to

$$
\bar{H} = \begin{bmatrix} 0.1363 & 0.1786 & 0.0134 & -0.1240 & -0.4645 & -0.2122 & -0.5258 \\ 0.0548 & 0.0169 & -0.1416 & -0.0499 & 0.0148 & -0.0793 & -0.1545 \\ 0.1895 & 0.2468 & 0.0533 & -0.4109 & -0.6244 & -0.5924 & -0.7069 \end{bmatrix}.
$$

(22)

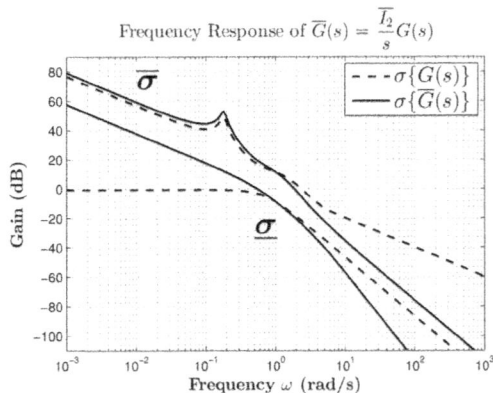

Fig. 3. Principal Gains of system $G(s)$ and the augmented system $\overline{G}(s)$

This $\overline{H}$ matrix plus $\overline{A}$ and $\overline{B}$ given in (21) defines a *Target Feedback Loop* $T_{ROL}(s)$ similar to the one that is given in (6), and its final form is exactly equal to the one that is given in (10). Therefore, the just designed *Target Feedback Loop* fulfills the basic requirements that it is expect to meet when it comes to design a robust LQG/LTR controller.

4   LQG/LTR Controller Design

It is already known that the system is controllable and observable and has no phase transmission zeros, thus it is now possible to design the classical LQG/LTR controller, which is represented here by $K(s)$.

It is important to assert that the integrators added to the system are actually a complement part of the controller $K(s)$. Therefore, the actual representation for the controlled system is shown in Fig. 4, where the two integrators added are located right before the block of $K(s)$ and the internal integrator is now an outward block of the system $G(s)$, which, for this reason, became $G_4(s)$. This representation is necessary since the outputs y_1, y_2 and y_3 will not be altered.

For the design of the LQG/LTR controller, this work also makes use of its conventional and well-known structure [6], shown in Fig. 5, which represents the state space equations of the augmented system $\overline{G}(s)$

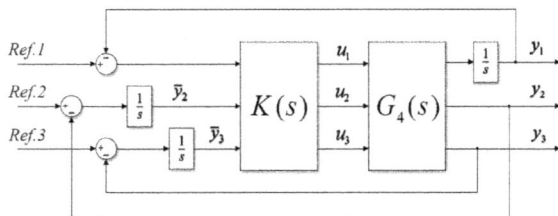

Fig. 4. Actual representation of the controlled system $K(s)\overline{G}(s)$

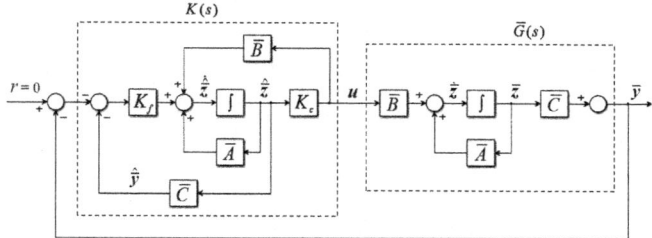

Fig. 5. Representation of the controlled system $K(s)\overline{G}(s)$ in the LQG/LTR structure

$$\begin{cases} \dot{\overline{z}} = \overline{A}\overline{z} + \overline{B}u \\ \overline{y} = \overline{C}\overline{z} \end{cases} , \tag{23}$$

and of the controller $K(s)$

$$\begin{cases} \dot{\hat{\overline{z}}} = (\overline{A} - \overline{B}K_c - K_f\overline{C})\hat{\overline{z}} + K_f\overline{y} \\ u = -K_c\hat{\overline{z}} + r \end{cases} . \tag{24}$$

The equations that represent the transfer matrices of the LQG/LTR controller and of the augmented system $\overline{G}(s)$ are, respectively

$$K(s) = -K_c(sI - \overline{A} + \overline{B}K_c + K_f\overline{C})^{-1}K_f , \tag{25}$$

$$\overline{G}(s) = \overline{C}(sI - \overline{A})^{-1}\overline{B} , \tag{26}$$

and the LQG control law is given by

$$u = -K_c\hat{\overline{z}} . \tag{27}$$

For the error, let us define a new variable $e = \overline{z} - \hat{\overline{z}}$, thus $\dot{e} = \dot{\overline{z}} - \dot{\hat{\overline{z}}}$, and hence

$$\begin{cases} \dot{e} = (\overline{A} - K_f\overline{C})e + \overline{B}r \\ \dot{\overline{z}} = (\overline{A} - \overline{B}K_c)\overline{z} + \overline{B}K_c e + \overline{B}r \end{cases} . \tag{28}$$

Therefore, the closed loop system $K(s)\overline{G}(s)$ is characterized by the following state space representation

$$\begin{bmatrix} \dot{\overline{z}} \\ \dot{e} \end{bmatrix} = \begin{bmatrix} \overline{A} - \overline{B}K_c & \vdots & \overline{B}K_c \\ \cdots & & \cdots \\ 0 & \vdots & \overline{A} - K_f\overline{C} \end{bmatrix} \begin{bmatrix} \overline{z} \\ e \end{bmatrix} + \begin{bmatrix} \overline{B} \\ \cdots \\ \overline{B} \end{bmatrix} r$$

$$\tag{29}$$

$$\overline{y} = \begin{bmatrix} \overline{C} & \vdots & 0 \end{bmatrix} \begin{bmatrix} \overline{z} \\ \cdots \\ e \end{bmatrix} .$$

Additionally, the closed-loop poles of $K(s)\overline{G}(s)$ are $\lambda(\overline{A} - \overline{B}K_c) \cup \lambda(\overline{A} - K_f\overline{C})$.

4.1 LQR Design

The optimal gains for the controller are defined by $K_c = R_c^{-1} \overline{B}^T P$, in which P is the unique solution of the CARE (Controller Algebraic Riccati Equation)

$$0 = P\overline{A} + \overline{A}P + Q_c - P\overline{B}R_c^{-1}\overline{B}^T P, \qquad P > 0, \tag{30}$$

where $R_c = \rho \times I$ is the positive definite matrix of weight for control effort and Q_c is the positive semi-definite matrix of weight for regulation of tracking performance, defined as: $Q_c = \overline{H}^I \overline{H}$.

The robust LQR regulator $T_{LQ}(s)$ for $K(s)\overline{G}(s)$ can now be defined. The relationship between the gains of $T_{LQ}(s)$ and $T_{ROL}(s)$, according to [4], is

$$\sigma[T_{LQ}(s)] \cong \frac{1}{\sqrt{\rho}} \sigma[T_{ROL}(s)], \tag{31}$$

and the matrix transfer function of $T_{LQ}(s)$ is by definition

$$T_{LQ}(s) = K_c(sI - \overline{A})^{-1}\overline{B}. \tag{32}$$

Now one may increase or decrease the *maximum crossover frequency* (where $T_{LQ}(s)$ crosses 0dB), taking into account that the gains of $T_{LQ}(s)$ increase as $\rho \to 0$.

4.2 Kalman Filter Design

The Kalman Filter for the augmented system $\overline{G}(s)$ is characterized by the following pair of equations:

$$\begin{cases} \dot{\hat{z}} = \overline{A}\hat{z} + \overline{B}u + K_f(\overline{y} - \hat{y}) \\ \hat{y} = \overline{C}\hat{z} \end{cases}. \tag{33}$$

The optimal gains for the Kalman Filter are defined by $K_f = \Sigma \overline{C}^T R_f^{-1}$, in which Σ is the unique solution of the FARE (Filter Algebraic Riccati Equation)

$$0 = \overline{A}\Sigma + \Sigma\overline{A}^T + Q_f - \Sigma\overline{C}^T R_f^{-1}\overline{C}\Sigma, \qquad \Sigma > 0, \tag{34}$$

where $R_f = \rho$, and Q_f is the modified covariance matrix of noise intensities, required for the LTR, and is defined as $Q_f = \overline{B}\,\overline{B}^T + q^2\overline{B}\,\overline{B}^T$, with $q^2 \to \infty$, according to [1], [6].

5 Results of Simulations

Fig. 6 shows the comparison between the gains of $T_{LQ}(s)$ and $K(s)\overline{G}(s)$. One can observe that the principal gains are fully equalized, until they reach the region of higher frequencies, where the uncertainties (not defined here) begin to affect the system. Intending to move the poles, near the origin, further to the left half-plane, we set $\rho = 0.5$, thus the *maximum crossover frequency* becomes $\omega_c = 1.383$ rad/s. In the presented example, for a good approximation of the gains of $K(s)\overline{G}(s)$ to the gains of the robust LQR

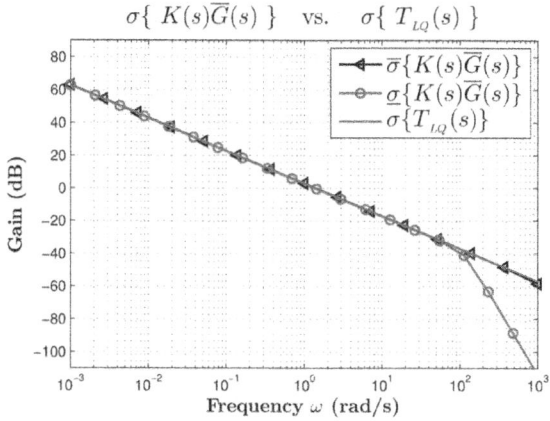

Fig. 6. Frequency response of the regulator $T_{LQ}(s)$ and the open-loop controlled system $K(s)\overline{G}(s)$

regulator $T_{LQ}(s)$, we chose $q^2 = 1 \times 10^{12}$. The gain margin and phase margin achieved are, respectively, GM = 44.54 and PM = 58.52, and the optimal gains of the LQR and the Kalman Filter are given in (35) and (36).

The poles and zeros of $K(s)\overline{G}(s)$ are shown in Table I. Note that the zeros of $K(s)$ and $K(s)\overline{G}(s)$ are the same.

$$K_c = \begin{bmatrix} 0.1927 & 0.2526 & 0.0189 & -0.1754 & -0.6569 & -0.3001 & -0.7436 \\ 0.0774 & 0.0240 & -0.2002 & -0.0705 & 0.0209 & -0.1121 & -0.2185 \\ 0.2681 & 0.3490 & 0.0754 & -0.5811 & -0.8831 & -0.8378 & -0.9997 \end{bmatrix} \tag{35}$$

$$K_f = \begin{bmatrix} -2.5556 & -7.7521 & 0.4842 & -5.8836 & 0.0018 & -0.0000 & 0.0000 \\ -0.0799 & 0.6722 & 6.6735 & 1.2226 & -0.0000 & 0.0014 & -0.0000 \\ -0.0388 & -2.4079 & 0.9860 & -3.7675 & 0.0000 & -0.0000 & 0.0002 \end{bmatrix}^T \times 10^6 \tag{36}$$

Table 1. Poles and Zeros of $K(s)\overline{G}(s)$

Open-loop Poles	Closed-loop Poles	Zeros
$-890.72 \pm 890.72i$	$-890.01 \pm 890.01i$	-
$-706.13 \pm 706.13i$	$-705.42 \pm 705.42i$	-
$-56.71 \pm 98.99i$	$-56.70 \pm 98.18i$	-
-114.806	-113.409	-
$-0.7801 \pm 1.0296i$	$-0.7801 \pm 1.0296i$	$-0.7801 \pm 1.0296i$
$-0.0176 \pm 0.1826i$	$-0.0176 \pm 0.1826i$	$-0.0176 \pm 0.1826i$
0	-1.414	-
0	-1.414	-
0	-1.414	-

Figs.7–9 show the output signals and the control signals in two time intervals, each, for a better analysis.

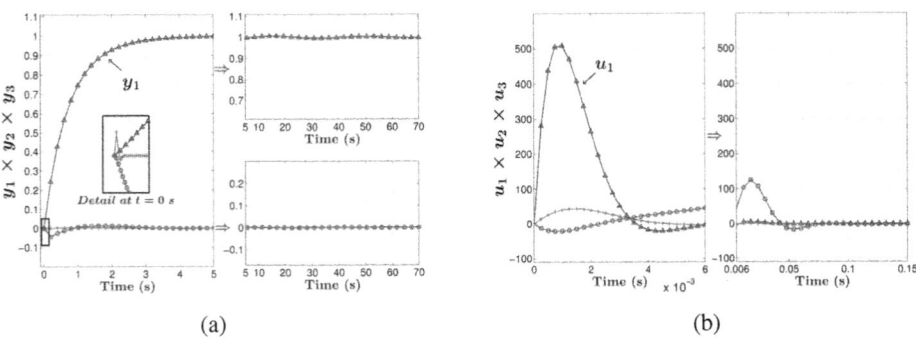

(a)   (b)

Fig. 7. (a) Response of the system to a step input to $Ref.1$; (b) Control signals u_1, u_2 and u_3 for a step input to $Ref.1$

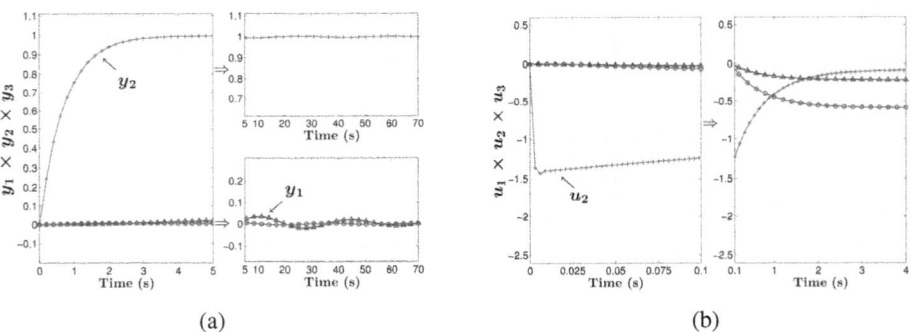

(a)   (b)

Fig. 8. (a) Response of the system to a step input to $Ref.2$; (b) Control signals u_1, u_2 and u_3 for a step input to $Ref.2$.

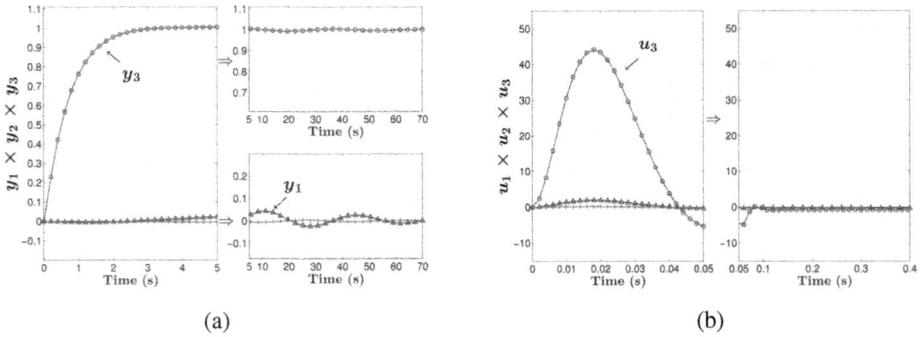

(a)   (b)

Fig. 9. (a) Response of the system to a step input to $Ref.3$; (b) Control signals u_1, u_2 and u_3 for a step input to $Ref.3$.

Analyzing the system's response one may observe that the output signal from a step input to $Ref.1$, which is y_1, requires a faster control action than y_2 or y_3. On the other hand, the outputs y_2 and y_3 remain quite stable from the point of view of $Ref.1$, due to the system decoupling.

By the point of view of $Ref.2$, the output signals y_1 and y_3 exhibit a loose coupling between themselves and the reference input. Similarly, it occurs to y_1 and y_2, analyzing from the point of view of $Ref.3$.

The loose coupling is not exactly an outcome of the mathematical procedure concerning the built-in integrator. But it occurs especially with y_1 (where the built-in integrator was placed), which drifts away from zero more than y_3 or y_2, as one can see in Figs. 8(a)–9(a), respectively. Hence, it becomes evident that in respect to the reference signals of the other channels, the output signal y_1 converges to zero, oscillating during a longer amount of time than the output signals y_2 or y_3. This alludes to the case of complex conjugate poles near the origin, analogously to what happens with the output signal of an underdamped system. In fact, the oscillation frequency of y_1 and its damping ratio are compatible with the open-loop poles of the system, situated next to the imaginary axis (seen in Table 1). However, it does not affect the stability of the system, since all output signals converge to a final value, be it zero, when no input signal is applied, or the reference, as the response to a step input.

6 Conclusion

This work presents the mathematical procedure for obtaining the equalization, in all frequencies, of the principal gains of a dynamic multivariable 3I3O system, with a single pole at the origin, and, the design of its, consequent, reduced order LQG/LTR controller. The procedure is based on the displacement of the natural built-in integrator from the internal structure of the system toward one of its output channels. After the procedure, the system $G(s)$ remains the same, but the representation of the basic structure of the system adopts a special partitioning, which purposefully produces a nonsingular state matrix. Once applied, the procedure allows the implementation of the formula given in (9), which originally only works for systems with $det(A) \neq 0$. An inherited advantage of this procedure is that, when using the natural built-in integrator, it reduces the number of necessary integrators to add to the system to decouple its channels. Consequently, the order of the LQG/LTR controller to be designed becomes smaller.

Acknowledgement. The authors acknowledge CNPq-Brazil for the financial support.

References

1. Doyle, J.C., Stein, G.: Robustness with observers. IEEE Trans. Automat. Contr. AC-24(4), 607–611 (1979)
2. Anderson, B.D.O., Moore, J.B.: Prentice-Hall. Inc, Englewood Cliffs, Englewood Cliffs (1990)
3. Doyle, J.C.: Guaranted margins for LQG regulators. IEEE Trans. Automat. Contr. AC-23(4), 756–757 (1978)

4. Doyle, J.C., Stein, G.: Multivariable feedback design: concepts for a classical/modern synthesis. IEEE Trans. Automat. Contr. AC-26(1), 4–16 (1981)
5. Athans, M.: A Tutorial for the LQG/LTR Method. In: IEEE ACC, Seattle, WA, USA, June 18-20, pp. 1289–1296 (1986)
6. Ridgely, D.B., Banda, S.S.: Introduction to Robust Multivariable Control, Flight Dynamics Laboratory, Wright-Patterson Air Force Base, USA, Final Rep. AFWAL-TR-85-3102 (1986)
7. Cruz, J.J.: Multivariable Robust Control original in portuguese: Controle Robusto Multivarivel, pp. 150–152. EDUSP, São Paulo (1996)
8. Glad, T., Ljung, L.: Control Theory: Multivariable and Nonlinear Methods, 1st edn., pp. 297–298. Taylor & Francis Group (2000)

Fuzzy Fractional PID Controller Tuned through a PSO Algorithm

Isabel S. Jesus and Ramiro S. Barbosa

GECAD – Knowledge Engineering and Decision Support Research Center
Institute of Engineering / Polytechnic of Porto (ISEP/IPP),
Dept. of Electrical Engineering, Porto, Portugal
{isj,rsb}@isep.ipp.pt

Abstract. The differentiation of non-integer order has its origin in the seventeen century, but only in the last two decades appeared the first applications in the area of control theory. In this paper we consider the development of an optimal fuzzy fractional $PD^{\beta}+I$ controller in which the parameters are tuned by a particle swarm optimization (PSO) algorithm. Several simulations are presented assessing the performance of the proposed controller.

Keywords: fractional calculus, fuzzy fractional PID control systems, PSO algorithm, optimization.

1 Introduction

Fractional calculus (FC) was introduced in science on a pure mathematical viewpoint. In fact, FC is a generalization of integration and differentiation to a non-integer order $\alpha \in C$, being the fundamental operator $_aD_t^{\alpha}$, where a and t are the limits of the operation [1, 2]. The FC concepts constitute a useful tool to describe several physical phenomena, such as heat, flow, electricity, magnetism, mechanics or fluid dynamics. Presently, the FC theory is applied in almost all areas of science and engineering, being recognized its ability in bettering the modelling and control of many dynamical systems. In fact, during the last years FC has been used increasingly to model the constitutive behavior of materials and physical systems exhibiting hereditary and memory properties. This is the main advantage of fractional-order derivatives in comparison with classical integer-order models, where these effects are simply neglected.

In this paper we investigate several control strategies based on fuzzy fractional-order algorithms. The fractional-order PID controller ($PI^{\alpha}D^{\beta}$ controller) involves an integrator of order $\alpha \in \Re^+$ and a differentiator of order $\beta \in \Re^+$. It was demonstrated the good performance of this type of controller, in comparison with the conventional PID algorithms. Recently, there have been a lot of researches in the application of fuzzy PID control [3, 4]. The fuzzy method offer a systematic procedure to design controllers for many kind of systems, that leads to a better performance than that of the conventional PID controller. It is a methodology of intelligent control that mimics human thinking and reacting by using a multivalent fuzzy logic and elements of artificial intelligence.

© Springer International Publishing Switzerland 2015   207
A.P. Moreira et al. (eds.), *CONTROLO'2014 - Proc. of the 11th Port. Conf. on Autom. Control*,
Lecture Notes in Electrical Engineering 321, DOI: 10.1007/978-3-319-10380-8_20

Bearing these ideas in mind, the paper is organized as follows. Section 2 gives the fundamentals of fractional-order control systems. Section 3 presents the control and optimization strategies. Section 4 gives some simulations results assessing the effectiveness of the proposed methodology. Finally, section 5 draws the main conclusions.

2   Fractional−Order Control Systems

In the following we provide a background for the remaining of the article by giving the fundamental aspects of the FC, and the discrete integer-order approximations of fractional-order operators.

2.1   Fundamentals of Fractional Calculus

The mathematical definition of a fractional-order derivative and integral has been the subject of several different approaches [1, 2]. One commonly used definition for the fractional-order derivative is given by the Riemann-Liouville definition ($\alpha > 0$):

$$_aD_t^\alpha f(t) = \frac{1}{\Gamma(n-\alpha)} \frac{d^n}{dt^n} \int_a^t \frac{f(\tau)}{(t-\tau)^{\alpha-n+1}} d\tau, \quad n-1 < \alpha < n \quad (1)$$

where $f(t)$ is the applied function and $\Gamma(x)$ is the Gamma function of x. Another widely used definition is given by the Grünwald-Letnikov approach ($\alpha \in \Re$):

$$_aD_t^\alpha f(t) = \lim_{h \to 0} \frac{1}{h^\alpha} \sum_{k=0}^{\left[\frac{t-a}{h}\right]} (-1)^k \binom{\alpha}{k} f(t-kh) \quad (2a)$$

$$\binom{\alpha}{k} = \frac{\Gamma(\alpha+1)}{\Gamma(k+1)\Gamma(\alpha-k+1)} \quad (2b)$$

where h is the time increment and $[x]$ means the integer part of x.

The "memory" effect of these operators is demonstrated by (1) and (2), where the convolution integral in (1) and the infinite series in (2), reveal the unlimited memory of these operators, ideal for modelling hereditary and memory properties in physical systems and materials.

Considering vanishing initial conditions, the fractional *differintegration* is defined in the Laplace domain, as:

$$L\{_aD_t^\alpha f(t)\} = s^\alpha F(s), \quad \alpha \in \Re \quad (3)$$

where $F(s) = L\{f(t)\}$.

2.2 Approximations of Fractional−Order Operators

In this paper we adopt discrete integer-order approximations to the fundamental element s^α ($\alpha \in \Re$) of a fractional-order control (FOC) strategy. The usual approach for obtaining discrete equivalents of continuous operators of type s^α adopts the Euler, Tustin and Al-Alaoui generating functions [5, 6].

It is well known that rational-type approximations frequently converge faster than polynomial-type approximations and have a wider domain of convergence in the complex domain. Thus, by using the Euler operator $w(z^{-1}) = (1-z^{-1})/T_c$, and performing a power series expansion of $[w(z^{-1})]^\alpha = [(1- z^{-1})/T_c]^\alpha$ gives the discretization formula corresponding to the Grünwald-Letnikov definition (2):

$$D^\alpha \left(z^{-1}\right) = \left(\frac{1 - z^{-1}}{T_c}\right)^\alpha = \sum_{k=0}^\infty \left(\frac{1}{T_c}\right)^\alpha (-1)^k \binom{\alpha}{k} z^{-k} = \sum_{k=0}^\infty h^\alpha (k) z^{-k} \quad (4)$$

where T_c is the sampling period and $h^\alpha(k)$ is the impulse response sequence.

A rational-type approximation can be obtained through a Padé approximation to the impulse response sequence $h^\alpha(k)$, yielding the discrete transfer function:

$$H\left(z^{-1}\right) = \frac{b_0 + b_1 z^{-1} + \ldots + b_m z^{-m}}{1 + a_1 z^{-1} + \ldots + a_n z^{-n}} = \sum_{k=0}^\infty h(k) z^{-k} \quad (5)$$

where $m \le n$ and the coefficients a_k and b_k are determined by fitting the first $m + n+1$ values of $h^\alpha(k)$ into the impulse response $h(k)$ of the desired approximation $H(z^{-1})$. Thus, we obtain an approximation that matchs the desired impulse response $h^\alpha(k)$ for the first $m + n+1$ values of k [5]. Note that the above Padé approximation is obtained by considering the Euler operator but the determination process will be exactly the same for other types of discretization schemes.

3 Control and Optimization Strategies

3.1 Fractional PID Control

The generalized PID controller, $G_c(s)$, has a transfer function of the form [7]:

$$G_c (s) = \frac{U(s)}{E(s)} = K_p + \frac{K_i}{s^\alpha} + K_d s^\beta, \qquad \alpha, \beta > 0 \quad (6)$$

where α and β are the orders of the fractional integrator and differentiator, respectively. The parameters K_p, K_i and K_d are correspondingly the proportional, integral, and derivative gains of the controller. Clearly, taking $(\alpha, \beta) = \{(1, 1), (1, 0), (0, 1), (0, 0)\}$ we get the classical $\{$PID, PI, PD, P$\}$ controllers, respectively [8]. Other PID controllers are possible, namely: PD^β controller, PI^α controller, PID^β controller, and so on. The fractional order controller is more

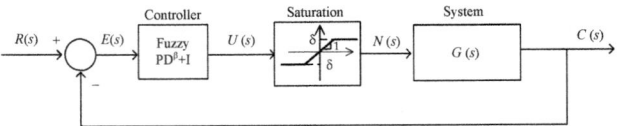

Fig. 1. Block diagram of the fuzzy control system

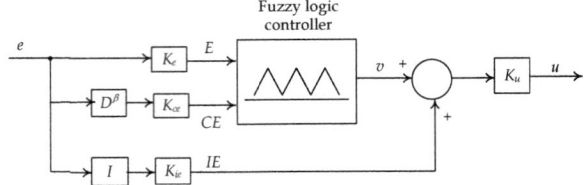

Fig. 2. Fuzzy PD$^\beta$ + I controller

flexible and gives the possibility of adjusting more carefully the closed-loop system characteristics [2].

In the time domain the PI$^\alpha$D$^\beta$ is represented by:

$$u(t) = K_p e(t) + K_i \, {}_0 D_t^{-\alpha} e(t) + K_d \, {}_0 D_t^{\beta} e(t) \tag{7}$$

where the fractional order differential operators may be implemented using the approximations (4) and/or (5).

3.2 Fuzzy Fractional PD+I Control

Fuzzy control emerged on the foundations of Zadeh's fuzzy set theory [3, 4]. This kind of control is based on the ability of a human being to find solutions for particular problematic situations. It is well know from our experience, that humans have the ability to simultaneously process a large amount of information and make effective decisions, although neither input information nor consequent actions are precisely defined. Through multivalent fuzzy logic, linguistic expressions in antecedent and consequent parts of IF-THEN rules describing the operator's actions can be efficaciously converted into a fully-structured control algorithm.

In the system of Fig.1, we apply a fuzzy logic control (FLC) for the PD$^\beta$ actions and the integral of the error is added to the output in order to find a fuzzy PD$^\beta$+I controller [4]. In fact, we can have a fuzzy PD$^\beta$+I$^\alpha$ controller, *i.e.* an added fractional integral action, as presented in [9]. The block diagram of Fig. 2 illustrates the configuration of the proposed fuzzy controller.

In this controller, the control actions are the error e, the fractional derivative of e and the integral of e. The U represents the controller output. Also, the controller has four gains to be tuned, K_e, K_{ie}, K_{ce} corresponding to the inputs and K_u to the output.

The control action U is generally a nonlinear function of error E, fractional change of error CE, and integral of error IE:

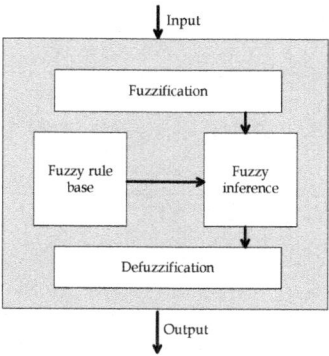

Fig. 3. Structure for fuzzy logic controller

$$U(k) = [f(E, CE) + IE]K_u =$$

$$\left[f\left(K_e e(k), K_{ce}D^\beta e(k)\right) + K_{ie}Ie(k)\right]K_u \tag{8}$$

where D^β is the discrete fractional derivative implemented as rational approximation (5) using the Euler scheme (4); the integral of error is calculated by rectangular integration:

$$I\left(z^{-1}\right) = \frac{T_c}{1 - z^{-1}} \tag{9}$$

To further illustrate the performance of the fuzzy PD^β+I a saturation nonlinearity is included in the closed-loop system of Fig.1, and inserted in series with the output of the fuzzy controller. The saturation element is defined as:

$$n(u) = \begin{cases} u, & |u| < \delta \\ \delta \, sign(u), & |u| \geq \delta \end{cases} \tag{10}$$

where u and n are respectively the input and the output of the saturation block and $sign(u)$ is the signum function.

Here we give an emphasis of the proposed FLC presented in Fig. 2. The basic structure for FLC is illustrated in Fig. 3. The fuzzy rule base, which reflects the collected knowledge about how a particular control problem must be treated, is one of the main components of a fuzzy controller. The other parts of the controller perform make up the tasks necessary for the controller to be efficient.

For the fuzzy PD^β+I controller illustrated in Fig.2, the rule-base can be constructed in the following form (see Table 1):

If E is NM and CE is NS Then u is NL

where NL, NM, NS, ZR, PS, PM, and PL are linguistic values representing "negative low", "negative medium" and so on, E is the error, CE is the fractional derivative of error and u is the output of the fuzzy PD^β controller. The membership functions for the premises and consequents of the rules are shown in Fig. 4 a).

Table 1. Fuzzy control rules

$E \setminus$ CE	NL	NM	NS	ZR	PS	PM	PL
NL	NL	NL	NL	NL	NM	NS	ZR
NM	NL	NL	NL	NM	NS	ZR	PS
NS	NL	NL	NM	NS	ZR	PS	PM
ZR	NL	NM	NS	ZR	PS	PM	PL
PS	NM	NS	ZR	PS	PM	PL	PL
PM	NS	ZR	PS	PM	PL	PL	PL
PL	ZR	PS	PM	PL	PL	PL	PL

With two inputs and one output the input-output mapping of the fuzzy logic controller is described by a non linear surface, presented in Fig.4 b).

The fuzzy controller will be adjusted by changing the parameter values of K_e, K_{ce}, K_{ie} and K_u. The fuzzy inference mechanism operates by using the product to combine the conjunctions in the premise of the rules and in the representation of the fuzzy implication. For the defuzzification process we use the centroid method.

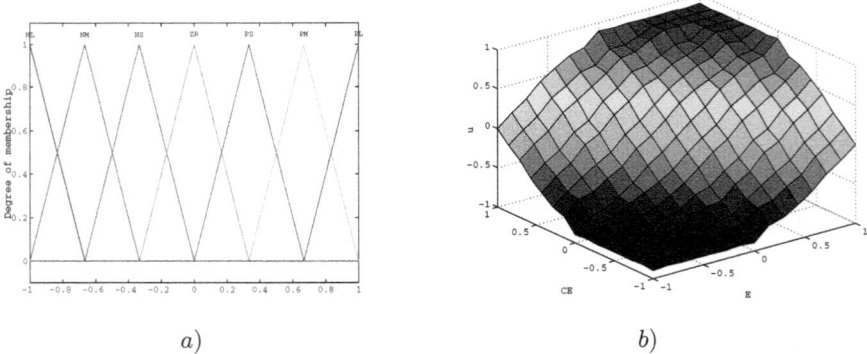

a)   b)

Fig. 4. a) Membership functions for E, CE and u, b) Control surface

3.3 Particle Swarm Optimization

The PSO is one of the latest evolutionary techniques developed by Dr. Eberhart and Dr. Kennedy in 1995, inspired by social behaviour of bird flocking or fish schooling. The PSO scheme optimizes searching by virtue of the swarm intelligence produced by the cooperation and competition among the particles of a species [10, 11]. The social system is discussed through the collective behaviours of simple individuals interacting with their environment and each other. Examples of this are the bird flock or fish school. Some applications of PSO are found in the field of nonlinear dynamical systems, data analysis, electrical engineering, function optimization, artificial neural network training, fuzzy control and many others in real world applications [11, 12].

It is important to mention that a reliable execution and analysis of a PSO usually requires a large number of simulations to provide that stochastic effects have been properly considered. In the PSO algorithm, all particles represent a potential solution to a problem, which is performed by adjusting their position taking into account both personal and group experiences [11, 12]. In each iteration, the velocity (v) is actualized by expression (11a). The new particle position $(cp_k + 1)$ is found by adding their actual position with the new velocity, as shown in (11b).

$$v_{k+1} = w * v_k + c_1 * r_1 (lbp - cp_k) + c_2 * r_2 * (gbp_k - cp_k) \tag{11a}$$

$$cp_{k+1} = cp_k + v_{k+1} \tag{11b}$$

where ω is the inertia weight, c_1 and c_2 are the individual and sociality weights coefficients for modelling attractive forces from the local and global best, respectively. Usually, $c_1 = c_2$ and ranges from $[0, 4]$ [13]. r_1 and r_2 are aleatory numbers between $[0, 1]$, lbp is the local best position, cp is the current position, and gbp is the global best position [10, 11]. The inertia term, forces the particle to move in the same direction as before by adjusting the old velocity. The cognitive term (personal best), forces the particle to go back to the previous best position. On the other hand, the social learning term (global best), forces the particle to move to the best previous position of its neighbors. The PSO optimizes an objective function by iteratively improving a swarm of solutions, called particles, based on special management of memory. Each particle is modified by referring to the memory of individual swarm's best information. Due to the collective intelligence of these particles, the swarm is able to repeatedly improve its best observed solution and converging to an optimum.

In this work we propose a fuzzy fractional $PD^\beta + I$ controller, where the gains will be tuned through the application of the PSO algorithm. The optimization fitness function corresponds to the minimization of the Integral Time Absolute Error (ITAE) criterion, that measures the response error as defined as [8]:

$$J(K_e, K_{ce}, K_{ie}, K_u) = \int_0^\infty t |r(t) - c(t)| \, dt \tag{12}$$

where $(K_e, K_{ce}, K_{ie}, K_u)$ are the $PD^\beta + I$ controller parameters to be optimized.

4   Simulations

In this section we analyze the closed-loop system of Fig. 1 with a fuzzy fractional $PD^\beta + I$ controller (Fig. 2). In all the experiments, the fractional order derivative

D^β in scheme of Fig. 2 is implemented by using a 4th order Padé discrete rational transfer function ($m = n = 4$) of type (5). It is used a sampling period of $T_c = 0.1$ s. The PD^β+I controller is tuned through the minimization of the ITAE (12) using a PSO. We use $\delta = 15.0$. Based in current studies we establish the following values for the PSO parameters: population number $PN = 25$, $c_1 = c_2 = 1$, $\omega = 0.9$ and a maximum number of iterations $I_{Max} = 50$. For guarantee that stochastic effects are properly considered, the experiments consist on executing the PSO 10 times, and we get the best result, *i.e.* the simulation that leads to the smaller error J.

In the first case, we compare a fuzzy fractional PD^β+I controller which leads to the lower error ($\beta = 0.9$), with a fuzzy integer PD+I controller ($\beta = 1$). Figure 5 a) shows the unit step responses of both controllers. The plant system $G_1(s)$ used is represented by a high-order transfer function [14]:

$$G_1(s) = \frac{1}{(s+1)(1+0.5s)(1+0.5^2s)(1+0.5^3s)} \tag{13}$$

The controller parameters, corresponding to the minimization of the ITAE index, lead to the values for the fuzzy integer PD+I controller: $\{K_e, K_{ce}, K_{ie}, K_u\} \equiv \{ 1.0705, 3.8194, 0.4206, 4.4784 \}$, with $J = 2.3763$, and for the fuzzy fractional PD^β+I controller to the following values:$\{K_e, K_{ce}, K_{ie}, K_u\} \equiv \{0.7193, 0.3551, 0.2702, 3.2029\}$, with $J = 0.8409$. These values lead us to conclude that the fuzzy fractional order controller produced better results than the integer one, since the transient response (namely, the settling time, rise time and overshoot) and the error J are smaller, as can be seen in Fig. 5 a).

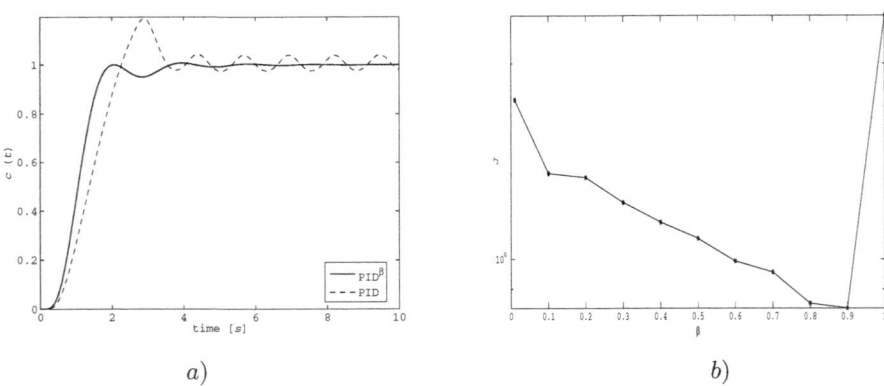

a)   b)

Fig. 5. a) Step responses of the closed-loop system, with a fuzzy PD+I and PD^β+I ($\beta = 0.9$) controllers, b) Error J versus β for $G_1(s)$

Figure 5 b) shows the ITAE error as function of β. The graph shows that the lowest error is produced for $\beta = 0.9$. We also see that the fractional controller is always better than its integer version considering $0 < \beta \leq 1$.

In a second experiment, we consider a fuzzy PD$^\beta$+I controller which leads for lower error to $\beta = 0.8$, applied to a process $G_2(s)$ represented by a non-minimum system given by the transfer function (14) [14].

$$G_2(s) = \frac{1 - 5s}{(s + 1)^3} \tag{14}$$

Once again, we consider for comparison the corresponding integer version ($\beta = 1$). Figure 6 a) shows the unit step responses of both controllers.

The controller parameters, corresponding to the minimization of the ITAE index, lead to the values for the fuzzy integer controller: $\{K_e, K_{ce}, K_{ie}, K_u\} \equiv$ $\{0.9953, 1.3058, 0.4360, 0.1991\}$, with $J = 70.9016$, and for the fuzzy fractional controller: $\{K_e, K_{ce}, K_{ie}, K_u\} \equiv \{0.0873, 0.0569, 0.0417, 2.5740\}$, with $J = 35.4189$. These values lead us to remain the previously conclusions drawn for $G_1(s)$, namely that the fuzzy fractional order controller produced better results than the integer one, since the transient response (in particular the settling time, rise time and overshoot) and the error J are smaller. Figure 6 b) shows the ITAE error as function of β. The graph shows a lowest error for $\beta = 0.8$. Also, the fractional controller (for $0 < \beta \leq 1$) is always better than the conventional fuzzy PID controller.

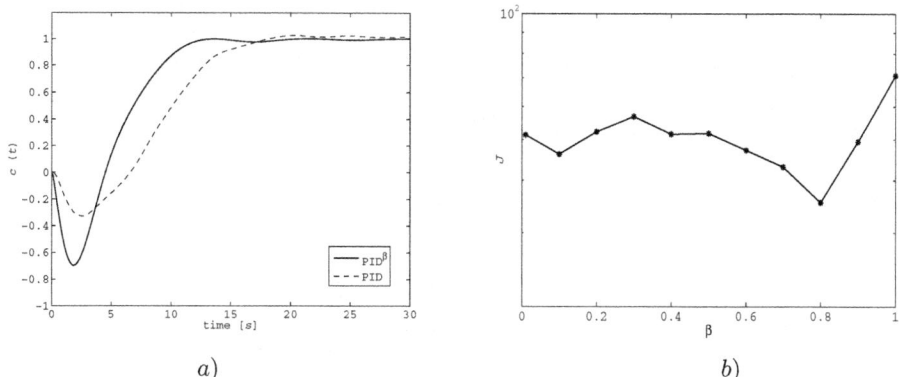

Fig. 6. a) Step responses of the closed-loop system, with fuzzy PD+I and PD$^\beta$+I ($\beta = 0.8$) controllers, b) Error J versus β for $G_2(s)$

In conclusion, with the fuzzy fractional PD$^\beta$+I controller we get the best controller tuning, superior to the performance revealed by the integer-order scheme.

5   Conclusion

This paper presented a fuzzy PD$^\beta$+I controller in which the parameters were tuned through a PSO algorithm. The control strategies presented, give better results than those obtained with conventional integer control structures, showing its effectiveness in the control of nonlinear feedback systems. These results points out the use of other values for the saturation element and/or the use of other type of systems and other $PI^\alpha D^\beta$ configurations which will be pursued in future research.

Acknowledgements. This work is supported by FEDER Funds through the "Programa Operacional Factores de Competitividade - COMPETE" program and by National Funds through FCT "Fundação para a Ciência e a Tecnologia".

References

[1] Oldham, K.B., Spanier, J.: The Fractional Calculus: Theory and Application of Differentiation and Integration to Arbitrary Order. Academic Press, New York (1974)

[2] Podlubny, I.: Fractional Differential Equations. Academic Press, San Diego (1999)

[3] Mizumoto, M.: Realization of PID controls by fuzzy control methods. Journal of Fuzzy Sets and Systems 70, 171–182 (1995)

[4] Barbosa, R.S.: On linear fuzzy fractional PD and PD+I controllers. In: The 4th IFAC Workshop Fractional Differentiation and its Applications - FDA 2010 (2010)

[5] Barbosa, R.S., Machado, J.A.T., Silva, M.F.: Time domain design of fractional differintegrators using least-squares. Signal Processing 86 (10), 2567–2581 (2006)

[6] Chen, Y., Vinagre, B.M., Podlubny, I.: Continued fraction expansion to discretize fractional order derivatives-an expository review. Nonlinear Dynamics 38 (1–4), 155–170 (2004)

[7] Podlubny, I.: Fractional-order systems and $PI^\lambda D^\mu$-controllers. IEEE Transactions on Automatic Control 44(1), 208–213 (1999)

[8] Jesus, I.S., Machado, J.A.T.: Fractional control of heat diffusion systems. Journal Nonlinear Dynamics 54(3), 263–282 (2008)

[9] Barbosa, R.S., Jesus, I.S.: A methodology for the design of fuzzy fractional PID controllers. In: 10th International Conference on Informatics in Control, Automation and Robotics - ICINCO 2013, pp. 276–281 (2013)

[10] Kennedy, J., Eberhart, R.C.: Particle swarm optimization. In: IEEE International Conference on Neural Networks IV, pp. 1942–1948 (1995)

[11] Eberhart, R.C., Shi, Y.: Particle swarm optimization: developments, applications and resources. Congress on Evolutionary Computation (2001)

[12] Jesus, I.S.: Application of PSO algorithm for the study of fractional order electrical potential. In: 18th IFAC World Congress - IFAC-WC 2011, pp. 10812–10817 (2011)

[13] Liu, Y., Sun, A., Loh, H.T., Lu, W.F., Lim, E.-P.: Advances of Computational Intelligence in Industrial Systems. Springer (2008)

[14] Ang, K.H., Chong, G.: PID control system analysis, design, and technology. IEEE Transactions on Control Systems Technology 13, 559–576 (2005)

Arduino Implementation of Automatic Tuning in PID Control of Rotation in DC Motors

Rodrigo Nuno Mendes Antunes[1], David Santinhos Ferreira[1],
Inês Isabel Gonçalves Santos[1], Ana Rita Das Neves Sousa[1],
and José Soares Augusto[1,2]

[1] Dep. de Física da Faculdade de Ciências da Universidade de Lisboa, Campo
Grande, 1749-016, Lisboa, Portugal
[2] Inesc-ID, Rua Alves Redol, 9, 1000-029, Lisboa, Portugal
jaaugusto@fc.ul.pt

Abstract. In this paper is described a laboratory control experiment[1]
that targets, through a practical "hands-on" approach, several important
control matters: PID control; digital control; controller design; automatic
tuning; nonlinear control and the describing function; and, above all,
the *practical* implementation issues which arise in real projects. Both
hardware and software scopes of the project are presented and discussed.
The inexpensive Arduino is used as the implementation platform, what
helps the interested readers in developing a similar experiment.

1 Introduction

The Arduino family of boards [1, 2] is a popular platform for implementing all
sorts of microcontroller based projects, both at the enthusiast, the academic,
and even the industrial levels. The reasons for its popularity are the low cost,
open engineering details, the fair performance versus cost ratio and the large
"ecosystem" of users and applications surrounding Arduino: there are many
available low-cost, or even free, add-ins (software and hardware) and, in forums,
the users' questions are promptly answered.

On the controller's front, the PID controller is pervasive in industrial environments (in [4] are cited studies reporting that the penetration of PID controllers
in mills was, a few years ago, well above 80 %), and is one of the most studied
"classical" controllers. Even today it is the support of involved automatic tuning
and/or adaptive techniques and algorithms [11, 12]. So, it is a natural choice for
being used in a laboratory experiment focusing digital controllers.

The standard control loop with unitary feedback is shown in Fig. 1. C is the
controller and G is the *plant* which encompasses the process to be controlled
itself as well as the actuator and sensor systems. The *reference input* is r, the
output is y, the quantity $e = r - y$ is the *error* and u is the *actuating signal*.

---

[1] This experiment was developed in the aim of the course "Controlo e Arquitecturas de
Sistemas de Instrumentação" (CASI) taught at the *Dep. de Física da Fac. de Ciências
da Un. de Lisboa* (DF-FCUL) to students engaged in the MSc in Engineering Physics:
http://www.fc.ul.pt/pt/cursos/mestrado-integrado/engenharia-fisica

© Springer International Publishing Switzerland 2015   217
A.P. Moreira et al. (eds.), *CONTROLO'2014 - Proc. of the 11th Port. Conf. on Autom. Control*,
Lecture Notes in Electrical Engineering 321, DOI: 10.1007/978-3-319-10380-8_21

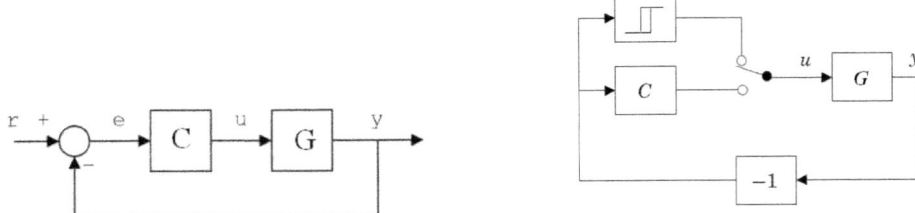

Fig. 1. <u>Left</u>: feedback control system with unitary feedback loop. C is the controller and G represents the plant. <u>Right</u>: architecture of a feedback control system equipped with automatic tuning (the input reference, $r(t) = 0$, is omitted from the picture). In normal operation, the controller C is in the loop, driving the plant G, and the switch is engaged in the lower position; when the switch is in the upper position (as shown) auto-tuning is in action.

Assuming linearity, the *transfer function* between the input r and the output y is easily calculated as

$$H(s) = \frac{Y(s)}{R(s)} = \frac{C(s)G(s)}{1 + C(s)G(s)} \tag{1}$$

Control practice has shown that often $G(s)$ is actually not known, what obviously prevents designing the controller. In this case, $G(s)$ is identified experimentally either explicitly or implicitly. Open loop experiments often are difficult to be performed: for instance, if there are integration effects in the signal path, parts of the system eventually saturate before measurements are finished. So, *closed loop experiments for identifying $G(s)$ are preferred* whenever possible.

The measurements either serve to *explicitly* identify the parameters of the system (after a reasonable model has been chosen as target) or instead are used to set the parameters of the controller without identifying the system (*implicit* identification). This last case applies to the Ziegler-Nichols rules [3, 4, 6, 7], a recipe which proposes values for the parameters of the PID controller based on results of an oscillation experiment.

The oscillation experiment proposed by Ziegler and Nichols, based in a proportional controller $C(s) = K_P$, is not very suitable for digital implementation. Here we use an effective autonomous oscillation alternative [3, 4, 8]: $C(s)$ is replaced by a relay (i.e. a nonlinear "hard" comparator, Fig. 1) while $r(t) = 0$, which induces auto-oscillations whose amplitude and frequency are measured. By changing the output saturation level, M, of the relay, several oscillation frequencies and amplitudes can be recorded. This information coupled with a predefined model for the plant $G(s)$ (see, as an example, (6) in page 222) allows designing the controller.

The organization of this paper is as follows: in the next section we present and discuss the design of the PID controller, in section 3 we discuss automatic tuning, in section 4 we present the experimental setup and the associated hardware, in section 5 we tackle concrete issues in practical identification and tuning, in section 6 programming details associated to Arduino implementation are discussed and, finally, we present results and some worthy conclusions.

2   PID Controllers

The classical "textbook" version of the PID controller is (Fig. 2):

$$u(t) = K_P\, e + K_I \int e + K_D\, \dot{e} = K\left(e(t) + \frac{1}{T_I}\int e(\tau)d\tau + T_D\frac{d\,e(t)}{dt}\right)$$

(where $K = K_P$, $K/T_I = K_I$ and $K\,T_D = K_D$). In the frequency domain, it is:

$$U(s) = K\left(1 + \frac{1}{s\,T_I} + s\,T_D\right)E(s) \tag{2}$$

In practice there are some variations upon (2): (i) a limit N_D is imposed to the high-frequency gain of the derivative (d/dt) action such that $T_D\,s$ becomes $T_D s/(1+T_D s/N_D)$; the derivative action is applied to $-y$ instead of to $e = r - y$; (iii) other safeguards are implemented, such as preventing windup caused by integration action. Details on these issues abound in the literature [5–7, 9].

There are many ways to convert the analogue PID into a digital PID [6, 7, 5, 9]. In this work we use the backward difference approximation of the derivative $s \Leftrightarrow \frac{1-z^{-1}}{T}$ on the analogue PID with gain limit in the derivative term. The transfer function in the $\mathcal{Z}$ domain is (where T is the sampling period):

$$C(z) = \frac{U(z)}{E(z)} = K\left[1 + \frac{(T/T_I)}{(1 - z^{-1})} + \frac{(T_D/T)(1 - z^{-1})}{1 + (T_D/T)(1 - z^{-1})/N_D}\right]$$

Since there are three free parameters in the PID controller, the use of design methods such as the root locus (RL) is not easy. Frequency based design using the phase (PM) or gain margins (GM) is possible, by making some assumptions on the relative magnitude of the parameters, but the quality of the design is strongly influenced by the type of the plant. Algebraic dominant pole

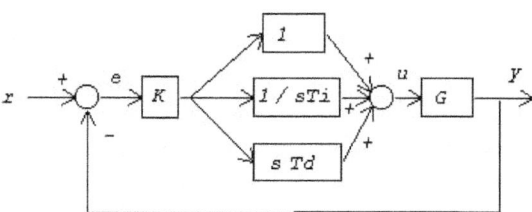

Fig. 2. Feedback control system with unitary feedback loop, showing a PID controller $C(s)$ with transfer function $C(s) = U(s)/E(s) = K(1 + 1/s\,T_I + s\,T_D)$ given by (2)). G represents the plant.

placement (2 dominant poles plus an extra pole), although less "elegant", gives more promising results [3, 4].

In practice $G(s)$ often is not known exactly and it must be measured, with either open loop or closed loop experiments. Opening the control loop can be difficult or impossible due to several reasons, e.g. $G(s)$ is changing "slowly" (a chemical reaction behaves like this). So, the ability of doing closed loop measurements and using these to "tune" the controller has a remarkable practical value.

The better known rules for tuning the PID controller with a closed loop experiment are the Ziegler-Nichols (ZN) rules [3, 4, 6, 7, 9]. They were designed to reduce the sensitivity of the feedback system to load (process) variations, and so they are not very suitable with respect to input (reference) variations: there is usually a large overshoot in the output and a small PM in the compensated system (around 25°). Nevertheless, ZN rules are used here as a typical example.

For performing automatic tuning of the PID controller one should know the plant $G(s)$ – a *System Identification Problem* (SIP). This issue, as well as automatic tuning itself, are presented in the next section. However, it shall be stressed that there is no need of identifying $G(s)$ when applying the ZN rules.

3   Automatic Tuning of PID Controllers

For tuning the PID controller with the ZN rules a proportional controller $C(s) = K_P$ is applied in the closed loop system (Fig. 1) [3–6, 8].

It is assumed that by increasing K_P the feedback system becomes unstable and oscillates with a period T_u – the *ultimate period*. This happens for $K_P = K_u$, where K_u is the *ultimate gain*. The ZN frequency response rules propose that $K = 0.6K_u$, $T_I = 0.5T_u$ and $T_D = 0.12T_u$ for a PID; $K = 0.4K_u$ and $T_I = 0.8T_u$ for a PI controller; and $K = 0.5K_u$ for a simple proportional controller [3, 4, 6].

Although easy to do manually (a knob is rotated and oscillation is monitored with an oscilloscope), the task of increasing K_P up to K_u is not very practical in digital implementations of the method. The *relay method* is preferred.

The Relay Method – The *relay method* [3, 4, 8], used in our experiment, allows either identifying $G(s)$ or apllying the ZN rules.

The linear controller $C(s)$ in Fig. 1 is replaced by the *relay* (Fig. 1), with saturation M, such that

$$u(t) = M \, \text{sign}[e(t)]$$

$u(t)$ is *insensitive* to the amplitude of $e(t)$, being sensitive only to its sign.

Oscillation in Linear Systems and the Barkhausen Criterion – If the transfer function in eq. (1) has a pair of conjugate imaginary poles (its denominator is zero when $s = \pm j\omega_0$) the system behaves as a linear oscillator. The Barkhausen criterion states that at the frequency of oscillation, ω_0, the loop gain is one: $-C(j\omega_0)G(j\omega_0) = 1$. The Barkhausen criterion doesn't tell the *amplitude* of the oscillations. This amplitude can be approximated by doing a nonlinear analysis based in the *describing function* [10, 9].

The Describing Function – If $C(s)$ in Fig. 1 is replaced by a nonlinear element $n(.)$, such as a relay, then $u(t) = n[e(t)]$. When $e(t)$ is a sinusoid $e(t) = A\sin(\omega t)$ and is feeding the relay, the control $u(t)$ has a Fourier series description [10]

$$u(t) = M\,\text{sign}[A\sin(\omega t)] = \sum_{n=1}^{\infty} a_n \cos(n\omega t) + \sum_{n=1}^{\infty} b_n \sin(n\omega t) \qquad (3)$$

The *describing function* $N(A)$, a kind of a "generalized" linear gain, is defined as the quotient $(b_1 + ja_1)/A$. The relay is a static odd nonlinearity, and the relevant first harmonic amplitudes are $a_1 = 0$ and $b_1 = 4M/\pi$. Then

$$N(A) = \frac{4M}{\pi A} \qquad (4)$$

Other nonlinearities give describing functions depending on ω, so $N \equiv N(A, \omega)$.

Nonlinear Oscillator – We make a further assumption which is a good approximation in real systems: the plant $G(s)$ has a low-pass nature and highly attenuates the higher harmonics ($n\omega$, for $n > 1$) in (3) and so $y(t) \approx e(t) = A\sin(\omega t)$. So, higher harmonics in the system only are relevant in $u(t)$ and are readily blocked by the plant. We apply the Barkhausen criterion to the first harmonic and obtain the oscillation condition for $\omega = \omega_0$

$$- N(A)G(j\omega_0) = 1 \qquad \Longleftrightarrow \qquad \frac{\pi A}{4M} = |G(j\omega_0)| \wedge \angle G(j\omega_0) = \pm\pi \qquad (5)$$

Identifying the Plant – In general, the user changes the amplitude of the relay, M, and each value M_i causes a certain oscillation amplitude A_i and frequency ω_i. By using least-squares fitting (LS) and the measured values, the unknown parameters of the model (such as (6)) can be found. Identification will be discussed below under the framework of the motor model in eq. (6).

4 The Experimental Setup

Electronics – A photo of the experiment, with named blocks, is shown in Fig. 3. The electronics system developed for the experiment is in Fig. 4.

The "pseudo-analogue" output of the Arduino is a Pulse Width Modulation (PWM) signal, with 0/5 Volt levels and 8 bits of precision. This signal is low-pass filtered with R1-C1 ($f_c \approx 43$ Hz), to get the average (or DC) analogue value, between 0 and 5 Volt. For driving the 12 V motor, a non-inverter amplifier with $G_V \approx 2.4$ and a custom power (1.5 A) amplifier with $G_V = 1$ are cascaded.

The circuit in the bottom of the schematic in Fig. 4 converts light pulses into electronic pulses whose frequency is proportional to the angular velocity of the motor. It comprises: a photo sensitive device: a high pass filter; and a LM311-based comparator with threshold at 0 Volt and a 0/5 Volt output. With this signal a tachometer is programmed in Arduino.

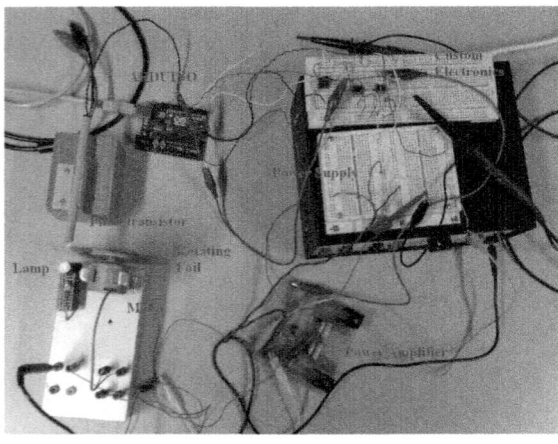 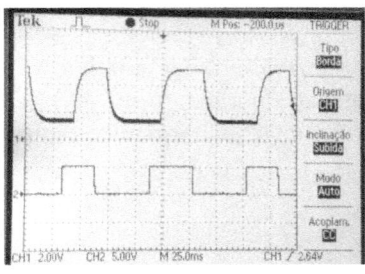

Fig. 3. Left: experimental setup (prototype) with the identified blocks. The "custom electronics" comprises the devices shown in Fig. 4. Right: the system operating in automatic tuning mode (oscillation with relay), where the frequency is approximately 12 Hz. The upper signal is the input of the LF356 (Fig. 4) – i.e. it is the PWM signal, switching between the two relay saturations, after being filtered by the low-pass R1-C1.

The DC Motor – A DC motor[2] rated with 3200 rpm @ 12 V, approximately, is used. The rotor has attached a circular foil with printed notches, such that $360°$ corresponds to 30 notches (Fig. 3). The curve of notch frequency (in Hertz) vs. the applied DC voltage is shown in Fig. 4. The approximated[3] coordinates of the endpoints are: $v_{min} \approx 3$ V; $v_{max} \approx 12$ V; $f_{min} \approx 315$ Hz; $f_{max} \approx 1580$ Hz. From now on, these values are used in the calculations. The motor doesn't work for voltages below v_{min} due to the static friction force.

The DC motor model (frequency vs. voltage, $\Omega(s)/V(s) \equiv Y(s)/U(s)$) is a first order low-pass multiplied by e^{-sL} for modelling apparent delays [3–5]:

$$G(s) = \frac{G_0}{1 + s\tau} e^{-sL} \tag{6}$$

5   Plant Identification and PID Tuning

Due to the pedagogical nature of the experiment, the simple averaging of several consecutive measurements (N_{meas} observations) is implemented in the experiment, in order to implement some robustness to noise. More elaborated methods of reducing variance in measurements are described in the literature [3–5].

A single value of M is used in the relay, since it is enough for guessing the parameters in (6).

---

[2] MMI-6S2RT, used in old tape recorders.

[3] It was verified that the f(v) curve of the motor was quite sensitive to temperature and time of operation: the presented curve is a typical one.

1. To get the static gain G_0 (the slope in Fig. 4), two DC voltages V_a and V_b are applied to the motor. They are kept constant until the frequency of the motor stabilizes, and the corresponding frequencies $\hat{f}_a$ and $\hat{f}_b$ are read. Then

$$G_0 = \frac{\hat{f}_b - \hat{f}_a}{V_b - V_a}$$

Note that this evaluation of G_0 is of the open-loop type.

2. Then, the relay is inserted in the closed loop and the frequency of oscillation, ω_u, is measured. This oscillation in the filtered PWM output of the Arduino is shown as the top signal in Fig. 3.

3. Still in self oscillation mode, the maximum $\hat{f}_{max}$ and minimum $\hat{f}_{min}$ of the frequency of the optical pulses measured by the tachometer are used to evaluate the amplitude A necessary for using the describing function (4). Indeed, $A_f = (\hat{f}_{max} - \hat{f}_{min})/2$ can be multiplied by 0.65 to get A (Fig. 5).

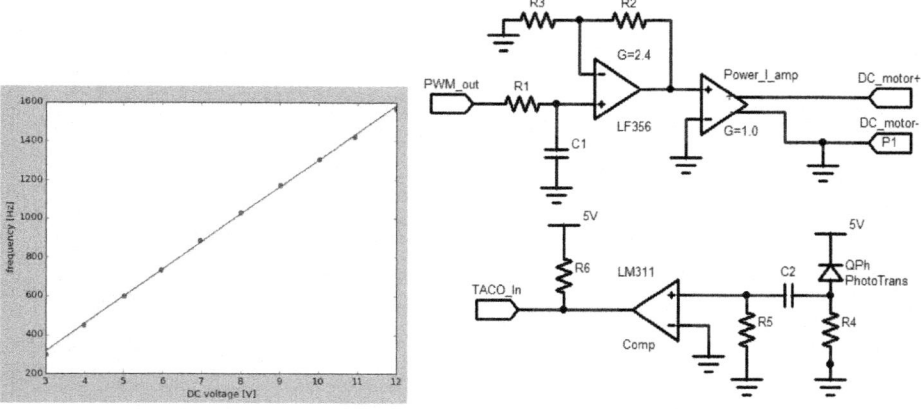

Fig. 4. Left: experimental $f(v)$ curve for the motor, where f is the output frequency, in Hertz, and v is the applied voltage, in Volt. The linear fit (the r coefficient is 0.9997) has an intercept of -105.0 and the slope is 140.5. Right: electronic system developed for the experiment. On top: Arduino PWM output driving the DC motor. At the bottom, the opto-electronics system associated to frequency measurement. The amplifiers and comparator are powered with ± 15 V.

Identification – At this point we can identify the parameters of the model (6). Using (5):

$$\frac{G_0}{\sqrt{1 + (\omega_u \tau)^2}} e^{-j[\arctan(\omega_u \tau) + \omega_u L]} = \frac{\pi A}{4M} e^{-j\pi}$$

from where it is easy to obtain the unknown parameters, τ and L:

$$\tau = \frac{1}{\omega_u} \sqrt{\left(\frac{4M G_0}{\pi A}\right)^2 - 1} \qquad L = [\pi - \arctan(\omega_u \tau)]/\omega_u \qquad (7)$$

PID Tuning – In the relay oscillation approach, the ZN rules are applied in a way similar to the "ultimate gain and ultimate period" approach. We set $T_u = 2\pi/\omega_u$, since this is the frequency of oscillation. On the other hand, under the ZN oscillation experiment with $K_P \equiv K_u$, the Barkhausen criterion indicates that the modulus of the loop gain is $K_u|G(j\omega_u)|$ while in the relay approach it is $N(A)|G(j\omega_u)| = \frac{4M}{\pi A}|G(j\omega_u)|$. The similarity leads to the conclusion that the ZN rules are applied with $K_u = \frac{4M}{\pi A}$ [3, 4, 8].

Certainly other design techniques are possible. The application of pole placement, refined ZN rules and several other techniques are described in [3, 4, 8].

Finally, it is important to notice that tuning can be reached without identification, which is a serious advantage of this approach when the model of the plant is either unknown or much more involved than (6).

6   Arduino Specifications and Programming

An Arduino UNO, rev. 3, (UNO-R3) board [1] was used in the experiment here reported. The next characteristics apply to this board, but many other boards are available. The UNO-R3 was not tweaked in any way: for instance, it is possible to increase the sampling frequencies of the internal ADCs and DACs [2] by writing directly on the registers of the processor ATmega328 (16 MHz clock), but that was not done here. That tweaking path can be followed, however, if faster systems are to be controlled with this board.

The sampling period of the digital controller was set to $T = 10$ ms (f_s=100 Hz), what conforms with the ratings of the ADCs and DACs of the board.

ADCs and DACs – The UNO-R3 has 6 analogue input channels. The 10 bit ADCs work at 10 KHz and accept input in a range from 0 up to 5 Volt.

The pseudo-analogue outputs are of the PWM type and were already described: precision is 8 bits, range is from 0 up to 5 Volt and output rate is around 1 KHz.

Interrupts – There are two interrupts available. One of them (INT0, attached to input pin 2) was used as the input for the algorithmic tachometer; in this way the error in frequency reading is minimized.

Numerics – In the implementation it is necessary to use constants to convert between the analogue and digital variables involved. A diagram of this issue is in Fig. 5. The reference frequency is set with a potentiometer, which applies a voltage between 0 and 5 Volt to one of the Arduino's analogue inputs. As soon as this voltage is read by Arduino, it is converted to an unsigned 10 bit scale, i.e. to a 0-1023 range. The same reasoning applies to the frequency in Hertz read by the software tachometer – it is converted for the same 10 bit range.

On the output front, the Arduino has a 8 bit PWM DAC, and so any calculated output of the controller which is out of the 0-255 range is immediately "saturated" for one of these limits. We present below snippets of the the code calling a PI controller (in accordance with expression (2) but with the last term, i.e. the derivative term, removed):

```
...
un = controller_pi(err);
analogWrite(pwmOut,(int) un);    // delivers PWM output
...
// PI controller
float controller_pi(float e) {
    float Tsecs=Ts/1000;    // from ms to seconds
    // satur() implements numeric saturation in the PWM output
    float usatur = satur(uant - Kp*eant + (Kp + Tsecs*Ki)*e );
    eant=e;
    uant=usatur;    // updates previous values, e(n-1) and u(n-1)
    return usatur;
}
...
// numeric output of Arduino is unsigned 8 bits: so 0 <= out < 256
float satur(float x) {
    if (x < 0) return 0;
    if (x > 255) return 255;
    return x;
}
...
```

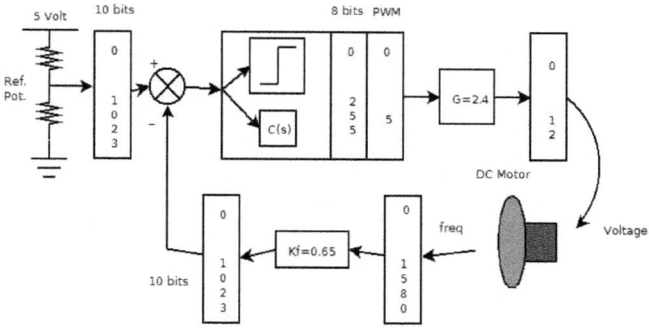

Fig. 5. Block diagram of numerics in digital controller implementation, showing conversion factors and acceptable ranges for the variables

Arduino's programming language is a subset of C, but it is flexible enough for most tasks. For instance, we did a strong use of global variables (the "effective" code is about a hundred lines), but we also used pointers (to pass arrays to functions by reference).

The complete code only uses 7.5% of the available Arduino memory, and so there is still a large room for the implementation of more complex algorithms, either at the controller design or at the system identification levels.

Results – The oscilloscope window in Fig. 3 shows that $\omega_u \approx 2\pi12$ rad/s with a relay level of $M = 95$ centred in a threshold of 160 (output units, in the 0-255 range). The observed extremal oscillation frequencies give $A_f = (\hat{f}_{max} - \hat{f}_{min})/2 = (1060 - 940)/2 = 60$, and so $A = 0.65\,A_f = 39$. The measured $G_0 = 4.25$ is in accordance with Fig. 4. Note that $G_0 = \Delta(\text{input units})/\Delta(\text{output units})$. Using (7) we calculate $\tau = 0.174$ s and $L = 22$ ms, concluding that the time constant τ of the motor is quite larger than L. The low-pass PWM filter, R1-C1, contributes with about 4 ms for this apparent delay L. Finally, we apply the ZN rules to the PID controller with $K_u = \frac{4M}{\pi A}$ and $T_u = 2\pi/\omega_u$ as already discussed.

7 Conclusion

PID controllers with automatic tuning capabilities have been very popular in the last decades. Several companies and models are available. Usually these are expensive devices, especially when prepared for industrial workloads and environments [3, 4]. It has been reported that, almost always, users keep the PID settings in their controllers as the factory defaults, despite the controlled plants being very diverse in practice, or else they tune the controllers in a wrong way. Thus, it can be concluded that automatic tuning can be a real help for the correct application of these instruments.

In this paper we have implemented a standard auto-tuning digital PID controller on an inexpensive platform using only a small fraction of the available memory resources. This frugal implementation opens several roads to future improvements: implementation of more complex identification algorithms (least squares, recursive least squares) targeting more complex models (analogue plants with more poles and zeros, ARMA Box-Jenkins digital models) and control of faster systems (by tweaking Arduino's ADC and DAC clock settings).

In the future we intend to interface the Arduino with a GUI in the host PC so that on-line interaction with the experiment becomes possible.

Above all, the development of this experiment has given to the students a strong familiarity with a broad range of subjects belonging to the control arena – PID, digital control, describing function, microcontroller programming, electronic actuators, just to mention the more relevant. And perhaps the most important skill passed to them was the experience of following a hands-on approach to controller development, from the ground up until the finished product – a much better experience than simply performing a few measurements in a "canned" commercial control experiment.

Acknowledgement. This work was supported by national funds through FCT, Fundação para a Ciência e Tecnologia, under projects PEst-OE/EEI/LA0021/ 2013 and PTDC/EEAELC/122098/2010 and by infrastructures of the *Dep. de Física da Fac. de Ciências da Un. de Lisboa* (DF-FCUL).

References

1. http://arduino.cc/ (consulted in January 2014)
2. Margolis, M.: Arduino Cookbook. O'Reilly (2011)
3. Ästrom, K.J., Hägglund, T.: Automatic Tuning of PID Controllers. Instrument Society of America (1988)
4. Ästrom, K.J., Hägglund, T.: PID Controllers: Theory, Design and Tuning, 2nd edn. Instrument Society of America (1995)
5. Ästrom, K.J., Wittenmark, B.: Adaptive Control, 2nd edn. Addison-Wesley (1995)

6. Chen, C.-T.: Analog and Digital Control System Design - Transfer-Function, State-Space and Algebraic Methods. Saunders (1993)
7. Nise, N.: Control Systems Engineering, 3rd edn. Wiley (2000)
8. Hägglund, T., Åstrom, K.J.: Method and an apparatus in tuning a PID-regulator, US Patent 4549123 (1983)
9. Phillips, C.L., Harbor, R.D.: Feedback Control Systems, 2nd edn. Prentice-Hall (1991)
10. Slotine, J.E., Li, W.: Applied Nonlinear Control. Prentice-Hall (1991)
11. Soltesz, K., Hägglund, T., Åstrom, K.J.: Transfer Function Parameter Identification by Modified Relay Feedback. In: 2010 American Control Conference, Baltimore, Maryland, USA (2010)
12. Soltesz, K., Hägglund, T.: Extending the Relay Feedback Experiment. In: 18th IFAC World Congress, Milano, Italy (2011)

Part VI
Control Applications

Guaranteed Estimation for Distributed Networked Control Systems

Ramón A. García[1], Luis Orihuela[2], Pablo Millán[2],
Manuel G. Ortega[1], and Francisco R. Rubio[1]

[1] Dpt. Ingeniería de Sistemas y Automática,
Universidad de Sevilla, Sevilla, Spain
{ramongr,mortega,rubio}@us.es
[2] Dpt. Matemáticas e Ingeniería,
Universidad Loyola Andalucía, Sevilla, Spain
{dorihuela,pmillan}@uloyola.es

Abstract. This paper proposes a distributed estimation and control scheme for discrete linear time-invariant systems. The plant is monitored and controlled by a network of agents that collect information about the system evolution and apply control inputs to drive its behavior. Bounded disturbances and measurement noises are considered through a zonotope-based approach. It is proven that the state of the plant belongs to the prediction and estimation zonotopes constructed by all the agents at every time instant. The performance of the proposed technique is tested by simulation using a classical system: the inverted pendulum on a cart.

Keywords: Distributed estimation and control, sensor networks, zonotopes.

1 Introduction

Networked Control Systems (NCS) consists of distributed dynamical systems, sensors, controllers and actuators that linked through real-time communication networks [10]. This architecture incorporates advantages such as reduction of wiring costs, more flexibility and easier maintenance. The advent of the so-called Sensor Networks (SN) has provided the control community with the technological solution to implement those distributed schemes.

Typically, these networks are not exclusively used by a unique control loop, but they are shared by a possibly large number of processes with real time requirements. Therefore, bandwidth limitation and congestion problems arise, which may result in delays, retransmissions, and packet dropouts, see [19].

This paper contributes with a novel solution in the field of distributed estimation and control. The final objective is to estimate the state of a plant and to stabilize it using a set of agents. These agents are geographically distributed in such a way that classical centralized solutions are neither feasible nor advisable. In the distributed control context, predictive control has been succesfully applied, see for instance [3,17]. As regards distributed estimation, different modifications based on Kalman filtering have been presented, see for example [6].

© Springer International Publishing Switzerland 2015   231
A.P. Moreira et al. (eds.), *CONTROLO'2014 - Proc. of the 11th Port. Conf. on Autom. Control*,
Lecture Notes in Electrical Engineering 321, DOI: 10.1007/978-3-319-10380-8_22

In an attemp to propose a new solution for the joint problem, this paper develops a estimation and control scheme based in guaranteed estimation techniques. These tecnhiques, also known as set membership estimation, guarantee that the plant state is contained in known compact sets at any time instant. Unlike the classical Kalman filter, statistical properties of the system disturbances and noises are not assumed, but their bounds are suppose to be known, see [8,13,18]. These methods have been applied to perform distributed estimation, see [11]. However, from the best of our knowledge, the distributed solutions in the literature lack from the control part. The main difficulty here is to guarantee that the plant state belongs to estimation zonotopes when unknown control signals are being applied by agents which may not be directly connected.

The paper is organized as follows. Section 2 presents some preliminaries on zonotopes and intruduces the notation. Section 3 describes the system and the agents dynamics. Next, Section 4 developes the proposed distributed algorithm. In Section 5, a simulation example is presented. Finally, Section 6 outlines the main conclusions.

2   Preliminaries: Zonotopes and Operations

Zonotopes are centrally symmetric convex polytopes that can be described as Minkowski sum of line segment. A zonotope can be represented by its center, $c \in \mathbb{R}^n$ and its generators $g_1, ..., g_m \in \mathbb{R}^n$:

$$Z = \left\{ c + \sum_{i=1}^{m} \zeta_i g_i : |\zeta_i| \leq 1 \right\} \tag{1}$$

Let $G = \begin{bmatrix} g_1 & \cdots & g_m \end{bmatrix} \in \mathbb{R}^{n \times m}$. Then, the zonotope can be written as $Z = c \oplus G[B]^m$, where $[B]^m$ is an unitary box of dimension m, i.e., it is a vector composed of m unitary intervals, $[-1, 1]$. An equivalent notation is $Z = \{c, G\}$. The order of the zonotope is given by the number of generators, that is, m.

Let $X = \{c_x, G_x\}$ and $Y = \{c_y, G_y\}$ be two zonotopes and R a matrix of appropriate dimensions, the following identities are verified:

$$RX = \{Rc_x, RG_x\} \tag{2}$$
$$X + Y = \{c_x + c_y, [G_x \ G_y]\} \tag{3}$$

A strip S is defined by the set $S = \{x \in \mathbb{R}^n \mid |cx - d| \leq \sigma\}$, where $c \in \mathbb{R}^{1 \times n}$ and $d, \sigma \in \mathbb{R}$.

In general, intersections between zonotopes and between zonotope and strip are not zonotopes. However, it is possible to find a zonotope containing the intersection. The interested reader may find the computational details in [1,5].

Finally, the reduction operator $\Diamond_p$ associates to Z the reduced zonotope $\Diamond_p Z$ of order p satisfying $Z \subseteq \Diamond_p Z$. The reduction operator is used to keep the order of the zonotopes bounded, at a cost of enlarging the size of the original zonotope, see [7].

3   System Description

Consider the distributed control and estimation scheme depicted in Fig 1. The system is modelled as a perturbed discrete linear time-invariant plant:

$$x(k+1) = Ax(k) + Bu(k) + w(k), \tag{4}$$

where $x \in \mathbb{R}^n$ is the plant state and A, B are well dimensioned matrices. The disturbance $w(k)$ may also account for unmodeled parts of the system and is assumed to remain in some known box $[W]$. The pair (A, B) is stabilizable.

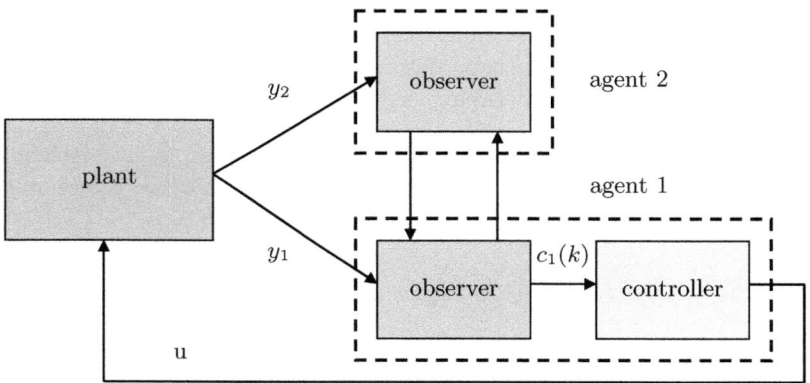

Fig. 1. Distributed scheme for control and estimation of the plant

This paper is concerned with the distributed observation and control through the use of sensor networks with limited communication bandwidth. As a first approximation to the problem, two agents with bidirectional communication are considered. Both agents receive incomplete information from the plant:

$$y_1(k) = C_1 x(k) + v_1(k) \in \mathbb{R}^{r_1}, \tag{5}$$
$$y_2(k) = C_2 x(k) + v_2(k) \in \mathbb{R}^{r_2}, \tag{6}$$

where $v_1(k), v_2(k) \in [V]$ are bounded measurement noises.

The pairs $(A, C_1), (A, C_2)$ are not necessarily observable nor detectable. However, we assume that the pair (A, C) is detectable, being C a matrix stacking both C_1 and C_2.

As Figure 1 illustrates, the estimation of agent 1 is used to generate the control signal. In particular, this paper proposes the following control law:

$$u(k) = Kc_1(k|k) \in \mathbb{R}^p, \tag{7}$$

where $c_1(k|k)$ is the center of the estimation zonotope of agent 1 at instant k. Matrix K represents a given controller such that $A + BK$ is stable. Next section details the algorithm to compute the estimation zonotopes.

4   Distributed Guaranteed Estimation Approach

This section presents the main contribution of the paper. Firstly, the algorithm to be implemented in the agents is given. Later, it is proven that the actual state of the plant is correctly estimated by both agents.

4.1   Control and Estimation Loops

Let $\hat{x}_i$ denote a zonotope with center c_i and generator matrix G_i. Then, prediction and estimation zonotopes can be defined:

Prediction: $\hat{x}_i(k|k-1)$ is a zonotope describing the prediction made by agent i of the state of the plant at time k, using the information available at $k-1$.

Estimation: $\hat{x}_i(k|k)$ is a zonotope describing the estimation made by agent i of the state of the plant at time k, using information available at k.

Table 1 presents the control and estimation loop that is executed in both agents at each time instant. It consists of 3 main steps: measure, communication and prediction.

Table 1. Control and estimation loop in both agents

Step	Agent 1	Agent 2
0. Initial	Know $\hat{x}_1(k|k-1)$	Know $\hat{x}_2(k|k-1)$
1. Measure	$y_1(k)$	$y_2(k)$
1.1. Strip	$S_1(k)$	$S_2(k)$
1.2. Intersection	$Z_1(k) = \hat{x}_1(k|k-1) \bigcap S_1(k)$	$Z_2(k) = \hat{x}_2(k|k-1) \bigcap S_2(k)$
1.3. Order Red.	$\Diamond Z_1(k)$	$\Diamond Z_2(k)$
2. Communication	Send $\Diamond Z_1(k)$	Send $\Diamond Z_2(k)$
	Receive $\Diamond Z_2(k)$	Receive $\Diamond Z_1(k)$
2.1. Intersection	$\hat{x}_1(k|k) = Z_1(k) \bigcap \Diamond Z_2(k)$	$\hat{x}_2(k|k) = \Diamond Z_1(k) \bigcap Z_2(k)$
2.2. Control	$u(k) = Kc_1(k|k)$	
3. Prediction	$\hat{x}_1(k+1|k) = A\hat{x}_1(k|k)$	$\hat{x}_2(k+1|k) = A\hat{x}_1(k|k)$
Control	$+Bu(k)$	$+BK\Diamond Z_2(k)$
Disturbances	$+[W]$	$+[W]$

Let us explain the main parts of the algorithm and the differences between the agents:

Measure: Once an agent measures the output from the plant, it calculates the strip that contains the system states consistent with the measured output. Then, it intersects the strip with its prediction zonotope. The exact intersection between a strip and a zonotope is not a zonotope, but there exist some methods to build a zonotope encompassing the intersection, see [1,5].

Communication: As it is assumed a limited bandwidth, the information to be transmitted through the network has to be wrapped in packets with a few bytes. In order to account of this limitation, the operator $\Diamond$ is employed to create a low-order zonotope. The received information at each agent is intersected with the available zonotopes to obtain the estimation zonotope $\hat{x}_i(k|k)$. Its center is used by agent 1 to compute the control action.

Prediction: The prediction step is different for each agent. As agent 1 knows the exact control action that is being applied to the plant, it uses it to compute its next prediction. On the contrary, agent 2 introduces additional uncertainty by considering a zonotope that contains the control action.

All the operations to be performed throughout the algorithm (finding the strip, intersections, order reduction, etc) are matrix operations which are not excessively complex, so they can be carried out in agents with relative low computational capabilities.

4.2   Distributed Guaranteed Estimation

The distributed guaranteed estimation problem has been previously studied in [11,14]. However, the problem becomes more difficult when the control part is introduced. The reason is obvious: the agents ignore the complete control signal that is being applied to the plant. In our particular scheme, agent 2 does not know $u(k)$, so it has to infer it and obtain a prediction $\hat{x}_2(k+1|k)$ that still contains the actual state of the plant.

The main theoretical result of this conference paper is presented now. It will be proven that the proposed algorithm results in a distributed guaranteed estimator, that is, the state of the system always belong to the zonotopes of both agents.

Lemma 1. Assuming that at instant k_0 the state of the system belongs to both prediction zonotopes, that is,

$$x(k_0) \in \hat{x}_1(k_0|k_0 - 1),$$
$$x(k_0) \in \hat{x}_2(k_0|k_0 - 1),$$

then, for all $k \geq k_0$, it will be satisfied that

$$x(k) \in \hat{x}_1(k|k), \ x(k+1) \in \hat{x}_1(k+1|k),$$
$$x(k) \in \hat{x}_2(k|k), \ x(k+1) \in \hat{x}_2(k+1|k).$$

Proof. The proof will demonstrate that, when a given state $x(k)$ is contained in the prediction zonotopes $\hat{x}_1(k|k - 1)$ and $\hat{x}_2(k|k - 1)$, then a) $x(k)$ will be contained in the estimation zonotopes $\hat{x}_1(k|k)$ and $\hat{x}_2(k|k)$; and b) $x(k+1)$ will be inside the prediction zonotopes $\hat{x}_1(k+1|k)$ and $\hat{x}_2(k+2|k)$. Then, as $x(k_0)$ is assumed to initially belong to $\hat{x}_1(k_0|k_0 - 1)$ and $\hat{x}_2(k_0|k_0 - 1)$, the result is proved by induction.

The proof of part a) is equivalent for both agents. Firstly, assume that $x(k) \in \hat{x}_i(k|k - 1)$, $i = 1, 2$. Furthermore, it is trivialy true that $x(k) \in S_i(k)$, since $S_i(k)$ is the strip including the states consistent with the measure $y_i(k)$. Hence,

$$\left. \begin{array}{c} x(k) \in \hat{x}_i(k|k - 1) \\ x(k) \in S_i(k) \end{array} \right\} \rightarrow x(k) \in \hat{x}_i(k|k - 1) \bigcap S_i(k) \rightarrow x(k) \in Z_i(k)$$

The reduction operator $\Diamond$ satisfies that $Z_i(k) \subseteq \Diamond Z_i(k)$, so $x(k) \in \Diamond Z_i(k)$. Finally,

$$\left. \begin{array}{l} x(k) \in \Diamond Z_i(k) \\ x(k) \in Z_j(k), j \neq i \end{array} \right\} \to x(k) \in \Diamond Z_i(k) \bigcap Z_j(k) \to x(k) \in \hat{x}_i(k|k).$$

Now, let us move to the prediction step, which is different in each agent. The actual state of the system at $k+1$ is given by

$$x(k+1) = Ax(k) + BKc_1(k|k) + w(k).$$

Agent 1 computes the prediction zonotope as $\hat{x}_1(k+1|k) = A\hat{x}_1(k|k) + BKc_1(k|k) + [W]$. Initially, it evolves the estimation zonotope with A. Therefore, as $x(k) \in \hat{x}_1(k|k)$, it turns out that $Ax(k) \in A\hat{x}_1(k|k)$. Secondly, it adds the vector $BKc_1(k|k)$ to all the points of the zonotope. Lastly, it takes into account all possible disturbances with $[W]$, ensuring that $x(k+1) \in \hat{x}_1(k+1|k)$.

Concerning agent 2, and for the same reasons, it holds that $Ax(k) \in A\hat{x}_2(k|k)$. In order to get a guaranteed prediction, agent 2 must add an uncertain zonotope containing the control actions. Although agent 2 ignores the value of $c_1(k|k)$, it holds that $c_1(k|k) \in \Diamond Z_2(k)$. Therefore, $BKc_1(k|k) \in BK\Diamond Z_2(k)$. Thus, $x(k+1) \in \hat{x}_2(k+1|k)$. $\square$

Please note that agent 2 considers the control signal as a bounded disturbance. With unrestricted communication, no reduction step would be needed, and both estimation zonotopes would be the same.

5   Simulation Example

In this section, the performance of the proposed distributed control and estimation scheme is tested by simulation. The results are compared with the method described in [15,16]. This method proposes the use of distributed Luenberger observers in combination with consensus techniques.

5.1   Plant Description and Modeling

We consider the classic problem of an inverted pendulum on a cart, described in [12]. The system and the control scheme can be observed in Figure 2a. The pivot of the pendulum is located on a cart, which can move back and forth powered by a motor that applies at instant t a force $u(t)$. This force is the control input of the system.

The linear position of the pivot at time t is denoted by $s(t)$, whereas the angular position is denoted by $\phi(t)$. The mass of the pendulum is m, L is the distance between the pivot and the centre of gravity of the pendulum, and J is the moment of inertia about its center of gravity. The cart has a mass M, and a friction coefficient with the ground F.

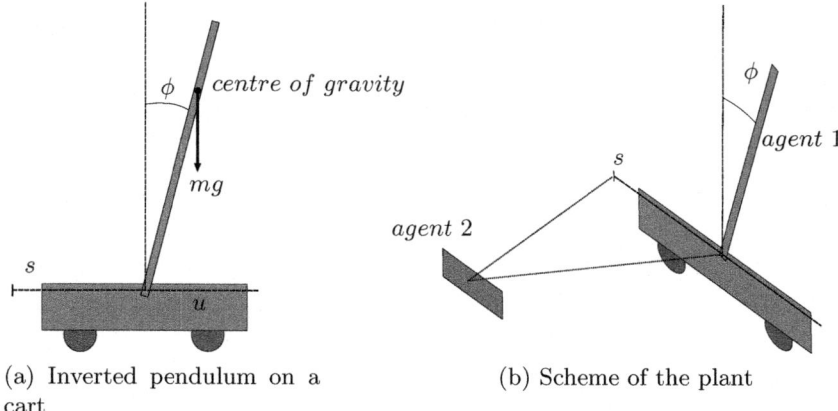

(a) Inverted pendulum on a cart

(b) Scheme of the plant

Fig. 2. Plant considered

As detailed in [12], if the mass of the cart M is assumed much greater than the mass of the pendulum m, the dynamics equations of the system are:

$$\ddot{s}(t) = \frac{1}{M}u(t) - \frac{F}{M}\dot{s}(t), \tag{8}$$

$$\ddot{\phi}(t) = \frac{g}{L'}\sin\phi(t) - \frac{1}{L'}\ddot{s}(t)\cos\phi(t), \tag{9}$$

where the constant value $L' = \frac{J+mL^2}{mL}$ can be understood as the *effective length of the pendulum*, because an ideal pendulum of length L' would have the dynamics described in (9).

By linearizing the equations around the point $u(t) = 0$, $s(t) = 0$, $\phi(t) = 0$ and defining the states as

$$x_1 = s(t), x_2 = \dot{s}(t), x_3 = s(t) + L'\phi(t), x_4 = \dot{s}(t) + L'\dot{\phi}(t),$$

the dynamics of the system can be written as follows:

$$\dot{x}(t) = \begin{bmatrix} 0 & 1 & 0 & 0 \\ 0 & -\frac{F}{M} & 0 & 0 \\ 0 & 0 & 0 & 1 \\ -\frac{g}{L'} & 0 & \frac{g}{L'} & 0 \end{bmatrix} x(t) + \begin{bmatrix} 0 \\ \frac{1}{M} \\ 0 \\ 0 \end{bmatrix} u(t) \tag{10}$$

The distributed control and estimation structure consists of two agents, depicted in Fig.2b, whose functions are:

- **Agent 1.** The first agent is physically located on the cart. It is responsible for measuring the angular position of the pendulum. It also applies the control signal to the motor. The output is:

$$y_1(t) = \begin{bmatrix} -\frac{1}{L'} & 0 & \frac{1}{L'} & 0 \end{bmatrix} x(t)$$

- **Agent 2.** The second agent is located in an external position. It is responsible for measuring the displacement of the cart. The output in this case is given by:

$$y_2(t) = \begin{bmatrix} 1 & 0 & 0 & 0 \end{bmatrix} x(t)$$

Finally, the model is discretized by using a sample time of 0.01 seconds. The parameters are $\frac{F}{M} = 1s^{-1}$, $\frac{1}{M} = 1kg^{-1}$, $\frac{g}{L'} = 11.65s^{-2}$ and $L' = 0.842m$.

5.2 Simulation Results

The simulations have been performed using the same controller as in [15], that is, an LQR with $Q = diag\{1, 1, 100000, 10\}$, $R = \{0.001\}$. We assume that the bounds for the plant disturbances and sensor noises are, respectively: $\|w(k)\| \leq 0.01$, $\|v_1(k)\| \leq 0.05$ and $\|v_2(k)\| \leq 0.08$.

Figure 3 represents the time evolution of the system outputs. Both linear and angular positions are shown with the application of the technique proposed in this paper (Figure 3a) and the results obtained in [15] (Fig. 3b).

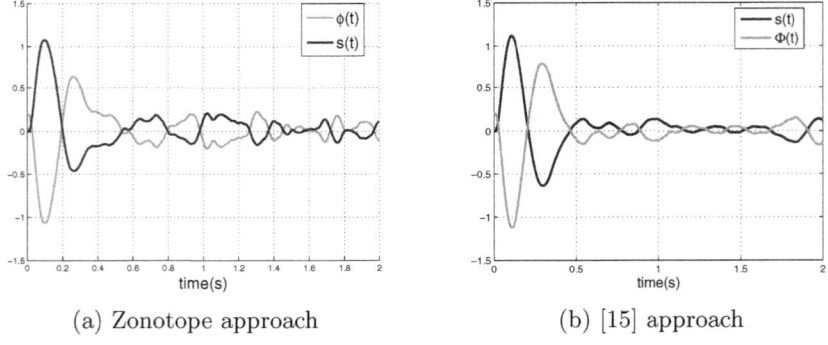

(a) Zonotope approach   (b) [15] approach

Fig. 3. Comparative

From Figure 3 it can be observed that the performance of the presented technique and the one proposed in [15] is very similar. However, the method developed in this paper can use pre-computed controllers to operate and it does not require the design of consensus matrices and Luenberger-like observers as in [15].

Additionally, the presented method guarantees that the plant state lies in the computed zonotopes at any time instant, which can be very interesting in applications where it is important to determine sets containing the plant evolution. This is the case, for instance, in robotic applications or in systems with process variables that compromise the security. For the same simulations, these sets are shown in Figure 4a and Figure 4b together with the evolution of the states 1 and 3, respectively.

Fig 4b shows that the estimates of agent 2 are less accurate than the one obtained by agent 1 for the third plant state. As agent 2 cannot measure this state, its estimations are only based on the information transmitted by agent 1.

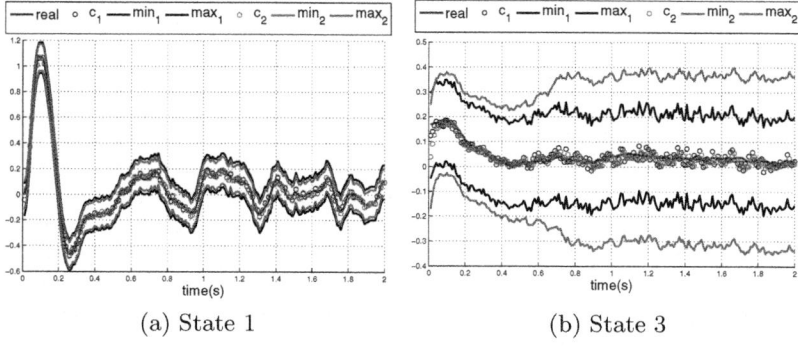

(a) State 1   (b) State 3

Fig. 4. Two states of the linearized system

6   Conclusions

The paper proposes a guaranteed distributed estimation and control technique making use of zonotopes. The method is of application in distributed networked control systems with limited communication bandwidth. In this first approach to the problem, the network contains two agents and only one has access to the actuator. None of the agents has complete observability of the plant. As it has been proven and shown in simulation, the developed technique guarantees that the estimation zonotopes encompass the actual system evolution.

Acknowledgments. The authors acknowledge MCyT (Grant DPI2010-19154) for funding this work.

References

1. Alamo, T., Bravo, J.M., Camacho, E.F.: Guaranteed state estimation by zonotopes. Automatica 41(6), 1035–1043 (2005)
2. Akyildiz, I.F., Weilian, S., Sankarasubramaniam, Y., Cayirci, E.: A survey on sensor networks. IEEE Communications Magazine 40(8), 102–114 (2002)
3. Alvarado, I., Limon, D., MuËœnoz de la PeËœna, D., Maestre, J.M., Ridao, M.A., Scheu, H., Marquardt, W., Negenborn, R.R., De Schutter, B., Valencia, F., Espinosa, J.: A comparative analysis of distributed MPC techniques applied to the HD-MPC four-tank benchmark. Journal of Process Control 21(5), 800–815 (2011)
4. Arampatzis, T., Lygeros, J., Manesis, S.: A Survey of Applications of Wireless Sensors and Wireless Sensor Networks. In: Proceedings of the 2005 IEEE International Symposium on Intelligent Control, Mediterrean Conference on Control and Automation, June 27-29, vol. 724, pp. 719–724 (2005)
5. Bravo, J. M.: Control Predictivo no lineal robusto basado en técnicas intervalares, Phd Thesis. Universidad de Sevilla, Spain (2004)
6. Carli, R., Chiuso, A., Schenato, L., Zampieri, S.: Distributed Kalman filtering based on consensus strategies. IEEE Journal on Selected Areas in Communications 26(4), 622–633 (2008)

7. Combastel, C.: A State Bounding Observer for Uncertain Non-linear Continuous-time Systems based on Zonotopes. In: Proceedings of the 44th IEEE Conference on Decision and Control, and the European Control Conference, Seville, Spain, pp. 7228–7234 (2005)

8. Raimondo, D.M., Braatz, R.D., Scott, J.K.: Active fault diagnosis using moving horizon input design. In: 2013 European Control Conference (ECC), July 17-19, pp. 3131–3136 (2013)

9. Estrin, D., Girod, L., Pottie, G., Srivastava, M.: Instrumenting the world with wireless sensor networks. In: 2001 IEEE International Conference on Proceedings Acoustics, Speech, and Signal Processing (ICASSP 2001), vol. 4, pp. 2033–2036 (2001)

10. Hespanha, J., Xu, Y.: A survey of recent results in neetworked control systems. Proceedings of the IEEE 95(1), 138–162 (2007)

11. Kieffer, M.: Distributed bounded-error state estimation. In: Proceedings of the International Symposium on System Identification, Saint-Malo, France, pp. 360–365 (2009)

12. Kwakernaak, H., Sivan, R.: Linear Optimal Control Systems. John Wiley and Sons, New York (1972)

13. Jaulin, L.: Robust set-membership state estimation; application to underwater robotics. Automatica 45(1), 202–206 (2009)

14. Leger, J.B., Kieffer, M.: Guaranteed robust distributed estimation in a network of sensors. In: Proceedings of the International Conference on Acousticas, Speech, and Signal Processing, Dallas, Texas, USA, pp. 3378–3381 (2010)

15. Orihuela, L., Millán, P., Vivas, C., Rubio, F.R.: Control y observación distribuida en sistemas de control a través de redes. In: Proceedings of the XXXII Jornadas de Automática, Seville, Spain (2011)

16. Millán, P., Orihuela, L., Vivas, C., Rubio, F.R., Dimarogonas, D.V., Johansson, K.H.: Sensor-network-based robust distributed control and estimation. Control Engineering Practice 21(9), 1238–1249 (2013)

17. Stewart, B.T., Venkat, A.N., Rawlings, J.B., Wright, S.J., Pannocchia, G.: Cooperative distributed model predictive control. Systems & Control Letters 59(8), 460–469 (2010)

18. Li, Y., Lin, Z.: Generalized criterions for determining the maximal ellipsoidal invariant set of linear systems under saturated linear feedback. In: 2013 32nd Chinese Control Conference (CCC), July 26-28, pp. 878–883 (2013)

19. Zampieri, S.: Trends in networked control systems. In: 17th IFAC World Congress, pp. 2886–2894 (July 2008)

Dual Mode Feedforward-Feedback Control System

Paulo Moura Oliveira[1], Damir Vrančić[2], and Hélio F. Freire[1]

[1] INESC TEC – INESC Technology and Science (formerly INESC Porto)
Department of Engineering, School of Sciences and Technology,
Universidade de Trás-os-Montes e Alto Douro, UTAD
5001–801 Vila Real, Portugal
[2] Department of Systems and Control,
Jožef Stefan Institute, Jamova cesta 39, Ljubljana, Slovenia
oliveira@utad.pt, damir.vrancic@ijs.si, freireh@gmail.com

Abstract. A dual mode control configuration involving open-loop feedforward control to deal with set-point tracking and proportional integrative and derivative control to deal with disturbance rejection is proposed. The feedforward controller is a Posicast pre-filter shaping the reference input using either two or three steps. The switching between the open-loop and closed-loop control is performed automatically. The particle swarm optimization algorithm is deployed as design tool both for the feedforward and PID controller. Simulation results are presented showing the merits of the proposed technique.

1 Introduction

Two classic control systems design objectives are commonly known as set-point tracking (SPT) and disturbance rejection (DR). For a great deal of open-loop stable linear systems deadbeat type of SPT responses can be achieved using feedforward control. Indeed, for some process control applications, such as pre-loading [1] open-loop feedforward control is used with the actuator set to the maximum output value, to reach the desired system output value as fast as possible, and then a feedback loop is closed to deal with DR. This switching between input-reference SPT, using an on-off open-loop feedforward approach and proportional integral and derivative (PID) feedback control, was termed as a dual mode control scheme by Shinskey [2]. However, it is important to state that this type of dual control methodology works well when the system is not frequently subject to disturbances in the tracking phase. The switching between the open-loop and the closed-loop control depends on the design methodology as it will be shown in the sequel.

An input reference signal open-loop control strategy, Posicast, was proposed by Smith [3], to deal with underdamped second-order dynamics. Posicast control is based on the modification of a step reference input to cancel system vibrations. Posicast originated most of the actual current vibration control theory [4,5], requiring shaping the input command signal in order to cancel vibration poles. In this paper a technique is proposed bridging Posicast control based on the input reference step input manipulation using two and three steps, and dual mode control using a PID controller within

© Springer International Publishing Switzerland 2015,
A.P. Moreira et al. (eds.), *CONTROLO'2014 - Proc. of the 11th Port. Conf. on Autom. Control*,
Lecture Notes in Electrical Engineering 321, DOI: 10.1007/978-3-319-10380-8_23

the feedback loop. Despite the development of more elaborated control techniques such as predictive control, PID control continues to be widely used within a great deal of practical applications attracting significant research efforts [6] deserving the organization of a specialized conference [7].

The problem under study here is to design dual mode control system, using a particle swarm optimization algorithm (PSO) [8] as a design tool. Some of the addressed issues involve the proper selection of step amplitudes within the on-off feedforward control, and PID controller gains.

2   Input Reference Feedforward Control

Considering the open-loop system represented in Figure 1 with: G_f representing a feedforward controller and G_p the system model, r the reference input, u the control signal and y the system output. The block in-between the feedforward control and the system represents the actuator saturation limits. If position control is considered, the reference input signal is a step, considered here of unitary amplitude for exposition convenience. The objective of applying open-loop feedforward control is to achieve deadbeat responses for set-point changes.

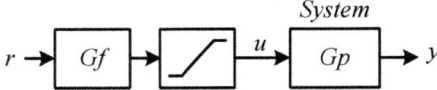

Fig. 1. Open-loop feedforward control with Posicast input command shaping

For underdamped second order dynamics:

$$G_p(s) = \frac{\omega_n^2}{s^2 + 2\zeta\omega_n s + \omega_n^2} \tag{1}$$

with: ζ representing the damping factor and ω_n undamped natural frequency, deadbeat responses can be accomplished by using a half-cycle Posicast controller [3]. This type of controller is governed by:

$$G_{fhc}(s) = A_1 + A_2 e^{-t_1 s} = A_1 + (1 - A_1)e^{-t_1 s} \tag{2}$$

with: A_1 and A_2 representing the two step amplitudes, applied respectively at time $t=0$ and $t=t_1$. The amplitude, A_1, and time, t_1, for the second step can be evaluated using the following expressions:

$$A_1 = \frac{1}{1 + M_p} = \frac{1}{1 + e^{-\frac{\zeta\pi\omega_n}{\omega_d}}} \tag{3}$$

$$t_1 = \frac{T_d}{2} \tag{4}$$

with: ω_d representing the damped natural frequency, T_d the undamped time period and M_p the first overshoot. The half-cycle shaped control signal, in the unit step input case, is in the interval: $0 \leq u(t) \leq 1$. Smith [3] proposed another Posicast controller requiring the use of three steps, governed by the following equation:

$$G_{fts}(s) = A_1 + A_2 e^{-t_1 s} + A_3 e^{-t_2 s} \quad 0 < t_1 < t_2 \tag{5}$$

subject to: $A_1 + A_2 + A_3 = 1$, with: A_1, A_2 and A_3 representing the three step amplitudes; t_1 and t_2 represent the time in which the second and third steps are applied to the system reference input. This type of three-step Posicast controller can be designated third-cycle, quarter-cycle, fifth-cycle, and so on, depending on the ratio used in relation to the undamped time period for obtaining the third step time, t_2. Thus, the time for the third step can be evaluated for the third-cycle (tc), quarter-cycle (qc) and fifth cycle (fc) Posicast using respectively:

$$t_{2tc} = \frac{T_d}{3}; \quad t_{2qc} = \frac{T_d}{4}; \quad t_{2fc} = \frac{T_d}{5} \tag{6}$$

The second step time, t_1, is always half of t_2. The three steps amplitudes can be evaluated using the following expressions:

$$A_1 = \frac{e^{2\zeta\omega_n t_1}}{1 - 2e^{\zeta\omega_n t_1}\cos(\omega_d t_1) + e^{2\zeta\omega_n t_1}} \tag{7}$$

$$A_2 = \frac{-2e^{\zeta\omega_n t_1}\cos(\omega_d t_1)}{1 - 2e^{\zeta\omega_n t_1}\cos(\omega_d t_1) + e^{2\zeta\omega_n t_1}} \tag{8}$$

$$A_3 = \frac{1}{1 - 2e^{\zeta\omega_n t_1}\cos(\omega_d t_1) + e^{2\zeta\omega_n t_1}} \tag{9}$$

An example of possible controller signals and respective system outputs is presented in Figures 2 and 3, respectively. As it can be observed, a possible limitation to this type of feedforward control can be the control signals amplitudes. Faster tracking responses require larger control signals magnitude which are actuator dependant.

The concept of using Posicast feedforward control can be used for system dynamics which are not second order. A similar approach has been proposed in [9,10], implemented in the discrete time domain, in which the control signal was set to the maximum value for a certain period (on), changed to the minimum value (off) and then set to the steady state value (u_{ss}). The type of control signal obtained is of the same type of the ones presented in Figure 2. The objective then [9,10] was to select the maximum control value (on), minimum control value (off), steady state control signal and transition times between the former states using genetic algorithms.

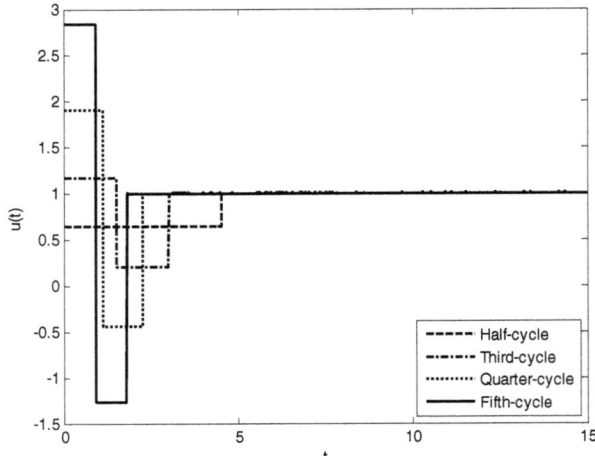

Fig. 2. Open-loop shaped signals, with half-cycle, third-cycle, quarter-cycle and fifth cycle Posicast

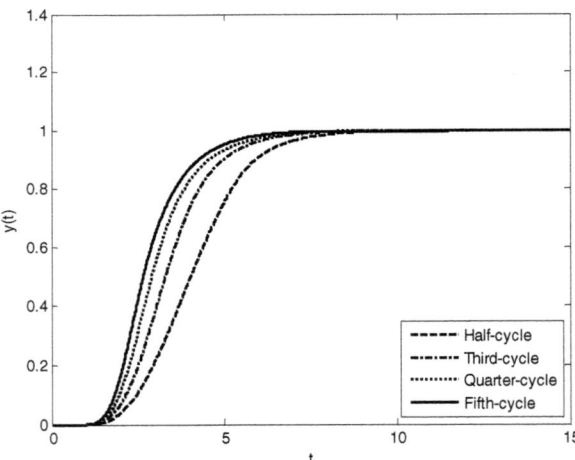

Fig. 3. Open-loop responses, with half-cycle, third-cycle, quarter-cycle and fifth cycle Posicast

3   Dual Mode Feedforward-Feedback PID Control

A dual mode PID control configuration shown in Figure 4 was proposed in [11,12] to control oscillatory systems. The objective of this control structure is to use the open-loop feedforward Posicast controller to accomplish set-point tracking, and then switch to PID feedback control in a pre-specified time t_s.

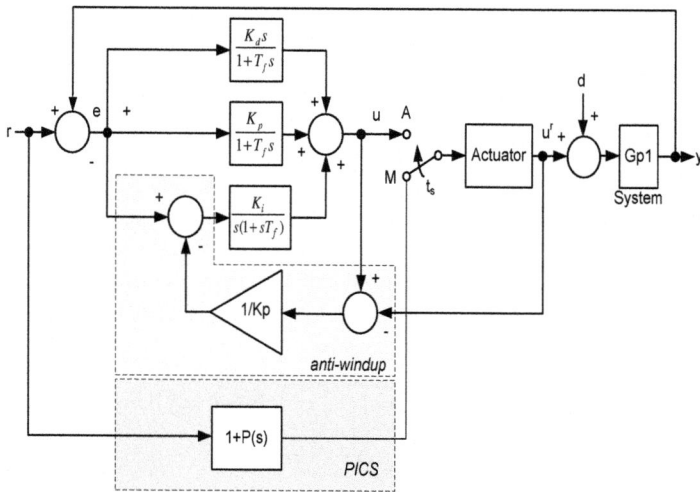

Fig. 4. Dual mode PID control structure with Posicast [11]

In this work, it is proposed to use the same principle. However, instead of making the transition between open-loop and closed-loop at a predefined transition time, this transition is performed based on the evaluation of the steady-state system output. A simplified diagram block of this dual mode control structure is presented in Figure 5.

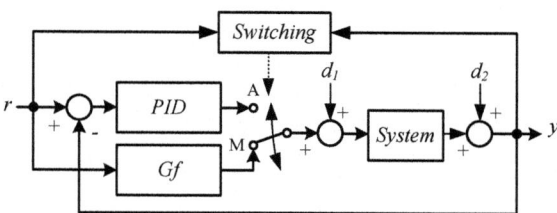

Fig. 5. Proposed dual mode control structure

The system output set-point tracking is evaluated and when it reaches the steady state-value the transition is performed. Assuming set-point changes can occur more frequently, it is also necessary to automatically change from closed-loop control to open-loop control. It is proposed here to monitor the reference signal to detect set-point changes and perform the transition from the closed-loop to open-loop, accordingly.

The expressions presented in (7-9) can only be applied for underdamped second order systems and do not allow specifying the controller signal magnitude. Thus a standard PSO algorithm is deployed here to design three-step Posicast feedforward controllers, to overcome this drawback.

Each swarm population element encodes a problem potential solution, representing in this case the following parameters $\{A_1, A_2, A_3, t_2, \beta\}$, with β representing the ratio factor which is used to obtain t_1:

$$t_1 = \frac{t_2}{\beta} \tag{10}$$

Each particle, i, is characterized by two fundamental 5-dimensional variables: a position (x) and velocity, (v), vectors. The first algorithm step consists in initializing the swarm position and velocity, within a predefined search space interval, using a uniform random procedure. Individual swarm elements performance is evaluated, within the problem to be solved, using a minimization objective function. In this case the criteria used is the well known IAE (integral of the absolute error value) for the SPT and DR system response. The swarm particle position and velocity are updated in each iteration t, by the following two equations:

$$v_{id}(t+1) = \omega v_{id}(t) + c_1\varphi_c.(b_{id}(t) - x_{id}(t)) + c_2\varphi_s.(g(t) - x_{id}(t)) \tag{11}$$

$$x_{id}(t+1) = x_{id}(t) + v_{id}(t+1) \tag{12}$$

taking into account that: $1 \le i \le m$ and $1 \le d \le 5$, with m and d representing the swarm size and particle dimension, respectively. In (11), b_i represents the best position achieved by particle i and g represents the best global position achieved within a specified neighborhood, which, in this paper, considers the entire swarm in a fully connected topology; In the same expression, c_1 and c_2 are known as the cognitive and social constants, respectively, and φ_c and φ_s represent uniformly randomly generated numbers within interval [0,1]. While the originally proposed value for c_1 and c_2 is 2, other values can be used. Parameter ω, known as inertia weight, is a modification to the original algorithm, aiming to control the PSO convergence rate. A high value assigned to the inertia weight (near one) promotes the search space global exploration, while a small value promotes local exploitation. Thus, it is common practice to gradually decrease this weight from a high initial value (ω_{init}) to a small value final value (ω_{fin}) throughout the search process following the expression:

$$\omega(t) = \omega_{init} - \frac{(\omega_{init} - \omega_{fin})t}{t_{fin}} \tag{13}$$

in which t_{fin} represent the final iteration.

4   Simulation Results and Discussion

The simulations presented here are conducted in the digital time domain. The first system considered is a standard underdamped second order system with a damping factor of $\zeta=0.35$ and a natural frequency $\omega_n=2$ rad/s. Assuming a sampling time of $T_s=0.01$s the discrete time domain model considering a sample and hold is represented by:

$$G_{p1}(z) = \frac{0.00019906z + 0.00019814}{z^2 - 1.9857z + 0.9861} \tag{14}$$

Implementing the PID controller using a standard digital incremental algorithm and considering the half cycle transfer function (2) represented in the discrete time domain as a difference equation:

$$u_{fhc}(k) = A_1 r(k_{fs}) + A_2 r(k_{fs} + k_{ss}) \tag{15}$$

where: r is the reference input signal, k_{fs} and k_{ss}, represent the samples when the first and second steps are applied, respectively. For system (14), $T_d=3.35s$, $t_1=1.68s$ and the half-cycle is represented by:

$$u_{fhc}(k) = 0.764 r(k_{fs}) + 0.236 r(k_{fs} + 168) \tag{16}$$

Figure 6 represents input references set-point tracking, load and output disturbance rejection responses for system (14), with the system controlled by the proposed dual mode control methodology, with the half-cycle controller represented by (16).

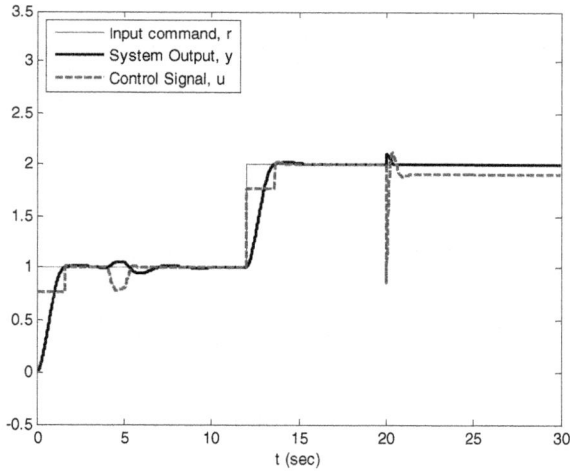

Fig. 6. Dual mode half-cycle PID input reference tracking and disturbance rejection for system (14)

Two unit steps were applied to the reference input at $t=0s$ and $t=12s$, a pulse load disturbance was applied in the period [4s,5s] with an amplitude of $d_1=0.2$ and a step output disturbance with an amplitude of $d_2=0.1$ was applied at t=20s. The PID controller gains used with $T_f=0.1s$ are: $K_p=1.62$, $K_i=4.72$, and $K_d=1.17$, obtained with the MOMI method described in [13]. As it can be observed from Figure 6, the dual mode controller is able to automatically detect input reference changes and apply open-loop feedforward control as well as to detect when the system reaches the steady state and change to PID control. Similarly, the three-step transfer function (5) can be represented in the discrete time domain as the following difference equation:

$$u_f(k) = A_1 r(k_{fs}) + A_2 r(k_{fs} + k_{ss}) + A_3 r(k_{fs} + k_{ts}) \tag{17}$$

where k_{ts} represents the sample in which the third step is applied. Evaluating a third-cycle three-step Posicast feedforward controller for system (14) results in the following transfer function, with $t_1 = 0.56s$ and $t_2 = 1.12s$.

$$u_{ftc}(k) = 1.280r(k_{fs}) - 0.866r(k_{fs} + 56) + 0.586r(k_{fs} + 112)$$ (18)

Similarly to the results presented in Figure 6, Figure 7 presents input references setpoint tracking, load and output disturbance rejection responses for system (14) when the system is controlled by the proposed dual mode control methodology, with the third-cycle controller represented by (18). As expected, the improvement of using the three-step controller can be observed in the SPT phase.

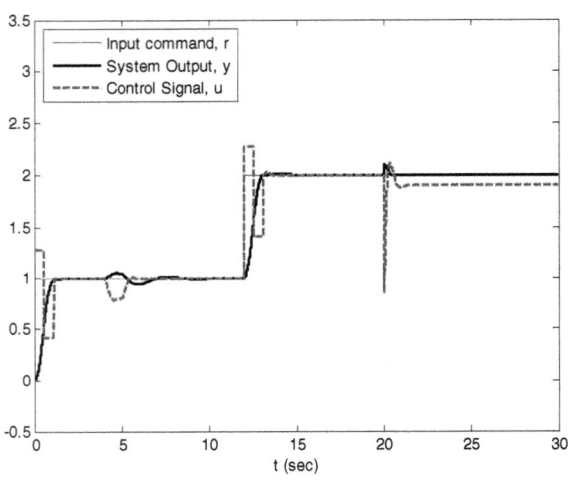

Fig. 7. Dual mode third-cycle PID input reference tracking and disturbance rejection for system (14)

The second system considered is a first order plus time delay with DC gain, $K=1$, time constant, $T=1s$, and time delay $L=1s$. Assuming a sampling time of $T_s=0.01s$ the discrete time domain model considering a sample and hold is represented by:

$$G_{p2}(z) = \frac{0.0099501662z}{z - 0.9900498337}$$ (19)

In this case the PSO was applied to obtain a deadbeat SPT response, assuming a three-step transfer function (5). As stated previously, each swarm element represents the following parameters $\{A_1,A_2,A_3,t_2,\beta\}$. The swarm size used was $m=30$, the algorithm was run for 100 iterations per simulation, $\omega_{init}=0.9$ and $\omega_{fin}=0.4$. The parameters obtained in this case were $\{2,1,0,0.829s,1.2\}$, resulting in $t_1=0.691s$. For these settings the feedforward controller expression (15) is represented by:

$$u_{ftc}(k) = 2r(k_{fs}) - r(k_{fs} + 69)$$ (20)

The step response obtained in the same conditions used for system (14), with $T_f=0.1s$ and PID controller gains: $K_p=0.97$, $K_i=0.70$, and $K_d=0.27$ obtained with the MOMI method described in [13], is presented in Figure 8. As it can be seen from the control signal presented in the feedforward tracking phase, the maximum value, 2, is used as the first step amplitude, and then the second negative step puts the system in the steady-state, thus the 0 value obtained for A_3. The system response presents a dead-beat behavior for SPT and is able to automatically change for PID control and vice-versa whenever necessary.

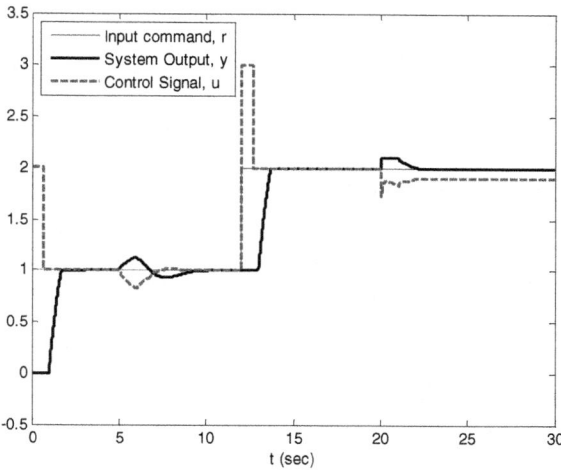

Fig. 8. Dual mode third-cycle PID input reference tracking and disturbance rejection for system (19)

5   Conclusion

In this paper a dual mode technique was proposed aiming the best of two-worlds regarding two of the most relevant control requirements: set-point tracking and disturbance rejection. To achieve dead-beat like tracking responses, half-cycle and three-step Posicast feedfoward controllers are deployed in open-loop. To deal with disturbance rejection the system is put in closed loop using a PID controller. The main contributions of this paper are: i) a new dual-mode control structure using Posicast and PID control is proposed; ii) an automatic switching methodology between the open-loop and closed-loop control, and vice-versa, is proposed; ii) the particle swarm optimization is used as a simple optimization tool to design three-step Posicast con-trollers. Simulation results showed that the dual mode control performs well both for tracking and regulation. As the drawback of the proposed technique is not being able to reject disturbances in the tracking phase, further research is necessary to provide this dual mode controller with rejection capabilities within tracking. The use of the proposed PSO designed based methodology to a wider set of plants will be the topic of future publications.

References

1. Dhanda, A., Franklin, G.: Vibration Control via Pre-loading. In: American Control Conference, Portland, OR, USA, June 8-10, pp. 3574–3579 (2005)
2. Shinkey, F.G.: Process Control Systems: Application, 4th edn. McGraw-Hill (1996)
3. Smith, O.J.M.: Posicast Control of Damped Oscillatory Systems. Proc. IRE 45(9), 1249–1255 (1957)
4. Singer, N.C., Seering, W.P.: Preshaping command inputs to reduce system vibration. Journal of Dynamic Systems Measurement and Control 112(3), 76–82 (1990)
5. Singhose, W.: Command Shaping for Flexible Systems: A Review of the First 50 Years. Int. Journal of Precision Eng. and Manufacturing 10(4), 153–168 (2009)
6. Åström, K.J., Hägglund, T.: The Future of PID Control. Control Engineering Practice 9(11), 1163–1175 (2001)
7. Vilanova, R., Visioli, A.: IFAC Conference on Advances in PID Control, PID 2012, Brescia, Italy, March 28-30. IFAC, S.l. (2012)
8. Kennedy, J., Eberhart, R.C.: Particle swarm optimization. In: Proc. IEEE Int'l. Conf. on Neural Networks, IV, pp. 1942–1948. IEEE Service Center, Piscataway (1995)
9. Jones, A.H., De Moura Oliveira, P.B.: Genetic Design of Dual Mode Controllers Through a Process of Co-evolution. In: 4th IEEE Mediterranean Symposium on New Directions in Control & Automation, Crete, Greece, June 10-14, pp. 794–798 (1996)
10. Jones, A.H., Lin, Y.-C., B, De Moura Oliveira P.B.: Auto-Tuning of Dual Mode Controllers using Genetic Algorithms. In: 2nd IEE Conference on Genetic Algorithms in Engineering Systems: Innovations and Applications, Glasgow, Scotland, September 2-4, vol. 446, pp. 516–521 (1997)
11. De Moura Oliveira, P.B., Vrančić, D.: Underdamped Second-Order Systems Overshoot Control. In: Vilanova, R., Visioli, A. (eds.) IFAC Conference on Advances in PID Control, PID 2012, Brescia, Italy, March 28-30, ThPS.12, 6 pages. IFAC, S.l. (2012)
12. De Moura Oliveira, P.B., Vrančić, D., Boaventura, C.J.: Posicast PID Control of Oscillatory Systems. In: 10th Portuguese Conference on Automatic Control, July 16-18, pp. 27–32 (2012)
13. Vrančić, D., Strmčnik, S., Juričić, Đ.: A magnitude optimum multiple integration method for filtered PID controller. Automatica 37, 1473–1479 (2001)

Computer Implementation of Robust Controllers for Anesthesia*

Daniela V. Caiado[1],**, João Miranda Lemos[2], and Bertinho Andrade Costa[2]

[1] INESC-ID, Rua Alves Redol 9, 1000-029 Lisboa, Portugal
daniela.caiado@ist.utl.pt
[2] INESC-ID/IST-UL, Rua Alves Redol 9, 1000-029 Lisboa, Portugal
jlml@inesc-id.pt,bac@inesc-id.pt

Abstract. The reversible state of a general anesthesia encompasses three components: hypnosis, paralysis and analgesia. In this paper design and implementation issues of the controllers used to maintain paralysis and hypnosis are presented. The controllers are designed from model databases of compartmental models and rely on the H_∞ method. In order to cope with the variability of the patient models, robust techniques are used to produce a controller that is able to withstand robust stability and robust performance. The controllers are designed in continuous and discretized for computer implementation. Clinical results during elective surgery are presented to illustrate the results obtained.

Keywords: Robust control, H_∞ design, Depth of anesthesia, Neuro-muscular blockade, Clinical trials.

1 Introduction

Closed-loop control of anesthesia has been the subject of several previous studies, either to identify models of the patient's response to drug infusion, or to design controllers based on several methodologies in order to achieve a proper anesthetic state. Like any other drug induced state, performing anesthesia is not a straightforward process, because the effect of a certain drug dose differs not only from patient to patient but over time of exposure to the drug. Nevertheless, the patient's response may be related to features such as age, weight, height, gender and ethnicity, that can be used to infer the most suitable dosage. The anesthetists use these informations to perform drug dosing and, as the surgery proceeds, they close the loop by adjusting the dosage relying on their experience. However, an automatic closed-loop mechanism, with the appropriate measured variables, can be a breakthrough in this medical process.

The reversible state of anesthesia comprehends three effects: hypnosis, paralysis and analgesia. In a general anesthesia, hypnosis or depth of anesthesia (DoA) is induced by the hypnotic drugs that act on the synaptic transmission on the

---

* This work was performed in the framework of contract PEst–OE/EEI/LA0021/2013.
** Corresponding author.

neuronal network, enabling the loss of consciousness during the surgical intervention. Several models of the hypnotic effect on patients have been described in [1–3], in particular for the effect of the drug *propofol*. Other studies identified the interaction of the analgesic drug *remifentanil* with the hypnotic effect, measured by the BIS index [4]. The BIS index is a processed data from the patient's electroencephalogram that can be related to the hypnotic state of the patient. The BIS index is a value between 0 and 100 that measures the "awareness" of the patient, being 100 for full awareness. During surgery, the BIS index is usually kept close to 50.

The effect of paralysis is achieved with muscle relaxant drugs that act on the neuromuscular junction between the neuron and the muscle fiber. This effect is also called the neuromuscular blockade (NMB) since the drugs, like *rocuronium*, block the neuromuscular signal transmission, inhibiting muscle contraction. Similarly to hypnosis, compartmental models have also been described for paralysis [5], relating the NMB index with the drug infusion rate. In this case, the NMB index is the train-of-four (TOF) ratio that measures the muscle contraction at the thumb after an electrical stimulation of the ulnar nerve, with four twitches. The contractions are measured by the sensor placed on the hand of the patient which result in a TOF ratio between 100 % and 0 %. During the induction of NMB the index drops from 100 %, that corresponds to no blockade, to (near) 0 %, where the paralysis is achieved. During the surgery, the NMB index is usually maintained near 10 % and should not be over 20 % to avoid muscle contractions that may interfere with the surgery procedure.

Control algorithms have been described for NMB and hypnosis, such as PID [6, 7] or adaptive controllers [8–10]. However, the robust approach to overcome uncertainty of the patient models, with the NMB index TOF ratio and the BIS index, is seldom in the literature. The uncertainty on these systems is very high, mostly due to the variability of the patient responses. The unmodelled dynamics and the errors introduced by the linear approximation of the nonlinear term of these Wiener structure models are also sources of uncertainty. The design of a controller that takes these uncertainties into account can be a great improvement on the automatic control of anesthesia, since the controllers are designed with the ability to stabilize a large group of patients and with appropriate control actions that provide appropriate clinical responses.

This paper presents two controllers, one for NMB and another for hypnosis, with the BIS index, that are designed based on the H_∞ methods. With the H_∞ approach, the loop shaping of the sensitivity and of the complementary sensitivity functions allow the desired performance and robust requirements to be fulfilled. The controllers are designed in continuous and then discretized for computer implementation. The contribution of the paper consists on the comparison of a reduced order controller that operates in discrete time and its assessment on clinical trial.

This paper is organized as follows: the control design based on H_∞ methods is briefly explained in section 2; the resulting controllers are set in the operating room during elective surgery and results are presented in section 3; conclusions are drawn in section 4.

2   Controller Design

The control design is performed based on the patient models described in Appendix. The nonlinear models are linearized around equilibrium values of 10 % for NMB index and 50 for BIS index. The corresponding incremental models are described in continuous time by a function $G(s)$, with the patient response being

$$y(s) = G(s) \ u(s), \tag{1}$$

where u is the drug infusion rate, s is the Laplace variable and y the increment of either BIS or NMB..

For the robust control design approach that is followed in the design of the controllers for NMB and for hypnosis, a database of models is considered for each one of the problems. The controller is designed for a nominal model and based on the uncertainty that is present in the model database. For the control of NMB a database of 50 models defined as $\mathcal{G}_{NMB} = \{G_i(s), i = 1, 2, ..., 50\}$ is considered, whereas the control of hypnosis considers a database of 18 models defined as $\mathcal{G}_{DOA} = \{G_i(s), i = 1, 2, ..., 18\}$. For each of the databases, the multiplicative uncertainty is denoted as $\Delta_i(j\omega)$, for frequency ω, by which each model G_i is related to the nominal model G_N by

$$G_i(j\omega) = G_N(j\omega)(1 + \Delta_i(j\omega)). \tag{2}$$

2.1   Robust Stability Condition

To design a robust controller K it is taken into account that the controller must be able to stabilize not only the nominal model, for which it is designed, but also the models of the class considered.

Defining an upper bound $l(\omega)$ for the uncertainty Δ_i

$$|\Delta_i(j\omega)| < l(\omega), \tag{3}$$

the robust stability condition can be written as

$$\frac{1}{l(\omega)} < |T_N(j\omega)|, \tag{4}$$

where T_N is the complementary sensitivity function defined by

$$T_N(j\omega) \triangleq \frac{KG_N(j\omega)}{1 + KG_N(j\omega)}. \tag{5}$$

Accordingly, the controller that yields condition (4), taking (3) into account, has robust stability. In this design, both controllers fulfill this requirement (data not shown).

2.2 Robust Performance Conditions

To design the controller for robust performance, loop-shaping methods are applied to the system, while assuring that the robust stability condition is fulfilled. For that sake, in order to design a controller K, interconnected with the plant as in Fig. 1, two weighting functions W_S and W_T are coupled to the system. The closed-loop response y is described then by

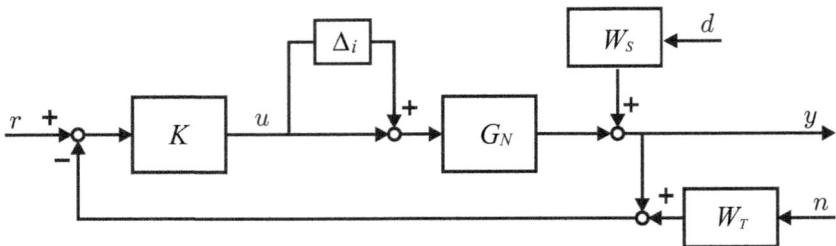

Fig. 1. Schematic representation of the control design system with use of the two weighting functions W_S and W_T and with the representation of the multiplicative uncertainty Δ_i of each model G_i with respect to the nominal model G_N; the resulting controller K filters the error between the reference r and the output y that is affected by output disturbances d and sensor noise n

$$y = \frac{W_S}{1 + KG_i} d + \frac{KG_i}{1 + KG_i} r + W_T \frac{KG_i}{1 + KG_i} n, \qquad (6)$$

and the control problem is to minimize the weighted sensitivity functions and the weighted complementary sensitivity functions such that,

$$|S_i \, W_S| < 1 \qquad \Leftrightarrow \qquad |S_i| < \frac{1}{|W_S|}, \qquad (7)$$

and

$$|T_i \, W_T| < 1 \qquad \Leftrightarrow \qquad |T_i| < \frac{1}{|W_T|}. \qquad (8)$$

Therefore, the inverse of the weighting functions W_S and W_T are selected as upper bounds of the sensitivity functions S_i and of the complementary sensitivity functions T_i, respectively.

The condition (7), with a high-pass stable function W_S^{-1} that bounds the low frequency range, assures that the output disturbance d is rejected and does not cause significant deviations since the gain of S_i is less than 1 (or 0 dB). On the other hand, the condition (8), with a low-pass stable function W_T^{-1} that bounds the high frequency range, guarantees that the high frequency sensor noise is also rejected since it shapes the functions T_i to have a gain less than 1. The conditions (7) and (8) are imposed for the control design along with the robust stability condition (4) and this is accomplished with the choice of the two weighting functions, shown in Fig. 2 for the controller of hypnosis (data not shown for the controller of NMB for lack of space).

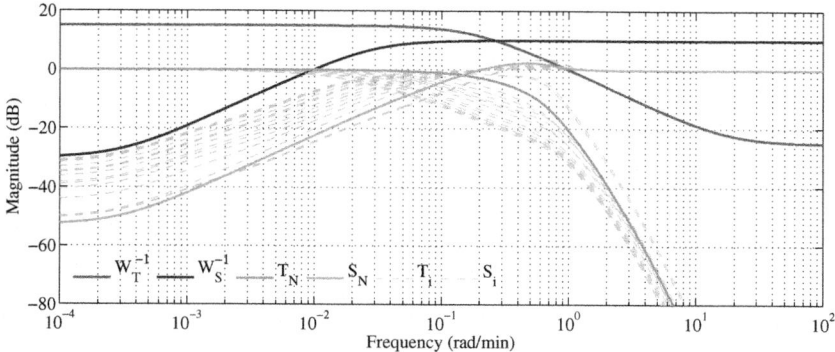

Fig. 2. Magnitude of the inverse of the weighting functions, W_S^{-1} and W_T^{-1}, and of the sensitivity functions S_i and of the complementary sensitivity functions T_i, for all models of the database $\mathcal{G}_{\mathcal{DOA}}$

The implementation of the design method is performed with the available software of MATLAB® Robust Control Toolbox™[11]. To avoid tracking errors integral action is forced in the controller. For appropriate implementation of the software used, the integrator is coupled with the model and, afterwards, it is coupled to the controller. The resulting controllers, before being enlarged with the integrator, are state-space systems of order 17 and 19 for NMB and hypnosis, respectively. These controllers are then approximated by systems of order 4 and 7, respectively, by using the method balanced truncation model order reduction implemented in the MATLAB® function **reduce** of the Robust Control Toolbox™. The frequency response of the complementary sensitivity and sensitivity functions of the nominal systems with the full order and with the reduced order controller are shown for comparison in Fig. 3 for hypnosis, and in Fig. 4 for NMB.

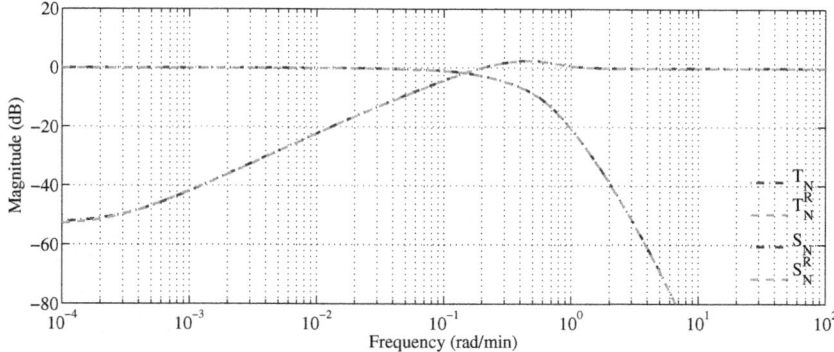

Fig. 3. Comparison between the full order and the reduced order controller for hypnosis: frequency response of the sensitivity functions of the nominal model, S_N and S_N^R respectively, and of the complementary sensitivity functions, T_N and T_N^R respectively

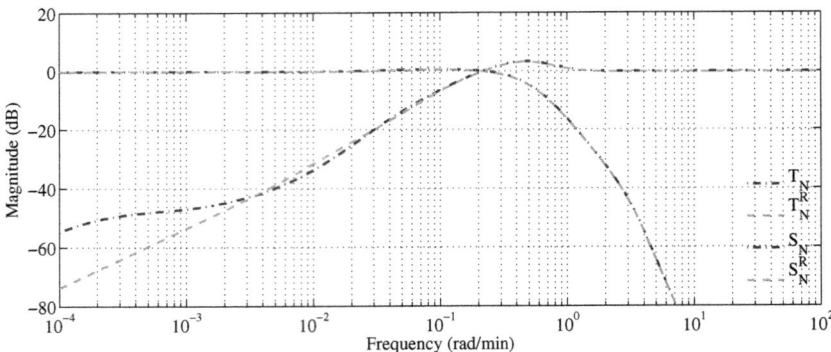

Fig. 4. Comparison between the full order and the reduced order controller for NMB: frequency response of the sensitivity functions of the nominal model, S_N and S_N^R respectively, and of the complementary sensitivity functions, T_N and T_N^R respectively

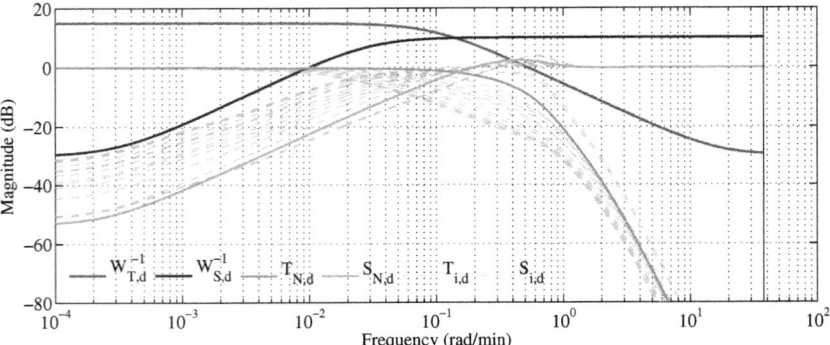

Fig. 5. Magnitude of the inverse of the weighting functions, W_S^{-1} and W_T^{-1}, and of the sensitivity functions S_i and of the complementary sensitivity functions T_i, for all models of the database $\mathcal{G}_{\mathcal{DOA}}$ in the discrete time domain

Since the controllers are designed in the continuous time domain, for computer implementation the controllers must be discretized. In this case, the controllers are discretized with the same sampling interval that the output is measured, which corresponds to 20 seconds for the controller of NMB and 5 seconds for the controller of hypnosis, and is performed with the ZOH method.

For the reduced complexity controllers in the discrete time domain, robust stability and robust performance conditions are checked. For both controllers, the robust stability condition is satisfied. The verification of the performance conditions is shown in Fig. 5 for the controller of hypnosis, in the discrete time domain. However, for the NMB controller not all the complementary sensitivity functions of $\mathcal{G}_{\mathcal{NMB}}$ fall bellow the upper bound W_T and, therefore, the noise rejection condition is not fulfilled in discrete time (data not shown). Nevertheless the controller is able to deliver appropriate performances, with high frequency noise rejection, both in simulations and in clinical trials.

3   Clinical Trials

The controllers have been evaluated in clinical trials. Several safety measures are taken in order to guarantee the patient's safety. The controller is equipped with an override command that enables the anesthetist to switch it off and also to adjust manually the dosage if required. In this paper, a clinical trial where both controllers are used in simultaneous is reported in Fig. 6. The controllers are switched on after an initial *bolus* of the drug that drives the patient state near the required levels. No override command is used, and the controllers are able to adjust the dosage for appropriate hypnosis and paralysis levels. The controllers are able to filter the high frequency noise.

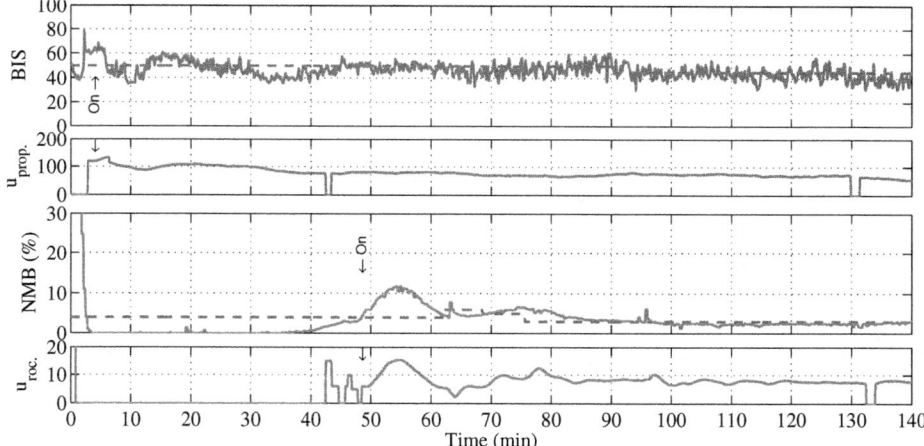

Fig. 6. Closed-loop control with the robust controllers used during a surgery in a 54 years old male patient with 76 kg and 1.76 m: the first couple plots are the BIS index measured and the drug dosage of *propofol*; the second couple plots are the NMB index measured and the drug dosage of *rocuronium*; the red dashed lines are the reference levels and the vertical arrows indicate the moment that the controllers are switched on; the first minutes of the surgery are omitted to simplify the exposition; the drug dosages are in $\mu g.kg^{-1}.min^{-1}$

4   Conclusions

In this paper two robust controllers designed for anesthesia, in order to automate the closed-loop control of hypnosis and paralysis are presented. The loop-shaping techniques applied allowed the design of robust controllers that are able to deliver adequate clinical performances. The controllers are able to stabilize the patient, both in simulation and in clinical trials, performing high frequency sensor noise rejection and output disturbance rejection. The order reduction and the discretization allowed an appropriate computer implementation of the closed-loop system and the results show the adequacy of the controllers designed.

Appendix: Dynamic Models of NMB and Hypnosis

The models of the patient's response to the drug infusion considered have a Wiener structure, where a static nonlinearity appears in series with a linear model. The patient models are pharmacokinetic/pharmacodynamic (PK/PD) models described as multi-compartmental models, where the drug is assumed to be uniformly distributed in each compartment, with constant transfer rates. The compartmental models are represented in Fig. 7.

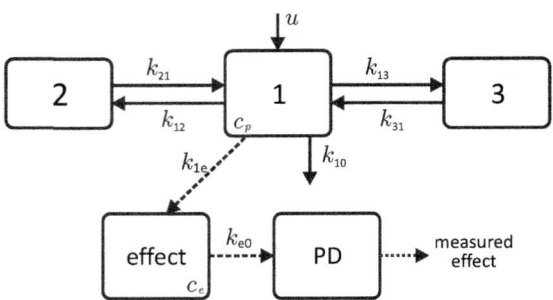

Fig. 7. Schematic representation of the patient response model: the drug u is perfused in the first compartment that is in equilibrium with two compartments; the drug concentration in plasma c_p is related with the drug concentration in the effect compartment c_e, which is related to the measured effect by the PD model; the compartment (3) is not considered in the NMB model

The linear model relates the effect concentration, which is the drug concentration that corresponds to the measured effect, with the drug infusion rate. The linear part of the patient model is described, in state-space representation, by

$$\begin{cases} \dot{\mathbf{x}}(t) = \mathbf{A}\,\mathbf{x}(t) + \mathbf{B}\,u(t) \\ c_e(t) = \mathbf{C}\,\mathbf{x}(t) \end{cases}, \tag{9}$$

where $\mathbf{x}$ and $\dot{\mathbf{x}}$ are the column vectors of the state-space variables x_i and of its derivatives, respectively. The state matrix $\mathbf{A}$ is a patient dependent matrix and matrices $\mathbf{B}$ and $\mathbf{C}$ are the input and output matrices of the system. The output of the linear system is the effect concentration c_e that is related to the measured NMB effect by a static nonlinearity modeled by the nonlinear Hill equation.

The NMB patient response model is represented by a 3-compartmental linear model (compartments (1), (2) and the effect compartment), that corresponds to the PD model, followed by the PD model described by the nonlinear Hill equation, as represented in Fig. 7. The linear part of the NMB model is described by (9), where the state matrix $\mathbf{A}$ is

$$\mathbf{A} = \begin{bmatrix} -(k_{10} + k_{12} + k_{1e}) & k_{21} & 0 \\ k_{12} & -k_{21} & 0 \\ k_{1e} & 0 & -k_{e0} \end{bmatrix}, \tag{10}$$

where k_{ij} is the constant rate from the i-th to the j-th compartment. The state is defined as $\mathbf{x} = [x_1 \ x_2 \ x_e]^T$, $\mathbf{B} = [1 \ 0 \ 0]^T$ and $\mathbf{C} = [0 \ 0 \ 1]$. The nonlinear Hill equation for the NMB model is defined as

$$y(t) = 100 \frac{C_{50}^\gamma}{C_{50}^\gamma + c_e^\gamma(t)}, \tag{11}$$

where $y(t)$ is the measured effect by the NMB index (TOF ratio), which is a value normalized between 0 % and 100 %, corresponding to full paralysis and no paralysis, respectively. The variables C_{50}, which corresponds to the effect concentration related to 50 % of the NMB index, and γ are patient dependent ones.

The hypnotic response model considered is represented by a 4-compartmental linear model, followed by the nonlinear Hill equation (Fig. 7). The linear part of the hypnotic model is described by (9), where the state matrix $\mathbf{A}$ is

$$\mathbf{A} = \begin{bmatrix} -(k_{10} + k_{12} + k_{1e}) & k_{21} & k_{31} & 0 \\ k_{12} & -k_{21} & 0 & 0 \\ k_{13} & 0 & -k_{31} & 0 \\ \frac{k_{1e}}{1000 \, V} & 0 & 0 & -k_{e0} \end{bmatrix}, \tag{12}$$

where k_{ij} is the constant rate from the i-th to the j-th compartment and V is the volume of the first compartment, and the state is defined as $\mathbf{x} = [x_1 \ x_2 \ x_3 \ x_e]^T$, $\mathbf{B} = [1000/60 \ 0 \ 0 \ 0]^T$ and $\mathbf{C} = [0 \ 0 \ 0 \ 1]$. The nonlinear part of the hypnotic model is described the Hill function described by

$$y(t) = E_0 + (E_{max} - E_0) \frac{c_e^\gamma(t)}{c_e^\gamma(t) + C_{50}^\gamma}, \tag{13}$$

where $y(t)$ is the BIS index that ranges between 100, when the patient is fully aware, and 0, E_0 is the baseline effect at zero drug concentrations, E_{max} is the peak drug effect, C_{50} is the concentration related with 50 % of the drug effect and γ is the steepness of the concentration-response relation. As described in [4], exists a synergetic effect between the hypnotic drug *propofol* and the analgesic drug *remifentanil*, shown in the electroencephalogram, for which the overall effect becomes described by

$$y(y) = \frac{97.7}{1 + \left((1 + \beta) \frac{c_e^p(t)}{C_{50}^p} + \frac{c_e^r(t)}{C_{50}^r} \right)^\gamma}, \tag{14}$$

where β is a patient dependent parameter and the super-indexes p and r refer to the variables associated with *propofol*, and the hypnotic model, and to the variables associated with *remifentanil*, and the analgesic model, respectively. In this work, the *remifentanil* dose that appears in the model is handled as an unaccessible disturbance.

References

1. Marsh, B., White, M., Morton, N., Kenny, G.N.C.: Pharmacokinetic model driven infusion of propofol in children. British Journal of Anaesthesia 67(1), 41–48 (1991)
2. Dyck, J.B., Shafter, S.L.: Effects of age on propofol pharmacokinetics. In: Implemented in the computer program Stanpump Seminars in Anesthesia, vol. 11, pp. 2–4 (May 1992)
3. Schnider, T.W., Minto, C.F., Shafer, S.L., Gambus, P.L., Andresen, C., Goodale, D.B., Youngs, E.J.: The influence of age on propofol pharmacodynamics. Anesthesiology 90(6), 1502–1516 (1999)
4. Bouillon, T.W., Bruhn, J., Radulescu, L., Andresen, C., Shafer, T.J., Cohane, C., Shafter, S.L.: Pharmacodynamic interaction between propofol and remifentanil regarding hypnosis, tolerance of laryngoscopy, bispectral index, and electroencephalographic approximate entropy. Anesthesiology 100(6), 1353–1372 (2004)
5. Lago, P., Mendonça, T., Gonçalves, L.: Online autocalibration of a PID controller of neuromuscular blockade. In: Proceedings of the 1998 IEEE International Conference on Control Applications, vol. 1, pp. 363–367 (September 1998)
6. Macleod, A.D., Asbury, A.J., Gray, W.M., Linkens, D.A.: Automatic control of neuromuscular block with atracurium. British Journal of Anaesthesia 63, 31–35 (1989)
7. Absalom, A., Sutcliffe, N., Kenny, G.N.: Closed-loop control of anesthesia using bispectral index: performance assessment in patients undergoing major orthopedic surgery under combined general and regional anesthesia. Anesthesiology 96(1), 67–73 (2002)
8. Struys, M., De Smet, T., Versichelen, L., Van de Vilde, S., Van der Brocke, R., Martier, E.: Comparison of closed-loop controlled administration of propofol using BIS as a controlled variable vesus standard practice controlled administration. Anesthesiology 95(1), 6–17 (2001)
9. Hui, Q., Haddad, W.M., Chellaboina, V., Hayakawa, T.: Adaptive control of mammilary drug delivery systems with actuator amplitude constraints and system time delay. European Journal of Control 11(6), 1–15 (2005)
10. Mendonça, T., Lemos, J.M., Magalhães, H., Rocha, P., Esteves, S.: Drug delivery for neuromuscular blockade with supervised multimodel adaptive control. IEEE Transactions on Control Systems Technology 17(6), 1237–1244 (2009)
11. Balas, G., Chiang, R., Packard, A., Safanov, M.: Robust Control Toolbox™ User's Guide. The MathWorks, Inc. (2012)

Auto Tuning Applied to Distillation Column Based on Armax Model Parameter Estimation

Nilton Silva, Heleno Bispo, Romildo Brito, and João Manzi

Department of Chemical Engineering - Federal University of Campina Grande,
Campina Grande/PB, Brazil
{nilton,heleno,romildo.brito,manzi}@deq.ufcg.edu.br

Abstract. An auto-tuning procedure for PID controllers based on model ARMAX is the main subject of this paper. Such a model has been used in the on-line controller parameters estimation through the relay method, resulting in a smaller deviation between the model and process variables when utilized a normalized regression matrix. The procedure was applied to an Aspen Plus Dynamics® model of Naphtha Treatment Unit. The Aspen model takes into account the main control structures and the principal observed disturbances. The procedure has been developed in Visual Basic for Applications programming language, making possible a real-time connection with the Aspen model. The on-line identification of the ARMAX model was made using the RELS (PR), thereby allowing the estimation of the controllers parameters from classical tuning procedures in closed loop. With the use of this strategy, the updating of the controllers parameters has been possible, promoting a superior performance of the control structure.

Keywords: Auto-tuning, PID, ARMAX, identification.

1 Introduction

In general, at the industrial practice, traditional controllers are designed to operate with fixed parameters control. Such a consideration can be inappropriate for stochastic processes, due the frequently changes in its values. The parameter changes are caused by variations or even disturbances, i.e., in product specification and material feed, being this main reason to avoid the fixing of parameters in this kind of process.

However, an alternative improvement can be acquired by means of the use of adaptive control systems [1]. A control system can be considered as adaptive when its parameters are continuously adjusted following the process variations [2]. Thus, to reduce the process analysis problem of stability and convergence, such parameters can be assumed constants but unknown. Thereby, an adaptive control system will converge for these values, even for an unknown process. Then, it is important to be noted that such control systems operates in a condition of self-tuning, aiming automatically the achievement of the best process performance. This control structure is known as self-tuning regulator (STR). Then, in accordance with the Figure 1, STR acts on the splitting of the controller parameters estimations from its basic design ([3],[4]).

© Springer International Publishing Switzerland 2015,
A.P. Moreira et al. (eds.), *CONTROLO'2014 - Proc. of the 11th Port. Conf. on Autom. Control*,
Lecture Notes in Electrical Engineering 321, DOI: 10.1007/978-3-319-10380-8_25

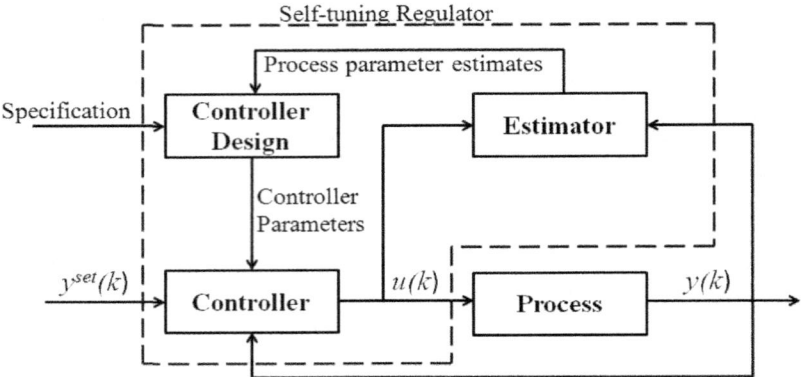

Fig. 1. Basic Self-tuning Regulator Diagram

The on-line estimation of the unknown controller parameters is made by a recursive estimation method, taking into account the uncertainty equivalence principle [5]. Different estimation structures can be used, such as, stochastic approximation, least squares, generalized and extended least squares, instrumental variable and maximum likelihood. The best estimation method will be defined in according to specifications of the system in closed-loop [3].

1.1   Objective

The main objective of this paper is the applying of an auto-tuning of PID controllers procedure based on auto-regressive model (ARMAX). Such a procedure will be used in an Aspen Plus Dynamics® model of Naphtha Treatment Unit, previously validated with data of a real industrial plant, as depicted by the Figure 2.

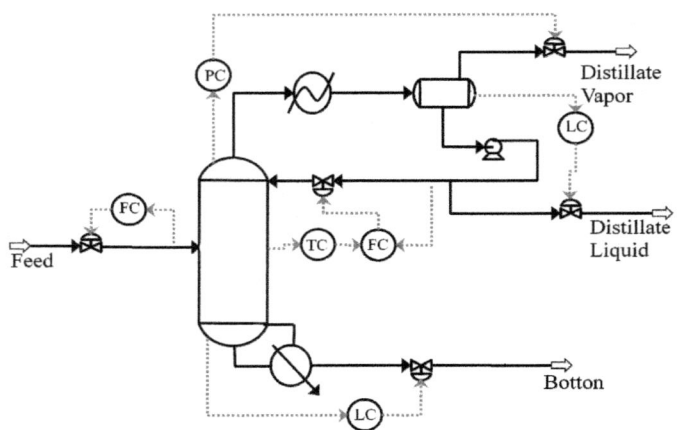

Fig. 2. Simplified Column Flowsheet and process control structures

2   On-Line Parameters Estimation

The mathematical modelling of processes is a tool of fundamental importance in science and engineering. From simulations of such models, it is possible to obtain valuable informations and conclusions about a specific process, or even, specific process conditions under analysis. However, sometimes, the development of a phenomenological modelling of a system is not possible, due i.e. its high complexity. In such situations, the system identification becomes an alternative and attractive option, allowing the off-line or on-line parameters estimation.

In some cases, the structural model of a process is known, while the time-varying parameters are not. This situation occurs basically due to changes in the operating conditions or disturbances during the process execution. Therefore, in such a condition, the off-line parameters estimation is inefficient. Moreover, on-line estimation structures are capable of a frequent process parameters estimation, promoting a correctly input/output data processing.

The estimation is performed by a comparison between the system response, y(k), and the parameterized model output, y(θ, k). Where k and θ are respectively the time-varing sample size and the model parameters. Thus, the parameters, θ(k), are continuously adjusted until y(θ, k) is close to y(k) with the increase of k. Then, under given conditions, when the model output, y_m, is near to the process output, y, it will imply that θ(k) will be close of the unknown process parameters, θ*.

In accordance to Bobal et al [1], the basic three steps for an on-line estimation are:

1. Selection of an appropriate model of parameterization for the process;
2. Determination of an adaptative law for generate and update θ(k);
3. Designation of process inlet to obtain θ(k) close to θ* with k → ∞.

However, in adaptive control, the first two steps are the most important.

3   The Models' Structures

The use of standard models is indicated when process cannot be defined by the first principles. Such a situation can happens due limited information about the system, or even when the physical relationships are very complex. It is well known that standard are able to handle many dynamic systems cases, and its most common class is the standard linear system. Then, generally, a discrete time model parameterized by θ can be written as:

$$y(k) = G(q^{-1}, \theta)u(k) + H(q^{-1}, \theta)e(k) \tag{1}$$

where, q^{-1}, is the *back shift* differential operator, $q^{-1}y(k) = y(k-1)$; $G(q^{-1}, \theta)$, process model; $H(q^{-1}, \theta)$, disturbance model; and, $e(k)$, white noise.

In the Black-Box Modelling with a prediction error method, the frequently structure used is the following:

$$A(q^{-1})y(k) = \frac{B(q^{-1})}{F(q^{-1})}u(k) + \frac{C(q^{-1})}{D(q^{-1})}e(k) \tag{2}$$

where,

$$A(q^{-1}) = 1 + a_1q^{-1} + a_2q^{-2} + \ldots + a_{na}q^{-na} \tag{3}$$

in analogous manner for other polynomials C, D e F, while:

$$B(q^{-1}) = b_1q^{-1} + b_2q^{-2} + \ldots + b_{nb}q^{-nb} \tag{4}$$

Some special cases for the Equation 2 general structure is shown in Table 1.

Table 1. Common Linear model structures

Polynomial Structure	Structure Identification
B	FIR (Finite Impulse Response)
A,B	ARX (Auto Regressive with eXogenous Inputs)
A,B,C	ARMAX (Auto Regressive with Moving Average with eXogenous inputs)
B,F	OE(Output Error)
B,C,D,F	BJ (Box-Jenkins)

4   Process Model

A statistical model can be used to represent the process. Such a model is commonly called convolution of random variables functions ([6],[7]). This discrete time model is a combination of the convolution and autoregressive ARMAX models, and can be expressed by:

$$y_{(k)} + a_1 y_{(k-1)} + a_2 y_{(k-2)} + \ldots + a_n y_{(k-n)} = b_1 u_{(k-d-1)} +$$
$$+ b_2 u_{(k-d-2)} + \ldots + b_m u_{(k-d-m)} + v_{(k)} \tag{5}$$

where, a_i and b_i are coefficients obtained by a regression, d is the time delay, $v(k)$ the combined noise effects in the sampling, $v^*(k)$ unmeasured disturbances at the sampling moment, and $\varepsilon(k)$ modeling errors. Thus, the combination of such effects can be given by: $v(k) = v^*(k) + \varepsilon(k-1)$.

Once, the least squares method is commonly used in the identification system, an unknown parameter of a mathematical model must be chosen minimization procedure. The precision degree can be obtained by the sum of squares of the difference between the observed and output values, the latter can be analytically predicted weight that measures the degree of precision. Knowing that, the criterion of least-squares is quadratic, an analytical solution is possible, while the measurement variable is linear in the unknown parameters.

Assuming that the process is described by a single input and a single output system (SISO), then:

$$A(q^{-1})y(k) = B(q^{-1})u(k-d) + C(q^{-1})e(k) \tag{6}$$

where,

$$A(q^{-1}) = 1 + a_1 q^{-1} + a_2 q^{-2} + ... + a_{na} q^{-na} \tag{7}$$

$$B(q^{-1}) = b_1 q^{-1} + b_2 q^{-2} + ... + b_{nb} q^{-nb} \tag{8}$$

$$C(q^{-1}) = 1 + c_1 q^{-1} + c_2 q^{-2} + ... + c_{nc} q^{-nc} \tag{9}$$

Since the Armax model is linear in relation to the parameters [8], it can be given in the following matricial form:

$$y(k) = \theta_{k-1}^T \varphi_{k-1} + e(k) \tag{10}$$

where, the parameter estimation matrix and the regression matrix are respectively given by:

$$\theta_{k-1} = \left[\hat{a}_1, \hat{a}_2, ..., \hat{a}_{na}; \hat{b}_1, \hat{b}_2, ..., \hat{b}_{nb}; \hat{c}_1, \hat{c}_2, ..., \hat{c}_{nc} \right]^T \tag{11}$$

$$\varphi_{k-1}^e = \left[-y_{k-1}, -y_{k-2}, ..., -y_{k-na}; u_{k-d-1}, u_{k-d-2}, ..., u_{k-d-nb}; e_{k-1}, e_{k-2}, ..., e_{k-nc} \right]^T \tag{12}$$

Once θ_{k-1} is available the Recursive Least Squares (RLS) Algorithm can be use in the recursive estimation of θ. However, all the φ_{k-1} terms are known except the last ones: nc. When the Recursive Extended Least Squares a Priori Prediction Errors RELS(PR) is used, such components are replaced using previously forecast errors: $\varepsilon_k = y_k - \theta_{k-1}^T \varphi_{k-1}$, where φ_{k-1} is the pseudo regressor, given by:

$$\varphi_{k-1} = \left[-y_{k-1}, -y_{k-2}, ..., -y_{k-na}; u_{k-d-1}, u_{k-d-2}, ..., u_{k-d-nb}; \varepsilon_{k-1}, \varepsilon_{k-2}, ..., \varepsilon_{k-nc} \right]^T \tag{13}$$

The linear regressor can be expressed in following matricial notation, taking into account the parameters and system data,

$$y_{(k)} = \theta^T \varphi_{(k)} + v_{(k)}^* \tag{14}$$

Thus, the model parameters are obtained by the classical method of recursive least squares (RLSM) ([7],[9]),

$$\hat{\theta}_k = \left[\sum_{k=1}^{N} \varphi_{(k)} \varphi_{(k)}^T \right]^{-1} \sum_{k=1}^{N} \varphi_{(k)} y_{(k)} \tag{15}$$

With the minimization of the difference between the model and the process, the resulting obtained parameters can be considered as the best predictions to the output variable in the sense of minimum variance, as presented by Silva et al [10].

5   Adapted Autotuning Procedure

The classical auto tuning strategy is performed with a relay directly connected to the process ([11],[12]), as depicted by Figure 3. Such a procedure replaces the PID controller, by the mentioned relay, stimulating the process as on-off controller. However, the main disadvantage in the use of this procedure is the fact of the process remains uncontrolled during the tuning time. Although, this problem can be solved with a slight modification in the procedure, as shown in Figure 4. In such a situation, the relay will stimulate the identified model parallel to the real process.

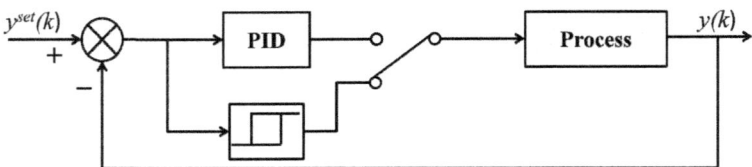

Fig. 3. Classical auto tuning structure

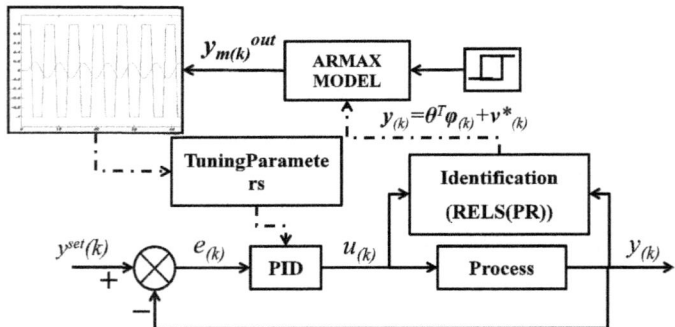

Fig. 4. Modified auto tuning structure

In this way, when the convolution model for the process identification is used associated to an auto tuning procedure, the strategy can be performed during the process operation, as well as, with the control system.

The detailed structure for the auto tuning parameters,applied to PID controllers, is shown by Figure 4. The identification block initially receives the historic of n input u(t) and output y(t) variables. Such information will be used to generate the parameters for the recursive convolution model, which will be stimulated by the relay, keeping the same stimulation amplitude observed in the real process.

The auto-tuning procedure implementation is performed with an introducing of an oscillatory impulse by the relay between $\pm\varepsilon$. Such an impulse is appropriately chosen to produce a controlled oscillation with a constant amplitude in the model output variable, $y_{(t)}^{m}$. From the generated model output, the ultimate gain *(Ku)* and period *(Tu)* can be determined. From the generated model output, the ultimate gain *(Ku)* and period *(Tu)* can be determined for each sampling cycle, and the controller parameters can be estimated using the classical tuning procedure. Thus, the conventional rules proposed by [11] can be used to the establishment of the PID controller parameters.

6   Procedure

A model of a Naphtha Treatment Unit in Aspen Plus Dynamics® platform has been used as a study case in this paper. The proposed auto tuning procedure has been made considering a depentanizer distillation column, as is shown by the simplified diagram given by the Figure 2. The column is designed with 37 stages, with an average feed stream of 57.55 ton/h, composed by paraffinic and aromatic compounds, which are the more representative of the real process load. The main control structures of the process under analysis are: level, flow, pressure and temperature. And main observed disturbances in the real process has been considered, i.e., inlet flow and composition variations, and disturbances on the heating and cooling duty.

Once the procedure has been developed in Visual Basic for Applications (VBA) programming language, the connection with the Aspen Plus Dynamics® platform was made by the Object Linking and Embedding (OLE) technology of communication, which is supported by both platforms. Thus, the ARMAX model parameters estimation and auto-tuning procedures are summarized by the following flowsheet given by Figure 5.

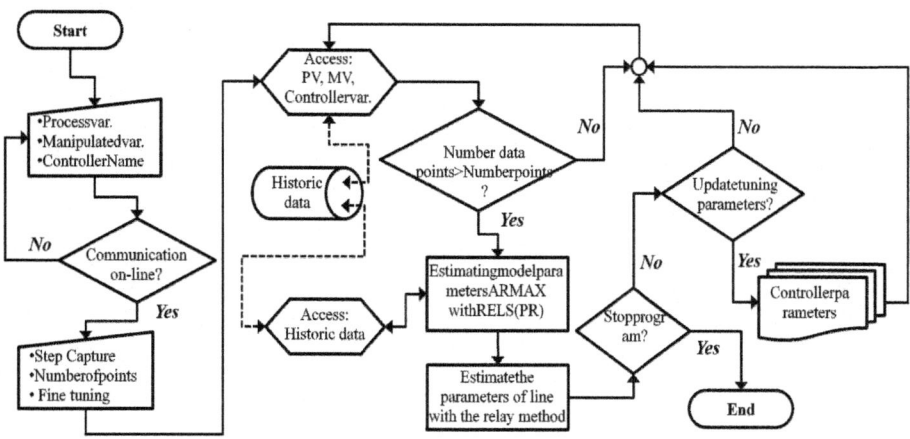

Fig. 5. Simplified auto tuning algorithm developed in VBA

7   Results and Discussion

7.1   Parameter Estimation and Identification

Better results can be obtained when input/output data are normalized. For such a normalization has been used as reference the average value of the input/output variable, u_{aver} and y_{aver}, respectively, as is shown by the following equations:

$$y'_{(k)} = \frac{(y_{(k)} - y_{aver})}{(y_{max} - y_{aver})} \qquad (16)$$

$$u'_{(k)} = \frac{(u_{(k)} - u_{aver})}{(u_{max} - u_{aver})} \qquad (17)$$

The normalization procedure has been used to avoid rounding errors during the matricial calculations for the parameters estimation. After the parameters determination, the following computations are carried out with the variables in the nominal values.

The identification method efficiency presented an appropriate functional behavior, when the model given by Equation 5, for n = 30, is considered. The results of the procedure applied to control structures of pressure, temperature and flow rate are shown in Figures 6, 7 and 8. And it is important to be mentioned that the parameters values have been found with a minimal computational effort.

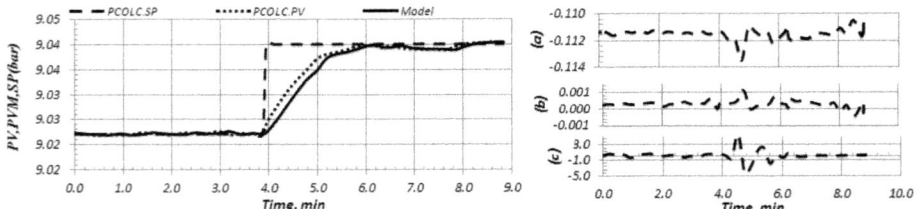

Fig. 6. Behavior of pressure control loop and the values of the model Armax parameters

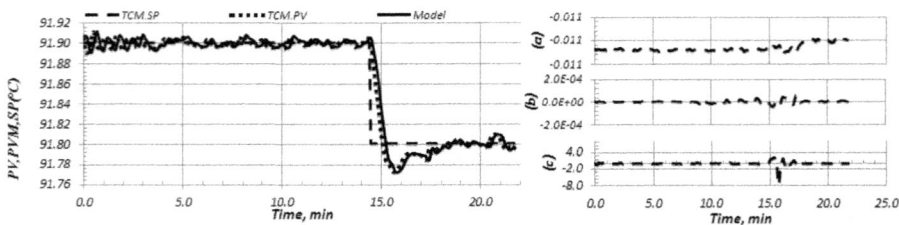

Fig. 7. Behavior of temperature control loop and the values of the model Armax parameters

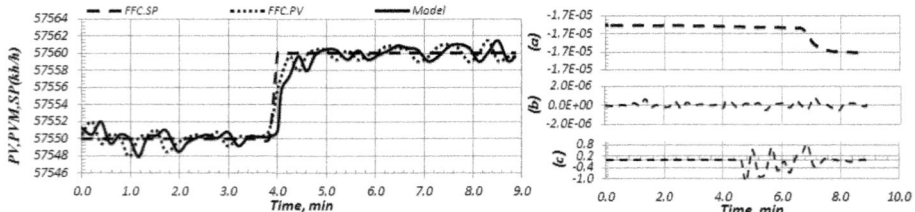

Fig. 8. Behavior offlow control loop and the values of the model Armax parameters

As can be observed, small setpoint changes for the controlled variables (pressure, temperature and flow), the model follows the process behavior. The online estimation introduces continuous corrections in the model parameters, given that the difference between the model and process can be evermore minimized. Then, such model can be used for estimation of controllers tuning parameters.

7.2   ARMAX Auto-Tuning

The online PID controller parameters adjustment for each control system has been established as shown in Figure 4, which is composed by a relay, the ARMAX model and an adjusting algorithm for the controllers tuning parameters.

The relay generated stimulation is based on an appropriate amplitude in the form of a square wave from $\pm\varepsilon$, which corresponds to a factor of 5% of the average normalized value of the manipulated variable. Knowing that the process stimulation range is used as reference to the recursive model, that will result in an output periodical wave with a constant amplitude.

Figure 9 shows the relationship between the input and output waves to the pressure control structure.

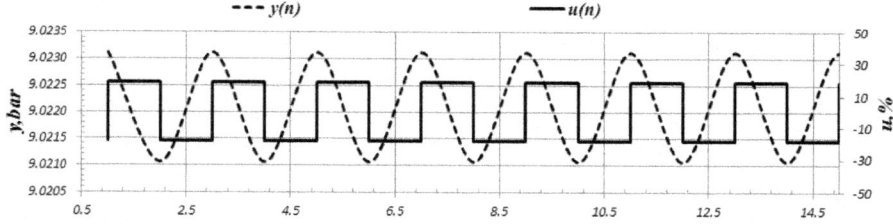

Fig. 9. Generated relay stimulation response and the output variable

From the periodical oscillation response of the identified Armax model, the ultimate period (Tu) and gain (Ku) can be determined. The ultimate period value is given by the length of wave period as shown in Figure 10, as well as the value of Ku, which is calculated by the two waves amplitude, in according to Equation 18 [13].

$$K_u = \frac{4\varepsilon}{\pi a} \tag{18}$$

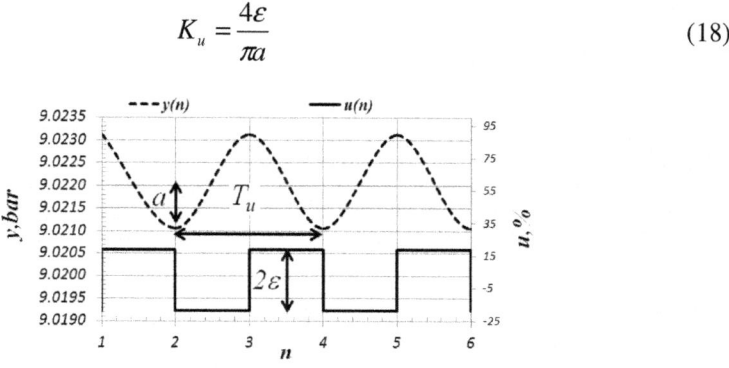

Fig. 10. Generated controller stimulation $u(n)$ and output variable $y(n)$

While the process is running, the PID controller parameters are automatically adjusted based on the rules of the Ziegler-Nichols method, as shown by Table 2. The P_{aj} parameter enables a controller fine-tuning, allowing to obtain the output response with different damping sizes.

Table 2. Tuning parameters equations

Controller	Parameters		
PID	$kc = 0.6 * P_{aj} * Ku$	$ti = 0.5 * Tu$	$td = 0.125 * Tu$

Therefore, a set of parameters for each sampling time have been obtained, resulting in an auto-tuning closed loop response. Figure 11 shows the pressure control structure, when submitted to a set point step change. The Integral of Time multiplied by the absolute value of error (ITAE) criterion performance has been used for both situations: without auto-tuning, Figure 11 (I) and with auto-tuning, Figure 11 (II). It can be observed a superior performance when the auto-tuning procedure is operating, where the ITAE (II) = 8.0458 has a more satisfactory result when compared to the process without the auto-tuning, ITAE (I) = 15.1659, even in the presence of large disturbances.

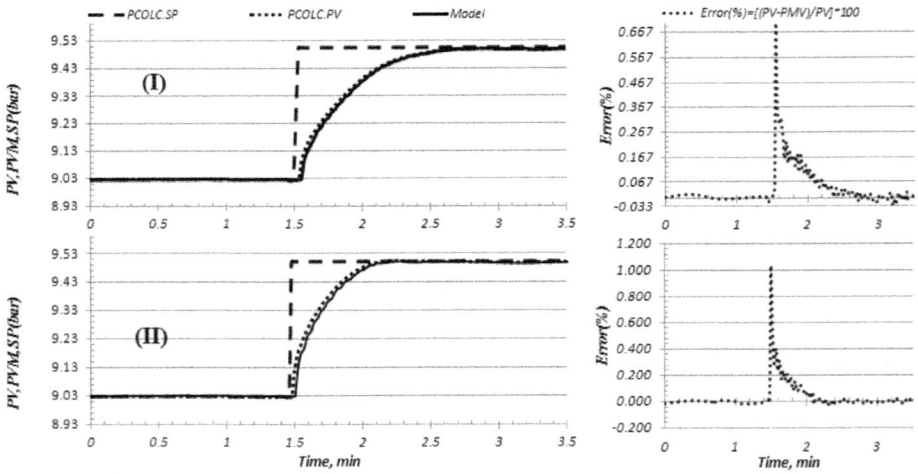

Fig. 11. Pressure control loop without auto-tuning(1), with auto-tuning(2)

Figure 12 shows the flow control structure, when submitted to a set point step change. The performance of two situations with and without auto-tuning, Figure 12 (I) and (II), are respectively given by: ITAE (I) = 692.79 and ITAE (II) = 11.37, where the control system operated with auto-tuning procedure has presented a superior performance.

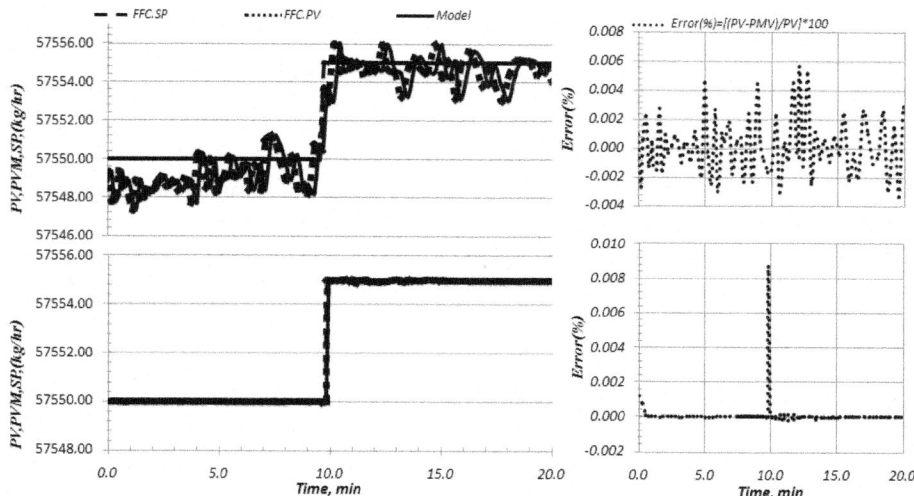

Fig. 12. Flow control loop without auto-tuning(1), with auto-tuning(2)

In Figure 13 the temperature control structure is presented. Such a structure is the master cascade structure, and similar to previous tests, these data are referent a step change in the set point. Once more, the performance in both two situations: with and without auto-tuning is considered, Figure 13 (I) and (II), respectively. A more satisfactory result can be obtained, in accordance with the following ITAE performance criterion: ITAE (I) = 3.532 and ITAE (II) = 0.3975, when the auto-tuning procedure is operating. It is important to be mentioned that for all presented structures a good agreement between the process and model can be observed.

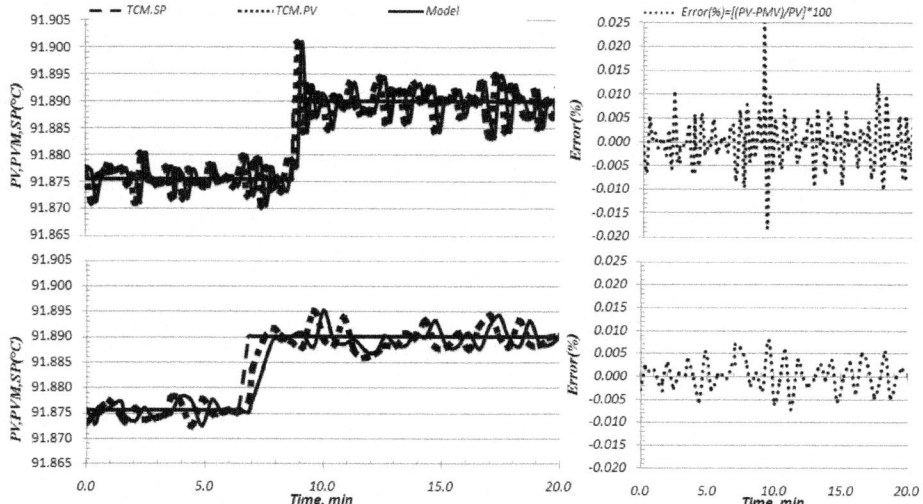

Fig. 13. Temperature control loop without auto-tuning(1), with auto-tuning(2)

8   Conclusions

The auto-tuning procedure presented and used in this study allowed the online controller parameters estimation and updating. Such a procedure uses in its structure the ARMAX model with recursive estimation as the process model. Finally, it must be observed that when there are changes in the operating point new values for the tuning parameters are of fundamental importance in the performance improvement of control structures.

References

1. Bobál, V., Böhm, J., Fessl, J., Machácek, J.: Digital Self-Tuning Controller – Algorithms. Implementation and Applications. Springer, London (2005)
2. Sastry, S., Bodson, M.: Adaptive Control: Stability, Convergence and Robustness. Prentice-Hall, Englewood Cliffs (1989)
3. Åström, K.J., Wittenmark, B.: Adaptive Control. Addison-Wesley Publishing Companny, Reading (1995)
4. Åström, K.J.: Tuning and Adaptation, IFAC 13th Triennial World Congress, San Francisco, USA, pp. 1–18 (1996)
5. Witsenhausen, H.S.: Separation of Estimation and Control for Discrete Time Systems. Proc. IEEE 59(11), 1557 (1971)
6. Soong, T.T.: Modelos Probabilísticos em Engenharia e Ciências. Trad. Alfredo Alves de Farias. LTC, Rio de Janeiro (1986)
7. Ljung, L., Söderström, T.L.: Theory and Practice of Recursive Identification. MIT Press, Cambridge (1983)
8. Moore, S.M., Lai, J.C.S., Shankar, K.: ARMAX Modal Parameter Identification in the Presence of Unmeasured Excitation - I: Theoretical Background. Mechanical Systems and Signal Processing 21, 1616–1641 (2007)
9. Holst, J.: Adaptive Prediction and Recursive Estimation. Department of Automatic Control. Lund Institute of Technology, Sweden (1977)
10. Silva, J.N., Bispo, H., Brito, R.P., Manzi, J.: Robust Stability Analysis Inspired by Classical Statistical Principles. Canadian Journal of Chemical Engineering 6, 82–89 (2013)
11. Åström, K.J., Hagglund, T.: PID Controllers: Theory, Design and Tuning, 2nd edn. Inst. Society of America, North Carolina (1995)
12. Rasmussen, H.: Automatic Tuning of PID Regulators. Aalborg University, Denmark (1993)
13. Åström, K.J., Hagglund, T.: Automatic Tuning of Simple Regulators with Specifications on Phase and Amplitude Margins. Automatica 20(5), 645–651 (1984)

Position Control Scheme for Induction Motors Using Sliding Mode Observers and Controller

Oscar Barambones and Patxi Alkorta

Dpto. Ingeniería de Sistemas y Automática, University of the Basque Country
Engineering School, Nieves Cano 12, 01006 Vitoria, Spain

Abstract. A robust position control for induction motors using sliding mode observers and controller is proposed. The proposed control scheme incorporates a sliding mode based flux and load torque estimator in order to avoid these sensors, that increases the cost and reduces the system reliability. The proposed control scheme presents a robust performance under system uncertainties and external load torque disturbances. Additionally, the proposed control scheme does not present a high computational cost and therefore can be implemented easily in high-performance real time applications for induction motors.

1 Introduction

Position control is often used in some applications of electrical drives like robotic systems, conveyor belts, etc. In these applications, traditionally DC motors are used due to their linear behavior. However, the squirrel cage induction motor (IM) presents some excellent constructional features such as reliability, high efficiency, ruggedness, low cost, and low maintenance, which make the use of an IM very attractive for some applications, but due to their highly coupled nonlinear structure, the IM control presents some drawbacks that should be solved using more sophisticated controllers [1]-[3].

On the other hand in the last decade remarkable efforts have been made to reduce the number of sensors in the control systems. The sensors increase the cost and also reduce the reliability of the control system because this elements are generally expensive, delicate and difficult to install [4],[5].

In the present work an adaptive robust position control for high-performance applications of induction motors is proposed. The design also incorporates a flux and load torque observers based on the sliding mode theory [6]. The overall control scheme does not involve a high computational cost and therefore can be implemented easily in a real time applications.

The controller incorporates an adaptation law for the sliding gain so that the sliding mode controller can adapt the sliding gain value that is necessary to overcome the system uncertainties. In this sense, the control signal of the proposed sliding mode control scheme will be smaller than the control signals of the traditional sliding mode control schemes, because in the latter, the sliding gain value should be chosen high enough to overcome all the possible uncertainties that could appear in the system over time.

© Springer International Publishing Switzerland 2015   273
A.P. Moreira et al. (eds.), *CONTROLO'2014 - Proc. of the 11th Port. Conf. on Autom. Control*,
Lecture Notes in Electrical Engineering 321, DOI: 10.1007/978-3-319-10380-8_26

The estimated rotor flux is used to calculate the rotor flux vector angular position, whose value is essential in order to apply the field oriented control principle.

Finally, the control scheme presented in this paper is validated in a real test, using a commercial IM, in order to demonstrate the real performance of this controller.

2   Sliding Mode Observer for the Rotor Flux

Using the well-known IM model dynamics in the stationary reference frame [7], [2] and taking into account the singular perturbation theory [8], it is obtained the singularly perturbed model of the IM, where the currents are the fast variables of the system and the fluxes are the slow variables:

$$
\begin{aligned}
\varepsilon \dot{i}_{ds} &= -L_m \alpha_r i_{ds} + \alpha_r \psi_{dr} + w_r \psi_{qr} + \frac{L_r}{L_m}(V_{ds} - R_s i_{ds}) \\
\varepsilon \dot{i}_{qs} &= -L_m \alpha_r i_{qs} - w_r \psi_{dr} + \alpha_r \psi_{qr} + \frac{L_r}{L_m}(V_{qs} - R_s i_{qs}) \\
\dot{\psi}_{dr} &= L_m \alpha_r i_{ds} - \alpha_r \psi_{dr} - w_r \psi_{qr} \\
\dot{\psi}_{qr} &= L_m \alpha_r i_{qs} + w_r \psi_{dr} - \alpha_r \psi_{qr}
\end{aligned}
\tag{1}
$$

where V_{ds}, V_{qs} are stator voltages; i_{ds}, i_{qs} are stator currents; ψ_{dr}, ψ_{qr} are rotor fluxes; w_r is motor speed; R_s, R_r are stator and rotor resistances; L_s, L_r are stator and rotor inductances; L_m, is mutual inductance; $\sigma = 1 - \frac{L_m^2}{L_s L_r}$ is leakage coefficient; $T_r = \frac{L_r}{R_r}$ is rotor-time constant; $\varepsilon = \frac{\sigma L_s L_r}{L_m}$ and $\alpha_r = \frac{1}{T_r}$.

The proposed sliding mode observer is based on the original system model, which has corrector terms with switching functions based on the system outputs. Therefore, considering the measured stator currents as the system output errors, the corresponding sliding-mode-observer can be constructed as follows:

$$
\begin{aligned}
\varepsilon \dot{\hat{i}}_{ds} &= -L_m \alpha_r i_{ds} + \alpha_r \hat{\psi}_{dr} + w_r \hat{\psi}_{qr} + \frac{L_r}{L_m}(V_{ds} - R_s i_{ds}) + k_1 e_{i_d} - g_{i_d} \operatorname{sgn}(e_{id}) \\
\varepsilon \dot{\hat{i}}_{qs} &= -L_m \alpha_r i_{qs} - w_r \hat{\psi}_{dr} + \alpha_r \hat{\psi}_{qr} + \frac{L_r}{L_m}(V_{qs} - R_s i_{qs}) + k_2 e_{iq} - g_{i_q} \operatorname{sgn}(e_{iq}) \\
\dot{\hat{\psi}}_{dr} &= L_m \alpha_r i_{ds} - \alpha_r \hat{\psi}_{dr} - w_r \hat{\psi}_{qr} - g_{\psi_d} \operatorname{sgn}(e_{id}) \\
\dot{\hat{\psi}}_{qr} &= L_m \alpha_r i_{qs} + w_r \hat{\psi}_{dr} - \alpha_r \hat{\psi}_{qr} - g_{\psi_q} \operatorname{sgn}(e_{iq})
\end{aligned}
\tag{2}
$$

where $\hat{i}$ and $\hat{\psi}$ are the estimations of i and ψ; k_1 and k_2 are positive constant gains; g_{i_d}, g_{i_q}, g_{ψ_d} and g_{ψ_q} are the observer gain matrix; $e_{i_d} = i_{ds} - \hat{i}_{ds}$ and $e_{i_q} = i_{qs} - \hat{i}_{qs}$ are de current estimation errors, and sgn() is the sign function.

Subtracting (2) from (1), the estimation error dynamics can be expressed in matrix form as:

$$
\begin{aligned}
\varepsilon \dot{e}_i &= +A e_\psi + K_i e_i + G_i \Upsilon_e \\
\dot{e}_\psi &= -A e_\psi + G_\psi \Upsilon_e
\end{aligned}
\tag{3}
$$

where $e_{\psi_d} = \psi_{dr} - \hat{\psi}_{dr}$, $e_{\psi_q} = \psi_{qr} - \hat{\psi}_{qr}$ are the flux estimation errors, $A = \alpha_r I_2 - w_r J_2$, $e_i = [e_{i_d} \; e_{i_q}]^T$, $e_\psi = [e_{\psi_d} \; e_{\psi_q}]^T$, $\Upsilon_e = [\mathrm{sgn}(e_{id}) \; \mathrm{sgn}(e_{iq})]^T$, $G_i = \begin{bmatrix} g_{i_d} & 0 \\ 0 & g_{i_q} \end{bmatrix}$, $G_\psi = \begin{bmatrix} g_{\psi_d} & 0 \\ 0 & g_{\psi_q} \end{bmatrix}$, $I_2 = \begin{bmatrix} 1 & 0 \\ 0 & 1 \end{bmatrix}$, $J_2 = \begin{bmatrix} 0 & -1 \\ 1 & 0 \end{bmatrix}$, $K_i = \begin{bmatrix} -k_1 & 0 \\ 0 & -k_2 \end{bmatrix}$

Following the two-time-scale approach, the stability analysis of the above system can be considered determining the observer gains G_i and K_i of the fast subsystem or measured state variables (i_{ds}, i_{qs}), to ensure the attractiveness of the sliding surface $e_i = 0$. Thereafter, the observer gain G_ψ of the slow subsystem or inaccessible state variables (ψ_{dr}, ψ_{qr}), are determined, such that the reduced-order system obtained when $e_i \cong \dot{e}_i \cong 0$ is locally stable [8].

The attractivity condition $e_i^T \dot{e}_i < 0$ of the sliding surface $(e_i = 0)$ for the fast reduced-order system is fulfilled with the following inequalities:

$$g_{i_d} < - \left| \alpha_r e_{\psi_d} + w_r e_{\psi_q} \right| + k_1 \left| e_{id} \right|$$
$$g_{i_q} < - \left| -w_r e_{\psi_d} + \alpha_r e_{\psi_q} \right| + k_2 \left| e_{iq} \right| \tag{4}$$

The proof is carried out using the following Lyapunov function candidate $V = \frac{1}{2}\varepsilon e_i^T e_i$ whose time derivative is:

$$\frac{dV}{d\tau} = e_i^T \varepsilon \dot{e}_i = e_i^T [A e_\psi + K_i e_i + G_i \Upsilon_e] \tag{5}$$

$$= \begin{bmatrix} e_{id} \left\{ g_{i_d} \, \mathrm{sgn}(e_{id}) + \alpha_r e_{\psi_d} + w e_{\psi_q} - k_1 e_{id} \right\} \\ e_{iq} \left\{ g_{i_q} \, \mathrm{sgn}(e_{iq}) - w e_{\psi_d} + \alpha_r e_{\psi_q} - k_2 e_{iq} \right\} \end{bmatrix}$$

Taking into account that all states and parameters of IM are bounded, then there exist sufficiently large negative numbers g_{i_d}, g_{i_q}, and positive numbers k_1 and k_2 so that the inequalities defined in (4) are verified.

Then, once the currents trajectory reaches the sliding surface $e_i = 0$, the observer error dynamics given by (3) behaves as a reduced-order subsystem governed only by the rotor-flux error e_ψ, assuming that $e_i = \dot{e}_i = 0$.

$$0 = +A e_\psi + G_i \Upsilon_e \quad \Rightarrow \quad e_\psi = -A^{-1} G_i \Upsilon_e$$
$$\dot{e}_\psi = -A e_\psi + G_\psi \Upsilon_e \quad \Rightarrow \quad \dot{e}_\psi = (G_i + G_\psi) \Upsilon_e \tag{6}$$

In order to demonstrate de stability of the previous system, the following Lyapunov function candidate $V = \frac{1}{2} e_\psi^T e_\psi$ is proposed, whose time derivative is:

$$\frac{dV}{dt} = e_\psi^T \dot{e}_\psi = -\Upsilon_e^T (G_i + G_\psi)^T A^{-1} G_i \Upsilon_e = -\Upsilon_e^T \left[(I_2 + G_\psi G_i^{-1}) G_i \right]^T A^{-1} G_i \Upsilon_e$$

$$= -\Upsilon_e^T G_i^T (A^{-1})^T A^T (I_2 + G_\psi G_i^{-1})^T A^{-1} G_i \Upsilon_e$$

$$= -(A^{-1} G_i \Upsilon_e)^T A^T (I_2 + G_\psi G_i^{-1})^T A^{-1} G_i \Upsilon_e$$

$$= -e_\psi^T A^T (I_2 + G_\psi G_i^{-1})^T e_\psi = -e_\psi (I_2 + G_\psi G_i^{-1}) A \, e_\psi^T \tag{7}$$

To ensure that $\dot{V}$ is negative definite the following sufficient condition can be requested:

$$(I_2 + G_\psi G_i^{-1}) A \geq \varrho I_2 \tag{8}$$

where ϱ is a positive constant

Solving the gain matrix G_ψ in (8) yields:

$$G_\psi \leq (\varrho A^{-1} - I_2)G_i \tag{9}$$

Therefore, the time derivative of the Lyapunov function will be negative definite if the observer gain G_ψ is chosen taking into account (9). As a result from (7) and (8) it is concluded the flux observer error converges to zero with exponential rate of convergence.

3   Sliding Mode Load Torque Observer

Using the field-orientation control principle, the current component i_{ds}^e is aligned in the direction of the rotor flux vector $\bar{\psi}_r$, and the current component i_{qs}^e is aligned in the perpendicular direction to it. At this condition, it is satisfied that $\psi_{qr}^e = 0, \psi_{dr}^e = |\bar{\psi}_r|$, and then the mechanical equation of an IM can be written as:

$$J\ddot{\theta}_m + B\dot{\theta}_m + T_L = T_e \tag{10}$$

where J and B are the inertia constant and the viscous friction coefficient of the IM respectively; T_L is the external load; θ_m is the rotor mechanical position, which is related to the rotor electrical position, θ_r, by $\theta_m = 2\,\theta_r/p$ where p is the pole numbers and T_e denotes the generated torque of an IM, defined as:

$$T_e = \frac{3p}{4}\frac{L_m}{L_r}\psi_{dr}^e i_{qs}^e = K_T\,i_{qs}^e \tag{11}$$

where ψ_{dr}^e is the rotor-flux linkages, with the subscript 'e' denoting that the quantity is referred to the synchronously rotating reference frame; i_{ds}^e and i_{qs}^e are the d-q stator current components.where K_T is the torque constant.

The load torque only changes at certain moments, and then it can be considered as a quasi-constant signal. Therefore, from (10) the system state space equations are:

$$\dot{w}_m = -\frac{B}{J}w_m + \frac{K_T}{J}i_{qs}^e - \frac{1}{J}T_L$$
$$\dot{T}_L = 0 \tag{12}$$

From singular perturbation theory [8], the stability of the above system can be demonstrated assuring the asymptotic stability of the fast component of this system (the rotor speed), and thereafter the convergence of the slow component (the load torque) for the reduced system, when the rotor speed estimation error is zero.

Then, from (12) the sliding mode observer can be constructed as:

$$\dot{\hat{w}}_m = \frac{-B}{J}w_m + \frac{K_T}{J}i_{qs}^e - \frac{1}{J}\hat{T}_L + k_{w_1}e_w + h_1\,\mathrm{sgn}(e_w)$$
$$\dot{\hat{T}}_L = -k_{w_2}e_w - h_2\,\mathrm{sgn}(e_w) \tag{13}$$

where $e_w = w_m - \hat{w}_m$, and k_{w_1}, k_{w_2}, h_1 and h_2 are a positive constants. Subtracting (13) from (12), the estimation error dynamic is obtained:

$$\dot{e}_w = -\frac{1}{J}e_T - k_{w_1}e_w - h_1 \operatorname{sgn}(e_w)$$
$$\dot{e}_T = k_{w_2}e_w + h_2 \operatorname{sgn}(e_w) \tag{14}$$

where $e_T = T_L - \hat{T}_L$

In order to demonstrate the stability of the fast component of the system the following Lyapunov function candidate is proposed $V = \frac{1}{2}e_w^2$, whose time derivative is:

$$\dot{V} = e_w \dot{e}_w = e_w \left(-\frac{1}{J}e_T - k_{w_1}e_w - h_1 \operatorname{sgn}(e_w) \right) = -h_1|e_w| - k_{w_1}e_w^2 - \frac{1}{J}e_w e_T$$

To ensure that $\dot{V}$ is negative definite the following sufficient condition can be requested:

$$h_1 \geq \left| \frac{1}{J}e_T \right| - k_{w_1}|e_w| + \eta_w \quad \eta_w > 0 \tag{15}$$

Therefore, the speed observation error reaches the equilibrium point, $e_w = 0$ and $\dot{e}_w = 0$, and then from (14) it is obtained that the observer error dynamics behaves as the reduced-order subsystem presented below:

$$0 = -\frac{1}{J}e_L - h_1 \operatorname{sgn}(e_w) \tag{16}$$
$$\dot{e}_T = h_2 \operatorname{sgn}(e_w) \tag{17}$$

From the previous equations it is obtained:

$$\dot{e}_T = -\frac{1}{J}\frac{h_2}{h_1}e_T \tag{18}$$

Consequently, the load torque estimation error tends exponentially to zero.

4   Adaptive Sliding Mode Position Control

The mechanical equation of an IM can be written as [2]:

$$\ddot{\theta}_m = -(a + \triangle a)\dot{\theta}_m - (f + \triangle f) + (b + \triangle b)i_{qs}^e \tag{19}$$

where the parameters are defined as:

$$a = \frac{B}{J}, \quad b = \frac{K_T}{J}, \quad f = \frac{\hat{T}_L}{J}; \tag{20}$$

and the terms $\triangle a$, $\triangle b$ and $\triangle f$ represents the uncertainties of the terms a, b and f respectively.

It should be noted that the load torque T_L has been replaced by the estimated load torque $\hat{T}_L$ and the difference between the real and the estimated value is taken as an uncertainty.

Let us define the position tracking error as follows:

$$e(t) = \theta_m(t) - \theta_m^*(t) \tag{21}$$

where θ_m^* is the rotor position command.

Taking the second derivative of the previous equation with respect to time yields:

$$\ddot{e}(t) = \ddot{\theta}_m - \ddot{\theta}_m^* = -a\,\dot{e}(t) + u(t) + d(t) \tag{22}$$

where the following terms have been collected in the signal $u(t)$,

$$u(t) = b\,i_{qs}^e(t) - a\,\dot{\theta}_m^*(t) - f(t) - \ddot{\theta}_m^*(t) \tag{23}$$

and the uncertainty terms have been collected in the signal $d(t)$,

$$d(t) = -\triangle a\,w_m(t) - \triangle f(t) + \triangle b\,i_{qs}^e(t) \tag{24}$$

Now, we are going to define the sliding variable $S(t)$ as:

$$S(t) = \dot{e}(t) + k\,e(t) \tag{25}$$

where k is a positive constant gain.

The sliding mode position controller is designed as:

$$u(t) = -(k-a)\,\dot{e}(t) - \hat{\beta}\gamma\,\mathrm{sgn}(S) \tag{26}$$

where k is the previously defined gain, β is the switching gain.

Finally, the switching gain $\hat{\beta}$ is adapted according to the following updating law:

$$\dot{\hat{\beta}}(t) = \gamma\,|S(t)| \qquad\qquad \hat{\beta}(0) = 0 \tag{27}$$

where γ is a positive constant that let us choose the adaptation speed for the sliding gain.

The stability proof of this position controller will be carried out using the following Lyapunov function: $V(t) = \dfrac{1}{2}S(t)S(t) + \dfrac{1}{2}\tilde{\beta}(t)\tilde{\beta}(t)$

where $\tilde{\beta}(t) = \hat{\beta}(t) - \beta$, and β is a constant that satisfies $\beta > |d(t)| + \eta \quad \forall\,t$, and $\eta > 0$.

The time derivative of $V(t)$ is:

$$\dot{V}(t) = S(t)\dot{S}(t) + \tilde{\beta}(t)\dot{\tilde{\beta}}(t) = S \cdot [\ddot{e} + k\dot{e}] + \tilde{\beta}\dot{\hat{\beta}} = S\,[(-a\,\dot{e} + u + d) + k\,\dot{e}] + \tilde{\beta}\gamma|S|$$

$$= S\left[-\hat{\beta}\gamma\,\mathrm{sgn}(S) + d\right] + (\hat{\beta} - \beta)\gamma|S| = d\,S - \beta\gamma|S| \tag{28}$$

$$\leq |d||S| - \beta\gamma|S| \leq |d||S| - (d_{max} + \eta)\gamma|S| = |d||S| - d_{max}\,\gamma|S| - \eta\gamma|S|$$

$$\leq -\eta\gamma|S| \tag{29}$$

Using the Lyapunov's direct method, since $V(t)$ is clearly positive-definite, $\dot{V}(t)$ is negative semidefinite and $V(t)$ tends to infinity as $S(t)$ and $\tilde{\beta}(t)$ tends to infinity, then the equilibrium at the origin $[S(t), \tilde{\beta}(t)] = [0, 0]$ is globally stable, and therefore the variables $S(t)$ and $\tilde{\beta}(t)$ are bounded. Since $S(t)$ is bounded then it is deduced that $e(t)$ and $\dot{e}(t)$ are bounded.

On the other hand, the derivative of (25) can be obtained as,

$$\dot{S}(t) = \ddot{e}(t) + k\dot{e}(t) = \dot{S}(t) = -a\dot{e}(t) + u(t) + d(t) + k\dot{e}(t) = d(t) - \hat{\beta}\gamma\,\mathrm{sgn}(S) \quad (30)$$

From equation (30) we can conclude that $\dot{S}(t)$ is bounded because $d(t)$, γ and $\hat{\beta}$ are bounded.

Now, from equation (28) it is deduced that

$$\ddot{V}(t) = d\,\dot{S}(t) - \beta\,\gamma\frac{d}{dt}|S(t)| \quad (31)$$

which is a bounded quantity because $\dot{S}(t)$ is bounded.

Under these conditions, since $\ddot{V}$ is bounded, $\dot{V}$ is a uniformly continuous function, so Barbalat's lemma let us conclude that $\dot{V} \to 0$ as $t \to \infty$, which implies that $S(t) \to 0$ as $t \to \infty$.

Finally, the torque current command, $i_{qs}^{e*}(t)$, can be obtained directly substituting (26) in (23):

$$i_{qs}^{e*}(t) = \frac{1}{b}\left[a\,\dot{\theta}_m^* + \ddot{\theta}_m^* + f(t) - (k - a)\,\dot{e} - \hat{\beta}\gamma\,\mathrm{sgn}(S)\right] \quad (32)$$

The global asymptotic stability of the closed-loop system with the proposed sliding mode observers, is provided by the separation principle, which requires the asymptotic stability of the observer fast enough, such that it brings the state estimate close enough to its real value in a short time and restores the stabilizing powers of the controller as a necessary and sufficient condition [9].

5   Experimental Results

In order to carry out the real experimental validation of the proposed control scheme, the control platform shown in figure 1 has been designed and constructed. This control platform allows us to verify the real time performance of the controllers in a commercial IM. The platform is formed by a PC with Windows XP in which it is installed MatLab7/Simulink R14 and ControlDesk 2.7 and the DS1103 Controller Board real time interface of dSpace. The power block is formed of a three-phase rectifier connected to 380 V/50 Hz AC electrical net and a capacitor bank of 27.200 μF in order to get a DC bus of 540 V. The platform also includes a three-phase IGBT/Diode bridge of 50A, and the M2AA 132M4 ABB induction motor of 7.5kW of die-cast aluminium squirrel-cage type. The platform also includes a 190U2 Unimotor synchronous AC servo motor of

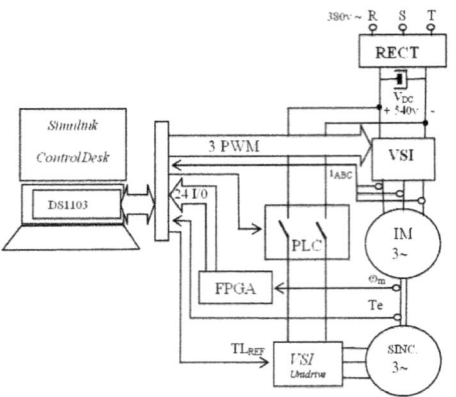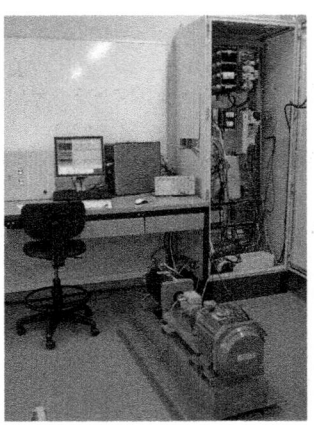

Fig. 1. Block diagram and photography of the induction motor experimental platform

10.6 kW connected to the IM to generate the load torque (controlled in torque). This servo motor is controlled by its VSI Unidrive inverter module.

In the experimental validation it is assumed that there is an uncertainty of 50% in the system parameters, that will be overcome by the proposed sliding mode control. The values for the controller parameters are: $k = 56$ and $\gamma = 10$, the values for the flux observer parameters are: $g_{i_d} = -44.5$, $g_{i_q} = -44.5$, $g_{\psi_d} = -50$, $g_{\psi_q} = -50$, $k_1 = 100$ and $k_2 = 100$, and the values for the load torque observer parameters are: $k_{w_1} = 25$, $k_{w_2} = 250$ $h_1 = 100$ and $h_2 = 100$.

Figure 2 shows the real test carried out in the experimental platform using the proposed observer-controller scheme.

In this real experiment the control scheme performance using various position reference changes and load torque variations, is presented. The load torque follows a ramp command until $t = 1\,s$, then it is maintained constant, then at time $t = 2\,s$ the load torque steps from $20\,Nm$ to $40\,Nm$ and finally steps from $40\,Nm$ to $60\,Nm$ at time $t = 3.25\,s$, which is a 20% above the nominal torque value $(49Nm)$.

The first graph shows the reference and the real rotor position, and the second graph shows the rotor position error. As it can be observed, after a transitory time, the rotor position tracks the desired position in spite of system uncertainties. Nevertheless, at time $t = 2\,s$ and $t = 3.25\,s$ a little position error can also be observed because there is a torque increment at this time. At these times, the controlled system lost the so called "sliding mode" because the actual sliding gain is too small in order to maintain the system in the sliding mode. However after a while, the sliding gain is adapted so that the new sliding gain value can lead the system to the sliding mode, and then the rotor position error is eliminated. The third graph shows the estimated rotor flux. The fourth graph shows the motor torque, the load torque and the estimated load torque. As it can be seen in this graph the proposed load torque observer provides a good estimation of the load torque value. Finally, the fifth graph presents the time evolution of

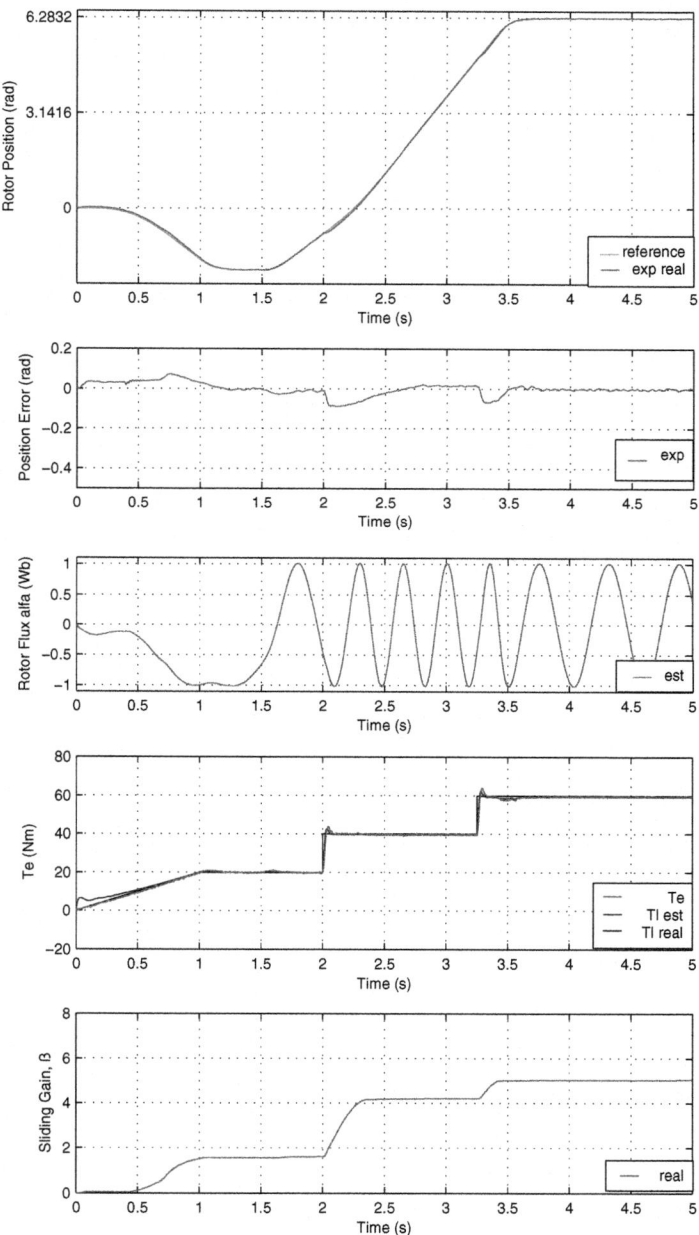

Fig. 2. Variable position tracking experimental results

the adaptive sliding gain. The sliding gain starts from zero and then it is increased until its value is high enough in order to lead the system to the sliding mode.

6 Conclusion

In this paper an IM position regulation using an adaptive sliding mode control for a real-time applications has been presented. The proposed design incorporates an adaptation law for the sliding gain, in order to calculate the sliding gain, that is necessary to overcome the system uncertainties and a rotor flux and load torque estimators based on sliding mode theory. The proposed observers and the proposed controller do not involve a high computational cost and therefore this control scheme can be easily implemented in a low cost DSP-processor. Finally, by means of real examples, it has been confirmed that the proposed position control scheme presents a good performance in practice, and that the position tracking objective is achieved under system uncertainties and also under load torque variations.

Acknowledgements. The authors are very grateful to the Basque Government by the support of this work through the projects S-PE12UN015 and S-PE13UN039, and to the UPV/EHU by its support through the projects GIU13/41 and UFI11/07.

References

1. Yazdanpanah, R., Soltani, J., Arab Markadeh, G.R.: Nonlinear torque and stator flux controller for induction motor drive based on adaptive input–output feedback linearization and sliding mode control. Energy Conversion and Management 49, 541–550 (2008)
2. Barambones, O., Alkorta, P.: A robust vector control for induction motor drives with an adaptive sliding-mode control law. Journal of the Franklin Institute 348, 300–314 (2011)
3. Sabanovic, A.: Variable Structure Systems With Sliding Modes in Motion Control. IEEE Trans. Indus. Informatics 7, 212–223 (2011)
4. Barambones, O., Garrido, A.J., Garrido, I.: Robust speed estimation and control of an induction motor drive based on artificial neural networks. Int. J. Adapt. Control Signal Process 22, 440–464 (2008)
5. Alanis, A.Y., Sanchez, E.N., Loukianov, A.G., Hernandez, E.A.: Discrete-time recurrent high order neural networks for nonlinear identification. Journal of the Franklin Institute 347, 1253–1265 (2010)
6. Spurgeon, S.K.: Sliding mode observers: a survey. International Journal of Systems Science 39, 751–764 (2008)
7. Bose, B.K.: Modern Power Electronics and AC Drives. Prentice Hall, New Jersey (2001)
8. Kokotovic, P.V., Khalil, H., O'Reilly, J.: Singular Perturbation Methods in Control: Analysis and Design. Academic Press, New York (1996)
9. Atassi, A.N., Khalil, H.K.: Separation results for the stabilization of nonlinear systems using different high-gain observer designs. Systems & Control Letters 39, 183–191 (2000)

Pulsed Power Current Source Inverter Based Active Power Filter

Fernando M. Camilo[1], Sonia F. Pinto[1,2], Jose Fernando Silva[1,2] and Duarte M. Sousa[1,2]

[1] Instituto Superior Técnico, U Lisbon, 1049-001 Lisbon, Portugal
{fernando.camilo,soniafp,
fernando.alves,duarte.sousa}@tecnico.ulisboa.pt
[2] INESC-ID, Lisbon, Portugal

Abstract. In this paper a novel current source inverter (CSI) based active power filter controlled by a space vector modulation (SVPWM) method is used to limit the peak level power consumption of a pulsed power load. Unlike conventional approaches, the CSI active power filter energy storage is inductor based. The approach to control the CSI is established using two controllers: one to control the AC three phase currents sourced by the electric grid, and the other to control the current on the L_{ind} inductor of the DC side of the CSI. The DC current controller will produce the reference values that will be further used in $\alpha\beta$ space vector modulation, to control the AC currents. Simulation results are presented, and discussed.

Keywords: Current source inverter, SVPWM control, active pulsed power filter compensation.

1   Introduction

Pulsed power loads may cause power quality issues, especially if they are connected to weak or isolated grids [1]. Pulsed power loads may absorb very high currents during very short time periods, introducing voltage disturbances in the electric grid [2]. To mitigate these issues active filtering may be used, in which electrical energy is stored for a relatively long time period followed by a subsequent energy release in very short pulses, either a single pulse or in a controlled repetitive sequence [3].

Therefore, this paper proposes a current source inverter (CSI) to behave as an active power filter to level the high power peaks absorbed by pulsed power loads. The CSI based filter stores energy in a DC inductor and has the potential to increase the quality of distribution of electrical power to preserve voltage waveform quality.

The proposed approach to control the CSI currents is established using two controllers: one to control the AC grid currents, and the other to control the DC current of the L_{ind} inductor on the DC side of CSI. The DC current controller will produce the reference values that will be further used in $\alpha\beta$ space vector modulation, to control the AC currents.

With the proposed control system, it is possible to improve the power quality of the grid voltage, minimizing the effect of the pulsed power load in the grid and guaranteeing nearly unitary power factor.

© Springer International Publishing Switzerland 2015,
A.P. Moreira et al. (eds.), *CONTROLO'2014 - Proc. of the 11th Port. Conf. on Autom. Control,*
Lecture Notes in Electrical Engineering 321, DOI: 10.1007/978-3-319-10380-8_27

2   Proposed System

2.1   Global Model

Figure 1 shows the circuit diagram of the global system for the proposed CSI filter.

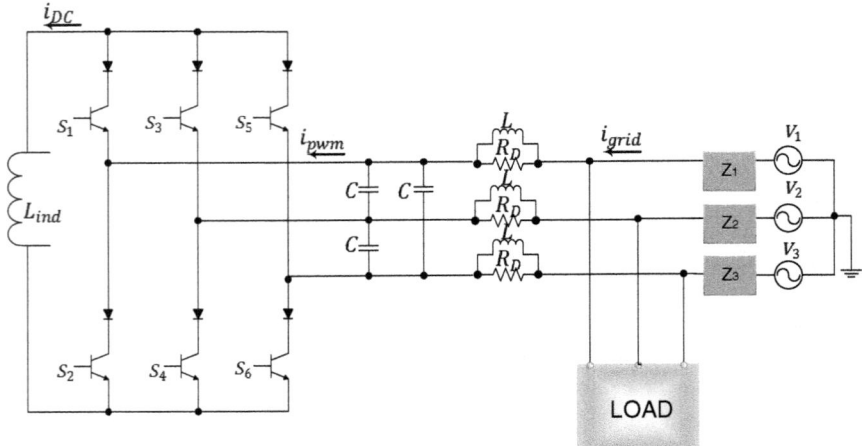

Fig. 1. Circuit diagram of the global system for proposed system

The system can be depicted as a CSI with an inductor L_{ind} current i_{DC} at the DC side, and a 2nd order low-pass filter at the AC side to reduce the harmonics created by the CSI modulation method. The CSI is connected to the grid, and is controlled to store energy in the DC inductor L_{ind}, which is then used to supply a pulsed power load, keeping the power supplied by the AC grid almost constant.

2.2   Current Source Inverter Modulation Method

The CSI is controlled using the space vector pulse width modulation (SVPWM) technique. This type of modulation guarantees high input/output transfer ratios and has the advantage of ensuring low harmonic distortion. The CSI presents nine possible combinations to connect its switches resulting in nine state space vectors, three of them being null. The spatial vectors can be obtained by applying the Concordia transformation matrix [4] to the AC currents that results from each one of these switching combinations. Knowing, in $\alpha\beta$ plane, the location of the desired AC currents, it is possible to synthesize the $\alpha\beta$ components of the AC current, using 2 vectors adjacent to the sector where the reference current vector is located. It is necessary to know the AC reference current vector angle relative to the sector where the vector is located, and the current modulation index. Considering that the switches states can be mathematically represented as "1" if they are closed and "0" if they are open, to prevent short circuits it must be ensured that the sum of the top three switches and the three lower switches at figure 1, must be equal to 1 [5]. The type of modulation used to control the CSI was symmetric commutation [6], presented in figure 2.

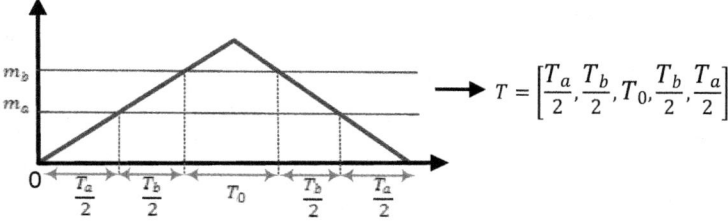

$$T = \left[\frac{T_a}{2}, \frac{T_b}{2}, T_0, \frac{T_b}{2}, \frac{T_a}{2}\right]$$

Fig. 2. Symmetric commutation

The current modulation index m (1), relates the amplitude of the AC grid current to the DC current.

$$m = \frac{I_{grid_{máx}}}{I_{DC}} \tag{1}$$

The parameters T_a T_b, T_0 for SVPWM modulation [7, 8, 9] are calculated from (2), where T_s is the switching period, ϕ_i defines the angle of the reference vector related to the sector where is located [5].

$$\begin{cases} T_a = m\,T_s\,sen\left(\frac{\pi}{3} - \phi_i\right) \\ T_b = m\,T_s\,sen(\phi_i) \\ T_0 = T_s - T_a - T_b \end{cases} \tag{2}$$

To reduce the harmonic content produced by the modulation, a symmetric commutation was selected.

2.3   Filter Design

In order to attenuate the high-frequency SVPWM harmonics a 2nd order low-pass filter is used [10, 11] (figure 1) to connect the CSI to the grid.

This filter is designed based on a single phase equivalent. The capacitor C (3), is calculated in order to maximize the power factor in the connection to the grid, guaranteeing a minimum power factor $P_{F_{min}}$. The capacitor value will depend on the grid voltage $V_{ac_{máx}}$, the mains frequency ω and on the minimum current $I_{ac_{min}}$ [11].

$$C = \frac{I_{ac_{min}}}{\omega\,V_{ac_{máx}}} \tan\left(\cos^{-1}\left(P_{F_{min}}\right)\right) \tag{3}$$

The inductance L of the filter (4) is calculated based on the filter cut-off frequency ω_p, which was assumed to be 500Hz.

$$L = \frac{1}{\omega_p^2\,C} \tag{4}$$

Due to the CSI constant DC current control, it is necessary to determine the negative incremental resistance r_i [11] (5), where η, V_0 and P_0 are the CSI efficiency, output voltage and output power, respectively.

$$r_i = -\eta \frac{V_0^2}{P_0} \tag{5}$$

The filter damping resistor (6), may be calculated from (5), and depends on the characteristic impedance Z_f.

$$R = \frac{r_i Z_f}{2 \xi r_i - Z_f} \tag{6}$$

The impedance Z_f is given by (7), where $H_p \leq 1/(2 \xi^2)$:

$$Z_f = \frac{2 \xi^2 H_p - 1}{\xi H_p} r_i \tag{7}$$

The damping factor was assumed to be $\xi = 0.7$.

2.4   Inductor L_{ind} Sizing

The inductor can be sized based on the variation (ΔI_{DC}) of the I_{DC} current.

$$2 \Delta I_{DC} \ll I_{DC} \tag{8}$$

Therefore, the inductor can be calculated from (9), where P_{ps} is the pulsed load power and Δ_t is the time duration that the pulsed load is connected.

$$L_{ind} = \frac{\Delta_t P_{ps}}{[0.5 \left[(I_{DC} + \Delta I_{DC})^2 - (I_{DC} - \Delta I_{DC})^2 \right]]} \tag{9}$$

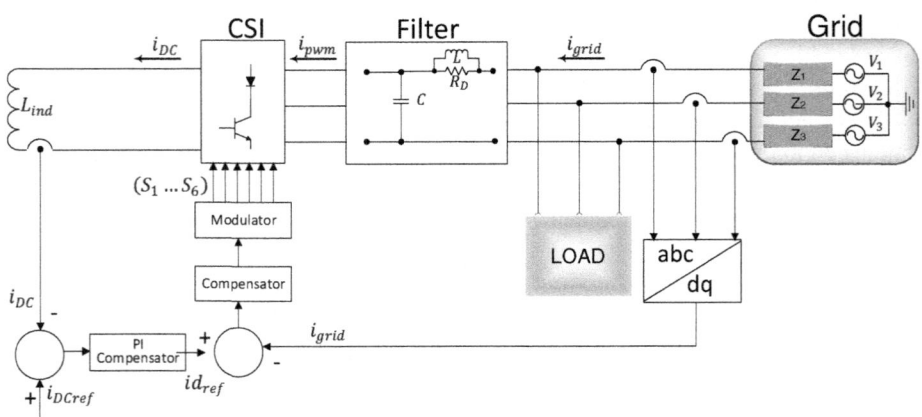

Fig. 3. Complete control scheme of the CSI active power filter

3   Controllers Design

The approach to control this CSI pulsed power filter is based in two controllers: one to control the AC current, and the other to control the DC current of the I_{DC} inductor. The figure 3 shows a final model of the controllers design.

3.1   Controller of the CSI AC Currents

With this controller it is aimed to control the AC grid currents, according to the block diagram presented in figure 4.

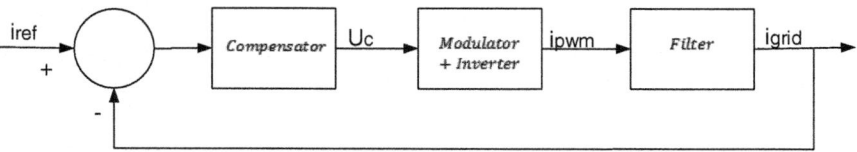

Fig. 4. Block diagram to control the CSI current

The block Filter is defined by the transfer function F(s) (10).

$$F(s) = \frac{i_{grid}}{i_{pwm}} = \frac{s\,L + R_D}{s^2\,L\,C\,R_D + s\,L + R_D} \tag{10}$$

The block Modulator+Inverter is usually defined by the transfer function M(s) [3], [12].

$$M(s) = \frac{k_D}{1 + s\,T_D} \tag{11}$$

The constant k_D represents the incremental conductance of the CSI, where I_{DC} is the current in the DC side of the CSI and $u_{cmáx}$ is the maximum value of the triangular carrier.

$$k_D = \frac{I_{DC}}{\sqrt{2}\,u_{cmáx}} \tag{12}$$

The parameter T_D represents the time delay caused by the CSI, where f_{sw} represents the switching frequency.

$$T_D = \frac{1}{2\,f_{sw}} \tag{13}$$

A polynomial compensator, with two zeros and one pole to cancel the dynamics of the 2nd order filter is used [12, 13]. The compensator can be defined by the transfer function C(s) (14), where k_l is a gain, that will be further calculated.

$$C(s) = \frac{k_l}{s}\,\frac{s^2\,L\,C\,R_D + s\,L + R_D}{s\,L + R_D} \tag{14}$$

Figure 5 shows the block diagram of the controlled system, considering that the filter dynamics is cancelled by the proposed compensator (14).

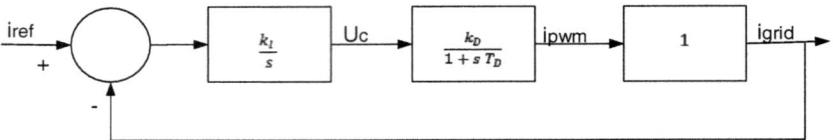

Fig. 5. Block diagram of the compensator of the CSI current, considering that the filter dynamics is cancelled by the polynomial compensator

The closed loop transfer function $G(s)$ of the controlled system can be expressed by (15):

$$G(s) = \frac{i_{grid}}{i_{ref}} = \frac{\frac{k_l k_D}{T_D}}{s^2 + s\frac{1}{T_D} + \frac{k_l k_D}{T_D}} \tag{15}$$

To equate the compensator (14) parameter k_l (16), the transfer function (15) is compared to the transfer function of a 2nd order system, considering a damping factor of $\xi = \frac{\sqrt{2}}{2}$.

$$k_l = \frac{1}{2 T_D k_D} \tag{16}$$

3.2   Control of the Current I_{DC}

This PI controller is based on dominant pole near the origin and the symmetrical optimum criterion is used [12]. It is intended to control the inductor DC current I_{DC}, according to the block diagram presented in figure 6.

The parameter T_{DL} represents the delay introduced by the converter and k_{DL} represents the gain, which depends on the voltage v_{id} on the AC side and the I_{DC} current on the DC side of the CSI, assuming that the power $P_{in} = P_{out}$.

$$k_{DL} = \frac{v_{id}}{I_{DC}} \tag{17}$$

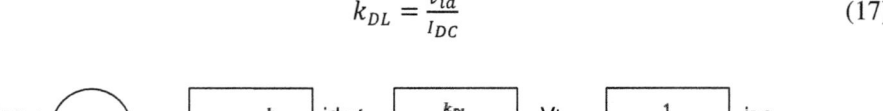
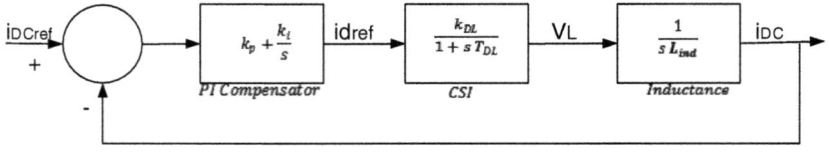

Fig. 6. Diagram block of the PI controller for The DC current

In accordance with dominant pole near the origin definition, the P and I gains can be expressed by (18) and (19), respectively, where the value of parameter a can be selected between $2 \leq a \leq 4$ [12].

$$k_p = \frac{L_{ind}}{a \, T_D \, k_D} \tag{18}$$

$$k_i = \frac{L_{ind}}{a^3 \, T_D{}^2 \, k_D} \tag{19}$$

4   Simulation Results

To test the performance of the CSI and the proposed control system, several simulations were done using Matlab/Simulink, and considering the following parameters: $L_{ind} = 1{,}84H$, $R_{load} = 1\Omega$, $L = 0{,}42mH$, $C = 80{,}9\mu F$, $R = 3\Omega$, $L_{load} = 10mH$, $I_{DC} = 240A$, $k_p = 2{,}3$, $k_i = 5{,}7$, $k_{DL} = 4$, $T_{DL} = 100ms$, $V_{ac_{rms}} = 230V$, $a = 2$, $S_{sc} = 10MVA$. The power of the pulsed load is 158,7kW.

Figure 7 represents the simulation of a pulsed load current. Without an active power filter, the pulsed load can produce a voltage drop in the power grid (figure 8), which will decrease the power quality.

Figures 9 to 11 show the simulation results obtained for the pulsed power CSI based APF. Figure 9 shows the DC I_{DC} current at the inductor. Is possible to see that the compensator guarantees that the I_{DC} current is nearly constant (240A) and with a value high enough to control the AC grid currents.

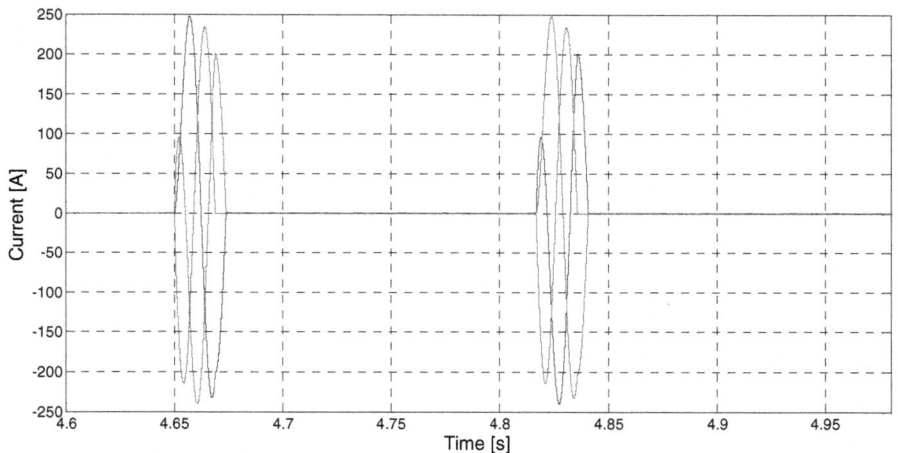

Fig. 7. Current waveforms of the three phase pulsed load

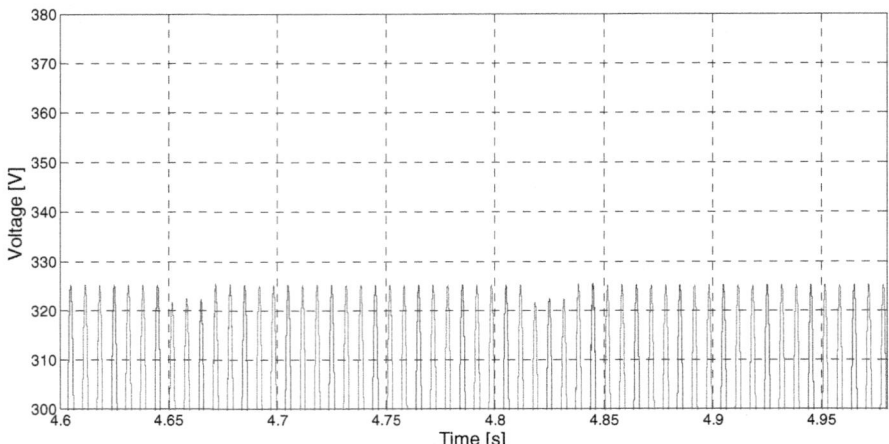

Fig. 8. Voltage drop at power grid caused by pulsed load

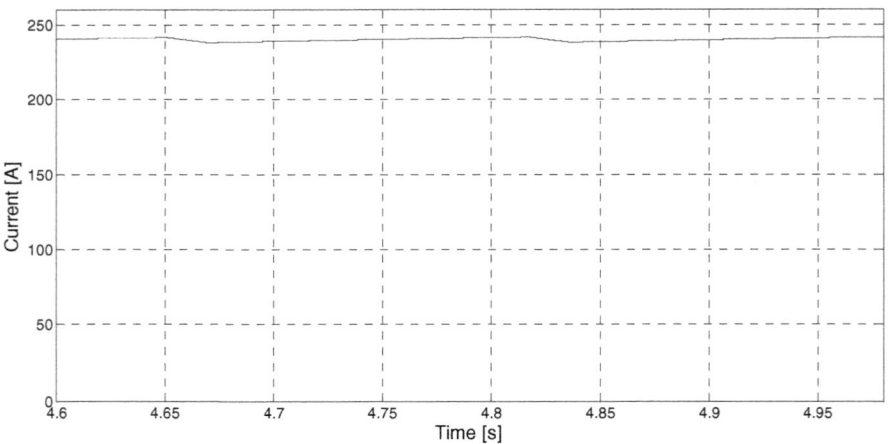

Fig. 9. Simulation of the closed loop control of the DC I_L current

From figure 10 it is possible to see that the CSI AC currents compensate the current required by the pulsed load, figure 7. As a result, and despite the pulsed power load, the amplitudes of the three phase grid currents are nearly constant, and significantly lower than the pulsed power currents, figure 11, which is a clear advantage in weak grids, minimizing the voltage drop.

As can be seen in figure 11 the peak current consumption is leveled, resulting, in this way, the quality improvement of the power grid voltage.

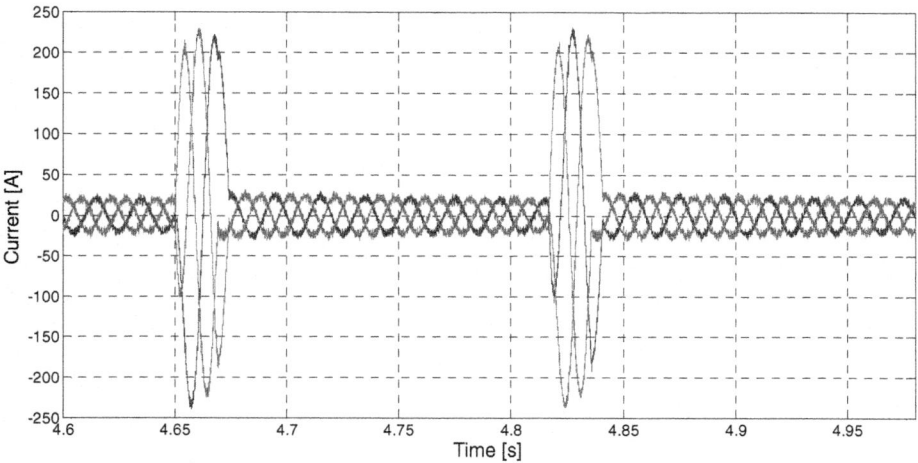

Fig. 10. CSI AC output currents

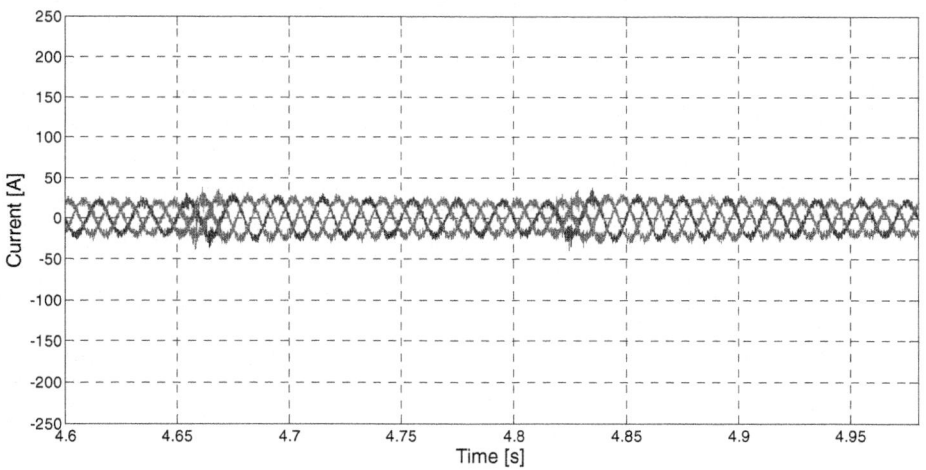

Fig. 11. Simulation of the AC currents at power grid side

5   Conclusions

A current source inverter active pulsed power filter controlled by SVPWM modulation was proposed. The CSI stores energy in the DC inductor and supplies it when needed by the pulsed power load, so that the amplitude of the AC currents supplied by the electrical network is nearly constant. Controllers were designed to actively compensate the current required by the pulsed load, allowing the tracking of the AC currents references. Simulation results were presented and allowed the validation of the proposed control system, showing that the proposed current source inverter active power filter has the capability to compensate pulsed power loads.

Acknowledgments. This work was supported by Portuguese national funds through FCT - Fundação para a Ciência e a Tecnologia, under project PEst-OE/EEI/LA0021/2013.

References

1. Xin-zhi, W., Li, W., Chao, Z., Shuang, W.: Investigating the Impact of Pulsed Load on Isolated Power System. In: International Conference on Modelling, Identification and Control, pp. 201–205 (June 2012)
2. Crider, J.M., Sudhoff, S.D.: Reducing Impact of Pulsed Power Loads on Microgrid Power Systems, Smart Grid. IEEE Transactions on Smart Grid 1(3), 270–277 (2010)
3. Silva, J.F., Pinto, S.F.: Advanced Control of Switching Power Converters. Cap. In: Rashid, M.H. (ed.) Power Electronics Handbook, 3rd edn., vol. 36, pp. 1037–1113. Butterworth-Heinemann, Burlington (2011), ISBN: 978-0-12-382036-5
4. Wu, B.: High-Power Converters and AC Drives, 2nd edn. John Wiley & Sons, New Jersey (2006)
5. Pinto, S., Silva, J.: Sliding Mode Direct Control of Matrix Converters. IET Electr. Power Appl. 1(3) (2007)
6. Lee, H., Jung, S., Sul, S.: A Current Controller Design for Current Source Inverter-fed PMSM Drive System. In: 8th International Conference on Power Electronics – ECCE. IEEE, Asia (2011)
7. Pham, D.C., Huang, S., Huang, K.: Modeling and Simulation of Current Source Inverters with Space Vector Modulation, pp. 978–971. IEEE (2010) ISBN: 978-1-4244-7720-3
8. Tan, L., Li, Y., Liu, C., Wang, P., Li, Z.: Advanced Voltage Control Methods for Current Source Inverters with Linear and Nonlinear Loads. In: IEEE International Symposium on Industrial Electronics, Seoul, Korea (2009)
9. Ronanki, D., Rajesh, K., Parthiban, P.: Simulation of SVPWM Based FOC of CSI Fed Induction Motor Drive. IEEE (2012) ISBN: 978-1-4673-0455-9
10. Pinto, S.F., Silva, J.F.: Input Filter Design for Sliding Mode Controlled Matrix Converters. In: IEEE Proc. PESC 2001, CD ROM, Vancouver, Canada (June 2001) ISBN 0-7803-7069-4
11. Silva, J.: Input Filter Design for Power Converters. IST, TULisbon (2013)
12. Silva, J.F.A.: Eletrónica Industrial: semicondutores e conversores de potência. IST, TULisbon (2013)
13. Pinto, S., Silva, J., Lopes, S.: Smart Microgeneration Systems for Power Quality Improvement, International Review of Electrical Engineering. Praize Worthy Prize 6(6), 2723–2735 (2011)

Control and Execution of Incremental Forming Using a Stewart Platform

João N.D. Torrão, Jorge A.F. Ferreira, and Ricardo J. Alves de Sousa

Universidade de Aveiro, Aveiro, Portugal

Abstract. The SPIF-A project consists on developing an entirely new and innovative machine for incremental sheet metal forming. It is mostly a team effort aiming to prove the concept, and covers various subjects from structural analysis to automation and control but also, kinematics, among others. This work focuses on machine development, namely on defining the platform's kinematics and devising its first position control system, as well as implementing a force measuring system.

The referred system is a fuzzy logic controller implemented via software on a real time target machine, that receives movement instructions from an operator station which interprets G-code.

Keywords: Stewart Platform Kinematics, Fuzzy Logic Controller, Incremental Forming, Industrial Robotics, Force Measurement System, Machine Development.

1 Forming Operations Requirements

Sheet metal parts range from simple bent and cut pieces to complex shapes, they are most commonly produced in press operations using specially made tools, in a process known as drawing. This is highly cost effective, when dealing with large batches, as they produce parts swiftly with every stroke, but since changing geometries implies costly tool and die modifications, versatility is very limited. To comply with this, many techniques have emerged, one such example is SPIF (Single Point Incremental Forming), an innovative yet simple process, where the forming tool is nothing more than a rod with a flat or spherical tip (Figure 1), which moves along a path pressing down on the sheet metal restrained on a blank holder, making it a die-less process [1].

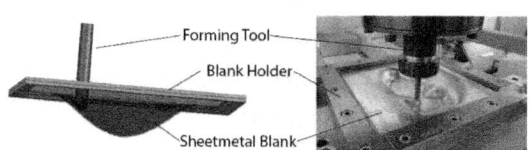

Fig. 1. SPIF components

© Springer International Publishing Switzerland 2015   293
A.P. Moreira et al. (eds.), *CONTROLO'2014 - Proc. of the 11th Port. Conf. on Autom. Control,*
Lecture Notes in Electrical Engineering 321, DOI: 10.1007/978-3-319-10380-8_28

It presents various advantages over more standard sheet metal processes such has high flexibility in terms of part size and shape, using the same tool, it is one of the few methods to produce metallic rapid prototypes, with it's main hindrance being its longer forming time [2].

While the necessary workload is significantly less than in press operations, the forming forces required are still in the 10 to 20 kN range [3,4] for a 1.6mm thick steel part, meaning that machinery must be able to provide the required power while preforming precise movements in order to properly form parts.

There are two main types of incremental forming machines, adapted and purpose built. While adapted NC machines and serial industrial manipulators may be the cheapest approach, they present various limitations [5], as both suffer from low structural stiffness and therefore are unable to handle the necessary work load for harder and/or thicker materials. Also, due to error propagations throughout their serial kinematic chain, they lack precision in same cases.

Purpose built machines, are designed specifically for incremental forming, and have the needed high stiffness and power for manufacturing of parts with complex geometries, while maintaining high accuracy and good surface finish. The Amino®Corp pioneered the market with their Dieless-NC machine [6] and has an entire range of products from small research models to large industrial ones. Another example is the Cambridge ISF capable of 26 kN work load allowing work on harder materials [5]. The only drawback of these is the fact that they have only 3 axis, meaning that the tool is always aligned with the Z-axis, while positioning the tool normal to the wall its forming proves advantageous [7].

Keeping in mind the previous requirements, and the observations of some authors [5,7], that parallel kinematic machines are a good solution for incremental forming, the SPIF-A research project [8] opted for Stewart platform. Its six parallel linear actuators grant it 6 DOF and its simple design makes it easy to build and grant the needed structural stiffness. Hydraulic cylinders are used in order to provide the necessary load to test harder and/or thicker materials that yet have to be researched.

2   SPIF-A's Kinematics and Coordinate Systems

When producing parts in machine tools, the input coordinates are derived from a point, on the work table, known as WCS (work coordinate system) and, since the base and platform geometries used in the kinematics use different references, its necessary to define the transformations between the different systems.

For a given point in the WCS, which implies position $P = (x, y, z)$ and orientation $R = (\phi, \theta, \psi)$, make up the affine transformation $^{Ws}[T]_{Pt}$.

The other transformations, from mobile platform to tool-tip/contact point $^{Mp}[T]_{Pt}$, from table to part reference point $^{Tb}[T]_{Ws}$, and from table to the fixed base on top $^{Tb}[T]_{Bp}$, can be obtained from the dimensions in Figure 2, and are used get the fixed base to mobile platform transformation $^{Bp}[T]_{Mp}$ needed to calculate the inverse kinematics, in this case the cylinder lengths.

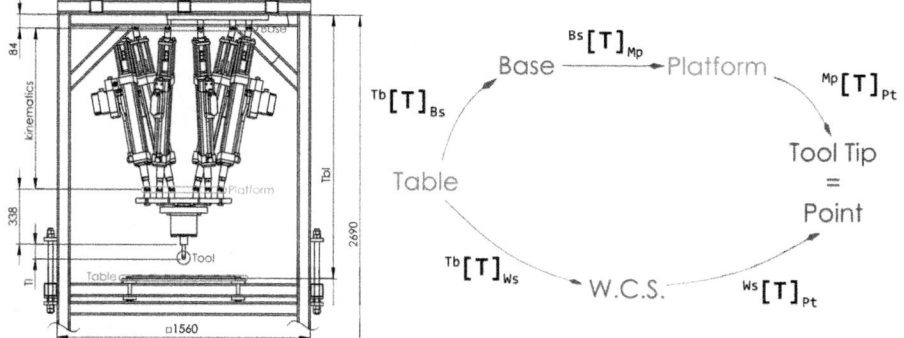

Fig. 2. SPIF-A Machine's dimensions and Kinematic reference points

2.1 Inverse Kinematics

The inverse or reverse kinematics of a Gough/Stewart platform are fairly simple and a single solution unlike it's serial manipulator counter parts.

Referring to Cauchy's model of the Articulated Octahedron, which describes a 3-3 type Stewart Platform, for any given configuration of the mobile face there is given length of the edges connecting it to the fixed face. The SPIF-A uses a 6-6 Platform, with a coplanar semi regular hexagon (SRH) base and plate geometry to retain high stiffness while still using standard u-joints avoiding the more complex and expensive joints for type 3-3 platforms [9].

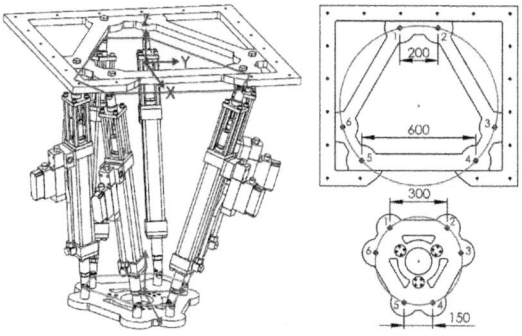

Fig. 3. SPIF-A's platform geometry and coordinate systems

The coordinates for both SRH can be calculated in relation to the length of their sides. For the coplanar base plate the vertex points $B_p(i) = [x_i, y_i, 0]^T$ with $i \in \{1, 2, 3, 4, 5, 6\}$:

$$
Bp = \begin{bmatrix} -404.145 & -404.145 & 115.470 & 115.470 & 288.675 & 288.675 \\ -100 & 100 & 400 & 300 & -300 & -400 \\ 0 & 0 & 0 & 0 & 0 & 0 \end{bmatrix}^T \quad (1)
$$

For the mobile plate the result is similar, for $M_p(i) = [p_i, q_i, 0]^T$ with $i \in \{1, 2, 3, 4, 5, 6\}$:

$$Mp = \begin{bmatrix} -173.205 & -173.205 & -43.301 & -43.301 & 216.506 & 216.506 \\ -150 & 150 & 225 & 75 & -75 & -225 \\ 0 & 0 & 0 & 0 & 0 & 0 \end{bmatrix}^T mm \quad (2)$$

When the platform is subjected to a spatial transformation T_f, either a translation, rotation or both, the euclidean distance between corresponding vertex points on the base and platform hexagons, corresponds to the 6 link lengths L_i required to achieve that particular position (x,y,z) and orientation (ϕ, θ, ψ):

$$L_i = \sqrt{(T_f \cdot Mp_i - Bp_i)^T \cdot (T_f \cdot Mp_i - Bp_i)} , i = \{1, 2, 3, 4, 5, 6\} \quad (3)$$

Using $T_f = {}^{Bs}[T]_{Mp}$ from Figure 2, the inverse kinematics can be solved to plan the forming toolpaths.

2.2   Forward Kinematics

The Forward or Direct Kinematic formulation of a Stewart platform is one of its bigger hindrances, but it is necessary to check if the machine is on target by reading the cylinders position (link lengths). While there is a single inverse kinematic solution, for a given set of link lengths, there are up to 64 complex solutions [10], this number can be cut short by optimizing the geometry, and when using SRH it drop to 24 [11]. To add to the problem the analytic equations to reach them (closed-form solutions) lead to non-linear systems of equations, that are difficult to solve[10]. One way around this is to use numerical iterative methods, which require less computational power but are dependent on the initial approximation, which should be the last know position. Mendes Lopes [12], proposed an algorithm using Newton's method based on ist fast convergence rate:

1. Select the initial approximation to the platforms position P=(x,y,z) and orientation R=(ϕ, θ, ψ) $Pw_0 \equiv Pw_k$ and the admissible error ϵ;
2. Determine the inverse Jacobian $J_E^{-1}(Pw_k)$;
3. Determine the error between the measured actuator lengths L and the length calculated using the reverse kinematics K^{-1} on the position/orientation candidate $(L - K^{-1}(Pw_k))$;
4. Solve the new candidate $Pw_{k+1} = Pw_k + J_E^{-1} \cdot (L - K^{-1}(Pw_k))$;
5. Calculate error for new candidate $\epsilon_{k+1} = |L - K^{-1}(Pw_{k+1})|$, if it surpasses the admissible error, return to step 2.

3   Control and Instrumentation

The geometry of parts made by the SPIF-A machine is described by G-code generated by CAM software and saved in a text file that is interpreted by a graphic interface in the operator's station. A Speedgoad™ SN1584 real time target machine is used to bridge the Operator and the actual hardware. It runs a preloaded Simulink™ model whose variables are read and/or changed by the operators interface via a TCP/IP connection, and used two I/O modules (analogue and digital) to interact with the valves, encoders and load cell amplifiers.

In the model there are several functions that convert the WCS movements to the required cylinder configurations, through the kinematic equations and check if the machine is on target, others that log the position and force values and also a motion controller for the actuators.

3.1   Fuzzy Logic Positional Controller

FLCs (Fuzzy Logic Controllers) have been used to some extent to control electro-hydraulic Stewart Platforms [13], having shown better behaviour than pure PID controllers, with faster response times and avoiding overshoot phenomena, having the advantage of being capable of being designed empirically by knowing how a system behaves without needing to define its complex dynamical model mathematically, for this reason, this type of control strategy was chosen as the first motion controller for the SPIF-A, in order to achieve proof of concept, and in doing so paving the way for future work on the prototype's controller. As stated parts are made using G-code to describe their geometry, to avoid a complex control scheme at this early stage, the method employed was to use a single input, single output controller for each cylinder, meaning that the positioning system would be controlled in the joint space and not the modal space, this is done by planning the path each G-code line describes the geometry and computing each cylinder's configuration every step of the way using the inverse kinematics. The forward kinematics are verified along this operation and when they verify that the target position as been reached it moves to the next line of code.

Based on Sulc's method [13], a Mandani type FLC was developed MatLab™'s Fuzzy Logic Toolbox, with the output signal dependent on the positional error and its derivative.

Seven rules were used for both the inputs and also for the output and are triangular in type with the exception being the extrema which are, the rule base was then converted into a lookup table [14], to be used in the Simulink™ model, there is the possibility of adjusting the gain for the error (K_e) and error derivative (K_{de}) inputs, this was meant for tuning the controller during the simulation stage, and also as a fine tuning mechanism in the itself to account for any discrepancies in behaviour any of the cylinders might have in comparison to its counter parts.

The controller was then tested using a transfer function from the manufacturers catalogue, following the path to form a truncated. Several gain values were tested for both the error and its derivative, as for the error the faster responses without losing stability happened using gain of 10, the derivative gain proved more sensitive therefore tuning was more challenging. What was discovered was that the gain needed to be very small, under 0.2, since the system would gradually start to oscillate around the target position, but discarding the derivative all

Table 1. Rule base for the SPIF-A's FLC

		nb	nm	ns	ze	ps	pm	pb

| | | error derivative | | | | | | |
		nb	nm	ns	ze	ps	pm	pb
	nb	nb	nb	nb	nm	nm	ns	ze
	nm	nb	nb	nm	nm	pns	ze	ps
	ns	nm	nm	ns	ns	ze	ps	pm
error	ze	nm	ns	ze	ze	ze	ps	pm
	ps	nm	ns	ze	ps	ps	pm	pm
	pm	ns	ze	ps	pm	pm	pb	pb
	pb	ze	ps	pm	pm	pb	pb	pb

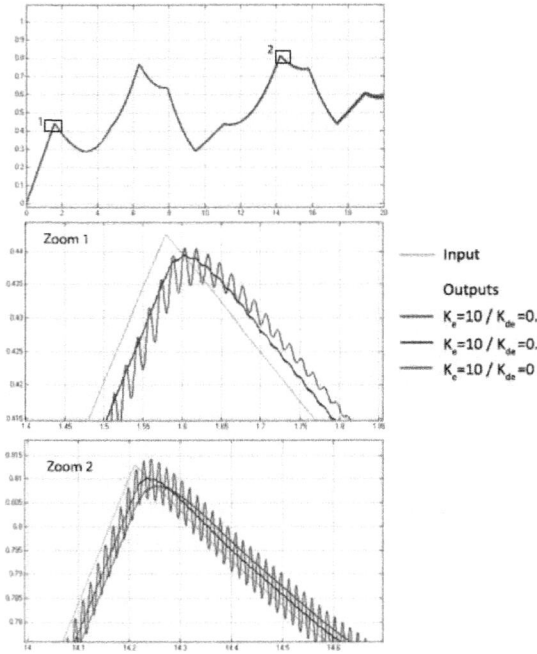

Fig. 4. Controller outputs for various tested gains

together (using a null gain) shown oscillations at the beginning of the planned path, therefore the most stable configurations tested used a 0.1 derivative gain.

3.2   Force Measuring System

Measuring the forming forces is essential to better understand the process, namely the vertical force F_v, normal to the sheet blank, and the horizontal force F_h tangent to the tool path [3]. To do so a force measuring system was devised using three load cells between the platform's moving plate and the spindle.

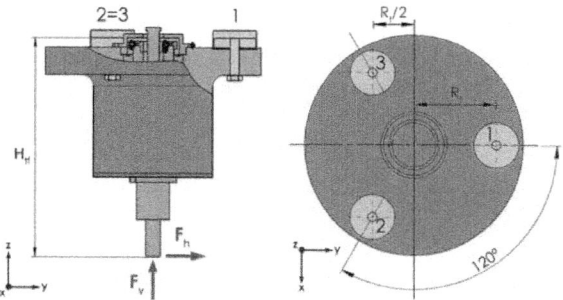

Fig. 5. Spindle forces and load cell configuration

Resolving the equilibrium equations in matrix form results in a system to convert the read load cell force values in the actual forming forces:

$$
\begin{bmatrix} F_{hx} \\ F_{hy} \\ F_v \end{bmatrix} = \begin{bmatrix} 0 & \dfrac{R_f \cdot \cos 30^\circ}{H_{tf}} & -\dfrac{R_f \cdot \cos 30^\circ}{H_{tf}} \\ \dfrac{R_f}{H_{tf}} & -\dfrac{R_f}{2 \cdot H_{tf}} & -\dfrac{R_f}{2 \cdot H_{tf}} \\ 1 & 1 & 1 \end{bmatrix} \cdot \begin{bmatrix} F_{1z} \\ F_{2z} \\ F_{3z} \end{bmatrix}
\tag{4}
$$

4   Results and Applications

The first tests involved simple trajectories or shapes shapes like lines and circles, which were concluded without discernible deviations.

Fig. 6. First path test performed

With the positioning system performing as expected, the next step was to test its part producing capabilities, and along the way to test its force measuring system. This was done by forming simple shapes such as truncated cones and pyramids, allowing to study the influence of various forming parameters and strategies, one such experiment was to compare the forming forces for different materials, namely AA1050 aluminium and DPS780 dual phase steel.

For the aluminium the results were consistent with the existing research [15], while the dual phase steel will the the focus of further studies since it has yet to be researched in incremental forming.

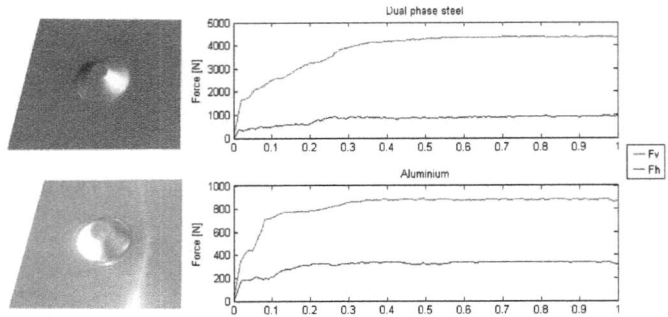

Fig. 7. Measured forces for truncated cones using different materials

Another batch of tests focused on complex geometries with specific applications, such cranial implants for medical applications [15], and to produce small batches of discontinued or custom parts for example for the automotive industry.

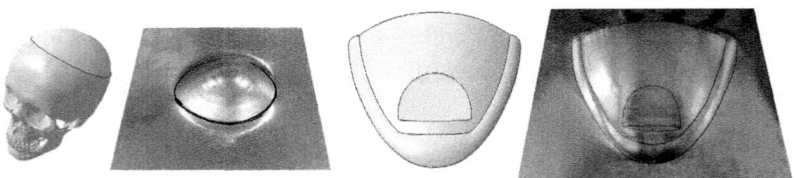

Fig. 8. Complex geometries formed by the SPIF-A machine

5   Conclusion

This work was done as part of a masters dissertation whose main objective was to prove the viability of using a hydraulic Stewart platform, validating Marabuto's proposal [8], and since this approach is unprecedented there are no comercial and/or viable solutions meaning all construction and component design was done in-house. As with all projects concerning the development of experimental hardware, sometimes setbacks occur, meaning that in such a tight schedule, some of the later task saw their intended work time reduced, nevertheless the proposed goals were met and proof of concept achieved, this section serving to explain how the several tasks were executed and to explore possibilities for future work. One main challenge was setting up the real time machine, to interface with the hardware, which required the development of an electrical control cabinet, to safely power the valves for the hydraulic circuit, and to house the I/O modules in order to implement the necessary control and data acquisition software. While this took time way from other tasks it was vital in ensuring that all safety precautions were met. While accounting for the schedule changes, the controller was kept simple in order to avoid further delays in part production. Even though it displays good results, now that the viability of this technology is starting to show, the controller will be subjected to further work namely in comparing different strategies for motion control and also by experimenting with other methods such as force control. Nevertheless the proposed controller and instrumentation software allowed for successful forming operations with satisfactory geometrical accuracy, a more in depth analysis is required using specialized measurement tools that were not available at the time, also since SPIF is prone to springback [15], implementation of correction algorithms that are still in development, this being another point to consider in future work. Overall this allowed research various forming parameters (tool size, vertical increment, aamong others), with the built machine serving as the first prototype for a possible industrial model.

References

1. Leszak, E.: Apparatus and Process for Incremental Dieless Forming. E. Patent US3342051A1 (1967)
2. Jeswiet, J., Micari, F., Hirt, G., Bramley, A., Duflou, J., Allwood, J.: Asymmetric-Single Point Incremental Forming of Sheet Metal. CIRP Manufacturing Technology 54, 88–114 (2005)
3. Allwood, J.M., King, G., Duflou, J.R.: Structured Search for Applications of Incremental Sheet Forming Process by Product Segmentation. Proceedings of the IMECH E, Part B Journal of Engineering and Manufacture 219, 239–244 (2004)
4. Decultot, N., Velay, V., Robert, L., Bernhart, G., Massoni, E.: Behaviour modeling of aluminium alloy sheet for Single Point Incremental Forming. International Journal of Material Forming (vol.1 Suppl.), 1151–1154 (2008)
5. Allwood, J.M., Houghton, N.E., Jackson, K.P.: The design of an Incremental Forming machine. In: 11th Conference on Sheet Metal, Erlangen, pp. 471–478 (2005)
6. Amino, H., Lu, Y., Maki, T., Osawa, S., Fukuda, K.: Dieless NC Forming, Prototype of Automotive Service Parts. In: Proceedings of the 2nd International Conference on Rapid Prototyping and Manufacturing (ICRPM), Beijing (2002)
7. Callegari, M., Amodio, D., Ceretti, E.: Sheet incremental forming: advantages of robotised cells vs. CNC machines. In: Huat, L.K. (ed.) Industrial Robotics: Programming, Simulation and Applications, pp. 493–514. ARS Publications (2007)
8. Marabuto, S.R., Afonso, D., Ferreira, J.A.F., Melo, F.Q.M., Martins, M.A.B.E., Sousa, R.J.A.: Finding the best machine for SPIF operations. Key Engineering Materials 473, 861–868 (2011)
9. Gao, X.S., Lei, D., Liao, Q., Zhang, G.F.: Generalized Stewart Platforms and Their Direct Kinematics. IEEE Transactions on Robotics 21, 141–151 (2005)
10. Dasgupta, B., Mruthyunjaya, T.S.: A Canonical Formulation of a Direct Position Kinematics Problem for a General 6-6 Stewart Platform. Mechanism and Machine Theory 31(6), 819–826 (1994)
11. Huang, X., Liao, Q., Wei, S.: Closed-form forward kinematics for a symmetrical 6-6 Stewart platform using algebraic elimination. Mechanism and Machine Theory 45, 327–334 (2010)

12. Lopes, A.M.F.M.: Um dispositivo robótico para controlo de força-impedância de manipuladores industriais, Doctoral Thesis, FEUP, Universidade do Porto (1999)
13. Sulc, B., Jan, J.A.: Non Linear Modelling and Control of Hydraulic Actuators. Acta Polytechnica 42, 41–47 (2002)
14. Omurlu, V.E., Yildiz, I.: Self-tuning fuzzy PD-based stiffness controller of a 3x3 Stewart platform as a man-machine interface. Journal of Electrical Engineering and Computer Science 19, 743–752 (2011)
15. Duflou, J.R., Lauwers, B., Verbert, J.: Medical application of single point incremental forming: cranial plate manufacturing. In: Proceedings of the 2005 VRAP Conference, Leiria, pp. 161–164 (2005)

Maximum Power Point Tracking of Matrix Converter Based Wind Systems with Permanent Magnet Synchronous Generator

Guilherme C. Fernandes[1], Sonia F. Pinto[1,2], and Jose Fernando Silva[1,2]

[1] Tecnico Lisboa, Universidade de Lisboa, Lisboa, Portugal
[2] INESC-ID, Lisboa, Portugal

Abstract. In this work, matrix converter based wind energy systems equipped with Permanent Magnet Synchronous Generators (PMSG) are proposed. To extract the maximum available power from the wind, the Matrix Converter is controlled using the Space Vector representation together with the Sliding Mode control technique, so that the converter supplies the generator with the required currents to provide the tracking of the established reference variables. The Maximum Power Point Tracking is achieved using two different approaches: speed control and torque control. To evaluate their performances, both control approaches are tested in MATLAB/SIMULINK. From the simulation results it is possible to confirm that with the Matrix Converter and adequate input filters it is possible to extract the maximum power from the wind with a nearly unitary power factor in the grid connection.

Keywords: Wind Energy, Matrix Converter, Space Vector Modulation, Maximum Power Point Tracking, Torque Control, Speed Control.

1   Introduction

Nowadays, wind power is one of the most promising renewable energies. In 1998 wind power capacity worldwide was around 10 GW and in the year 2012 it reached a record of 280 GW [1].

Modern wind turbines are usually equipped with a generator that may be a Variable-Speed Synchronous Generator or a Doubly-Fed Induction Generator. Both solutions use AC/AC power electronic converters to connect the generator to the grid. Usually, the AC/AC power converter is a rectifier-inverter pair structure with an intermediate DC-link with electrolytic capacitor banks, which result in increased losses, weight, size and costs of the equipment, as well as in decreased lifetime [2].

Due to these disadvantages, Matrix Converters [3,4] have become an alternative solution to the DC-link power converters, as they are single stage AC/AC bidirectional power converters capable of establishing the desired output voltages and currents with variable frequency, and nearly unitary power factor in the connection to the grid. Matrix Converters have a simple topology, are made almost exclusively by semiconductors and have nearly no energy storage components.

© Springer International Publishing Switzerland 2015,   303
A.P. Moreira et al. (eds.), *CONTROLO'2014 - Proc. of the 11th Port. Conf. on Autom. Control*,
Lecture Notes in Electrical Engineering 321, DOI: 10.1007/978-3-319-10380-8_29

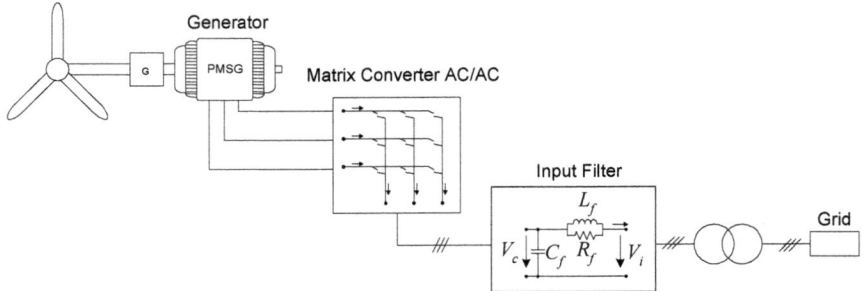

Fig. 1. Proposed wind generation system

The main goal of this work is to evaluate matrix converter based wind energy systems equipped with Permanent Magnet Synchronous Generators (PMSG) connected to the grid through matrix converters. Maximum Power Point Tracking (MPPT) is ensured using two different approaches: Speed Control and the Torque Control.

2   Maximum Point Power Tracking

2.1   Wind Turbine Model

The wind turbine should be controlled in order to extract the maximum available power from the wind. Usually the MPPT approach sets the pitch angle to zero and is applied when the wind speed u is between the cut-in speed and nominal speed of the wind generator. The application of this control strategy can be made by controlling the generator torque or the generator speed [5,6]. Both methods are described and compared in this paper. The power P_e extracted from the wind can be obtained from (1), where the power coefficient C_p is given by (2), ρ is the air density (Kg/m³), A is the swept area (m²) of the turbine rotor and u is the wind velocity [5,6].

$$P_e = \frac{1}{2} C_p(\beta, \lambda) \rho A u^3 \tag{1}$$

$$C_p = 0.22 \left(\frac{116}{\lambda_i} - 0.4\beta - 5 \right) e^{-\frac{12.5}{\lambda_i}} \tag{2}$$

The power coefficient C_p depends on the pitch angle β and on the tip-speed ratio λ, defined as the ratio between the linear velocity of the blade tip ($R\ \omega_T$) and the wind velocity (u) (3):

$$\lambda = \frac{R\ \omega_T}{u} \tag{3}$$

$$\lambda_i = \frac{1}{\dfrac{1}{\lambda + 0.08\beta} - \dfrac{0.035}{1 + \beta^3}} \tag{4}$$

The torque applied to the wind turbine rotor is then obtained from (5):

$$T_T = \frac{P_e}{\omega_T} \tag{5}$$

2.2   Optimum Speed

In order to obtain the optimum speed value, that guarantees the MPPT, it is required to determine the maximum available mechanical power supplied by the wind turbine (6):

$$\frac{dP_e}{d\omega_T} = 0 \tag{6}$$

From (1) and solving (6), for β=0 the generator speed reference is then given by (7), where G is the wind turbine gearbox ratio.

$$\omega_{opt} = G\,\frac{6.32497\,u}{R} \tag{7}$$

This optimum speed will be used as a reference for the speed control strategy.

2.3   Optimum Torque

Replacing (7) in (1), it is possible to determine the maximum power extracted from the wind. From (7), the torque that guarantees the MPPT is given by (8) [5,6].

$$T_{REF} = T_{MPPT} = \frac{0.843213\,R^3\,u^3}{G\,\omega_{Topt}} \tag{8}$$

Further, the matrix converter reference currents will be established from the reference torque (8) in order to extract the maximum power from the wind.

3   Permanent Magnet Synchronous Generator

The proposed wind generation system is equipped with a permanent magnet synchronous generator. In the rotating dq frame the voltages u_{ds}, u_{qs} applied to the stator windings are given by (9), where i_{ds}, i_{qs} are the stator currents, and ψ_{ds}, ψ_{qs} are the stator fluxes [7,8,9].

$$\begin{cases} u_{ds} = r_s\,i_{ds} + \dfrac{d\psi_{ds}}{dt} - \omega_e\,\psi_{qs} \\[2mm] u_{qs} = r_s\,i_{qs} + \dfrac{d\psi_{qs}}{dt} + \omega_e\,\psi_{ds} \end{cases} \tag{9}$$

The relation between the stator currents i_{ds}, i_{qs} and stator fluxes ψ_{ds}, ψ_{qs} is given by (10), where ψ_{f0} is the permanent magnet flux and L_{ds}, L_{qs} are the stator inductances:

$$\begin{cases} \psi_{ds} = \psi_{f0} + L_{ds}\,i_{ds} \\ \psi_{qs} = L_{qs}\,i_{qs} \end{cases} \tag{10}$$

The electromagnetic torque T_{em} of the generator is given by (11), where p is the PMSG number of pole pairs.

$$T_{em} = p\,(\psi_{ds}\,i_{qs} + \psi_{qs}\,i_{ds}) \tag{11}$$

3.1   Stator Flux Oriented Control

The Stator Flux Oriented Control approach is used to control the generator currents. The dq frame is aligned with the stator linkage flux ψ_{ds}, resulting in $i_{ds}=0$. Then, from (10) the stator flux will equal the permanent magnet flux $\psi_{ds} = \psi_{f0}$ and from (11) it is possible to establish a linear relation between the electromagnetic torque and the i_{qs} current (12).

$$i_{qs} = \frac{T_{em}}{p\,\psi_{f0}} \tag{12}$$

If T_{em} is the same as the reference torque, i_{qs} will be the reference current to the matrix converter current controller.

3.2   Speed Controller

The model of the speed controller is presented in figure 2 [10].

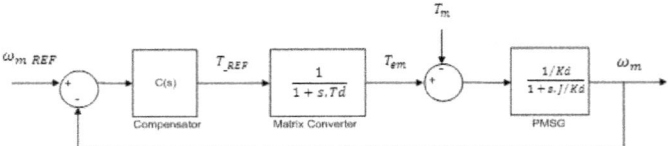

Fig. 2. Block diagram block of the speed controller

The reference torque establishes the currents that will be applied to the PMSG stator so that the generator speed can track its reference torque. The sizing of C(s) compensator is done considering that the open-loop chain is a second order system with two real poles at $-1/T_d$ and $-K_d/J$. In order to minimize the effect of the T_m disturbance and to guarantee fast response times and zero tracking error to the step response, a PI controller is used [10]:

$$C(s) = K_p + \frac{K_i}{s} = \frac{1 + s\,T_z}{s\,T_p} \tag{13}$$

To cancel the effect of the low frequency pole $-K_d/J$, the zero T_z of the PI compensator is given by (14):

$$T_z = \frac{J}{K_d} \tag{14}$$

From the closed-loop transfer function of the system and then comparing it to the transfer function of a second order system it is possible to obtain T_p (15):

$$T_p = \frac{1}{\omega_0^2 K_d T_d} \tag{15}$$

4   Matrix Converter

The three phase matrix converter is an array of nine bidirectional switches [3] S_{ij} (16) that allow the connection of any output phase to any input phase.

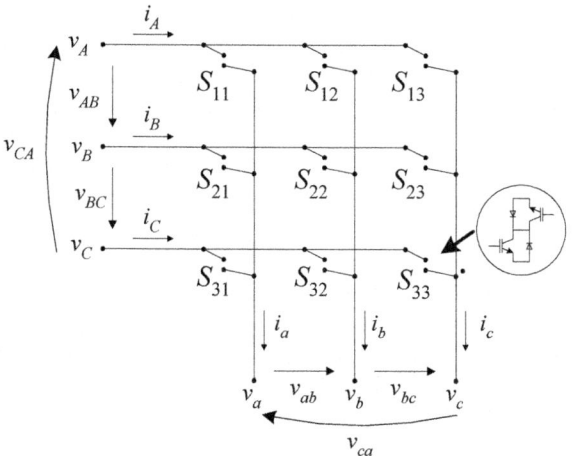

Fig. 3. Matrix converter

The power switches S_{ij} are usually represented as two possible logical states: $S_{ij} = 1$ if the switch is ON and $S_{ij} = 0$ if the switch is OFF. Due to topological restrictions, there are only 27 possible switching states.

$$\mathbf{S} = \begin{bmatrix} S_{11} & S_{12} & S_{13} \\ S_{21} & S_{22} & S_{23} \\ S_{31} & S_{32} & S_{33} \end{bmatrix} \tag{16}$$

$$[v_a \quad v_b \quad v_c]^\mathrm{T} = \mathbf{S}\,[v_A \quad v_B \quad v_C]^\mathrm{T}$$
$$[i_A \quad i_B \quad i_C]^\mathrm{T} = \mathbf{S}^\mathrm{T}\,[i_a \quad i_b \quad i_c]^\mathrm{T} \tag{17}$$

The application of Concordia transformation to the 27 voltages and currents (17) that result from these switching states allow the representation of the input currents and output voltages as vectors in the $\alpha\beta$ plane [4], [11]. This vector representation is useful to control the matrix converter output currents, which will be the PMSG currents, and the input power factor in the connection to the grid. The connection to the grid is made using a second order filter [4], [11] that minimizes the harmonics of the currents injected in the grid.

4.1   Control of Matrix Converter Output Currents

The reference currents are obtained from (12) and result from the PMSG speed or torque controller. The sliding surfaces used to control the matrix converters output currents are represented in (18) [11]-[13].

$$\begin{cases} S_\alpha(e_\alpha, t) = K_\alpha \ (i_\alpha^{REF} - i_\alpha) \\ S_\beta(e_\beta, t) = K_\beta \ (i_\beta^{REF} - i_\beta) \end{cases} \quad K_\alpha > 0, K_\beta > 0 \tag{18}$$

To assure that the system slides along the defined surfaces it is required that they verify the stability conditions (19):

$$\begin{cases} S_\alpha(e_\alpha, t) \cdot \dot{S}_\alpha(e_\alpha, t) < 0 \\ S_\beta(e_\beta, t) \cdot \dot{S}_\beta(e_\beta, t) < 0 \end{cases} \tag{19}$$

The Space Vector selection criterion is presented in table 1 and is based in equations (18) and (19).

Table 1. Criteria to choose the vector to control the PMSG currents

$S_{\alpha,\beta}$	Space Vector Choice	Value of $S_{\alpha,\beta}$
$S_{\alpha,\beta} > \Delta$	Vector that increases $i_{\alpha,\beta}$ value	+1
$S_{\alpha,\beta} > -\Delta$ and $S_{\alpha,\beta} < \Delta$	Vector that does not change $i_{\alpha,\beta}$	0
$S_{\alpha,\beta} < -\Delta$	Vector that decreases $i_{\alpha,\beta}$ value	-1

According to [11], in each time instant there are always two different vectors to apply in order to control the currents.

4.2   Power Factor Control in the Connection to the Grid

To control the Matrix Converter input currents and to obtain a nearly unitary power factor, the reactive power (20) in the connection to the grid should be nearly zero.

$$Q_{dq} = v_d i_q - v_q i_d \tag{20}$$

Choosing a dq reference frame synchronous with the grid voltage, then $i_{q\_ref} = 0$. To control the input power factor, the sliding surface is established [11].

$$S_{iq}(e_{iq}, t) = K_{iq}(i_{q_{ref}} - i_q) \tag{21}$$

The sliding surface also has to guarantee the stability condition (22).

$$S_{iq}(e_{iq}, t) \cdot \dot{S}_{iq}(e_{iq}, t) < 0 \tag{22}$$

Table 2. Criteria to choose the vector to control the grid currents

S_{iq}	Space Vector Choice	Value of S_{iq}
$e_{iq} > \Delta$	Vector that increases i_q	+1
$e_{iq} < -\Delta$	Vector that decreases i_q	-1

The input current controller defines a criterion (table 2) that allows to choose which of the two available space vectors is the most adequate to apply [11]. The right choice of space vectors guarantees that the controlled variables track their references.

5   Simulation Results

The proposed system was tested in MATLAB/SIMULINK platform. The simulation was based on the Siemens wind turbine SWT-1.3-113 (table 3), PMSG parameters presented in table 4, and the speed controller parameters are presented in table 5.

Table 3. Characteristics of Siemens wind turbine SWT-1.3-113

$S_N[MVA]$	$V_N[V]$	Blade Length [m]	Gear Box ratio
2.3	690	55	77

Table 4. Permanent Magnet Synchronous Generator parameters

p	r_s [mΩ]	L_d[mH]	L_q[mH]	B_m	$J_{in}[Kgm^3]$	$\psi_{f0}[Wb]$
4	2	0.09	0.09	0.05	100	0.865

Table 5. Speed Controller parameters

K_t	$T_d[ms]$	ξ	$T_z[s]$	$\omega_0[rad/s]$	$T_p[s]$	K_p	K_i
1	1	$\sqrt{2}/2$	2000	707.1068	0.04	50000	25

The wind chart of Fig. 4 was obtained based on average measurements of wind speed. Regarding the available simulation computer, the time span to be simulated had to be reduced to thirty seconds and the inertia of the turbine plus generator was also reduced in order to obtain scaled results.

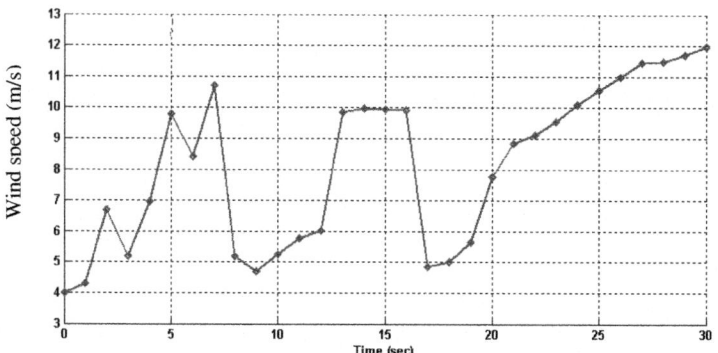

Fig. 4. Wind Speed Chart

From the obtained results, the MPPT speed controller can track the wind speed variations with adequate response times. As can be seen from figure 5, the speed reference established by the controller is tracked by the generator. Figure 6 represents the electromagnetic torque obtained for the speed control case.

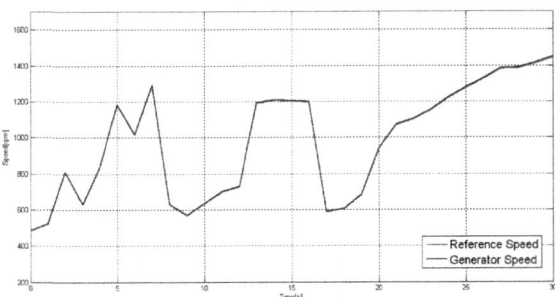

Fig. 5. Generator speed tracking the reference

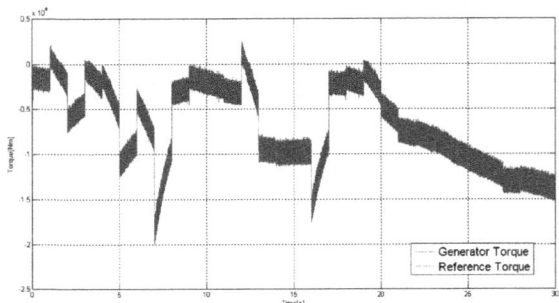

Fig. 6. Generator torque tracking the reference (speed control case)

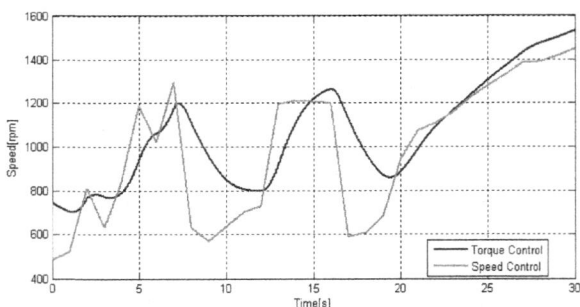

Fig. 7. Speed comparison of Torque Control and Speed Control

In contrast, the MPPT torque controller, although following the optimum torque slowly enforces the MPPT speed. As expected, the MPPT speed controller is much faster than MPPT torque controller. The speed controller allows the generator to accurately track the wind speed variation. The torque controller considers that the generator is always rotating at its optimal speed, as can be seen in figure 7, which can be a serious limitation when the wind speed changes fast.

Figure 8 represents the electromagnetic torque obtained for the torque control case. The electromagnetic torque produced by the generator tracks accurately the reference torque established by the torque control. As can be seen, the values of the electromagnetic torque obtained in this case are different from the ones obtained with the speed control.

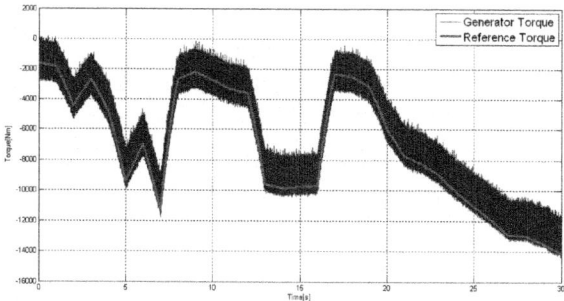

Fig. 8. Generator torque tracking the reference (torque control case)

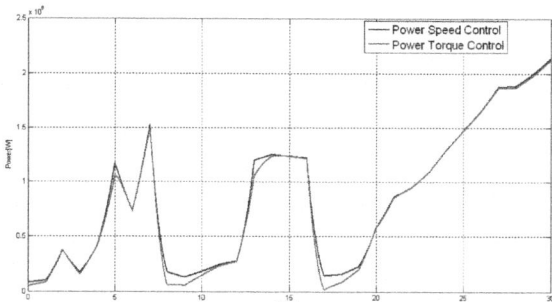

Fig. 9. Electric power in both approaches

Figure 9 represents the electric power generated using both approaches. It is possible to see that the speed controller is able to extract more power from the wind than the torque controller when the wind speed is between 4 and 7 m/s. The torque controller assumes the generator speed is at its optimal value and defines a torque reference to the generator controller. However, as can be seen from figure 7 when the wind speed values are low, the generator speed in the torque control case is much higher than the optimal speed. As a result it produces less power than the speed control approach.

Calculations were made in order to estimate the differences in the power extraction. The torque control case produced 5 kWh with a power coefficient of 42% and the speed control case produced 7.4 kWh with a power coefficient of 44%.

6   Conclusions

Both Maximum Power Point Tracking techniques were tested and produced different results. The speed controller extracts more power from the wind than the torque controller because the last does not perform well when wind speed values are near the cut-in speed limit. However, the reference for the electromagnetic torque to be produced by the generator in the speed control case has discontinuities when wind speed values change fast, and it reaches very high torque values when the wind speed is close to the nominal speed of the wind generator. These high values of torque result in overcurrents which are undesirable. In spite of extracting less power from the wind, the electromagnetic torque produced in the torque control case has lower values of torque and it is continuous and more stable than in the speed control approach.

From these results one might conclude that when wind speeds are closer to the wind generator cut-in speed, the best solution is to control the generator speed, because this controller extracts more power from the wind.

Acknowledgements. This work was supported by Portuguese national funds through FCT - Fundação para a Ciência e a Tecnologia, under project PEst-OE/EEI/LA0021/2013.

References

[1] GWEC, Global wind Statistics, Global Wind Energy Council (2013)

[2] Ristow, A., Begovic, M., Pregelj, A., Rohatgi, A.: Development of a Methodology for Improving Photovoltaic Inverter Reliability. IEEE Transactions on Industrial Electronics, Vo 55(7) (2008)

[3] Wheeler, P., Rodriguez, J., Clare, J., Empringham, L., Weinstein, A.: Matrix Converters: A Technology Review. IEEE Trans. on Industrial Electronics 49(2), 276–288 (2002)

[4] Monteiro, J., Silva, J.F., Pinto, S.F., Palma, J.: Linear and Sliding Mode Control Design for Matrix Converter Based Unified Power Flow Controllers. IEEE Transactions on Power Electronics 29(7), 3357–3367 (2014)

[5] Anaya-Lara, O., Jenkins, N., Ekanayake, J., Carthwright, P., Hughes, M.: Wind Energy Generation: Modeling and Control. John Wiley and Sons, Cardiff (2009)

[6] Pinto, S.F., Aparício, L., Esteves, P.: Direct Controlled Matrix Converters in Variable Speed Wind Energy Generation Systems. In: IEEE International Conference on Power Engineering, Energy and Electrical Drives, POWERENG 2007, Setúbal, Portugal, pp. 654–659 (May 2007)

[7] Jones, C.V.: The Unified Theory of Electrical Machines. Plenum Press (1967)

[8] Rolna, A., Luna, A.: Modelling of a Variable Speed Wind Turbine with a Permanent Magnet Synchronous Generator. In: IEEE Symposium in Industrial Electronics, ISIE 2009, Seoul Olympic Parktel, Seoul, Korea (July 2009)

[9] Ming, Y., Li, G., Zhou, M., Zhao, C.: Modeling of the Wind Turbine with a Permanent Magnet Synchronous Generator for Integration. In: Power Engineering Society General Meeting, Florida, USA (2007)

[10] Afonso, L., Pinto, S.F., Silva, J.F.: Maximum Power Point Tracker for Wind Energy Generation Systems Using Matrix Converters. In: Proc. POWERENG 2013, Istanbul, Turley (May 2013)

[11] Pinto, S., Silva, J.: Sliding Mode Direct Control of Matrix Converters. IET Electr. Power Appl. 1(3) (May 2007)

[12] Silva, J.F., Pinto, S.F.: Advanced Control of Switching Power Converters, Cap. In: Rashid, M.H. (ed.) Power Electronics Handbook, 3rd edn., vol. 36, pp. 1037–1113. Butterworth-Heinemann, Burlington (2011) ISBN: 978-0-12-382036-5

[13] Silva, J.F., Pires, V.F., Pinto, S.F., Barros, J.D.: Advanced Control Methods for Power Electronics Systems. Mathematics and Computers in Simulation 63(3), 281–295 (2003)

Control of the Platform
of an Offshore Wind Turbine*

Celso Monteiro and José Sá da Costa

IDMEC, Instituto Superior Técnico, Universidade de Lisboa
Av. Rovisco Pais, 1049-001 Lisboa, Portugal
celso.pinto.monteiro@ist.utl.pt, sadacosta@dem.ist.utl.pt

Abstract. This paper addresses the control of the platform of a floating wind turbine. The floating wind turbine consists in a semi-submersible type platform and a 5 MW wind turbine. The platform pitch was identified as the platform critical motion. Since a non null pitch can displace the rotor plane from its perpendicular position in relation to wind direction, control mechanisms were introduced in the platform in order to control the platform pitch. Two controllers were developed, a proportional-integral-derivative (PID) controller and a pole placement controller. With both controllers it was possible to guarantee a null pitch and so maintain the rotor plane in its perpendicular position in relation to wind direction.

Keywords: Offshore, DeepCwind, NREL offshore 5-MW baseline wind turbine, pitch, PID, pole placement.

1 Introduction

Nowadays wind energy has become a mature technology that can not be ignored in the socio-economic context. This form of producing clean energy is seen by the main community as an alternative to fossil fuels. Because vast offshore wind resources are available, offshore wind farms become a sustainable alternative for onshore wind farms. Currently, most of the offshore wind farms are located in Europe, using wind turbines installed in shallow water on bottom-mounted substructures[1]. These substructures include monopiles and gravity bases used in water about 30 meters (m) depth and lattice frames, such as jacket used in water to about 60 m. Nevertheless, because of the abundance of wind resources in depths greater than 60 m, floating support platforms will be the most economical type of support structure to use.

The present paper concerns the study of a floating wind turbine, in particular the control of the floating platform. The floating wind turbine consists in a semi-submersible type platform, called DeepCwind[10] and a 5-megawatt(MW) wind turbine called NREL offshore 5-MW base line wind turbine[6]. The aim of this

---
* This work was supported by FCT, through IDMEC, under LAETA. Project: PEst - OE/EME/LA0022/2011-LAETA.

paper is to address the controller design for the platform most critical motion, the pitch. The other motion modes of the platform, namely: surge, sway, heave, roll and yaw are also controlled but due to the lack of space these controllers designs are not addressed here.

Since a non null platform pitch removes the rotor plane of the wind turbine from its perpendicular position regarding the wind direction, the platform pitch is considered the platform critical motion. To avoid this issue, control action will be applied to the platform pitch in order to maintain a null pitch. The performance of the controller is verified by simulating the wind turbine using FAST[5], an aero-hydro-servo-elasto dynamic modeling tool, as the wind turbine emulator.

In the next section (Section 2) we present the nonlinear model of the floating wind turbine, where a theoretical formulation for the dynamic of the wind turbine is formulated. In Section 3, we will present the control mechanisms which will allow us to control the platform motions. As mentioned before, only the pitch motion will be controlled, being its controller designed in Section 4. In Section 5, the performance of the pitch controller will be tested through simulations. In the last section (Section 6), conclusions regarding this work will be presented.

2   Nonlinear Model of the Floating Wind Turbine

In this section a theoretical formulation of the nonlinear model of the floating wind turbine is presented. The work of J. Jokman [4] provides the hydrodynamic formulation needed to establish the dynamics of the support platform and the work of P. Moriarty and A. Hasen [8] provides the aerodynamic formulation needed to establish the dynamic of the wind turbine.

The nonlinear time-domain equation of motion of the fully coupled wind turbine and support platform is given by [4]:

$$M_{ij}(q, u_{\mathrm{c}}, t)\ddot{q} = f_i(q, \dot{q}, u_{\mathrm{c}}, t) \ , \tag{1}$$

where M_{ij} is the (i, j) component of the inertia mass matrix, u_{c} is the set of control inputs, f_i is the component of the forcing function associated with degree of freedom (DOF) i, t is time, q is the DOFs, $\dot{q}$ is the first time derivative of the DOFs and $\ddot{q}$ is the second time derivative of the DOFs.

2.1   Hydrodynamic Formulation

Jonkman[4] establishes that the total external load on the support platform, F_i^{Platform}, is:

$$F_i^{\mathrm{Platform}} = A_{ij}\ddot{q}_j + F_i^{\mathrm{Hydro}} + F_i^{\mathrm{Lines}} \ , \tag{2}$$

where A_{ij} is the (i, j) component of the impulsive hydrodynamic-added-mass, F_i^{Hydro} is the hydrodynamic load applied on the support platform and F_i^{Lines}

is the load on the support platform due to all mooring lines. The hydrodynamic load (F_i^{Hydro}) can be split into three independent and simpler components, namely radiation, diffraction, and hydrostatics [4]. The radiation component refers to the load on the floating platform when the platform is forced to oscillate in its various modes of motion and no incident surface waves are present [4]. The wave-radiation load includes contributions from hydrodynamic added mass and damping, both obtained numerically with WAMIT®[7]. The diffraction component is related to the loads on a floating platform when the platform is fixed at is mean position (no motion) and incident waves are present and scattered by the platform [4]. This load is closely related to the wave elevation effect. The hydrostatic load is elementary but nevertheless crucial in the overall behavior of a floating platform [4] and it is related to the buoyancy of the supporting platform. Additional description of the hydrodynamic formulation can be found in [4].

2.2   Aerodynamic Formulation

In Moriarty an Hansen[8] a aerodynamic formulation is established based on blade element momentum (BEM) theory. The BEM originates from two different theories: the blade element theory and the momentum theory [8]. The blade element theory assumes that blade can be divide into small elements that act independently of surrounding elements and operate aerodynamically as two-dimensional airfoils whose aerodynamic forces can be calculated based on the local flow conditions [8]. In a wind turbine these elements are considered along the span of the blade and total forces and moments exerted on the wind turbine are calculated by summing the forces and moments of all elements. The momentum theory, assumes that the loss of pressure or momentum in the rotor is caused by the work done by the airflow passing through the rotor plane on the blade elements [8]. Using this theory, one can calculate the induces velocities from momentum lost in the flow in the axial and tangential directions. These induced velocities affect the inflow in the rotor plane and therefore also affect the forces calculated by blade element theory. From the BEM theory a set of equations, based on the angle of attack, α, is available to solve iteratively in order to obtain the induced velocities and the forces on each blade element. This set of equations, two from blade element theory and two from momentum theory, consists in equations for the dominant forces on the turbine design which are the thrust and torque. Factors of corrections of the BEM theory and additional information can be found in [8].

3   Control Mechanisms

The gross motion of the platform is restricted by the mooring lines attached to the sea bottom. However, to fine control locally the 6 DOFs platform motion two main control actuators are used that consists in two sets of pumps. The first one, located at the bottom of the offset columns, is capable to project water to counterbalance surge, sway and yaw motion. The other set, located at the top of offset columns, is capable to transfer water from one column to another in order to counterbalance pitch and roll, and heave by adding or removing water from the columns. The first set of pumps correspond to a propulsion mechanism and the second one to an active ballast mechanism. Schematically the propulsion mechanism consist of horizontal forces applied on the bottom of the offset columns and the active ballast mechanism consist in vertical forces applied on the upper part of the offset columns (Fig. 1). FAST [5] don't allow control of the platform. To implement it, we recompile the FAST source code in order to introduce the control mechanisms that allow us to perform control on the platform. Details on how to recompile the FAST source code can be found in [3].

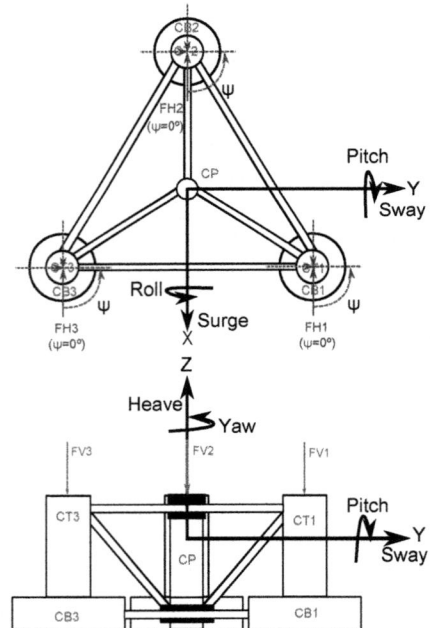

Fig. 1. Control mechanisms scheme

4   Control Design

Suitable linear models were obtained from the nonlinear model by applying system identification techniques, particularly the logarithmic decrement method, on the platform motions responses. These responses consists of steps responses wherein the steps were applied by means of the control actuators described in previous section. Those linear models are used to control design purposes. Here, only the platform pitch controller design will be presented, because it is the critical platform motion when the concern is power production. The controlled system must fulfil certain requirements. First, the controlled system must guarantee a nearly zero platform pitch in order to maintain the rotor plane in its perpendicular position in relation to wind direction. Second, the resonant frequency of the controlled system must be away from the exciting frequency of the sea states, in order to avoid resonance and guarantee an overall stability of the system. These will be the two significant requirements that the pitch controlled system must fulfil.

The impossibility of modelling the actual pump dynamics, drive us to adopt the referred on reference [11] as the actuator model. In [11] it is mentioned that the active ballast of the WindFloat is capable to transfer approximately 200 ton of water in approximately 30 minutes. By assuming this approximation as the approximation for DeepCwind active ballast, it is possible to modelling the water transfer as a first order model. Thus, the process to control ($G_\mathrm{p}(s)$) become:

$$G_\mathrm{p}(s) = G_\mathrm{a}(s)G_\mathrm{Pitch}(s) \ , \tag{3}$$

where $G_\mathrm{a}(s)$ is the first order transfer function of the actuator and $G_\mathrm{Pitch}(s)$ is the identified second order transfer function of the pitch motion of the platform.

$$G_\mathrm{p}(s) = \frac{8.4E - 5}{s^3 + 0.00512s^2 + 0.06s + 9.998E - 5} \ . \tag{4}$$

Two controllers are proposed to control the platform pitch. One controller based on the transfer function representation and the other controller based on the state space representation. The first one consists in a proportional-integral-derivative (PID) controller and the second one in a state space pole placement controller. In the PID controller the proportional parameter (K_p), the integral parameter (T_i) and the derivative parameter (T_d) were obtained iteratively in order to fulfill the requirements. The second method of Ziegler and Nichols [9] provide the starting parameters of the iterative process. Despite the fact that we obtain a null pitch with these parameters, obtained by the Ziegler Nichols method, we could not use them because they induce the system to response with high overshoot. To deal with the overshoot we perform iteratively some fine tuning. For the state space controller design the state feedback gain matrix (K) was obtained by the Ackermanns's formula [9] at each step of an iterative process, consisting in placing the poles in different locations until the requirements were fulfilled. In this iterative process all the states variables were considered measurable and available for feedback. This assumption was made for the time being because the inclusion of a state observer, despite of being stable, cause the entire system to be unstable, which occurs since the state observer was synthesized from a approximate linear model and applied to a non linear system. This robustness design is under study and will be not addressed here.

5   Results

Validation results obtained by simulation of the nonlinear plant with the pitch controllers, designed using a linear model approximation of the nonlinear plant, will be presented. Both controllers will be tested for several environmental conditions, where the wind state is characterized by the Kaimal spectrum[2], and the sea state is characterized by the JONSWAP spectrum [4]. Two irregular wind states are considered one with mean value of 11.7 miles-per-second (mps) and the other with mean value of 25 mps, and six periodic sea states as depicted in Table 1. The total simulation time is 5400 second (s) where the first 1800 s will be discarded because it is the time necessary to dissipate the effects of the initial conditions. For a better visualization of the results we only present the results for sea state 5. For the other sea states similar results are obtained.

Table 1. Periodic sea state, peak period – T_p, significant wave height – H_s

Sea State	T_p (s)	H_s (m)
1	2.0	0.09
2	4.8	0.67
3	8.1	2.44
4	9.7	3.66
5	10	6.00
6	17	15.24

5.1   Irregular Wind with 11.7 mps Mean Value

Figure 2 depict the platform pitch response, for periodic sea state 5 and 11.7 mps irregular wind speed, when the platform pitch is under control of a PID (Fig. 2a) and state space controller (Fig. 2b).

Figure 2 shows that with both controllers, was possible to maintain a null platform pitch. Assuming a hub height of 90 m, like we have in this wind turbine [6], one can observe that the displacement at the hub height will be approximately 16 cm, when the platform pitch is 0.1°, which is considered negligible regarding the dimension of the structure. For the remaining sea states it was also possible to guarantee a null platform pitch.

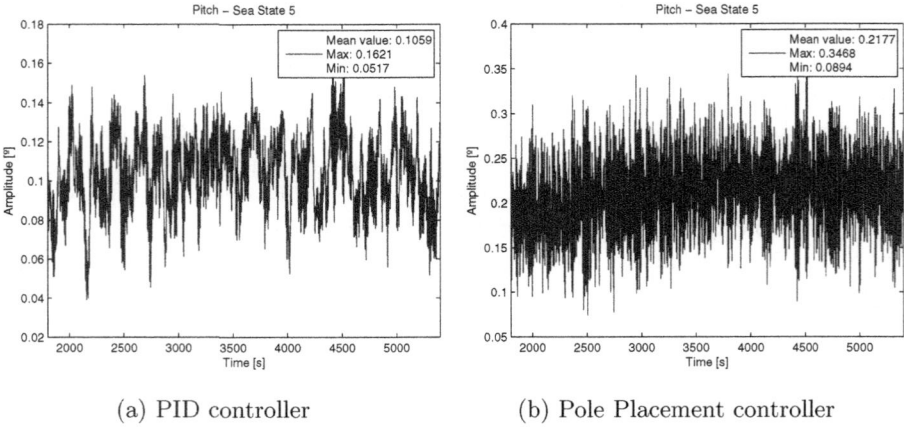

(a) PID controller   (b) Pole Placement controller

Fig. 2. PID controller and Pole Placement controller (11.7 mps irregular wind)

5.2   Irregular Wind with 25 mps Mean Value

Figure 3 depict the platform pitch response for periodic sea state 5 and 25 mps irregular wind speed, when the platform pitch is under control of a PID (Fig. 3a) and state space controller (Fig. 3b).

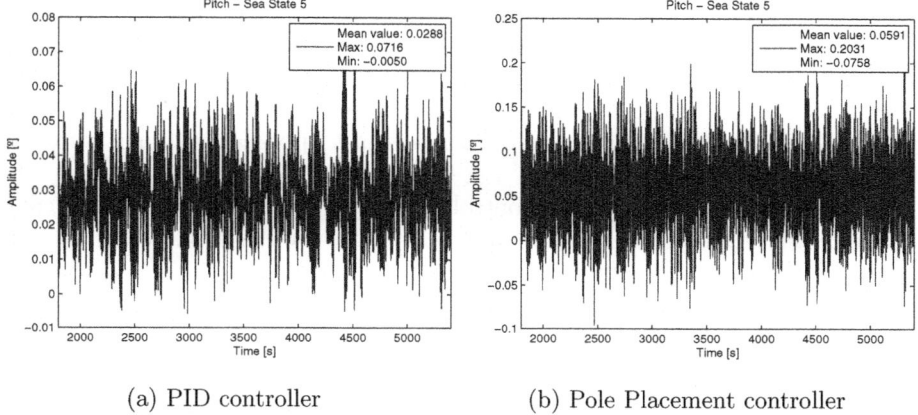

(a) PID controller   (b) Pole Placement controller

Fig. 3. PID controller and Pole Placement controller (25 mps irregular wind)

When a irregular wind with 25 mps mean value is considered, both controllers demonstrate good performance, by being capable to guarantee a null platform pitch (Fig. 3). The platform pitch mean value obtained, when a wind with 25 mps mean value is considered is less than when a wind with 11.7 mps mean value is considered. This happens because a wind with 25 mps mean value is the cut-out wind speed of this wind turbine [6], which implies more effort from the blade pitch controller, offering less resistance to wind flow, in order to preserve the integrity of the wind turbine, causing the platform pitch to be even lower, when both, blade pitch and platform pitch controllers, are combined. Additional details about the blade pitch controller can be found in [6]. Similar results were obtained when others sea states were considered.

6   Conclusions

In this paper, we present the control design of a platform pitch of an offshore floating wind turbine. The platform pitch was identified as the platform critical motion when the concern is power production. To counterbalance the pitch motion, control actuators were introduced in the platform in order to control pitch motion. Two controllers have been developed, one based on a transfer function representation, the PID controller and the other based on a state space representation, the pole placement controller.

Both controllers provide good results by being capable to maintain a null platform pitch and guarantee an overall stability of the system for all simulation condition. Despite of the fact that the linear model of the platform pitch be obtained by an approximate method, logarithmic decrement method, the controllers demonstrated robustness, being capable to fulfil the requirements for the nonlinear plant.

When compared, the PID controller demonstrate a better performance than the pole placement controller. However the difference between them is insignificant, being the mean value obtained with the PID controller, in stationary regime, half of the mean value obtained with the pole placement controller.

These results are promising and can be used to evaluate the need of introducing these control mechanisms in this type of plants. Because of its simplicity, the PID controller is the suitable controller to use instead of the pole placement controller. Note that observer design for this last approach was not addressed here.

In future we propose to apply control action in yaw motion, in order to counterbalance the gyroscopic moment induced by the nacelle yaw, using the same control mechanisms presented in this paper.

References

[1] The European Offshore Wind Industry - Key Trends and Statistics 2012. Technical report, European Wind Energy Association (January 2013)

[2] Jonkman, B., Kilcher, L.: TurbSim User's Guide. In: NREL/TP - Draft Version. National Renewable Energy Laboratory, Golden (September 2012)

[3] Jonkman, B., Michalakes, J., Jonkman, J., Buhl, M., Platt, A., Sprague, M.: NWTC Programmer's Handbook: A Guide for Software Development Within the FAST Computer-Aided Engineering Tool. In: NREL/TP - Draft Version. National Renewable Energy Laboratory, Golden (August 2012)

[4] Jonkman, J.: Dynamics Modeling and Loads Analysis of an Offshore Floating Wind Turbine. NREL Technical Report No. TP-500-41958 (November 2007)

[5] Jonkman, J., Buhl, M.: Fast User's Guide. In: NREL/EL-500-38230. National Renewable Energy Laboratory, Golden (August 2005)

[6] Jonkman, J., Butterfield, S., Musial, W., Scott, G.: Definition of 5-MW Reference Wind Turbine for Offshore System Development. In: NREL/TP-500-38060. National Renewable Energy Laboratory, Golden (February 2009)

[7] Lee, C., Newman, J.: WAMIT® User Manual, Version 7. PC, WAMIT, Inc., Chestnut Hill (2006)

[8] Moriarty, P., Hansen, A.: AeroDyn Theory Manual. In: NREL/EL-500-36881, National Renewable Energy Laboratory, Golden (December 2005)

[9] Ogata, K.: Modern Control Engineering, 5th edn. Prentice-Hall (2010)

[10] Robertson, A., Jonkman, J., Masciola, M., Song, H., Goupee, A., Coulling, A., Luan, C.: Definition of the Semisubmersible Floating System for Phase II of OC4. NREL Technical Report (2012)

[11] Roddie, D., Cermelli, C., Aubaul, A., Weinstein, A.: WindFloat: A Floating Foundation for Offshore Wind Turbines. Journal of Renewable and Sustainable Energy (May 2010)

Temperature Control of a Solar Furnace with Exact Linearization and Off-Line Identification[*]

Bertinho Andrade Costa and João Miranda Lemos

INESC-ID/IST, Univ. Lisboa, Rua Alves Redol, 9 1000-029 Lisboa Portugal
{bac,jlml}@inesc-id.pt

Abstract. The ability to predict the behavior of materials is crucial in several industries due to safety aspects, such as in high temperature furnaces for the production of glass, in the nuclear energy industry or in the high concentrated thermal solar systems. Solar furnaces use concentrated solar radiation that can be used to perform material stress tests with high temperatures. These devices have nonlinear dynamics that are caused by the actuator, the shutter, and by the interaction between the solar energy and the properties of the material. The contribution of this paper consists in the design of a controller for practical use based on the exact linearization method with off-line identification. The aim is to improve the controller performance to track the temperature profile. In this context it is assumed that the thermodynamic properties of the sample material to be tested are unknown.

Keywords: Nonlinear Control, Exact linearization, Off-line identification , Solar furnaces.

1 Introduction

Solar furnaces are devices that concentrate solar energy in a focus and are used to perform material tests, fig.1, [1],[2], [3].

The Solar Energy Laboratory (Odeillo, southern France) and the Plataforma Solar de Almeria (southern Spain) are two sites where solar furnaces are used for material stress tests. Experimental results obtained with the solar furnace of Odeillo, considered in this paper, are described in [4], where a PI controller is used to command the shutter position as a function of the temperature tracking error. To tune the controller a local characterization of the dynamics using off-line identification was used. An enhanced temperature control architecture was proposed in [6], the aim was to explore a cascade control architecture with two loops to decouple the shutter nonlinearity from the temperature dynamics. The inner control loop, fig.2 compensates the nonlinear static behavior of the shutter and sun power variability caused by atmospheric perturbations such as clouds.

---

[*] This work was developed under the SFERA II project (http://sfera.sollab.eu) and was partial supported by the program PEst-OE/EEI/LA0021/2013 of *Fundação para a Ciência e a Tecnologia*.

© Springer International Publishing Switzerland 2015   321
A.P. Moreira et al. (eds.), *CONTROLO'2014 - Proc. of the 11th Port. Conf. on Autom. Control*,
Lecture Notes in Electrical Engineering 321, DOI: 10.1007/978-3-319-10380-8_31

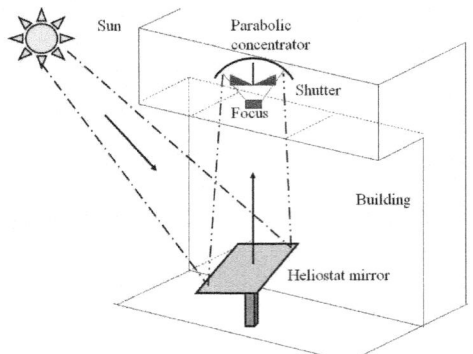

(a) General view of a solar furnace.

(b) Shutter of the 6kW solar furnace (top). Test tube in the focus with a sample (center).

Fig. 1. Schematic of the solar furnace subsystems at Odeillo PMSE Laboratory

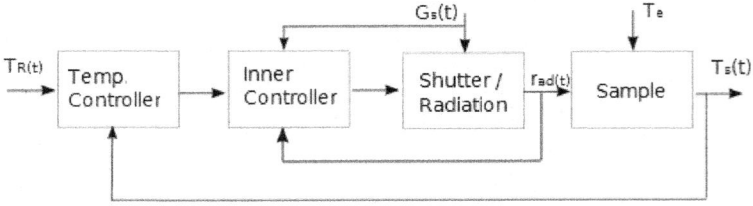

Fig. 2. Control system architecture for the solar furnace based in a cascade structure

It receives the solar flux signal and the reference signal for the flux signal and adjusts the shutter's angle. The outer control loop controls the temperature of the material sample. This controller receives the temperature reference and the sample's temperature signal and computes a control signal that is used by the inner control loop as the reference signal for the flux at the focus of the furnace.

In [7] a predictive adaptive controller was used in parallel with a PI controller to control the temperature of the sample. The aim was to use the adaptive controller to improve the performance of the PI controller in situations where the PI controller was not well tuned. However in this approach the use of online adaptation may cause stability problems.

The contribution of this paper it to explore the exact linearization method and on off-line identification, that explores the structure of the nonlinear dynamics of the temperature process, to enhance the controller performance each time a test is performed. In this case the knowledge about the thermodynamic properties of the material can be improved each time a test is performed.

This paper is organized as follows, section 2 describes the dynamic model of solar furnace. The control architecture and the design methodology is presented

in section 3. The evaluation of the control methodology using computer simulations are presented in section 4. Conclusions are described at the end of the
paper.

2   Solar Furnace Model

A solar furnace system comprises, fig.1, a heliostat mirror to track the sun and to
guide the sun's light to a parabolic mirror. A parabolic mirror that concentrates
the energy of the sun at the focus and a shutter that is used to change the
amount of energy in the focus by moving a set of blades. Figure 1(b) show an
example of solar furnaces with a shutter[1]. The solar furnace model comprises
two dynamic models: a dynamic model that describes the interactions between
the concentrated solar radiation at the focus and the temperature of the sample,
and the model that describes the behaviour of the shutter. These models are
presented hereafter.

2.1   Shutter Model

In practice, the shutter operates in closed loop using a servo mechanism. In this
work it is assumed that the shutter has a dynamics much faster than the thermal
closed-loop subsystem [4] [6], only the static function of the shutter is considered,

$$s_{fs}(u_s(t)) = 1 - \frac{\cos(\theta_0 + u_s(t)(90° - \theta_0)/100)}{\cos(\theta_0)} \qquad (1)$$

with the shutter command being limited to, $0 \le u_s(t) \le 100$ and $\theta_0 = 25°$. The
controller of the shutter is able to move the blades to the target angle in less
than $0.5s$.

2.2   Temperature Model of the Sample

An energy balance is used to model the temperature of the sample, such as the
one made in [1]. The sample has a circular shape with a diameter of $2cm$ and a
height of $2mm$. The temperature of the sample, $T_s(t)$ [K] is described by

$$\frac{dT_s(t)}{dt} = -\alpha_1[T_s^4(t) - T_e^4(t)] - \alpha_2[T_s(t) - T_e(t)] + \alpha_3 G_s(t) s_{fs}(u_s(t)) . \qquad (2)$$

T_e [K] represents the temperature of the environment that contributes to losses
by radiation and convection. The factors α_1 , α_2 and α_3 are defined by the
following equations

$$\alpha_1 = \frac{\epsilon(T_s)\sigma A_{sr}}{C_p(T_s)m}; \quad \alpha_2 = \frac{h_{conv}(T_s, T_e)A_{sc}}{C_p(T_s)m}; \quad \alpha_3 = \frac{\alpha_s A_{si} g_f}{C_p(T_s)m} . \qquad (3)$$

---

[1] At Odeillo Processes Materials and Solar Energy Laboratory (Oriental Pyrenees in
the South of France).

Table 1. Thermal model parameters

Parameter : Description
ρ $[kgm^{-3}]$: Density of the material
C_p $[Jkg^{-1}K^{-1}]$: Material Specific Heat
m $[kg]$: Mass of the sample
ϵ : Emissivity of the material
σ $[Wm^{-2}K^{-4}]$: Stefan-Boltzmann const.
A_{sr} $[m^2]$: Sample's loss radiation area
A_{sc} $[m^2]$: Sample's convection area
A_{si} $[m^2]$: Sample's incident area
L_c $[m]$: Characteristic length
h_{conv} $[Wm^{-2}K^{-1}]$: Convection factor
α_s : Sample's solar absorption factor
g_f : Furnace gain
G_s $[W/m^2]$: Max. Solar Flux

The parameters in 3 are described in Table 1. The thermodynamic data of three materials SiC, Al_2O_3 and steel $AISI1010$ were obtained from [5]. It is assumed that maximum value of the solar flux $(max(G_s(t)))$ is $1000Wm^{-2}$ and the furnace gain is $g_f = 1000$. The losses due to convection are described by the convection factor

$$h_{conv}(T_s, T_e) = 1.91((T_s - T_e)/L_c)^{0.25} , \tag{4}$$

where laminar flow of air is assumed to occur [5].

3   Control Strategy

If a new material must be tested, there is an initial lack of information about its thermal properties (C_p, ϵ, α_s are unknown) and the parameters α_1, α_2 and α_3 that enter in equation of the temperature can not be computed. As a consequence, if a first controller is used to track the temperature reference profile with a small error then there is a risk that the sample will be melt due to temperature overshoots.

To solve the problem, data from the process for temperatures below the melting point of the material sample must be collected. This can be done by performing open loop tests, where the shutter is adjusted by an human operator, or a simple (proportional) controller can be used. In this case the human operator can adjust the controller gain by "small increments". The proposed approach to solve the problem is based on the definition of a virtual control input such that $u_r(t) = G_s(t)s_{fs}(u_s(t))$. The aim is to impose $u_r(.)$ and to invert the nonlinearity such that $u_s(.)$ is computed and applied to the process input, note that $s_{fs}()$ is known and $G_s(t)$ is measured. The virtual input is generate according with

$$u_r(t) = K_P e_s(t) \tag{5}$$

where the K_P parameter is adjusted online and the tracking error $e_s(t)$ is defined by

$$e_s(t) \overset{\Delta}{=} T_R(t) - T_s(t) \tag{6}$$

where $T_R(t)$ represents the temperature reference. With this approach the non-linear effect of the shutter is compensated.

In order to characterize the process dynamics, it is assumed that the signals in eq.(2), $\frac{dT_s(t)}{dt}$, $[T_s^4(t) - T_e^4(t)]$, $[T_s^4(t) - T_e^4(t)]$ and $u_r(t)$ can be computed and used with off-line Least Square Method (LMS) to compute estimates of the parameters α_1, α_2 and α_3. The estimates of α_1, α_2 and α_3 are used to design a controller based on the exact linearization method. At this stage it is important to highlight the main difficulties that the thermal process poses. Equation 2 that described the temperature dynamic behaviour is a nonlinear first order differential equation that depends on the 4th power of the temperature. This means that some parameters of a local model obtained using the (local) linearization will show a large variation with temperature. Considering the tracking error dynamics $\dot{e}_s(t) \overset{\Delta}{=} \dot{T}_R(t) - \dot{T}_s(t)$, it can be written as

$$\dot{e}_s(t) = \dot{T}_R(t) + \alpha_1[(T_R(t) - e_s(t))^4 - T_e^4(t)] $$
$$+ \alpha_2[T_R(t) - e_s(t) - T_e(t)] - \alpha_3 u_r(t) . \tag{7}$$

Expanding the nonlinear term $(T_R(t) - e_s(t))^4$ as $T_R^4(t) - 4T_R^3(t)e_s(t) + 6T_R^2(t)e_s^2(t) - 4T_R(t)e_s^3(t) + e^4(t)$ and assuming that $e_s(t)$ is small enough such that $(T_R(t) - e_s(t))^4 \approx T_R^4(t) - 4T_R^3(t)e_s(t)$, then eq.7 can be approximated by

$$\dot{e}_s(t) = -[\alpha_1 4T_R^3(t) + \alpha_2]e_s(t) + \dot{T}_R(t) + \alpha_1[T_R^4(t) - T_e^4(t)] $$
$$+ \alpha_2[T_R(t) - T_e(t)] - \alpha_3 u_r(t) . \tag{8}$$

The parameters of the local model, $4 * \alpha_1 T^3$, α_2 and α_3 are shown in the fig.3 for the three material types, where $4 * \alpha_1 T^3$ has the largest temperature dependence. Figures 4 and 5 show the data that are used to identify the process dynamics, that is, the solar power, the temperature profile $T_R(t)$, the temperature of the sample $T_s(.)$, and the command of the shutter $u_s(.)$, where a sampling time of $0.5s$ is used.

3.1   Off-Line Identification of the Temperature Model

The identification methodology assumes that the parameter α_1 , α_2 and α_3 are constant. Two important points must be tackled, the first one it to avoid the computation of the numeric time derivative of $T_s(.)$, the order point is to use sampled data.

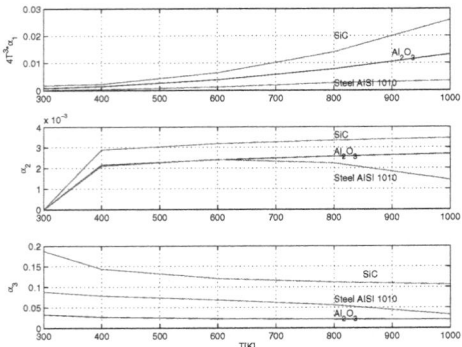

Fig. 3. Parameters of the linearized temperature model and their temperature dependence

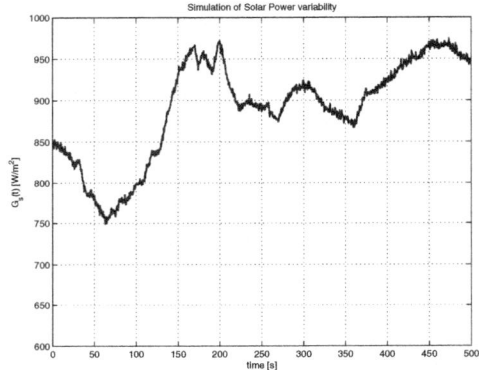

Fig. 4. Simulation of $G_s(t)$. Its behavior is based on experimental data

To tackle the problem, one first define $\zeta_1(t) = [T_s^4(t) - T_e^4(t)]$, $\zeta_2(t) = [T_s(t) - T_e(t)]$ and $\zeta_3(t) = u_r(t) = G_s(t)s_{fs}(u_s(t))$ and apply a stable low-pass filter $O(s) = a/(s+a)$ with unitary static gain to eq. (2), yielding

$$\frac{as}{(s+a)}T_s(s) = -\alpha_1\frac{a\zeta_1(s)}{(s+a)} - \alpha_2\frac{a\zeta_2(s)}{(s+a)} + \alpha_3\frac{a\zeta_3(s)}{(s+a)}. \tag{9}$$

Parameter a is selected according with the level of noise present on $T_s(.)$ and with the dynamics of $T_s(.)$. The general rule is that filter must be much faster than the temperature dynamics. It follows that equation (9) admits the continuous time representation,

$$\zeta_{f0}(t) = -\alpha_1\zeta_{f1}(t) - \alpha_2\zeta_{f2}(t) + \alpha_3\zeta_{f3}(t) \tag{10}$$

with

$$\zeta_{f0}(t) = a(T_s(t) - T_{sf}(t)) \tag{11}$$

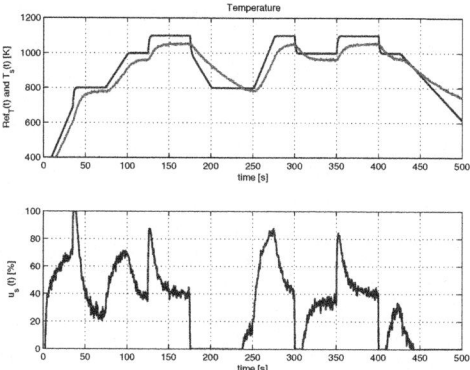

Fig. 5. Sim1: Obtained data for process identification with a not well tuned P controller. Top: Temperature $T_s(t)$ and reference $T_R(t)$. Bottom: Control signal $u_s(t)$.

$$\frac{dT_{sf}(t)}{dt} = -aT_{sf}(t) + aT_s(t) \tag{12}$$

$$\frac{d\zeta_{f1}(t)}{dt} = -a\zeta_{f1}(t) + a\zeta_1(t) \tag{13}$$

$$\frac{d\zeta_{f2}(t)}{dt} = -a\zeta_{f2}(t) + a\zeta_2(t) \tag{14}$$

$$\frac{d\zeta_{f3}(t)}{dt} = -a\zeta_{f3}(t) + a\zeta_3(t) \,. \tag{15}$$

In order to solve the equations of the filters (11) in discrete time, the first order hold (FOH) method is used with the sampling time $h = 0.5s$. The estimation of parameters α_1, α_2 and α_3 are computed using the Least Mean Square (LMS) method with discrete time signals $\zeta_{f0}[t]$, $\zeta_{f1}[t]$, $\zeta_{f2}[t]$, $\zeta_{f3}[t]$.

Considering eq. 10 and the data at each time sample, it can be organized in the following form,

$$\begin{bmatrix} y[t] \\ y[t-h] \\ \cdots \\ y[t-nh] \end{bmatrix} = \begin{bmatrix} \zeta_{f1}[t] & \zeta_{f2}[t] & \zeta_{f3}[t] \\ \zeta_{f1}[t-h] & \zeta_{f2}[t-h] & \zeta_{f3}[t-h] \\ \cdots & \cdots & \cdots \\ \zeta_{f1}[t-nh] & \zeta_{f2}[t-nh] & \zeta_{f3}[t-nh] \end{bmatrix} \begin{bmatrix} \alpha_1 \\ \alpha_2 \\ \alpha_3 \end{bmatrix}, \tag{16}$$

that can be represented by $Y = \Phi\alpha$. The parameters are obtained from

$$\alpha = (\Phi^T \Phi)^{-1} \Phi Y \tag{17}$$

assuming that the matrix $\Phi^T \Phi$ has inverse. Note that in the present problem there is a huge difference between the values of $\zeta_{f1}[t]$ $\zeta_{f2}[t]$ $\zeta_{f3}[t]$, which are related with $T^4(.)$, $T(.)$ and $G_s(t)s_{fs}(u_s(t))$. This can cause numerical problems. To mitigate this problem the matrix Φ must be scaled by a diagonal matrix such that $Y = \Phi\Lambda\alpha_L$, with $\alpha_L = \Lambda^{-1}\alpha$.

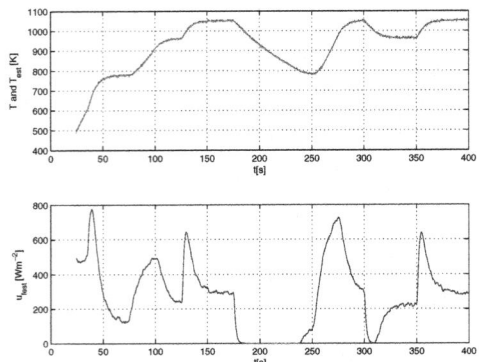

Fig. 6. Off-line process identification. Top: Temperature $T_s(t)$ (green colour) and prediction (blue colour) one-step ahead of the process output. Bottom: Filtered control signal $u_r(t)$.

The results of the off-line identification using the data from fig.5 is shown in fig.6 where a one-step ahead prediction of the process output is shown. The scale matrix was selected as $diag(\Lambda) = [1.0 \times 10^{-12} \ 1.0 \times 10^{-3} \ 1.0]$, the estimates of α_1, α_2 and α_3 are respectively, 3.4801×10^{-12} 3.0×10^{-3} 0.0224, with $\sigma_{\alpha_1} = 1.9501 \times 10^{-13}$ $\sigma_{\alpha_2} = 2.8784 \times 10^{-4}$ and $\sigma_{\alpha_3} = 1.7681 \times 10^{-4}$.

3.2 Improving the Temperature Controller

Having the estimates $\hat{\alpha}_i$ of the process parameters α_i and estimates of the error bounds, such that $\alpha_i = \hat{\alpha}_i + \Delta\alpha_i$, the control signal is defined as $u_r(t) = \overline{u}_r(t) + \delta_r(t)$ with

$$\overline{u}_r(t) = \frac{\dot{T}_r + \hat{\alpha}_1(T_R^4(t) - T_0^4) + \hat{\alpha}_2(T_R(t) - T_e)}{\hat{\alpha}_3}. \tag{18}$$

In this approach the exact linearization is used. The dynamics of the tracking error can now be written as

$$\dot{e}_s(t) = \frac{\Delta\alpha_3}{\hat{\alpha}_3}\dot{T}_R(t) + (\Delta\alpha_1 + \hat{\alpha}_1\frac{\Delta\alpha_3}{\hat{\alpha}_3})(T_s^4(t) - T_e^4(t)) + \tag{19}$$

$$+(\Delta\alpha_2 + \hat{\alpha}_2\frac{\Delta\alpha_3}{\hat{\alpha}_3})(T_s(t) - T_e(t)) - \alpha_3\delta_r(t).$$

The input $\delta_r(t)$ is used to compensate small parameter errors. Defining $\delta_r() = K_l/\hat{\alpha}_3(e_s(t) + K_p \int(e_s(\tau)d\tau)$ and considering that the $\Delta\alpha_i$ are small, the dynamics of the tracking error can be written as

$$\dot{e}_s(t) = -\frac{\alpha_3}{\hat{\alpha}_3}K_l(e_s(t) + K_i \int(e_s(\tau)d\tau). \tag{20}$$

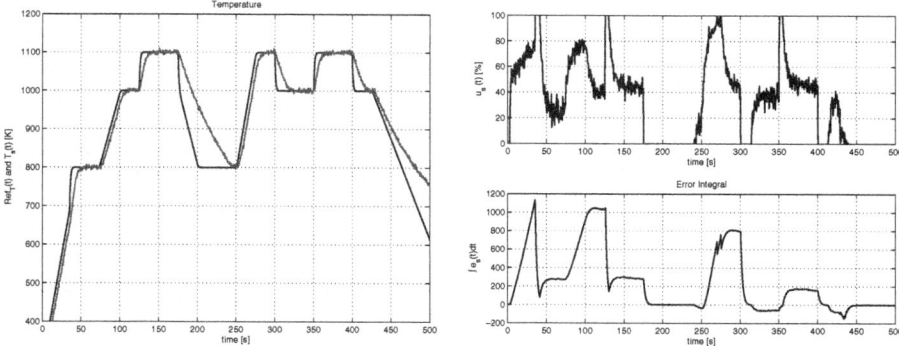

(a) Temperature $T_s(t)$ and reference $T_R(t)$. (b) Top: Control signal $u_s(t)$ that has a large noise amplitude. Bottom: Evolution of the signal at the integrator.

Fig. 7. Simulation n.2: Temperature closed loop control

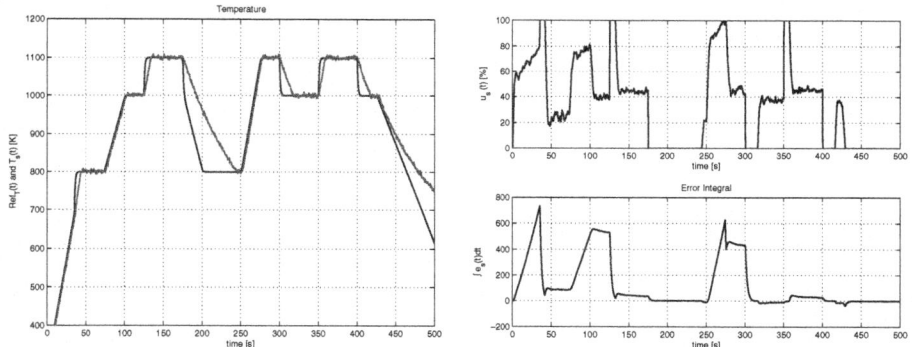

(a) Temperature $T_s(t)$ and reference $T_R(t)$. (b) Top: Control signal $u_s(t)$ has a small noise amplitude. Bottom: Evolution of the signal at the integrator.

Fig. 8. Simulation n.3: Improving the temperature control by using a filter to attenuate the noise present in $T_s(t)$.

The parameter of the PI controller can now be selected such the the dynamics of the tracking error eq.(20) is stable. Note that the terms of eq.(19) that depend on $\Delta\alpha_i$ can be evaluated for the temperature reference profile and bounds can be computed and used to evaluate the robustness of the controller.

4 Evaluation of the Temperature Controller

In order to evaluate the controller, the estimates obtained with the LM method are used. The PI controller parameters are $K_l = 0.006$ and $Ki = 0.25 * K_l$, in the ideal case, this selection imposed a real double pole on the tracking error dynamics. The results are shown in fig.s 7(a) and 7(b). The control signal $u_s(t)$ has a high level of noise due to the amplification of noise that is present in the $T_s(t)$ signal.

To attenuate the noise in the control signal a filter with the transfer function $(1/(s+1))$ was used to filter the temperature of the sample $T_s(t)$. This allows the gain K_l to be increased to 0.01 which improves the tracking error without incurring in overshoots. The results are shown in fig.s 8(a) and 8(b).

5   Conclusions

The present paper describes the design of a control system for solar furnaces which are used to perform temperature stress test in material samples. The control methodology that is envisage explores the structure of the nonlinear dynamics of the process and uses off-line identification to improve the characterization of the nonlinear dynamics of the process with the experiment performed to evaluate the material samples. This methodology avoids the use of online adaptation mechanisms that may cause stability problems during stress test. The information that is obtained with the off-line identification can also be applied to improve and to assess the robustness of the temperature controller.

References

1. Berenguel, M., Camacho, E.F., Garcia-Martin, F.J., Rubio, F.R.: Temperature control of a solar furnace. IEEE Control Systems 19(1), 8–24 (1999)
2. Fernandes, J.C., Oliveira, F.A.C., Granier, B., Badie, J.M., Rosa, L.G., Shohoji, N.: Kinetic aspects of reaction between tantalum and carbon material (active carbon or graphite) under solar radiation heating. Solar Energy, Science Direct 80, 1553–1560 (2006)
3. Fernandes, J.C., Amaral, P.M., Rosa, L.G., Shohoji, N.: Weibull statistical analysis of flexure breaking performace for alumina ceramic disks sintered by solar radiation heating. Ceramics International 26, 203–206 (2000)
4. Costa, B.A., Lemos, J.M., Guillot, E., Olalde, G., Rosa, L.G., Fernandes, J.C.: Temperature control of a solar furnace for material testing. In: 8th Portuguese Conference on Automatic Control, CONTROLO 2008 (2008)
5. Çengel, Y.A.: Heat Transfer - A pratical approach, 2nd edn. McGraw-Hill
6. Costa, B.A., Lemos, J.M.: Singular Perturbation Stability Conditions for Adaptive Control of a Solar Furnace with Actuator Dynamics. In: European Control Conference 2009, Budapest,Hungary (2009)
7. Andrade, C.B., Lemos, J.M.: Predictive Adaptive Temperature Control in a Solar Furnace for Material Stress Tests. In: 2012 IEEE Multi-conference on Systems and Control 2012, Dubrovnik, Croatia (2012)

Part VII
Decision

An Hp-Adaptative Pseudospectral Method for Conflict Resolution in Converging Air Traffic

Santiago Vera, Jesé Antonio Cobano, Guillermo Heredia, and Anibal Ollero

Robotics, Vision and Control Group
University of Seville
Seville, Spain
{svera,jcobano,guiller,aollero}@us.es

Abstract. This paper presents a conflict resolution method in converging air traffic based on an hp-adaptative pseudospectral method. The technique generates conflict free trajectories in scenarios with multiple aircraft. This method computes an optimal solution numerically. It assigns a speed profile to each aircraft sequencing the order of arrival to the TMA entry point such that the separation between them is greater than a minimum safety distance and the total deviation from the initial trajectories is minimized. The paper also presents simulations in converging scenarios by considering unlocked and locked aircraft.

1 Introduction

This paper addresses the problem of converging air traffic in terminal area by including unlocked and locked aircraft. The trajectories of unlocked aircraft could change to solve the conflicts, while the trajectories of locked aircraft are known and fixed, so they can not change. The proposed conflict resolution method is based on speed planning to sequence the arrival of aircraft. The method computes an optimal solution and uses an hp-adaptive pseudospectral method whose main characteristic is the low computational load. Also, the generated trajectories deviate as little as possible of the initial trajectories..

An hp-adaptive pseudospectral method divides the problem into temporal segments and polynomially approximates the state in each segment considered. The method computes the solution increasing the number of segments and the degree of the polynomial within a segment to achieve an error less than the allowed tolerance error [1] [2]. This method overcomes the limitations of the direct and pseudospectral methods [3]. This flexibility to increase collocation points (degree of the polynomial) and segments makes it a suitable method to address the conflict resolution in converging air traffic.

Collocation methods have been widely used to solve optimal control problems in aerospace, including optimal trajectory generation. However, most of published work on pseudospectral and direct collocation methods consider trajectory generation for one aircraft [4] [5]. This paper addresses simultaneous conflict-free trajectory generation for multiple aircraft flying in the same area. Although it is posed as an optimal control problem, the collocation-based method proposed

© Springer International Publishing Switzerland 2015   333
A.P. Moreira et al. (eds.), *CONTROLO'2014 - Proc. of the 11th Port. Conf. on Autom. Control,*
Lecture Notes in Electrical Engineering 321, DOI: 10.1007/978-3-319-10380-8_32

to solve it is a numerical method, which guarantees that the computed solution computed is valid in the collocation points, so the minimum separation among aircraft is maintained in these points. Therefore, an evaluation considering a model of the vehicle should be carried out in order to ensure that the minimum separation is not violated during the flight.

The paper is organized into six sections. Section 2 presents works on this problem. The problem formulation is described in Section 3. The proposed method is explained in Section 4. Section 5 shows the simulations performed, and the conclusions are detailed in Section 6.

2   State of the Art

Aircraft conflict detection and resolution (CDR) have been studied extensively and many methods for the assistance of ground operators [6] have been published. An overview of proposed models on CDR in ATM can be found in [7], where the CDR methods are characterized depending on the following factors: dimensions of the state information, technique for dynamic state propagation, conflict detection threshold, conflict resolution technique, maneuvering dimensions, and management of multiple aircraft conflicts.

Among the different methods can include integer programming [8], non-linear programming (NLP) [9], collocation methods reducing the number of dimensions of the problem [4] [5], particle swarm optimization [10], evolutionary computation methods [11], ant colony optimization methods [12], among many others.

The method presented in [13] is devoted only to pair wise conflicts and involve computational studies to forecast future traffic levels. The geometric approach presented in [14] is also for pair wise non-cooperative aircraft collision avoidance. A CDR method based on a mixed-integer linear program (MILP) optimizes the total flight time by modifying velocity or heading [8]. However, this method only allows one speed change for each aircraft. The method described in [15] is based on MILP to solve pair wise conflicts by changing speeds for multiple aircraft. Another approach [16] involves the application of a hybrid system model that generates safe maneuvers. The stochastic method described in [17] is based on the Monte Carlo approach and the computing time is again high. Other collision detection and resolution methods based on speed planning are presented in [18]. The main drawback of genetic algorithms [11] is that the computation time is not predictable and the convergence to a solution is not ensured in a finite time interval. The aircraft trajectory optimization is considered in [19] by applying a $5th$ degree Gauss-Lobatto direct collocation method and then solved using a nonlinear programming based branch-and-bound algorithm.

3   Problem Formulation

The problem considered in this paper concerns conflict resolution among multiple aircraft in converging air traffic by including unlocked and locked aircraft.

In order to maintain the safety distance and sequencing the order of arrival of the aircraft, speed changes are allowed so the initial spatial trajectory is maintained.

The trajectory of each aircraft is given by an initial waypoint, the TMA entry point and a final waypoint. Each waypoint is defined by: 2D coordinates (x,y) and speed from that waypoint (v). It is assumed that all aircraft trajectories are known. Aircraft maintain the safety separation if they are separated by a minimum distance, D.

An optimal control problem is considered. The cost is defined by the changes of velocity.

The inputs of the method are the enitial trajectory of each aircraft up to the TMA entry point and the aircraft model.

The objective is to find conflict-free trajectories to sequence the order of arrival in a TMA that minimizes the probability of having a conflict while minimizing the changes of speed for each aircraft.

4   Hp-Adaptive Pseudospectral Method

The hp-adaptive pseudospectral method numerically solves optimal control problems. The basic approach is to transform the optimal control problem into a sequence of nonlinear constrained optimization problems by discretizing the state and control variables.

This method determines the number of temporal segments and the degree of the polynomial in each segment that provides an accurate approximation to the solution of the optimal control problem.

The optimal control problem is considered in Bolza form. The following cost function should be minimized:

$$J = \phi(x(-1), t_0, x(+1), t_f) + \frac{t_f - t_0}{2} \int_{-1}^{1} \mathcal{L}(x(\tau), u(\tau), \tau) d\tau \tag{1}$$

subject to the dynamic constraints:

$$\frac{dx}{d\tau} = \frac{t_f - t_0}{2} f(x(\tau), u(\tau), \tau) \tag{2}$$

Boundary conditions are considered:

$$\phi(x(-1,), t_0, x(+1), t_f) = 0 \tag{3}$$

and the inequality path constraints:

$$C(x(\tau), u(\tau), \tau, t_0, t_f) \leq 0 \tag{4}$$

where $x(\tau)$ is the state, $u(\tau)$ is the control, and τ is time. The variable $\tau \epsilon [-1, 1]$ and $t \epsilon [t_0, t_f]$ are related as

$$t = \frac{t_f - t_0}{2} \tau + \frac{t_f - t_0}{2} \tag{5}$$

The hp-adaptive pseudospectral method can increase the number of segments and the degree of the polynomial within a segment in each iteration to achieve an error less than the tolerance error allowed [1]. The method uses a two-tiered strategy:

1. If the error across a particular segment has a uniform-type behavior, then the number of collocation points is increased.
2. If the error at isolated points is significantly larger than errors at other points in a segment, then a segment is subdivided at these isolated points.

The solution is determined by the number of segments, the width of each segment and the polynomial degree (number of collocation points) required in each segment.

Therefore, the method computes an accurate solution at the collocation points. Moreover, it should compute an accurate solution between the collocation points in order to ensure that the constraints are met in the whole problem.

The main advantage of this method is that it leads to higher accuracy solutions with less computational load than is required in a global pseudospectral method.

The method could address the 3D problem but in order to clarify the results, the 2D problem has been considered.

4.1   Model

One of the main advantages of using pseudospectral collocation methods for trajectory generation is that dynamic models of different complexity can be used to model the aircraft dynamics [20]. A simple model is used in the simulations.

In this work straight paths segments are considered for all aircraft involved in the converging traffic and the conflicts are resolved with speed changes. Without loss of generality, the altitude is assumed to be constant (the formulation would be similar for 3D). The state vector is defined by the position of the aircraft x_i and the speed of the aircraft v_i. The input control is the speed reference u_i.

The model considered is:

$$\dot{x}_i = v_i \cdot cos(\Theta_i) \tag{6}$$

$$\dot{v}_i = \frac{-1}{\tau}(v_i - u_i) \tag{7}$$

where, Θ_i is the slope of the straight line that defines the trajectory between two waypoints and τ is the time needed by the aircraft to reach the speed reference u_i.

A system with multiple aircraft can be defined by concatenating the state of all them. Therefore, the state vector and control vector are defined as follows:

$$X = [x_1, v_1, x_2, v_2, ..., x_n, v_n] \tag{8}$$

$$A = [a_1, a_2, ..., a_n] \tag{9}$$

where n is the number of aircraft.

The solution should satisfy constraints taking into account the physical limitations of each aircraft and the separation between them. The speed of the aircraft will be constrained:

$$v_{min} < v_{cruise} < v_{max} \tag{10}$$

and the separation between a_i and a_j should meet:

$$distance(a_i, a_j) \geq D \tag{11}$$

where D is the safety distance and y is obtained from x:

$$y_i = tan(\Theta_i)x_i + B_i, \ with \ \theta \neq 0 \tag{12}$$

Finally, the speed in the final waypoint should be the cruise speed, v_{cruise}.

5   Simulations

Simulations have been carried out to validate the proposed method in converging air traffic in a TMA. The problems have been solved by using the open-source pseudospectral optimal control software GPOPS [3] [2]. The algorithms have been run in a PC with a CPU Intel Core i7-3770 @ 3.4 Ghz and 16 GB of RAM. The operating system used in the simulations was Kubuntu Linux 12.10 OS and the code has been implemented in Matlab.

Two different scenarios are considered (see Figure 1):

- **Scenario 1:** Only unlocked aircraft are considered. The number of aircraft is from two to seven.
- **Scenario 2:** One locked aircraft is considered and the rest are unlocked aircraft.

The distance from the initial waypoint to the TMA entry point is the same for each aircraft. The considered values of safety distance and speeds are: $D = 500m$, $v_{min} = 50$, $v_{cruise} = 100$ and $v_{max} = 200$.

Twenty simulations are performed for each number of aircraft. Table 1 shows the minimum computation time, its standard deviation, σ, and the aircraft considered from Figure 1. Obviously, the computation time depends on the number of aircraft. Note that the time does not exponentially increase because the numerical problem is bounded. This is due to that the number of constraints increases as the number of aircraft increases. Each computation time is suitable to compute the conflict-free trajectories in this kind of scenario. Moreover, the scalability is proved.

A simulation experiment is presented to illustrate the solutions obtained in each case. Concretely, a simulation with four aircraft is considered. Figure 2 shows the number and localization of each collocation point. Figure 3 depicts the speed profile computed. Note that every speed profile is different for each

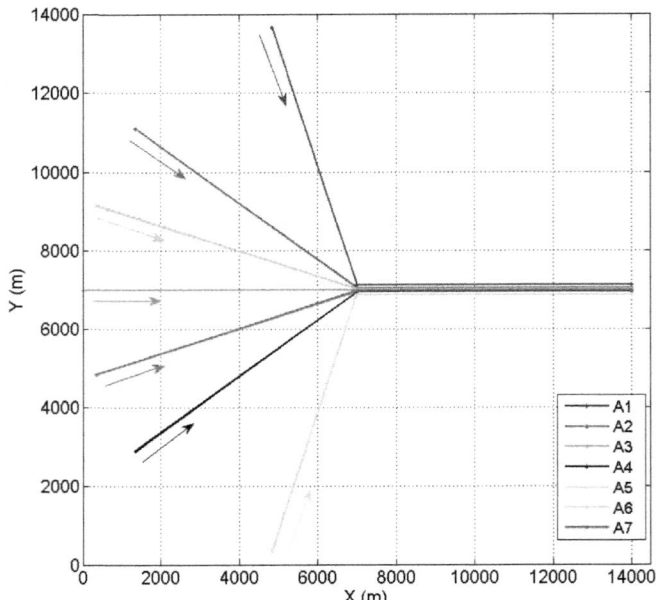

Fig. 1. Scenario considered in the simulations

Table 1. Computation time obtained depending on the number of aircrafts

UAVs	Time (s)	σ (s)	Aircraft considered
2	3.017	0.013	A1,A2
3	3.683	0.004	A1,A2,A6
4	4.009	0.016	A1,A2,A3,A6
5	9.968	0.038	A1,A2,A3,A6,A7
6	12. 641	0.065	A1,A2,A3,A4,A6,A7
7	25.082	0.236	A1,A2,A3,A4,A5,A6,A7

aircraft, and there is a coordination between them in order to met the constraints. Morevoer, the initial and final speed are equal, $v_{cruise} = 100m/s$.

The separation between aircraft is presented in Figure 4. Note that the safety distance is met during the whole flight, so the trajectories are safe.

Simulation with different separations, $D = 600, 700$ and $800m$, in this same scenario with four aircraft are analyzed to show the behaviour of the method when the constraint (11) is changed. The results show that the trajectories computed are always safe and the computation time is similar. The order of arrival is the same for $D = 500m$, $D = 700m$ and $D = 800m$: $A1, A2, A4, A3$. The order for $D = 600m$ is: $A1, A4, A2, A3$.

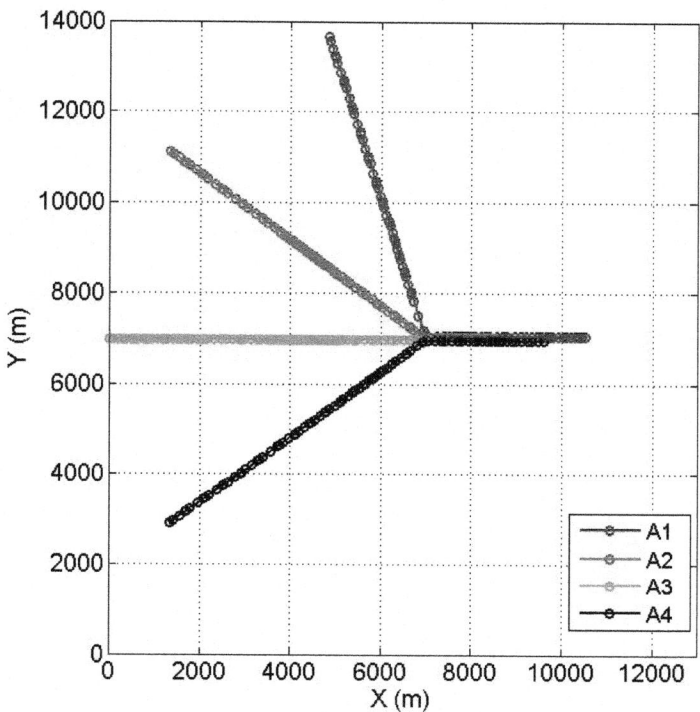

Fig. 2. Collocation points computed in each trajectory considering four aircraft

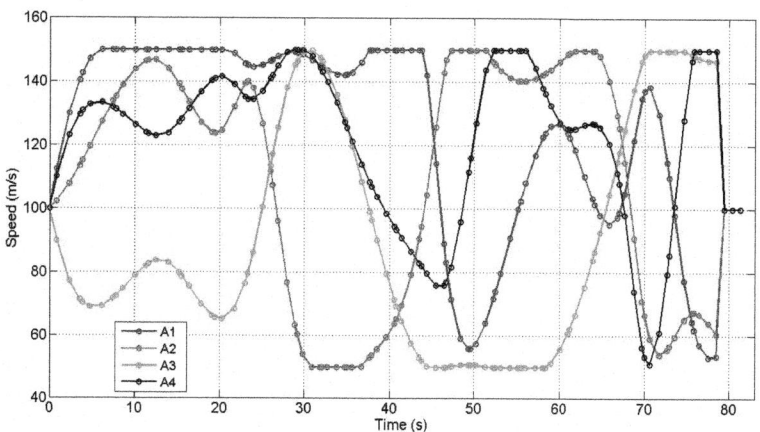

Fig. 3. Speed profile of each aircraft considering four aircraft

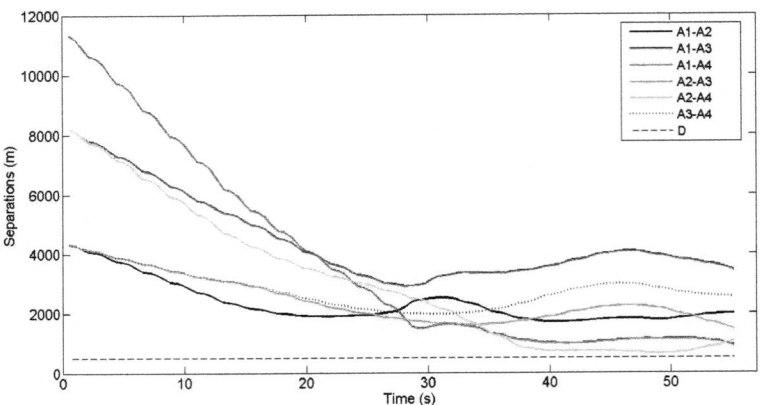

Fig. 4. Separation between aircraft considering four aircraft and D=500m

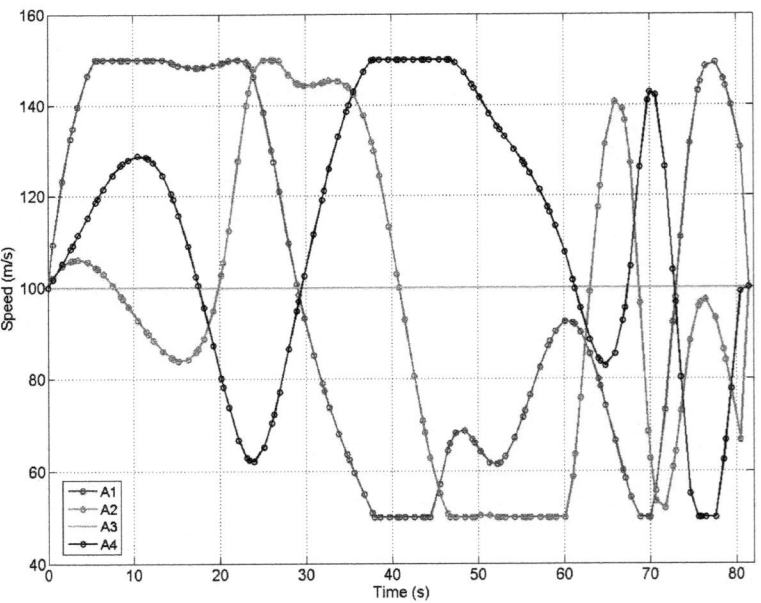

Fig. 5. Speed profile of each aircraft considering three unlocked aircraft and one locked aircraft

The Scenario 2 considers three unlocked aircraft, $A1$, $A2$ and $A6$, and a locked aircraft, $A3$, (see Figure 1). The speed profile of the aircraft are shown in Figure 5. Note that the speed of $A3$ is constant and the rest of aircraft change their speeds. In this case the arrival order is $A1$, $A4$, $A3$, $A2$.

Finally, Figure 6 depicts the separation between aircraft showing that the trajectories are safe.

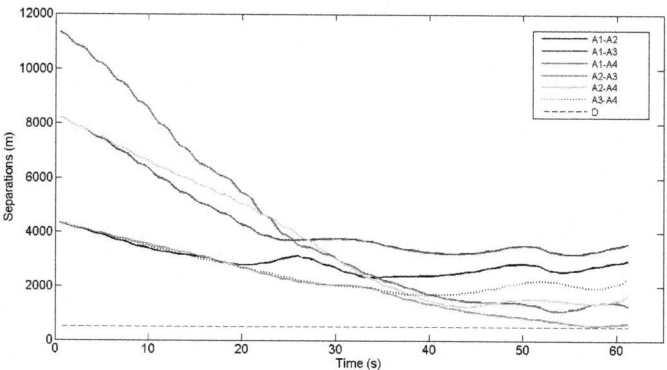

Fig. 6. Separation between aircraft considering three unlocked aircraft and one locked aircraft.

6   Conclusion

The proposed conflict resolution method in this paper solves conflicts for multiple aircraft in converging air traffic in a TMA by changing the speed profiles of each aircraft. The method is based on an hp-adaptive pseudospectral method. The scalability and low computational load are its main advantages.

It should be highlighted that the method considers multiple aircraft compared to most of works published. Other important characteristic is that several changes of speed can be carried out as in [18] but the proposed method considerably improves the computation time and ensures the optimal solution.

Furthermore, unlocked and locked aircraft are considered. It has been demonstrated that the method is suitable in both scenarios.

Acknowledgment. This work was supported by the European Commission FP7 ICT Programme under the EC-SAFEMOBIL project (288082) and the CLEAR project (DPI2011-28937-C02-01) funded by the Ministerio de Ciencia e Innovacion of the Spanish Government.

References

1. Darby, C.L., Hager, W.W., Rao, A.V.: An hp-adaptive pseudospectral method for solving optimal control problems. Optimal Control Applications and Methods 32(4), 476–502 (2010)
2. Patterson, M.A., Rao, A.V.: GPOPS-II: A MATLAB Software for Solving Multiple-Phase Optimal Control Problems Using hp-Adaptive Gaussian Quadrature Collocation Methods and Sparse Nonlinear Programming. ACM Transactions on Mathematical Software 39, 1–41 (2013)
3. Rao, A.V., Benson, D.A., Darby, C., Francolin, C., Patterson, M., Sanders, I., Huntington, G.T.: Algorithm: GPOPS, a Matlab software for solving multiple-phase optimal control problems using the gauss pseudospectral method. ACM Transactions on Mathematical Software 41, 81–126 (2010)

4. Geiger, B.R., Horn, J.F., Sinsley, G.L., Ross, J.A., Long, L.N.: Flight testing a real time implementation of a UAV path planner using direct collocation. In: Proceedings of the AIAA Guidance, Navigation and Control Conference and Exhibit, South Carolina, USA (2007)

5. Basset, G., Xua, Y., Yakimenkob, O.A.: Computing short time aircraft maneuvers using direct methods. Journal of Computer and Systems Sciences International 49, 481–513 (2010)

6. G. Chaloulus, G. Roussos, J.L., Kyriakopoulos, K.: Ground assisted conflict resolution in self-separation airspace. In: AIAA Guidance, Navigation and Control Conference and Exhibit, Reston, VA, USA (2008)

7. Kuchar, J.K., Yang, L.C.: A review of conflict detection and resolution modeling methods. IEEE Transactions on Intelligent Transportation Systems 1, 179–189 (2000)

8. Pallottino, L., Feron, E.M., Bicchi, A.: Conflict resolution problems for air traffic management systems solved with mixed integer programming. IEEE Transactions on Intelligent Transportation Systems 3, 3–11 (2002)

9. Prasanna, H., Ghosey, D., Bhat, M., Bhattacharyya, C., Umakant, J.: Interpolation-aware trajectory optimization for a hypersonic vehicle using nonlinear programming. In: AIAA Guidance, Navigation, and Control Conference and Exhibit, San Francisco, California, USA (2005)

10. Cobano, J.A., Alejo, D., Heredia, G., Ollero, A.: 4d trajectory planning in atm with an anytime stochastic approach. In: Proceedings of the 3rd International Conference on Application and Theory of Automation in Command and Control Systems, Naples, Italy (2013)

11. Vivona, R., Karr, D., Roscoe, D.: Pattern-based genetic algorithm for airborne conflict resolution. In: AIAA Guidance, Navigation and Control Conference and Exhibit, Keystone, Colorado (2006)

12. Durand, N., Alliot, J.: Ant colony optimization for air traffic conflict resolution. In: Proceedings of the Eighth USA/Europe Air Traffic Management Research and Development Seminar (ATM2009), Napa,CA, USA (2009)

13. Erzberger, H.: Automated conflict resolution for air traffic control. In: Proceeding International Congress Aeronautical Sciences, pp. 179–189 (2006)

14. Carbone, C., Ciniglio, U., Corraro, F., Luongo, S.: A novel 3d geometric algorithm for aircraft autonomous collision avoidance. In: 2006 45th IEEE Conference on Decision and Control, pp. 1580–1585 (2006)

15. Vela, A., Solak, S., Singhose, W., Clarke, J.-P.: A mixed integer program for flight-level assignment and speed control for conflict resolution. In: Proceedings of the 48th IEEE Conference on Decision and Control, 2009 held jointly with the 28th Chinese Control Conference, CDC/CCC 2009, pp. 5219–5226 (2009)

16. Tomlin, C., Mitchell, I., Ghosh, R.: Safety verification of conflict resolution manoeuvres. IEEE Transactions on Intelligent Transportation System 2, 110–120 (2001)

17. Lecchini, A., Glover, W., Lygeros, J., Maciejowski, J.: Monte carlo optimization for conflict resolution in air traffic control. IEEE Transaction on Intelligent Transportation System 7, 470–482 (2006)

18. Cobano, J.A., Alejo, D., Ollero, A., Viguria, A.: Efficient conflict resolution method in air traffic management based on the speed assignment. In: Proceedings of the 2nd International Conference on Application and Theory of Automation in Command and Control Systems, ATACCS 2012, pp. 54–61. IRIT Press, Toulouse (2012)

19. M., O.A.S., E., S.: Multiphase mixed-integer optimal control applied to 4d trajectory planning in air traffic management. In: Proceedings of the 3rd International Conference on Application and Theory of Automation in Command and Control Systems (2013)

20. Base of Aircraft Data (BADA) website,
http://www.eurocontrol.int/eec/public/standard_page/proj_BADA.html

An Evolutionary Multiobjective Optimization Approach for HEV Energy Management System

Alberto Pajares Ferrando, Xavier Blasco Ferragud,
Gilberto Reynoso-Meza, and Juan Manuel Herrero Dura

Instituto de Automática e Informática Industrial,
Universitat Politècnica de València, Spain
{alpafer1,xblasco,gilreyme,juaherdu}@upv.es

Abstract. Hybrid vehicles have become a promising solution to mitigate the negative effects of pollution and fossil fuel dependency, consequences (among other causes) of an increasing demand on mobility of people and goods. A hybrid vehicle is integrated by many subsystems, where one of the most important is the energy management system, which coordinates when to switch between energy sources to give a desired output. The energy management system needs to take into account several objectives and specifications, most of the times in conflict, to guarantee an acceptable vehicle's performance. This situation makes it a complex system to control and design. In this context, multiobjective optimization could play a significant role as a design tool, since it enables the designer to analyse the tradeoff among design alternatives. In this paper we present a multiobjective optimization design procedure by means of evolutionary multiobjective optimization in order to tune the energy management system of hybrid vehicles. To this aim, new meaningful objectives are stated and optimized. The presented results validate this approach as viable and useful for designers.

Keywords: Energy Management System, Hybrid Electrical Vehicle, Multiobjective Optimization Problem.

1   Introduction

Nowadays, there is an increasing demand for mobility of goods and people. Nevertheless, this demand has been covered at expense of more emissions and fossil fuel dependency. In order to mitigate the negative effects of both of them, different alternatives have been proposed; among them, hybrid vehicles rise as a practical and valuable alternative.

These vehicles are very efficient and several car makers are already producing different models. These vehicles do not require special maintenance and the hybrid system could last the same or more than the conventional car [1]. So, in the medium term they would be a good economic solution [2]. Nevertheless, there is research to be done on advanced control strategies to improve their performance and durability.

Hybrid electrical vehicles (HEV) [3] are perhaps the most popular of the hybrid vehicles. They combine two or more energy sources (at least one electrical) that must

© Springer International Publishing Switzerland 2015

A.P. Moreira et al. (eds.), *CONTROLO'2014 - Proc. of the 11th Port. Conf. on Autom. Control*,
Lecture Notes in Electrical Engineering 321, DOI: 10.1007/978-3-319-10380-8_33

be managed effectively to assure a good performance. The energy sources have to be carefully operated by the energy management system (EMS) [4] satisfying different requirements and specifications such as: the minimization of the fuel consumed, the maximization of the state of charges of the batteries (SOC), durability improvement, among others. This problem might be stated as an optimization problem where all the requirements are represented by design objectives that must be optimized. In order to obtain new and better solutions important efforts should be oriented towards looking for the appropriated indicators (objectives) focusing on pertinent aspects of the problem.

Recently, a variety of control strategies for energy management systems have been proposed for hybrid vehicles. Fuzzy control is one of the most used for this purpose [5-6]. However, in these works the proposed control strategies are not considering the health of the battery as well as dynamic properties of hybrid vehicles. This paper aims to design an EMS including all the objectives that could be relevant and appropriate for the designer. Therefore, it is necessary to make a study of the typical performance indicators for system power management and develop new indicators.

After the definition of these indicators the problem becomes a multiobjective optimization problem (MOP) [7]. There are several alternatives to solve a problem like this. In particular, multiobjective optimization techniques can deal with problems with multiple design objectives by optimizing all simultaneously. This kind of optimization might be used to have a close embedment of the designer with the design process, since it enables the designer to analyse the exchange between design objectives and select the most convenient design alternative. In this case, it is required an optimization algorithm to find the set of optimal solution (Pareto Set) and decision support tools to assist the decision maker (DM) selecting the most preferred solution, according to the designer's preferences. In this work, it is offered a preliminary approach to incorporate such tools in the EMS tuning problem.

This work is organized as follows: Section 2 the multiobjective optimization statement is presented. In section 3 the multiobjective optimization problem of the EMS is describe and explained. In section 4 the calculations and results are discussed. Finally, in section 5 some concluding remarks and future trends are commented.

2   Multiobjective Optimization Statement

The multiobjective optimization problem, without loss of generality, can be stated as follows:

$$\min_{\theta \in \Re^n} J(\theta) = [J_1(\theta), \dots, J_m(\theta)] \in \Re^m \tag{1}$$

where $\theta \in \Re^n$ is defined as the decision vector and J as the objective vector.

In general, it does not exist a unique solution because there is not a solution better than other in all objectives. Therefore a set of solutions, the Pareto set Θ_p is defined and its projection into the objective space J_p is known as the Pareto front. Each point in the Pareto front is said to be optimal Pareto (see definition 1) and non-dominated solution (see definition 2).

Definition 1. Pareto optimality: an objective vector $J(\theta^1)$ is optimal Pareto if not exists other objective vector $J(\theta^2)$ such that $J_i(\theta^2) \leq J_i(\theta^1)$ for all $i \in [1,2,...,m]$ and $J_j(\theta^2) < J_j(\theta^1)$ for at least one $j, j \in [1,2,...,m]$

Definition 2. Dominance: given a solution θ^1 with cost function value $J(\theta^1)$ dominates a second solution θ^2 with cost value $J(\theta^2)$ if and only if:

$$\forall_i \in [1,2,...,m], J_i(\theta^1) \leq J_i(\theta^2)\} \wedge \{\exists q \in [1,2,...,m]: J_q(\theta^1) \leq J_q(\theta^2) \qquad (2)$$

which is denoted as $\theta^1 \prec \theta^2$

Multiobjective optimization techniques search for the best discrete approximation $\Theta_p{}^*$ of the Pareto set wich generates the best description for the Pareto front $J_p{}^*$. In this way, the DM has a set of solutions for a given problem and more flexibility to choose a particular or desired solution.

Three elements are (at least) required to introduce the aforementioned ideas into a multiobjective optimization design (MOOD) procedure. First of all, the definition of a multiobjective problem; afterwards, the optimization process, in order to obtain a set of Pareto optimal solutions; finally, the multi-criteria decision-making (MCDM) step, in order to select the solution that will be implemented. The aforementioned steps are explained below, in the HEV control context.

3   Multiobjective Optimization Problem

For this work, the "HEV_ParallelSeries" model [8] has been chosen as vehicle hybrid model. It contains three operating modes: acceleration, cruise and braking. It also includes the relevant subsystems and devices of a HEV: Control, Engine and Electrical subsystems; Power Split Device, Vehicle Dynamics and speed demand (input profiles). In addition, the EMS is easy to understand and modifiable according to the requirements of this work.

3.1   Multiobjective Problem Definition

In the HEV model, the battery supplies initially the power to the vehicle, up to a certain threshold; after that, the battery is no longer required and the HEV starts using the engine to supply the necessary power. Later, when the power demand decreases below a certain threshold, the EMS recommends using only the battery, in order to reduce fuel consumption. The selection of such thresholds is essential to guarantee the proper functioning of the vehicle. On the one hand, low threshold values would cause the fuel consumption to increase unnecessarily; on the other hand, high values, even when they ensure low fuel consumption, could cause an unsatisfactory response to the user's demand. Therefore it is necessary to reach a compromise between them and the design objectives to be described below. For these reasons it was decided to use the values of start (θ1) and end (θ2) of the engine ignition as decision variables. These thresholds are measured in rpm according to the velocity profile used in the model. It is important to notice that θ1 (threshold on) must be greater or equal than θ2 (threshold off).

In the MOP it is important determining the design concept to be used, the design objectives and the constraints. The MOP statement shall specify the objectives and decision variables of the problem. The objectives to design the EMS will be mainly related with the performance of the vehicle and the battery.

With regard to the performance of the vehicle, two objectives are clearly defined; the state of charge of the batteries (SOC), Eq. (3), and fuel used, Eq. (4).

$$J1 = SOC(\theta_1, \theta_2) \tag{3}$$

$$J2 = FUEL(\theta_1, \theta_2) \tag{4}$$

where θ_1, θ_2 are the decision variables of the problem (inputs). The SOC is calculated with the current balance in the battery. Total fuel is calculated by accumulation of the fuel used at each instant of the simulation. These two objectives depend on the decision variables already detailed ($\theta1, \theta2$), and are calculated by model simulation.

Additionally, a new objective for performance estimation is designed: the difference between the demanded and the real speed of the vehicle. It shows how quickly and efficiently the vehicle responds to the user's request, as shown in Eq. (5). This response is vital to the proper functioning of the vehicle.

$$J3 = \int_0^T \frac{|V_{dem} - V_{real}(\theta_1, \theta_2)|}{|V_{dem}|} \tag{5}$$

Where T is the simulation time. Besides the performance, it is vital to take into account the health and life of the batteries [9]. These elements are subject to a very large level of demand. They should be able to contain a high load with the least possible mass (energy density). On the other hand, must withstand wide temperature ranges, possible accidents and thousands of recharge cycles.

Different studies [10-11] conclude with some recommendations to extend the life of the LI-ION batteries, typically used for this type of vehicle, SOC has to be maintained between 20% and 80%. This requirement is included as a constraint of the problem. Additionally it is interesting maintaining the state of charge around the 40 % when stored for a long time.

The batteries usually have a limited number of operation cycles, so it is important not to increase the number of cycles unnecessarily. Discharge cycles (objective J4), gets the number of SOC discharges higher than 10%. Continuous discharge bigger that 10% is considered as a single cycle.

$$J4 = \sum(Discharges(\theta_1, \theta_2) > 10\%) \tag{6}$$

In addition, it is calculated the percentage of utilization of the charge. This objective (J5) calculates the charge finally stored in the battery. So, it shows what percentage of charge has been consumed in the journey.

With these two objectives, the health and useful life of the battery could be monitored and, if previous experiments are known, the remaining battery life can be obtained.

$$J5 = 100 \frac{SOC(\theta_1, \theta_2)_{final} - SOC_{initial}}{Charge} \tag{7}$$

Therefore the MOP must maximize the SOC of the battery and the utilization of the charge, and minimize fuel, difference between real speed and demand speed and discharge cycles. The MOP is stated as:

$$J = \min_\theta [-J1 \quad J2 \quad J3 \quad J4 \quad -J5] \tag{8}$$

3.2   Multiobjective Optimization Algorithm

There are several algorithms to obtain the set of optimal solutions for a MOP. In any case some characteristics desirable for these algorithms are: convergence, diversity, pertinency, among others. As detailed in [12], it is more important to have an algorithm covering the desirable features, than worry about the *optimization engine* behind the algorithm. Multiobjective evolutionary algorithms (MOEAs [13]), has shown to be effective in treatment complex design problems. In this work sp-MODE[1] [14] algorithm is used because it provides the basic features for this problem: convergence, diversity and pertinency. It has shown good performance in the design of complex problems [7, 15].

3.3   Decision Making Stage

The MCDM stage consists in analysing the set of solutions in order to select the most preferable design alternative for the designer. This task is not trivial, since the analysis of a Pareto front with 3 or more objectives leads to a multi-dimensional analysis.

It is known that visualization tools are valuable and provide designers a meaningful way to analyse the Pareto front in order to select the best solutions for them. While visualizing a problem with 2-3 dimensions could be considered an easy task to perform, appreciating trade-offs with a problem with more objectives becomes more difficult. Therefore is important to have mechanisms or tools to analyse multi-dimensional data and, if possible, visualize them.

In this paper Level Diagrams [16] are used to perform the graphical analysis. For this purpose, ti is used a LD-Tool[2] available for designers. In addition, it is used a tool for multi-data analysis[3] [17], with filtering capacity.

To appreciate the tradeoff among solutions, the tool LD-Tool will be used interactively with the tool designed for the analysis of multi-criteria data. In this way, it is possible to visualize optimal solutions simultaneously with the filtering process in order to select the best solution according to the designer specifications.

4   Simulation Procedure and Results

A set of simulations will be carried out to evaluate the proposal contained in this paper. In all these experiments the input is the vehicle speed profile, which emulates a

---

[1] Available in Matlab Central.

[2] Available in http://cpoh.upv.es/es/investigacion/software.html

[3] *Ibidem.*

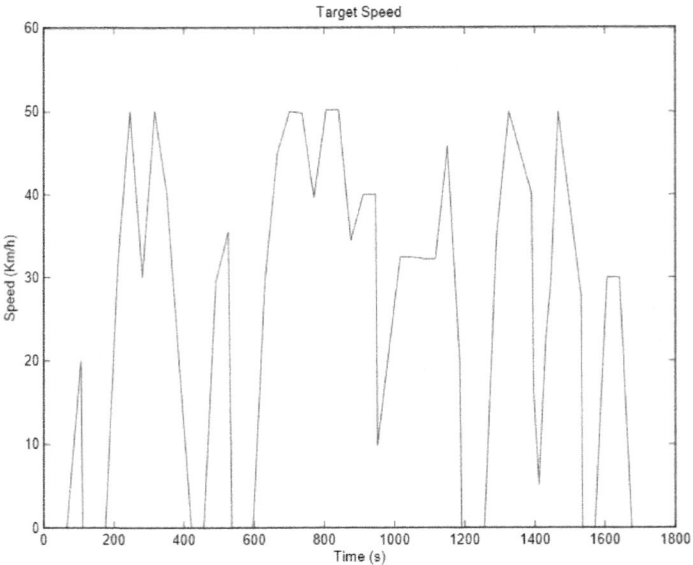

Fig. 1. City speed profile used in simulation test

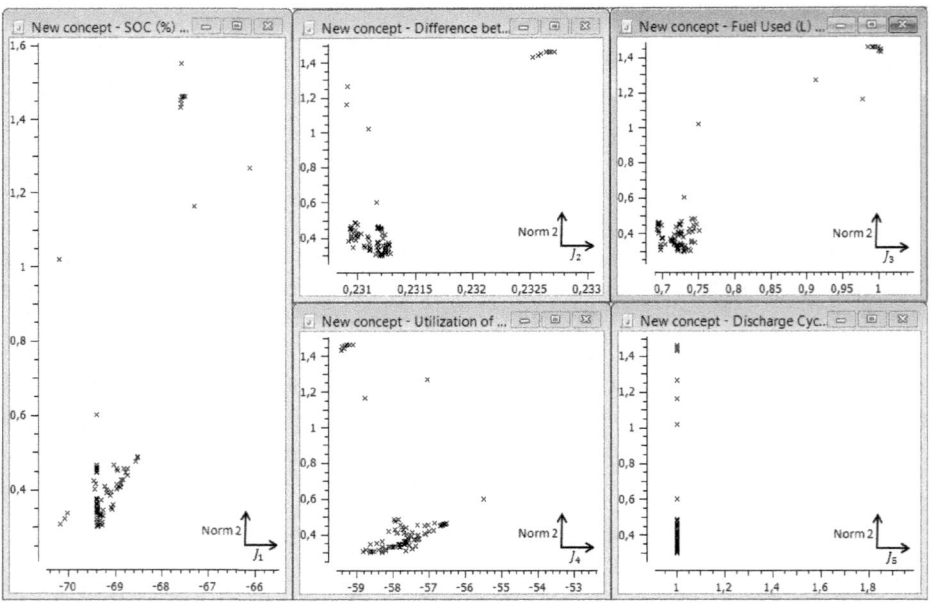

Fig. 2. Pareto Front with LD-Tool

real path. Due to the limitation of space, only one profile is presented (Fig. 1); it is a typical city profile of about half an hour (low speed, several accelerations and stops because of traffic lights, stops, etc.).

The other two inputs are the decision variables that are set by the optimization algorithm during its evolution. When optimization is finished, a Pareto front and set is available for the decision making stage. In Fig. 2 the Level Diagram representation of the Pareto Front obtained is depicted.

The first decision (for this profile), after analysing the approximated Pareto front, has been to discard two of the objectives. The objective J3 (difference between real speed and demanded speed) has a practically uninteresting variation in all solutions. Additionally, the discharge cycles (objective J4) is the same in all the design alternatives. Therefore both J3 and J4 have not been used in the selection phase.

Only three objectives (J1, J2 and J5) seem meaningful for the decision making process. Assuming that the designer is interested especially in solutions with low fuel consumption, two more solutions (the ones consuming more than 0.8l of fuel) are discarded. After this first filtering phase 87 solutions are now under consideration.

As a next step in the decision process, an advanced Physical Programming [18] filter is introduced. This filter classifies each solution according to meaningful physical ranges. For each objective the designer has to decide which are the values highly desirable (HD), desirable (D), tolerable (T), undesirable (U) and highly undesirable (HU) (see table 1). With this information the filter can order each solution from most preferable to less preferable (see [16] for Physical Programming details).

Table 1. Physical Programming filter

Objectives	$\leftarrow$ HD $\rightarrow$	$\leftarrow$ D $\rightarrow$	$\leftarrow$ T $\rightarrow$	$\leftarrow$ U $\rightarrow$	$\leftarrow$ HU $\rightarrow$	
$J1(\theta_1, \theta_2)$	-75	-70	-69	-68.8	-68.5	-66
$J2(\theta_1, \theta_2)$	0.65	0.7	0.73	0.75	0.85	1
$J3(\theta_1, \theta_2)$	0	1	1	1	1	1
$J4(\theta_1, \theta_2)$	0	2	2	2	2	2
$J5(\theta_1, \theta_2)$	-60	-59	-58	-57.5	-56	-50

Finally, the Pareto front is visualized through LD-Tool, and it is concluded that the best solution for the current specifications is the squared (red) solution shown in Fig. 3. As observed, the selected solution is the one which minimizes the indicator calculated by the Physical Programming filter. This solution is obtained with ignition (threshold on) and shutdown (threshold off) values of engine usage equal to 1070.6 rpm. So, the battery does not provide energy to vehicle when the number of rpm of the vehicle exceeds this value. The selected design alternative has a final SOC 69.39 %, as shown in Fig. 4a. The fuel consumed is 0.7223l. Additionally, the difference between demanded and real speed is very similar, as shown in Fig. 4b.

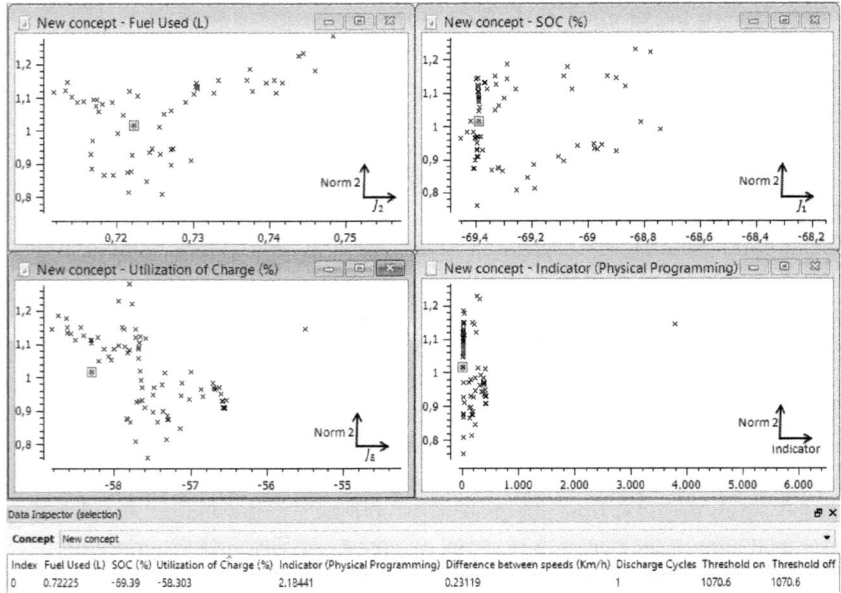

Fig. 3. Pareto Front with LD-Tool and selected solution obtained (squared in red)

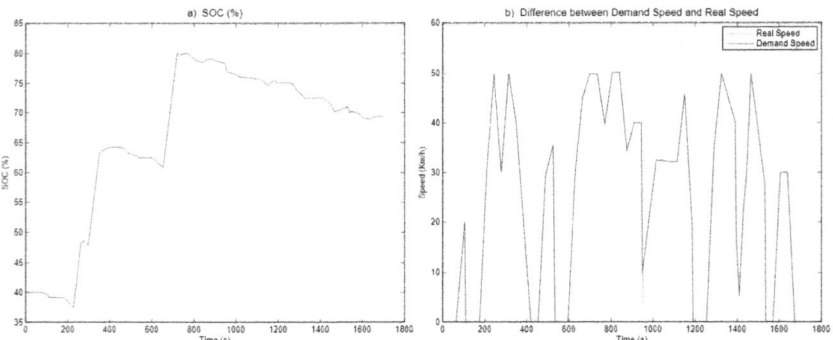

Fig. 4. SOC (%) (a) and Difference between demand speed and real speed (b)

5   Conclusions and Future Work

A proposal for energy management system tuning in hybrid vehicles has been presented. This approach is based on the multiobjective (and simultaneous) optimization of meaningful criteria. The methodology used a model to obtain estimations of vehicle variables in order to use them in the optimization phase.

A fundamental step of the approach was the selection and/or development of meaningful design objectives. For this first approach, five criteria have been selected taking into account the vehicle performance and battery health. After the multiobjective optimization process a set of optimal solutions is available and a decision making step is needed. The decision phase has required using graphical tools and the application of an adapted version of Physical Programming advance filtering. As it has been shown in the simulation results, the methodology seems to be an adequate approach in order to give a higher degree of flexibility to the control engineer at the decision making stage. Other vehicle speed profiles have been tested but, because of the limitation of space, they have not been included in this paper.

This work is a first approach to validate the possibilities of a MOOD procedure in EMS for HEV. Further research is required, for example, to seek for better and meaningful criteria to evaluate the performance. For instance, it can be seen in section 4, that the objective difference between speed real and demanded speed, has not produced the expected results. This objective provides very similar results for all the solutions obtained. This may be because there is no aggressive speed changes in the input profiles.

Other optimization criteria and or decision variables could be relevant for efficient energy management, for instance, it would be interesting to have some parameter that report the emission from the vehicle or the battery temperature (operating in an inadequate range can shorten considerably the battery life).

Additionally, because the high variability of speed profiles and driving situation it is necessary a statistical validation of all the result.

Acknowledgments. This work was partially supported by the ATOM project (PAID-05-12, number SP20120460) Universitat Politècnica de València (Spain) and the project ENE2011-25900 from the Spanish Ministry of Economy and Competitiveness.

References

1. Feng, A.N., Santini Danilo, J.: Mass impacts on fuel economies of conventional vs. hybrid electric vehicles. Argonne National Laboratory, France (2004)
2. Miller, J.M.: Propulsion Systems for Hybird Vehicles, vol. 45 (2004)
3. Chan, C.C.: The state of the art of electric and hybrid vehicles. Proc. IEEE 90(20), 247–275 (2002)
4. Lai, J.-S., Nelson, D.J.: Energy Management Power Converters in Hybrid Electric and Fuel Cell Vehicles. Proceedings of the IEEE 95(4) (2007)
5. Li, Q., Chen, W., Li, Y., Liu, S., Huang, J.: Energy management strategy for fuel cell/battery/ultracapacitor hybrid vehicle based on fuzzy logic. International Journal of Electrical Power & Energy Systems 43(1), 514–525 (2012)
6. Solano Martínez, J., Mulot, J., Harel, F., Hissel, D., Péra, M.C., John, R.I., Amiet, M.: Experimental validation of a type-2 fuzzy logic controller for energy management in hybrid electrical vehicles. Engineering Applications of Artificial Intelligence (2013)
7. Reynoso-Meza, G., Blasco, X., Sanchis, J., Martínez, M.-.: Algoritmos evolutivos y su empleo en el ajuste de controladores del tipo PID: Estado actual y perspectivas. Revista Iberoamericana de Automática e Informática Industrial 10(3), 251–268 (2013)
8. Millar, S.: Hybrid-Electric Vehicle Model in Simulink. Matlab central. Agosto 2010, actualizado Marzo (2013)
9. Scrosati, B., Garche, J.: Lithium batteries: Status, prospects and future. Journal of Power Sources 195(9), 2419–2430 (2010)
10. Battery university. How to Prolong Lithium-based Batteries (2010)

11. Battery university. Charging Lithium-ion (2010)
12. Reynoso-Meza, G., Blasco, X., Sanchis, J., Martínez, M.: Controller tuning using evolutionary multi-objective optimisation: Current trends and applications. Control Engineering Practice 28, 58–73 (2014)
13. Das, S., Suganthan, P.N.: Differencial evolution: A survey of the state-of-the-art. IEEE Transactions on Evolutionary Computation 15(1), 4–31 (2010)
14. Reynoso-Meza, G., Sanchis, J., Blasco, X., Martínez, M.: Multiobjective design of continuous controllers using differencial evolution and spherical pruning. In: Di Chio, C., et al. (eds.) EvoApplicatons 2010, Part I. LNCS, vol. 6024, pp. 532–541. Springer, Heidelberg (2010)
15. Reynoso-Meza, G., Blasco, X., Sanchis, J.: Optimización evolutiva multi-objetivo y selección multi-criterio para la ingeniería de control. X Simposio CEA de Ingeniería de Control.
16. Blasco, X., Herrero, J.M., Sanchis, J., Martínez, M.: A new graphical visualization of n-dimensional Pareto front for decision-making in multiobjective optimization. Information Sciences 178(20), 3908–3924 (2008)
17. Pajares, A., Blasco, X., Reynoso-Meza, G., Herrero, J.M.: Desarrollo de una herramienta para el análisis de datos multi-criterio. Aplicación en el ajuste de controladores del tipo PID. XXXIV Jornadas de Automática. Terrassa – Barcelona, 4-6 de Septiembre (2013)
18. Messac, A., Mattson, C.: Generating well-distributed sets of Pareto points for engineering design using Physical Programming. Optimization and Engineering 3, 431–450 (2002), doi:10.1023/A:1021179727569

A Tabu Search Algorithm for the 3D Bin Packing Problem in the Steel Industry

Joaquim L. Viegas, Susana M. Vieira,
Elsa M.P. Henriques, and João M.C. Sousa

LAETA, IDMEC, Instituto Superior Técnico, Universidade de Lisboa, Portugal
{joaquim.viegas,susana.vieira,elsa.h,jmsousa}@tecnico.ulisboa.pt

Abstract. This paper presents a tabu search and best-fit decreasing (BFD) algorithms to address a real-world steel cutting problem from a retail steel distributor. It consists of cutting large steel blocks in order to obtain smaller tailored blocks ordered by clients. The problem is addressed as a cutting & packing problem, formulated as a 3-dimensional residual bin packing problem for minimization of stock variation. The performance of the proposed approaches is compared to an heuristic and ant colony optimization (ACO) algorithms. The proposed algorithms were able to reduce the stock variation by up to 179%. The comparison of results between the tabu search and BFD algorithm shows that a multiple order joint analysis benefits the optimization of the addressed objective.

Keywords: Cutting & Packing, Steel Cutting, 3D Residual Bin Packing Problem, Real-world Application.

1 Introduction

The Cutting & Packing (C&P) class of problems is present in a wide number of industry and services applications such as the cutting of paper, metal, glass, basic pallet loading and container loading. A C&P problem is structured by a set of large objects (input, supply) and a set of small items (output, demand), which are defined in n geometric dimensions. The objective of this type of problem is to assign the small items to the large objects in order to optimize a given objective function.

The addressed problem consists in cutting a set of boxes (orders from clients) from a set of bins (steel stock pieces), where both sets are strongly heterogeneous. The objective is to minimize the stock growth in order to reduce growing stock costs. According to the most recent C&P typology [1], this problem can be considered a 3D residual bin packing problem with the specificity of the use of a timeline. There is a variable daily set of boxes to be cut from a set of bins that suffers changes resulting from the cuts.

Silveira et al. [2] approached this problem with the objective of scrap minimization, using different scrap conditions and without considering chip generation. Good results were achieved using a first-fit decreasing (FFD) type heuristic and a ant colony optimization (ACO) algorithm.

© Springer International Publishing Switzerland 2015   355
A.P. Moreira et al. (eds.), *CONTROLO'2014 - Proc. of the 11th Port. Conf. on Autom. Control,*
Lecture Notes in Electrical Engineering 321, DOI: 10.1007/978-3-319-10380-8_34

The literature presents a variety of approaches used to address C&P problems. Heuristic approaches like first-fit and best-fit decreasing are popular due to their simplicity and ability to achieve good results. These have also been used as the basis for more complex approaches based on metaheuristics. [3, 4, 5]

Metaheuristic approaches as tabu-search [3, 5, 6], ACO [7], genetic [8] and branch-and-bound [9] algorithms have also been used with success. Mixed integer programming using column-based approaches achieve very good results [4].

According to the most recent typology of C&P problems [1], from 2007, only two studies are found for the residual bin packing problem (RBPP). More recently, a 3D residual bin packing problem was studied by Bang-Jensen [10], proving the increasing need of more complex algorithms, i.e. for three-dimensional problems with multiple bins, in order to solve real-world problems.

A tabu search and a best-fit decreasing (BFD) type algorithms for the 3D residual bin packing problem with guillotine constraint are proposed in this work. These approaches search for the best bin and packing point to place each box and use an heuristic algorithm (3DHA) as the basis to construct feasible solutions. The tabu search approach searches for benefits of the combined analysis of the packing of multiple boxes. The algorithms are tested using real data from a company of the steel industry. The performance is compared to the 3DHA and ACO algorithm presented in [2].

2   3D Steel Cutting Problem

The addressed problem, which arises from a steel retailer, consists on the cutting of large steel blocks coming from steel plants in order to obtain the smaller, tailored blocks, that are made on demand to supply different industries like the tooling industry.

This problem consists in executing orthogonal cuts in order to obtain a set of boxes from a set of bins, where both sets are rectangular-shaped. Each box j is characterized by length l, width w and depth d, where $l \geq w \geq d$. The bins are characterized by length L, width W and depth D, where $L \geq W \geq D$. All orthogonal orientations of the boxes are allowed. The coordinates origin is located at the bottom-left-behind of the bin and (x_j, y_j, z_j) represents the point where the bottom-left-behind point of the box j is placed.

In order to have a feasible arrangement, the following 3-dimensional packing constraints have to be met: each box is placed completely within the bin; a box i may not overlap another box j; each box is placed parallel to the side walls of the bins (orthogonal packing).

The clients orders (boxes) may have an associated quantity greater than one. The steel cutting process is characterized by the guillotine cut constraint, which means that all the cuts have to be edge-to-edge.

Chips Generation: When processing a saw steel cutting operation, part of the material is lost in the form of chips. The chips have approximately the dimension of the cut area with a 2 millimeter thickness. When a cut is made, a volume of $A_c \times 2$ (mm^3) is considered to be lost (A_c being the cut area). Figure 1 exemplifies

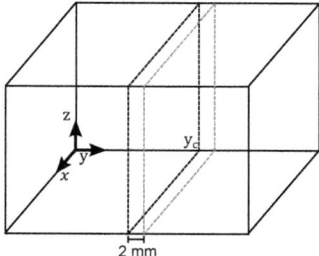

Fig. 1. Chip representation during a cutting operation at y_c position

the chip volume during a cutting operation between the areas limited by dashed lines.

Scrap Metal Definition: After a cutting operation is finished, the leftover pieces (residual bins) that are not considered useful for future cuts and do not compensate to be kept in stock are sold back to the steel plants for recycling. A residual bin, this is, a volume of material obtained after the cutting of a bin, is considered scrap metal, and not kept in stock (P_{stk}), if it complies to any of the conditions presented in equations 1, 2 and 3.

$$L \leq 375 \wedge 150 \leq W \leq 300 \wedge 151 \leq D \leq 300 \quad (mm) \tag{1}$$

$$L \leq 1950 \wedge W \leq 225 \wedge D \leq 120 \quad (mm) \tag{2}$$

$$weight \leq 35 \quad (kg) \tag{3}$$

Objective: The optimization objective used is the minimization of stock growth. On a first analysis it can seem that scrap minimization would be the most important objective to consider but, from the analysis of stock data and contact with the retail steel cutting company, scrap minimization leads to a high stock growth and generation of small pieces that, even if they do not comply to any of the scrap conditions, have a small probability of being used in the future. The weight of chips generated from cuts is used to tie-break solution with equal stock growth.

The considered problem is divided into several small daily problems according to the daily demand, in which the final stock of one day becomes the initial stock of the next day.

The cutting decision system presented in figure 2 represents the problem. With the orders and stock as input and the cuts to execute in order to minimize the objective as the desired output. Its function is to make the piece association for clients' orders in order to minimize the generation of non-scrap residual bins. The proposed decision system is developed in order to simultaneously decide on the cutting of various orders. The joint analysis of multiple orders instead of one at a time was made to research the possible benefits of joint order optimization.

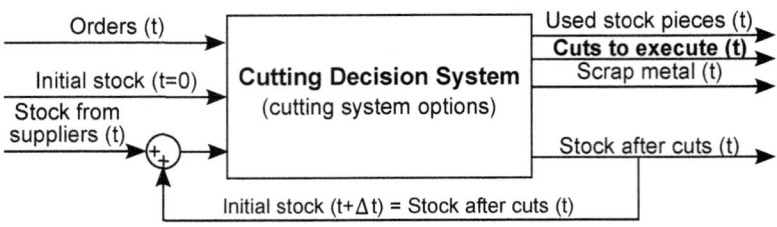

Fig. 2. Cutting decision system

2.1   Problem Characterization

The following presents the steel cutting problem characteristics classification according Wäscher's typology [1].

- Dimensionality - Three dimensions are used to characterize the problem.
- Kind of assignment - As all orders have to be fulfilled in order to minimize a function related to the bins used; the problem is considered of the "input minimization" type.
- Assortment of boxes - The assortment of all boxes can be included in the "strongly heterogeneous assortment" type. Of the total 582 boxes there are 322 different boxes. The assortment of boxes changes depending on the day.
- Assortment of bins - The assortment of bins is included in the "strongly heterogeneous assortment of large objects" (1064 different bins of a total of 1068)

According to the above presented criteria, the problem is comparable to re-fined problem type of the 3-dimensional residual bin packing (3D RBPP), which is described by a strongly heterogeneous set of small items that have to be assigned to a strongly heterogeneous assortment of large objects in order to minimize a function related to the bins.

The guillotine cut constraint, the assortment of boxes' characteristics that varies significantly in each day and the assortment of bins that has almost the double of items of the assortment of boxes are characteristics that make this problem quite unique in comparison to C&P problems found in the literature

3   Constructive Heuristic

The 3D heuristic algorithm (3DHA) developed by Silveira [2], which is used as the basis for the approaches presented in this work, was inspired by the first-fit decreasing (FFD) type algorithms developed by Crainic [11]. The heuristic was designed for the same problem, with a simpler scrap condition, no chip generation and the objective of scrap minimization.

This base heuristic uses the Extreme Points (EPs) [11] concept for the definition of packing points, box sorting and EP sorting modules, a packing and direction algorithm and two cutting decision algorithms. The open guillotine cut (OGC) algorithm is used after the placement of each box in order to find the cutting pattern which maximizes each extreme point's dimensions and, at the end of the day, a closed guillotine cut (CGC) algorithm is used to obtain the final cutting configurations. In algorithm 1, a simplification of the used heuristic is presented. A more detailed explanation of the heuristic in presented in [2].

Algorithm 1. Simplified 3DHA

---

initialize
 define the box sorting method
 define the EP sorting method

sort list of boxes
for all boxes **do**
 sort the list of EPs
 select the first feasible EP of the sorted list
 apply the packing direction and orientation algorithm in order to place the box
 update the list of EPs
 apply the OGC algorithm to generate virtual cuts
end for
apply the CGC algorithm to obtain the final cuts

---

4 ACO Algorithm

The ACO algorithm proposed by Silveira [2], inspired by the work of Levine and Ducatelle [7] for the 1-dimensional bin packing and cutting stock problems was applied to the problem. This algorithm searches for the optimal order of orders and makes use of the constructive heuristic in order to generate solutions.

This algorithm was used in order to compare its performance with the proposed approaches. It manipulates the box selection process in contrast to the tabu search and BFD approaches that control the packing point selection.

5 Proposed Tabu Search Algorithm

The proposed tabu search algorithm, described by the pseudo-code of algorithm 2, makes use of the constructive heuristic and was designed to explore multiple packing point options for the different boxes (orders) in order to achieve better results. This algorithm works by changing the packing point selection of the constructive heuristic: instead of choosing the first available extreme-point (EP) the tabu search algorithm will search, for each box, the best EP to use in order to optimize the complete solution.

The algorithm starts by generating the initial solution and its element range. The element range is the method used for the algorithm to bound the solutions generated by the neighborhood generation algorithm to feasible solutions.

Having the initial solution, the algorithm generates a large neighborhood and reduces it to a fixed size without having solutions pertaining to the tabu list. The objective function value of the neighborhood solutions is then obtained and the one with best (lower) fitness is chosen as the next iteration's solution and, if its value is higher than the best solution's fitness, the best solution is changed to be equal to the iteration's solution. The cycle finishes by updating the tabu list and starts again generating the next neighborhood.

Algorithm 2. TS Algorithm

N_b: Number of Boxes
s, s^*: current and best solution
$f(s)$: fitness of solution s
n, n_b: stopping condition
m, m_b: Neighborhood size multiplier
T: Tabu List
ER: Element range (Range for each solution element)
$GoBack$: Option to make the algorithm go back to s^*

initialize
   1. generate initial solution s and obtain ER
   2. $s^* = s$
   3. $T = \emptyset, T \leftarrow s$
repeat
     generate s neighborhood: $LN(s)$
     $N(s) \leftarrow LN(s)[1 : N_b \times m]$
     compute $f(s') \quad \forall s' \in N(s)$
     $s = s', f(s') = min(f(s') \quad \forall s' \in N(s))$
     update ER
     $T \leftarrow s$
     if $f(s) < f(s^*)$ **then**
        $s^* = s$
     end if
until $f(s^*)$ stays unchanged for n iterations
if GoBack **then**
     $s = s^*, n = n_b, m = m_b, GoBack = 0$
     go to the start of **repeat** cycle
else
     stop
end if

The stopping criterion used is a fixed number of iterations without improvement. The *GoBack* option for the algorithm to go back to the best solution and continue for another fixed number of iterations is included as a way for the algorithm to further explore the best solution's neighborhood.

5.1 Solution Encoding

The encoding used for the tabu search algorithm guides the EP selection process of the constructive heuristic; it represents, for each box (box type or order) the index of the EP, in a sorted list of possible packing points, to be used. This approach was inspired by the work of Bortfeldt el al. [12], in which a tabu search for the CLP that uses encoded solutions related to the used packing points in an ordered list was proposed.

For a daily instance with N_b number of boxes, an encoded solution s is defined by a vector of N_b integers, encoded solution elements s_i, equal or greater than one. The solution obtained by the constructive heuristic alone, which is used as the initial tabu search solution is then equal to a vector of ones with size N_b.

$$s = (s_1, s_2, ..., s_{N_b})$$

$$s_{initial} = (1, 1, ..., 1)$$

5.2 Tabu List

The tabu list is used to escape going back to already selected solutions. The used tabu list is a simple list of the full solutions selected at each iteration. This list is defined with no fixed size due to the fact that the algorithm was designed to run for a low number of iterations.

5.3 Neighborhood Structure

The neighborhood algorithm generates all the possible solutions with exactly one different element from the parent solution and is bounded by the element range vector ER. The neighborhood is then limited to the first $N_b \times m_n$ elements. The neighborhood is ordered by the number incremented or subtracted to the initial element, from left to right element changes and increments followed by subtractions.

Example: If $s = (1, 2, 3)$ and $ER = (2, 3, 3)$

$$N = \begin{bmatrix} 2 & 2 & 3 \\ 1 & 3 & 3 \\ 1 & 1 & 3 \\ 1 & 2 & 2 \\ 1 & 2 & 1 \end{bmatrix}$$

5.4 Best-Fit Decreasing Algorithm

A best-fit decreasing type algorithm that uses the tabu-search encoding was also developed. It starts by obtaining the initial solution; the algorithm then optimizes each one of the solution's elements at a time, from first to last. Having the initial solution, it generates a Max number of solutions obtained by incrementing the initial solution's first element from 1 to Max, the objective function of the solutions is obtained and, the solution with the minimum fitness, if lower than the initial solution, is set as the new initial solution; after this the algorithm goes to the next solution element and so on.

6 Problem Data and Results

The addressed problem is from an anonymous company from the steel industry with the objective of minimizing the stock variation. The used data corresponds to 3 months of operation and is composed by the list of stock pieces at the start of the operation period, the stock pieces obtained from suppliers added to the stock at a determined day and the daily orders that need to be satisfied.

6.1 Data

Table 1 presents the most important characteristics of the problem's data. Volumetric data is presented in mm^3. $\overline{V}$ represents the average volume. The standard deviation and relative standard deviation (RSD) are also presented. It is possible to verify the heterogeneity of the boxes and bins by observing the deviations that are greater than 250%. The presented data is proportional to the real data due to confidentiality issues.

Table 1. Characteristics of the boxes and bins

	Types	Number	Min.Vol	Max. Vol.	$\overline{V}$	$\delta(V)$	$\frac{\delta}{\overline{V}}$
Boxes	322	582	1.5×10^5	7×10^9	2.3×10^8	6.5×10^8	2.87
Bins	1064	1068	1.7×10^6	1.7×10^{10}	7×10^8	2.4×10^9	3.45

6.2 Results

This section presents the results of the heuristic, ACO, tabu search and best-fit decreasing algorithms. The tabu search algorithm was parametrized with the following parameters: $n = 5$, $n_b = 2$, $m = 5$, $m_b = 30$, $GoBack = 1$. The ACO algorithm uses the following parametrization: 20 ants, 80 iterations or 5 without improvement, $\rho = 0.2$, $\alpha = 1$, $\beta = 2$, $p_{best} = 0.05$, $q = 1$, $p = 2$ and iteration best pheromone update. The best-fit decreasing algorithm uses a range of 30. The EP ordering method obtains the lists of increasing d and volume and combines them by setting the odd entries to the d decreasing list and the even entries to the volume decreasing list.

In table 2, the total stock variation, stock reduction in comparison to the 3DHA, total weight of scrap, scrap from residual bins (kg) and the chips generated (kg) are presented. The tabu search algorithm achieved the best performance, generating a solution with a -125 stock variation, reducing the stock growth by 179% while generating an 40% increase in scrap metal and reducing the chips' scrap (proportional to the area cut) by 26%. The BFD algorithm obtained slightly worse results in comparison to the tabu search algorithm, obtaining an greater stock variation and scrap generated. The ACO algorithm wasn't able to achieve good performance, obtaining an equal stock variation in comparison to the 3DHA.

The increase of performance obtained using the tabu search algorithm in comparison with the BFD approach shows that, for this problem, the joint analysis of the cutting of multiple orders brings some benefit and, in comparison with the ACO results, the individual box to bin assignment, with a flexible choice of packing points, is very important. The ACO algorithm works in the box sorting space and the BFD and tabu search algorithms manipulate the packing point selection.

Table 2. Results for each algorithm

Algorithm	Stock Variation	Stock Variation Reduction	Total Scrap	Scrap	Chips
3DHA	158	-	49520	26585	22935
ACO	158	0%	48990	26375	22615
BFD	-119	175.3%	70365	52385	261925
Tabu Search	-125	179.1%	69345	52485	16860

7   Conclusions

The proposed algorithms were able to significantly reduce the stock variation generated in the addressed steel cutting problem. In comparison with the 3DHA and ACO algorithm, the stock variation was reduced by up to 179%. The chips generated were also significantly reduced, resulting from a reduction of the total cutting area.

Having obtained significant better results, using algorithms that are flexible in terms of the choice of packing point and bin (BFD and TS), in comparison to the ACO algorithm that worked in the box selection space, it can be concluded that the correct box to bin/packing point assignment is of great importance in the stock variation minimization.

The slight performance increase obtained using the tabu search in comparison to the BFD algorithm shows that this problem benefits from a joint analysis of multiple orders. The tabu search is able to combine orders and, when a bin is the best choice for different boxes, choose the one that minimizes the most the optimization's objective.

Acknowledgments. This work was supported by FCT, through IDMEC, under LAETA Pest-OE/ EME/LA0022 and by a FCT post-doctoral grant SFRH/BPD/65215/2009, Ministério da Educação e Ciência, Portugal. This work was developed within the ToolingEdge project, partly funded by the Incentive System for Technology Research and Development in Companies, under the Competitive Factors Thematic Operational Programme, of the Portuguese National Strategic Reference Framework, and EUs European Regional Development Fund.

References

[1] Wäscher, G., Hauner, H., Schumann, H.: An improved typology of cutting and packing problems. European Journal of Operational Research 183(3), 1109–1130 (2007)

[2] Silveira, M.E., Vieira, S.M., Da Costa Sousa, J.M.: An ACO algorithm for the 3D bin packing problem in the steel industry. In: Ali, M., et al. (eds.) IEA/AIE 2013. LNCS, vol. 7906, pp. 535–544. Springer, Heidelberg (2013)

[3] Crainic, T.G., Perboli, G., Tadei, R.: Teodor Gabriel Crainic, Guido Perboli, and Roberto Tadei. TS2PACK: a two-level tabu search for the three-dimensional bin packing problem. European Journal of Operational Research 195(3), 744–760 (2009)

[4] de Queiroz, T.A., Miyazawa, F.K., Wakabayashi, Y., Xavier, E.C.: Algorithms for 3d guillotine cutting problems: Unbounded knapsack cutting stock and strip packing. Computer & Operations Research 39, 200–212 (2012)

[5] Lodi, A., Martello, S., Vigo, D.: TSpack: A unified tabu search code for multidimensional bin packing problems. Annals of Operations Research 131(1-4), 203–213 (2004)

[6] Allen, S.D., Burke, E.K., Kendall, G.: A hybrid placement strategy for the three-dimensional strip packing problem. European Journal of Operational Research 209, 219–227 (2011)

[7] Levine, J., Ducatelle, F.: Ant colony optimisation and local search for bin-packing and cutting stock problems. Journal of the Operational Research Society 55(7), 705–716 (2004)

[8] Wang, H., Chen, Y.: A hybrid genetic algorithm for 3D bin packing problems. In: 2010 IEEE Fifth International Conference on Bio-Inspired Computing: Theories and Applications (BIC-TA), pp. 703–707 (2010)

[9] Martello, S., Pisinger, D., Vigo, D.: The three-dimensional bin packing problem. Operations Research 48(2), 256–267 (2000)

[10] Bang-Jensen, J., Larsen, R.: Efficient algorithms for real-life instances of the variable size bin packing problem. Computers & Operations Research 39(11), 2848–2857 (2012)

[11] Crainic, T.G., Perboli, G., Tadei, R.: Extreme point-based heuristics for three-dimensional bin packing. INFORMS Journal on Computing 20(3), 368–384 (2008)

[12] Bortfeldt, A., Gehring, H., Mack, D.: A parallel tabu search algorithm for solving the container loading problem. Parallel Computing 29(5), 641–662 (2003)

Fuzzy Decision Tree to Predict Readmissions in Intensive Care Unit

Cláudia Silva, Susana M. Vieira, and João M.C. Sousa

IDMEC/LAETA
Instituto Superior Técnico, Universidade de Lisboa
claudia.silva@dem.ist.utl.pt, {susana.vieira,jmsousa}@tecnico.ulisboa.pt

Abstract. Patients readmissions to Intensive Care Unit (ICU) are introduced as a problem associated with increased mortality, morbidity and costs, which complicates the performance of a good clinical management and medical diagnosis. The aim of this work is to use the fuzzy decision tree using the axiomatic fuzzy set (AFS) theory in a type of coherence membership function to apply in the risk of readmission. Three fitness functions are used to obtain the threshold using different assessment measures: accuracy; area under the curve ROC (AUC); and Cohen's kappa coefficient. The results for this problem data demonstrated that the model using fitness function with Cohen's kappa coefficient obtains better performance than the others fitness functions.

Keywords: Fuzzy Decision Tree, Readmissions, Axiomatic Fuzzy Set.

1 Introduction

Intensive care unit (ICU) readmissions are a commonly used as quality measure but despite decades of research, these adverse events continue to occur. Of particular concern is that readmitted patients during the same hospitalization have much worse prognoses than those not readmitted [1]. Patients readmitted have a increased risk of death, length of stay and higher costs [2]. Previous studies have demonstrated overall readmission rates of $4 - 14\%$ [3, 4], of which nearly a third can be attributed to premature discharge from the critical care setting [3, 5]. Increasing pressures on managing care and resources in ICUs is one explanation for strategies seeking to rapidly free expensive ICU beds. Faced with this scenario, a clinician may elect to discharge a patient currently in the ICU who has already had the benefits of stabilization and intensive monitoring, to make room for more acute patients allocated in the emergency department, exposing the outwardly transferring patients to the risk of readmission in the short term [6]. Moreover, besides the existence of morbidity and mortality issues around readmission, the Centers for Medicare & Medicaid Services already reduced funding for specified avoidable conditions, and it is quite possible avoidable readmission to an ICU will also receive attention. Previous studies have examined different variables that are assessed at discharge, and that are considered to be predictive of readmission. In [7], a wrapper method based on tree search feature selection was implemented to obtain the six most predictive variables of readmissions. The present works consists on the fact that knowledge can be extracted from medical data collected in the last 24 hour period before patients discharge of the ICU, for medical diagnosis and particularly to predict readmissions outcome.

Decision trees are one among interesting and commonly encountered architectures used for learning, reasoning and organization of datasets. The incorporation of fuzzy sets [8] into decision trees enables us to combine the uncertainty handling and approximate reasoning capabilities of the former with the comprehensibility and ease of application of the latter. This trend gave rise to the name of fuzzy decision trees and has resulted in a series of development alternatives [9]. The axiomatic fuzzy set (AFS) decision tree [10, 11] uses the theoretical findings of the axiomatic fuzzy set theory in a new type of coherence membership function.

The main goal is to use a AFS decision tree to classify patients as readmitted or no-readmitted. Three fitness functions are used to choose the threshold: based on accuracy; based on area under the curve ROC (AUC); and based on Cohen's kappa coefficient. Considering that kappa is a more conservative measure and the dataset is unbalanced, our hypothesis is that the AFS decision tree using fitness function based on Cohen's kappa coefficient have better results than the others.

The outline of the work is as follows: in section II the basis for AFS decicion tree is introduced, followed by the model performance measures description. An introduction to the medical database is given in section III. Main results are presented in section IV, followed by the drawing of conclusions and future work in section V.

2   Methods

2.1   AFS Decision Tree

The AFS decision tree uses the theoretical findings of the AFS theory in a new type of coherence membership function. This type of membership function embraces both the factor of fuzziness (by capturing subjective imprecision) and randomness (by referring to the objective uncertainly) and treats both of them in a consistency manner [11]. The AFS classifier algorithm consists of the six steps.

Fuzzification of Data: In order to properly define the membership function of a fuzzy concept, we need to define AFS structure of data. Let X, M be sets and 2^M be the power set of M. Let $\tau : X * X \to 2^M$. (M, τ, X) is called an AFS structure if τ satisfies the following axioms:

$$AX1 : \forall (x_1, x_2) \in XX, \tau(x_1, x_2) \subseteq \tau(x_1, x_1) \tag{1}$$

$$AX2 : \forall (x_1, x_2), (x_2, x_3) \in XX, \tau(x_1, x_2) \cap \tau(x_2, x_3) \subseteq \tau(x_1, x_3) \tag{2}$$

X is called universe of discourse, M is called a concept set and τ is called structure [11].

The algorithm assigns three fuzzy terms (simple concepts) to each attribute. The first cut point is the minimal value of the attribute, the second one is the mean and the third one is the maximal value of the attribute. The membership functions of fuzzy terms are defined by

$$\mu_\xi = sup_{i \in I} inf_{y \in A_i} \frac{\sum_{u \in A_i^\tau(x)} \rho_y(u) N_u}{\sum_{u \in x} \rho_y(u) N_u}, \forall x \in X \tag{3}$$

where N_u is the number of times $u \in X$ is observed and ρ_y is the weight function for a simple concept $y \in M$ (set of simple concept) [11].

Construction of AFS Decision Tree: The node splitting criterion, that is the growth process of the decision tree is guided by the maximum information gain [12]. P^N and I^N denote the total example count and information measure for node N, where I^N is the standard information content described as

$$I^N = -\sum_{v_c \in c} \frac{P_{v_c}^N}{P^N} . log(\frac{P_{v_c}^N}{P^N}) \tag{4}$$

The information gain is denote by $G_{V_i}^N = I^N - I^{S_{v_i}^N}$ when using the fuzzy attribute V_i to split N, where

$$I^{S_{v_i}^N} = \sum_{v_i^P \in D_i^N} \frac{P^{N|v_P} . I^{N|v_i^P}}{\sum_{v_i^P \in D_i^N} P^{N|v_P}} \tag{5}$$

is the weighted information content. The $N|v_p^i$ is a particular child node N created by the use of the fuzzy set attribute V_i to split N.

There are three stopping criteria (the first two are a sort of precondition and the third one come in a form of some postcondition): a given node N can be expanded if the samples in the threshold cut set of fuzzy set are not in the same class; the current node N can be expanded if $V_N \neq V$, the set V_N is the attribute applied from root to the node N, V is the set of all attributes; and when the maximum information gain at the current node N is negative or the set of Ns children is empty, we stop expanding nodes [10, 11].

Rule Extraction: Each path starting from the root traversing down to a classification node (terminal node) is converted to a rule. Suppose the rule $r_1, r_2,..., r_i$ is extracted from decision tree, the antecedent part of the rule r_i is a fuzzy concept [11].

Pruning the Rule-Base: first, it removes each rule from the rule-base, and classify the training data using the remaining rules. Then, it delete the rule, whose corresponding remaining rules have the maximal increase of accuracy or AUC or Coehn's kappa coefficient on training data. Repeat this until the resulting pruned rule-base becomes worse than the original one. Finally, it is used

logic operation $\vee$, sum all fuzzy concepts representing the antecedents of the rules with the same consequent. For a training data set with l classes, a rule can be obtained and read as: If x is ξ (fuzzy concepts) then x belongs to the class l [11].

Determining Optimal Threshold: different values of threshold (δ) can produce different trees. Thus, the performance of the tree depends of the threshold δ. It is computed the sub-optimal value of the δ using the Fitness Function $F(\delta)$ with Genetic Algorithm. The sub-optimal δ maximizes the following three fitness functions:

$$F(\delta) = \|X\|.ACC - \delta.NN \tag{6}$$

$$F(\delta) = \|X\|.AUC - \delta.NN \tag{7}$$

$$F(\delta) = \|X\|.K - \delta.NN \tag{8}$$

where $\|X\|$ is the number of training samples, ACC, AUC and K is the classification accuracy, area under the curve ROC and Cohens Kappa, respectively, reported on the training samples for the rule-base obtained by threshold δ via the algorithm of building AFS decision tree and NN is the total nodes of the pruned tree.

Inference of Decision Assignment and Associated Confidence Degrees. In order to predict the class label of the new samples x which are not included in X, it is necessary calculate the membership degree $\mu_\xi^L(x)$ and x belongs to the class $argmax_{v_k^c \in D_c} \mu_{\xi_k}^L(x), k = 1, 2, ..., l$. The confidence degree of the membership degree of x is defined as follows:

$$C_\xi(x) = 1 - (\mu_\xi^U(x) - \mu_\xi^L(x)) \tag{9}$$

For sample x, the closer the upper bound of membership function of ξ is to the lower bound, the larger value of confidence degree we obtain. The larger value of $C_\xi(x)$ advises us to trust $\mu_\xi^L(x)$ as the membership degree of x belonging to ξ [11].

2.2   Model Assessment

In describing the performance of binary classiers, the accuracy of classication cannot be considered alone [13]. Specially in medical applications, this measure shows limitations, where the positive class (disease-positive) is often underrepresented and misclassifications in this class will not have great impact in the accuracy value. Both the sensitivity, or hit rate, and specificity, or true rejection rate, must also be analyzed. From the clinical perspective, achieving the highest sensitivity and specificity simultaneously is often desirable [14].

In this work, the measures used to assess the quality of the obtained classifier were: specificity (10), which represented the cases where the patient was correctly classified as not readmitted (class 0):

$$\text{specificity} = \frac{TN}{TN + FP} \tag{10}$$

The sensitivity (11), which represented the cases where the patient was correctly classified as readmited (class 1):

$$\text{sensitivity} = \frac{TP}{TP + FN} \tag{11}$$

The accuracy (12), which consisted on the proportion of true classifications over the total number of observations:

$$\text{accuracy} = \frac{TN + TP}{TN + TP + FP + FN} \tag{12}$$

where, TN and TP are patients that were correctly identified as belonging to class 0 and class 1, respectively; FN and FP are patients that were incorrectly identified as belonging to class 0 and class 1, respectively.

In medical diagnosis and in the machine learning community, one of the methods for combining sensitivity and specificity measures into the evaluation task is the analysis of the area under the ROC curve (AUC), with ROC being the receiver operating curve, a graphical plot which illustrates the performance of a binary classifier system as its discrimination threshold is varied [15]. A value of 0.50 is achieved through random predictions and 1 represents a perfect discrimination.

Cohens kappa coefficient is a statistical measure of inter-rater agreement for qualitative items. It is generally thought to be a more robust measure than simple percent agreement calculation, since it takes into account the agreement occurring by chance. The equation for Cohens kappa is:

$$K = (Pr(a) - Pr(e))/(1 - Pr(e)) \tag{13}$$

where $Pr(a)$ is the total agreement probability, and $Pr(e)$ is the hypothetical probability of chance agreement, using the observed data to calculate the probabilities of each observer randomly saying each category [16].

3   Data

Multi-parameter Intelligent Monitoring in Intensive Care (MIMIC II) Database was used, acquired from the Beth Israel Deaconess Medical Center in the United States: the MIMIC II (Multi-parameter Intelligent Monitoring for Intensive Care) Database. MIMIC II data was collected from 2001 to 2006, and it has been de-identified by removal of all protected health information. The database is formed by 25 549 patients, of which 19 075 are adults (15 years old at time of admission). It includes high frequency sampled data of bedside monitors, clinical data (laboratory tests, physicians and nurses notes, imaging reports, medications and other input/output events related to the patient) and demographic data [17]. Variables included for modeling purposes are shown in table 1

Table 1. Physiological variables used

Type of variable	Variable name(units)	Average sampling frequency (samples per day)
Monitoring signals	Heart rate (beats/min)	27.25
	Respiratory rate (breaths/min)	26.54
	Temperature (C)	9.63
	SpO2 (%)	26.71
	Non-invasive arterial Blood pressure (systoloc) (mmHg)	25.72
	Non-invasive arterial Blood pressure (mean) (mmHg)	25.56
Laboratory tests	Red blood cell count ($cellsx10^3/\mu L$)	3.15
	White blood cell count ($cellsx103/\mu$)	3.04
	Platelets (cellsx103/μL)	2.83
	Hematocrit (%)	2.78
	BUN (mg/dL)	2.16
	Sodium (mg/dL)	4.37
	Potassium (mg/dL)	3.28
	Calcium (mg/dL)	2.56
	Chloride (mg/dL)	1.68
	Creatinine (mg/dL)	2.94
	Magnesium (mg/dL)	1.66
	Albumin (g/dL)	1.86
	Arterial pH	1.72
	Arterial base excess (mEq/L)	1.68
	Lactic acid (mg/dL)	2.28
Other	Urine output (mL/h)	22.25

The outcome was defined as binary: one if patient was readmitted and zero if not. The distribution of classes in dataset is as follow: 87% are patients not readmitted and 13% are patients readmitted.

4   Results

The dataset was randomly divided into two equal parts: one for feature selection (FS subset) and the other for model assessment (MA subset). The datasets were balanced, MA and FS subsets contained approximately the same percentage of classes. Feature selection was not performed and two dataset was used, one using all variables and the other using only the most predictive variables selected by the sequential forward selection approach were found in [7] and included

Table 2. Results for full dataset with different fitness functions (FF)

	FF with ACC	FF with AUC	FF with kappa
AUC	0.58 ± 0.07	0.59 ± 0.07	0.61 ± 0.08
ACC	0.62 ± 0.12	0.61 ± 0.12	0.62 ± 0.11
Sensitivity	0.46 ± 0.25	0.48 ± 0.24	0.51 ± 0.26
Specificity	0.65 ± 0.15	0.65 ± 0.15	0.66 ± 0.14

Table 3. Results for six variables with different fitness functions (FF)

	FF with ACC	FF with AUC	FF with kappa
AUC	0.56 ± 0.05	0.58 ± 0.07	0.60 ± 0.08
ACC	0.59 ± 0.09	0.60 ± 0.11	0.46 ± 0.08
Sensitivity	0.47 ± 0.25	0.47 ± 0.24	0.52 ± 0.23
Specificity	0.59 ± 0.13	0.64 ± 0.15	0.67 ± 0.14

(ordered by predictive power): mean heart rate, mean temperature, mean platelets, mean non-invasive arterial blood pressure (mean), mean spO2, and mean lactic acid. A 10-fold cross-validation was performed to assess the model, by using a 90% of the MA subset as the training set, and the remaining 10% as the validation set. To reduce variability, 10 rounds of this validation process were performed always using different partitions. The model performance measures AUC, ACC, sensitivity and specificity results in this validation stage were averaged over the rounds. The threshold found to be the optimal for fuzzy decision tree varied between a minimum value of 0.30 and a maximum value of 0.80. Three fitness functions are used (6 7 8). In the following tables the results obtained for the models performance after 10 fold cross-validation are presented. In table 2 the three fitness functions based on ACC, AUC and Cohen's kappa coefficient were used in full dataset. As it possible to observe, the overall AFS decision tree with fitness function based on Cohen's kappa results are better than the results obtain with the other fitness functions. The ACC presents approximately the same values, as also the specificity, although this one is slightly lower for the fitness function with AUC. There is an increasing in the sensitivity, comparatively to the the fitness function with ACC.

In table 3 the three fitness functions based on ACC, AUC and Cohen's kappa coefficient are used in the six predictive variables. Once more time, the overall AFS decision tree with fitness function based on Cohen's kappa results are better than the results obtain with the other fitness functions. Comparing the results obtained in both table 2 and table 3, it is possible to observe that have similar results, so it can mean that no information was lost in the choose of this six variables.

5   Conclusions

This work applies a fuzzy decision tree to predict ICU readmissions. A method based on AFS decision tree was introduced and different fitness functions were performed. The proposed fitness function based in Cohen's kappa coefficient was able to slightly increase the performance of fuzzy decision tree in comparison with the others fitness function. The increase of performance obtained using the fitness function with cohen's kappa coefficient in comparison with the fitness function with ACC and AUC shows that, for this specific problem, is a important model assessment measure.

As future work, the AFS decision tree with fitness function based on cohen's kappa coefficient should be further applied and confirmed to other applications beyond clinical data. Future work should also involve the improvement of the AFS decision treee.

Acknowledgments. This work was supported by FCT, through IDMEC, under LAETA Pest-OE/EME /LA0022 and partially supported by the project PTDC/EMS-SIS/3220/2012 - IC4U, IDMEC, and by a FCT post-doctoral grant SFRH/BPD/65215/2009, Ministério da Educação e Ciência, Portugal.

References

[1] Elliott, M., Worrall-Carter, L., Page, K.: Intensive care readmission: A contemporary review of the literature. Intensive and Critical Care Nursing (2013)

[2] Boudesteijn, E., Arbous, S., van den Berg, P.: Predictors of intensive care unit readmission within 48 hours after discharge. Critical Care 11(Suppl. 2), 475 (2007)

[3] Baigelman, W., Katz, R., Geary, G.: Patient readmission to critical care units during the same hospitalization at a community teaching hospital. Intensive Care Medicine 9(5), 253–256 (1983)

[4] Andrew, L., Rosenberg, T.P., Hofer, R.A.: Hayward, Cathy Strachan, and Charles M Watts. Who bounces back? physiologic and other predictors of intensive care unit readmission. Critical Care Medicine 29(3), 511–518 (2001)

[5] Charles, J.R., Durbin, G.: Robert F KopelL. A case-control study of patients readmitted to the intensive care unit. Critical Care Medicine 21(10), 1547–1553 (1993)

[6] Chalfin, D.B., Trzeciak, S., Likourezos, A., Baumann, B.M., Dellinger, P.: Impact of delayed transfer of critically ill patients from the emergency department to the intensive care unit*. Critical Care Medicine 35(6), 1477–1483 (2007)

[7] Fialho, A.S., Cismondi, F., Vieira, S.M., Reti, S.R., Sousa, J.M., Finkelstein, S.N.: Data mining using clinical physiology at discharge to predict icu readmissions. Expert Systems with Applications 39(18), 13158–13165 (2012)

[8] Zadeh, L.A.: Zadeh. Fuzzy sets. Information and Control 8(3), 338–353 (1965)

[9] Janikow, C.Z.: Fuzzy decision trees: issues and methods. IEEE Transactions on Systems, Man, and Cybernetics, Part B: Cybernetics 28(1), 1–14 (1998)

[10] Liu, X., Pedrycz, W.: The development of fuzzy decision trees in the framework of axiomatic fuzzy set logic. Applied Soft. Computing 7(1), 325–342 (2007)

[11] Liu, X., Feng, X., Pedrycz, W.: Extraction of fuzzy rules from fuzzy decision trees: An axiomatic fuzzy sets (afs) approach. Data & Knowledge Engineering (2012)

[12] Yuan, Y., Shaw, M.J.: Induction of fuzzy decision trees. Fuzzy Sets and systems 69(2), 125–139 (1995)

[13] Swets, J.A.: Measuring the accuracy of diagnostic systems. Science 240(4857), 1285–1293 (1988)

[14] Grzybowski, M., Younger, J.G.: Statistical methodology: Iii. receiver operating characteristic (roc) curves. Academic Emergency Medicine 4(8), 818–826 (1997)

[15] Fawcett, T.: An introduction to roc analysis. Pattern Recogn. Lett. 27(8), 861–874 (2006)

[16] Vieira, S.M., Kaymak, U., Sousa, J.M.: Cohen's kappa coefficient as a performance measure for feature selection. In: 2010 IEEE International Conference on Fuzzy Systems (FUZZ), pp. 1–8. IEEE (2010)

[17] Saeed, M., Lieu, C., Raber, G., Mark, R.G.: Mimic ii: a massive temporal icu patient database to support research in intelligent patient monitoring. In: Computers in Cardiology, pp. 641–644. IEEE (2002)

On the Current Payback Time for Small Investors in the Photovoltaic Systems in the Region of Madeira

Sandy Rodrigues Abreu[1], Marco Leça[1], Xiaoju Chen[2], and Fernando Morgado-Dias[1]

[1] Madeira Interactive Technologies Institute and Centro de Competências de
Ciências Exactas e da Engenharia, Universidade da Madeira
Campus da Penteada, 9000-039 Funchal, Madeira, Portugal
morgado@uma.pt
[2] Civil and Environmental Engineering Department,
Carnegie Mellon University, 5000 Forbes Avenue, Pittsburgh, PA 15213, USA

Abstract. Following a period of strong investment in renewable energy, Portugal is now facing a huge reduction in the support for such clean energy sources, namely through the reduction of the feed-in tariffs and removal of tax incentives. The region of Madeira benefits from a very good solar exposition and has strong fossil fuel based products dependency, which make it a favorable place to invest in photovoltaic. But unfortunately these recent changes made the feed-in tariffs too low for any informed investor to put money there. Throughout this paper we analyze the payback time for these investments, we select the best regime for selling energy and forecast the future for this area of activity along with the next decisions that will be important in this area.

Keywords: Photovoltaic systems, Feed-in tariff, Payback time, Micro Production, Mini Production, Self-consumption.

1 Introduction

Following a period of strong investment in renewable energy, Portugal is now facing a huge reduction in the support for such clean energy sources, namely through the reduction of the feed-in tariffs and removal of tax incentives. The scenery change has been forced by a deep financial crisis but the payback time (PT) for small investors through the photovoltaic micro production regime has been getting higher and is no longer an attractive investment, in spite of a huge reduction of the necessary investment that is now below half of the necessary amount 8 years ago.

The Region of Madeira (RoM), situated almost one thousand kilometers southwest from the mainland, benefits from a good solar exposition and has strong fossil fuel based products dependency. It is potentially an ideal location for investing in photovoltaic production for small investors.

According to the Portuguese legal framework, Photovoltaic (PV) installations can be under several different regimes: micro production (up to 3.68KW), mini production (up to 250 KW) and general regime. The first two benefits from special feed-in tariffs (mini production only up to 20KW and after that falls into a reverse auction

A.P. Moreira et al. (eds.), *CONTROLO'2014 - Proc. of the 11th Port. Conf. on Autom. Control*,
Lecture Notes in Electrical Engineering 321, DOI: 10.1007/978-3-319-10380-8_36

for prices) and in the latter energy is bought at the same price the consumers pay, until a recent change described in section 2.3.

During the last decade a large number of small companies were created to deal with the great number of requests for new micro production installations. These companies have now an unattractive product to offer to their clients. In this paper we analyze this situation and propose the most favorable regime for selling energy. We also forecast the future of this market and point the next decisions that will be important in this area.

1.1   Energy Situation of RoM

In energetic terms, RoM has a high dependence on fossil fuel based products, although this has been mitigated by recent minor investments in renewable energy, specifically in the areas of photovoltaic and wind energy. This effort, though small, is commendable but is still in its infancy.

In the regional context we verify that there are many good examples of the application of renewable energy, with tremendous success in several points, especially in hydro, wind and photovoltaic energies, being these in descending order, from highest to lowest energy produced. The production of electricity comes mainly from diesel and natural gas plants, representing 74.5% of the total energy consumed, followed by 9% in hydropower, wind power is standing around 9.7%, solar energy at 3.8% and finally, energy from solid municipal waste (SMW) at 3%.

Figure 1 presents a graph of the energy situation of the RoM during the last 7 years.

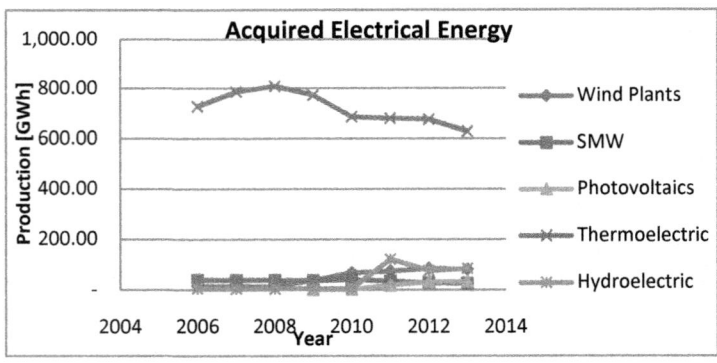

Fig. 1. Total acquired electrical energy by source in Madeira, 2013

The electricity from solar power had an amazing growth over the past three years, with an extraordinary 200% increase in installed capacity. Already at the same level of wind energy, there has been a 120% increase in investment in this area.

1.2   Solar Radiation in Madeira

Solar energy is an inexhaustible, abundant energy source on the planet. In Europe, Portugal is the country with more solar exposure, registering an annual average of approximately 8 hours/day with a reception above 1200W/m². The daily solar energy that reaches Portugal is on average 430000 GWh, which is equivalent to the energy the country consumes in approximately 1000 days [1] [2].

For the study of global radiation of RoM, we used data from various meteorological stations, in particular the Institute of Meteorology (IM), the Regional Laboratory of Civil Engineering (LREC), NASA Surface Meteorology and Solar Energy and World Meteorological Organization (WMO) [3]. Figure 2 shows a graph of the monthly variation of the daily solar radiation average, from 2002 to 2005.

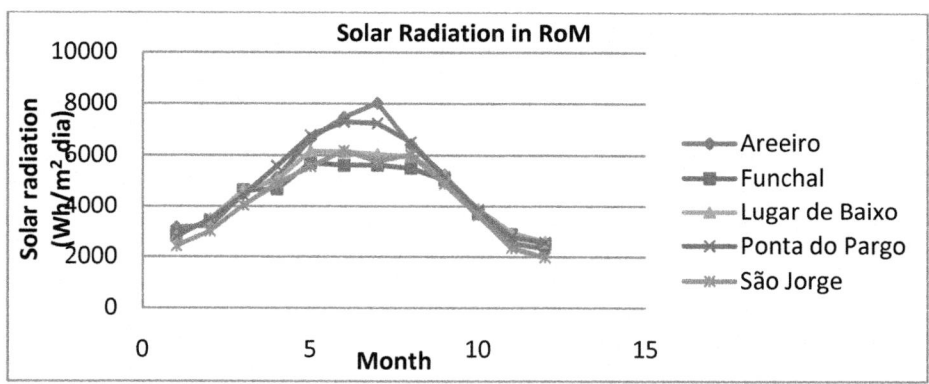

Fig. 2. Monthly variation of daily average solar radiation of RoM [4]

From figure 2, it can be concluded that the highest annual mean values were recorded in Areeiro (4873 Wh/m²/day), followed closely by Ponta do Pargo (4868 Wh/m²/day) in Calheta. The highest values recorded for the monthly average solar radiation, were for the months of June and July for the Areeiro and Ponta do Pargo stations. This period corresponds to the beginning of the summer season. The maximum monthly value was registered in July (8023 Wh/m²/day).

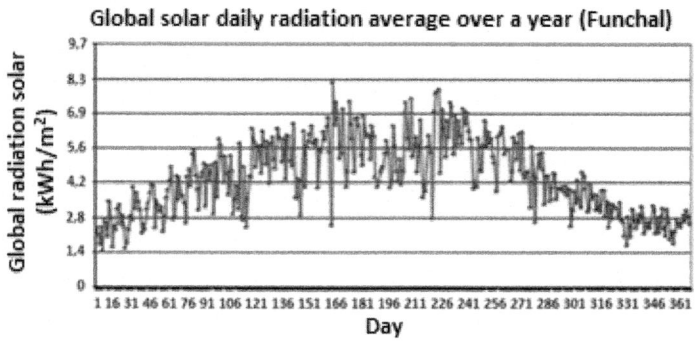

Fig. 3. Average daily solar radiation through the year for Funchal [5]

Figure 3 shows the daily average of solar radiation for Funchal over a period of five years (1999-2003) [5]. Through the data of NASA Surface Meteorology and Solar Energy System, the average measured values for monthly solar radiation on the horizontal plane was 5.63 kWh/m^2/day and the number of hours of sunshine for Madeira are around 2600 hours/year. These values are based on data prior to March 2008.

Figure 4 shows the radiation map of Madeira, Porto Santo and Desertas islands.

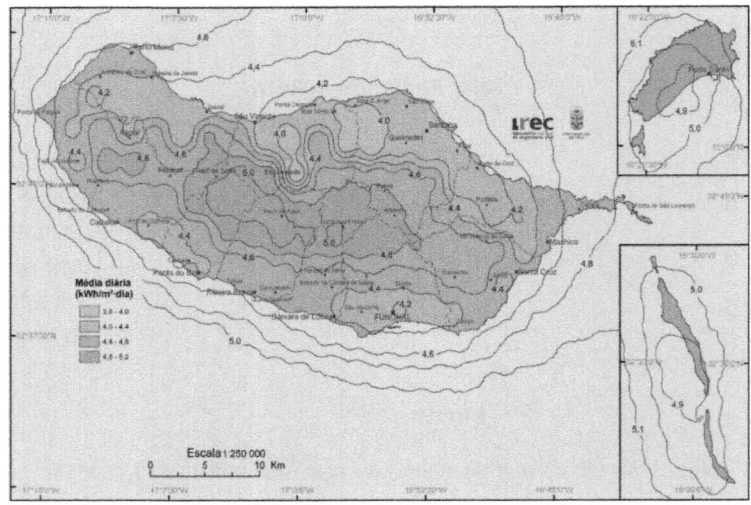

Fig. 4. Solar radiation map RoM (Vázquez et al, 2008)

Considering figures 2-4, it can be concluded that Madeira has very favorable conditions for the installation of photovoltaic panels as has been happening in the last decade.

2   Legislative Framework

For small investors currently there are two regimes of selling the energy to the local electrical energy company: Micro production (McP) and Mini production (MnP).

These regimes have changed over time both in terms of feed-in tariffs and regulation. A short oversight is given in the next section.

2.1   Micro Production Regime

The McP regime has an upper limit of 3.68 KW of installed power, provided that the installed power is less than half of the hired from the electrical company.

Reviewing the prices since this regime was created, in 2008 the feed-in tariff was 0.65 €/kWh for the first 5 years. After that, for the next 5 years, the tariff is reduced annually, according to the ordinance definition to approximately 77 % of the value of previous year's tariff. Afterwards, the paid value goes down to the value of the cost per kWh paid by the consumers (general regime) [6]. At that time, in 2008, the cost of each kWh consumed was around 0.11 €.

In 2010, the buying regime is composed of a contract with duration of 15 years divided into two periods, the initial with 8 years and the subsequent of 7. At the end the producer is forced to join general regime where energy is bought at the same price the local company sells the energy to consumers [7]. The feed-in tariff is 0.40 €/kWh for the first period and 0.24 €/kWh for the second period, and the value of both rates is decreased by 0.02 €/kWh a year for new installations [7].

For 2014 the new values defined by government ordinance are 0.066 €/kWh for the first 8 years and 0.145 €/kWh for the remaining 7 years.

2.2   Mini Production Regime

These facilities produce electricity from renewable resources, based on a single production technology (e.g. wind or PV) and their maximum power supply is 250 kW, provided that the installed power is less than half of the hired from the electrical company. This type of scheme has originally been launched in 2011, with a 15 year contract with a fixed rate of 0.25 €/kWh produced, with its value reduced annually by 7% [9]. Nowadays (2014), the reference tariff for MnP is 0.106 €/kWh with feed-in tariff and 0.142 €/kWh for the general regime.

2.3   General Regime for Micro Production

The general regime was characterized by energy being bought at the same price consumers pay, but in 2013 a new tariff was introduced according to the following equation:

$$\operatorname{Re} m_m = W_m \times P_{ref} \times \frac{IPC_{n-1}}{IPC_{ref}} \tag{1}$$

Where Rem_m is the remuneration of the month m(€), Wm is the energy produced in month m(kWh), Pref is the value of the share of energy from simple rate between 2.3 and 20.7 kVA applied in the year 2012 to the consumer; IPC_{ref} is the index of prices at the consumer, excluding housing, for the month of December 2011 (published by the National Statistics Institute), and IPC_{N-1} is the same index for the month of December of the year n-1 [8]. Currently the values for this regime are 0.142€/KWh.

3   Price of Energy

As can be seen from the previous section the values of the feed-in tariffs have been changing over time. Also the prices for energy at the consumer level have been increasing based on inflation and the costs to produce energy, but were also subjected to changes in taxes, namely a change in VAT value. The VAT value was initially of 4% (2008) and was increased to 16% (2011) and is now of 22% (since 2013), but since these values were changed during fiscal years the average value in 2011 was 7% and in 2012 20.5%.

Figure 5 compares these values, considering the prices of energy at the consumer level under the Low Voltage Normal regime (BTN) up to 20.7 kVA, with bi-hourly rate, which is the most common contractual arrangement in RoM, including taxes.

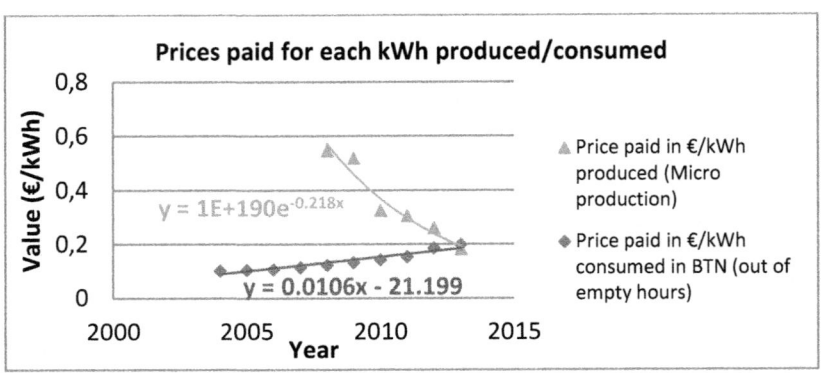

Fig. 5. Prices paid for each kWh produced in micro production and prices paid for each kWh consumed in the public network

Analyzing this graph it can be seen that since 2010, the average value per kWh produced with McP feed-in tariff decreased significantly (from 0.547 €/kWh to 0.182 €/kWh, considering the average values along the 15 year contracts), continuing a descending trend that is approximated by $y = 1E+ 190e^{-0.218x}$.

In 2013, the amount paid for each kWh consumed exceeded the value received for each kWh produced under the McP feed-in tariff. This is a new situation where it just started to compensate to produce energy for self-consumption, instead of applying to obtain a feed-in tariff.

Figure 6 shows the feed-in value paid for each kWh in MnP and the amount paid for each kWh consumed (including taxes) during peak and full hours, under the Special Low Voltage regime (BTE) exceeding 41.4 kW, with tetra-hourly rate, which is the most common contractual arrangement in RoM for public buildings.

The values used for figure 6 were approximated by straight lines: the price paid for each kWh produced in MnP regime ($y = 0.1579$) and for each kWh consumed during peak hours ($y = 0.0082x - 16.422$) intersected in the year 2012, i.e., from this date on it compensates to consume the energy produced by the photovoltaic system for self-consumption rather than to sell it to the local energy company, since the amount received per kWh produced is lower than the price paid for each kWh consumed.

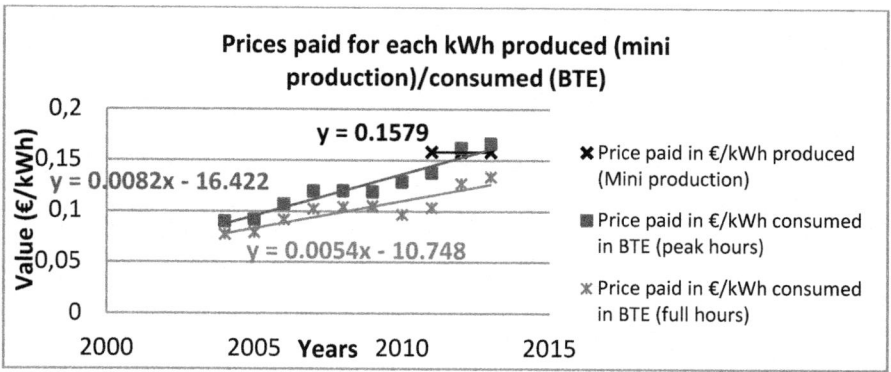

Fig. 6. Prices paid for each kWh produced in mini production and prices paid for each kWh consumed in public network

With the current increasing trend in the price paid for each kWh consumed at full hours (y = 0.0054 x − 10.748) and the stable value paid for each kWh in MnP, it is expected that in late 2018 the former will exceed the latter.

Also note that, the amount paid for each kWh consumed in Portugal, is 13.6% higher than the average EU price and with a further increase of 2.8% proposed by the Energy Services Regulatory Authority (ERSE), added on January 1st, 2014 [10] [11].

4   Cost of Photovoltaic Installations in RoM

This section analyses the prices of photovoltaic installations under the McP regime in RoM. This type of installation was selected because it was the first to be regulated, and prices could be obtained from several companies. This is a regional analysis since the prices of solar modules vary by country and region in which they are acquired. The objective is not only to calculate the PT but to predict trends for future prices of these installations. The prices used are for a maximum output power of 3.68 kW, as defined in the legislation but the installing companies found out that to achieve a better payback time it would compensate to install a higher power so on average these installations are of 4.05 kW. The prices shown in figure 7 include VAT.

Since 2007 until early 2013, with the market stimulated by the presence of various manufacturers of solar cells, competition between companies has increased considerably, causing a steady decline of about 15 % per year in the prices. From mid-2013 to date, there has been a small slowdown on this decrease of prices, anticipating future price stabilization.

Nowadays, a PV installation of this size costs around 10500€. It can be seen that there is a steady fall in prices of PV installations. To characterize this behavior we have chosen an exponential function with a negative exponent. Notice that there is not enough data to build a detailed model, but this function will be much more appropriate for the behavior that can be expected in this situation. Analyzing the function of the trend line (y = 2E+155e$^{-0.173x}$) it is observed that in the short term (≈ 4 years), will reach up the minimum value of 5000€, from which the price of the PV installations will remain roughly stable.

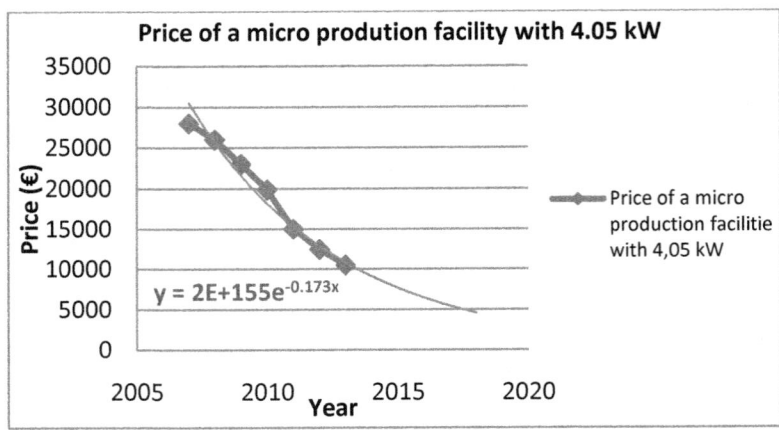

Fig. 7. Historic of prices of micro productions facilities with 4.05 kW of installed power

5   Payback Time

The graph in Figure 8 represents the number of years necessary for PT to be reached for a McP plant with an installed capacity of 4.05 kW located in Funchal, under different energy buying regimes. The value of installed capacity was calculated using the average installed power of micro production in the Funchal area. From these installations we have selected a set of 10 with available data and no failures over 2 years to calculate an average annual production of approximately 5800 kWh. These values were used in the calculations that follow.

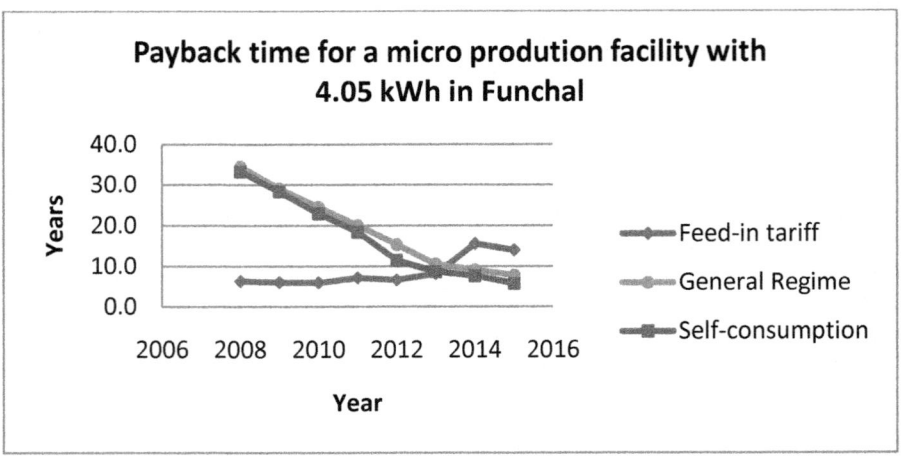

Fig. 8. Comparison between different types of regimes

PT is a key issue to attract investment in this area. As can be seen from figure 8 the initial feed-in tariffs were designed so that this solution is commercially attractive with a PT around 6 years. Gradually these feed-in tariffs got worst and even though the installation prices reduced a lot, the PT is now over 10 years.

At the same time the energy prices for consumers increased an astonishing 209.3% over the last ten years and also had to face a VAT increase from 4% to 22%. As a result it is now more attractive for investors the use of energy for self-consumption or to go to the general regime than to apply for the feed-in tariffs. It should be noted that the self-consumption always had better PT than the general regime. Nevertheless self-consumption requires that there is enough consumption all the time to use the energy produced which will not be the case for most of the households.

It is anticipated that the future trend regarding the PT for the feed-in tariffs will be increasing, thus favoring the general regime and self-consumption.

One should note that these conclusions only apply to McP facilities, since, for MnP a different formula is applied. Moreover, the forecast for the price of electricity was determined using the average value of the index of prices at the consumer from the previous years.

6   Conclusion

From the average annual solar radiation map, we can say that Madeira has very favorable general conditions for using this resource since, in general, values above 4kWh/m^2/day are attractive for investment in PV or even heat based systems. It was shown that the areas with higher annual average solar radiation are Areeiro and Calheta. It can be stated that in most of RoM (exceeding 75% of the area), solar energy is an endogenous renewable energy resource with great interest in exploration.

The price of PV installations decreased 59.6 % over the last 5 years and the price paid to the local company per kWh consumed rose approximately 186 % in BTE scheme and 209.3 % in the BTN regime in the past 10 years.

The cost of McP facilities has decreased in the last eight years, to the current values near 10500€ and it is expected that it will stabilize at values close to 5,000€ in 2018, if the verified trend continues.

From the analysis of the different PV production regimes, it was found that currently they are not very attractive to investors, especially the McP. This is because the PT is high, between 9-10 years and, especially, the price paid per kWh produced is relatively low and less than the amount paid for each kWh consumed.

To change this situation and make it appealing again, the average price paid per kWh produced should be greater or equal to 0.275€, in order to reduce the PT to less than 6 years, as seen in 2008. If we consider a continuous decrease in the cost of PV, the PT will decrease, making the regime more attractive.

However, with the progressive increase of energy prices (BTN clients) and considering the past downwards trend of installation costs for McP, we estimate that, in 2016, the PT will be less than 6 years.

Currently, the most attractive regime is the self-consumption though it needs the associated electrical installation to be able to use the energy produced. An interesting alternative could come from the expected legislation for this regime if it contemplates measuring production and consumption using the prices for consumers.

Acknowledgements. The authors would like to acknowledge the Portuguese Foundation for Science and Technology for their support through project PEst-OE/EEI/LA0009/2011.

The authors would also like to acknowledge the regional renewable energy companies in particular Intelsol and AMV-renewables, and LREC, Dr. Carlos Magro for their support in developing this article.

References

1. Clima, S.: http://www.suaveclima.pt/ficheiros/galeria_92_1.pdf (accessed July 12, 2013)
2. Castro, R.: Uma introdução às energias renováveis: Eólica, Fotovoltaica e Mini-hídrica. IST Press, Lisboa (2011)
3. NASA Surface meteorology and Solar Energy: RETScreen Data, https://eosweb.larc.nasa.gov/cgi-bin/sse/retscreen.cgi?email=rets%40nrcan.gc.ca&step=1&lat=32&lon=16&submit=Submit (accessed July 12, 2013)
4. Magro, C., Vásquez, M., Belmonte, P.: Navarro, José e Cerqueira, Maria, Atlas da Radiação Solar do Arquipélago da Madeira. Funchal, LREC (2008)
5. Ervilha, Ana e Pereira, José, "Avaliação do Potencial Energético Solar na Região Autónoma da Madeira". ERAMAC – Maximização da Penetração das Energias Renováveis e Utilização Racional da Energia nas Ilhas da Macaronésia; e AREAM – Agência Regional da Energia e Ambiente da Região Autónoma da Madeira, Funchal (2005)
6. Ordinance n.º 363/2007, consultado a, http://www.edpsu.pt/pt/PRE/Microproducao/RegulamentaoDocs/Decreto-Lei%2025_2013.pdf (July 12, 2013)
7. Ordinance n.º 118-A/2010, http://www.edpsu.pt/pt/PRE/Microproducao/RegulamentaoDocs/Decreto-Lei%2025_2013.pdf (accessed July 12, 2013)
8. Ordinance n.º 25/2013, http://www.edpsu.pt/pt/PRE/Microproducao/RegulamentaoDocs/Decreto-Lei%2025_2013.pdf (accessed July 12, 2013)
9. Ordinance n.º 34/2011, http://www.dre.pt/pdf1s/2011/03/04700/0131601325.pdf (accessed July 12, 2013)
10. Rosa, E.: Preços da Energia em Portugal, http://www.dn.pt/inicio/economia/interior.aspx?content_id=3587239&seccao=Dinheiro+Vivo (accessed July 12, 2013)
11. Aumento aprovado: Eletricidade vai subir 2,8% em 2014, Economia - DN, http://www.dn.pt/inicio/economia/interior.aspx?content_id=3587239&seccao=Dinheiro+Vivo (accessed February 13, 2013)

Part VIII
Estimation

Unsupervised Learning of Gaussian Mixture Models in the Presence of Dynamic Environments
A Multiple-Model Adaptive Algorithm*

Abdolrahman Khoshrou and António Pedro Aguiar

Faculty of Engineering, University of Porto, Portugal
{majid.khoshrou,pedro.aguiar}@fe.up.pt

Abstract. This paper tackles the on-line unsupervised learning problem of Gaussian mixture models in the presence of uncertain dynamic environments. In particular, we assume that the number of Gaussian components (clusters) is unknown and can change over time. We propose a multi-hypothesis adaptive algorithm that continuously updates the number of components and estimates the model parameters as the measurements (sample data) are being acquired. This is done by incrementally maximizing the likelihood probability associated to the estimated parameters and keeping/creating/removing in parallel a number of hypothesis models that are ranked according to the *minimum description length* (MDL), a well-known concept in information theory. The proposed algorithm has the additional feature that it relaxes "the sufficiently large data set" restriction by not requiring in fact any initial batch of data. Simulation results illustrate the performance of the proposed algorithm.

Keywords: On-line learning, Adaptation and learning, Gaussian mixture models.

1 Introduction

Over the years, the research on identifying and classifying unknown number of components in a dynamic environment has been an important topic in computer vision and pattern recognition communities. In particular, for data clustering, mixture models, where each component density of the mixture represents a given set of individuals/samples in the total population, has been applied in a widespread of applications. Mixture models are able to represent arbitrarily complex probability density functions (pdfs). This makes them an excellent choice for representing complex class-conditional pdfs (e.g. likelihood functions) in Bayesian supervised learning scenarios or prior probabilities for Bayesian parameter estimation [1]. For off-line clustering, and more precisely to compute

---

* This work was partially supported by project CONAV/FCT-PT [PTDC/EEACRO/ 113820/2009].

the parameters that define the mixture model given a finite data set, a widely used procedure is to apply the *expectation-maximization* (EM) algorithm that incrementally converges to a maximum likelihood estimate of the mixture model. Notice however that the basic EM algorithm is not able to deal with on-line data since it is an iterative algorithm that requires all the batch of data in each iteration. Another important restriction is the number of components of the mixture which is to be fixed (does not change) and has to be known a-priori.

To solve some of the above problems, several approaches have been developed. In [2], starting from a fixed number of clusters in a batch of data, a split-and-merge approach together with a dissimilarity index concept is presented that adaptively updates the number of mixture models. Z. Zivkovic et al. [3] inspired by [4] proposed an on-line (recursive) algorithm that estimates the parameters of the mixture and simultaneously selects the number of components by starting with a high number of components in a small batch and searching for the *maximum a posteriori* (MAP) solution, and discarding the irrelevant components. A. Declercq and J. H. Piater et al. [5] presents a method to incrementally learning Gaussian mixture models (GMMs) based on a new fidelity criterion for splitting and merging mixture components.

In this paper we address the on-line unsupervised learning problem of GMMs in the presence of uncertain dynamic environments, i.e., we assume that the number of Gaussian components (clusters) is not only unknown but it also can change over time. Inspired by the work in [4], namely the use of the *minimum description length* (MDL) concept, we propose a multi-hypothesis adaptive algorithm that continuously updates the number of components and estimates the model parameters as the measurements (sample data) are being acquired. The proposed algorithm has the additional feature that it relaxes "the sufficiently large data set" restriction by not requiring in fact any initial batch of data. Simulation results illustrate the performance of the proposed algorithm where it shows that indeed it is able to continuously adapt to the dynamic changes of the number of clusters and estimate the parameters of the mixture model.

2   Problem Statement

Let $\{\mathbf{Y}_n, \ n = 0, 1, 2, \ldots\}$ be a discrete-time random process where for each particular time n, $\mathbf{Y}_n$ follows a K-component mixture of d-dimensional Gaussian with probability density function (pdf) given by

$$p(\mathbf{y}|\theta, w) := \sum_{k=1}^{K} w^{[k]} \, p^{[k]}(\mathbf{y}|\theta^{[k]}), \qquad (1)$$

where $\mathbf{y}$ represents one particular outcome of $\mathbf{Y}_n$ and $w := \{w^{[1]}, \ldots, w^{[K]}\}$ is the mixing weight set that satisfies

$$\sum_{k=1}^{K} w^{[k]} = 1, \quad w^{[k]} > 0, \qquad (2)$$

K denotes the number of components of the mixture, $\theta^{[k]} := \{\mu^{[k]}, \, \Sigma^{[k]}\}$ is the mean and covariance matrix of the kth component, with $\theta := \{\theta^{[1]}, \ldots, \theta^{[K]}\}$, and

$$p^{[k]}(\mathbf{y}|\theta^{[k]}) := \frac{1}{(2\pi)^{d/2}|\Sigma^{[k]}|^{1/2}} \exp\left(-\frac{1}{2}(\mathbf{y} - \mu^{[k]})^T (\Sigma^{[k]})^{-1}(\mathbf{y} - \mu^{[k]})\right). \quad (3)$$

Note that for simplicity of notation we have omitted in the parameters K, w, θ their explicit dependence on the time n.

We can now formulate the problem addressed in the paper: *Given a sequence of observations $Y_0, Y_1, \ldots$, find on-line, as the samples are arriving, a sequence of estimates for the parameters K, w, θ that is most likely to be in some sense close to the correct characterization of the random process $\{\mathbf{Y}_n, \; n = 0, 1, 2, \ldots\}$.*

3 Preliminaries and Basic Results

This section presents several background results, starting with the EM algorithm, that are needed to understand the proposed on-line unsupervised learning algorithm.

3.1 The Basic Expectation-Maximization (Off-line) Algorithm

For finite mixture models, given a set of n independent and identically distributed samples $Y = \{Y_1, \ldots, Y_n\}$, the log-likelihood corresponding to a K-component mixture where all the components are d-dimensional Gaussian is [1]

$$\ell := \log p(Y|\theta, w) = \log \prod_{i=1}^{n} p(Y_i|\theta, w) = \sum_{i=1}^{n} \log \sum_{k=1}^{K} w^{[k]} p(Y_i|\theta^{[k]}) \quad (4)$$

It is well-known that the *maximum likelihood* (ML) or *maximum a posteriori* (MAP) estimates cannot be found analytically [1, Ch. 9]. An elegant and powerful method for finding ML or MAP solutions for models with latent variables is called the *expectation-maximization* or EM algorithm [6], [1, Ch. 9]. The EM is an easily implementable algorithm that iteratively increases the posterior density or likelihood function. In order to describe the EM, we need to introduce for each observation Y_i, a discrete unobserved indicator vector $Z_i = [Z_i^{[1]}, \ldots, Z_i^{[K]}]$. This vector specifies from which component the observation Y_i was drawn, i.e., if $Z_i^{[k]} = 1$ and $Z_i^{[p]} = 0$ for $k \neq p$, then this means that the sample Y_i was produced by the kth component. Hence, the complete log-likelihood function (i.e. the one from which we could estimate θ, w if the *complete* data $X = \{Y, Z\}$ was observed [7]) can be written as a product

$$\log p(Y, Z|\theta, w) := \sum_{i=1}^{n} \sum_{k=1}^{K} Z_i^{[k]} \log \left[w^{[k]} p(Y_i|\theta^{[k]}) \right] \quad (5)$$

The EM algorithm runs over the whole data set Y and until some convergence criterion is met, iteratively produces a sequence of estimates $\hat{\theta}_m, \hat{w}_m, m = 0, 1, 2, \ldots$ by alternatively applying two steps:

E-Step: Given Y and the current estimates $\hat{\theta}_m, \hat{w}_m$ and by considering the fact that $\log p(Y, Z|\theta, w)$ is linear with respect to the missing Z, the so-called Q-function computes the conditional expectation of the complete log-likelihood function as

$$Q(\theta, \hat{\theta}_m) := E\left[\log p(Y, Z|\theta, w)\big|Y, \hat{\theta}_m, \hat{w_m}\right] = \log p(Y, \Gamma|\theta, w), \quad (6)$$

where $\Gamma \equiv E[Z|Y, \hat{\theta}_m, \hat{w}_m]$ is the a conditional expectation that each observation is generated by which component. Since the elements of Z are binary, as mentioned in [4], their conditional expectations are given by

$$\Gamma_i^{[k]} := E\left[Z_i^{[k]}|Y, \hat{\theta}_m, \hat{w}_m\right] = \Pr[Z_i^{[k]} = 1|Y_i, \hat{\theta}_m, \hat{w}_m] = \frac{\hat{w}_m^{[k]}p(Y_i|\hat{\theta}_m^{[k]})}{\sum\limits_{k=1}^{K} \hat{w}_m^{[k]}p(Y_i|\hat{\theta}_m^{[k]})},$$

$$(7)$$

where $\hat{w}_m^{[k]}$ corresponds to the a priori probability that $Z_i^{[k]} = 1$ in the m-th iteration of the basic EM algorithm over Y, while $\Gamma_i^{[k]}$ is the a posteriori probability that $Z_i^{[k]} = 1$, after observing Y_i.

M-Step: Maximizing Q by constructing a Lagrangian function to update the parameter estimation

$$\hat{\theta}_{m+1} = \arg\max_{\theta} Q(\theta, \hat{\theta}_m), \quad (8)$$

for the ML estimation. In the case of MAP criterion, instead of $Q(\theta, \hat{\theta}_m)$, we need to maximize $\{Q(\theta, \hat{\theta}_m) + \log p(\theta)\}$.

Since in many real world applications, the number of components is unknown and may change over time, and we may also have memory and time constraints, the above EM algorithm, for that type of applications, has to be modified to accommodate those issues and also to be applicable in an on-line context. Before we introduce the proposed algorithm, in the next section, first we briefly describe the criterion that is used in [4] in order to find the number of components in a batch of data. Later, we explain how to use this criterion in real time applications.

3.2   The Minimum Description Length (MDL) Principle

The MDL principle is rooted on the fact that any regularity in a given set of data can be used to compress the size of the data. This principle can be also used for inductive inference to the model selection problem [8,9]. Given a set of hypotheses $\mathcal{H} = \{\mathcal{H}_1, \mathcal{H}_2, \ldots\}$ and a data set Y, the goal is to find the hypothesis or combination of hypotheses in $\mathcal{H}$ that most compress Y. For the particular case of a data set $Y = \{Y_1, \ldots, Y_n\}$, that has been generated according to

Eq.(1)-Eq.(3), which has to be encoded and transmitted, the description length can be obtained as follows [4,10]:

$$\mathcal{L}(\theta, w, Y) = \frac{N}{2} \sum_{k=1}^{K} \log(\frac{nw^{[k]}}{12}) + \frac{K}{2} \log \frac{n}{12} + \frac{K(N+1)}{2} - \ell, \tag{9}$$

where N is a constant that grows quadratically with the dimension d of the data, K is the number of components, n is the total number of samples, $w^{[k]}$ is the mixing weight of the kth component, and $-\ell$ can be viewed as the code-length of the data, given by Eq.(4).

3.3 Titterington's On-Line Algorithm for a Multivariate Normal Mixture

As mentioned earlier, the original EM algorithm works in a batch manner. In contrast to the traditional version of the EM, on-line EM variants can flexibly update the parameters of Y_n as soon as a new sample is observed. In [11], the application of an on-line EM algorithm proposed by Titterington [12] for estimating the multivariate normal mixture in computer vision tasks is investigated. In the proposed on-line algorithm the *Titterington-type on-line parameter recursion* for multivariate normal mixtures are given by

$$\mu_{n+1}^{[k]} = \mu_n^{[k]} + \frac{1}{n} \frac{\Gamma_{n+1}^{[k]}}{w_n^{[k]}}(Y_{n+1} - \mu_n^{[k]}) \tag{10}$$

$$\Sigma_{n+1}^{[k]} = \Sigma_n^{[k]} + \frac{1}{n} \frac{\Gamma_{n+1}^{[k]}}{w_n^{[k]}} \left[(Y_{n+1} - \mu_n^{[k]})(Y_{n+1} - \mu_n^{[k]})^T - \Sigma_n^{[k]} \right] \tag{11}$$

$$w_{n+1}^{[k]} = w_n^{[k]} + \frac{1}{n}(\Gamma_{n+1}^{[k]} - w_n^{[k]}) \tag{12}$$

where Y_{n+1} is the new observation, $\Gamma_{n+1}^{[k]}$ is the a posteriori probability in Eq.(7), n is the time, $w_{n+1}^{[k]}$ is the mixing weight of kth component at time $n+1$, $\mu_{n+1}^{[k]}$ is the updated mean of kth component and $\Sigma_{n+1}^{[k]}$ is the updated covariance of kth component. For more details and the derivation of the formulas see [11].

3.4 Gaussian Mixture Reduction

In this work, we chose a pairwise merging of components method that measures the dissimilarity between the post-merge mixture with respect to the pre-merge mixture based on an easily-computed upper bound of the *Kullback-Leibler* (KL) discrimination measure presented in [13].

Given a mixture of two Gaussian components m, p with the parameters θ and w, where $\theta^{[i]} \equiv \{\mu^{[i]}, \Sigma^{[i]}\}$, $i \in \{m, p\}$ and $w^{[m]} + w^{[p]} = 1$, we can obtain the parameters of merging of these two as follow

$$\mu^{[mp]} = w^{[m]}\mu^{[m]} + w^{[p]}\mu^{[p]} \tag{13}$$

$$\Sigma^{[mp]} = w^{[m]}\Sigma^{[m]} + w^{[p]}\Sigma^{[p]} + w^{[m]}w^{[p]}(\mu^{[m]} - \mu^{[p]})(\mu^{[m]} - \mu^{[p]})^T$$

The KL dissimilarity measure B, between two components m and p can be obtained according to (see [13] for details):

$$2B\left((\mu^{[m]}, \Sigma^{[m]}, w^{[m]}), (\mu^{[p]}, \Sigma^{[p]}, w^{[p]})\right) = tr(\Sigma^{[mp]^{-1}}\breve{\Sigma}^{[mp]})$$

$$+(w^{[m]} + w^{[p]})\log\det(\Sigma^{[mp]}) - w^{[m]}\log\det(\Sigma^{[m]}) - w^{[p]}\log\det(\Sigma^{[p]}) \quad (14)$$

where

$$\breve{\Sigma}^{[mp]} = w^{[m]}\Sigma^{[m]} + w^{[p]}\Sigma^{[p]} - (w^{[m]} + w^{[p]})\Sigma^{[mp]} + \frac{w^{[m]}w^{[p]}}{w^{[m]} + w^{[p]}}(\mu^{[m]} - \mu^{[p]})(\mu^{[m]} - \mu^{[p]})^T$$

Algorithm 1. On-line EM-based Clustering

Input: Sample data: $Y_0, Y_1, \ldots$,
Mean and covariance of the initial component $\theta^{[0]}:(\mu^{[0]}, \Sigma^{[0]})$,
Maximum number of hypotheses: $\aleph_{max}$,
Dissimilarity measure threshold: B_{max}
Output: Number of the components: K,
Mean and covariance of the components: $\{\theta^{[1]}, \ldots, \theta^{[K]}\}$,
Mixing weights: $\{w^{[1]}, \ldots, w^{[K]}\}$
1.Update the current components in each hypothesis $\mathcal{H}_1, \ldots, \mathcal{H}_\aleph$:
Find the a posteriori probabilities Γ_{n+1} as Eq.(7)
Update the current components by Eq.(10)-Eq.(12)
Update the log-likelihood ℓ as in Eq.(16)
Update the description length $\mathcal{L}$ as in Eq.(9)
2.Add a new hypothesis: $\mathcal{H}_{\aleph+1}$
Create a new component according to Eq.(15)
Define the log-likelihood of this new hypothesis: $\ell_{\aleph+1} = \ell_1$
Obtain the description length of this new hypothesis: $\mathcal{L}_{\aleph+1}$ in Eq.(9)
3.Check if we can add another hypothesis: $\mathcal{H}_{\aleph+2}$
Find B for every pair of components in $\mathcal{H}_1$ according to Eq.(14)
if $min(B) < B_{max}$ **then**
    Merge two components according to Eq.(13)
    Set the log-likelihood of $\mathcal{H}_1$ as the log-likelihood of new hypothesis: $\ell_{\aleph+2} = \ell_1$
    Obtain the description length for this new hypothesis $\mathcal{L}_{\aleph+2}$ in Eq.(9)
end if
4. Refresh the model
Re-order incrementally the hypotheses according to their description length $\mathcal{L}$
Keep the first $\aleph_{max}$ hypotheses
Acquire the next sample and go to 1

---

4   The Proposed On-Line Unsupervised Learning Algorithm

In this section we describe the proposed on-line unsupervised learning algorithm, which is composed of several models as it will become clear later. Algorithm 1 describes the pseudo-code for one model and its rational is as follows:

- Start with one single observation and build the first hypothesis $\mathcal{H}_1$ described by a single Gaussian distribution with mean $\mu^{[0]}$ at the point itself, and some predefined covariance $\Sigma^{[0]}$. Then, calculate the log-likelihood of this hypothesis $\ell_1 = -\log\sqrt{(2\pi)^d|\Sigma|}$ and find the corresponding description length $\mathcal{L}_1$ according to Eq.(9).

- The second acquired sample, updates the first hypothesis $\mathcal{H}_1$ according to Eq.(10)-Eq.(12), and builds the second hypothesis $\mathcal{H}_2$ which contains two components: the first updated component and a second component with mean $\mu^{[K+1]}$ at the point itself, with some predefined covariance $\Sigma^{[K+1]}$

$$\mu^{[K+1]} = Y_{n+1} \quad , \quad w^{[K+1]} = \frac{1}{n+1} \tag{15}$$

where K is the number of components at the time (being $K = 1$ for the case of the second sample).

- The third point will update the two current hypotheses and build another one by adding a new component to $\mathcal{H}_1$ and so on and so forth. For the sake of computational speed and memory, the number of hypotheses has to be bounded. Thus, after reaching the limit of maximum hypotheses $\aleph_{max}$, we rank the hypotheses in an increasing order according to their description length and keep only the first $\aleph_{max}$ hypotheses and discard the rest.

- As explained above, in each iteration we add a new hypothesis by assuming that the new arriving point is a new component, according to Eq.(15), beside the current Gaussian mixture in $\mathcal{H}_1$. Thus, it is likely that we face the very common problem of over fitting. To avoid that, in each iteration after updating the current components in all hypotheses, we check the possibility of adding another hypothesis by merging two most similar components in $\mathcal{H}_1$ (the hypothesis with minimum description length), according to the dissimilarity measure B_{max}. For example at time n, if there were 5 components in $\mathcal{H}_1$, by receiving a new point Y_{n+1}, first we would update the components in $\mathcal{H}_1$ as we do in all other hypotheses; then if there were two similar components according to a threshold in $\mathcal{H}_1$, we would merge them and add another hypothesis $\mathcal{H}_{\aleph+2}$ (see Algorithm 1) composed by the post-merge mixture. For this new hypothesis the log-likelihood is set to be the same as the log-likelihood of the pre-merge mixture in $\mathcal{H}_1$, since it is assumed that the two components were very similar to each other.

- The dissimilarity measure threshold B_{max} is an important quantity since a very small value would not be helpful in tackling the over-fitting problem and setting a very high threshold can cause under-fitting of the components. To address this problem, we propose to run different set of models in parallel, that is, several processes using Algorithm 1 but with different values of B_{max}. For each time n the model with MDL is selected as output.

Another point that needs to be taken into consideration is the fact that the computation of the log-likelihood has to be done in a recursive on-line format. Thus, after updating θ_n and w_n, the log-likelihood ℓ from Eq.(4) is updated as

$$\ell_{n+1} = \ell_n + \log p(Y_{n+1} | \theta_n, w_n) \tag{16}$$

where $\theta_n = \{\theta_n^{[1]}, ..., \theta_n^{[K]}\}$ and $w_n = \{w_n^{[1]}, ..., w_n^{[K]}\}$.

For some practical reasons, in Eqs. (10), (11), (12), we changed the *learning rate* $\frac{1}{n}$ to a faster decaying envelope, i.e. we added a sufficiently large enough constant to n in order to reduce the problem of instability as proposed in [3].

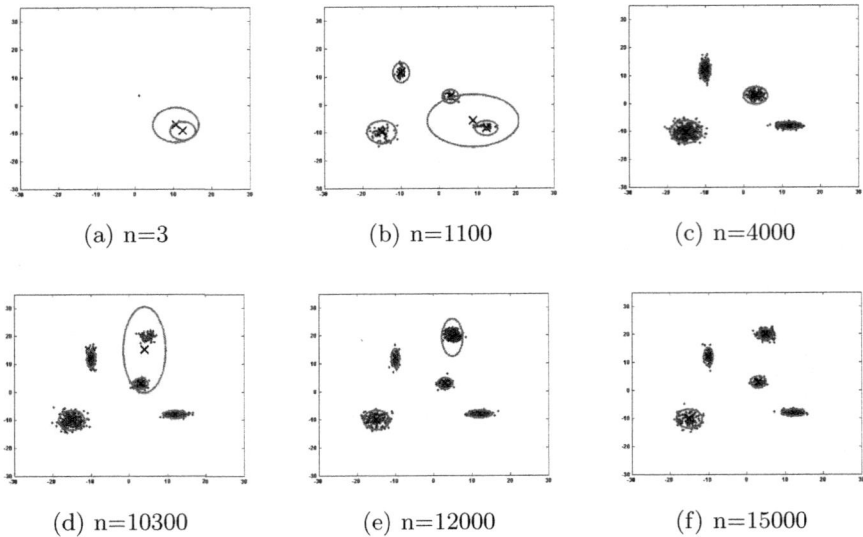

(a) n=3   (b) n=1100   (c) n=4000

(d) n=10300   (e) n=12000   (f) n=15000

Fig. 1. An example of the execution behaviour of the proposed algorithm

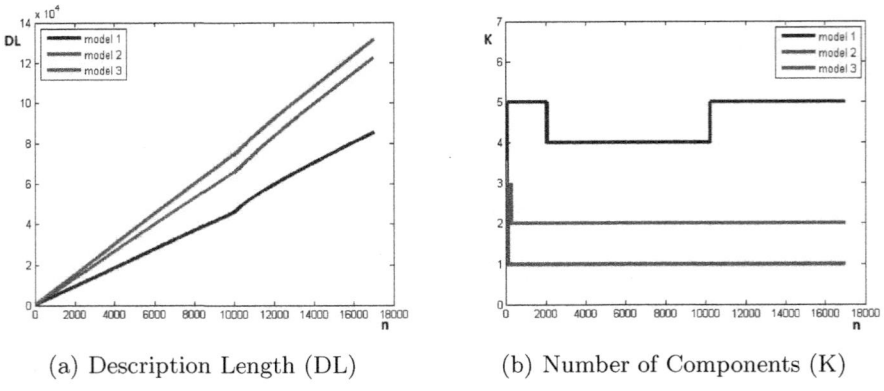

(a) Description Length (DL)   (b) Number of Components (K)

Fig. 2. Time evolution of the description length (DL) and K

5   Simulation Results

This section illustrates the behaviour of the proposed algorithm for two types of experiments: a synthetic Gaussian mixture data set and the Iris data set.

5.1   A Gaussian Mixture Data Set

Fig.1 shows an example of 3 models running in parallel in order to find a mixture of well separated synthetic Gaussian components in real time starting with one

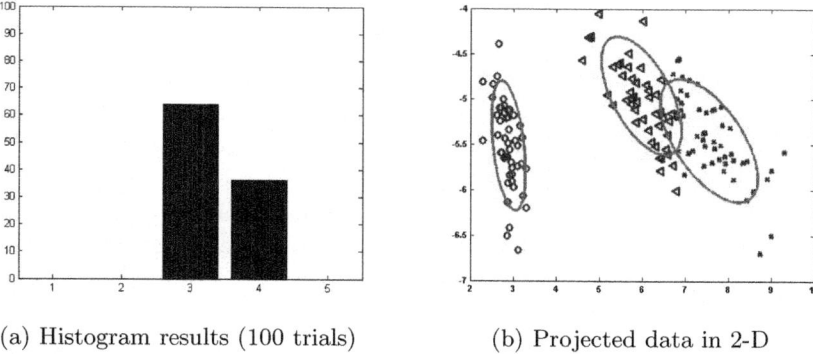

(a) Histogram results (100 trials)   (b) Projected data in 2-D

Fig. 3. Iris Data Set Results

single observation. The maximum number of hypotheses was set to $\mathcal{N}_{max} = 10$ and the merging threshold B_{max} to 0.008, 0.08, and 0.8, respectively. The experiment was split in two phases. For $n < 10000$, the observed data were randomly extracted, according to Eq.(1) with 4 components (K=4) and mixing weights (from left to right) $w = [0.35, 0.25, 0.15, 0.25]$. It can be seen, after some transient situation, that the algorithm merged the two most similar components and were able to correctly determine the 4 components. Then, for $n \geq 10000$, we started to extract data from another component beside those previous ones. The algorithm was able to converge to the solution rapidly. Fig.2 shows the output of 3 different models running in parallel. Intuitively we can say that setting a higher threshold in models 2 and 3, led to early merging in the components and in turn smaller estimation for K. This problem which is known as under-fitting, can cause reduction in the log-likelihood and increase the description length in turn.

5.2   The Iris Data Set

We used the well-known 3-component 4-dimensional "Iris" data set [14]. This data set has only 150 samples, and therefore we had to randomize and repeat them 60 times. We set the maximum number of hypotheses $\mathcal{N}_{max} = 50$ in 10 different models with the merging threshold starting from $B_{max} = 0.002$. Fig.3(a) shows that in 64 out of 100 trials the 3 components were correctly identified. By visual inspection we could observe that the linearly separated component (iris setosa) could almost perfectly be identified. On the other hand, the properly identification of the other two non-linear separable components (iris versicolor and iris virginca) was more challenging since the order in which the data is presented can influence the recursive solution. The typical solution is shown in Fig.3(b) by projecting the 4-dimensional data to the first two principal components.

6   Conclusion

This paper proposed an on-line unsupervised learning of GMMs algorithm in the presence of uncertain dynamic environments. The algorithm relies on a multi-hypothesis adaptive scheme that continuously updates the number of components and estimates the model parameters as the measurements (sample data) are being acquired. The hypothesis models are ranked according to the MDL. In general, we could conclude that the algorithm has a good performance specially when the components are well separated. However, it is worth to mention that a critical issue is the initial selection of the covariance when a new component is created. This has to be done carefully because choosing a very small covariance can be experimentally problematic since in the process of calculating the a posteriori probability in Eq.(7), the result in Eq.(3) could be zero due to finite precision. On the other hand, choosing an extremely large covariance can lead to the "under-fitting" problem. This is something that deserves further investigation.

References

1. Bishop, C.M.: Pattern recognition and machine learning. Springer (2006)
2. Greggio, N., Bernardino, A., Victor, J.S.: A practical method for self adapting gaussian expectation maximization. In: ICINCO, pp. 36–44 (2010)
3. Zivkovic, Z., van der Heijden, F.: Recursive unsupervised learning of finite mixture models (2004)
4. Figueiredo, M.A.T., Jain, A.K.: Unsupervised learning of finite mixture models 24, 381–396 (2000)
5. Declercq, A., Piater, J.H.: Online learning of gaussian mixture models − a two level approach, pp. 605–611 (2008)
6. Dempster, A.P., Laird, N.M., Rubin, D.B.: Maximum likelihood from incomplete data via the EM algorithm 39, 1–38 (1977)
7. McLachlan, G., Krishnan, T.: The EM Algorithm and Extensions. John Wiley & Sons, New York (1997)
8. Lanterman, A.D.: Schwarz wallace and rissanen: Intertwining themes in theories of model selection (2000)
9. Grünwald, P.D.: The minimum description length principle. The MIT Press (2007)
10. Raudys, S.J., Jain, A.K.: Small sample size effects in statistical pattern recognition: Recommendations for practitioners. IEEE Transactions on Pattern Analysis and Machine Intelligence 13(3), 252–264 (1991), cited By (since 1996)462
11. Li, D., Xu, L., Goodman, E.: On-line EM variants for multivariate normal mixture model in background learning and moving foreground detection (2012)
12. Titterington, D.M.: Recursive parameter estimation using incomplete data 46, 257–267 (1984)
13. Runnalls, A.R.: A kullback-leibler approach to gaussian mixture reduction 43, 989–999 (2007)
14. Iris data set, http://archive.ics.uci.edu/ml/datasets/Iris

Prediction of Solar Radiation
Using Artificial Neural Networks

João Faceira[1], Paulo Afonso[2], and Paulo Salgado[1]

[1] Departamento de Engenharias - ECT,
Universidade de Trás-os-Montes e Alto Douro, Portugal
[2] ESTGA/IT, Universidade de Aveiro, Portugal
`joao_faceira14@hotmail.com, pafnaa@ua.pt, psal@utad.pt`

Abstract. Solar radiation data is needed by engineers, architects and scientists in the framework of studies on photovoltaic or thermal solar systems. A stochastic model for simulating global solar radiation is useful in reliable power systems calculations. The main objective of this paper is to present an algorithm to predict hourly solar radiation in the short/medium term, combining information about cloud coverage level and historical solar radiation registers, which increased the performance and the accuracy of the forecasting model. The use of Artificial Neural Networks (ANN) model is an efficient method to forecast solar radiation during cloudy days by one day ahead. The results of three statistical indicators - Mean Bias Error (MBE), Root Mean Square Error (RMSE), and t-statistic (TS) - performed with estimated and observed data, validate the good performance accuracy of the proposed three indicators.

Keywords: Forecasting model, Neural networks, Solar radiation.

1 Introduction

Hourly solar radiation data forecasting has significant consequences in most solar applications such as energy system sizing and meteorological estimation. Accurate forecasting of solar radiation improves the efficiency of the outputs of many applications.

The global solar radiation shows not only regular yearly and daily variations but also a random behavior. The yearly and daily variations can be described in a deterministic way while the random behavior has a high correlation with the state of the atmosphere. Solar Radiation Outside of Atmosphere (SROA) can accurately be determined [1] and the atmosphere will induce the randomness [2]. The transmissivity of solar radiation in the atmosphere depends on various factors, like humidity, air pressure and cloud type, but the factor with major impact is the cloud coverage. By assuming a deterministic relation between cloud coverage and hourly global solar radiation, the need for measurement of the latter disappears.

The issue of transforming the measured solar radiation data into radiation data useful for energy applications is very important, and often dealt with various time scales (monthly average values, daily or hourly data) in the scientific literature. There are

© Springer International Publishing Switzerland 2015

A.P. Moreira et al. (eds.), *CONTROLO'2014 - Proc. of the 11th Port. Conf. on Autom. Control*,
Lecture Notes in Electrical Engineering 321, DOI: 10.1007/978-3-319-10380-8_38

two approaches that allow the quantification of solar radiation: the "physical modeling" based on physical processes occurring in the atmosphere and influencing solar radiation and the "statistic solar climatology" mainly based on time series analysis [3]. Several authors have also developed empirical models for solar radiation prediction [4, 5].

Classically, the solar radiation data can be regarded as a random time series produced by a stochastic process, and its prediction depends on accurate mathematical modeling of the underlying stochastic process. By using an accurate model, the prediction is mathematically defined by the conditional expectation of the data given from the present and the past data samples (historical behavior).

On the other hand, the computation of such conditional expectation requires the knowledge of the distribution of the samples including higher order statistics. Since the available or recorded data is finite, such distributions can be estimated or fit into pre-set stochastic models such as Auto-Regressive (AR) [6], Auto-Regressive Moving-Average (ARMA) or Markov [7]. Another approach is the use of an Artificial Neural Networks (ANN) model [8,9,10,21] that is trained to predict results from available historical data capable of extracting nonlinear modeling relationships. This model can be used for forecasting or modeling, given that the structure of the ANN is adequate (in number, type and links of elements - Neurons) and the training process is sufficiently efficient to extract the relevant relationships from relevant data set. By combining the solar radiation model with an ANN model of cloud coverage the forecasting method could be even more suitable.

The level of cloudiness or transmissivity is expressed in Oktas [11], a unit of measurement used to describe the amount of cloud cover at any given location such as a weather station. Sky conditions are estimated in terms of eighths of the sky covered in cloud, ranging from 0 Oktas (completely clear sky) through 8 Oktas (completely overcast). By using cloud observations as input for the simulations, the local solar radiation can be estimated with accuracy. Also, with a good cloudiness forecasting values, from the medium term to the long run horizon, the prediction of solar radiation can be forecasted. This information is currently provided by official sources of weather forecasting, meteorological satellites and ground weather stations [12]. Here, the proposed model will adjust the cloud coverage information taking into account the geographic specificities of the solar farm as well as the recorded historical weather data.

In this paper, three ANN models are proposed to predict solar radiation for a period of one day ahead based on the knowledge of cloudiness forecast provided for the day, the calculated extraterrestrial radiation (ER) and the measured past values of global solar irradiation. With this solution, we hope to increase the quality of management of solar panels systems and thus contribute for a best electric energy efficiency production. This can be obtained by including a stochastic model for the short-term variations: three ANN structures are proposed, whose errors and limitations are estimated and discussed.

2   Solar Data

The term of extra-atmospheric solar irradiation, employed for denoting the solar irradiation that reaches the extra atmospheric zones of the earth, can be accurately computed [13]. Its magnitude depends only on the Solar Constant, geographical latitude and the solar time and when this radiation penetrates in the atmosphere, it suffers a reflection and attenuation due to clouds, dusty particles, etc., therefore, only a fraction of the total irradiation reaches the earth surface. However, the magnitude of this global solar irradiation is also affected by climatological conditions, which are unpredictable, and its value is not now deterministic computed. The atmospheric transmittance or clearness index, C_k, for hourly events is a measure of this fraction and is defined by Beer-Lambert law as:

$$C_k = \frac{I_k}{I_{0,k}} \tag{1}$$

where:

I_k is the global solar irradiation (GSI) on the earth surface,

$I_{0,k}$ is the extraterrestrial radiation (ER) on the geographic local, in the kth hour.

C_k is the atmospheric transparency that induces randomness to the solar irradiation measured on earth, and its probability distribution behaves in a quasi-universal manner [14]. For atmospheric science applications, $I_{0,k}$ can be given as

$$I_{0,k} = e^{-\tau/\mu} \tag{2}$$

where

τ is the optical depth, $\mu = |\cos(\theta)|$ and θ is the angle of propagation of the ray obtained from the normal of the surface.

Given a coefficient of extinction within an atmospheric column, β, the optical depth between sunrise and sunset times is defined by:

$$\tau = \int_{\theta_1}^{\theta_2} \beta(\theta) \cdot d\theta \tag{3}$$

where θ is the hour angle and θ_1 and θ_2 correspond respectively to the sunrise and sunset hour angles.

In this paper, the value of clearness is computed between the ratio of sensor measurement value of the global solar irradiation and the calculated theoretical value of the extraterrestrial radiation [15]. Figure 1a) shows the normalized values of C_k for a sub-period of data used in this work, while figure 1b) exhibits the sky conditions rounded in eighths levels of sky covered clouds (SCC), ranging from 0 to 8 Oktas. In both figures, during the night period, the values are assumed to be zero.

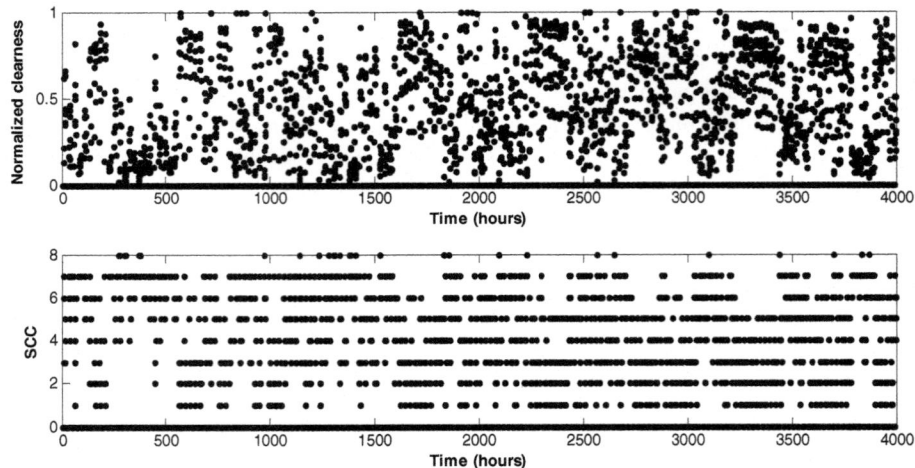

Fig. 1. First 4000 hours of the training data set: a) Normalized clearness index, C_k; b)   Level of the sky covered clouds (SCC) in Oktas

The progression of C_k values could be described via a sampling procedure of a random variable upon a given distribution probability function, if such hourly events were to be independent.

Our study proposes to analyze the radiation time series (Wh.m^{-2}) measured at the meteorological station of Ponte da Barca, Portugal (41.80300°N, -8.42001°W). Data representing the global solar radiation were measured on a daily basis from January 2010 to December 2011.

3   Time Series Prediction Using Neural Networks

Time series prediction or forecasting takes an existing series of data $x_{k-n}, \cdots, x_{k-2}, x_{k-1}$ and forecasts the $x_k, \cdots, x_{k+m}$ data values. The goal is to observe or model the existing data series to enable future unknown data values to be forecasted accurately. Thus a prediction x_k can be expressed as a function of the recent history of the time series, $x_k = f(x_{k-1}, \cdots, x_{k-n})$. As the time series prediction is based on their past values, it is necessary to obtain a data record. When obtaining a data set, the objective is to maximize the information data and an adequate number of records for prediction purposes. Hence, future values of a time series x_k can be predicted as a function of past values. The problem of time series prediction becomes a problem of system identification; the unknown system to be identified is the function f which inputs are the past values of the time series.

Artificial neural network is an interconnected group of artificial neurons that uses a mathematical or computational model for information processing based on a connectionist approach to computation. ANNs represent a class of distinct mathematical models originally motivated by the information processing in biological neural networks, many of them applicable to forecasting tasks and modeling nonlinear functions f. The MultiLayer Perceptron (MLP) [8], [9], [10] [11] represents a well-researched non-recurrent ANN paradigm, which offers great flexibility in forecasting through flexibility in the number of input and output variables. Thus MLPs offer large degrees of freedom towards the forecasting model design. A fixed number n of past values is fed to the input layer of the MLP and the output is required to predict a future value of the time series. The problem of time series prediction now becomes a problem of system identification; the unknown system to be identified is the function f with the past input values of the time series. This method is often called the sliding window technique as the N-tuple input slides over the full training set. Figure 2 represents the structure of the forecasting model, where the inputs are the extraterrestrial radiation $I_{0,k}$, level expected of cloudiness SCC and the solar global radiation past values. The $I_{0,k}$ values are previously estimate by the SROA, known the geographic localization and the date and time of the forecasting task.

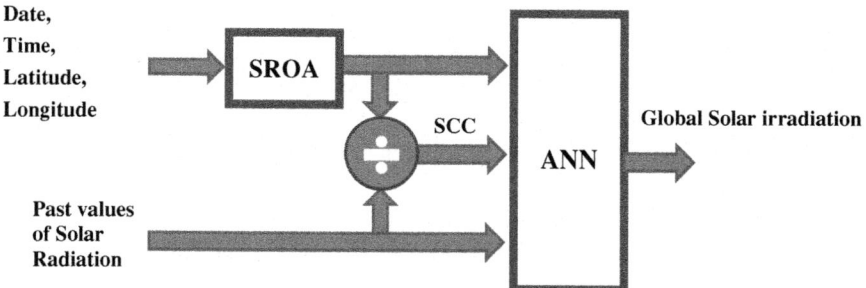

Fig. 2. Structure of the forecasting ANN Model

The ANN implementation is made in several stages:

1) ANN structure: we used a MultiLayer Perceptron (MLP) using feedforward back-propagation. This structure is the most used in the literature for the estimation of solar radiation with the best results [16,17].

Figure 3 represents the three models that were used for forecasting purposes. The input numbers of the ANN are two for models (I) and (II) and five for the last model. In the last model a feedback structure was used. These configurations are tested in order to give the best results with a small error. Through the compromise between the number of hidden layers and the number of neurons in each layer, it is possible to obtain a fast and robust network with the best results. A number of 20 neurons were selected in the hidden layers of the three models.

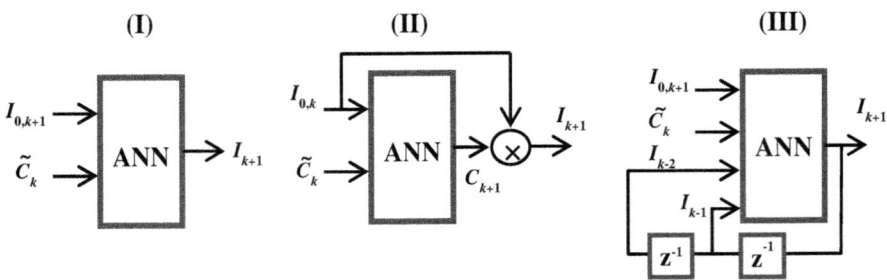

Fig. 3. Three MPL's for forecasting GSI values: (I) $\tilde{I}_{k+1} = f\left(I_{0,k+1}, \tilde{C}_k\right)$; (II) $\tilde{I}_{k+1} = f\left(I_{0,k+1}, \tilde{C}_k\right) \times I_{0,k+1}$; (III) $\tilde{I}_{k+1} = f\left(I_k, I_{k-1}, I_{0,k+1}, \tilde{C}_k\right)$, where $\tilde{C}$ is SCC (Oktas)

2) ANN transfer function: the combination of a sigmoid and a linear transfer functions allows a good approximation of several types of functions, as well as the estimation of solar data [16,17];

3) learning algorithm: Lavenberg–Marquardt back propagation algorithm (LM algorithm) was used for training the ANN model. LM algorithm is a variation of Newton's method. This algorithm provides a nice compromise between the speed of Newton's method and guaranteed convergence of steepest descent algorithm [16,17];

4) training/test set: among all the available data, we must use a percentage for the training and the rest will serve for the testing. Here 75% of random choice data is used in the training phase and the remaining data in the test phase.

The accuracy of the different models is compared using four statistical indicators: the root mean square error (RMSE), the mean bias error (MBE), the mean absolute error (MAE) and the mean absolute percentage error (MAPE) which are defined as:

$$RMSE = \sqrt{\frac{1}{N}\sum_{k=1}^{N}(d_k - y_k)^2} \tag{4}$$

$$MBE = \frac{1}{N}\sum_{k=1}^{N}(d_k - y_k) \tag{5}$$

$$MAE = \frac{1}{N}\sum_{k=1}^{N}|d_k - y_k| \tag{6}$$

$$MAPE = \frac{100}{N}\sum_{k=1}^{N}\left|\frac{d_k - y_k}{d_k}\right| \quad (\%) \tag{7}$$

where d_k and y_k are the ith measured and computed values of radiation quantities, respectively, while N is the number of measurements taken into account.

The prediction is found to be accurate with quality measured data [18,19]. It is evaluated with mean absolute percentage error (MAPE), where MAPE<10% means high prediction accuracy, 10%<MAPE<20% means good prediction, 20%<MAPE<50% means reasonable prediction and MAPE>50% means inaccurate forecasting. The test of MBE provides information on the long term performance of the models. A positive MBE gives the average amount of overestimation in the predicted values and a negative MBE value gives the average amount of underestimation in the predicted values. One drawback of this test is that overestimation in one observation will cancel the underestimation in another observation. The test on RMSE provides information on the short term performance of the models and it is always a positive value. Apart from these statistical tests, the t-statistic test proposed by Stone [20] was also employed to assess the proposed models. This method has an additional advantage of telling whether the models estimates are statistically significant or not at a particular confidence level. The t-statistic (TS) is as follows:

$$TS = \sqrt{(N-1)\frac{MBE^2}{RMSE^2 - MBE^2}} \qquad (8)$$

A two tailed t-test was carried out at the 95% confidence level. TS should lie within the interval defined by $-T_c$ and $+T_c$ where T_c, the critical TS value is obtained from the Student's distribution at desired confidence level with $(N-1)$ degrees of freedom (Daniel and Terrell 1992). The smaller the value of TS, the better the performance of the model.

4   Results

This section presents the main steps undertaken for the development of the algorithm in Matlab® for the prediction of solar radiation over two years (2010 and 2011). The data was collected in the Meteorological Station located at Ponto da Barca [9] and Matlab neural network toolbox was used. After data collection and analysis, the Backpropagation method is used to train three ANN weights in order to predict the cloud coverage level, or directly the GSI values. In this process, half of the available data were involved, while the remaining data was employed to promote the model prediction validation. Jointly with the SROA model, the forecasting model is tested for computing ER values, $I_{0,k}$.

Three ANN's with their own structures were used to predict the same series of data. Firstly, the search for the ideal network structure is a complex and crucial task. In order to determine the best network configuration, all the parameters available in this network architecture were studied. As a result of this iterative process, the selected network has three neuron layers: input, hidden and output layer. There was no significant difference in the use of 1, 2 and 3 hidden layer architectures. One hidden layer was used in order to minimize the complexity of the proposed ANN model. The network has the following characteristics: in the input layer the number of neurons are equal to the number of input variables, twenty neurons on the hidden layer and one neuron on the output layer. Concerning the transfer functions and the training algorithm, the best results were obtained with the Sigmoidal (hidden layer) and linear (output layer) function and the Levenberg–Marquardt second-order algorithm.

In Figure 4, the measured global solar radiation and prediction values from SROA model are plotted as a function of time. It can be observed that the actual radiation is very close to the theoretical radiation for some days, but in another set of days the curves are no longer close. This results mainly from the influence of the clouds that can absorb solar radiation. To solve this problem, ANN's were implemented to predict the cloudiness, C_k, or GSI values, I_k. In the 1st and 3rd proposed model, an ANN structure is used to predict future I_k values while in the second model the ANN structure is used to estimate next clearness C_k values. As inputs, we have the ER values, $I_{0,k}$, and the measured SCC data (*i.e.* $\tilde{C}$, in Oktas), for the time horizon forecasting. The 3rd model is a second order dynamic ANN, where the past output values of the ANN are also used as inputs. Here the $\tilde{C}$ values are quantified in 8 discrete levels and normally this information can be found in the National Weather forecasting institute for a period of various days.

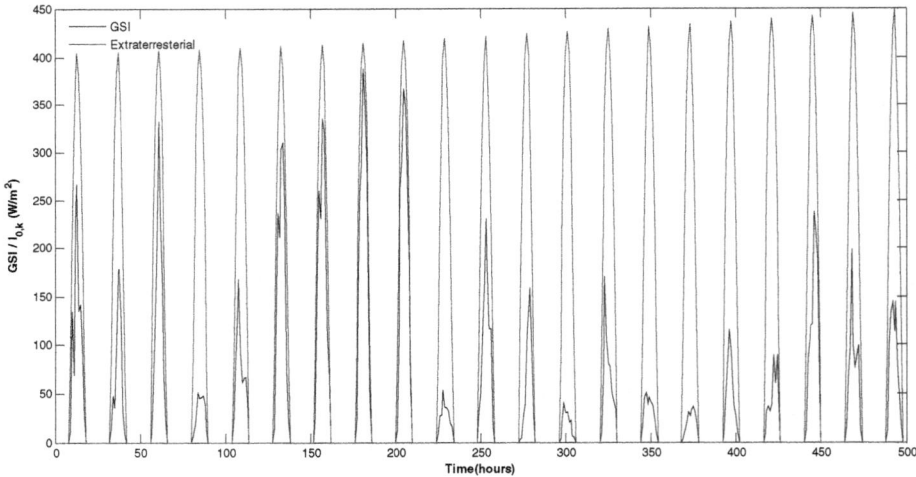

Fig. 4. Global Solar Radiation (black line) and extraterrestrial radiation (blue line)

The performance of the proposed three models was evaluated using the *t*-statistic (TS), Root Mean Square Error (RMSE) and Mean Bias Error (MBE). These indicators are mainly employed for adjustment of solar radiation data (see table 1). The forecasting results of 3rd model in displayed in the figure 5, for the firsts 500 hours.

Table 1. Forecast model *versus* performance statistical indicator

Model	RMSE	MAE	MBE	MAPE	TS
(I)	12.10	6.459	0.0419	11,2%	0.772
(II)	12.11	6.433	-0.0978	18,9 %	1.066
(III)	11.68	6.172	-0.0614	5,1%	0.695

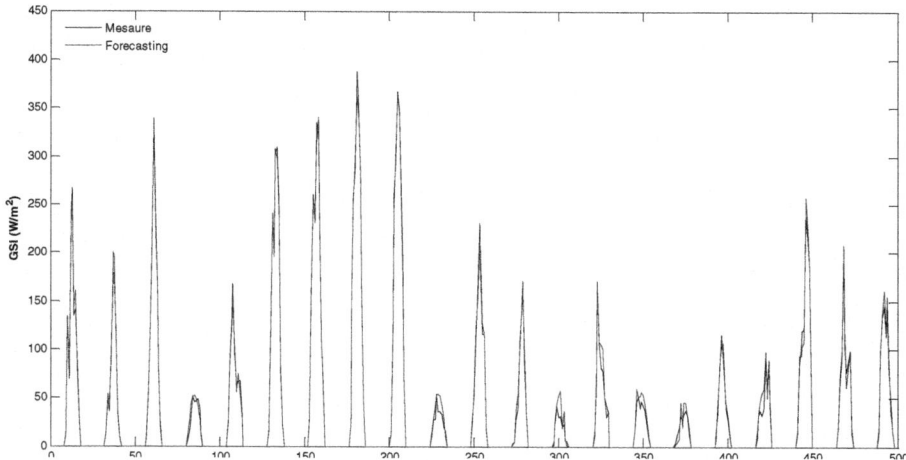

Fig. 5. Global Solar Radiation: Measure values (black line); forecasting values (blue line)

Based only on MAPE criterion, models (I) and (II) are considered as good prediction (10% < MAPE < 20%), while model (III) achieves high prediction accuracy (MAPE <10%). The test of MBE provides information on the long-term performance of the proposed model. With the exception of model (I), that has a slight overestimation, the other models have a slight underestimation. The test of RMSE provides information on the short-term performance of the proposed model. The lower RMSE and TS values of the third model indicate that this proposed model has the best short-term performance. The TS measure is used to achieve a more reliable statistical result, as it is possible to have experimental results with simultaneously large RMSE values and very small MBE. Small values of TS mean better model performance, therefore the low t-value of the proposed models (0.695–1.066) demonstrate the excellent performance of all models.

With these results, it can be concluded that forecasting values of solar radiation are very close to the measured ones. Their extrapolation can then be used to predict the energy production of photovoltaic panels in a local farm.

5   Conclusion

In this work, a new model for daily solar radiation forecasting is comprehensively evaluated through the combination of the SROA model with the ANN prediction model. This model predicts the solar radiation for 24 hours. The values obtained with the use of neural networks (tuned by historical data values) show that the developed tools allow the estimation of values of radiation with great precision. With this information, the power output of a photovoltaic production center can be forecasted, allowing the implementation of more efficient strategies for energy management.

References

1. Stull, R.B.: Meteorology Today For Scientists and Engineers: A Technical Companion Book. West Publishing Company, Minneapolis/St. Paul (1995)
2. Graham, V., Hollands, K.: A method to generate synthetic hourly solar radiation globally. Int. J. Solar Energy 44, 333–341 (1990)
3. Badescu, V.: Modelling Solar radiation at the earth surface. Springer (2008) ISBN 978-3-540-77454-9
4. Khatib, T., Mohamed, A., Sopian, K.: A review of solar energy modeling techniques. Renewable and Sustainable Energy Reviews 16, 2864–2869 (2012)
5. Batlles, F.J., Rubio, M.A., Tovar, J., Olmo, F.J., Alados-Arboledas, L.: Empirical modeling of hourly direct irradiance by means of hourly global irradiance. Energy 25, 675–688 (2000)
6. Aguiar, R., Collares-Pereira, M.: A time dependent autoregressive Gaussian model for generating synthetic hourly radiation. Solar Energy 49, 167–174 (1992)
7. Maafi, A.A.A.: A two state Markovian model of global irradiation suitable for photovoltaic conversion. Solar and Wind Technology 6, 247–252 (1989)
8. Mellit, A., Benghanemb, M., Hadj Arabc, A., Guessoumd, A.: A simplified model for generating sequences of global solar radiation data for isolated sites: using artificial neural network and a library of Markov transition matrices approach. Solar Energy 79, 469–482 (2005)
9. Cao, J.C., Cao, S.H.: Study of forecasting solar irradiance using neural networks with preprocessing sample data by wavelet analysis. Energy 31, 3435–3445 (2006)
10. Capizzi, G., Napoli, C., Bonanno, F.: Innovative Second-Generation Wavelets Construction With Recurrent Neural Networks for Solar Radiation Forecasting. IEEE Trans. on Neural Networks and Learning Syst. 23(11), 1805–1815 (2012)
11. Jones, P.: Cloud-cover distribution and correlations. J. Appl. Meteor. 31, 732–741 (1992)
12. Hourly solar radiation, http://snirh.pt/index.php?idMain=2&idItem=1
13. Paulescu, M., et al.: Weather Modeling and Forecasting of PV Systems Operation. Green Energy and Technology, pp. 17–42 (2013), doi:10.1007/978-1-4471-4649-0_2
14. Aguiar, R., Collares-Pereira, M.: TAG: A time-dependent, autoregressive, Gaussian model for generating synthetic hourly radiation. Solar Energy 49(3), 167–174 (1992)
15. Kandilli, C., Ulgen, K.: Solar Illumination and Estimating Daylight Availability of Global Solar Irradiance. Energy Sources
16. Ahmad, M.J., Tiwari, G.N.: Solar radiation models – review. International Journal of Energy and Environment 1(3), 513–532 (2010)
17. Bakirci, K.: Models of solar radiation with hours of bright sunshine: A review. Renewable and Sustainable Energy Reviews 13, 2580–2588 (2009)
18. Myers, D.R.: Solar radiation modeling and measurements for renewable energy applications: data and model quality. Energy 30, 1517–1531 (2005)
19. Muneer, T., Younes, S., Munawwar, S.: Discourses on solar radiation modeling. Renewable and Sustainable Energy Reviews 11, 551–602 (2007)
20. Stone, R.J.: Improved statistical procedure for the evaluation of solar radiation estimation models. Solar Energy 51(4), 289–291 (1993)
21. Ferreira, P.M., Gomes, J.M., Martins, I.A.C., Ruano, A.E.: A Neural Network based Intelligent Predictive Sensor for Cloudiness, Solar Radiation, and Air Temperature. Sensors 12, 15750–15777 (2012)

Body Fall Detection with Kalman Filter and SVM

Paulo Salgado[1] and Paulo Afonso[2]

[1] Departamento de Engenharias – ECT,
Universidade de Trás-os-Montes e Alto Douro, Portugal
[2] ESTGA/IT, Universidade de Aveiro, Portugal
psal@utad.pt, pafnaa@ua.pt

Abstract. In this paper, an approach for human body fall detection is presented that can be supported with a modern smartphone equipped with accelerator sensors. Falling is one of the most significant causes of injury, mainly for elderly citizens, and is one of the reasons why many individuals are forced to leave the comfort and privacy of their homes and live in an assisted-care environment. The acceleration measured by the embedded tri-accelerometer sensor, was utilized to collect the information about the body motion and was used to develop a robust algorithm to accurately detect a fall. This data is incorporated by a real-time Pose Body Model (PBM) which is identified by an Extended Kalman Filter (EKF) algorithm. Moreover, a Support Vector Machine (SVM) performs a binary classification of the observed data, allowing the detection of fall incidents. This fall detection system is tailored for mobile phones and has an important application in the field of safety and security, but can also be used in motion analysis of body moving and live style monitoring. Experimental results showed that this methodology can detect most types of single human falls quite accurately.

Keywords: Body fall detection, Kalman filter, Support Vector Machine.

1 Introduction

Human falls are one of the major health problems for elderly people. Falls are dangerous and often cause serious injuries that may even lead to death. In fact, fall related injuries have been among the five most common causes of death amongst the elderly population. Statistics show that every year, 30% of the senior citizen population suffers harmful accidents from falling [1], which are the leading cause for hospital admission in emergency and long-stay hospitalizations. It is the sixth leading cause of death for people over 65, the second leading cause for people between 65 and 75, and the first for people over 75 [2, 3]. Early detection of a fall is therefore an important issue.

An automatic fall detection system can help to address this problem by reducing the time between the fall and the arrival of required assistance. This type of system can generally be classified under one or a combination of the following three broad categories [4,5,6]:

© Springer International Publishing Switzerland 2015

A.P. Moreira et al. (eds.), *CONTROLO'2014 - Proc. of the 11th Port. Conf. on Autom. Control*,
Lecture Notes in Electrical Engineering 321, DOI: 10.1007/978-3-319-10380-8_39

1. Video monitoring-based fall detectors, based on a set of cameras placed in every room where the occurrence of falls needs to be detected [7]. A set of processing algorithms is running in a computer designed to identify unusual inactivity and recognize a fall episode by analyzing on-time the captured images. This recognition process is passive in the sense that it generally does not require the user to wear any device.
2. Floor vibration-based fall detectors, based on the assumption that it is possible to detect human falls by observing the vibration patterns in the floor through the analysis of the vibration signature of sensor signals [8]. The vibration signature of the floor generated by a human fall is significantly different from that generated by falling objects and is significantly different from that generated by normal daily activities such as walking, tapping, *etc.*
3. Automatic wearable fall detectors, where the falls are generally characterized by an impact on the floor followed by the near horizontal orientation of the faller. Wearable sensor based fall detection means embedding some micro sensors into the clothes, girdle, *etc.*, to monitor the movement parameters of human body in real-time, and determine whether a fall has occurred based on the analysis of these parameters.

Accelerometers are employed in the design to detect an impact whereas tilt sensors and/or gyroscopes are used to determine the orientation of the faller after the impact [9,10,11,12]. The impact of a human body fall on the floor generates sound and can be listened by microphones for analysis and recognition tasks. The sound of an impact is very distinctive and can be easily distinguished from others dropped items. However, its analysis is very difficulty due to its dependency on the floor kind, the ground impact angle and the worn clothing.

These technological devices are now embedded in the majority of the smartphone devices through Microelectromechanical systems (MEMS). With real time acquired acceleration, vibration and tilt signals, a fall detection system will be set up. Typically, the recognition methodology is based on a few thresholds decision making of these signals [13,14,15]. But there are some drawbacks concerning this kind of algorithms, including lack of adaptability, accurate classification deficiency, *etc.* One possible solution to overcome these problems may involve the sensor fusion and an estimation process, through a robust and accurate kinematic model of the pose movement of the body [16].

In this paper, a tri-axial accelerometer is used to capture the movement signals of the human body. With the combined information from these three-accelerometer, a Pose Body Model (PBM) is built. This is based on a 4th order model with the following state variables: angular position, angular rate, angular acceleration and radius curvature. A PBM is proposed together with an identification method based on the EKF algorithm, from which it is possible to estimate the body position, stated as the vertical angle pose value. From this observation model, a SVM algorithm is able to scrutinize a body fall episode. The results show that this method can detect the falls effectively.

2   Pose Body Model

Traditionally, tracking and identifying human dynamics involve exact modeling of limbs as well as the torso, to infer the dynamic events [7,17]. Our approach uses a simpler and robust representation, which focus just on the torso (see figure 1); the smartphone with the accelerator sensors is attached to a human body segment, which rotates with respect to a global coordinate system, called the navigation frame N.

From the streaming data continually provided by the fall sensor embedded in the Smartphone, the challenge was how to use it with the higher effectiveness in order to detect real body fall episodes. In a regular standing position, the only value noticeable is the effect of the gravity vector g on the z-axis. In theory, if the user is wearing the device correctly, the x-axis and y-axis will be zero. Once this initial position has been provided by the calibration procedure, it is then possible to use the dot product of the z vector and the g vector to determine how the user is oriented in comparison to the upright initial position. Furthermore, as the orientation of Smartphone is connected to the body, the frontal and lateral components of accelerations can be aggregated into a single acceleration variable, and therefore, the problem is reduced to a 2D kinematic space, as illustrated in figure 1.

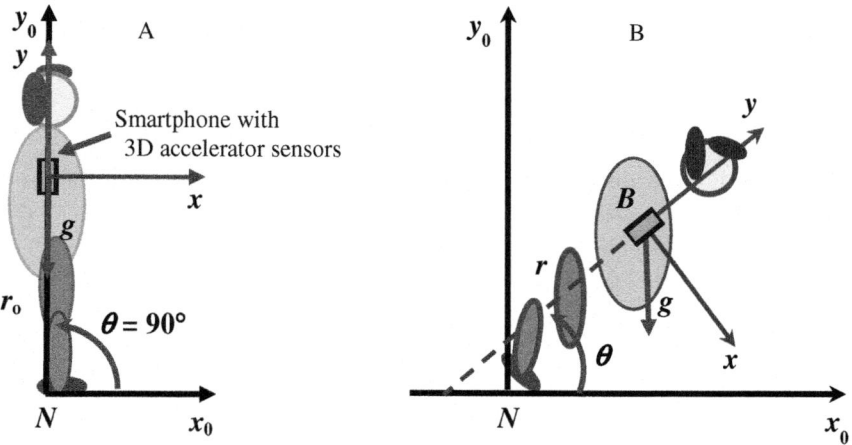

Fig. 1. Body (A) and navigation frames (B)

The system model is simply an integrator described by a set of equations representing a 4th order state model. Vector $\bar{x}=[\theta \ w \ \alpha \ r]^{T}$ represents the state variables, whose components are the angular position, θ, the angular velocity, $w = \dot{\theta}$, the angular acceleration, $\alpha = \ddot{\theta}$ and the radius of the curvature, r. The system outputs z represent the information available on the system, namely the measurement of the bi-axis-acceleration a_x and a_y. These are the measures of the body trunk coming from the accelerometers.

For each frame, corresponding to a time instant, θ is measuring the angle of the trunk with the horizontal plan (floor). In the erect mode, θ is about $\pi/2$ rad and r has a value equal to the height from the floor to the smartphone, r_0. The discrete kinematic equations of frame B are then:

$$\theta_{k+1} = \theta_k + w_k T + \frac{1}{2}\alpha_k T^2 + n_{\theta,k}$$
$$w_{k+1} = w_k + \alpha_k T + n_{w,k} \tag{1}$$
$$\alpha_{k+1} = \alpha_k + n_{\alpha,k}$$
$$r_{k+1} = r_k \cdot \varepsilon + (1-\varepsilon) r_0 + n_{r,k}$$

The sample time T is $100ms$, $\varepsilon = 0.9$ is a forgetting parameter of radius r. The measure model is given by:

$$a_{x,k} = \alpha_k r_k + g \cos(\theta_k) + \eta_{a_x,k}$$
$$a_{y,k} = w_k^2 r_k - g \sin(\theta_k) + \eta_{a_y,k} \tag{2}$$

The model error n and the observed model noise η are assumed to be white Gaussian, with zero mean. The covariance of the noise angle, $n_{\theta,k}$, and the radius variables $n_{r,k}$ are q_θ and q_r, respectively. The nature of the measurement errors η reflects the error of the observed model and the sensor signal noise; it is also assumed to be Gaussian noise and uncorrelated, with zero mean and covariance matrix R. Values of R_k are given by $R_k = \gamma^2 \cdot I$, where γ is the *rms* of the measured acceleration signals.

The PBM model estimates the angular variable, the angular velocity and the angular acceleration and should be of 3rd order at least. The part of the model associated with the radius of curvature is assumed to be of 1st order, for a radius r that converges to r_0. Equations (1) and (2) can be rewritten in the matrix form of the state space, as:

$$x_k = A_{k-1} x_{k-1} + B_{k-1} r_0 + n_{k-1}$$
$$z_k = g_k(x_k) + e_k \tag{3}$$

where $B_k = \begin{bmatrix} 0 & 1-\varepsilon \end{bmatrix}^T$ and $A_k = \begin{bmatrix} A_{rot,k} & 0;0 & 0 & 0 & \varepsilon \end{bmatrix}$ with the sub-matrix of angular variables:

$$A_k^{rot} = \begin{bmatrix} 1 & T & T^2/2 \\ 0 & 1 & T \\ 0 & 0 & 1 \end{bmatrix} \tag{4}$$

The observed model estimates the accelerations of the body trunk in the referential xOy, taking into account the centripetal and tangential acceleration, as well as the projection of the gravity acceleration, by the vector function:

$$g_k(x_k) = \begin{bmatrix} \alpha_k r_k + g\cos(\theta_k) \\ w_k^2 r_k - g\sin(\theta_k) \end{bmatrix} \tag{5}$$

The state model (1) covariance of the model error is:

$$Q_k = \begin{bmatrix} Q_k^{rot} & 0 \\ 0 & Q_k^{transl} \end{bmatrix} \tag{6}$$

with $Q_k^{transl} = q_r$ and $Q_k^{rot} = q_\theta E\left[W^T W\right]$, for $W = \left[T^3/3!\ T^2/2!\ T\right]$.

As the state equations are linear and the observation equations are nonlinear, the PBM model is also nonlinear.

The Kalman Filter (KF) is a recursive data-processing algorithm that combines the available measurements with prior knowledge of the system, to produce a state estimate x_k in such a manner that the mean square is statistically minimized. Thus, KF performs the above tasks for linear systems and linear measurements in which the driving and measurement noises are assumed to be mutually uncorrelated, with zero-mean and Gaussian distribution. In nature, however, most physical problems or processes are nonlinear. Consequently, the nonlinear systems must be linearized before the linear filter theory can be applied.

Using an Extended Kalman Filter (EKF), a sampling period T is chosen and maintained constant during the study period. The actual state estimation and prediction equations are given by:

$$\begin{aligned} x_{k|k-1} &= A_{k-1} x_{k-1|k-1} \\ x_{k|k} &= x_{k|k-1} + K_k e_k \\ e_k &= z_k - g_k\left(x_{k|k-1}\right) \\ S_k &= H_k P_{k|k-1} H_k^T \end{aligned} \tag{7}$$

where $H_k = dg_k/dx$; e_k is the measurement residue, a Gaussian random variable with zero mean and covariance S_k, and R_k is the covariance of the measurement noise w.

The equations for the filter gain, K, and the covariance matrix, P, of the state prediction are then:

$$\begin{aligned} K_k &= P_{k|k-1} H_k^T \left(H_k P_{k|k-1} H_k^T + R_k\right)^{-1} \\ P_{k|k-1} &= A_{k-1} P_{k-1|k-1} A_{k-1} + Q_{k-1} \end{aligned} \tag{8}$$

where Q_k is the covariance of the process noise n.

3   Support Vector Machine

The support vector machine (SVM) is a statistical classification method originally designed for binary classification. The SVM constructs a hyperplane or a set of hyperplanes in a high or infinite-dimensional space in order to be used in classification, regression, or other tasks. The basic idea is to find a hyperplane which separates the m-dimensional data $x_k \in \mathbb{R}^m$ perfectly into two classes, through the label class variable $y_k \in \{-1;1\}$, where x_k represents the kth training data of set $S = \{x_1, \cdots, x_n\}$. A good separation between classes is achieved by the hyperplane $f(x_k) = w^T \phi(x_k) - b$ that produces the largest distance to the nearest training data points of any class (the so-called functional margin), i.e. $2/\|w\|$. In general, the maximization of the margin leads to a small error of the classifier. Due to the fact that in many cases, the data is not linearly separable, the SVM uses the concept of "kernel induced feature space", where a set of Kernel functions ϕ's casts the data into a higher dimensional space, where it is separable. The maximum hyperplane problem can be formulated as the following Quadratic Programming (QP) optimization problem [18][19]:

$$J = \arg\min_{w,\xi,b} \left\{ \frac{1}{2} w^T w + C \sum_{k=1}^{n} \xi_k \right\} \tag{9}$$

subject to (for any $k = 1,\ldots, n$)

$$y_k \cdot \left(w^T \phi(x_k) + b + \xi_k \right) \geq 1 \tag{10}$$

where ξ_k is a non-negative slack variable that measures the degree of misclassification of data x_k. The objective function is then increased by a function which penalizes non-zero ξ_k, and the optimization becomes a trade-off between a large margin and a small error penalty. Computations involving ϕ are handled using the kernel function $K(x_i, x_j) = \phi(x_i)\phi(x_j)$. This work deals with a Gaussian radial basis function, $K(x_i, x_j) = e^{-\gamma\|x_i - x_j\|^2}$ which implements a nonlinear classifier.

The space variables x_k of this problem, are the state variables θ_k, w_k and α_k used to classify the region on the space with the SVM algorithm. These variables may correspond or not to situations of a body fall situation.

4   Experimental Results

A set of experiments was carried out by a human equipped with a Smartphone (Huawei U8800, IOS Android 2.3.5) placed in the shirt pocket, subject himself to various normal activities that ended in a fall incident. The smartphone data is collected with a sampling time of $T=0.01$s. The 3D accelerations data are transformed in two acceleration components (i.e., ax and ay components, of body frame B) and then used by EKF filter with PBM to estimate the unobserved state variables, x's. Figure 2 shows the time behavior of the state variables for one fall experiment where the fall occurs around the 7th second.

The joint performance of EKF algorithm and PBM model gives promising results of state and measured variables, which agree well with the analytic arguments.

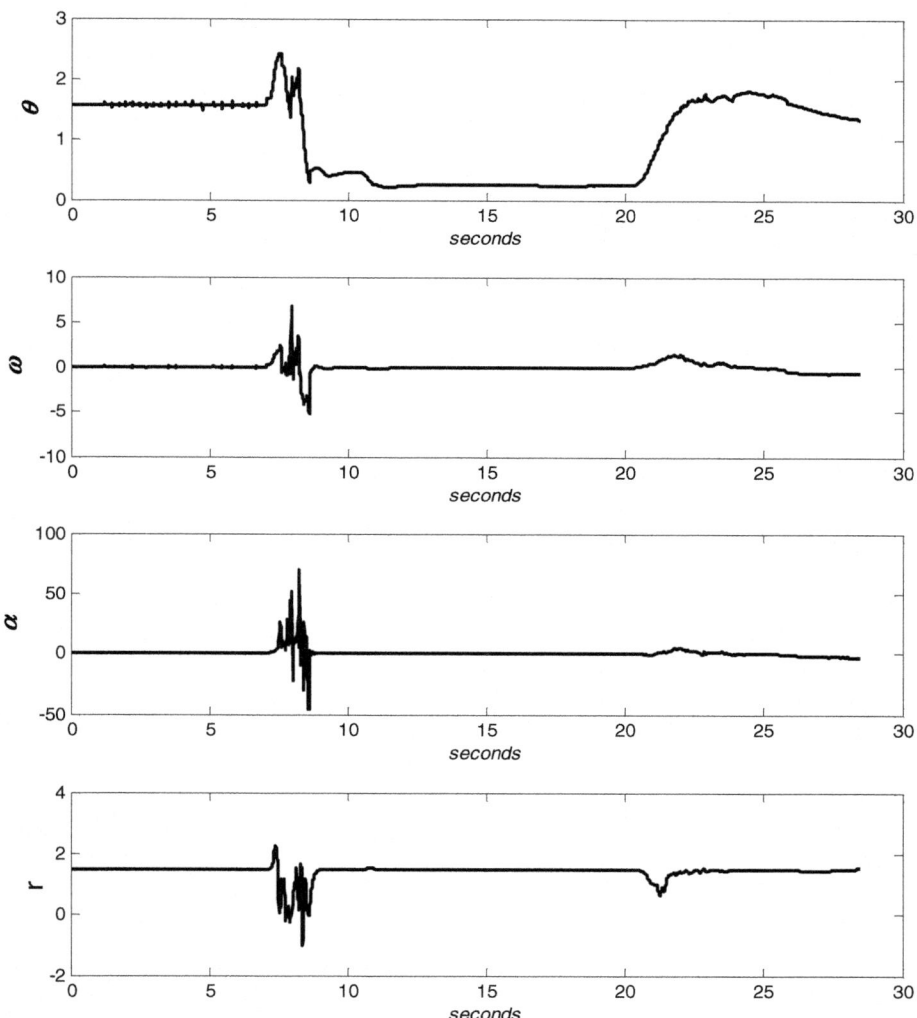

Fig. 2. State-space variable of PBM model

A SVM with Gaussian kernel function is used for detecting the fall event. The inputs are the x's variables and all experimental data was used to train the SVM. The third graphic of figure 3 shows the result of detection by the SVM classifier of a fall experiment. New tests validated its performance with a success detection rate of 96%.

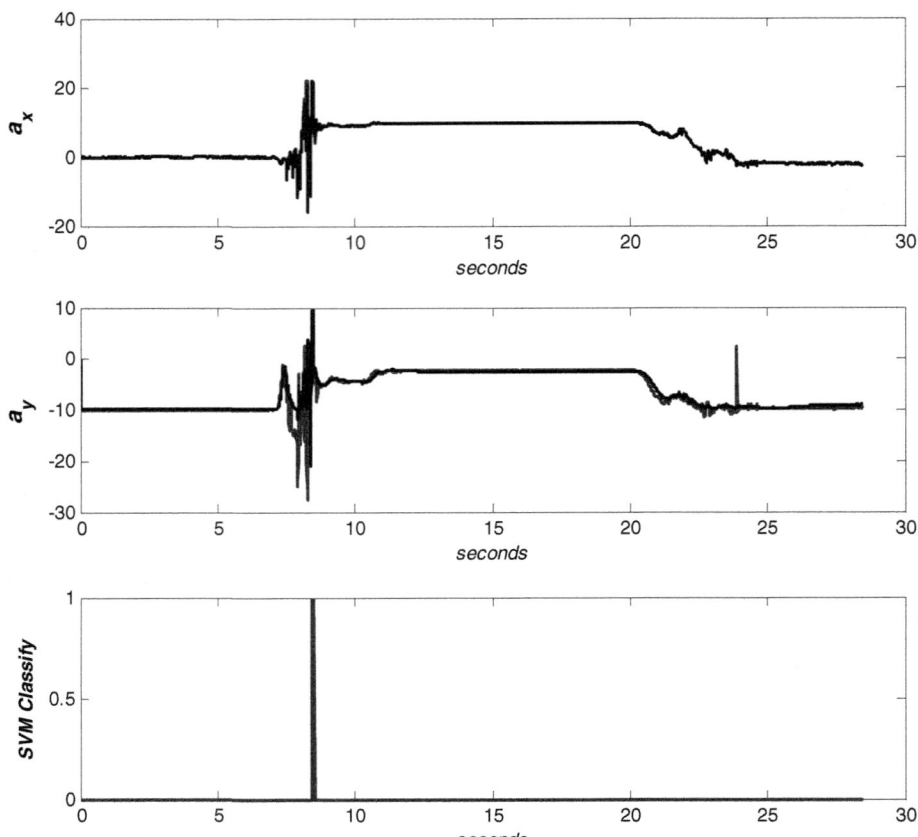

Fig. 3. Measured and PBM estimated accelerations, a_x and a_y. (two upper graphics); the result of SVM classify module (third graphic below).

5   Conclusions

Human motion is an important neuro-musculo-skeletal event which incorporates mechanical, physiological and anatomical factors with many underlying elements such as movement planes, kinematics, biomechanics, *etc*, that must be understood when quantifying and defining human movement. Accelerometry has proven to be an appropriate and viable mean of determining the movement of human body, provided by the widespread technology incorporated in the modern smartphones. As a result, accelerometry has become a relatively non-intrusive mean of assessing ambulatory movement, posture, and detection of body fall episodes.

In this paper, a PBM model combined with an EKF filter provides a robust framework for tracking human motion through a series of discontinuities, namely for the body fall episodes. By focusing on the motion of the human torso and developing a simple representation of this part of the body, it is possible to develop a tracking system that can follow through extremely sudden discontinuities. The SVM algorithm is used with success to detect the fall events.

References

1. Sixsmith, A., Johnson, N.: A smart sensor to detect the falls of the elderly. IEEE Pervasive Computing 3(2), 42–47 (2004)
2. Vellas, B., Cayla, F., Bocquet, H., De Pemille, F., Albarede, J.L.: Prospective study of restriction of activity in old people after falls. Age Ageing 16(3), 189–193 (1987)
3. Bourke, A.K., Van de Ven, P., Gamble, M., O'Connor, R., Murphy, K., et al.: Evaluation of waist-mounted tri-axial accelerometer based fall-detection algorithms during scripted and continuous unscripted activities. J. Biomech. 43, 3051–3057 (2010)
4. Welch, G.F., Foxlin, E.: Motion tracking: no silver bullet, but a respectable arsenal. IEEE Comput. Graph. Appl. 22, 24–38 (2002)
5. Rajendran, P., Corcoran, A., Kinosian, B., Alwan, M.: Falls, Fall Prevention, and Fall Detection Technologies. In: Eldercare Technology for Clinical Practitioners Aging Medicine, ch. 8, pp. 187–202 (2008)
6. Mubashir, M., Shao, L., Seed, L.: A survey on fall detection: Principles and approaches. Neurocomputing 100, 144–152 (2013)
7. Hazelhoff, L., Han, J., de With, P.H.N.: Video-Based Fall Detection in the Home Using Principal Component Analysis. In: Blanc-Talon, J., Bourennane, S., Philips, W., Popescu, D., Scheunders, P. (eds.) ACIVS 2008. LNCS, vol. 5259, pp. 298–309. Springer, Heidelberg (2008)
8. Alwan, M., et al.: A Smart and Passive Floor-Vibration Based Fall Detector for Elderly. In: 2nd IEEE International Conference on Information and Communication Technologies (2006)
9. Chen, K., Bassett, D.: The technology of accelerometry-based activity monitors: current and future. Med. Sci. Sports Exerc. 37(11), 490–500 (2005)
10. AN-1023 APPLICATION NOTE, Fall Detection Application by Using 3-Axis Accelerometer ADXL345. Analog Device, http://www.analog.com
11. Chen, K., Bassett, D.: The technology of accelerometry-based activity monitors: current and future. Med. Sci. Sports Exerc. 37(11), 490–500 (2005)
12. Li, Q., Stankovic, J.A., Hanson, M.A., Barth, A.T., Lach, J., Zhou, G.: Accurate, fast fall detection using gyroscopes and accelerometer-derived posture information. In: BSN 2009: Proceedings of the 2009 Sixth International Workshop on Wearable and Implantable Body Sensor Networks, pp. 138–143. IEEE Computer Society, Washington, DC (2009)
13. Bourke, A., O'Brien, J., Lyons, G.M.: Evaluation of a threshold-based tri-axial accelerometer fall detection algorithm. Gait Posture 26(2), 194–199 (2006)
14. Bourke, A.K., O'Donovan, K.J., ÓLaighin, G.M.: The identification of vertical velocity profiles using an inertial sensor to investigate pre-impact detection of falls. Med. Eng. Phys. 30(7), 937–946 (2008)
15. Culhane, K.M., Lyons, G.M., Hilton, D., Grace, P.A., Lyons, D.: Long-term mobility monitoring of older adults using accelerometers in a clinical environment. Clin. Rehabil. 18(3), 335–343 (2004)
16. Salgado, P., Afonso, P.: Fall body detection algorithm based on tri-accelerometer sensors. In: IEEE 14th International Symposium on Computational Intelligence and Informatics (CINTI), Budapest, Hungary, pp. 355–358 (2013)

17. Chapman, A.: Biomechanical Analysis of Fundamental Human Movements, Human Kinetics, United States (2008)
18. Cortes, C., Vapnik, V.N.: Support-Vector Networks. Machine Learning 20 (1995)
19. Steinwart, I., Christmann, A.: Support Vector Machines. Springer, New York (2008) ISBN 978-0-387-77241-7

System Identification Methods for Identification of State Models

Marta Sofia Esteves[1], Teresa P. Azevedo Perdicoúlis[1],
and Paulo Lopes dos Santos[2]

[1] Universidade de Trás-os-Montes e Alto Douro, Vila Real, Portugal
[2] Faculdade de Engenharia da Universidade do Porto, Porto, Portugal

1 Introduction

System Identification (SI) is a methodology for building mathematical models of dynamic systems from experimental data, i.e., using measurements of the system input/output (IO) signals to estimate the values of adjustable parameters in a given model structure. The process of SI requires some steps, such as measurement of the IO signals of the system in time or frequency domain, selection of a candidate model structure, choice and application of a method to estimate the value of the adjustable parameters in the candidate model structure, validation and evaluation of the estimated model to see if the model is right for the application needs, which should be done preferably with a different set of data, [PS] and [Lj1].

State space models of linear time invariant (LTI) systems play an important role in modern control theory and applications. The methods of identification of classical systems were developed by mid-year 1980. However, the introduction of the concept of state is recent and led to the development of many methods of subspace, based on the theory of classical stochastic realization, [Ka]. The minimal realization is commonly referred to as fitting given Markov parameters (or impulse response of a system) with a state space model being determined from observed IO data. For example, in stochastic SI, Markov parameters are identified first as sample covariance and then a model is derived through a minimal realization algorithm, [Ka]. In particular, the general problem of subspace identification consists in determining, from a set of available samples of the IO signals, the order n of the system matrices respective to each type of subspace method. The subspace system identification (SSI) algorithms are based on concepts withdrawn from system theory, numerical linear algebra and statistics. A key element of the subspace methods is to understand how the state vectors of the Kalman filter and the extended observability matrix are obtained by the use of tools of numerical linear algebra, based on either singular value (SVD) or QR decomposition, [Lj1], with no need for any optimization techniques or to write the system in a canonical form. This implies that the subspace algorithms can also be applied to all systems. Subspace methods are characterized by a conceptual simplicity and algorithms with good results. The above mentioned advantages justify the use of these methods over the traditional ones. It is important to note the difference between conventional methods and the

© Springer International Publishing Switzerland 2015   417
A.P. Moreira et al. (eds.), *CONTROLO'2014 - Proc. of the 11th Port. Conf. on Autom. Control*,
Lecture Notes in Electrical Engineering 321, DOI: 10.1007/978-3-319-10380-8_40

SSI. When using classical methods, the transfer function is first identified and the state space model is obtained by means of a particular realization. Then through the state space model, we can calculate the state vectors of Kalman filters. In subspace methods, we begin by constructing estimates of the states through the data IO using tools of numerical linear algebra and a state space model is then obtained by solving a problem of least squares (LS), [Ka] and [Es1].

SI Education is an essential part of control engineering and requires simple software tools, [Lj2]. Subspace Identification Tool (SIT) is a new interactive tool, that contains different types of identification, following a sequence which should be performed to identify a particular system, with fast execution and excellent facilities for interactive graphics. This toolbox is expected to have an impact on education in SSI, as this type of interactive tools provides a real time connection between the decisions taken during the design stage, and the results obtained in the analysis phase, for all type of systems, i.e., SISO, SIMO, MISO and MIMO systems.

The paper is organized as follows: the theoretical methods behind SSI are presented in Section 2, for the definition of LTI system, in Section 3 for the explain a matrix IO equation and the Section 4 is for Subspace Identification. The functionalities of SIT are presented in Section 5.

2   LTI System

In LTI systems the parameters are not changing over time and, therefore, the same entry will give the same result at every instance, which is simple and advantageous for a first stage in the study of subspace algorithms, [Ka]. The choice of a model set is a difficult issue in SI. Usually several class of LTI systems are used, because they are simple and they can describe accurately the dynamics of a great number of systems from a control point of view. Usually, identified models show the relation between the IO signals and they don't reflect the knowledge of the first principles of the system. Such kind of models are referred as black-box models, [PS].

Consider an innovation representation of a discrete-time LTI system, [OM2], with m inputs and l outputs, of the form

$$x(k+1) = Ax(k) + Bu(k) + w(k),$$
$$y(k) = Cx(k) + Du(k) + v(k) \tag{1}$$

where, $x(k) \in \mathrm{R}^n$ is the state vector, $w(k) \in \mathrm{R}^m$ is the process noise, $u(k) \in \mathrm{R}^m$ the input vector, $y \in \mathrm{R}^l$ is the output vector, and $v(k) \in \mathrm{R}^l$ is the measurement noise, which $w(k)$ and $v(k)$ are vectors signals, that are sequences of white noise, zero mean and stationary and $A \in \mathrm{R}^{m \times n}$, $B \in \mathrm{R}^{n \times m}$, $C \in \mathrm{R}^{l \times n}$ and $D \in \mathrm{R}^{l \times m}$ are the parameters.

3   A Matrix IO Equation

In this section, we derive some algebric properties of a LTI system described in equation (1). Define the Markov parameters as $h(k) = \begin{cases} D, se & k = 0, \\ CA(k-1)^B, se & k > 0 \end{cases}$ where D, C, A and B refers to (1). Let $Y \in \mathbb{R}^{2li \times j}$ and $U \in \mathbb{R}^{2mi \times j}$ be block Hankel matrices, defined as in [Es1].

Let Γ_i the extended observability matrix of order i ($i > n$) which is defined as: $\Gamma_i = \begin{bmatrix} C \ CA \ CA^2 \ldots \ldots CA^{i-1} \end{bmatrix}^T \in \mathbb{R}^{li \times n}$. Also, the extended controllability matrix ($i > n$), $\mathcal{C}_i$ (where the index i represents the number of columns of the matrix) is defined as: $\mathcal{C}_i = \begin{bmatrix} A^{i-1} \ B \ A^{i-2}B \ldots AB \ B \end{bmatrix} \in \mathbb{R}^{n \times mi}$.

H_i is the lower triangular Toeplitz matrix which contains a Markov parameters can then be defined as $H_i =$

$$\begin{bmatrix} D & 0 & 0 & \ldots & 0 \\ CB & D & 0 & \ldots & 0 \\ CAB & CB & D & \ldots & 0 \\ CA^2B & CAB & CB & \ldots & 0 \\ \vdots & \vdots & \vdots & \vdots & \vdots \\ CA^{i-2}B & CA^{i-3}B & CA^{i-4}B & \ldots & D \end{bmatrix}_{li \times mi}.$$

Let $X \in \mathbb{R}^{n \times j}$ be the matrix of the consecutive states from the initial state to the final state, [Es1].

The matrix IO equations play a very important role in the SSI, because its permits implicit and explicit states estimates of the system. These equations are derived directly from the state space representation

$$Y_i(i,j) = \Gamma_i X_i + H_i U_i(i,j), \tag{2}$$

where the index j represents the number of rows of the matrix. The singular values of the block Hankel matrix U are used as a quantification of the concept of persistency of excitation.

4   Subspace Identification

In this section, we present the algorithms implemented in the toolbox for the different types of identification.

4.1   Deterministic Identification

Deterministic models work with exact relationships between the measured variables and are expressed without uncertainty, therefore, deterministic algorithms estimate deterministic models (A, B, C, D) (with $w(k) = v(k) = 0$).

Under the following assumptions, [OM2]:

- The input $u(k)$ is persistently excited.
- The space spanned by the intersection of rows U_f and the row space of X_p should be a empty set.

– The user defines the weight matrices $W_1 \in \mathbb{R}^{li \times li}$, $W_2 \in \mathbb{R}^{j \times j}$ and that W_1 have full rank W_1 obeys: $r(W_1) = r(W_p.W_2)$, where W_p is the block Hankel matrix with the past IO.
   Indeed, we assume now, $\mathcal{O}_i$ as oblique projection $\mathcal{O}_i := Y_f / _{U_f} W_p$, and the decomposition SVD is:

$$W_1 \mathcal{O}_i W_2 = \begin{bmatrix} U_1 & U_2 \end{bmatrix} \begin{bmatrix} S_1 & 0 \\ 0 & 0 \end{bmatrix} \begin{bmatrix} V_1^T \\ V_2^T \end{bmatrix} = U_1 S_1 V_1^T \tag{3}$$

We then so that:

1. The matrix $\mathcal{O}_i$ is equal to the product of the observability matrix, Γ_i, and the sequence of states, X_f, i.e., $\mathcal{O}_i = \Gamma_i X_f$.
2. The order of system is equal to the number of singular values not null of equation (3).
3. The extended observability matrix is equal to $\Gamma_i = W_1^{-1} U_1 S_1^{\frac{1}{2}} T$, when considering a similarity matrix T.
   when considering a similarity matrix T.
4. The part of the sequence of states that is contained in the space spanned by the columns of W_2 can be recovered by $X_f W_2 = T^{-1} S_1^{\frac{1}{2}} V_1^T$.
5. Finally it is said that the equation of state X_f is equal to

$$X_f = T_i^\dagger \mathcal{O}_i, \tag{4}$$

where $T_i^\dagger$ is the pseudo-inverse of T_i. X_f is also located in row space of the future IO Hankel matrices, U_f and Y_f. Moreover, it can be proved that X_f is the intersection of the row spaces of W_p and $W_f = \begin{bmatrix} U_f \\ Y_f \end{bmatrix}$, where p is relative to past and f is relative to future.

Implementation of Deterministic Algorithms. Three algorithms have been implemented, namely Ho-Kalman, Intersec and Project, because are the most classical in literature, [Es1].

Ho-Kalman Algorithm

– Calculate the impulse response, $h(k), k = 0, \ldots, i+j-1$ with $i > n$ e $j > n$, by the LS method.

– Generate the Hankel matrix $H_{ij} = \begin{bmatrix} h(1) & h(2) & \cdots & h(j) \\ h(2) & h(3) & \cdots & h(j+1) \\ h(1) & h(2) & \cdots & h(j) \\ \vdots & \vdots & \vdots & \vdots \\ h(i) & h(i+1) & \cdots & h(i+j-1) \end{bmatrix} = \Gamma_i \mathcal{C}_j.$

– Perform the SVD $H_{ij} = USV^T$. Through singular values, it can be possible to determine the order n of system.
– Assume that $\Gamma_i = U(:, 1 : n) S^{\frac{1}{2}}(1 : n, 1 : n)$ e $\mathcal{C}_j = S^{\frac{1}{2}}(1 : n, 1 : n) V^T(:, 1 : n)$
– Let C is the l first rows of Γ_i and $A = \Gamma_{i-1}^\dagger \overline{\Gamma_i}$, where $\Gamma_{i-1}^\dagger$ is the pseudo-inverse of Γ_{i-1} and $\Gamma_{i-1} = \Gamma_i(1 : (i-1)l, :)$, $\overline{\Gamma_i} = \Gamma_i(l : il, :)$. Let B is the block constituted by the m first rows of Γ_i, and let $D = h(0)$.

Intersec Algorithm

- Through the equation (3) and the intersection by equation (4), is possible to analyze the singular values, and, consequently, the order of system.
- Determine the matrices A, B, C e D by $\begin{bmatrix} X_{i+1}^d \\ Y_{i|i} \end{bmatrix} = \begin{bmatrix} A & B \\ C & D \end{bmatrix} \begin{bmatrix} X_i^d \\ U_{i|i} \end{bmatrix}$,
  where $X_i^d = [x(i)\ x(i+1)\ x(i+2)\ \dots\ x(i+j-1)]$ is the states sequence, where i is the first element of states sequence. $U_{i|i}$ and $Y_{i|i}$ are blocks Hankel matrices with m and l lines, respectively.

Project Algorithm

- As it happens for Intersec algorithm, the order of system is determined by (3);
- $A = \Gamma_{i-1}^{\dagger} \overline{\Gamma_i}$ and C is calculated by the first l rows of Γ_i. The matrices B and D can be determined trough two different ways, see [OM2, page 54].

4.2 Stochastic Identification

Stochastic model concepts can contain uncertainty and probability. Contains quantities that are described by stochastic variables. And so, stochastic algorithms estimate innovation stochastic models (A, C, K) with $u(k) = 0$. Assuming that, [OM2]:

- The process noise $w(k)$ and the measurement noise $v(k)$ are not null.
- The number of measurements tends to infinity $j \to \infty$.
- The weight matrices $W_1 \in \mathbb{R}^{li \times li}$ and $W_2 \in \mathbb{R}^{j \times j}$ are such that W_1 has full rank and W_2 obeys to $r(Y_p) = r(Y_p.W_2)$, where Y_p is the block Hankel matrix, which contain the past outputs.

And $\mathcal{O}_i$ is defined as $\mathcal{O}_i := Y_f/Y_p$ and SVD is defined as it was for the deterministic case. So we have so:

1. The matrix $\mathcal{O}_i$ is equal to the product of the extended observability matrix for the sequence of states estimated by the Kalman filter $\mathcal{O}_i = \Gamma_i \hat{X}_i$.
2. The order of system is equal to the number of singular values not null of equation (3).
3. The observability and controlability extended matrix are equal to $\Gamma_i = W_1^{-1} U_1 S_1^{1/2} T$ and $\mathcal{C}_i^c = \Phi_{[Y_f, Y_p]}$, where Y_p and Y_f are the past and future output, respectively, defined as: $Y_p = \Gamma_i X_p + H_i, Y_f = \Gamma_i X_f + H_i U_f$
4. The part of the sequence of states $\hat{X}_i$ which lies in the column space and W_2 can be obtained from $\hat{X}_i W_2 = T^{-1} S_1^{1/2} V_1^T$.
5. The sequence of forward state, $\hat{X}_i$, is $\hat{X}_i = \Gamma_i^{\dagger} \mathcal{O}_i$.

Implementation of Stochastic Algorithms. One algorithm has been implemented, because it is simple algorithm and explains all steps of the stochastic identification theorem.

Stochastic Algorithm

- In this case, we are in the presence of noise. We focus with a definite positive real sequence for forward subsystem.
- The order of the system is calculated through analysis of the singular values of the equation (3).
- The extended observability and controllability matrices are equal to equations $\Gamma_{i]}$ and $\mathcal{C}_i$.
- Through reasoning that was enunciated in this subsection can be shown to contain the following $\mathcal{O}_i = Y_{f-}/Y_{p+} = \Gamma_{i-1}/\hat{X}_{i+1}$.
- $\mathcal{O}_i$ can be easily calculated using the output data. It is also easily verified that if we take the last l lines of Γ_i, we find Γ_{i-1}, as $\Gamma_{i-1} = \underline{\Gamma}_i$, where $\underline{\Gamma}_i$ represents the matrix Γ_i without l last rows. Now $\hat{X}_{i+1}$ can be calculated with output data, through $\hat{X}_{i+1} = \Gamma_{i+1}^{\dagger}\mathcal{O}_{i-1}$.
- To resolve A e C by $\begin{bmatrix} \hat{X}_{i+1} \\ Y_{i|i} \end{bmatrix} = \begin{bmatrix} A \\ C \end{bmatrix}\hat{X}_i + \begin{bmatrix} \rho_w \\ \rho_v \end{bmatrix}$. Intuitively, once the residues of the Kalman filter, ρ_w and ρ_v (i.e., innovations) are not correlated with $\hat{X}_i$, it will solve also the set of equations using the method LS. And so, the solution of this method resolve one estimate of A and C as $\begin{bmatrix} A \\ C \end{bmatrix} = \begin{bmatrix} \hat{X}_{i+1} \\ Y_{i|i} \end{bmatrix}\hat{X}_i^{\dagger}$.

- To resolve G and Λ_0 is by $E_j\left[\begin{pmatrix} \rho_w \\ \rho_v \end{pmatrix}(\rho_w^T\ \rho_v^T)\right] = \begin{bmatrix} Q_i & S_i \\ S_i^T & R_i \end{bmatrix}$ where the superscript i which indicates that this estimate covariance is not a steady state as in the previous equation, but it is not a state of stable covariance matrices of the state not stable of Kalman filter $P_{i+1} = AP_iA^T + Q_i$, $G = AP_iC^T + S_i$, $\Lambda_0 = CP_iC^T + R_i$, where i goes to ∞, $Q_i \to Q, S_i \to S$ and $R_i \to R$. However, this process introduces a bias in the estimate when $i \neq \infty$. However, in exchange bias, in this estimate, we have a sequence of positive real covariance. This is easily observed from the fact that (by construction): $\begin{bmatrix} Q & S \\ S^T & R \end{bmatrix} = E_j\left[\begin{pmatrix} \rho_w \\ \rho_v \end{pmatrix}(\rho_w^T\ \rho_v^T)\right] > 0$. Thus, the matrix G and Λ_0 may be calculated based on the Lyapunov equation for $\sum^s$, in other words $\sum^s = A\sum^s A^T + Q$, and then, $G = A\sum^s C^T + S$, $\Lambda_0 = C\sum^s C^T + R$. Finally, the model can be converted to a forward innovation model through the Riccati equation for the matrix P, such as $P = APA^T + (G - APC^T)(\Lambda_0 - CPC^T)^{-1}(G - APC^T)^T$, and then the Kalman gain, K can be calculated as $K^f = (G - APC^T)(\Lambda_0 - CPC^T)^{-1}$, where f is the superscript for the forward model.

4.3   Combined Deterministic-Stochastic Identification

Model deterministic-stochastic combined the properties of deterministic and stochastic case. And so deterministic-stochastic algorithms estimate all the matrices of system (1). Assuming that, [OM2]:

- The deterministic input, $u(k)$, is not correlated with noises $w(k)$ and $v(k)$.
- The input should be persistently excited to order $2i$.
- The number of measurements tends to infinity.
- The weight matrices $W_1 \in \mathbb{R}^{li \times li}$ and $W_2 \in \mathbb{R}^{j \times j}$ are such that W_1 has full rank, r and W_2 obeys to $r(Y_p) = r(Y_p.W_2)$, where Y_p is the block Hankel matrix, which contais the past IO.

Indeed, if we assume now that $\mathcal{O}_i$ is defined as the oblique projection $\mathcal{O}_i := Y_f / U_f W_p$, and thus the SVD will be given by the same expression in (3). And so,

1. The matrix $\mathcal{O}_i$ is equal to the product of Γ_i, and the sequence of states of the Kalman filter, $\hat{X}_i$ $\mathcal{O}_i := \Gamma_i \hat{X}_i$,
   with $\hat{X}_i := \hat{X}_{i_{[\hat{X}_0, P_0]}}$, $\hat{X}_0 = X_p^d / U_f U_p$, $P_0 = -[\sum^d - S^{uu}(R^{uu})^{-1}(S^{xu})^T]$.
2. The order of the system for this type of identification is equal to the number of non-zero singular values of equation (3).
3. The extended observability matrix is equal to:

$$\Gamma_i = W_i^{-1} S_1^{\frac{1}{2}} T. \tag{5}$$

4. The part of the sequence of states $\hat{X}_i$, which is contained in the space generated by the columns of W_2, can be recovered by $\hat{X}_i W_2 = T^{-1} S_1^{\frac{1}{2}} V_1^T$ And so, the equation of states $\hat{X}_i$ is equal to $\hat{X}_i = T_i^{\dagger} \mathcal{O}_i$.

Implementation of Deterministic-Stochastic Algorithms. Two algorithms have been implemented, mainly N4SID and MOESP, because are more classical of literature, [OM2] and [Ka].

N4SID Algorithm

- Calculate the oblique projection, $\mathcal{O}_i$ and $\mathcal{O}_{i+1}$ by: $\mathcal{O}_i = Y_f / U_f W_p = \Gamma_i \hat{X}_i$, $\hat{X}_i = \Gamma_i^{\dagger} \mathcal{O}_i$, and $\mathcal{O}_{i+1} = Y_f^- / \bar{U}_f W_p^+ = \Gamma_{i-1} \hat{X}_{i+1}$, $\hat{X}_{i+1} = \Gamma_{i-1}^{\dagger} \mathcal{O}_{i+1}$.
- Calculate SVD by (3).
- Calculate the sequences of state $\hat{X}_i$ and $\hat{X}_{i+1}$ by the equations $\hat{X}_i$ and $\hat{X}_{i+1}$. The following set of linear equations allows to reach the estimation of A, B, C and D: $\begin{bmatrix} \hat{X}_{i+1} \\ \hat{Y}_{i|i} \end{bmatrix} = \begin{bmatrix} A & B \\ C & D \end{bmatrix} \begin{bmatrix} \hat{X}_i \\ U_{i|i} \end{bmatrix} \begin{bmatrix} \rho_w \\ \rho_v \end{bmatrix}$.

MOESP Algorithm

- This algorithm is based on the LQ decomposition by

$$\begin{bmatrix} U_f \\ W_p \\ Y_f \end{bmatrix} = \begin{bmatrix} L_{11} & 0 & 0 \\ L_{21} & L_{22} & 0 \\ L_{31} & L_{32} & L_{33} \end{bmatrix} \begin{bmatrix} Q_1^T \\ Q_2^T \\ Q_3^T \end{bmatrix}.$$

- Calculate the SVD (3) to examine the order n, and determine the observability matrix by (5).

- Likewise, the order of the system can be calculated using the singular values of S_1^m, i.e. $\Gamma_i = U_i^m (S_i^m)^{\frac{1}{2}}$, where m is the superscript related to MOESP. The weight matrices are $W_1 = I_{li}$ and $W_2 = \Pi_{U_f \perp}$.
- The sequences $\mathcal{Z}_i$ and $\hat{X}_{i+1}$ are : $\hat{X}_i = \hat{X}_{i_{[\hat{x}_0, P_0]}}$, $\hat{X}_{i+1} = \hat{X}_{(i+1)_{[\hat{x}_0, P_0]}}$. $\mathcal{Z}_i = Y_f / \begin{bmatrix} W_p \\ U_f \end{bmatrix} = \Gamma_i \hat{X}_i + H_i^d U_f$, $\mathcal{Z}_{i+1} = Y_f^- / \begin{bmatrix} W_p^+ \\ U_f^- \end{bmatrix} = \Gamma_{i-1} \hat{X}_{i+1} + H_{i-1}^d U_f^-$.
- Get A and C by $\begin{bmatrix} \Gamma_{i-1}^\perp \mathcal{Z}_i \\ Y_{i|i} \end{bmatrix} = \begin{bmatrix} A \\ C \end{bmatrix} \Gamma_i^\perp \mathcal{Z}_i + K U_f + \begin{bmatrix} \rho_w \\ \rho_v \end{bmatrix}$, where ρ_w and ρ_v are residues of Kalman filter. Get B and D by different way that matrices A and C, see [Ka, page 159].

5 SIT Description

This section describes the functionalities of the toolbox. SIT contains all the features of subspace identification mentioned in the previous sections, namely it allows the user to choose what kind of system wants to analyze, to import the IO signals, to choose the algorithms, to estimate the model parameters and, finally, to validate the estimated model, [Es1]. When developing a tool of this kind, one of the most important things that the developer needs to keep in mind is the organization of the main windows and menus, in order to facilitate the user to understand the subspace identification technique as a whole, [Lj1] and [Lj2]. This toolbox contains help menus to assist new users to realize all features of the software and analyze all the steps needed through an example. In the same folder of SIT, examples of SISO and MIMO systems are also available.

Using SIT. On the left side of the window, under the caption "Signal Analysis", one can import IO data and represent it graphically. At the top center, under the caption of "Estimation", one can estimate the parameters of the model by choosing a convenient algorithm according to the nature of the system. Once an algorithm is selected, a window pops up that prompts the use to select the order of the model through the visualization of the singular values of the system. Once this is done, one may see the system matrices on the very right center of the window. At the bottom center, different ways to validate the estimated model are available. Next we detail the functionalities of SIT, according to the order that the user should follow, [Es1].

1. To open the SIT, the user should evoke the SIT command from the "command window" of MATLAB. Also, the IO data should be available in .mat format in a predefined folder.
2. To import the data, under section "Signal Analysis", the user is asked through a pop-up menu what type of system wants to analyze, i.e., to use deterministic, stochastic or combined deterministic-stochastic identification.
3. Once this selection is done, the system allows the user to select the data and additionally shows the path files to the folder where the data is stored. The user can then verify that the correct files have been selected.

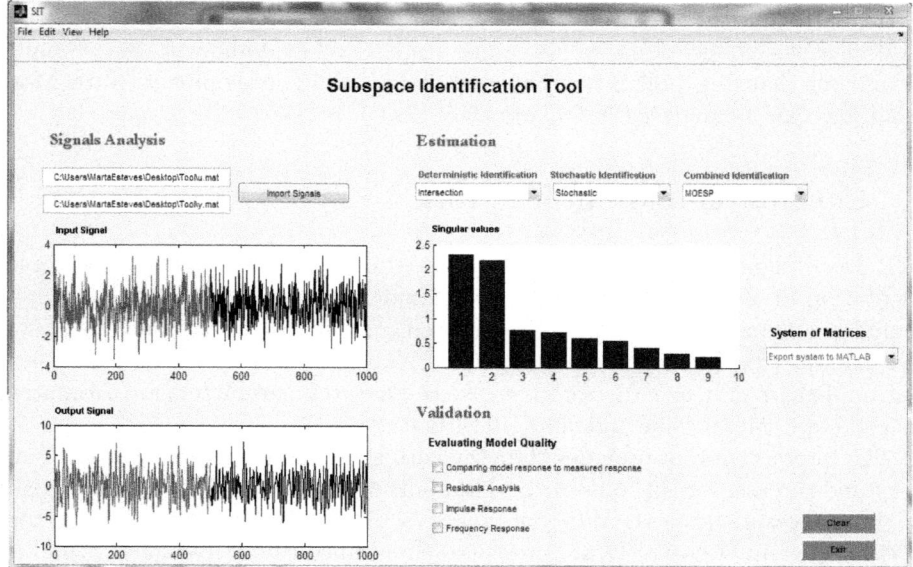

Fig. 1. Graphical layout of the various steps in IS in subspaces in SIT, with a SISO example for deterministic identification -Intersec algorithm, [Es1]

4. The IO signals are divided into two sets: an estimation (identification) set in red color and a validation data set in the black color. After data import, the system is automatically classified as SISO, SIMO, MISO or MIMO.
5. Once the IO signals have been analyzed, the system is able to proceed to the estimation of the order of the system and its parameters.
6. To estimate the parameters, under section "Estimation", the user selects the algorithm according to the type of identification that has been chosen. Three different menus exist, relating to the different types of identification, that is, for **Deterministic Identification**, the algorithm Intersection, Projection and Ho-Kalman, for **Stochastic Identification**, the algorithm Stochastic and for **Combined Deterministic-stochastic**, the algorithms N4SID and MOESP.
7. The singular values are shown graphically and the user is prompted to select the order n of the system and the calculated system matrices.
8. To view the estimated system matrices, the user has options in the menu displayed: "Export system to MATLAB" or "Export system to a text file".
9. The last step is the validation of the model through the study of: "Comparing model response to measured response", "Residuals Analysis", "Impulse Response", "Frequency Response". By checking an option, new windows open to perform the required validation.

Figure 1 illustrates a simple example, that show us the complexity of SIT. For this case, we can see the layout of SIT, representation of IO data, the selection of algorithm and analysis the order of system trough the graphically representation of singular values, interactively. I.e., it is a simple demonstration, but much other examples, with educational value, can be studies and explored with SIT. Finally, we assume that this toolbox is very recent and is still in development (Note: More examples can be analysed in Chapter 5 of [Es1]).

6   Conclusion and Future Work

SIT was designed to serve as a support for education in SI, through the implementation of algorithms for constructing models of linear state space and also includes some additional options to be used, mainly, for educational purposes. For example, the instructor can import the IO signals and then select the methods implemented in accordance to his needs. Once the parameters are estimated, he can export it to a file and share it with students.

The interactive tool provides the user with several degrees of freedom to understand the theoretical concepts and in addition gives sensitivity to the impact of the choices made in the different stages of IS. A main advantage of this tool is that different stages of IS are presented simultaneously in a single screen, i.e., IO data, choice of different models, estimation of parameters and validation of the estimated models. These interactive features of the tool allows the user to understand and experience the relationships between these different stages and its meaning.

References

AM. Antsaklis, P.J., Michel, A.N.: A Linear Systems Primer. Birkhauser, Boston (2007)

Es1. Esteves, M.: Identificacão de sistemas - Métodos para identificacão de modelos de estado. Master thesis. UTAD. MEEC (2013), https://repositorio.utad.pt/browse?type=author&value=Esteves%2C+Marta+Sofia+Alves (consulted on February 24, 2014)

Es2. Esteves, M.: ITSIE: An Interactive Software Tool for System Identification Education (2011), http://aer.ual.es/ITSIE/tools/manual_ITSIE (consulted on February 1, 2014)

Ka. Katayama, T.: Susbspace Methods for System Identification. Springer (2005) ISBN: 1852339810

Lj1. Ljung, L.: System identification: Theory for the User, 2nd edn. Prentice-Hall, Englewood Cliffs (1999)

Lj2. Ljung, L.: Matlab System identification toolbox users guide (2011), http://www.mathworks.com/help/pdf_doc/ident/ident.pdf (consulted on January 8, 2014)

Mo. De Bart, M.: Mathematical concepts and techniques for modeling of static and dynamic systems. PhD thesis. Department of Electrical Engineering. Katholieke Universiteit Leuven. Belgium (1988)

OM1. Overschee, P. and M., Bart: Available software for subspace identification, http://homes.esat.kuleuven.be/~smc/sysid/software/ (consulted on September 1, 2012)

OM2. Overschee, P., Bart, M.: Subspace identification for linear systems (1996) ISBN: 0-7923-9717-7

PS. Pintelon, R., Schoukens, J.: System Identification, A frequency Domain Approach. Institute of Electrical and Electronics Engineers. United States of America (2012)

Adaptively Tuning the Scaling Parameter of the Unscented Kalman Filter

Leonardo Azevedo Scardua[1,2] and José Jaime da Cruz[1]

[1] Automation and Control Laboratory, University of São Paulo, São Paulo, SP, Brazil
leonardo.scardua@usp.br, jaime@lac.usp.br
[2] Federal Institute of Espírito Santo, Serra, ES, Brazil
lascardua@ifes.edu.br

Abstract. This paper describes a new approach to adaptively tuning the scaling parameter of the unscented Kalman filter. The proposed algorithm is based on the idea of moment matching and is computationally inexpensive, allowing it to be executed online. Two nonlinear filtering problems are used to numerically compare the performance of the proposed algorithm with the performances of recently published adaptive unscented Kalman filters. An unscented Kalman filter enhanced with the proposed adaptive algorithm outperformed the other adaptive filters in both numerical problems.

Keywords: unscented transform, unscented Kalman filter, scaling parameter, automatic tuning, nonlinear state estimation.

1 Introduction

Let us consider the problem of estimating the current state of a nonlinear stochastic dynamic system described by the following discrete-time equations

$$\mathbf{x}_k = \mathbf{f}_k(\mathbf{x}_{k-1}) + \mathbf{v}_{k-1} \tag{1}$$
$$\mathbf{y}_k = \mathbf{h}_k(\mathbf{x}_k) + \mathbf{w}_k \tag{2}$$

where

1. $\mathbf{x}_k \in \Re^{n_x}$ is the state vector at discrete time k and $\mathbf{y}_k \in \Re^{n_y}$ is the measurement at discrete time k;
2. $\mathbf{f}_k : \Re^{n_x} \to \Re^{n_x}$ is the process model equation and $\mathbf{h}_k : \Re^{n_x} \to \Re^{n_y}$ is the measurement model equation;
3. $\mathbf{v}_{k-1} \sim \mathcal{N}(0, \mathbf{Q}_{k-1})$ e $\mathbf{w}_k \sim \mathcal{N}(0, \mathbf{R}_k)$ are independent Gaussian distributed noises.

Under the Bayesian framework, there is a set of recursive equations that optimally solve this problem [2]. Unfortunately, closed-form solutions to this set of equations can only be found in particular situations, such as when the dynamic

© Springer International Publishing Switzerland 2015   429
A.P. Moreira et al. (eds.), *CONTROLO'2014 - Proc. of the 11th Port. Conf. on Autom. Control*,
Lecture Notes in Electrical Engineering 321, DOI: 10.1007/978-3-319-10380-8_41

system is linear and Gaussian. The case of nonlinear dynamic systems demands resorting to approximate solutions. One popular approach to obtaining such approximate solutions are Gaussian filters. Gaussian filters provide approximate solutions to the set of recursive equations by assuming that the filtering distribution is Gaussian. In other words, it is assumed that

$$p(\mathbf{x_k}|\mathbf{y_{1:k}}) \simeq \mathcal{N}(\mathbf{x}_k|\mathbf{m}_k, \mathbf{P}_k) \; , \tag{3}$$

where the mean $\mathbf{m}_k$ and the covariance $\mathbf{P}_k$ are approximated via moment matching. The general additive-noise Gaussian assumed density filter can be written as [11]:

Prediction

$$\mathbf{m}_k^- = \int \mathbf{f}_k(\mathbf{x}_{k-1})\mathcal{N}(\mathbf{x}_{k-1}|\mathbf{m}_{k-1}, \mathbf{P}_{k-1})d\mathbf{x}_{k-1}$$

$$\mathbf{P}_k^- = \int (\mathbf{f}_k(\mathbf{x}_{k-1}) - \mathbf{m}_k^-)(\mathbf{f}(\mathbf{x}_{k-1}) - \mathbf{m}_k^-)^T \mathcal{N}(\mathbf{x}_{k-1}|\mathbf{m}_{k-1}, \mathbf{P}_{k-1})d\mathbf{x}_{k-1}$$

$$+\mathbf{Q}_{k-1} \; . \tag{4}$$

Update

$$\mathbf{K}_k = \mathbf{C}_k\mathbf{S}_k^{-1}$$
$$\mathbf{m}_k = \mathbf{m}_k^- + \mathbf{K}_k(\mathbf{y}_k - \boldsymbol{\mu}_k)$$
$$\mathbf{P}_k = \mathbf{P}_k^- - \mathbf{K}_k\mathbf{S}_k\mathbf{K}_k^T \tag{5}$$

where

$$\boldsymbol{\mu}_k = \int \mathbf{h}_k(\mathbf{x}_k)\mathcal{N}(\mathbf{x}_k|\mathbf{m}_k^-, \mathbf{P}_k^-)d\mathbf{x}_k$$

$$\mathbf{S}_k = \int (\mathbf{h}_k(\mathbf{x}_k) - \boldsymbol{\mu}_k)(\mathbf{h}_k(\mathbf{x}_k) - \boldsymbol{\mu}_k)^T \mathcal{N}(\mathbf{x}_k|\mathbf{m}_k^-, \mathbf{P}_k^-)d\mathbf{x}_k + \mathbf{R}_k$$

$$\mathbf{C}_k = \int (\mathbf{x}_k - \mathbf{m}_k^-)(\mathbf{h}(\mathbf{x}_k) - \boldsymbol{\mu}_k)^T \mathcal{N}(\mathbf{x}_k|\mathbf{m}_k^-, \mathbf{P}_k^-)d\mathbf{x}_k \; . \tag{6}$$

According to [13], all Gaussian filters can be seen as a family of filters that rely on the common structure given by (4), (5) and (6). Each member of this family is characterized by the integration strategy adopted to tackle the multidimensional integrals involved in the calculation of the moments (4) and (6). It can thus be seen that the quality of the estimates produced by a Gaussian filter is dependent on the filter's ability to correctly estimate the mean and covariance of $\mathbf{x}_k$ and $\mathbf{y}_k$.

The unscented Kalman filter (UKF) [9] is a Gaussian filter that uses the unscented transform (UT) [6] to approximately calculate the moments (4) and (6). For the UT, the accuracy to which it can predict the mean and covariance of a nonlinearly transformed random variable $\mathbf{z} = g(\mathbf{x})$ is a direct function of the accuracy to which it can capture the mean and covariance of the random variable **x**. In this paper, we work with the commonly used second-order UT[8], which correctly captures the sample mean, covariance and all higher odd-ordered central moments of a Gaussian random variable **x**. The UT's scaling parameter κ can be used to correct the representation of the fourth and higher order moments of **x**[7]. Thus, proper tuning of κ can improve the estimates yielded by the UT and ultimately improve the estimates produced by the filter. Based on this idea, the adaptive algorithm proposed in this paper aims at improving the performance of the UKF by finding the value of κ that, at each discrete time instant, improves the ability of the UT to correctly estimate the mean and covariance of the measurement $\mathbf{y}_k$.

The rest of this paper is organized as follows. Section 2 details the role of the scaling parameter on the results produced by the UT and ultimately by the UKF. Section 3 describes the proposed approach to adaptively tuning the scaling parameter. Section 4 compares the performance of the proposed approach with the performances of the standard UKF and of three versions of a recently published adaptive UKF ([3],[4],[12]). Finally, section 5 presents the conclusions and future directions for research.

2   The Unscented Transform

The UT is a method aimed at capturing various types of information about different kinds of distributions[7]. As we are dealing with the UKF, our interest lies in estimating the mean and covariance of a random variable $\mathbf{y} = g(\mathbf{x})$ resulting from applying a nonlinear transformation $g(\cdot)$ to a Gaussian random variable **x**. Let us now describe how the UT accomplishes this task.

The n_x-dimensional random variable **x**, with mean $\boldsymbol{\mu}_x$ and covariance $\mathbf{P}_x$, is approximated by the following set of weighted samples, called *sigma points*

$$\mathcal{X}_0 = \boldsymbol{\mu}_x \qquad\qquad w_0 = \kappa/(n_x + \kappa)$$
$$\mathcal{X}_i = \boldsymbol{\mu}_x + (\sqrt{(n_x + \kappa)P_x})_i \qquad w_i = 1/2(n_x + \kappa)$$
$$\mathcal{X}_{i+n_x} = \boldsymbol{\mu}_x - (\sqrt{(n_x + \kappa)P_x})_i \qquad w_{i+n_x} = 1/2(n_x + \kappa) \quad (7)$$

where $(\sqrt{(n_x + \kappa)P_x})_i$ is the ith row or column of the matrix square root of $((n_x + \kappa)P_x)$. The sigma points have the same sample mean, covariance, and higher odd-ordered central moments as the distribution of **x**, with the scaling parameter κ affecting the fourth and higher order sample moments of the sigma points[7]. The mean and covariance of the transformed random variable **y** are predicted by the UT as

$$\mathcal{Y}_i = g(\mathcal{X}_i)$$

$$\boldsymbol{\mu}_y = \sum_{i=0}^{2n_x} w_i \mathcal{Y}_i$$

$$P_y = \sum_{i=0}^{2n_x} w_i (\mathcal{Y}_i - \boldsymbol{\mu}_y)(\mathcal{Y}_i - \boldsymbol{\mu}_y)^T . \tag{8}$$

It can be shown that to correctly approximate the mean and covariance of $\mathbf{y}$ up to the m-th order term of the Taylor series expansions of these moments, it is necessary that the sigma points correctly approximate the moments of $\mathbf{x}$ up to the mth order moment [8]. As the fourth and higher order sample moments of the sigma points are functions of κ, correctly tuning κ improves the predictions (8). As the UKF is a Gaussian filter, the overall performance of the filter is also improved.

3   The Proposed Approach

In section 2, we have seen that the performance of the UKF is dependent on the quality of the approximations yielded by the UT. We have also seen that proper tuning of the scaling parameter κ can improve the quality of such approximations. It is thus natural to think that the criterion for the automatic tuning of κ must be the improvement of the approximations yielded by the UT. To this end, the algorithm proposed in this section is based on the idea of finding the value of κ that minimizes the difference between the first two moments of the noisy measurements received by the UKF and the approximations to these moments produced by the UT. Such approximations are obtained when the UT algorithm is fed with the current estimate of the state of the system and the nonlinear function $g(\cdot)$ is the measurement equation (2). Considering that the main source of new information is the innovation sequence[10], we execute the optimization of the scaling parameter before the measurement update step of the UKF. At each time step k, the current implementation of the proposed optimization approach comprises the following steps:

1. To each κ_j in the previously chosen n_k-element set $K = \{\kappa_1, \cdots, \kappa_n\}$:
   (a) Using measurement equation (2), apply the UT to the current state estimate $(\hat{\mathbf{x}}_k, \hat{\mathbf{P}}_{x_k})$, yielding the estimate $(\hat{\mathbf{y}}_k, \hat{\mathbf{P}}_{y_k})$ for the next measurement to be received by the filter.
   (b) Upon receiving the new measurement $\mathbf{y}_k$, assume that the mean of this measurement is $\mathbf{y}_k$. The covariance $\mathbf{P}_{y_k}$ of this measurement can be given by the covariance of the measurement noise, $\mathbf{R}_k$, or by the measurements' sample covariance. In the end of this section, it is discussed how to choose between these two alternatives.
   (c) Compute the distance between $(\hat{\mathbf{y}}_k, \hat{\mathbf{P}}_{y_k})$ and $(\mathbf{y}_k, \mathbf{P}_{y_k})$ as in

$$\kappa_k^{\text{KMM}} = \arg\min_{\kappa \in K} \left[(\hat{\mathbf{y}}_k - \mathbf{y}_k)^T (\hat{\mathbf{y}}_k - \mathbf{y}_k) + \sum_{i=1}^{n_x} \sum_{i=1}^{n_x} \sqrt{(\hat{P}_{y_k}^{i,i} - P_{y_k}^{i,i})^2} \right] \tag{9}$$

where $P_{y_k}^{i,i}$ is the element in line i and column i of matrix $\mathbf{P}_{y_k}$ and $\hat{P}_{y_k}^{i,i}$ is the element in line i and column i of matrix $\hat{\mathbf{P}}_{y_k}$.

2. The best κ_j for time step k is the one that yields the smallest distance (9).

The adaptive algorithm just described follows the same basic steps of the adaptive algorithm described in [3,4]. Namely, it uses a grid-based optimization strategy to find the value of a parameter that minimizes a certain cost function. In both cases, the parameter is κ and the UT is used to generate κ-dependent predictive estimates of the mean and covariance of the measurement, with such estimates being used to calculate the value of a cost function. Nevertheless, the proposed adaptive algorithm is built around a new moment-based optimization criterion for κ, which seeks to minimize the distance between $(\hat{\mathbf{y}}_k, \hat{\mathbf{P}}_{y_k})$ and $(\mathbf{y}_k, \mathbf{P}_{y_k})$. To our knowledge, this criterion, which involves an innovative way of guessing the covariance $\mathbf{P}_{y_k}$ of the current measurement, has not yet been proposed.

We now discuss how to choose a value for the covariance of the current measurement $(\mathbf{P}_{y_k})$. In this paper we experimented with two possible tuning procedures. The first is setting $\mathbf{P}_{y_k}$ equal to the covariance of the measurement noise $(\mathbf{R}_k)$. The second is setting $\mathbf{P}_{y_k}$ equal to the sample covariance of the measurements received by the filter (In section 4, this sample covariance is calculated using only the current and the last measurements).

To choose between such tuning procedures, we adopted a strategy that seeks to assess the sensitivity of the sample covariance of the measurements to the level of $\mathbf{R}_k$. The basic idea consists in using the measurement model to generate sequences of measurements for different levels of the covariance $\mathbf{R}_k$ (such levels are obtained by multiplying the covariance matrix $\mathbf{R}_k$ by a scalar number). For each level, the sample covariance of the sequence of measurements is calculated (the calculation uses all measurements generated). If the sample covariance of the measurements changes significantly with the changes in the level of $\mathbf{R}_k$, the first tuning procedure is adopted, otherwise, the second tuning procedure is chosen.

4   Numerical Results

In this section, the classic vertical falling object tracking problem [1] and a general nonlinear filtering problem [3] are used to compare the proposed algorithm, the adaptive UKF described in [3,4] and the regular UKF. In the remainder of this paper, the proposed algorithm is called KMM and the adaptive UKF proposed in [3,4] is called by the same name of the optimization criteria it adopts, which is ML. Let us now describe how the scaling parameters of such filters are tuned.

The scaling parameter of the regular UKF is always set to the standard recommendation $\kappa = 3 - n_x$[7]. For the ML, the tuning is more sophisticated. The algorithm seeks to maximize the likelihood function $p(\mathbf{y}_k|\mathbf{y}_{k-1})$. As this function is generally unknown, it is replaced by the approximation

$$\hat{p}(\mathbf{y}_k|\mathbf{y}_{k-1}, \kappa) = \mathcal{N}\{\mathbf{y}_k : \hat{\mathbf{y}}_{k|k-1}(\kappa), \mathbf{P}_{y,k|k-1}(\kappa)\} \tag{10}$$

where $\hat{\mathbf{y}}_{k|k-1}(\kappa)$ and $\mathbf{P}_{y,k|k-1}(\kappa)$ are simply the (κ dependent) UT estimates for the mean and covariance of the measurement $\mathbf{y}_k$. Thus, ML chooses the best value of κ at discrete time k as

$$\kappa_k^{\text{ML}} = \arg\max_{\kappa \in K}(\hat{p}(\mathbf{y}_k|\mathbf{y}_{k-1}, \kappa)) \ . \tag{11}$$

Reference [4] stresses that other optimization criteria could be used instead of (11). To choose the other criteria, we took into consideration the fact that increases in the dimension of the state vector make the values of the likelihood in (11) approach zero [12]. Conceding that such behavior can make the ML criterion (11) sensitive to numerical errors, we replaced it by the moment-based criteria (12) [12] and (13) [12]. As we did for the ML algorithm, we chose to label the algorithms resulting from such replacements with the same names of the optimization criteria they adopt, which are MMPE and MEPE.

Criterion MMPE (12) is based on minimizing the square of the Euclidean norm of the measurement prediction error $\tilde{\mathbf{y}}_{k|k-1}(\kappa) = \mathbf{y}_k - \hat{\mathbf{y}}_{k|k-1}(\kappa)$ and is given as

$$\kappa_k^{\text{MMPE}} = \arg\min_{\kappa \in K}(\tilde{\mathbf{y}}_{k|k-1}(\kappa)^T \tilde{\mathbf{y}}_{k|k-1}(\kappa)) \ . \tag{12}$$

Criterion MEPE (13) adds to (12) the minimization of the covariance of the measurement prediction error, yielding

$$\kappa_k^{\text{MEPE}} = \arg\min_{\kappa \in K} \left(\tilde{\mathbf{y}}_{k|k-1}(\kappa)^T \tilde{\mathbf{y}}_{k|k-1}(\kappa) + \text{tr}(\mathbf{P}_{y,k|k-1}(\kappa)) \right) \ . \tag{13}$$

To compare the quality of the estimates produced by the different filters, we use the mean-squared error(MSE)

$$\text{MSE} = \frac{\sum_{m=1}^{M} \sum_{k=0}^{N} \sum_{i=1}^{n_x} (x_{i,k}^m - \hat{x}_{i,k}^m)^2}{MNn_x} \tag{14}$$

where M is the number of Monte Carlo runs of the simulation process, N is the fixed number of samples of each simulation, $x_{i,k}^m$ is the i-th component of the true state in the m-th simulation at discrete time k and $\hat{x}_{i,k}^m$ is the corresponding filter estimate.

The other performance index, labeled VAR, is the variance of the estimates produced by the filter. Such index is calculated as the variance of the mean squared errors that result from the M Monte Carlo runs of the filter. The idea is to verify if the overall performance of a filter is characterized by usually good results or by sequences of very good and very bad results.

The algorithms UKF, ML, MMPE and MEPE used in the numerical tests are implemented with the software package denominated Nonlinear Estimation Framework (NEF) [5], which was used by the authors of such algorithms in [3]. Nevertheless, it is important to stress that, while KMM used the system's measurement data that was generated by NEF, its implementation did not use the software functions provided by the package. This makes it difficult to compare the execution times of different algorithms. Nonetheless, the execution times are presented to support the claim that KMM is a fast algorithm, suitable for online execution on modern computers. The tests were run on a notebook with Intel® Core™ i5 processor, 4GB of RAM and Microsoft® Windows 7.

4.1   Ballistic Fall Tracking

The ballistic fall tracking problem deals with the estimation of the altitude, velocity, and constant ballistic coefficient of a vertically falling body. It is considered to be a difficult filtering problem because of the strong nonlinearities exhibited by the forces which act on the vehicle[9]. The measurements are taken at discrete time instants by a radar that measures range in the presence of white Gaussian noise. The continuous time process model is

$$\dot{x}_1(t) = -x_2(t)$$
$$\dot{x}_2(t) = -e^{-\alpha x_1(t)}x_2^2(t)x_3(t)$$
$$\dot{x}_3(t) = 0 \tag{15}$$

where $x_1(t)$ is the altitude, $x_2(t)$ is the velocity, $x_3(t)$ is the ballistic coefficient and $\alpha = 5 \times 10^{-5}$. The discrete-time measurement model is

$$z_k = \sqrt{M^2 + (x_1(k) - Z)^2} + v_k \tag{16}$$

where the variance of the zero mean Gaussian noise v_k is 1×10^4 and $M = Z = 1 \times 10^5$. The evaluation was executed on filtering data produced by 500 Monte Carlo runs. Each run simulated 30 seconds of falling, with the Fourth-order Runge-Kutta method used for the integration of the process equations (15). The integration step was 0.05 seconds and the sampling interval was 0.1 seconds. For the filters, the integration step was made equal to the sampling step, making the filtering problem more difficult. It was assumed complete knowledge of the system's model, including of the process and measurement noises. The initial state for the dynamic system was $x_0 = \begin{bmatrix} 3 \times 10^{+5} & 2 \times 10^{+4} & 1 \times 10^{-3} \end{bmatrix}^T$. The initial state for the filters was

$$p(\hat{x}_0) \sim \mathcal{N} \left(\begin{bmatrix} 3 \times 10^{+5} \\ 2 \times 10^{+4} \\ 3 \times 10^{-5} \end{bmatrix} ; \begin{bmatrix} 1 \times 10^{+6} & 0 & 0 \\ 0 & 4 \times 10^{+6} & 0 \\ 0 & 0 & 1 \times 10^{-4} \end{bmatrix} \right) \tag{17}$$

and the resulting data is shown in Table 1, where the last line displays the mean execution time per sample of each algorithm.

4.2   Nonlinear Dynamics

The second nonlinear filtering problem is

$$x_{k+1} = (1 - 0.05\Delta T)x_k + 0.04\Delta T x_k^2 + w_k \tag{18}$$
$$z_k = x_k^2 + x_k^3 + v_k \tag{19}$$

Table 1. Vertical Fall Tracking

	UKF $\kappa = 3 - n_x$	KMM $\kappa \in \{0 : 1 : 4\}$	ML $\kappa \in \{0 : 1 : 4\}$	MMPE $\kappa \in \{0 : 1 : 4\}$	MEPE $\kappa \in \{0 : 1 : 4\}$
MSE	10283.305	9115.228	9648.832	9527.664	9864.423
VAR	5.246×10^7	3.679×10^7	4.308×10^7	3.977×10^7	4.464×10^7
t[s]	0.005	0.002	0.012	0.011	0.013

Table 2. Nonlinear Dynamics

	UKF $\kappa = 3 - n_x$	KMM $\kappa \in \{0 : 1 : 4\}$	ML $\kappa \in \{0 : 1 : 4\}$	MMPE $\kappa \in \{0 : 1 : 4\}$	MEPE $\kappa \in \{0 : 1 : 4\}$
MSE	0.134	0.078	0.087	0.796	0.713
VAR	0.080	0.009	0.005	3.982	2.347
t[s]	0.003	0.001	0.006	0.006	0.006

where $\Delta T = 0.01$, $k = 0, 1, \ldots, N$, $N = 150$, $Q_k = 0.5$ and $R_k = 0.09$, $\forall k$. The initial state for the dynamic system and also for the filters was $p(x_0) = \mathcal{N}(2.3; 0.01)$. The evaluation was executed on filtering data produced by 500 Monte Carlo runs for each set of κ values. For the filters, it was assumed complete knowledge of the system's model, including of the process and measurement noises. The resulting data is shown in Table 2.

4.3   Analysis

Considering the MSE criterion, KMM outperformed the other filters. Considering the VAR criterion, KMM was the second best in the nonlinear dynamics case, but was the best in the vertical fall case. To obtain such performance, it was necessary to use both strategies described in section 3 to set the target value for the predicted covariance of the current measurement (P_{yk}). For each of the numerical examples, fig 1 displays the evolution of the measurements's covariance (using all measurements available), as a result of a gradual decrease of the covariance (intensity) of the measurement noise. In the nonlinear dynamics case, the almost straight line shows that the measurement's covariance is practically unaffected by the increase in the noise reduction factor, prompting the adoption of the sample measurements' covariance as the target value for P_{yk}. In the vertical fall case, the measurements' covariance decreases rapidly with the increase in the noise reduction factor, justifying the use of the measurement noise covariance as the target value for P_{yk}.

Another important aspect to analyse is the resemblance between KMM and MEPE criteria. Though the first terms of both equations (13) and (9) are the same and the second terms are similar, criteria KMM and MEPE apply the moment matching idea to different extents. For criterion MEPE, the second term seeks to minimize the covariance of the measurement prediction error. This can be seen as maximizing the confidence the UT has on the predicted measurement it outputs. Thus, MEPE limits the moment matching process to the first moment of the current measurement. For criterion KMM, the second

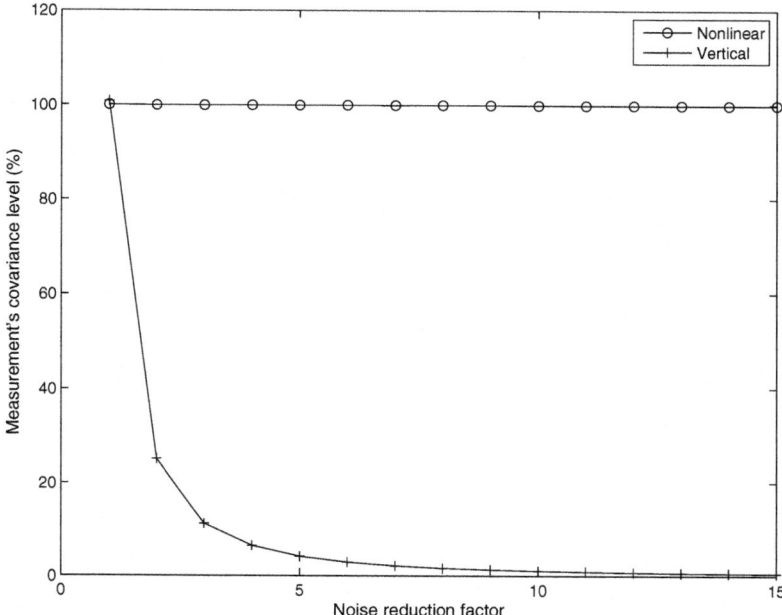

Fig. 1. Sensitivity of the measurement's covariance

term seeks to maximize the similarity between the covariance of the received measurement and the covariance of the predicted measurement, performing a more complete moment matching process than MEPE does. The results obtained in our numerical tests seem to indicate that this difference is meaningful.

5   Concluding Remarks

Motivated by the fact that the estimation performance of the UKF can be greatly affected by the tuning of the UT scaling parameter, a computationally efficient algorithm aimed at the online adaptation of such parameter is proposed. The new algorithm is based on the theoretically sound principle that states that the quality of the estimates produced by a Gaussian filter is dependent on the filter's ability to correctly estimate the first two moments of nonlinearly transformed random variables. In the tests performed, the new algorithm outperformed the fixed parameter UKF and three variations of a recently published adaptive unscented Kalman filter (ML, MMPE and MEPE). Further research includes investigating ways of dealing with imperfect noise knowledge and with modelling errors.

Acknowledgments. The first author is grateful to Capes and the second author is grateful to Fapesp and CNPq (grants 2010/11113-2 and 303627/2012-3).

References

1. Athans, M., Wishner, R., Bertolini, A.: Suboptimal state estimation for continuoustime nonlinear systems from discrete noisy measurements. IEEE Transactions on Automatic Control 13(5), 504–514 (1968)
2. Candy, J.V.: Bayesian Signal Processing: Classical, Modern and Particle Filtering Methods. Wiley-Interscience, New York (2009)
3. Dunik, J., Simandl, M., Straka, O.: Adaptive choice of scaling parameter in derivative-free local filters. In: 2010 13th Conference on Information Fusion (FUSION), pp. 1–8 (2010)
4. Dunik, J., Simandl, M., Straka, O.: Unscented kalman filter: Aspects and adaptive setting of scaling parameter. IEEE Trans. Automat. Contr. 57(9), 2411–2416 (2012)
5. Flidr, M., Straka, O., Havlik, J., Simandl, M.: Nonlinear estimation framework: A versatile tool for state estimation. In: 2013 18th International Conference on Methods and Models in Automation and Robotics (MMAR), pp. 490–495 (2013)
6. Julier, S.J., Uhlmann, J.K.: A general method for approximating nonlinear transformations of probability distributions. Tech. rep. (1996)
7. Julier, S.J., Uhlmann, J.K., Durrant-Whyte, H.F.: A new method for the nonlinear transformation of means and covariances in filters and estimators. IEEE Transactions on Automatic Control 45(3), 477–482 (2000)
8. Julier, S.J., Uhlmann, J.K.: Consistent debiased method for converting between polar and cartesian coordinate systems. In: Proc. SPIE, vol. 3086, pp. 110–121 (1997), http://dx.doi.org/10.1117/12.277178, acquisition, Tracking, and Pointing XI
9. Julier, S.J., Uhlmann, J.K.: A new extension of the kalman filter to nonlinear systems. In: Proc. of AeroSense: The 11th Int. Symp. on Aerospace/Defense Sensing, Simulations and Controls (1997)
10. Mohamed, A.H., Schwarz, K.P.: Adaptive kalman filtering for INS/GPS. Journal of Geodesy 73(4), 193–203 (1999), http://dx.doi.org/10.1007/s001900050236
11. Sarkka, S.: Bayesian Filtering and Smoothing. Cambridge University Press (2013)
12. Straka, O., Dunik, J., Simandl, M.: Gaussian sum unscented kalman filter with adaptive scaling parameters. In: 2011 Proceedings of the 14th International Conference on Information Fusion (FUSION), pp. 1–8 (2011)
13. Wu, Y., Hu, D., Wu, M., Hu, X.: A numerical-integration perspective on gaussian filters. IEEE Transactions on Signal Processing 54(8), 2910–2921 (2006)

Part IX
Modelling

DC Motors Modeling Resorting to a Simple Setup and Estimation Procedure

José Gonçalves[1,3], José Lima[1,3], and Paulo Gomes Costa[2,3]

[1] Polytechnic Institute of Bragança, Department of Electrical Engineering,
Bragança, Portugal
{goncalves,jllima}@ipb.pt
[2] Faculty of Engineering of the University of Porto,
Department of Electrical Engineering and Computers, Porto, Portugal
paco@fe.up.pt
[3] INESC-TEC (formerly INESC Porto), Portugal

Abstract. This paper describes a procedure applied to model DC motors. An example of the procedure apply is shown for a 12V brushed DC motor, equipped with a 29:1 metal gearbox and an integrated quadrature encoder. It is described the developed setup applied to obtain the experimental data and the developed algorithm applied to estimate the actuator parameters. It was obtained an electro-mechanical dynamical model that describes the motor, its gear box and the encoder. The motivation to develop a simple and easy to assemble procedure that allows to model DC motors is due to the fact that these actuators are intensively used in mobile robotics, being realistic simulation, based in accurate sensor and actuator models, the key to speed up Robot Software developing time.

Keywords: Actuators, Sensors, Modeling, Simulation.

1 Introduction

This paper describes a procedure applied to model DC motors. An example of the procedure apply is shown for a 12V brushed DC motor, equipped with a 29:1 metal gearbox and with an integrated quadrature encoder. A detailed description of the referred DC Motor can be found in [1].

It is described the developed setup applied to obtain the experimental data and the developed algorithm applied to estimate the actuator parameters. It was obtained an electro-mechanical dynamical model that describes the motor, its gear box and the encoder. The goal of the actuator model parameters estimation is to provide more models that can be used in SimTwo, which is illustrated in Figure 1, being a realistic simulation software that can support several types of robots. Its main purpose is the simulation of mobile robots that can have wheels or legs, although industrial robots, conveyor belts and lighter-than-air vehicles can also be defined. Basically any type of terrestrial robot definable with rotative joints and/or wheels can be simulated in this software [2] [7] [8].

© Springer International Publishing Switzerland 2015
A.P. Moreira et al. (eds.), *CONTROLO'2014 - Proc. of the 11th Port. Conf. on Autom. Control*,
Lecture Notes in Electrical Engineering 321, DOI: 10.1007/978-3-319-10380-8_42

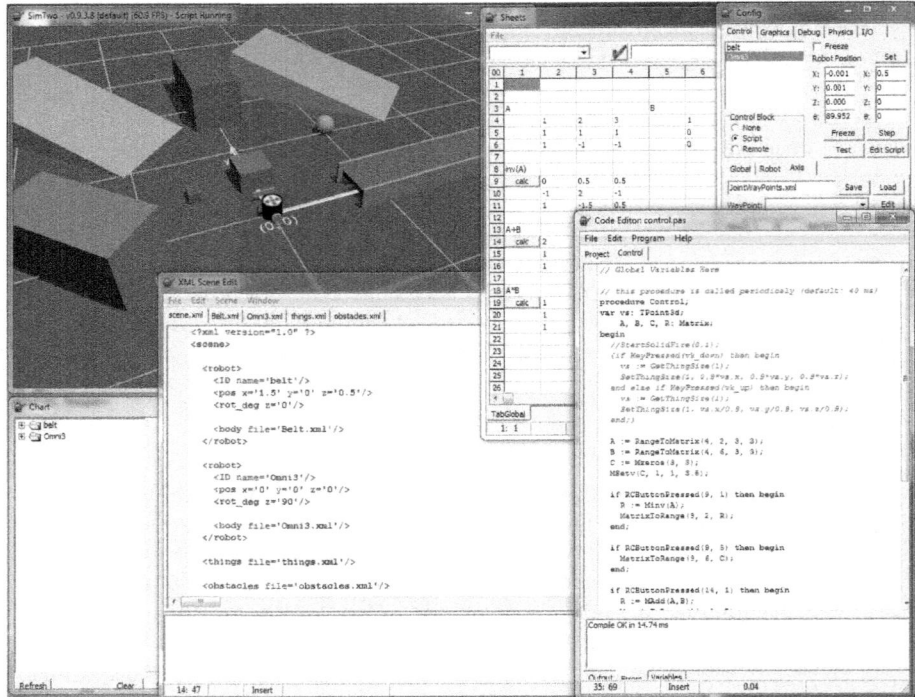

Fig. 1. SimTwo 3d View

The motivation to develop a simple and easy to assemble procedure that allows to model and simulate DC Motors is due to the fact that these actuators are are intensively used in mobile robotics and because realistic simulation, based on accurate sensor and actuator models is the key to speed up Robot Software developing time [6].

The paper is organized as follows: After a brief introduction it is described the developed setup applied to obtain the experimental data and the actuator parameters estimation. Finally some conclusions and future work are presented.

2   DC Motor Experimental Setup

DC motors are actuators worldwide popular in the mobile robotics domain [5]. The fact that a DC motor is equipped with encoders is an important feature because it provides important data to obtain the closed loop velocity control and to obtain relative measurements based on the odometry calculation [3]. The chosen DC Motor is a 12V brushed DC motor, equipped with a 29:1 metal gearbox and an integrated quadrature encoder, being shown in Figure 2.

In order to obtain experimental data the setup, shown in Figures 3 and 4, was implemented. The experimental setup is based on an Arduino Uno programmed in C, a PC software application developed in Free Pascal (Remote),

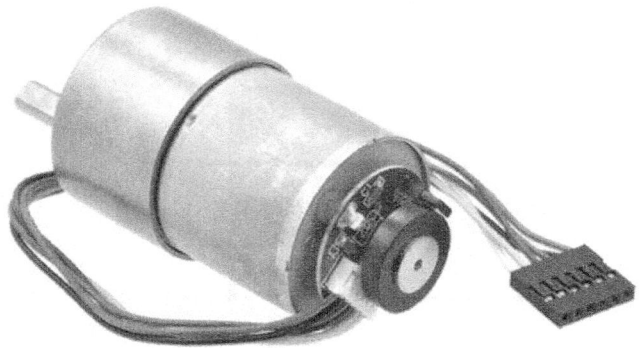

Fig. 2. 12 V DC Geared Motor [1]

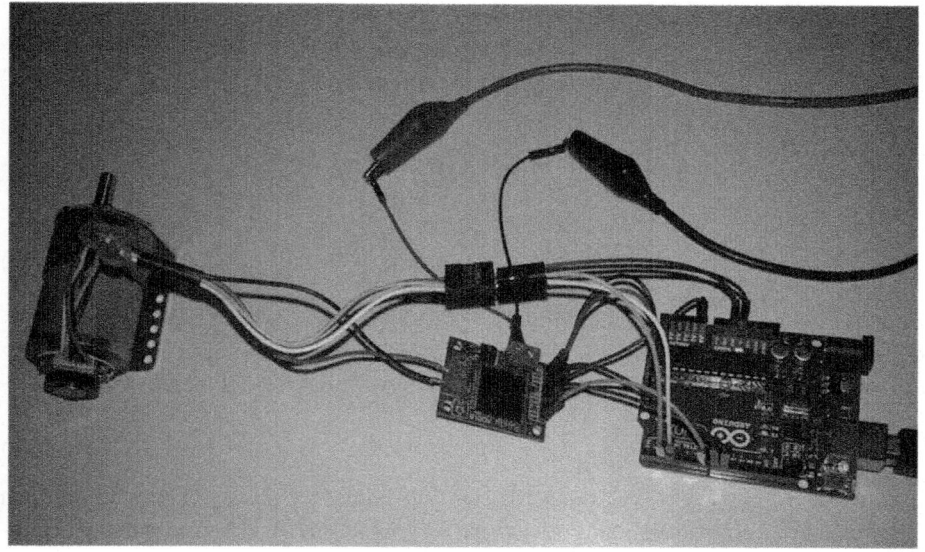

Fig. 3. Experimental setup

the VNH3SP30 Drive, a DC Power source and a DC motor without Load. The registered data is the load angular velocity and the input voltage. Two tests were performed, the first was to obtain the step response for a 12 Volt input (transitory response data) and the second test was the steady state response for several input voltages (steady state data).

3   DC Motor Parameters Estimation

The DC Motor model can be defined by the following equations, where U_a is the converter output, R_a is the equivalent resistor, L_a is the equivalent inductance and e is the back emf (electromotive force) voltage as expressed by equation (1).

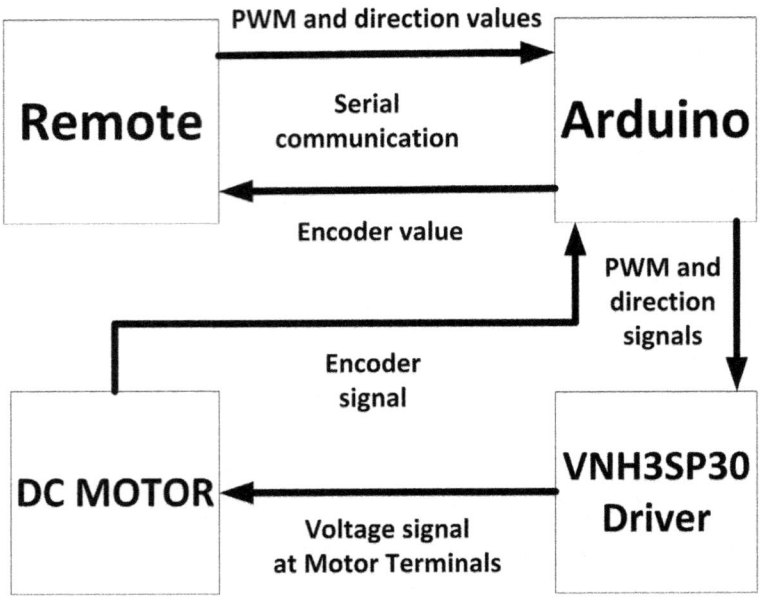

Fig. 4. Experimental setup block diagram

$$U_a = e + R_a i_a + L_a \frac{\partial i_a}{\partial t} \tag{1}$$

The motor can provide a torque T_L that will be applied to the load, being the developed torque (T_d) subtracted by the friction torque, wich is the sum of the static friction (T_c) and viscous friction $(B\omega)$, as shown in equation 2.

$$T_L = T_d - T_c - B\omega \tag{2}$$

Current i_a can be correlated with the developed torque T_d through equation (3), the back emf voltage can be correlated with angular velocity through equation (4) and the load torque T_L can be correlated with the moment of inertia and the angular acceleration through equation 5 [4].

$$T_d = K_s i_a \tag{3}$$

$$e = K_s \omega \tag{4}$$

$$T_L = J\dot{\omega} \tag{5}$$

Resorting to equation 2, equation 3 and equation 5, equation 6 was obtained.

$$\dot{\omega} = \frac{K_s i_a - T_c - B\omega}{J} \tag{6}$$

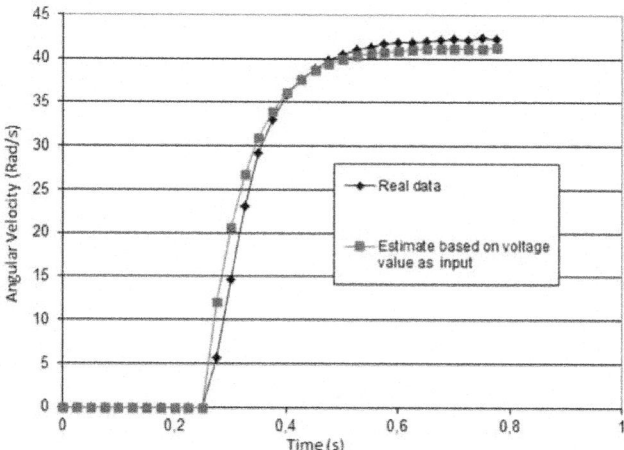

Fig. 5. Motor transitory response data

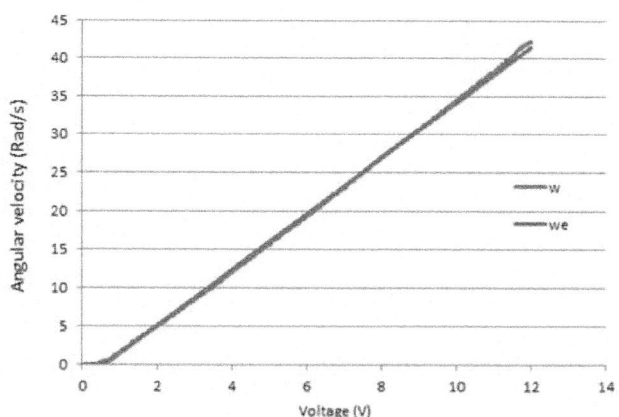

Fig. 6. Motor steady state response data

Table 1. DC Motor estimated parameters

Parameters	Value
K_s	2.64E-1
L_a	3.4E-3
R_a	3.5486
B	7.06E-4
T_c	4.6604E-2
J	1.47E-3

After discretizing equation 6, applying the Euler method, equation 7 is obtained, where ΔT is the sampling time (25 ms).

$$\omega[k] = \omega[k-1] + \Delta T \frac{K_s i_a[k-1] - T_c - B\omega[k-1]}{J} \tag{7}$$

By minimizing the sum of the absolute error between the estimated (equation 7) and the real transitory response data (assuming initial know values for T_c and K_s), parameters B and J were estimated. Then using equations 1, 2, 3, 4 and 5 and assuming that voltage drop due to L_i is negligible, equation 8 is obtained.

$$J\dot{\omega} = \frac{K_s}{R_a}(U_a - K_s\omega) - B\omega - T_c \tag{8}$$

Solving the first order differential equation, equation 9 is obtained:

$$\omega(t) = \frac{a}{b}(1 - e^{-bt}) \tag{9}$$

where:

$$a = \frac{K_s U_a - R_a T_c}{R_a J} \tag{10}$$

$$b = \frac{K_s^2 + R_a B}{R_a J} \tag{11}$$

In steady state $\omega = \frac{a}{b}$, resulting in equation 12.

$$\omega = \frac{K_s}{K_s^2 + R_a B}U_a - \frac{R_a T_c}{K_s^2 + R_a B} \tag{12}$$

By minimizing the absolute error between estimated and the steady state data, assuming an initial value for R_a, parameters K_s and T_c are estimated. Finally resorting to equation 9, by minimizing the absolute error between the estimated data and the transitory response data, R_a is estimated. The described optimization process must be repeated until the estimated parameters converge to their true values. Parameters such as T_c, R_a and K_s that are initially assumed as known are replaced by the estimated ones, every time the estimate process is repeated. The estimated and the real transitory and steady state responses are shown in Figures 5 and 6 respectively.

The estimated parameters, in SI Units, are shown in Table 1, where the equivalent inductance values is measured.

4   Conclusions and Future Work

This paper describes a simple procedure applied to model DC Motors. As an example of the procedure apply it is shown the modeling of a 12V brushed DC motor, equipped with a 29:1 metal gearbox and with an integrated quadrature encoder. It was described the developed setup applied to obtain the experimental data and the algorithm applied to estimate the actuator parameters. It is obtained an electro-mechanical dynamical model that describes the motor, its gear box and the encoder. The motivation to develop a simple and easy to assemble procedure that allows to model DC motors is due to the fact that these actuators are intensively used in mobile robotics, being realistic simulation, based in accurate sensor and actuator models the key to speed up Robot Software developing time.

As future work the authors intend to test the presented procedure in the modeling of different DC motors, whenever a new robot prototype is assembled.

Acknowledgements. The present work is funded by Best Case - Cooperation Project "NORTE-07-0124-FEDER-000060", which is financed by the North Portugal Regional Operational Programme (ON.2 O Novo Norte), under the National Strategic Reference Framework (NSRF), through the European Regional Development Fund (ERDF), and by national funds, through the Portuguese funding agency, Foundation for Science and Technology (FCT).

References

1. DC Motor (2014), http://www.pololu.com/product/1443
2. Costa, P., Gonçalves, J., Lima, J., Malheiros, P.: Simtwo realistic simulator: A tool for the development and validation of robot software. International Journal of Theory and Applications of Mathematics Computer Science (2011)
3. Borenstein, J., Everett, H., Feng, J.: 'Where am I?' Sensors and Methods for Mobile Robot Positioning. Technical Report, The University of Michigan (1996)
4. Bishop, R.: The Mechatronics Handbook. CRC Press, New York (2002)
5. Conceição, A., Moreira, A., Costa, P.: Dynamic Parameters Identification of an Omni-directional Mobile Robot. In: Proceedings of the International Conference on Informatics in Control, Automation and Robotics (2006)
6. Browning, B., Tryzelaar, E.: UberSim: A Realistic Simulation Engine for RobotSoccer. In: Proceedings of Autonomous Agents and Multi-Agent Systems (2003)
7. Michel, O.: Webots^{TM}: Professional Mobile Robot Simulation. International Journal of Advanced Robotic Systems 1(1) (2004) ISSN 1729-8806
8. Michel, O.: Khepera Simulator version 2.0, User Manual (1996), http://www2.deec.uc.pt/pm/RM/manual.pdf

Fractional Calculus in Economic Growth Modelling: The Spanish Case

Inés Tejado[1,2], Duarte Valério[1], and Nuno Valério[3]

[1] IDMEC, Instituto Superior Técnico, Universidade de Lisboa, Portugal
{ines.tejado,duarte.valerio}@tecnico.ulisboa.pt
[2] Industrial Engineering School, University of Extremadura, Spain
itejbal@unex.es
[3] Instituto Superior de Economia e Gestão, Universidade de Lisboa, Portugal
valerio@iseg.ulisboa.pt

Abstract. A variety of fractional order models have been proposed in the literature to account for the behaviour of financial processes from different points of view. The objective of this work is to model the growth of national economies, namely, their gross domestic products (GDPs), by means of a fractional order approach. The particular case of Spain is addressed, and results show that fractional models have a better performance than the other alternatives considered and proposed in the literature.

Keywords: Fractional Calculus, Dynamic models, Gross Domestic Product, Economic Growth.

1 Introduction

Fractional derivatives, which are a generalisation of usual, integer-order derivatives, are non-local, and thus suitable for constructing dynamic models for long series where a memory effect can be found — more so than models using integer derivatives and integrals alone [25]. This is the reason why fractional differential equations possess large advantage in describing economic phenomena over large time periods.

A variety of fractional order models have been proposed in the literature to account for the behaviour of financial processes from different points of view. For example, as diffusion or stochastic processes by means of Lévy models [1,4,5,6,17] or continuous time random walks for the movement of log-prices (e.g. [10,16,18,19,22,23]), respectively. A modified fractality concept was applied in [12] to describe the stochastic dynamics of the stock and currency markets. Likewise, a macroeconomic state space model was proposed in [24] for national economies consisting of a group of fractional differential equations. A similar form was used in [27] but with variable orders.

The objective of this work is to model the growth of national economies, namely, their gross domestic products (GDPs), using a dynamic model of fractional order. In the literature many models for the evolution of the GDP have been published, among which the classical papers [3,21]. Yet, to the best of our knowledge, no fractional model of GDP as a function of a vector of inputs has yet been found.

In this paper, the GDP of a national economy was modelled as function of a vector with nine variables. The particular case of the economy of Spain along the last five decades is studied. The paper is organised as follows. Section 2 introduces fractional derivatives for reference purposes. Section 3 briefly describes the considered model to account for the behaviour of national economies. In Section 4, the obtained results after fitting are given. Finally, Section 5 draws the concluding remarks and future works.

2   Fractional Calculus

Let us define differential operator D as ${}_cD_t^n f(t) = \frac{\mathrm{d}^n f(t)}{\mathrm{d}t^n}$ and ${}_cD_t^{-n}f(t) = \underbrace{\int_c^t \cdots \int_c^t f(\tau)\,\mathrm{d}\tau \cdots \mathrm{d}\tau}_{|n| \text{ integrations}}$. It can be shown by mathematical induction that

$$
{}_cD_t^n f(t) = \lim_{h \to 0} \frac{\sum_{k=0}^{n}(-1)^k \binom{n}{k} f(t - kh)}{h^n}, \quad n \in \mathbb{N} \tag{1}
$$

where combinations of a things, b at a time are given by $\binom{a}{b} = \frac{a!}{b!(a-b)!}$. This can be generalised using the Gamma function, which verifies $\Gamma(n) = (n-1)!, n \in \mathbb{N}$ and is defined in $\mathbb{C}\backslash\mathbb{Z}^-$, as

$$
\binom{a}{b} = \begin{cases} \dfrac{\Gamma(a+1)}{\Gamma(b+1)\Gamma(a-b+1)}, & \text{if } a, b, a - b \notin \mathbb{Z}^- \\[2mm] \dfrac{(-1)^b\Gamma(b-a)}{\Gamma(b+1)\Gamma(-a)}, & \text{if } a \in \mathbb{Z}^- \wedge b \in \mathbb{Z}_0^+ \\[2mm] 0, & \text{if } [(b \in \mathbb{Z}^- \vee b - a \in \mathbb{N}) \wedge a \notin \mathbb{Z}^-] \vee (a, b \in \mathbb{Z}^- \wedge |a| > |b|) \end{cases} \tag{2}
$$

Using (2), it is reasonable to generalise (1) for non-integer orders as

$$
{}_cD_t^\alpha f(t) = \lim_{h \to 0^+} \frac{\sum_{k=0}^{\lfloor \frac{t-c}{h} \rfloor}(-1)^k \binom{\alpha}{k} f(t - kh)}{h^\alpha} \tag{3}
$$

Values c and t are called terminals. The upper limit of the summation in (3) is diverging to $+\infty$. When $\alpha \in \mathbb{N}$, all terms with $k > \alpha$ will be zero; thus (3) reduces to (1) when $h > 0$. This is the only case in which the summation has a finite number of terms and the result does not depend on terminal c.

The upper limit $\lfloor \frac{t-c}{h} \rfloor$ was set so that, if $\alpha = -1, -2, -3, \ldots$, (3) becomes a Riemann integral (calculated from c to t).

For more details on operator D, properties, alternative definitions, and Laplace transforms, see [25,26].

3 Economic Growth Model

Consider a simple model of a national economy in the following form:

$$y(t) = f(x_1, x_2 \ldots) \tag{4}$$

where the output model y is the GDP (in 2012 euros) and the x_k are the variables on which the output depends. The inputs considered and their rationale are the following:

- natural resources are represented by x_1 (land area, km^2), and their quality by x_2 (arable land, km^2);
- human resources are represented by x_3 (population), and their quality by x_4 (average years of school attendance);
- manufactured resources are represented by x_5 (gross capital formation (GCF), in 2012 euros);
- external impacts in the economy are represented by x_6 (exports of goods and services, in 2012 euros);
- internal impacts in the economy are represented as follows: budgetary impacts by x_7 (general government final consumption expenditure (GGFCE), in 2012 euros), monetary impacts by x_8 (money and quasi money (M2), in 2012 euros), and investment by x_5. Rather than having x_5 play two roles, we will rather use a ninth variable $x_9 \equiv x_5$ to represent the impact of investment in the economy.

This choice of variables joins those traditionally considered in growth accounting [8,13,14] to those acknowledged by Keynesian models having short-term inputs related to impacts in the economy.

Thus, the following integer and fractional order models were considered:

$$y(t) = C_1 x_1(t) + C_2 x_2(t) + C_3 x_3(t) + C_4 x_4(t) + C_5 \int_{t_0}^{t} x_5(t)\mathrm{d}t +$$

$$+ C_6 x_6(t) + C_7 x_7(t) + C_8 \frac{\mathrm{d}x_8(t)}{\mathrm{d}t} + C_9 \frac{\mathrm{d}x_9(t)}{\mathrm{d}t}, \tag{5}$$

$$y(t) = \sum_{k=1}^{9} C_k D^{\alpha_k} x_k(t), \tag{6}$$

where C_k and α_k are constant weights and the differentiation orders for each of the variables, respectively, and t_0 is the first year considered. Notice that in the integer model the accumulated gross capital formation ($\int_{t_0}^{t} x_5(t)\mathrm{d}t$) is used as a measure of manufactured resources; the variation of M2 ($\frac{\mathrm{d}x_8(t)}{\mathrm{d}t}$) is used as a measure of the monetary impacts in the economy; and the variation of the gross capital formation ($\frac{dx_9(t)}{dt} = \frac{dx_5(t)}{dt}$) is used as a measure of the impact of investment in the economy.

4   Results for the Spanish Economy

Using the models above, the economy of Spain was modelled in the period between 1960 to 2012. This period was considered not only because it is the one for which reliable data can be easily obtained, but also because this is the period where modern economic growth consistently took hold of the Spain economy. (See data in the Appendix.) The goal of the fitting is to calculate parameters α_k and C_k of the dynamic models (5) and (6) for this particular national economy. The fitting procedure was implemented in MATLAB, using Nelder-Mead's simplex search method as implemented in function *fminsearch*, by minimising the mean square error (MSE). To evaluate the goodness-of-fit of the obtained models, the following performance indices were also calculated: (i) mean absolute deviation (MAD); (ii) coefficient of determination (R^2); (iii) $t-$ and $p-$values for each variable.

The obtained models are shown in Fig. 1, with the values of the orders α and the coefficients C given in Table 1. It can be seen that the differentiation orders obtained for x_1, x_2, x_3, x_6 and x_7 of fractional order model (6) are zero (or almost zero), which leads us to consider a simpler model, in which only variables x_4, x_5, x_8 and x_9 are assumed to have fractional order influence, as follows:

$$y(t) = \sum_{k=1,2,3,6,7} C_k x_k(t) + \sum_{k=4,5,8,9} C_k D^{\alpha_k} x_k(t). \tag{7}$$

What this means is that not all economic indicators have the same influence over time on the GDP: for some (those with $\alpha = 0$) only the current value matters. The results related to this model were also included in Fig. 1 and Table 1.

The performance indices calculated for models (5), (6) and (7) are summarized in Table 2, where $t-$values corresponding to variables which are necessary for the model, assuming a 5% significance level, are in bold. As observed, the population (x_3) and the variation of GCF (x_9) have a considerable effect on the integer model, whereas the remaining variables have low influence. In contrast, for the fractional model, it is clear that the arable land (x_2) and the GGFCE (x_7) are variables without much influence in the GDP. Likewise, the fractional model has a clearly better performance, at the expense of needing more variables (7 against 2 for the integer model).

Taking into account the low influence of variables x_2 and x_7 in the model, let us consider a simpler model with only 7 inputs, with an integer form given by

$$y(t) = C_1 x_1(t) + C_3 x_3(t) + C_4 x_4(t) + C_5 \int_{t_0}^{t} x_5(t) \mathrm{d}t +$$
$$+ C_6 x_6(t) + C_8 \frac{\mathrm{d}x_8(t)}{\mathrm{d}t} + C_9 \frac{\mathrm{d}x_9(t)}{\mathrm{d}t}, \tag{8}$$

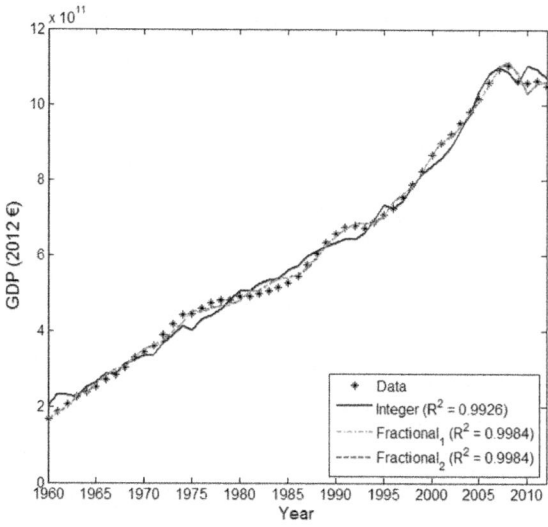

Fig. 1. Fitting results for the Spanish case (models with 9 variables): the integer model is given by (5), the fractional₁ model by (6), and the fractional₂ model by (7)

Table 1. Fitting results: orders of the fractional operator and coefficients

	α_1	α_2	α_3	α_4	α_5	α_6	α_7	α_8	α_9
Integer (5)	0	0	0	0	-1	0	0	1	1
Fractional (6)	0	0	0	0.068	0.860	0	1.250×10^{-4}	-1.020	-0.834
Fractional (7)	0	0	0	0.066	0.855	0	0	-1.016	-0.822
Integer (8)	0	—	0	0	-1	0	—	1	1
Fractional (9)	0	—	0	0.198	-0.809	0	—	-0.995	0.988

	C_1 ($\times10^5$)	C_2 ($\times10^6$)	C_3 ($\times10^4$)	C_4 ($\times10^{10}$)	C_5 ($\times10^{-2}$)	C_6 ($\times10^{-2}$)	C_7 ($\times10^{-1}$)	C_8 ($\times10^{-2}$)	C_9 ($\times10^{-1}$)
Integer (5)	-3.209	-3.137	2.365	2.974	1.385	-4.541	8.841	14.711	10.315
Fractional (6)	8.760	1.616	-1.767	-3.431	645.925	46.860	-5.935	-2.310	4.788
Fractional (7)	8.789	1.579	-1.735	-3.742	647.620	49.818	-3.536	-2.393	4.917
Integer (8)	-12.09	—	1.841	5.111	4.067	13.23	—	204.30	7.070
Fractional (9)	10.31	—	-1.153	-3.991	41.643	38.34	—	-20.98	829.228

and a fractional form as

$$y(t) = \sum_{k=\ 1,3,6} C_k x_k(t) + \sum_{k=4,5,8,9} C_k D^{\alpha_k} x_k(t). \qquad (9)$$

The obtained models consisting of 7 variables are shown in Fig. 2. The values of the orders α and the coefficients C are also given in Table 1, and the performance indices in Table 2.

Table 2. Performance indices

Index / Statistic	Variable	Models with 9 variables			Models with 7 variables	
		Integer (5)	Fractional (6)	Fractional (7)	Integer (8)	Fractional (9)
MSE ($\times 10^{20}$)		5.610	1.228	1.241	6.084	1.320
R^2		0.9926	0.9984	0.9984	0.9920	0.9983
MAD ($\times 10^{10}$)		2.033	0.912	0.920	2.0820	0.9257
t−values	x_1	−0.425	**3.953**	**3.831**	−2.150	**5.190**
	x_2	−1.836	2.036	2.044	—	—
	x_3	**3.276**	**−4.117**	**−3.962**	2.917	**−3.634**
	x_4	0.724	**−8.277**	**−7.355**	1.879	**−8.121**
	x_5	0.385	**11.977**	**10.731**	1.339	**17.764**
	x_6	−0.113	**4.019**	**4.008**	0.474	**3.669**
	x_7	0.719	−1.489	−0.936	—	—
	x_8	2.237	**−16.560**	**−16.236**	3.437	**−15.678**
	x_9	**2.736**	**12.264**	**9.508**	2.093	**12.359**
p−values	x_1	0.673	2.762×10^{-4}	4.013×10^{-4}	3.682×10^{-2}	4.634×10^{-6}
	x_2	7.903×10^{-2}	4.777×10^{-2}	4.695×10^{-2}	—	—
	x_3	3.993×10^{-3}	1.662×10^{-4}	2.688×10^{-4}	5.441×10^{-3}	7.001×10^{-4}
	x_4	0.480	1.622×10^{-10}	3.465×10^{-9}	6.657×10^{-2}	1.964×10^{-10}
	x_5	0.720	2×10^{-15}	7.3×10^{-14}	0.187	0
	x_6	0.912	2.252×10^{-4}	2.330×10^{-4}	0.637	6.315×10^{-4}
	x_7	0.502	0.143	0.354	—	—
	x_8	3.334×10^{-2}	0	0	1.257×10^{-3}	0
	x_9	8.941×10^{-3}	1×10^{-15}	3.090×10^{-12}	4.190×10^{-2}	0

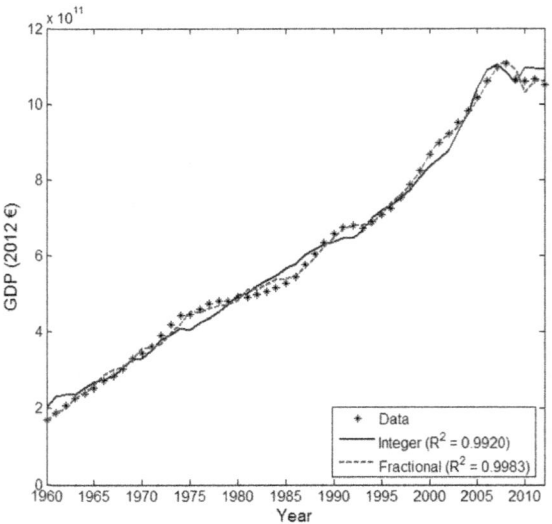

Fig. 2. Fitting results for the Spanish case (models with 7 variables): the integer model is given by (8), and the fractional model by (9)

There is, of course, a slight deterioration of performance, but the results obtained with fractional model (9) remain highly satisfactory. Again, this is achieved at the expense of the model needing more independent variables than its integer counterpart, as seen from the $t-$ and $p-$values.

Furthermore, fractional orders of x_8 and x_9 appearing in Table 1 are nearly ± 1. It is worth mentioning that the sign of α_8 is different in the integer and fractional models (8) and (9). This is particularly significant since it shows that M2 has an effect over a long time (a derivative of order almost -1 is not a local operator). On the other hand, variables x_1, x_3, x_6 and x_9 turn out to have influence in the present only. Finally, we can note some similarity between the fractional models obtained and those of fractional diffusion processes [15]. We can thus hypothesize that such diffusion models (useful in areas such as bioengineering or soil dynamics) can also explain how these variables affect the economy; this hypothesis can only be checked when more countries are studied.

5   Conclusions

This paper investigated modelling of national economic growth, namely, the gross domestic product (GDP), using models from Fractional Calculus. Nine macroeconomic indicators, chosen according to the practice established in the literature, were used to account for the behaviour of this financial process. The particular case of Spain was studied for the period 1960–2012, and results show that fractional models have a better performance than the other alternatives considered and proposed in the literature. In the end, a simplified model with only seven inputs was obtained. External and internal impacts, manufactured resources, and the quality of the natural and human resources are seen to be important factors.

Our future efforts will focus on study other economies of the European zone, and verifying our results using other criteria, such as the AIC or the BIC [11].

Appendix

Sources for the economic data in Table 3 are as follows:

- x_1 is taken from [2]. The data concerns what is currently the territory of Spain only, and not what are now Equatorial Guinea and Western Sahara, which were always separate national economies. Slight variations in area, found in the database, which are spurious, since the territory of Spain did not change in the period considered, were discarded. This input is thus constant.
- x_2 and x_3 are taken from [2].
- x_4 is taken from [7]. As the data has a 5-year sampling time (starting in 1960), a third-order spline interpolation was used for intercalary years.
- x_5, x_6 and x_7 are taken from [2], in current euros. The price index mentioned below was used to convert values to 2012 euros.

Table 3. Spanish economic data for years 1960–2012. GDP, x_5, x_6, x_7 and x_8 in 2012 euros, x_1 in km^2, x_2 in % of x_1, x_3 in people and x_4 in years

Year	GDP ($\times 10^{11}$)	x_1	x_2	x_3	x_4	GCF ($\times 10^{10}$)	x_6 ($\times 10^{10}$)	x_7 ($\times 10^{10}$)	x_8 ($\times 10^{10}$)
1960	1.69	499780	32.51	30455000	4.7	13.5	1.41	1.52	9.91
1961	1.89	499780	32.51	30739250	4.74	15.2	1.50	1.66	11.33
1962	2.07	499780	32.61	31023366	4.77	1.58	1.72	1.80	12.34
1963	2.27	499780	32.42	31296651	4.79	14.5	1.75	2.04	13.42
1964	2.39	499780	31.85	31609195	4.81	13.7	2.11	2.10	15.28
1965	2.54	499780	31.95	31954292	4.82	13.3	2.08	2.29	16.46
1966	2.73	499780	31.03	32283194	4.83	12.4	2.43	2.55	17.27
1967	2.85	499780	31.49	32682947	4.85	10.8	2.43	2.87	18.34
1968	3.03	499780	31.40	33113134	4.87	10.2	3.21	2.96	20.48
1969	3.30	499780	32.18	33441054	4.91	10.6	3.74	3.24	23.18
1970	3.45	499780	31.39	33814531	4.95	9.51	4.29	3.49	25.37
1971	3.61	499780	32.69	34191678	5.01	9.15	4.81	3.72	29.16
1972	3.90	499780	32.59	34502705	5.07	10.4	5.34	3.98	32.99
1973	4.20	499780	32.12	34817071	5.15	11.7	5.74	4.28	36.82
1974	4.44	499780	31.85	35154338	5.22	13.8	6.01	4.71	38.07
1975	4.47	499780	31.66	35530725	5.3	13.1	5.67	4.50	38.78
1976	4.61	499780	31.34	35939437	5.37	12.8	5.95	5.58	39.61
1977	4.74	499780	31.29	36370050	5.44	12.2	6.45	5.84	38.18
1978	4.81	499780	31.31	36872506	5.51	11.3	6.85	6.14	37.81
1979	4.82	499780	31.18	37201123	5.58	11.0	6.77	6.40	38.33
1980	4.92	499780	31.15	37439035	5.66	11.7	7.22	6.88	39.55
1981	4.92	499780	31.17	37740556	5.75	10.9	8.21	7.34	41.16
1982	4.98	499780	31.16	37942805	5.85	10.9	8.67	7.52	42.40
1983	5.07	499780	31.22	38122429	5.95	10.7	9.92	7.89	43.74
1984	5.16	499780	31.34	38278575	6.06	10.3	11.27	7.90	45.36
1985	5.28	499780	31.16	38418817	6.17	10.7	11.29	8.28	47.26
1986	5.45	499780	31.16	38535617	6.28	11.5	10.17	8.38	48.38
1987	5.75	499780	31.20	38630820	6.38	13.0	10.45	9.14	52.48
1988	6.04	499780	31.19	38715849	6.49	14.9	10.72	9.51	56.20
1989	6.33	499780	31.06	38791473	6.61	16.4	10.78	10.30	60.39
1990	6.57	499780	30.70	38850435	6.73	17.2	10.60	10.97	62.88
1991	6.74	499780	30.55	38939049	6.86	17.1	10.89	11.71	65.46
1992	6.80	499780	30.44	39067745	6.7	15.9	11.29	12.44	64.49
1993	6.73	499780	29.99	39189400	7.14	14.1	12.23	12.68	67.90
1994	6.89	499780	29.64	39294967	7.28	14.5	14.36	12.57	69.98
1995	7.08	499780	28.12	39387017	7.42	1.55	15.86	12.81	72.81
1996	7.25	499780	28.93	39478186	7.56	15.7	17.14	13.05	75.56
1997	7.53	499780	28.60	39582413	7.69	16.6	19.82	13.17	77.00
1998	7.87	499780	27.40	39721108	7.83	18.5	20.99	13.63	75.94
1999	8.24	499780	26.96	39926268	7.97	20.7	21.99	14.16	79.58
2000	8.66	499780	26.85	40263216	8.13	22.8	25.17	14.85	84.71
2001	8.98	499780	26.20	40720484	8.29	23.7	25.63	15.29	89.47
2002	9.22	499780	25.87	41313973	8.47	24.6	25.20	15.82	92.14
2003	9.51	499780	26.07	42004522	8.64	26.1	25.02	16.46	100.5
2004	9.82	499780	26.09	42691689	8.81	27.8	25.46	17.44	115.3
2005	10.17	499780	25.87	43398143	8.97	30.0	26.10	18.27	143.3
2006	10.58	499780	25.49	44116441	9.11	32.7	27.83	19.02	175.1
2007	10.95	499780	25.22	44878945	9.23	33.9	29.46	20.07	202.2
2008	11.05	499780	25.04	45555716	9.32	32.2	29.28	21.54	214.6
2009	10.63	499780	25.05	45908594	9.37	2.55	25.43	22.69	223.2
2010	10.60	499780	25.12	46070971	9.39	2.42	28.82	22.69	224.0
2011	10.65	499780	25.08	46174601	9.47	2.29	32.22	22.30	214.6
2012	10.49	499780	25.04	46217961	9.56	2.06	33.80	21.14	199.5

- x_8 is taken from [9] in current euros in the 1999–2012 period. In the 1962–1968 period, it is taken from [2] also in current euros. These two series are clearly coherent. [20] has data for 1941–1970 in current pesetas; values for 1962–1970 are consistently 60% of those in [2]: and so for 1960–1961 we used the values of [20] converted to euros and divided by 0.6. The price index mentioned below was used to convert values to 2012 euros.
- The price index mentioned several times above is the one implicit in [2], that for several variables provides values in current euros and in constant euros.

Acknowledgments. Inés Tejado would like to thank the Portuguese Fundação para a Ciência e a Tecnologia (FCT) for the grant with reference SFRH/BPD/81106/2011. This work was partially supported by Fundação para a Ciência e a Tecnologia, through IDMEC under LAETA, and under the joint Portuguese–Slovakian project SK-PT-0025-12.

References

1. Baeumer, B., Meerschaert, M.M.: Fractional diffusion with two time scales. Physica A: Statistical Mechanics and its Applications 373, 237–251 (2007)
2. World Bank. World development indicators (September 2013), http://data.worldbank.org/
3. Barro, R.J.: Economic growth in a cross section of countries. The Quarterly Journal of Economics 106(2), 407–443 (1991)
4. Blackledge, J.: Application of the fractal market hypothesis for modelling macroeconomic time series. ISAST Transactions on Electronics and Signal Processing 2(1), 89–110 (2008)
5. Blackledge, J.: Application of the fractional diffusion equation for predicting market behaviour. International Journal of Applied Mathematics 40(3), 130–158 (2010)
6. Cartea, Á., del Castillo-Negrete, D.: Fractional diffusion models of option prices in markets with jumps. Physica A: Statistical Mechanics and its Applications 374(2), 749–763 (2007)
7. de la Fuente, Á., Doménech, R.: Educational attainment in the OECD, 1960–2010. Technical report, BBVA (2012)
8. Denison, E.F.: Why Growth Rates Differ. Brooking Institutions, Washington (1967)
9. Eurostat. Statistics (2013), http://epp.eurostat.ec.europa.eu/portal/page/portal/statistics/themes
10. Gorenflo, R., Mainardi, F., Scalas, E., Raberto, M.: Fractional Calculus and Continuous-Time Finance III: the Diffusion Limit. In: Mathematical Finance Trends in Mathematics, pp. 171–180. Birkhäuser Basel (2001)
11. Konishi, S., Kitagawa, G.: Information Criteria and Statistical Modeling. Springer, Berlin (2008)
12. Laskin, N.: Fractional market dynamics. Physica A: Statistical Mechanics and its Applications 287, 482–492 (2000)
13. Lucas, R.E.: On the mechanics of economic development. Journal of Monetary Economics 22, 3–42 (1988)

14. Maddison, A.: Explaining the economic performance of nations, 1820–1989. In: Baumol, W., et al. (eds.) Convergence of Productivity, pp. 20–61. Oxford University Press, Oxford (1994)
15. Magin, R.L.: Fractional Calculus in Bioengineering. Begell House (2004)
16. Mainardi, F., Raberto, M., Gorenflo, R., Scalas, E.: Fractional calculus and continuous-time finance II: The waiting-time distribution. Physica A: Statistical Mechanics and its Applications 287, 468–481 (2000)
17. Marom, O., Momoniat, E.: A comparison of numerical solutions of fractional diffusion models in finance. Nonlinear Analysis: Real World Application 10, 3435–3442 (2009)
18. Meerschaert, M.M., Scalas, E.: Coupled continuous time random walks in finance. Physica A: Statistical Mechanics and its Applications 370, 114–118 (2006)
19. Meerschaert, M.M., Sikorskii, A.: Stochastic Models for Fractional Calculus. Studies in Mathematics, vol. 43. Walter de Gruyter & Co. (2012)
20. Argandoña, A.: La demanda de dinero en España, 1901–1970. Cuadernos de Economía 3(6), 3–49 (1975)
21. Sala-I-Martin, X.X.: I just ran two million regressions. The American Economic Review 87(2), 178–183 (1997)
22. Scalas, E.: The application of continuous-time random walks in finance and economics. Physica A: Statistical Mechanics and its Applications 362, 225–239 (2006)
23. Scalas, E., Gorenflo, R., Mainardi, F.: Fractional calculus and continuous-time finance. Physica A: Statistical Mechanics and its Applications 284(1-4), 376–384 (2000)
24. Skovranek, T., Podlubny, I., Petráš, I.: Modeling of the national economies in state-space: A fractional calculus approach. Economic Modelling 29(4), 1322–1327 (2012)
25. Valério, D., da Costa, J.S.: An Introduction to Fractional Control. IET, Stevenage (2013) ISBN 978-1-84919-545-4
26. Valério, D., da Costa, J.S.: An introduction to single-input, single-output Fractional Control. IET Control Theory & Applications 5(8), 1033–1057 (2011)
27. Xu, Y., He, Z.: Synchronization of variable-order fractional financial system via active control method. Central European Journal of Physics 11(6), 824–835 (2013)

LuGre Friction Model: Application to a Pneumatic Actuated System

João Falcão Carneiro and Fernando Gomes de Almeida

IDMEC, Faculdade de Engenharia, Universidade do Porto,
Rua Dr. Roberto Frias, s/n, 4200-465, Porto, Portugal
{jpbrfc,fga}@fe.up.pt

Abstract. Pneumatic systems offer a wide range of load capacity and velocities, present advantageous power to weight ratios and do not generate thermal or magnetic fields. However, their broader use in industry has been hindered by the difficulties found in fine motion or force control. Amongst the several phenomena that contribute to this scenario, friction in piston and rod seals is probably the most important one. Industrial pneumatic cylinders present friction forces with several undesired features for control purposes like highly nonlinear behavior, hysteresis and dwell time dependence.

The LuGre friction model presents a good trade-off between complexity and accuracy. In this work the LuGre model of an industrial pneumatic cylinder is identified and validated using experimental data. Static and dynamic parameters are estimated using a procedure described in literature and a new procedure to estimate one of the dynamic parameters is proposed. A comparison between both methods is provided.

Keywords: Pneumatic systems, Friction models, LuGre model.

1 Introduction

Pneumatic systems are an actuation solution with several advantages. They have low acquisition and maintenance costs, provide a large range of working loads and offer a high power to weight ratio. Despite these advantages, they are typically only used in simple motion tasks since complex position or tracking tasks are very hard to achieve. One of the most important causes of the complexity in pneumatic systems control is the friction force caused by piston and rod seals. Several features concur to this, as friction force is highly nonlinear, presents hysteresis and is dependent on the time the piston spent previously still. These characteristics cause several undesired behaviors in a pneumatic cylinder motion like, for instance, the well-known stick-slip phenomena [1].

There are several friction models in literature (see for example the study performed by Armstrong-Hélouvry *et al.* in [2]), and a particular choice is typically made keeping in mind a trade-off between low complexity and good friction prediction. It is well known that static friction models, like for instance the Karnopp model [3] are incapable of correctly reproducing several relevant phenomena. Examples of this

© Springer International Publishing Switzerland 2015   459
A.P. Moreira et al. (eds.), *CONTROLO'2014 - Proc. of the 11th Port. Conf. on Autom. Control*,
Lecture Notes in Electrical Engineering 321, DOI: 10.1007/978-3-319-10380-8_44

phenomena are, among others, the friction force dependence on the previously followed trajectory and on the time moving parts were previously still [4]. In order to be able to reproduce these behaviors, dynamic models are required. Amongst the several models available, the Dahl [5] and the LuGre [4] ones are perhaps the most referenced in literatures as providing a good compromise between accuracy and complexity. Notice that there are more evolved dynamic models like, for instance, the Generalized Maxwell-Slip model [6]. However, these models include a large number of parameters and states, compromising its use in control applications.

The Dahl model is not directly applicable to pneumatic cylinders since it does not include the viscous phenomenon. Regarding the LuGre model, it has already been successfully used in several pneumatic systems studies. Its parameters are typically determined either i) as part of the unknown system parameters to be estimated by a control law adaptation scheme [7] or ii) with experimental data fitting [8]. This study follows this last approach, using the experimental procedure presented in [9] to determine both static and dynamic parameters. Furthermore, this work proposes an alternative procedure to the one presented in [9] to estimate one of the dynamic parameters. This alternative method is compared with the previously mentioned one. A further contribution of this work is to provide preliminary results regarding the experimental validation of the LuGre model for pneumatic cylinders in the pre-sliding regime. This paper is organized as follows. Next section presents the experimental setup and its mechanical model. Section 3 and Section 4 are devoted to the experimental estimation of the LuGre model static and dynamic parameters, respectively. Both sections include a description of the experimental procedures followed as well as an experimental validation of the results obtained. Finally, section 5 presents the main conclusions drawn from this work.

2   Experimental Setup: Description and Mechanical Model

The experimental setup used in this work comprises an industrial actuator from ASCO-JOUCOMATIC, with a piston diameter $\phi_p = 32$ mm, rod diameter $\phi_r = 12$ mm and a stroke length $l = 400$ mm. This actuator is connected to a carriage that moves along a mechanical guide. The total moving mass is $M = 2.86$ kg. The system is driven by two servovalves (FESTO MPYE-5-1/8-HF-010-B) to control the mass flow inserted in each chamber separately, as this configuration may be advantageous in control applications [10]. Alternatively, two pressure reduction valves, (ASCO-JOUCOMATIC Sentronic 609060111) may be used, in order to easily enforce pressure profiles in each cylinder chamber.

Regarding variable measurement, pressure in each chamber is determined using two pressure transducers (Druck PTX 1400). The position of the piston is measured using two transducers: for "large" motions, a digital encoder embedded in the mechanical guide (5 μm resolution) is used. For "small" motions, an external LVDT, with 1 μm resolution and ± 30 μm stroke, is available. The system also comprises an accelerometer manufactured by FGP with a ± 50G range. Data provided by the accelerometer, along with position data x, is used to estimate velocity using a classic reduced order observer.

The mechanical model of the system is presented in equation (1). In this equation, F_p is the pneumatic force developed by the cylinder, F_{fr} is the friction force and F_{ext} are external forces, considered to be null in this work.

$$M\ddot{x} = F_p - F_{fr} - F_{ext} \tag{1}$$

The pneumatic force developed by the cylinder is dependent on the pressure in chambers A (P_A) and B (P_B) and is defined by equation (2)

$$F_p = P_A.A_A - P_B.A_B - P_{atm}.(A_A - A_B) \tag{2}$$

where P_{atm} is the atmospheric pressure and A_A and A_B are chambers A and B areas, respectively. The friction force in this work is estimated using the LuGre model [4] defined by equations (3) to (5).

$$\hat{F}_{fr} = \sigma_0.z + \sigma_1.\frac{dz}{dt} + f(\dot{x}) \tag{3}$$

$$\frac{dz}{dt} = \dot{x} - \sigma_0.\frac{|\dot{x}|}{g(\dot{x})}z \tag{4}$$

$$f(\dot{x}) = \alpha_2\dot{x} \tag{5}$$

In these equations z represents the average bristle deformation and σ_0 and σ_1 the stiffness and damping coefficients of the bristle motion. Function $f(\dot{x})$ is used to include the viscous friction parcel (cf. equation (5)) and $g(\dot{x})$ is an arbitrary positive function with strictly negative slope. The choice of $g(\dot{x})$ allows the inclusion of static characteristics like Coulomb friction and the Stribeck effect [4], [11-13] . The typical choice found in literature is presented in equation (6)

$$g(\dot{x}) = \alpha_0 + \alpha_1 e^{-(\dot{x}/v_s)^2} \tag{6}$$

In equation (6) α_0 represents Coulomb friction, α_1 represents a coefficient related to the difference between the maximum static friction and the Coulomb friction and v_s represents the Stribeck velocity. In steady state conditions ($\dot{z}, \ddot{x} = 0$) the LuGre model reduces to equation (7)

$$\hat{F}_{fr} = \left(\alpha_0 + \alpha_1 e^{-(\dot{x}/v_s)^2}\right) sgn(\dot{x}) + \alpha_2\dot{x} \tag{7}$$

The LuGre model requires therefore the estimation of a total of six parameters: four static (α_0, α_1, α_2, v_s) and two dynamic (σ_0, σ_1). In order to fit the model to experimental data, the following sections will present methodologies to estimate both static and dynamic parameters.

3   Static Parameters Experimental Estimation

The friction force was determined using experimental measured pressures and accelerations and by applying equation (1), with $F_{ext} = 0$, to each experimental sample obtained.

3.1   Experimental Procedures

It is possible to find several studies in literature providing estimation methods to determine static parameters [9], [13], [14]. Distinct ways of collecting data are used but the background methodology is essentially the same in all of them. This methodology is based on fitting equation (7) to experimental data obtained at different constant velocities. A broad range of velocities should be used in the fitting process in order to capture not only low velocity features (like the static and Coulomb friction or the Stribeck velocity) but also higher velocity features (like the viscous coefficient).

The approach followed in this work to collect experimental data comprehends two different sets of experiments: i) experiments at constant velocity and ii) experiments where the piston was forced to leave the stick condition. For constant velocity experiments, the servovalves where used in open loop control by imposing symmetrical voltages. These experiments where performed for different voltages and two distinct pressure levels (4 and 7 bar, absolute source pressure). Given the asymmetric behavior of the system friction, these conditions where repeated for both outward and inward rod movements. The friction force and velocity considered in each trial of experiment i) was the average friction and velocity of the regions where friction force is nearly constant. For each source pressure, velocity and motion direction considered, four trials were performed.

Experiments ii) were performed around the piston central position and were performed by incrementing pneumatic force up until the piston moved. The criterion used to detect the motion of the piston was a sudden drop in the pneumatic force applied. In order to change the pressure inside each chamber, the pressure reducing valves were used. Since there is some dependence of the break-away force on the pneumatic force gradient applied, different force gradients were tested, with an average pressure of 4 bar in the cylinder chambers. Furthermore, trials ii) were also performed at different mean pressures inside chambers, with a 1 bar step, from 3 to 6 bar, using a low force gradient value (0.12 Ns^{-1}) that allows us to neglect inertia forces. Each experiment ii) was repeated 6 times and the average results obtained are presented in Fig. 1.

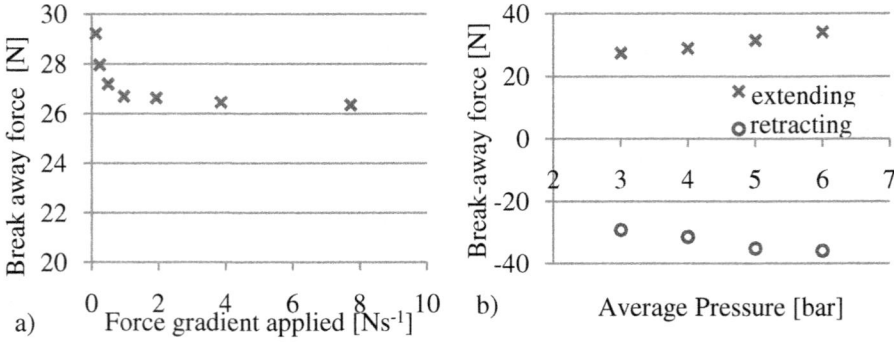

Fig. 1. Experiment ii) a) break-away force vs a) force gradient applied and b) average pressure

The static friction force retained to perform LuGre model parameters estimation was the one corresponding to an average pressure of 4 bar, with the slowest force rate. The choice of the average pressure is justified by the fact that it has been shown [10] that, for a source pressure of 7 bar, that average pressure can lead to a maximization of the available pneumatic force of the actuator. A set of six further trials under the above conditions was made. No significant dispersion was found in the results, as confirmed by the low standard deviation obtained in these six trials: $F_S^+=28{,}86$ N, $\sigma = 0{,}43$ N and $F_S^- = -31{,}50$ N, $\sigma = 0{,}34$ N.

3.2   Parameter Estimation and Model Validation

The Matlab Curve Fitting Toolbox was used to minimize the difference between the friction force estimate provided by equation (7) and the experimental data presented in the previous section.

The parameters obtained with the above mentioned procedure are resumed in Table 1, where superscripts represent either positive (+) or negative (-) velocity parameters. Fig. 2 presents a comparison between the experimental results and the ones obtained using equation (7) with the parameters presented in Table 1. A good fitting can be observed.

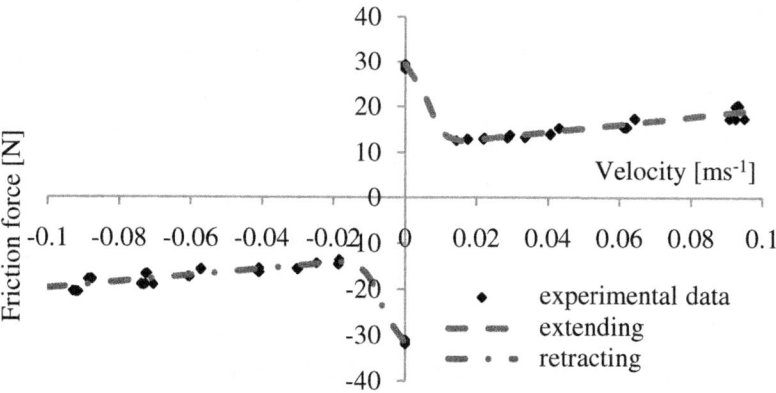

Fig. 2. Experimental values and static friction model prediction

Table 1. Estimated static parameters

Parameter	Value	Parameter	Value
$\hat{\alpha}_0^-$	12,92 N	$\hat{\alpha}_0^+$	11,05 N
$\hat{\alpha}_1^-$	18,75 N	$\hat{\alpha}_1^+$	18,24 N
$\hat{\alpha}_2^-$	69,97 Nsm^{-1}	$\hat{\alpha}_2^+$	83,29 Nsm^{-1}
$\hat{v}_s^-$	0,008 ms^{-1}	$\hat{v}_s^+$	0,0074 ms^{-1}

4   Dynamic Parameters Experimental Estimation

4.1   Experimental Procedure

As previously mentioned, there are two dynamic parameters in the LuGre model, σ_0 and σ_1. In this work the damping coefficient σ_1 was determined algebraically in order to ensure the model passivity [15]. The stiffness coefficient σ_0, on the other hand, was determined using two different procedures: one proposed by C. Canudas de Wit *et al.* in [9] and a new one presented in this paper. Experimental data for dynamic parameters fitting was determined in experiments performed in both directions, at an average pressure of 4 bar, with the pressure reducing valves imposing an increasing force, starting from zero, with a rate of $0,12$ Ns^{-1}. In order to obtain a higher resolution in motion measurement, the piston position signals were obtained with both the LVDT (for strokes up to 60 μm) and the integrated encoder (for higher strokes).

4.2   Parameter Estimation

4.2.1   Stiffness Parameter σ_0

Method 1

Canudas de Wit *et al.* [9] proposed a methodology based on the assumption that the applied force gradient is constant and sufficiently small, so that inertial and damping forces may be neglected. With this assumption, it is possible [9] to obtain equations (8) and (9) for $\dot{x} > 0$.

$$\hat{F}_{fr} = F_p = \sigma_0 z \tag{8}$$

$$\frac{dz}{dt} = \dot{x} - \frac{1}{\alpha_0 + \alpha_1} \cdot \frac{dF_p}{dt} \cdot t \cdot \dot{x} \tag{9}$$

Numerical integration of equation (9) provides the evolution of z in a pre-sliding experiment. A least squares estimate of σ_0 can then be found by [9]:

$$\hat{\sigma}_0 = \frac{\mathbf{z}^T \mathbf{F}_p}{\mathbf{z}^T \mathbf{z}} \tag{10}$$

where $\mathbf{Z}$ and $\mathbf{F}_p$ represent the vectors containing, respectively, z and F_p experimental values.

Method 2

The procedure proposed in this work is based on the observation that equation (9) can be re-written as (11):

$$\frac{dz}{dt} = \frac{dx}{dt}\left(1 - \frac{F_p(t)}{\alpha_0 + \alpha_1}\right) \tag{11}$$

As the piston position evolution in time is known, $F_p(t)$ can be written as as function of x, leading to

$$\frac{dz}{dx} = \left(1 - \frac{F_p(x)}{\alpha_0 + \alpha_1}\right) \tag{12}$$

Re-writing equation (8) in a differential form and replacing in equation (12) leads to:

$$\sigma_0 = \frac{d\hat{F}_{fr}}{dz} = \frac{dF_p}{dz} = \frac{dF_p}{dx} \cdot \frac{dx}{dz} = \frac{dF_p}{dx}\left(1 - \frac{F_p(x)}{\alpha_0 + \alpha_1}\right)^{-1} \tag{13}$$

Equation (13) can be rewritten as:

$$\frac{dF_p}{dx} = \sigma_0 - \frac{\sigma_0}{\alpha_0 + \alpha_1} F_p(x) \tag{14}$$

The solution to equation (14) is given by:

$$F_p(x) = (\alpha_0 + \alpha_1) \cdot \left(1 - e^{-\frac{\sigma_0}{\alpha_0 + \alpha_1} \cdot x}\right) \tag{15}$$

The stiffness parameter σ_0 is determined by fitting equation (15) to the data obtained experimentally in the pre-sliding domain.

Comparison between Methods 1 and 2

Using both methodologies presented above, the stiffness parameter σ_0 obtained in each of the three trials performed is shown in Fig. 4 a) and b). These figures show a clear discrepancy between the results obtained by method 1 and method 2. An also noticeable fact is the difference between extending and retracting motions, which may be attributable to the asymmetry of the cylinder.

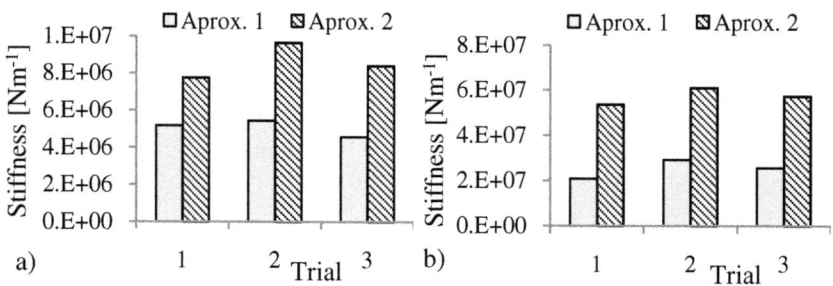

Fig. 3. Stiffness parameter σ_0 determined by method 1 and 2: results obtained when, respectively, a) extending and b) retracting the piston

Fig. 4 a) and b) represents the mean square error obtained in each method. Clearly, the method proposed in this work leads to a smaller mean square error, thus presenting an advantageous alternative to the one proposed by [9]. The average of the σ_0

parameters determined with method 2, in both extending and retracting motions ($\bar{\sigma}_0$ = 32862500 Nm^{-1}), will therefore be retained for subsequent tests of the LuGre model.

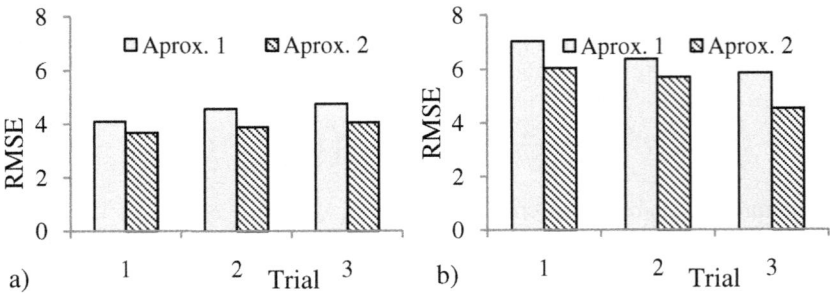

Fig. 4. Stiffness parameter σ_0 determined by method 1 and 2: mean square errors obtained when, respectively, a) extending and b) retracting the piston

4.2.2   Damping Parameter σ_1

In this study the damping parameter σ_1 was determined using the approach presented in [15]. This approach considers a damping parameter σ_1 that varies with velocity using equation (16), where σ_1', defined in equation (17), may be considered a pre-sliding damping parameter and v_c is defined in equation (18)

$$\sigma_1(\dot{x}) = \sigma_1' . e^{-(\dot{x}/v_c)^2} \tag{16}$$

$$\sigma_1' = 2.\zeta\sqrt{\sigma_0 . M} - \alpha_2 \tag{17}$$

$$v_c < 4.\sqrt{2.e}.\alpha_0/\sigma_1' \tag{18}$$

Using equation (17), the previously determined $\hat{\alpha}_2^-$, $\hat{\alpha}_2^+$ and $\bar{\sigma}_0$ parameters and considering $\zeta = 1$ [15], leads to σ_1'= 41540 Nsm^{-1} for the retracting motion and to σ_1' = 41553 Nsm^{-1} for the extending motion. As these values are quite similar, in this study the average value was considered: $\bar{\sigma}_1'$ = 41546,5 Nsm^{-1}.

The value of parameter v_c was calculated for both motion directions, using equation (18), leading to v_c^-< 0,0024 ms^{-1} and v_c^+< 0,0029 ms^{-1}. In order to ensure the passivity of the model under all circumstances, in this work the value of v_c has been chosen as v_c= 0,0024 ms^{-1}.

4.3   Model Validation in the Pre-sliding Regime

Given the importance of the friction pre-sliding regime for control applications, this section is devoted to a preliminary validation of the LuGre model in the pre-sliding regime. In order to do so, equations (1) and (3) were implemented in Matlab-Simulink and its results were then compared with the experimentally obtained. Fig. 5 presents simulation and experimental results obtained when F_p is a sinusoidal based signal

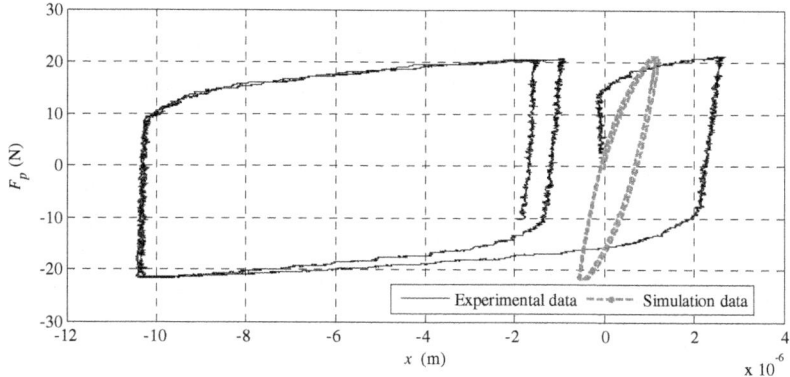

Fig. 5. Experimental and simulation results obtained in cyclic force experiment

with a frequency of 0.0075 rads^{-1} and amplitude of ca. 21 N, clearly below the static friction force. Given the low frequency of F_p, inertia forces may be neglected, so $F_{fr} \approx F_p$.

Fig. 5 experimental results show a clear difference between the piston displacement in the first cycle and the other ones. This may be due to the accommodation of the seals and guiding elements to their respective cavities so this first cycle will be faced as a transient one. Regarding the other two cycles, an elasto-plastic behavior may be observed with a small elastic recovery when unloading the applied force. Two additional notable behaviors are the symmetry of the piston displacement between positive and negative forces and the negative drift between the second and the third cycles. Regarding simulation results, it is noticeable that the experimental behavior is not correctly reproduced. This may be attributed to the inability of the LuGre model to capture the observed pre-sliding plastic behavior. Nevertheless, the elastic one is well reproduced. A small drift effect is present, although at a very smaller scale than the experimental one.

5   Conclusions

In this work the friction of a pneumatic actuated system was described using the LuGre model. Static and dynamic LuGre model parameters were experimentally estimated using a procedure available in literature. Furthermore, a novel estimation procedure for one of the dynamic parameters was proposed. This procedure led to lower fitting errors, thus being advantageous. Finally, this work also presented preliminary experimental results regarding the behavior of the LuGre model in the pre-sliding region. These results showed that the LuGre model does not capture the experimental evidence in the pre-sliding region, mainly due to its inability to describe the plastic behavior. Further studies in the pre-sliding domain will therefore be pursued.

468 J.F. Carneiro and F.G. de Almeida

Acknowledgments. This work has been supported by Portuguese State Funding through FCT – Fundação para a Ciência e a Tecnologia within project number PEst-OE/EME/LA0022/2013.

References

1. Carneiro, J.F., Almeida, F.G.: Undesired oscillations in pneumatic systems. In: Tenreiro Machado, J.A., Machado, J.A.T., Luo, A.C.J., Barbosa, R.S., Silva, M.S., Figueiredo, L.B. (eds.) Nonlinear Science and Complexity, pp. 229–243. Springer, Heidelberg (2011)
2. Armstrong-Hélouvry, B., Dupont, P., Canudas de Wit, C.: A Survey of Models, Analysis Tools and Compensation Methods for the Control of Machines with Friction. Automatica 30(7), 1083–1138 (1994)
3. Karnopp, D.: Computer Simulation of Stick-Slip Friction in Mechanical Dynamic Systems. ASME J. Dyn. Syst., Meas., Control 107(1), 100–107 (1985)
4. Canudas de Wit, C., Olsson, H., Astrom, K., Lischinsky, P.: A new model for control of systems with friction. IEEE Transactions on Automatic Control 40(3), 419–425 (1995)
5. Dahl, P.: A solid friction model. Technical Report, Aerospace Corporation (1968)
6. Al-Bender, F., Lampaert, V., Swevers, J.: The Generalized Maxwell-Slip Model: A Novel Model for Friction Simulation and Compensation. IEEE Transactions on Automatic Control 50(11), 1883–1887 (2005)
7. Schindele, D., Aschemann, H.: Adaptive Friction Compensation Based on the LuGre Model for a Pneumatic Rodless Cylinder. In: 35th Annual Conference of IEEE on Industrial Electronics, IECON 2009, Porto, Portugal, November 3-5, pp. 1432–1437 (2009)
8. Meng, D., Tao, G., Chen, J., Ban, W.: Modeling of a pneumatic system for high-accuracy position control. In: International Conference on Fluid Power and Mechatronics (FPM), August 17-20, pp. 505–510 (2011)
9. Canudas de Wit, C., Lischinsky, P.: Adaptive friction compensation with partially known dynamic friction model. International Journal of Adaptive Control and Signal Processing 11(1), 65–80 (1997)
10. Carneiro, J.F., Almeida, F.G.: Using two servovalves to improve pneumatic force control in industrial cylinders. Int. J. of Adv. Manuf. Technol. 66(1-4), 283–301 (2013)
11. Åström, K.J.: Control of Systems with Friction. In: Proceedings of the 4th International Conference on Motion and Vibration Control, pp. 25–32 (1998)
12. Olsson, H., Åström, K.J., Canudas de Wit, C., Gafvert, M., Lischinsky, P.: Friction Models and Friction Compensation. European Journal of Control 4(3), 176–195 (1998)
13. Abebe, W.N.: Simulated and Experimental Sliding Mode of a Hydraulic Positioning System Control. MSc thesis, University of Akron (2006)
14. Meng, D., Tao, G., Chen, J., Ban, W.: Modeling of a Pneumatic System for High-Accuracy Position Control. In: International Conference on Fluid Power and Mechatronics, pp. 505–510 (2011)
15. Åström, K.J., Canudas de Wit, C.: Revisiting the LuGre Model - Stick-Slip Motion and Rate Dependence. IEEE Control Systems Magazine 28(6), 101–114 (2008)

Part X
Robotics

Transporting Hanging Loads
Using a Scale Quad-Rotor

Cesáreo Raimúndez and José Luis Camaño

Universidad de Vigo, Spain
{cesareo,cama}@uvigo.es

Abstract. In this paper, we develop a control strategy that allows hanging load delivery, using a quad-rotor as controller device. The pendular load is hanging from a rope attached to a quad-rotor. The feasibility of this experiment is due to the great flexibility of the quad-rotor maneuverability, possessing a wide and rich dynamic range. The ultimate aim of this work is to show that the procedure can be used in scenarios such as those arising in collaborative work environments.

Keywords: scale quad-rotors, adaptive control, Neural Networks.

1 Introduction

The purpose of this paper is to show a way of integration between two dynamic models, in this case a scale quad-rotor and a hanging spherical pendulum, composing a more complex system which is a load delivery problem using a quad-rotor as control device. The integration with an inverted pendulum was also focused in [16] using a controller issued from Controlled Lagrangian's formulation. In [22] is considered an integration procedure using a Lagrangian approach in which the controller inherits the actuator dynamics resulting in elaborate formulations. The main objective of the technique developed in this publication is to facilitate formulating participatory transportation problems with several agents like in [19] and [20], but within a simpler formal framework.

The large dynamic range of the quad-rotor is amply known, being capable of aggressive maneuvers: [12], [5],[21]. Supposing that the pendulum dynamics are slow enough regarding the quad-rotor's, the adopted technique is proper as can be concluded from the simulation results. Regarding the adaptive tracking controller used in this paper we refer the reader to publications [17] and [18].

The structure of this paper is as follows: In Section 2 we present a controller that globally stabilizes the spherical hanging pendulum, using a proportional plus derivative controller. In Section 3 is presented a simplified modeling of a scale quad-rotor according to a Lagrangian paradigm. In Section 4 is shown how to ensemble both systems: the hanging pendulum and de quad-rotor linked as a whole system, according also to a Lagrangian procedure. In this section is introduced the main result of this paper: stating an equivalent dynamic system to an holonomic link which models the quad-rotor locked to the pendulum. In

© Springer International Publishing Switzerland 2015   471
A.P. Moreira et al. (eds.), *CONTROLO'2014 - Proc. of the 11th Port. Conf. on Autom. Control*,
Lecture Notes in Electrical Engineering 321, DOI: 10.1007/978-3-319-10380-8_45

Section 5 is presented the control methodology to ascertain the convergence of the equivalence holonomic relation, treated as an error tracking problem. In Section 6 an adaptive augmentation technique is presented to cope with modeling errors and finally in Section 7 the simulation results are shown, and Section 8 is for the conclusions.

2 Modeling and Stabilization of the 3D Hanging Pendulum

The standard Euler-Lagrange equations are given as

$$\frac{d}{dt}\left(\frac{\partial L}{\partial \dot{q}}\right) - \frac{\partial L}{\partial q} = \tau \tag{1}$$

where $q = (q_1, \cdots, q_n)$ are generalized configuration coordinates for the n degrees of freedom. The Lagrangian is $L = T - V$, where T is the kinetic energy and V the potential energy; and $\tau = (\tau_1, \cdots, \tau_n)$ is the vector of generalized forces. In standard mechanical systems the kinetic energy is given by $T = \frac{1}{2}\dot{q}^\top M(q)\dot{q}$, where $M(q)$ is the $n \times n$ inertia matrix, which is symmetric and definite positive for all q. Now according to figure 1, considering the platform coordinates as $P_c = (x_r, y_r, 0)$ and the hanging load coordinates as $P_b = P_c + \left(x_s, y_s, -\sqrt{l_p^2 - x_s^2 - y_s^2}\right)$ supposing a massless rope hanging the load, the kinetic and potential energy can be modeled as

$$T_p = \frac{1}{2}\left(m_c <\dot{P}_c, \dot{P}_c> + m_b <\dot{P}_b, \dot{P}_b>\right), \quad V_p = -m_b g\sqrt{l_p^2 - x_s^2 - y_s^2} \tag{2}$$

By hypothesis, the platform with coordinates P_c will evolve in an horizontal plane which in this formulation will be the $z = 0$ plane. Considering $\tau = u_p = (u_x, u_y)$, $q = (x_r, y_r, x_s, y_s)$ and developing 1 we obtain the equations of motion as:

$$M_p(q)\ddot{q} + C_p(q, \dot{q})\dot{q} + G_p(q) = u_p \tag{3}$$

with

$$M_p(q) = \begin{pmatrix} m_c + m_p & 0 & m_p & 0 \\ 0 & m_c + m_p & 0 & m_p \\ m_p & 0 & m_p\left(\frac{x_s^2}{l_p^2 - x_s^2 - y_s^2} + 1\right) & \frac{m_p x_s y_s}{l_p^2 - x_s^2 - y_s^2} \\ 0 & m_p & \frac{m_p x_s y_s}{l_p^2 - x_s^2 - y_s^2} & m_p\left(\frac{y_s^2}{l_p^2 - x_s^2 - y_s^2} + 1\right) \end{pmatrix} \tag{4}$$

and

$$
C_p(q,\dot{q}) = \begin{bmatrix} 0 & 0 & 0 & 0 \\ 0 & 0 & 0 & 0 \\ 0 & 0 & \frac{m_p x_s\left(\left(l_p^2-y_s^2\right)\dot{x}_s+x_s y_s \dot{y}_s\right)}{\left(l_p^2-x_s^2-y_s^2\right)^2} & \frac{m_p x_s\left(x_s y_s \dot{x}_s+\left(l_p^2-x_s^2\right)\dot{y}_s\right)}{\left(l_p^2-x_s^2-y_s^2\right)^2} \\ 0 & 0 & \frac{m_p y_s\left(\left(l_p^2-y_s^2\right)\dot{x}_s+x_s y_s \dot{y}_s\right)}{\left(l_p^2-x_s^2-y_s^2\right)^2} & \frac{m_p y_s\left(x_s y_s \dot{x}_s+\left(l_p^2-x_s^2\right)\dot{y}_s\right)}{\left(l_p^2-x_s^2-y_s^2\right)^2} \end{bmatrix}, \; G_p(q) = \begin{bmatrix} 0 \\ 0 \\ \frac{m_p g x_s}{\sqrt{l_p^2-x_s^2-y_s^2}} \\ \frac{m_p g y_s}{\sqrt{l_p^2-x_s^2-y_s^2}} \end{bmatrix}
$$
$$(5)$$

Now, supposing the platform restricted to a $z = C^{te}$, this system can be point stabilized using a proportional plus derivative controller with the structure

$$u_p = \sigma(k_p(q - q_o) + k_d \dot{q}) = \mathcal{K}(q, \dot{q}) \tag{6}$$

Here $k_p \succ 0$, $k_d \succ 0$ are positive definite matrixes, $q_o = (x_o, y_o, 0, 0)^\top$ is the desired rest point, and $\sigma(\cdot)$ is a suitable saturation function to avoid actuation excesses prone to linear controllers.

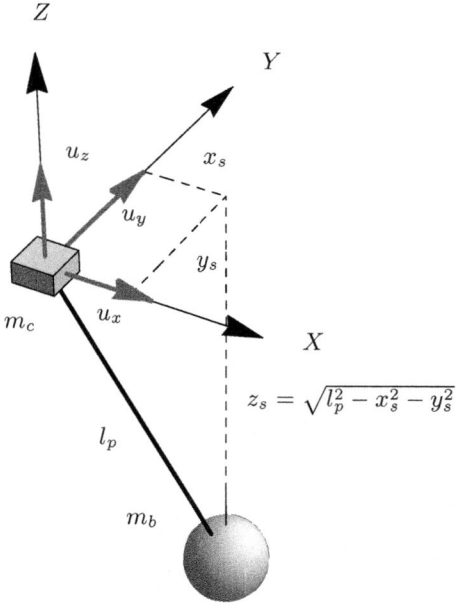

Fig. 1. Pendulum arrangement

3   Quad-Rotor: Simplified Modeling and Lagrangian Formulation

The generalized coordinates for the quad-rotor are $q = (\xi, \eta)$ where $\xi = (x, y, z)$, denote the position of the center of mass concerning the inertial frame and

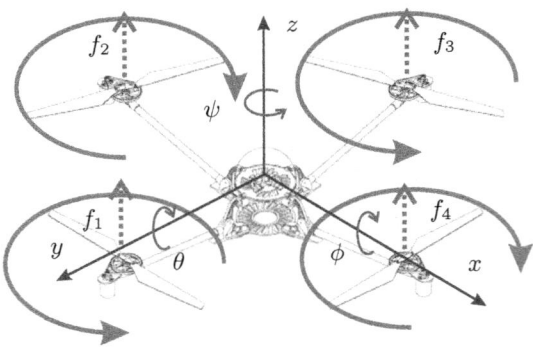

Fig. 2. quad-rotor representation

$\eta = (\psi, \theta, \phi)$ are the three Euler angles (yaw, pitch and roll) representing the quad-rotor pose. The total quad-rotor kinetic energy is given by T_q and the potential energy is given by V_q with the corresponding lagrangian $L_q = T_q - V_q$, with

$$T_q = \frac{1}{2}m\dot{\xi}^\top\dot{\xi} + \frac{1}{2}\omega_b^\top J\omega_b, \; V_q = mgz, \; \omega_b = Q(\eta)\dot{\eta} \tag{7}$$

Here m denotes the mass of the quad-rotor. Also

$$\omega_b = R(\eta)\dot{R}^\top(\eta) = Q(\eta)\dot{\eta} \Rightarrow Q(\eta) = \begin{pmatrix} -s_\theta & 0 & 1 \\ c_\theta s_\phi & c_\phi & 0 \\ c_\theta c_\phi & -s_\phi & 0 \end{pmatrix} \tag{8}$$

where $R(\eta)$ is the transformation matrix representing the quad-rotor pose.

$$J = \begin{pmatrix} J_1 & J_{12} & J_{13} \\ J_{12} & J_2 & J_{23} \\ J_{13} & J_{23} & J_3 \end{pmatrix}, \; I(\eta) = Q(\eta)^\top JQ(\eta) \tag{9}$$

Here $I(\eta)$ is the inertia matrix regarding the inertial frame. The change from an inertial to a local frame for the quad-rotor is done according to

$$R(\eta) = \begin{pmatrix} c_\theta c_\psi & s_\theta s_\psi & -s_\theta \\ c_\psi s_\theta s_\phi - s_\psi c_\phi & s_\psi s_\theta s_\phi + c_\psi c_\phi & c_\theta s_\phi \\ c_\psi s_\theta c_\phi + s_\psi s_\phi & s_\psi s_\theta c_\phi - c_\psi s_\phi & c_\theta c_\phi \end{pmatrix} \tag{10}$$

where c_θ, s_θ stand for $\cos\theta, \sin\theta$, respectively.

4   Quad-Rotor Pendulum Whole Dynamics

Calling now $L = L_p + L_q + \lambda^\top r$ with $\lambda = (\lambda_1, \lambda_2, \lambda_3)$ and $r = (x-x_r, \, y-y_r, \, z-z_r^o)$ with $z_r^o = C_o$ a suitable constant, and using the Lagrange-d'Alembert principe [2]

$$\delta \int (L + <\delta\eta, \tau> + <\delta\xi, uR.e_3>) \, dt = 0 \tag{11}$$

give the set quad-rotor-pendulum movement equations submitted to the holo-nomic constraint $r = 0$. The complete set of movement equations is:

$$\dot{\zeta} = \nu$$
$$M_p(\zeta)\dot{\nu} = -C_p(\zeta,\nu)\nu - G_p(\zeta) + E_p f_p$$
$$\mathcal{K}(\zeta,\nu) = f_p$$
$$\dot{\xi} = v$$
$$m\dot{v} = -f_0 - f_p + R(\eta)f_b$$
$$\omega_b = Q(\eta)\dot{\eta}$$
$$J\dot{\omega}_b = -\omega_b \wedge J\omega_b + \tau \qquad (12)$$
$$r(\xi,\zeta) = 0$$
$$m_p g \frac{x_s}{\sqrt{l^2 - x_s^2 - y_s^2}} = \lambda_1$$
$$m_p g \frac{y_s}{\sqrt{l^2 - x_s^2 - y_s^2}} = \lambda_2$$
$$-mg = \lambda_3$$

and

$$f_b = \begin{pmatrix} 0 \\ 0 \\ u \end{pmatrix}, \quad f_0 = \begin{pmatrix} 0 \\ 0 \\ mg \end{pmatrix}, \quad f_p = \begin{pmatrix} u_x \\ u_y \\ u_z \end{pmatrix} \qquad (13)$$

Here u represents the main thrust, u_x, u_y, u_z represent the reaction forces to the pendular load, whose fulcrum is located at the quad-rotor mass center and $\tau = (\tau_\psi, \tau_\theta, \tau_\phi)$ represent the moments regarding the local reference frame. The component u_z can be easily calculated as:

$$u_z = m_p \frac{d^2}{dt^2}\sqrt{l^2 - x_s^2 - y_s^2} \qquad (14)$$

The problem of stabilizing the pendular load via quad-rotor, is solved deter-mining controls $\{u, \tau\}$ that verify the conditions of the equations of motion. The f_p forces are the pendulum control forces according to (6) and are link contact internal forces. The external controlling inputs are u and τ. Our contribution is to replace the holonomic constraint $r = 0$ using instead an equivalent dynamic condition.

$$\ddot{r} + k_d \dot{r} + k_p r = 0 \quad (\equiv) \ r = 0, \ \{k_d \succ 0, \ k_p \succ 0\} \qquad (15)$$

The $\{u, \tau\}$ control inputs can be modeled in a first degree of approximation, without considering rotor dynamics, as:

$$u = \sum_{i=1}^{4} f_i$$
$$f_i = B_o \omega_i^2$$
$$\tau_\psi = (D_o/B_o)(f_2 + f_4 - f_1 - f_3) \qquad (16)$$
$$\tau_\theta = l(f_4 - f_2)$$
$$\tau_\phi = l(f_3 - f_1)$$

where f_i are the lifting forces in each rotor, ω_i the corresponding angular velocities, l the diagonal distance between axes of the respective rotors, and D_o, B_o
are *drag* and *thrust* factors, respectively. The relationship between $\{u, \tau\}$ and
the rotations ω_i is straightforward.

$$\omega_1^2 = \frac{1}{4B_oD_ol}(D_olu - 2D_o\tau_\phi - B_ol\tau_\psi)$$
$$\omega_2^2 = \frac{1}{4B_oD_ol}(D_olu - 2D_o\tau_\theta + B_ol\tau_\psi)$$
$$\omega_3^2 = \frac{1}{4B_oD_ol}(D_olu + 2D_o\tau_\phi - B_ol\tau_\psi)$$
$$\omega_4^2 = \frac{1}{4B_oD_ol}(D_olu + 2D_o\tau_\theta + B_ol\tau_\psi)$$

$$(17)$$

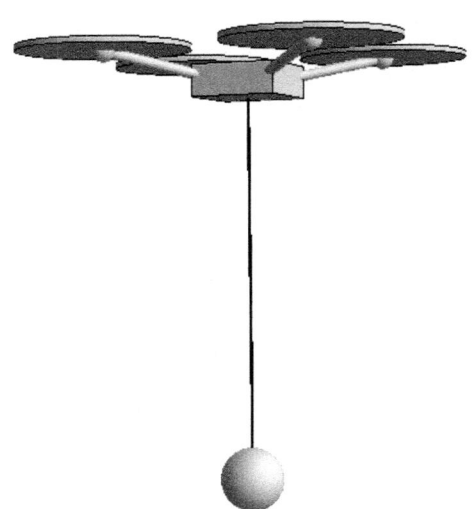

Fig. 3. Quad-Rotor-Hanging Pendulum setup

5   Error Tracking Controller

The error equivalence condition (15) is embedded in a more general tracking error condition, involving now the quad-rotor configuration space. Being
$P_r = (\xi_r, \eta_r), P = (\xi, \eta)$ the reference and actual trajectories and defining
$e = (e_\xi, e_\eta) = P_r - P$ is obvious that regarding error convergence,

$$\ddot{e} + K_d\dot{e} + K_pe = 0 \Rightarrow \quad \ddot{r} + k_d\dot{r} + k_pr = 0,$$
$$\{k_d \subset K_d \succ 0,\ kp \subset K_p \succ 0\}$$

$$(18)$$

or simply $|e| \leq c_o \Rightarrow |r| \leq c_o$ being c_o a convenient constant, because r is embedded in e. Establishing now

$$
\begin{aligned}
\ddot{e}_\xi &= \ddot{\xi}_r + m^{-1} f_0 - m^{-1} R(\eta) f_b \\
&= \nu_\xi \\
&= -k_{p\xi} e_\xi - k_{d\xi} \dot{e}_\xi \\
\ddot{e}_\eta &= \ddot{\eta}_r - I^{-1}(\eta) \left(\frac{1}{2} \frac{\partial}{\partial \eta} \left(\dot{\eta}^\top I(\eta) \dot{\eta} \right) - \dot{I}(\eta) \dot{\eta} + \tau \right) \\
&= \nu_\eta \\
&= -k_{p\eta} e_\eta - k_{d\eta} \dot{e}_\eta
\end{aligned}
\tag{19}
$$

with (ν_ξ, ν_η) the pseudo-control components and $k_{p\xi}$, $k_{p\eta}$, $k_{d\xi}$, $k_{d\eta}$ positive matrices. The expression for $\ddot{e}_\eta$ is obtained substituting (7) and (9) in (15). From

$$
\ddot{\xi}_r + m^{-1} f_0 - m^{-1} R(\eta_o) f_b = \nu_\xi
\tag{20}
$$

follows that

$$
\eta_o = \begin{cases}
\psi_o = \psi_r \\
\theta_o = \arcsin \left(\dfrac{\ddot{x}_r - \nu_{\xi_x}}{u} \right) \\
\phi_o = \arctan \left(\dfrac{\ddot{y}_r - \nu_{\xi_y}}{\ddot{z}_r + g - \nu_{\xi_z}} \right)
\end{cases}
\tag{21}
$$

$$
u = m \left\| \ddot{\xi}_r + m^{-1} f_0 - \nu_\xi \right\|
$$

Calling now

$$
\Delta \eta = \begin{pmatrix} 0 \\ \theta_r - \theta_o \\ \phi_r - \phi_o \end{pmatrix}
\tag{22}
$$

the error on the η coordinates is corrected, resulting in $\nu_\eta = -k_{p\eta}(e_\eta - \Delta\eta) - k_{d\eta} \dot{e}_\eta$ which defines the control law

$$
\begin{aligned}
\tau &= \dot{I}(\eta) \dot{\eta} - \frac{1}{2} \frac{\partial}{\partial \eta} \left(\dot{\eta}^\top I(\eta) \dot{\eta} \right) \\
&\quad + I(\eta) \left(\ddot{\eta}_r + k_{p\eta}(e_\eta - \Delta\eta) + k_{d\eta} \dot{e}_\eta \right)
\end{aligned}
\tag{23}
$$

which will stabilize the P_r trajectory tracking with a bounded error. In figure (4) the structure of the controller is shown, consisting of two proportional-derivative terms, namely PD_ξ, PD_η where S_ξ, S_η represent the operations described in equations (20) and (23) respectively. QR represents the plant (quad-rotor) and C the generator of trajectory commands.

6 Adaptive Augmentation

In order to cancel the presence of unmodeled dynamics, two corrective components are added to the control loops presented in figure (4), which are generated by the single hidden layer neural network adaptive element defined by SHL-NN

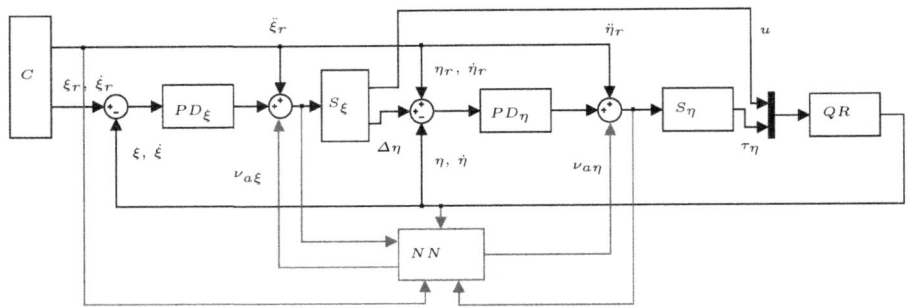

Fig. 4. Augmented Linear Controller with an Adaptive SHL-NN

$= (NN_\xi, \ NN_\eta)$. Let $\Delta = (\Delta_\xi, \Delta_\eta)$ be the vector of modeling errors. Equations (19) can be written as:

$$
\begin{aligned}
\ddot{e}_\xi &= \ddot{\xi}_r - (\ddot{\xi} + \Delta_\xi) \\
\ddot{e}_\eta &= \ddot{\eta}_r - (\ddot{\eta} + \Delta_\eta)
\end{aligned}
\tag{24}
$$

By adding to the control effort the adaptive terms $\nu_{a\xi}, \nu_{a\eta}$ the following representation of the error dynamics is obtained:

$$
\begin{aligned}
\ddot{e}_\xi + k_{p\xi} e_\xi + k_{d\xi}\dot{e}_\xi + \nu_{a\xi} - \Delta_\xi &= 0 \\
\ddot{e}_\eta + k_{p\eta} e_\eta + k_{d\eta}\dot{e}_\eta + \nu_{a\eta} - \Delta_\eta &= 0
\end{aligned}
\tag{25}
$$

which can also be written as

$$
\frac{d}{dt}\begin{pmatrix} e \\ \dot{e} \end{pmatrix} = \begin{pmatrix} O & I \\ -K_p & -K_d \end{pmatrix}\begin{pmatrix} e \\ \dot{e} \end{pmatrix} + B(\nu_a - \Delta)
\tag{26}
$$

with

$$
\begin{aligned}
K_p &= \begin{pmatrix} k_{p\xi} & O \\ O & k_{p\eta} \end{pmatrix}, \quad K_d = \begin{pmatrix} k_{d\xi} & O \\ O & k_{d\eta} \end{pmatrix} \\
B &= \begin{pmatrix} O \\ I \end{pmatrix}, \quad \nu_a = \begin{pmatrix} \nu_{a\xi} \\ \nu_{a\eta} \end{pmatrix}, \quad \Delta = \begin{pmatrix} \Delta_\xi \\ \Delta_\eta \end{pmatrix}
\end{aligned}
\tag{27}
$$

and with $e = (e_\xi, e_\eta)$. Here again, O, I are suitable null and identity matrices respectively. If the SHL-NN output signal ν_a perfectly cancels Δ, then we have asymptotically stable error dynamics. ν_a has the structure

$$
\nu_a = \left(W_\xi^\top \bar{\sigma}(V_\xi^\top \bar{\xi}), W_\eta^\top \bar{\sigma}(V_\eta^\top \bar{\eta})\right)
\tag{28}
$$

Weight propagation for $W_{\{\xi,\eta\}}, V_{\{\xi,\eta\}}$, is done according to the adaptation laws

$$
\begin{aligned}
\dot{W}_i &= -[(\bar{\sigma} - \bar{\sigma}' V_i^\top \bar{q})r^\top + \kappa \|e\| W_i]\Gamma_{W_i} \\
\dot{V}_i &= -\Gamma_{V_i}[\bar{q}(r^\top W_i^\top \bar{\sigma}') + \kappa \|e\| V_i]
\end{aligned}
\tag{29}
$$

with $r = (e^\top PB)^\top$, and $i = \{\xi, \eta\}$. The representation of $\bar{\sigma}(V_\xi^\top \bar{q})$ as $\bar{\sigma}$, as well as that of $\bar{\sigma}'$, is done for the sake of clarity. $\Gamma_{V_i} \succ 0$, $\Gamma_{W_i} \succ 0$ are definite positive matrices and $\kappa > 0$ is a real constant, being $\bar{q}$ the extended input vector, this is, $\bar{q} = (1, q)$ where q is the input vector.

6.1   Obtaining the Adaptation Laws

Let us consider the Lyapunov function

$$
\mathbf{V}(e, \tilde{V}, \tilde{W}) = \frac{1}{2}\left(e^\top P e \right.
$$
$$
\left. + \operatorname{tr}\left(\tilde{W}^\top \Gamma_W^{-1}\tilde{W}\right) + \operatorname{tr}\left(\tilde{V}^\top \Gamma_V^{-1}\tilde{V}\right)\right)
\tag{30}
$$

where P solves the equation

$$
A^\top P + PA + Q = 0, \quad A = \begin{pmatrix} O & I \\ -K_p & -K_d \end{pmatrix}
\tag{31}
$$

with $-Q$ and P definite positive. In order to obtain the adaptation equations (29) we must follow the steps required to proof that, on the error orbits, the condition $\dot{\mathbf{V}} \leq 0$ is satisfied as explained in [8]. The following steps are given in order to show the parameters regarding an adequate tuning of the controller. The details of the proof of convergence follow the above mentioned reference. Let us consider

$$
\epsilon = \nu_a^* - \Delta = W^{*\top}\bar{\sigma}(V^{*\top}\bar{q}) - \Delta
\tag{32}
$$

where W^*, V^* are the optimum values that best approximate Δ. The error dynamics is

$$
\dot{e} = Ae + B\left(W^{*\top}\bar{\sigma}(V^\top\bar{q}) - W^\top\bar{\sigma}(V^{*\top}\bar{q}) + \epsilon\right)
\tag{33}
$$

Defining now $\tilde{W} = W - W^*$, $\tilde{V} = V - V^*$ and using the Taylor series expansion of σ with respect to V in the neighborhood of V^*, which is the optimum value, we obtain

$$
\dot{e} = Ae + B\left(\tilde{W}^\top(\sigma - \sigma'V^\top\bar{q}) + W^\top\sigma'\tilde{V}^\top\bar{q} + w\right)
\tag{34}
$$

with

$$
w = \epsilon - W^{*\top}\left(\sigma^* - \sigma + \sigma'\tilde{V}^\top\bar{q}\right) + \tilde{W}^\top\sigma'V^{*\top}\bar{q}
\tag{35}
$$

Substituting now (29) and (34) in the expression of $\dot{\mathbf{V}}$ we have

$$
\dot{\mathbf{V}} = -\frac{1}{2}e^\top Q e + e^\top PBw - \kappa\|e\|\operatorname{tr}\left(\tilde{Z}^\top Z\right)
\tag{36}
$$

where

$$
Z = \begin{pmatrix} V & 0 \\ 0 & W \end{pmatrix}, \quad \tilde{Z} = Z - Z^*
\tag{37}
$$

Using $\operatorname{tr}(\tilde{Z}^\top Z) \leq \left\|\tilde{Z}\right\|\|Z^*\| - \left\|\tilde{Z}\right\|^2$ and following [8] there exist $a_0, a_1, c_3, \kappa > \|PB\|\, c_3$ such that

$$
\dot{\mathbf{V}} = -\frac{1}{2}\lambda_{min}(Q)\|e\|^2 - (\kappa - \|PB\|\, c_3)\|e\|\left\|\tilde{Z}\right\|^2 +
$$
$$
+ a_0\|e\| + a_1\|e\|\left\|\tilde{Z}\right\|
\tag{38}
$$

and, with $Z_m = \frac{a_1 + \sqrt{a_1^2 + 4a_0(\kappa - \|PB\|c_3)}}{\kappa - \|PB\|c_3}$,

$$\|e\| \geq \frac{a_0 + a_1 Z_m}{\frac{1}{2}\lambda_{min}(Q)} \Rightarrow \dot{\mathbf{V}} \leq 0 \tag{39}$$

Thus for convenient initial conditions, the tracking error e is ultimately uniformly bounded.

7 Simulations

For the purpose of stabilizing the inverted pendulum in the origin of coordinates, the setup was simulated for the pendulum parameters $\{l_p = 0.5, m_p = 0.35\}$ and quad-rotor parameters $\{J_1 = 0.5, J_2 = 0.5, J_3 = 0.2, n_1 = 12, n_2 = 3, n_3 = 3, \Gamma_{V_\xi} = 2I, \Gamma_{W_\xi} = 2I, \Gamma_{V_\eta} = 1I, \Gamma_{W_\eta} = 1I, k_{p\xi} = k_{d\xi} = 1I, k_{p\eta} = 7I, k_{d\eta} = 2I, \kappa_\xi = 2, \kappa_\eta = 1\}$.

As can be depicted through the pictures, the adaptive augmentation contributes additional dissipation to the stabilization process.

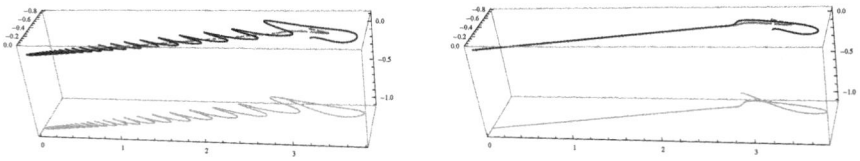

Fig. 5. Tracking during pendulum stabilization without adaptation (left) and with adaptation (right). Thick continuous: actual quad-rotor GC, Thick dashed desired quad-rotor GC, Thin green: load path.

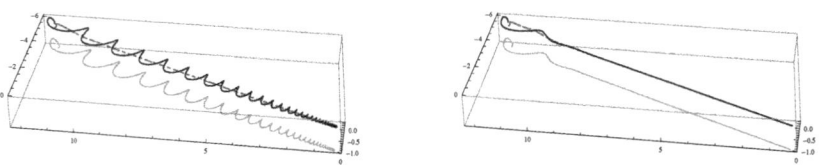

Fig. 6. Tracking during pendulum stabilization without adaptation (left) and with adaptation (right). Thick continuous: actual quad-rotor GC, Thick dashed desired quad-rotor GC, Thin green: load path.

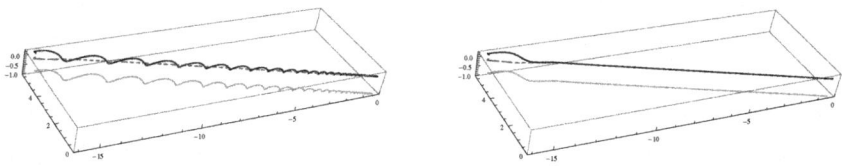

Fig. 7. Tracking during pendulum stabilization without adaptation (left) and with adaptation (right). Thick continuous: actual quad-rotor GC, Thick dashed desired quad-rotor GC, Thin green: load path.

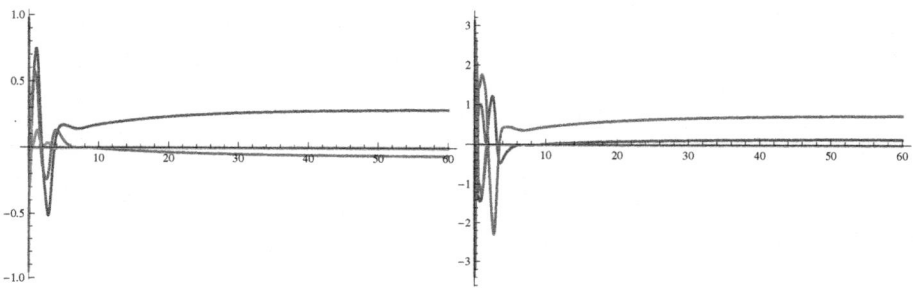

Fig. 8. Typical Neural Network weights evolution: $W_\xi(t)$ and $W_\eta(t)$

8   Conclusions

The method of connecting two dynamic systems affected by a holonomic link, is robust enough as shown through simulations, for succeed in experimental implementation. The system integration process can be used in many practical cases as those dealing with collaborative work.

References

1. Avila-Vilchis, J.C., Brogliato, B., Dzul, A., Lozano, R.: Nonlinear Modeling and Control of Helicopters. Automatica 39, 1583–1596 (2003)
2. Bullo, F., Lewis, A.D.: Geometric Control of Mechanical Systems. Texts in Applied Mathematics. Springer (2005)
3. Hehn, M., D'Andrea, R.: A Flying Inverted Pendulum. In: 2011 IEEE International Conference on Robotics and Automation, Shanghai, China (2011)
4. Hornik, K., Stinchcombe, M., White, H.: Multilayer feedforward networks are universal approximators. IEEE Transactions on Neural Networks 2, 359–366 (1989)
5. Huang, H., Hoffmann, G.M., Waslander, S.L., Tomlin, C.J.: Aerodynamics and Control of Autonomous Quadrotor Helicopters in Aggressive Maneuvering. In: 2009 IEEE International Conference on Robotics and Automation, Japan (2009)
6. Johnson, E.: Limited Authority Adaptive Flight Control. PhD Thesis. Georgia Institute of Technology (2000)
7. Isidori, A.: Nonlinear control systems. Springer (1995)

8. Kannan, S.K., Johnson, E.N.: Adaptive Flight Control for an Autonomous Unmanned Helicopter. In: AIAA Guidance, Navigation, and Control Conference and Exhibit, pp. 1–11 (2002)

9. Kim, N.: Improved methods in neural network based adaptive output feedback control, with applications to flight control. Ph. D. thesis. Georgia Institute of Technology (2003)

10. Koo, T.J., Ma, Y., Sastry, S.S.: Nonlinear Control of a Helicopter Based Unmanned Aerial Vehicle Model. Tech. report, UC Berkeley (2001)

11. Liu, G., Challa, S., Yu, L.: Revisit Controlled Lagrangians for Spherical Inverted Pendulum. International Journal of Mathematics and Computers in Simulation 2(1), 209–214 (2007)

12. Mellinger, D., Michael, N., Kumar, V.: Trajectory generation and control for precise aggressive maneuvers with quadrotors. International Journal of Robotics Research 31, 664–674 (2012)

13. Nardi, F.: Neural Network Based Adaptive Algorithms for Nonlinear Control. Ph. D. thesis. Georgia Institute of Technology (2000)

14. Shin, Y.: Neural Network Based Adaptive Control for Nonlinear Dynamic Regimes. Ph. D. Thesis. Georgia Institute of Technology (2005)

15. Sreenath, K., Michael, N., Kumar, V.: Trajectory Generation and Control of a Quadrotor with a Cable-Suspended Load. A Differentially-Flat Hybrid System. In: IEEE International Conference on Robotics and Automation, ICRA 2013 (2013)

16. Raimundez, C., Camaño, J.L., Barreiro, A.: Stabilizing an Inverted Spherical Pendulum using a Scale Quad-Rotor. In: IEEE CYBER 2014, 4th Annual IEEE International Conference on CYBER Technology in Automation, Control, and Intelligent Systems, Hong Kong, China (2014)

17. Raimundez, C., Camaño, J.L.: Tracking in Scale Quad-Rotors Through Adaptive Augmentation. Neural Computing and Applications, pp. 941–643. Springer ISSN 0941-0643

18. Raimundez, C., Villaverde, A.: Adaptive Tracking Control for a Quad-Rotor. In: 6th EuroMech Nonlinear Dynamics Conference (ENOC 2008) (2008)

19. Ritz, R., D'Andrea, R.: Carrying a Flexible Payload with Multiple Flying Vehicles. IEEE/RSJ International Conference on Intelligent Robots and Systems, 3465–3471 (2013)

20. Michael, N., Fink, J., Kumar, V.: Cooperative Manipulation and Transportation with Aerial Robots University of Pennsylvania, Philadelphia, Pennsylvania

21. Brescianini, D., Hehn, M., D'Andrea, R.: Quadrocopter Pole Acrobatics. In: IEEE/RSJ International Conference on Intelligent Robots and Systems, pp. 3472–3479 (2013)

22. Hehn, M., D'Andrea, R.: A flying inverted pendulum. In: 2011 IEEE International Conference on Robotics and Automation (ICRA), pp. 763–770 (2011)

Planar Modeling of an Actuated Camera Onboard a MAV*

Jesús G. Villagómez, Manuel Vargas,
Manuel G. Ortega, and Francisco R. Rubio

Dpt. of Systems Engineering and Automation, Avda. de los Descubrimientos, s/n.
University of Seville, Spain
{villagomez,mvargas,mortega,rubio}@us.es

Abstract. This work presents the modeling and control of a multi-body air vehicle composed of a miniature aerial vehicle (MAV) and a camera positioner subsystem. The goal is to get an actuated system to freely position the camera in the space, disengaging dynamic couplings between the underactuated aircraft and the camera frame due to typical setups, to improve the current operational profile of visual sensors onboard MAVs. The Newton-Euler formalism is applied aiming to obtain coupling terms between the aerial and the camera positioner systems and a latter linearization is computed as a first approach for the control tasks.

Keywords: Multibody mechanics, Fly-the-Camera, MAV, Linearization.

1 Introduction

The development and application of Miniature Air Vehicles (MAV) has been growing during last years, specially among hobbyists and civil environments. This kind of vehicles provides an excellent alternative due to its operational functionalities as vertical take-off and landing (VTOL), maneuverability and hovering [1]. Its low cost profile with respect to traditional manned aircraft and a wide range of applicability shows that are very useful tools to provide an extension of the environment perception to the operator, by means of exteroceptive sensors or cameras, for data collecting or aerial photography and filmography. But due to its inherent limitations, i.e. the payload is quite constrained, reducing drastically the range of extra devices that can be attached to the frame to enhance its current functional profile. Works where the modeling of a system compounded by an industrial manipulator and an aerial platform have been recently presented [2], [3]. Whereas this concept is quite promising, and let the system to interact with the environment for cargo transportation or aerial grasping, represents new challenges such as dynamic couplings or time varying grasped weight.

Computer vision has been widely used in MAVs, where most of previous works considered only the visual information provided by a camera rigidly attached to

---
* This work was partially supported by the spanish Ministry of Education (MECD) under national research projects $DPI2012-37580-C02-02$ and $DPI2010-19154$.

© Springer International Publishing Switzerland 2015   483
A.P. Moreira et al. (eds.), *CONTROLO'2014 - Proc. of the 11th Port. Conf. on Autom. Control*,
Lecture Notes in Electrical Engineering 321, DOI: 10.1007/978-3-319-10380-8_46

the airframe [4], [5]. In contrast, this work presents a setup which comprises the addition of a 2 degrees of freedom (DOF) positioner camera jointly with a camera to an underactuated MAV, to achieve the 6 DOF control in orientation subsystem of the camera frame. Mechanically there are two ways in which it can be done: the former consists on a *tilt − roll* camera attached to the front of the MAV, and the later consists on adding a *pan − tilt* camera to the bottom of the airframe, which is considered in this proposal. Most of the works concerning the use of an actuated camera include at least one camera operator to control the visual sensor attitude [6]. Typically the camera positioning problem onboard a MAV is divided into two interconnected control tasks. Once the aircraft gets to its initial position, hovering or translating avoiding obstacles, guidance should be achieved in such a way that the camera could gets its initial reference frame and "see" the object scene. Concurrently, the camera operator must compensate for the motion of the aircraft and reject these perturbations while keeps trying to track the interest object. By combining the 3 translational DOF plus the yaw angle provided by the quad-rotor with the 2 rotational DOF supplied by a camera positioner, or camera *gimbal*, it is possible to freely position the six DOF of the onboard camera. On this operational mode, also known as "fly-the-camera", references will be given from the visual system, which tracks and keeps the interest object's path with commands generated in the camera field-of-view (FOV) to move both devices. Previous works already considered an approach to this alternative, but only to improve the camera operator experience, regarding aerial platform and camera gimbal as two independently controlled devices [7]. What we propose is the full automation of the composite system, where references will be given from the camera frame's perspective.

The outline of this work is organized as follows: section 2 describes the modeling of a camera positioner onboard a MAV through the motion in a plane of the quad-rotor. In section 3 a linearization of the model obtained in previous section is proposed and some assumptions for the control task are introduced. Section 4 presents a cascade controller scheme to control the inertial camera position and orientation. Final remarks and future perspectives are given in section 4.

2   Modeling

Our approximation for modeling the mechanics of the system is focused on the Planar Vertical Take Off and Landing (PVTOL) aircraft with the camera positioner in configuration *pan − tilt*, in the vertical plane. This model will be called tPVTOL (*tilt* PVTOL), since the *pan* angle is fixed by definition. Usually it is assumed the camera to be an exteroceptive sensor for the quad-rotor platform. As a consequence, the complete system modelling has been typically assumed considering the aircraft and its extra hardware as a free solid rigid evolving in the space, influenced by external forces. In contrast, in this work the camera is evolving in space, actuated by the motion of the quad-rotor and the camera positioner, hence equations of motion of the hole composite system will be given from the camera perspective.

Consider an 1 DOF camera positioner system evolving within the plane, thanks to the flight of a miniature rotorcraft, to which it is rigidly mounted. This camera actuator system is intended to position a visual sensor in an absolute angular position, i.e., *tilt* positioned with no dependence of the current quad-rotor's attitude. Such positioning tasks are performed during near-hovering or in-motion maneuvers. The inertial camera angular position and rates will be measured using a low cost Inertial Measurement Unit (IMU), properly attached to the camera body. This setup could be considered as a multi-body mechanical system, where the simultaneous operation generates dynamic couplings and perturbations. Accordingly, based on weight of both the camera and the camera actuator and involved links lengths, the in-flight camera positioning task could shift the rotorcraft center of mass (CM_B), generating reaction forces and consequently disturbing the inertial camera orientation.

2.1 Reference Frames and Notation

The following dynamic model of the aircraft with the camera positioner is presented for the simple case where the camera is pointing forwards (*tilt* $\sim 0^o$), while the model proposed is also valid for other setups, such as camera pointing downward (*tilt* $\sim 90^o$). The kinematics of the flying system comprises three right-handed reference coordinate systems [8]. Let $(\hat{x}_W, \hat{y}_W, \hat{z}_W)$, which defines the fixed inertial frame W, whose origin O_W is located at the Earth surface. For the planar case, the vector basis becomes $(\hat{x}_W, 0, \hat{z}_W)$. Let $(\hat{x}_B, \hat{y}_B, \hat{z}_B)$ be the body-fixed frame B, whose origin O_B corresponds to the geometric center of the quadrotor. Similarly, for the longitudinal case, the vector basis becomes $(\hat{x}_B, 0, \hat{z}_B)$. Let $(\hat{x}_C, \hat{y}_C, \hat{z}_C)$ be the body-fixed frame C, whose origin O_C corresponds to the center of the camera frame. The vector basis becomes $(\hat{x}_C, 0, \hat{z}_C)$ for the longitudinal case. The orientation of the rigid body is given by a rotation $\mathbf{R} : CM_B \rightarrow W$, where $\mathbf{R} \in SO(2)$ is an orthogonal rotation matrix.

According to figure (Fig. 1(a)), θ_B is the quad-rotor's pitch angle, θ_{CB} is the angle of the camera positioner respect to (w.r.t.) $\hat{y}_C$, and θ_C is the angle describing how far is the camera frame orientation w.r.t. the inertial horizontal plane because of the quad-rotor's attitude. ℓ_{12} is the distance from the quad-rotor's center of mass to the camera actuator's joint, and ℓ_2 the distance between the center of this actuator and the position of the camera frame.

2.2 Newton-Euler Model

In order to characterize the behaviour and the motion of the composite system, the equations of motion are computed through Newton-Euler formulation. The advantage of this approach is that forces and torques applied to the vehicle are clearly displayed, and dynamic couplings between both subsystems are obtained in a natural way. Aerodynamic effects and additional torques, such as propellers rotating, are not considered in this first approach. The composite system can be first seen as two different bodies subject to external forces, applied to the center of mass of the quad-rotor and the camera frame respectively, expressed in the

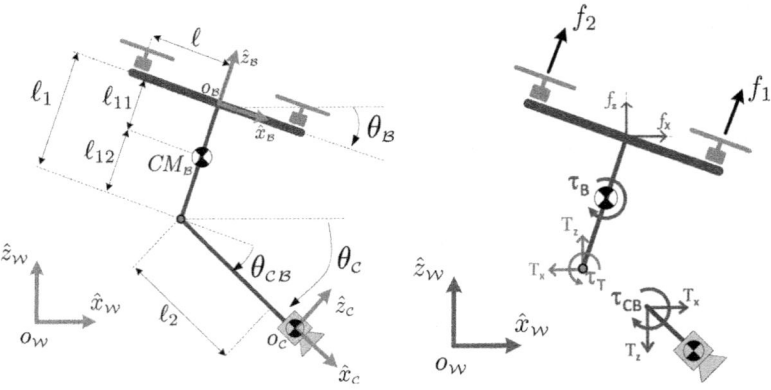

Fig. 1. Freebody diagram: (a) tPVTOL Frames of reference and (b) depiction of forces and moments

body-fixed frame. Couplings are obtained according to Newton's third law, which states that for every action force provided by the camera positioner, there is an equal and opposite reaction force applied directly to the quad-rotor subsystem. Similarly, forces applied in the quad-rotor frame will be directly passed through the kinematic chain till the camera frame. Consequently, Newton's equations of motion yield the following dynamic model for que quad-rotor airframe:

$$\begin{cases} m_\mathcal{B}\ddot{x}_{CM_B} = f_x - T_x \\ m_\mathcal{B}\ddot{z}_{CM_B} = f_z + T_z - m_\mathcal{B}g \\ I_\mathcal{B}\dot{\omega}_\mathcal{B} = \tau_\mathcal{B} - \tau_T + T_x\ell_{12}\cos(\theta_\mathcal{B}) + T_z\ell_2\sin(\theta_\mathcal{B}) \end{cases} \quad (1)$$

where $\tau_\mathcal{B}$ is the applied torque provided by the difference between upward thrust f_1 and f_2 supplied by frontal and rear rotors, respectively, with components f_x and f_z. $\tau_{C\mathcal{B}}$ is the applied torque provided by the camera positioner rotational joint, generating reaction forces T_x and T_z and a reaction torque labelled as τ_T in the quad-rotor frame. Excluding the total mass of the assembly comprising the camera and the positioner (henceforth Camera Composite System - CCS), the total mass of the quad-rotor is denoted as $m_\mathcal{B}$ with $I_\mathcal{B}$ as the inertia mass-moment, w.r.t. its center of mass. $\dot{\omega}_\mathcal{B}$ is the angular acceleration of the aircraft expressed in the inertial frame, length ℓ_{12} is depicted in figure (Fig. 1(a)) and g is the gravitational constant. Since the dynamic model is computed from the camera frame perspective, the relationship is stated through the following coordinates

$$x_{CM_B} = x_c + \ell_{12}\sin(\theta_\mathcal{B}) - \ell_2\cos(\theta_c)$$
$$z_{CM_B} = z_c + \ell_{12}\cos(\theta_\mathcal{B}) + \ell_2\sin(\theta_c) \quad (2)$$

Similarly, the dynamic model of the camera positioner is as follows:

$$\begin{cases} m_c \ddot{x}_c = T_x \\ m_c \ddot{z}_c = -T_z - m_c g \\ I_c \dot{\omega}_{CB} = \tau_{CB} + T_x \sin(\theta_c) \ell_2 - T_z \cos(\theta_c) \ell_2 \end{cases} \tag{3}$$

with

$$\tau_T = \tau_{CB} \tag{4}$$

where $\boldsymbol{\xi} = (x_c, z_c)^\top$ represents the position of the camera frame, τ_{CB} is considered as $\tau_{CB} = \tau_C - \tau_B$, I_c is the inertia mass-moment of the kinetic chain compounded by frame ℓ_2 with total mass m_c, including visual sensor, and $\dot{\omega}_{CB}$ is the angular acceleration exerted by the camera frame expressed in the inertial frame.

2.3 Equations of Motion

Through the Newton-Euler formulation, equations modeling the overall motion of the tPVTOL are derived. Since we are interested in the system dynamic behaviour from the camera frame outlook, eqs. (2) are used in (1) to get the absolute motion of the camera frame given by (3), considering translational and rotational motion of both systems as two interconnected dynamics [10]. Hence, the translational subsystem

$$f_x = m\ddot{x}_c + m_B \ell_{12}(c\theta_B \ddot{\theta}_B - s\theta_B \dot{\theta}_B^2) + m_B \ell_2(s\theta_c \ddot{\theta}_c + c\theta_c \dot{\theta}_c^2) \tag{5}$$

$$f_z = m(\ddot{z}_c + g) - m_B \ell_{12}(s\theta_B \ddot{\theta}_B + c\theta_B \dot{\theta}_B^2) + m_B \ell_2(c\theta_c \ddot{\theta}_c - s\theta_c \dot{\theta}_c^2) \tag{6}$$

and the rotational motion, with the following form:

$$\tau_B = \tau_{CB} + I_B \ddot{\theta}_B + m_c \ell_{12}(\sin(\theta_B)(\ddot{z}_c + g) - \cos(\theta_B)\ddot{x}_c) \tag{7}$$

$$\tau_{CB} = I_c \ddot{\theta}_B - m_c \ell_2(\cos(\theta_c)(\ddot{z}_c + g) + \sin(\theta_c)\ddot{x}_c) \tag{8}$$

The total mass of the system is depicted as $m = m_B + m_c$. The control input $\mathbf{f} = [f_x \ f_z]^\top = \mathbf{R}(\theta_B) \cdot [0 \ f]^\top$ is the thrust vector, where $f = f_1 + f_2$ is the total upward thrust in the body frame provided by frontal and rear motors. s$(\cdot)$ and c$(\cdot)$ stands for $sin(\cdot)$ and $cos(\cdot)$, respectively.

3 Problem Statement

The goal is getting the tPVTOL to follow a predefined trajectory, given in the camera frame. Specifically, we are interested in controlling the camera frame position (x_c, z_c) and its *tilt* angle (θ_c), w.r.t. the inertial frame. For control purposes and complexity simplification, coupling terms in the rotational subsystem, i.e, sideward accelerations among the quad-rotor and the CCS, will be given as function of the angular variables using eqs. (5-8), which yield finally the motion equations, for the translational motion:

$$f \cdot s\theta_B = m\ddot{x}_c + m_B \ell_{12}(c\theta_B \ddot{\theta}_B - s\theta_B \dot{\theta}_B^2) + m_B \ell_2(s\theta_c \ddot{\theta}_c + c\theta_c \dot{\theta}_c^2) \tag{9}$$

$$f \cdot c\theta_B = m(\ddot{z}_c + g) - m_B \ell_{12}(s\theta_B \ddot{\theta}_B + c\theta_B \dot{\theta}_B^2) + m_B \ell_2(c\theta_c \ddot{\theta}_c - s\theta_c \dot{\theta}_c^2) \tag{10}$$

The rotational subsystem is slightly different from eqs. (7-8), which is not directly related with $\ddot{x}_C$ and $\ddot{z}_C$, as follows:

$$\tau_B = \tau_{CB} + (I_B + \tfrac{m_B m_C}{m}\ell_{12}^2)\ddot{\theta}_B + \tfrac{m_B m_C}{m}\ell_{12}\ell_2(s\theta_{CB}\ddot{\theta}_C + c\theta_{CB}\dot{\theta}_C^2) \tag{11}$$

$$\tau_{CB} = (I_C - \tfrac{m_B m_C}{m}\ell_2^2)\ddot{\theta}_C + \tfrac{m_B m_C}{m}\ell_{12}\ell_2(s\theta_{CB}\ddot{\theta}_B + c\theta_{CB}\dot{\theta}_B^2) - \tfrac{m_C}{m}\ell_2 c\theta_{CB} \tag{12}$$

3.1   System Model Linearization

In hover mode, for the purpose of control design we will use the linearized version of the nonlinear model obtained in previous section, since in general, for this first approach the vehicle operates in regions where $\theta_B \simeq 0^\circ$. The linearized model is found by means of Taylor approximation in the vicinity of the nominal operating point, given by the small value increment of the generalized coordinates:

$$\begin{array}{lll}
\tau_B(t) = \tau_B^{eq} + \tau_B^\delta(t) & x_C(t) = x_C^{eq} + x_C^\delta(t) & \theta_B(t) = \theta_B^{eq} + \theta_B^\delta(t) \\
\tau_{CB}(t) = \tau_{CB}^{eq} + \tau_{CB}^\delta(t) & z_C(t) = z_C^{eq} + z_C^\delta(t) & \theta_C(t) = \theta_C^{eq} + \theta_C^\delta(t) \\
f(t) = f^{eq} + f^\delta(t) & &
\end{array} \tag{13}$$

Considering the camera positioner angle around a fixed value ($\theta_C = 0^\circ$), the following equilibrium conditions are obtained:

$$\begin{array}{lll}
\tau_B^{eq} = \tau_{CB}^{eq} + m_C g \ell_{12} s\theta_B^{eq} & x_C^{eq} = 0 & \theta_B^{eq} = 0; \\
\tau_{CB}^{eq} = -m_C g \ell_2 c\theta_C^{eq} & z_C^{eq} = 0 & \theta_C^{eq} = 0 \\
f^{eq} = \tfrac{mg}{c\theta_B^{eq}} & &
\end{array} \tag{14}$$

This assumption leads to the following time-invariant linearized representation of the system, i.e. the translational:

$$\ddot{x}_C^\delta = \tfrac{f^{eq}}{m}\theta_B^\delta - \tfrac{m_B \ell_{12}}{m}\ddot{\theta}_B^\delta \tag{15}$$

$$\ddot{z}_C^\delta = \tfrac{f^\delta}{m} - \tfrac{m_B \ell_2}{m}\ddot{\theta}_C^\delta \tag{16}$$

and the rotational subsystem:

$$\tau_B^\delta - \tau_{CB}^\delta = \alpha_1 \ddot{\theta}_B^\delta + \alpha_3 s\theta_{CB}^{eq} \ddot{\theta}_C^\delta \tag{17}$$

$$\tau_{CB}^\delta = \alpha_2 \dot{\theta}_C^\delta + \alpha_3 s\theta_{CB}^{eq} \ddot{\theta}_B^\delta - \tfrac{m_C}{m}\ell_2 c\theta_{CB}^{eq} f^\delta + m_C g \ell_2 s\theta_{CB}^{eq}\theta_{CB}^\delta \tag{18}$$

with

$$\alpha_1 = I_B + \tfrac{m_B m_C}{m}\ell_{12}^2 \qquad \alpha_2 = I_C - \tfrac{m_B m_C}{m}\ell_2^2 \qquad \alpha_3 = \tfrac{m_B m_C}{m}\ell_{12}\ell_2$$

Based on equations (13 - 18), the application of Laplace transformation yield the following transfer functions, which are considered for the control tasks:

$$x_C(s) = \frac{1}{ms^2}(f^{eq} - m_\mathcal{B}\ell_{12}s^2)\theta_\mathcal{B}(s) \tag{19}$$

$$z_C(s) = \frac{1}{ms^2}f(s) - \frac{m_\mathcal{B}}{m}\ell_2\theta_C(s) \tag{20}$$

$$\theta_\mathcal{B}(s) = \frac{1}{\alpha_1 s^2}(\tau_\mathcal{B}(s) - \tau_{C\mathcal{B}}(s)) \tag{21}$$

$$\theta_C(s) = \frac{1}{\alpha_2 s^2}(\tau_{C\mathcal{B}}(s) + \frac{m_C\ell_2}{m}f(s)) \tag{22}$$

where, e.g., $x_C(s) = \mathscr{L}[x_C^\delta(t)]$.

4   Control Strategy

In order to evaluate more sophisticated control structures, it is commonly introduced as a first approach linear control laws around a specific operating point. Consequently, the proposed scheme to control the camera position and orientation is carried out by a classical cascade controller, based on proportional derivative (PD) laws for the rotational and translational subsystems close to the hovering condition (nearly zero roll and pitch angle values). Both dynamics are partially decoupled and controlled by an attitude and position controller, as depicted in figure (Fig. 2).

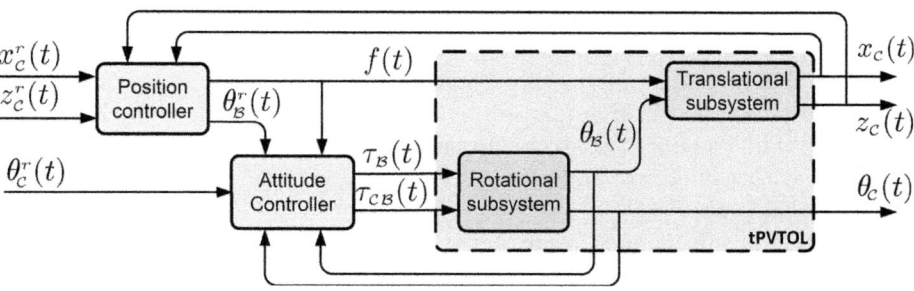

Fig. 2. Control scheme overview consisting on cascade PD controllers: rotational subsystem is partially decoupled by means of feed-forward compensation. Translational subsystem provides reference for the $\theta_\mathcal{B}$ angle.

A description of the system parameters is as follows:

1. The proposed control strategies will be tested by simulation, considering system parameter values (mechanically suitable for the dimension and weight of the proposed platform) collected in table (Tab. 1).

2. It is considered propellers forces to be limited, with a value not exceeding $f_{max} = mg = 13.29$ N. This means the four rotor set is able to give peak thrust forces equivalent to four times the multirotor system's weight. Consequently, the applied thrust in the quad-rotor frame is bounded. Applied torques in the quad-rotor frame defined as $\tau_B = (f_2 - f_1)\ell$ are also bounded, where ℓ is the length between the position of each propeller and the geometric center of the airframe (considered symmetric).

3. The maximum torque provided by the camera positioner rotational joint is also bounded, with a value $|\tau_{CB}| \leq 4m_c g\ell_2 = 1.56$ N·m.

Table 1. System model parameter values

Parameter	Value	Unit	Parameter	Value	Unit	Parameter	Value	Unit
g	9.81	$\frac{m}{s^{-2}}$	I_B	0.43	$kg \cdot m^2$	ℓ_{11}	0.03	m
m_B	0.955	kg	I_C	0.25	$kg \cdot m^2$	ℓ_2	0.10	m
m_C	0.4	kg	ℓ_1	0.10	m	ℓ	0.25	m

4.1 Attitude Control

Since both coupled dynamics are different, in terms of system response, a design specification for the attitude control law is that the rotational subsystem in closed loop must be about 10 times faster than the response of the translational subsystem. More specifically, the cascade control scheme presents good results when the rotational subsystem controller is designed with a time response about $t_s = 0.15$ s. A feed-forward term is introduced in order to pre-compensate the effects produced by changing thrust, which is considered as a disturbance for the angular position (τ_{CB}) controller. In figure (Fig. 3), it is depicted the evolution of θ_B and θ_C angles with changing references, related with the corresponding applied torques τ_B and τ_{CB} and input thrust f.

Mainly due to coupling dynamics, changes in θ_c reference produces additional torques to be applied in the quad-rotor frame, depicted in the behaviour of θ_B in $t = 3$ s. The computed control commands from the attitude controller must be fed with the corresponding equilibrium values (τ_{CB}^{eq} and τ_B^{eq}) before final control signal is applied.

Aerodynamic effects and, in particular, additional coupling forces are neither considered in the model given in eqs. (1 - 3) nor in the proposed control laws. This means that the rotational subsystem closed-loop response should be fast enough in comparison to the specified translation closed-loop dynamics.

4.2 Position Control

The translational subsystem is based on PD controllers, where output signals will be used as input references in the rotational subsystem, as depicted in figure (Fig. 2). As stated before, the controller for the camera frame position is designed in way that the translational subsystem time response in closed loop must be at least 10 times higher than the rotational loop, in such a way that unmodeled

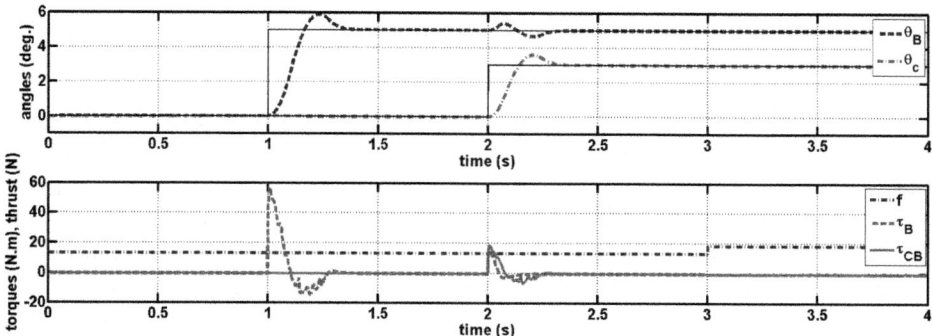

Fig. 3. (a) Angular airframe and camera position: step responses of the closed loop rotational subsystem and (b) applied torques and thrust

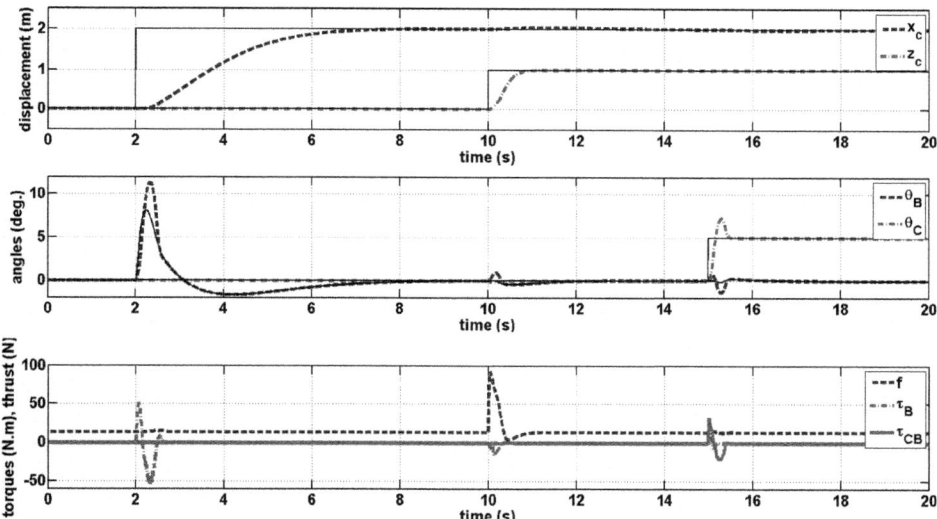

Fig. 4. (a) Evolution of the camera frame position: step responses of the closed loop translational subsystem and (b) applied torques and thrust

forces could not disturb the translational subsystem response. In figure (Fig. 4) it is depicted the evolution of the camera frame position x_C and z_C, and applied torques and thrust, for input changing references.

The x_C control position is quite constrained for the evidence of a non-minimum phase zero in (eq. 19), being underactuated by the θ_B angle and input thrust on its equilibrium value (eq. 14). Nevertheless, the height of the camera frame can be directly controlled since there is direct action over the variables which z_C has dependence on. Similarly, the translational controller control output must be increased with the equilibrium values obtained in (14).

5   Conclusions and Future Perspectives

This paper is mainly focused on the modeling of a multi-body platform consisting of an aerial platform (quad-rotor) and camera positioner, where the novel aspect is the dynamic behaviour of the system from the camera perspective. The aim is to use the two additional degrees of freedom provided by a camera positioner in combination with an underactuated quad-rotor aircraft to get an actuated system from the camera frame, allowing free arbitrary rotations of the camera in the space, but constrained for the task space where the aerial platform could fly to. The intrinsic relationships resulting from this kind of multi-body air system, i.e. the modeling of the composite system, were obtained through the Newton-Euler formalism of the system evolving in a longitudinal plane. Future work will cover the full 3D modeling of the system, evolving in the space. Simple linear control techniques are proposed as a first approach to the control task. Control techniques proposed in this work are suitable to be compared and extended with more sophisticated techniques, e.g., *Backstepping, Sliding-Mode* or LQR control laws [1], which could be subsequently applied in order to control the camera frame position as well as to improve some of the results obtained in this work.

References

1. Raffo, G.V.: Robust Control Strategies for a Quadrotor Helicopter. An Underactuated Mechanical System. Ph.D. Thesis, Engineering School. University of Seville (2011)
2. Escareno, J., Rakotondrabe, M., Flores, G., Lozano, R.: Rotorcraft Mav Having an Onboard Manipulator: Longitudinal Modeling and Robust Control. In: European Control Conference (ECC), Zürich, Switzerland, pp. 267–269 (2013)
3. Lippiello, V., Ruggiero, F.: Exploiting Redundancy in Cartesian Impedance Control of UAVs Equipped With a Robotic Arm. In: IEEE/RSJ International Conference on Intelligent Robots and Systems, pp. 3768–3773 (2012)
4. Chriette, A., Hamel, T., Mahony, R.: Visual Servoing for a Scale Model Autonomous Helicopter. In: Proceedings ICRA, IEEE International Conference on Robotics and Automation, vol. 2, pp. 1701–1706 (2001)
5. Corke, P.: Robotics, Vision and Control. Fundamental Algorithms in MATLAB®. Springer (2011)
6. Lee, D., Chitrakaran, V., Burg, T., Dawson, D., Xian, B.: Control of a Remotely Operated Quadrotor Aerial Vehicle and Camera Unit Using a Fly-the-Camera Perspective. In: 46th IEEE Conference on Decision and Control, pp. 6412–6417 (2007)
7. Neff, A.E., Lee, D., Chitrakaran, V.K., Dawson, D.M., Burg, T.C.: Velocity Control for a Quad-rotor UAV Fly-By-Camera Interface. In: IEEE Proceedings of SoutheastCon, pp. 273–278 (2007)
8. Etkin, B., Reid, L.D.: Dynamics of Flight, Wiley, New York (1959)
9. Fantoni, I., Lozano, R.: Nonlinear Control for Underactuated Mechanical Systems. In: Communications and Control Engineering Series. Springer (2002)
10. Castillo, P., Lozano, R., Dzul, A.E.: Modelling and Control of Mini-Flying Machines. Springer (2005)

Appendix A

The nonlinear model of the composite system, obtained in (9-12), can also be clustered in compact matrix form, where $\mathbf{q} = [x_c \ z_c \ \theta_B \ \theta_c]^\top$ is the generalized coordinates vector, as follows:

$$\mathbf{M}(\mathbf{q})\ddot{\mathbf{q}} + \mathbf{v}(\mathbf{q}, \dot{\mathbf{q}})\dot{\mathbf{q}} + \mathbf{g}(\mathbf{q})\dot{\mathbf{q}} = \mathbf{u} \tag{23}$$

Inertia terms are gathered in $\mathbf{M}(\mathbf{q})$. Vector $\mathbf{v}(\mathbf{q}, \dot{\mathbf{q}})$ collects coriolis, centrifugal and tangential forces provided by the relationship between the quad-rotor and the kinematic chain which the camera is attached to, exerted in the camera frame.

$$\mathbf{M} = \begin{bmatrix} m & 0 & m_B\ell_{12}c_{\theta_B} & m_B\ell_2 s_{\theta_c} \\ 0 & m & -m_B\ell_{12}s_{\theta_B} & m_B\ell_2 c_{\theta_c} \\ -m_c(\ell_{12}c_{\theta_B} + \ell_2 s_{\theta_c}) & m_c(\ell_{12}s_{\theta_B} - \ell_2 c_{\theta_c}) & I_{CM_B} & I_{CM_C} \\ -m_c\ell_2 s_{\theta_c} & -m_c\ell_2 c_{\theta_c} & 0 & I_{CM_C} \end{bmatrix} \tag{24}$$

$$\mathbf{v} = \begin{bmatrix} m_B(\ell_2 c_{\theta_c}\dot{\theta_c}^2 - \ell_{12}s_{\theta_B}\dot{\theta_B}^2) \\ -m_B(\ell_2 s_{\theta_c}\dot{\theta_c}^2 + \ell_{12}c_{\theta_B}\dot{\theta_B}^2) \\ 0 \\ 0 \end{bmatrix} \tag{25}$$

The effects of the gravity on the system are collected in vector $\mathbf{g}(\mathbf{q})$, while the force vector represents all external forces applied to the system, as follows:

$$\mathbf{g} = \begin{bmatrix} 0 & mg & m_c g(\ell_{12}s_{\theta_B} - \ell_2 c_{\theta_c}) & -m_c g\ell_2 c_{\theta_c} \end{bmatrix}^\top \tag{26}$$

$$\mathbf{u} = \begin{bmatrix} f_x & f_z & \tau_B & \tau_{CB} \end{bmatrix}^\top \tag{27}$$

Optimization Based Control for Target Estimation and Tracking via Highly Observable Trajectories*

An Application to Motion Control of Autonomous Robotic Vehicles

Andrea Alessandretti[1,2], António Pedro Aguiar[2,3], and Colin N. Jones[1]

[1] École Polytechnique Fédérale de Lausanne (EPFL), Lausanne, Switzerland
[2] Institute for Systems and Robotics (ISR), Instituto Superior Tecnico (IST), Lisbon, Portugal
[3] Faculty of Engineering, University of Porto (FEUP), Porto, Portugal

Abstract. This paper proposes a Model Predictive Control (MPC) scheme to solve the target estimation and tracking problem. The objective is to derive a feedback law that drives an autonomous robotic vehicle to follow a target vehicle using an on-line estimate of the target's state. In this scenario, when the target is observed through a nonlinear observation model, e.g., bearing only or range only sensors, it is possible to show that solving the tracking problem independently from the estimation problem can lead to an unsatisfactory result where the follower-target system is driven by the controller through unobservable or weakly observable trajectories and, as result, the state of the target vehicle cannot be recovered or cannot be recovered with high accuracy leading to the failure of the control strategy. In this paper, we propose an optimization based scheme that embeds, in a seamless way, an index of observability in the design of the target tracking controller resulting in a closed loop behavior that balances the objective of target tracking with the competing objective of maintaining a good estimate of the state of the target. Numerical results are presented that illustrate this type of behavior.

Keywords: Target tracking, target estimation, model predictive control, highly observable trajectories, observability

1 Introduction

This paper addresses the design of a continuous time output feedback sample-data MPC controller for target estimation and tracking.

* This work was supported by projects CONAV/FCT-PT [PTDC/EEACRO/113820/2009], MORPH [EU FP7 ICT 288704], and FCT [PEst-OE/EEI/LA0009/2011]. The first author benefited from grant SFRH/BD/51073/2010 of the Foundation for Science and Technology (FCT), Portugal.

One of the main challenges in the design of a target tracking algorithm steams from the unavailability of the state of the target vehicle which has to be estimated on-line using measurements from the sensors with often highly nonlinear models. Such nonlinearities often define a set of unobservable state and input trajectories that should be avoided in order to maintain a good estimate of the state of the target vehicle, which is crucial for the success of the target tracking algorithm.

To face this problem, [4,9] propose solutions based on suitable maneuvers designed to keep the state of the target vehicle observable. Although some of the results are promising, within the framework proposed, it is difficult to asses the quality of the overall estimate obtained during the all vehicle trajectory, and moreover the resulting system may result in artificial closed loop position trajectories.

In this work we exploit the potentiality of optimization based control (see, e.g., [5,8,7] and, more recently, [10,6] for an overview) to design a control law that jointly minimizes the distance target-follower and an index of observability specifically designed to penalize weakly observable trajectories. Related works are [2] and [3] where an economic index is embedded in a MPC controller to influence the transient and asymptotic closed loop behavior, respectively.

The organization of the paper is as follows. Section 2 contains the main result of the paper. In Section 3, we illustrate the potentials of the proposed method by solving the target estimation and tracking problem first in a decoupled manner, i.e., without introducing the observably index, and then using the proposed scheme. Section 4 closes the paper with some conclusions.

2   Follower-Target MPC Design

In this section a solution the simultaneous target estimation and tracking problem for autonomous robotic vehicles is proposed. We start with the problem statement description of the follower-target system. Then, a generic MPC control law for target tracking is presented in combination with an observability index to obtain the proposed control law strategy.

2.1   Follower-Target System Description

Let the follower and target vehicles be described by

$$\dot{x}_f(t) = f_f(x_f(t), u_f(t), w_f(t)), \qquad u_f(t) \in \mathcal{U}_f, \qquad t \geq 0$$
$$\dot{x}_t(t) = f_t(x_t(t), u_t(t), w_t(t)), \qquad u_t(t) \in \mathcal{U}_t, \qquad t \geq 0$$

where $(x_f(t), u_f(t), w_f(t)) \in \mathbb{R}^{n_f} \times \mathbb{R}^{q_f} \times \mathbb{R}^{p_f}$ and $(x_t(t), u_t(t), w_t(t)) \in \mathbb{R}^{n_t} \times \mathbb{R}^{q_t} \times \mathbb{R}^{p_t}$ denote the triples of state, input, and disturbance vectors of the follower vehicle and target vehicle, respectively, evaluated at time t. For sake of simplicity, the dependence on time and parameters is dropped whenever clear from the context. We consider the follower and target input vectors to be constrained in the sets $\mathcal{U}_f$ and $\mathcal{U}_t$, respectively. The follower-target system is defined by

$$\dot{x} = f(x, u_f, u_t, w_f, w_t) := \begin{pmatrix} f_f(x_f, u_f, w_f) \\ f_t(x_t, u_t, w_t) \end{pmatrix}, \qquad x := \begin{pmatrix} x_f \\ x_t \end{pmatrix} \qquad (1)$$

where the state of the follower vehicle is measured through the observation model

$$y_f = h_f(x_f, v_f),$$

and, moreover, the follower vehicle takes measurements of the target vehicle through the observation model described by

$$y_t = h_t(x_f, x_t, v_t),$$

where $(y_f, v_f) \in \mathbb{R}^{m_f} \times \mathbb{R}^{b_f}$ and $(y_t, v_t) \in \mathbb{R}^{m_t} \times \mathbb{R}^{b_t}$ are the pairs of measurement and measurement noise of the follower and target observation models, respectively. Thus, the observation model of the follower-target system is defined by

$$y = h(x, v) := \begin{pmatrix} h_f(x_f, v_f) \\ h_t(x_t, v_t) \end{pmatrix}, \qquad y := \begin{pmatrix} y_f \\ y_t \end{pmatrix} \qquad v := \begin{pmatrix} v_f \\ v_t \end{pmatrix}. \qquad (2)$$

The function $h_f(\cdot)$ is related to the sensors on-board of the follower vehicle, e.g., Global Positioning System (GPS), Inertial Measurement Unit (IMU), and magnetic compass, and the function $h_t(\cdot)$ with the sensor used to observe the target vehicle, e.g., a bearing only sensor, like a camera, or a range only sensor, like a sonar.

In order to perform target tracking it is convenient to estimate the future control input signal of the target vehicle, and thus its future position. Toward this goal, we choose a finite dimensional smooth parameterization $p_u \in \mathbb{R}^{n_p}$ of such signal, with the associated time derivative $\dot{p}_u = f_u(p_u)$. The parameter p_u is then estimated, together with the state of the follower-target system, by the observer defined by

$$\dot{x}_o = f_o(x_o, u_f, y), \qquad (3a)$$

$$\begin{pmatrix} \hat{x} \\ \hat{p}_u \end{pmatrix} = h_o(x_o) \qquad (3b)$$

where $x_o \in \mathbb{R}^{n_o}$ denotes the internal state vector of the observer and $\hat{x} \in \mathbb{R}^{n_f + n_t}$ and $\hat{p}_u \in \mathbb{R}^{n_p}$ denote the estimates of the state vector x and parameter p_u, respectively. We denote with $\bar{u}_t(\cdot; p_u)$ the predicted input signal of the target associated with the parameter p_u.

As an example, a possible choice of nonlinear observer could be the well known Extended Kalman Filter (EKF) and a possible parameterization of the input of the target vehicle could be $p_u = u_f$ and $\dot{p}_u = f_u(p_u) = 0$, which captures the set of constant (and in practice slowly varying) target inputs, i.e., $\bar{u}_t(\tau; p_u) = p_u$, for all $\tau > 0$.

2.2 MPC for Target Tracking

Using the notation introduced above, in this section we describe a generic MPC controller for target tracking that uses as input the estimate provided by the observer (3). For a generic trajectory $x(\cdot)$, $x([t_1, t_2])$ denotes the trajectory considered in the time interval $[t_1, t_2]$ and we use the notation $x(\cdot; z_1, \dots)$ whenever we would like to make explicit the dependence of the trajectory $x(\cdot)$ on the generic optimization problem parameters $z_1, \dots$.

The MPC optimization problem $\mathcal{P}(z, p)$, with $(z, p) \in \mathbb{R}^{n_f + n_t} \times \mathbb{R}^{n_p}$ is defined as follows:

Definition 1. *(MPC problem) Given the pair of vectors $(z, p) \in \mathbb{R}^{n_f + n_t} \times \mathbb{R}^{n_p}$, with the associated parametric target input signal $\bar{u}_t(\cdot; p)$, and a horizon length $T \in \mathbb{R}_{>0}$, the open loop MPC optimization problem $\mathcal{P}(z, p)$ consists of finding the optimal control signal $\bar{u}_f^*([0, T])$ that solves*

$$
\begin{aligned}
J_T^*(z, p) = \min_{\bar{u}_f([0,T])} &\int_0^T l_{tt}(\bar{x}(\tau), \bar{u}_f(\tau), \bar{u}_t(\tau; p))d\tau + m_{tt}(\bar{x}(T)) \\
s.t. \quad &\dot{\bar{x}} = f(\bar{x}, \bar{u}_f, \bar{u}_t(\tau; p), 0, 0) \\
&\bar{u}_f \in \mathcal{U}_f, \ \bar{x}(0) = z, \ \bar{x}(T) \in \mathcal{X}_f
\end{aligned}
$$

□

The *finite horizon cost* is composed of the *stage cost* $l_{tt} : \mathbb{R}^{n_f + n_t} \times \mathbb{R}^{q_t} \times \mathbb{R}^{n_p} \to \mathbb{R}_{\geq 0}$ and the *terminal cost* $m_{tt} : \mathbb{R}^{n_f + n_t} \times \mathbb{R}^{n_p} \to \mathbb{R}_{\geq 0}$, which is defined over the *terminal set* $\mathcal{X}_f \subseteq \mathbb{R}^{n_f + n_t}$. The subscript tt is used to emphasize that the stage and terminal costs are designed for target tracking.

In a sample-data receding horizon strategy, the control input is computed at discrete sample times $\mathcal{T} := \{t_0 = 0, t_1, \dots\}$, and the MPC control law is defined as

$$
u_f(t) := \bar{u}^*(t - \lfloor t \rfloor; \hat{x}(\lfloor t \rfloor), \hat{p}_u(\lfloor t \rfloor)), \tag{4}
$$

where $\lfloor t \rfloor$ is the maximum sampling time $t_i \in \mathcal{T}$ smaller or equal than t, i.e., $\lfloor t \rfloor = \max_{i \in \mathbb{N}_{\geq 0}} \{t_i \in \mathcal{T} : t_i \leq t\}$. Since the system is not time varying, the space of trajectories over which we optimize are considered, without loss of generality, in the interval $[0, T]$ and t_0 is chosen to be the time zero. Note that (3)-(4) is an output dynamic feedback control law since it uses the estimates of the state and target control parameter provided by the observer.

2.3 Observability Index

In this work we propose a strategy to modify the MPC optimization problem of the previous section in order to avoid weakly observable/non observable closed loop trajectory resulting in an effective target estimation and tracking controller. In this section we propose an index of observability.

Consider the observability matrix of the system (1) associated to the output y defined in (2), both considered in the nominal case (i.e., $v_f = 0$, $v_t = 0$, $w_f = 0$, $w_t = 0$), i.e.,

$$\mathcal{O}(x, u_f, u_t) = \frac{\partial}{\partial x} \left(y' \; \dot{y}' \; \ddot{y}' \; \ldots \; y^{\{r\}'} \right)'$$

where $y^{\{r\}}$ denotes the rth derivative of the output y with respect of time. From the properties of the observability matrix, given $r \in \mathbb{N}_{>0}$ the state of system (1) is locally observable at a given state and input pair $(\bar{x}, \bar{u}_f)$ if the matrix $\mathcal{O}(\bar{x}, \bar{u}_f, \bar{u}_t)$, for a given estimated target input vector $\bar{u}_t$, is full rank. For general nonlinear systems the number of derivatives r to be considered is not known a priori. An intuitive procedure to select r consists in increasing it until the observability matrix becomes full rank for some values of the state and input vectors. Then, driving the system through those values is enough to guarantee observability.

Let $\sigma_{min}(A)$ and $\sigma_{max}(A)$ denote the minimum and maximum singular value of a generic matrix A. To obtain a measure of the degree of observability, one possibility is to use the index $1/\sigma_{min}(\mathcal{O})$, which increases as $\mathcal{O}$ gets close to singularity and becomes infinity when $\mathcal{O}$ loses rank. Another index of interest is the condition number of $\mathcal{O}$, i.e., $\kappa(\mathcal{O}) := \sigma_{max}(\mathcal{O})/\sigma_{min}(\mathcal{O})$, which broadly speaking, provides a measure of the difference of the "quality" of observability of the state components, where $\kappa(\mathcal{O}) = 1$ if all the state components have the same "quality" of observability. Prompted by these observations, we select the following obeservability index :

$$l_o(x, u_f, u_t) = k \arctan \left(\frac{1}{k} \left(\frac{\alpha_1}{\sigma_{min}\mathcal{O}(x, u_f, u_t)} + \alpha_2(\kappa(\mathcal{O}(x, u_f, u_t)) - 1)^2 \right) \right) \tag{5}$$

for some positive constants $\alpha_1 > 0$ and $\alpha_2 > 0$, where the positive constant $k > 0$ defines the width of the region where the nonlinearity $\arctan(\cdot)$, used as smooth saturation-like function, behaves almost linearly.

Note that the observability matrix is not the only method to define the index of observability and other mathematical tools, e.g., the determinant or the trace of the Fisher Information Matrix (FIM), can be exploited in a similar fashion.

2.4 Proposed Controller

Using the observability index suggested in the previous section, the proposed controller is obtained from the controller of Section 2.2 but redefining the stage cost as

$$l(x, u_f, u_t) = l_{tt}(x, u_f, u_t) + l_o(x, u_f, u_t) \tag{6}$$

Note that from (5) the value of the function $l_o(\cdot)$, responsible to keep the closed loop trajectories observable, is saturated to ensure that it does not constantly dominate the costs $l_{tt}(\cdot)$ and $m_{tt}(\cdot)$, which are in charge of driving the follower toward the target. By varying the value of k in (5), it is possible to regulate the importance of the observability of the closed loop trajectories with the conflicting goal of perfect target tracking.

3   Simulation Results

In this section we consider the target estimation and tracking problems for both the target and the follower unicycle-like vehicles.

Let $\{I\}$ be an inertial coordinate frame and $\{B_f\}$ a body coordinate frame attached to the follower vehicle. The pair $(p_f(t), R(\theta_f(t))) \in SE(2)$ denotes the configuration of the follower vehicle, position and orientation, where $R(\theta_f(t))$ is the rotation matrix, from body to inertial coordinates, associated with the angle $\theta_f(t)$. For a unicycle-type vehicle, the kinematic model of a vehicle satisfies

$$\dot{p}_f(t) = R(\theta_f(t)) \begin{pmatrix} v_f(t) \\ 0 \end{pmatrix}, \qquad\qquad \dot{\theta}_f(t) = \omega_f(t),$$

where $\omega_f(t)$ denotes the angular velocity express in the body frame. For this example we consider the case where the control input of the follower vehicle

$$u_f(t) = \big(v_f(t)\ \omega_f(t)\big)' \in \mathcal{U}_f,$$

is constrained in the set

$$\mathcal{U}_f = \left\{ \begin{pmatrix} v_f \\ \omega_f \end{pmatrix} : -2 \le v \le 2, -\pi \le \omega \le \pi \right\}.$$

We assume that the target vehicle is of the same kind of the follower vehicle, but with unconstrained input. Thus, the follower-target system is defined as follows:

$$\dot{x} = \begin{pmatrix} \dot{p}_f \\ \dot{\theta}_f \\ \dot{p}_t \\ \dot{\theta}_t \end{pmatrix} = \begin{pmatrix} R(\theta_f(t)) \begin{pmatrix} v_f(t) \\ 0 \end{pmatrix} \\ \omega_f(t) \\ R(\theta_t(t)) \begin{pmatrix} v_t(t) \\ 0 \end{pmatrix} \\ \omega_t(t) \end{pmatrix}, \qquad x = \begin{pmatrix} p_f \\ \theta_f \\ p_t \\ \theta_t \end{pmatrix}.$$

The input signal of the target vehicle, which is unknown to the follower, is defined as $v_t(t) = 0.5$ and $\omega_t(t) = (\pi/200)\cos(0.05t)$.

The position and heading of the follower vehicle is continuously measured, i.e,

$$y_f(t) = h_f(x_f, v_f) = \begin{pmatrix} p_f(t) \\ \theta_f(t) \end{pmatrix}$$

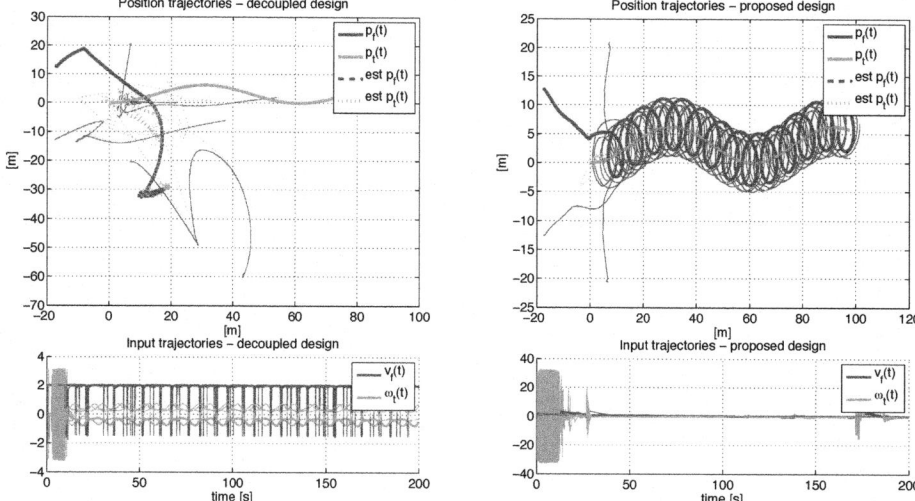

Fig. 1. The closed loop system trajectories associated with the decoupled (left) and coupled scheme (right) for different initial conditions. The ticker lines are associated to the same initial condition. The figure on the top displays the follower and target position trajectories, with solid blue and green lines, and the associated estimates, with dashed lines of the same color. The input signals of the target tracking controller is plotted in the bottom figures.

and we consider that the position of the target vehicle is continuously observed by an omnidirectional camera centered at the position of the follower vehicle. In this case, the observation model can be described as

$$y_t(t) = h_t(x_f, x_t, v_t) = \frac{p_t(t) - p(t)}{\|p_t(t) - p(t)\|},$$

where $y_t(t) \in \mathbb{R}^2$ is a bearing only observation, which provides information about the relative direction of the vehicle but not about the distance (in fact, $\|y_t(t)\| = 1$).

The future input of the follower is parameterized by the parameter $p_u = u_t$, with associated time derivative $\dot{p}_u = 0$, and an Extended Kalman Filter is used to jointly estimate the state of the follower-target system and of the parameterization p_u using $x_o = (x', p'_u)'$ and $h_o(x_o) = x_o$.

The target-tracking controller is designed using the MPC for trajectory-tracking proposed in [1]. In that work, the trajectory to be followed is defined by specifying a desired position, its derivative, and a bound on the value of the derivative. In our case we wish to track the position of the target vehicle, thus we can use the target position, the target velocity, which are components of our state vector, and a bound on the target velocity, which is considered to be 0.5. Using the same

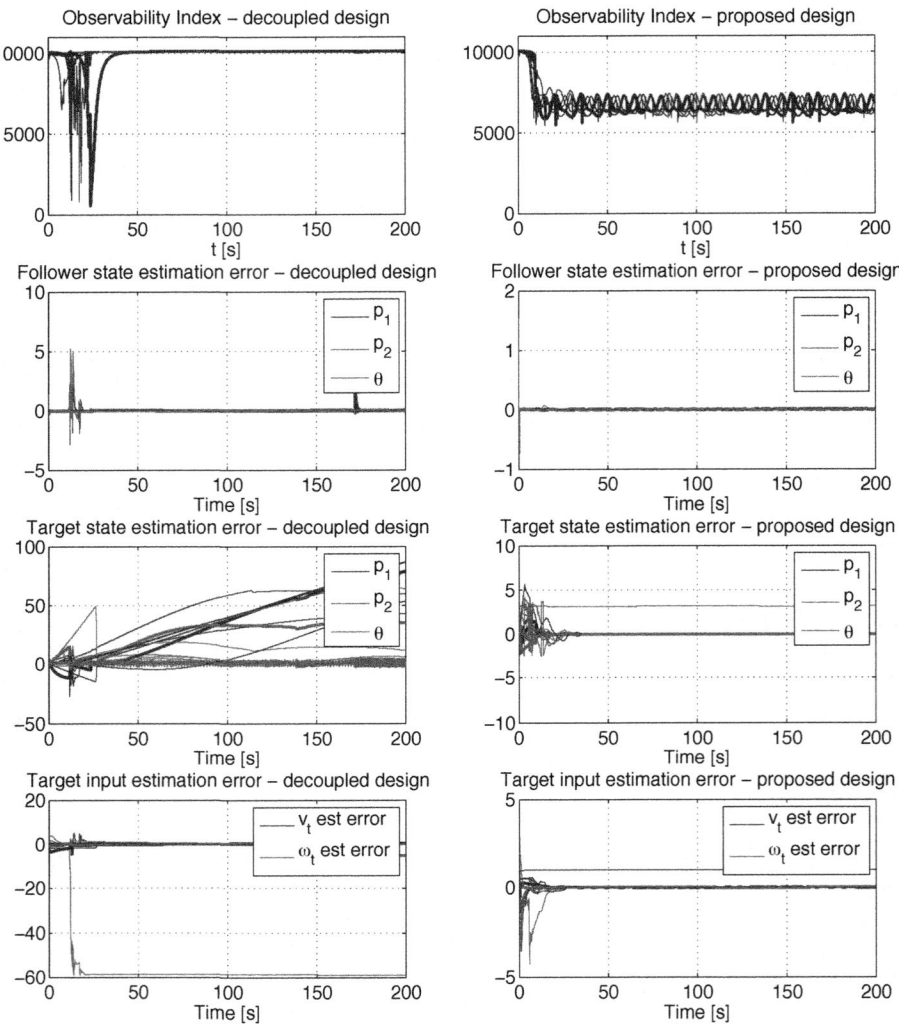

Fig. 2. The closed loop observability index (first row) and the estimation errors for the follower state (second row), target state (third row), and target inputs (bottom row) for the decoupled (left column) and coupled (right column) design

notation of [1], the MPC controller parameters are $\epsilon = \begin{pmatrix} -0.2 \\ 0 \end{pmatrix}$, $K = 0.1I_{2\times2}$, $O = 0.1I_{2\times2}$, and $Q = 10I_{2\times2}$, where $I_{2\times2}$ denotes an identity matrix of size 2×2. The resulting terminal set is $\mathcal{E}_f = \{e'e \leq 26.4^2\}$.

For simulation purposes the system is discretized with discretization step of 0.1, the set of sampling times is $\mathcal{T} = \{t_i = 0.1i, \ i \in \mathbb{N}_{\geq0}\}$, and the horizon length is chosen to be $T = 0.3$ s.

As expected using a decoupled design, i.e., using the target tracking controller in closed loop with the observer without embedding the observability index, the system is driven through unobservable/weakly observable trajectory and, as consequence, we have an increase of the estimation error along time and eventually a failure of the target tracking algorithm. Fig.1 displays the trajectories of the position of the follower and target together with the associated estimates provided by the Kalman filter (top) and the associated input signal (bottom) for the case of decoupled (left) and coupled (right) design. Fig.(2)(top) displays the observability index evaluated along the closed loop trajectories. It can be seen that for the decoupled design (left) the system is driven through unobservable/weakly observable state-input trajectories resulting in a bad estimate of the state of the follower-target system.

In order to avoid this problem, we use the proposed method adding in the stage cost of the MPC control the observability index (5) with $k = 10^5, \alpha_1 = 10^3, \alpha_2 = 10^2$. To reduce the computational complexity, the observability matrix, with $r = 3$, was computed considering the vehicles to be single integrators and using the real linear velocity to evaluate it. In this case we notice from Fig. 1 that the follower do not reach the vehicle but approaches it and start orbiting around it, so reducing the observability index. After an initial transient behavior, along the closed loop state and input trajectories the system is always observable, i.e., the observability index never saturates and, as effect, the estimation error converges to zero, Fig. 2 (right column). Note that, for one initial condition, the estimation error does not converge to zero, although it converges to an indistinguishable configuration (coherent with the observation) where the target moves backward (negative velocity).

4   Conclusion

A systematic procedure to design an optimization based target estimation and tracking controller was presented. The main idea consist in embedding in the stage cost of the MPC controller an observability index that reward highly observable trajectories resulting in this way a good estimate of the state of the follower-target system and, thus, to a successful and accurate tracking of the target.

References

1. Alessandretti, A., Aguiar, A.P., Jones, C.N.: Trajectory-tracking and path-following controllers for constrained underactuated vehicles using Model Predictive Control. In: 2013 European Control Conference (ECC), pp. 1371–1376 (2013)
2. Alessandretti, A., Aguiar, A.P., Jones, C.: A Model Predictive Control Scheme with Additional Performance Index for Transient Behavior. In: Proceedings of the 52nd IEEE Conference on Decision and Control (2013)
3. Alessandretti, A., Aguiar, P., Jones, C.N.: Ultimately Bounded MPC with Economic Performance Index. Technical report
4. Cochran, J., Krstic, M.: Nonholonomic Source Seeking With Tuning of Angular Velocity. IEEE Transactions on Automatic Control 54(4), 717–731 (2009)

5. Fontes, F.A.C.C.: A general framework to design stabilizing nonlinear model predictive controllers. Systems & Control Letters 42(2), 127–143 (2001)
6. Grune, L., Pannek, J.: Nonlinear model predictive control: theory and algorithms (2011)
7. Mayne, D.Q., Rawlings, J.B., Rao, C.V., Scokaert, P.O.M.: Constrained model predictive control: Stability and optimality. Automatica 36(6), 789–814 (2000)
8. Morari, M., Lee, J.H.: Model predictive control: Past, present and future. Computers & Chemical Engineering 23(4), 667–682 (1999)
9. Namaki-Shoushtari, O., Pedro Aguiar, A., Khaki-Sedigh, A.: Target tracking of autonomous robotic vehicles using range-only measurements: a switched logic-based control strategy. International Journal of Robust and Nonlinear Control 22(17), 1983–1998 (2012)
10. Rawlings, J.B., Mayne, D.Q.: Model Predictive Control Theory and Design. Nob Hill Pub. (2009)

Comprehensive Review of the Dispatching, Scheduling and Routing of AGVs

Kelen C.T. Vivaldini[1], Luís F. Rocha[2], Marcelo Becker[1], and António Paulo Moreira[3]

[1] EESC-USP, Mechatronics Group - Mobile Robotics Lab. S. Paulo, Brazil
[2] INESC TEC - INESC Technology and Science (formerly INESC Porto), Porto, Portugal
[3] INESC TEC - INESC Technology and Science (formerly INESC Porto), Porto, Portugal
and FEUP Faculty of Engineering, University of Porto, Porto, Portugal
{kteixeira,becker}@sc.usp.br,
luis.f.rocha@inescporto.pt, amoreira@fe.up.pt

Abstract. Automated Guided Vehicle System (AGVS) has become an important strategic tool for automated warehouses. In a very competitive business scenario, they can increase productivity and reduce costs of FMS (Flexible Manufacturing System) transportation systems. The AGV System provides efficient material flow and distribution among workstations at the right time and place. To attend such requirements, AGVS involves dispatching and scheduling of tasks and routing of AGVs. Some studies have approached such procedures in a similar form, although they have different functionalities. This paper reviews the literature related to the dispatching, scheduling and routing of AGVs (Automated Guided Vehicles) and highlights their main differences in comparison with the common management of vehicles transportation systems. To obtain a theoretical base, the definitions of dispatching, routing and scheduling procedures for materials handling applications are presented and the main methods to solve them are discussed.

Keywords: Automated Guided Vehicles, Routing, Dispatching, Scheduling.

1 Introduction

Materials handling is an essential activity in any production process and its efficiency has severe impacts on the production costs. A materials handling system is applied from the start of the mass production and by using mainly manual (for lighter loads) or mechanical (forklifts, conveyors, etc.) solutions. However, in more recent years and due to the need for increasing flexibility in production systems (small series production and customization), fully automated systems (Automated Guided Vehicle Systems - AGVS or Automatic Storage / Retrieval Systems - AS / RS) have been considered the best alternative.

AGVs (Automated Guided Vehicles) are autonomous vehicles widely used to transport materials between workstations in flexible manufacturing systems and perform a variety of tasks that involve automation in industrial environments [1]. Automated logistic systems, such as warehouses, cross docking centers and container terminals frequently use AGVs to optimize their materials handling tasks [2].

© Springer International Publishing Switzerland 2015
A.P. Moreira et al. (eds.), *CONTROLO'2014 - Proc. of the 11th Port. Conf. on Autom. Control,*
Lecture Notes in Electrical Engineering 321, DOI: 10.1007/978-3-319-10380-8_48

AGVs are among the classes of materials handling equipment that has grown most rapidly in quantity of equipment installed in industries and become a common tool for such activities. The equipment can quickly respond to transportation patterns that frequently change and can be integrated into fully automated control systems. AGVs have been increasingly used to transfer a wide variety of application materials in the area of distribution logistics.

The system that controls the AGVs is AGVS and consists of several AGVs operating concurrently. According to Kalinovcic et al [3] and Vivaldini et al [4], the design of an AGV System requires decision making regarding the best strategies to solve the several problems associated with its functioning, such as dispatching, routing and scheduling. Therefore, the solutions that best adapt to the existing requirements imposed by the materials handling system for AGVS must be chosen.

An AGVS must be developed to ensure an efficient flow of materials during the production process and provide the necessary materials to the appropriate workstation, at the right time and in the right quantity. This control system must guarantee the coordination among AGVs in terms of both scheduling and routing, ensuring an efficient operation of the system (conflict-free, collisions and deadlocks) independently of the number of vehicles used [4-5]. If these requirements are not guaranteed, the overall performance of the production system will decline, become less efficient and generate less profit or operate at higher costs [6]. However, if these requirements are carried out as per the needs of the productive system, the control system is a viable option for the increase in enterprise competitiveness.

We emphasize that such issues, as the selection of a route, dispatching and scheduling of tasks influence the overall performance of the AGVs management system, and must be addressed for the creation of an effective AGVS. If the decision on receiving an order is taken, both dispatching routing and scheduling will be planned so that the AGV is capable of transporting pallets from the origin to their destination.

This paper contributes to the clarification of AGVs dispatching, scheduling and routing and provides a comprehensive review of the main existent methods to solve them. Furthermore, the comparison of this problem with the general transportation systems applied to common vehicles outside the industrial area will be presented in the following section.

2   AGVs Challenges

The problems inherent to Industrial Logistics based on AGVS are similar to the ones studied in the field of urban Transportation Systems (TS). In TS, the main objective is to assure that vehicles efficiently arrive at the desired destinations within a highly dynamic environment, where traffic jams and unpredicted events that may occur are either avoided or prevented. Moreover, the reduction in transport costs is a constant pressure in the logistic area.

One of the main focuses regarding Intelligent Transportation System research and urban TS is the vehicle routing problem (VRP). VRP has high applicability because it reflects the decisions that need to be taken daily by enterprises. It deals mainly with both the computation of an efficient route taking into account the environment information and each vehicle as an entity whose main objective is independent of the other vehicles. At this point, the industrial transportation system becomes differentiable.

In the case of industrial logistics based on AGVS, the AGVs work together to attend all transportation requests in both time and correct sequence, delivering products to workstations so that the production process can follow the desired production flow. Several constraints are faced by system modeling and may be related to the total time of the route and time window, at which the service should be started. If these transportation tasks are not performed efficiently, the production system is severely affected. Therefore, such objectives must not be independent of each other, but establish some relation of precedence or synchronization. Other constraints generally negligible to general transportation, such as length of the vehicle, path capacity, network layout (normally a reduced size network with uni or\and bi-directional routes), and priority of the tasks and AGVs (e.g. priority defined considering their battery level) must also be taken into account. For instance, a fleet of vehicles must be efficiently controlled with hard restrictions that can be relaxed in more general transportation systems problems. These characteristics require complex control architectures, in which the routing algorithm proposed in the urban transportation system must be complemented by dispatching and task scheduling algorithms. These three control layers must be closely related, however the integrated problem becomes considerably more complex.

3   Dispatching, Routing and Scheduling Problem

The control policy of AGVS aims to attend transport demands as fast as possible within the deadline and without the occurrence of conflicts. Co and Tanchoco [7] and Lagevin [8] defined the function of AGVS management in Dispatching, Routing and Scheduling: "Dispatching is the process of selecting and assigning tasks to vehicles, Routing is the selection of the specific paths that each vehicle will execute to accomplish its transportation tasks, and Scheduling is the determination of the arrival and departure times of vehicles at each segment along their routes to ensure collision-free travel". Vis et al [5] affirm that at least the activities of dispatching of loads to AGVs, route selection, scheduling of AGVs and dispatching of AGVs to parking locations must be performed by a controller of the AGVS. According to Le-Ahn [9], the control of AGVs requires an online scheduling or dispatching systems. Therefore, dispatching, routing and scheduling decisions can be made either simultaneously or separately.

3.1   AGVs Dispatching

In general, the dispatching rules are divided into two types of operation decisions: workstation-initiated and vehicle-initiated, depending on whether the system has idle vehicles (vehicle-initiated) or queued transportation requests (workcenter-initiated) [10]. The vehicle-initiated dispatching determines the load to be assigned to a vehicle when the vehicle is ready for the next task, whereas the workcenter-initiated dispatching determines the vehicle to be selected when loads initiate transportation requests [11].

Le-Ahn [9] presented a dispatching system as a scheduling with zero planning horizon. A dispatching decision is made when a vehicle has dropped a load, a vehicle has reached its parking location or a new load has arrived.

A dispatching system uses dispatching rules to control vehicles. Online dispatching rules are simple and can be easily adapted to automated guided vehicle management systems. The common objectives are minimization of load waiting time, maximization of the system throughput, minimization of queue length, and guarantee of a certain service level at stations.

3.2   AGVs Scheduling

Scheduling defines the allocation process of AGVs for tasks considering the time and cost operations [12] and guaranteeing conflict-free routes. Typically, the goals of scheduling are related to the processing time of tasks or use of resources (number of AGVs involved, system throughput or total travel time of all vehicles) under certain constraints, such as deadlines, priorities, etc. However, unviable results can be achieved if the functioning of the transport system does not consider the scheduling limitations [5][13].

According to Le-Ahn [9], the main goal of most scheduling problems is to transport loads (products, pallets or containers) as quickly as possible to satisfy time-window constraints. Other criteria can be the minimization of the maximum load waiting time and maximum number of items in critical queues.

However, the scheduling can be divided into two key factors: is a predictive mechanism that determines the planned start and completion time of labor operations and a reactive mechanism that monitors the progress of the schedule and deals with unexpected events (failures, breakdowns, cancellations, date changes, etc.) [13].

The scheduling system decides when and where a vehicle must perform its tasks. If all tasks are known prior to planning a period of work, the problem can be solved off-line. In practice, the changes in the tasks information after route planning complicate the off-line scheduling. In these cases, an online scheduling becomes essential. The off-line and on-line approaches are clearly presented in [9].

In the off-line scheduling case, all available tasks are scheduled at once. This previously generated scheduling must be reviewed and updated when necessary throughout the production cycle. In the on-line scheduling, the task scheduling decisions are taken in a dynamic way, i.e. decisions are made over time, according to changes in the system state [14]. Therefore, any unforeseen event in the system can be controlled automatically and efficiently by the scheduling.

Real-time scheduling is a process in which short-term decisions are made, i.e., the scheduling is based on the current state of the system and its general needs and can be used by an off-line and/or on-line method. If the scheduling uses an off-line method, the process is reprogrammed, whereas in the on-line method, the task scheduling decision is made when a change occurs in the system status.

3.3   AGVs Routing

The routing problem in the AGV system can be compared with the Vehicle Routing Problem (VRP). It has been extensively addressed and an overview of the literature in this area was presented by [15-18].

Vehicle Routing refers to the process of determining one or more paths or sequences of path to be performed by a fleet of vehicles. The purpose is to visit a set of nodes geographically dispersed in predetermined locations that must be attended. According to Vivaldini et al. [2], the task of AGVs routing in an industrial environment is to find a route for each AGV from its current position to the desired destination ensuring a conflict-free travel along the selected path. In recent years, several algorithms have been proposed to solve routing problems [5][19-23]. They are classified into two categories: static routing algorithms and dynamic routing algorithms.

The difference between static and dynamic vehicle routings is indicated in the routing problem definition. In the static routing problem, the input data do not change during the path execution or in the execution of the routing algorithm. The route from node i to node j is previously determined and always used if the load   must be transported from i to j. A simple solution is to choose the route of shortest distance from i to j. However, static algorithms cannot adapt to changes in the AGVs traffic system. On the other hand, in dynamic routing, the routing decision is based on real-time information (i.e., the inputs can be either changed or updated during the execution of the algorithm or even while running the route), therefore, multiple paths between i and j can be chosen [5]. Presently, the routing algorithm and execution of the path are processes that evolve simultaneously in dynamic situations, in contrast to static situations, in which the first process clearly precedes the second without overlapping [24].

In static routing, all data are known prior to the calculation of the route and do not change during its execution. The route is calculated without taking into account collision avoidance, which can dramatically affect the system performance due to possible deadlocks (Fig.1-a) and traffic jams (Fig. 1-b).

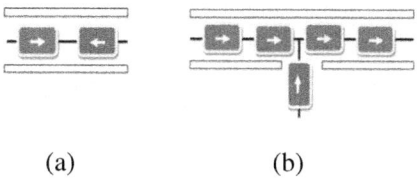

(a)   (b)

Fig. 1. Static routing problem: in (a) deadlocks, both AGVs use the same route in different directions, and in (b) traffic jam, a AGV is blocked by AGVs [2][22]

In this case, an additional system must be used for collision avoidance. This change results in different traffic times which cannot be foreseen at the time of their computation, since the problems are handled only upon the execution of the routes. This is the main disadvantage of applying static routing to systems that depend on previous knowledge on the arrival time [2].

4   Revision of the State-of-the-Art

This section provides a literature review on the main methods developed for solving the problems of dispatching, routing and scheduling. The idea is not to present an exhaustive state-of-the-art review, but rather some of the most used approaches, addressing their main goals and constraints.

4.1   AGVs Dispatching

As previously mentioned, dispatching consists in allocating a task list for the transportation system (AGVs) taking into account some restrictions, like deadlines, priorities etc. Its ultimate goal is associated with the optimization of resources, respecting production times, while minimizing the number of AGVs required to maintain or maximize the production ratio [4].The main approaches are dispatching rules, meta-heuristics, and integer/mixed programming.

Using dispatching rules, Hwang and Kim [25] proposed a new task dispatching algorithm for AGVs based on restrictive rules. The information from the WIP (Work-In-Progress) and the travel time of AGVs (considered deterministic) is used for the algorithm operation. The proposed solution divides the problem of allocating a task to an AGV into a combination of 3 functions. The first takes into account the distance from a vehicle to the workstation   and requests the execution of a task transportation; the other two functions weigh the urgency of that task execution, taking into account the state of the input buffer of the destination station and the output buffer of the station that requests the transportation.   To assess the algorithms performance, the authors in [25] compared the proposed algorithm with some commonly used dispatching restrictions, namely modified-first-come first-serve (MFCFS) and shortest-travel-time-distance (STTD). The developed algorithm provided better results due to the shop-floor used. However, any type of trajectory concern, routing and scheduling was taken. In other words, the routes for each task were pre-defined at the beginning of the algorithm and any collision or deadlock monitoring algorithm was considered. This is a common practice in the use of dispatching rules, however neglecting deadlocks in the routing may cause severe impacts in the transportation system.

Udhayakumar and Kumanan [12] compared the genetic algorithm and the ant-colony for the AGV task dispatching problem considering Meta-Heuristics. The algorithms had a double objective function: to balance the number of tasks given to the AGVs (only two were considered) and minimize the total transportation time so that the utilization ratio of AGVs could be maximized. Other meta-heuristic functions reported in the literature are particle swarm optimization, Tabu search, etc.

For integer/mixed programming, Kasilingan [26] proposed a model to solve the problem of dispatching tasks and determining the minimum number of AGVs required. The model minimizes the total cost of the system, i.e., the sum of the total operation time of the AGVs and the cost of transporting parts between workstations and capacity of vehicles. Moreover, the model underestimates the number of vehicles required.

Rajotia, Shanker and Batra [27] presented a combination of an analytical dispatching model and a simulation approach to determine the optimum number of AGVs. The mixed integer programming model (MIP) considers time handling, load, empty travel time, waiting time and congestion. The objective of the model is to minimize the travel time of the vehicle unloaded.

In addition to the vast dispatching articles in the area of FMS (Flexible Manufacturing Systems) and port terminals (container terminals), dispatching has also been widely reported in the literature, in part due to its complexity in the management of task deadlines and minimization of the number of vehicles used applied to a complex environment and with large-scale work [28-29].

4.2   AGVs Scheduling and Routing

Vehicle routing and scheduling of tasks aim to minimize costs and ensure conflict-free operations and order fulfillment.

The vehicle routing problem consists basically in determining m vehicle routes and minimizing the total distance of all routes by taking into account constraints, such as assignment of each pallet to one AGV, and the cargo capacity of the AGV, which must not beexceed . In real applications, a given number of constraints, such as restrictions on task scheduling, complicates the model. The problem refers to the Vehicle Routing Problem with Time Windows (VRPTW), in which a time window [s, t] is defined for each AGV to perform its tasks [30]. Rajotia, Shanker and Batra [31] added time window to the nodes to represent the arrival and departure times which the AGVs will occupy. Other AGVs can travel through a specific node at a time point not included in one of the time window. Desrochers et al. [32] provided an overview of methods to solve routing with time-window constraints. Several studies on VRPTW have been developed: Lagrangian relaxation [17], Branch and bound methods [33], insertion heuristics[34], constrained shortest path relaxation [35], survey of approximation and optimal approaches [36], dynamic routing method [37] and genetic algorithm [38].

To solve traffic jams and congestion of AGVs, the development of conflict-free routes has emerged in the literature. Broadbent et al. [41] introduced the first concept of conflict-free and shortest-time AGV routing. The routing procedure described employs Dijkstra's shortest path algorithm for the generation of a matrix, which describes the path occupation time of vehicles. Krishnamurthy, Batta and Karwan [42] developed a column generation method for the static routing problem in which an AGV has to move in a bidirectional conflict-free network and minimize the makespan. Maza and Castagna [21] proposed a robust predictive method of routing without conflicts. Möhring et al. [22] extended the approaches of Huang, Palekar and Kapoor [44] and Kim and Tanchoco [20] and proposed an algorithm for the problem of routing AGVs without conflicts at the time of the route computation. Klimm et al. [23] presented an efficient algorithm to cope with the problem of congestion and detours, avoiding potential deadlock situations by using a so-called static approach. Vivaldini et al. [2] proposed an algorithm based on Dijkstra's shortest path and time-window approaches to solve traffic jams and generate optimized conflict-free paths.

Chen et al.[45] also combined the methods used in [2] for a shortest and conflict-free path planning.

In the scheduling literature, Zaremba et al. [47], Veeravalli, Rajesh and Viswanadham [48] and Bing [49] proposed analytical models for the AGVs scheduling. Hartmann [50] introduced a general model for scheduling materials handling equipment (AGVs) in a container terminal such that the average lateness of a job and the average set up time could be minimized. Several types of methods have been employed to conflict-free routes of AGVs [20-23]. We can conclude the issues of routing and scheduling are often studied separately and their integration is a challenging problem.

5   Conclusions

This article has reviewed the state-of-the-art of the AGVs high-level controlling system, addressing their main challenges in comparison with other non-industrial intelligent transportation systems. Dispatching, routing and scheduling were theorically clarified and the main approaches in this area were presented.

Despite all the scientific work developed, the number of automated vehicles installed in industries over the past few years is below what is expected due to high installation costs and the difficulty in taking full advantage of the system. This fact is in part explained by the difficulty in maintaining a close interoperability among dispatching, routing, scheduling and the enterprise MRP (Manufacturing Resource Planning). This is a very complex task which is generally treated as separated problems in the scientific sphere. Furthermore, the difficulty in developing a solution sufficiently generic to be applied to a significant amount of industrial problems has also contributed to the reduced utilization of AGVs in the industrial area.

Acknowledgments. The authors would like to acknowledge Project "NORTE-07-0124-FEDER-000057" financed by the North Portugal Regional Operational Programme (ON.2 – O Novo Norte), under the National Strategic Reference Framework (NSRF), through the European Regional Development Fund (ERDF), and by national funds, through the Portuguese funding agency, Fundação para a Ciência e a Tecnologia (FCT).

References

1. Kalinovcic, L., Petrovic, T., Bogdan, S., Bobanac, V.: Modified Banker's algorithm for scheduling in multi-agv systems. In: IEEE - CASE, pp. 351–356 (2011)
2. Vivaldini, K.C.T., et al.: Automatic Routing System for Intelligent Warehouses. In: IEEE Int. Conference on Robotics and Automation, pp. 93–98 (2010)
3. Vivaldini, K.C.T., Tamashiro, G., Martins Junior, J., Becker, M.: Communication infrastructure in the centralized management system for intelligent warehouses. In: Neto, P., Moreira, A.P., et al. (eds.) WRSM 2013. CCIS, vol. 371, pp. 127–136. Springer, Heidelberg (2013)

4. Qiu, L., Hsu, W., Huang, S., e Wang, H.: Scheduling and routing algorithms for AGVs: A survey. International Journal of Production Research 40(3), 745–760 (2002)
5. Vis, I.F.A.: Survey of research in the design and control of automated guided vehicle systems. EJOR 170(3), 677–709 (2006)
6. Rocha, R.P.P.: Desenvolvimento de um Sistema de Gestão de AGVs, 648 p. Dissertação (Mestrado) – Faculdade de Engenharia da Universidade do Porto, Porto (1998)
7. Co, C.G., Tanchoco, J.M.: A A review of research on AGVS vehicle management. Engineermg Costs and Production Economics 21, 35–42 (1991)
8. Lavegin, A., Lauzon, D., Riopel, D.: Dispatching, Routing, and scheduling ot two automated Guided Vehicles in a Flexible Manufacturing System. Int. J. of Flexible Manufacturing Systems 8, 247–262 (1996)
9. Le-Ahn, T.: Intelligent Control of Vehicle-Based Internal Transport Systems. ERIM Ph.D. Series Research in Management 51. Erasmus University Rotterdam (2005)
10. Egbelu, P.J., Tanchoco, J.M.: A Characteriaztion of automatic guided vehicle dispatching rules. Int. J. Prod. Res. 22(3), 359–374 (1984)
11. De Koster, R.B.M., Le-Ahn, T., Van der Meer, R.: Testing and classifying vehicle dispatching rules in three real-world settings. J. of Op. Managemente 22, 369–386 (2004)
12. Udhayakumar, P., Kumanan, S.: Task scheduling of AGV in FMS using non-traditional optimization techniques. Int. J. Simul. Model. 9, 28–39 (2010)
13. Akturk, M.S., e Yilmaz, H.: Scheduling of automated guided vehicles in a decision making hierarchy. Int. J. Prod. Res. 32, 577–591 (1996)
14. Sabuncuoglu, I., Bayiz, M.: Analysis of reactive scheduling problems in a job shop environment. EJOR 126, 567–586 (2000)
15. Bodin, L.D., et al.: Routing and scheduling of vehicles and crews: the state of the art. Int. J. of Computers and Operations Res. 10(2), 63–211 (1983)
16. Laporte, G.: The vehicle routing problem: an overview of exact and approximate algorithms. EJOR 59(3), 345–358 (1992)
17. Fisher, M.L., Jörnsten, K.O., Madsen, O.B.G.: Vehicle Routing with Time Windows: Two Optimization Algorithms. Operations Research 45, 487–491 (1997)
18. Desrosiers, J., et al.: Time constrained routing and scheduling. Handbooks in Operations Research and Management Science 8, 35–139 (1995)
19. Walker, S.K., et al.: Free-ranging AGV and scheduling system. AGVS, 301–309 (1987)
20. Kim, C.W., Tanchoco, J.M.A.: Conflict-free shortest-time bidirectional AGV routing. International Journal of Production Research 29(12), 2377–2391 (1991)
21. Maza, S., e Castagna, P.: Conflict-free AGV routing in bi-directional network. In: IEEE Int. Conf. On Emerging Tech. and Factory Automation, New York, pp. 761–764 (2001)
22. Möhring, R.H., et al.: Conflict-free real-time AGV routing. In: Hein, F., Dic, H., Peter, K. (eds.) Operations Research Proc. 2004, pp. 18–24. Springer, Heidellberg (2004)
23. Klimm, M., et al.: Conflict-free vehicle routing: load balancing and deadlock prevention.. (2007), http://www.matheon.de/preprints/
5137_preprint-static-routing.pdf
24. Psaraftis, H.N.: Dynamic vehicle routing problems. In: Golden, B.L., Assad, A.A. (eds.) Vehicle Routing: Methods and Studies, pp. 223–248. Elsiever, North-Holland (1988)
25. Hwang, H., Kim, S.H.K.: Development of Dispatching Ruler for Automated Guided Vehicles Systems. IEEE Journal of Manufacturing Systems, 137–143 (1998)
26. Kasilingam, R.G.: Mathematical modeling of the AGVS capacity requirements planning problem. Engineering Costs and Production Economics 21, 171–175 (1991)
27. Rajotia, S., Shanker, K., Batra, J.L.: Determination of optimal AGV fleet size for an FMS. Int. J. Prod. Res. 36(5), 1177–1198 (1998)

28. Grunow, M., Gunther, H., Lehmann, M.: Strategies for dispatching AGVs at automated seaport container terminals. Container Terminals and Cargo Systems 7, 155–178 (2007)
29. Cheng, Y., Sen, H., Natarajan, K.: Dispatching Automated Guided Vehicles in a Container Terminal. Supply Chain Optimization. Applied Optimization 98, 355–389 (2003)
30. Larsen, A.: The dynamic Vehicle Routing Problem. Institute of Mathematical Modeling, Bookbinder Hans Meyer, Lyngby, Technical University of Denmark (2000)
31. Rajotia, S., Shanker, K., Batra, J.L.: A semi-dynamic time window constrained routeing strategy in an AGV system. Int. J. Prod. Res. 36(1), 35–50 (1998)
32. Desrochers, M., et al.: Vehicle routing with time windows: optimization and approximation. In: Vehicle Routing: Methods and Studies, pp. 65–84. Elsevier Science (1988)
33. Kolen, et al.: Vehicle routing with time windows. Op. Res. 35(2), 266–273 (1987)
34. Solomon, M.M.: Algorithms for the vehicle routing and scheduling problems with time window constraints. Operations Research 35(2), 254–265 (1987)
35. Desrochers, M., Desrosiers, J., Solomon, M.: A New optimization algorithm for the vehicle routing problem with time windows. Operations Research 40, 342–354 (1992)
36. Cordeau, J.-F., et al.: VRP with time windows. In: Vehicle Routing Problem, ch. 7, pp. 157–193. SIAM Monographs on Discrete Mathematics and Applications, Philadelphia (2002)
37. Smolic-Rocak, N., et al.: Time windows based dynamic routing in multi-AGV systems. IEEE Transactions on Aut. Sc. Eng. 7(1), 151–155 (2010)
38. Ulrich, C.A.: Integrated machine scheduling and vehicle routing with time windows 227(1), 152–165 (2013)
39. Dumas, Y., Desrosiers, J., Soumis, F.: The pickup and delivery problem with time windows. EJOR 54(1), 7–22 (1991)
40. Solomon, M.M., Desrosiers, J.: Time window constrained routing and scheduling problems. Transportation Science 22, 1–13 (1988)
41. Broadbent, A.J., et al.: Free-ranging AGV and scheduling system. Automated Guided Vehicle Systems 43, 301–309 (1987)
42. Krishnamurthy, N.N., Batta, R., Karwan, M.H.: Developing conflict-free routes for automated guided vehicles. Operations Research 41(6), 1077–1090 (1993)
43. Qiu, L., Hsu, W.J.: A Bi-directional path layout for conflict-free routing of AGVs. International Journal of Production Research 39(1), 2177–2195 (2001)
44. Huang, J., Palekar, U.S., Kapoor, S.: A labeling algorithm for the navigation of automated guided vehicles. Journal of Engineering for Industry 115, 315–321 (1993)
45. Chen, T., et al.: On the shortest and conflict-free path planning of multi-agv systembased on dijkstra algorithm and the dynamic time-window method. Advanced Materials Research 645, 267–271 (2013)
46. Shuhei, E., et al.: Petri net decomposition approach for bi-objective conflict-free routing for AGV systems. In: IEEE Int. Conf. on Systems, Man and Cybernetics, pp. 820–825 (2011)
47. Zaremba, M.B., et al.: A max-algebra approach to the robust dis-tributed control of repetitive AGV systems. Int. J. Prod. Res. 35(10), 2667–2687 (1997)
48. Veeravalli, B., Rajesh, G., Viswanadham, N.: Design and analysis of optimal material distribution policies in flexible manufacturing systems using a single AGV. Int. J. Prod. Res. 40(12), 2937–2954 (2002)
49. Bing, W.X.: The Application of analytic process of resource in an AGV scheduling. Computers and Industrial Engineering 35(1), 169–172 (1998)
50. Hartmann, S.: A General framework for scheduling equipment and manpower at container terminals. OR Spectrum 26, 51–74 (2004)

Coordination for Multi-robot Exploration Using Topological Maps[*]

Tiago Pereira[1,2,3], António Paulo Moreira[1,2], and Manuela Veloso[3]

[1] INESC TEC - INESC Technology and Science (formerly INESC Porto), Porto, Portugal
[2] FEUP - Faculty of Engineering, University of Porto, Portugal
{tiago.raul,amoreira}@fe.up.pt
[3] Carnegie Mellon University, Pittsburgh, PA 15213, USA
tpereira@andrew.cmu.edu, mmv@cs.cmu.edu

Abstract. This paper addresses the problem of decentralized exploration and mapping of unknown environment by a multiple robot team. The exploration methodology relies on individual decision rules and communication of topological maps to achieve efficient and fast mapping, minimizing overlap of explored space. This distributed solution allows scalability of the proposed methods.

Each robot broadcasts a graph representing the topological map, with information of exploration status of each region. Therefore, this kind of information can be transmitted to robots that are not in the communication range, through other robots in a multi-hop network.

This work has been tested in simulation, and the results demonstrate the performance improvements and robustness that arise from our multi-robot approach to exploration.

Keywords: multi-robot, exploration, coordination, topological maps.

1 Introduction

One of the central tasks in robotics is the localization and mapping. Without those, it is impossible to execute most tasks, such as navigation, path generation and motion control.

When doing SLAM, robots have also to explore the environment: plan and decide where to go in order to get the whole map. Particularly in SLAM with occupancy grid maps, the frontier-based exploration and information gain maximization has been used as exploration methodologies.

Being an established solution for single robots scenarios, it is still a research challenge for multi-robot teams when using distributed systems. The solution involves exchanging messages between robots, in order to coordinate the team

---

[*] This work is financed by the ERDF – European Regional Development Fund through the COMPETE Programme (operational programme for competitiveness), by National Funds through the FCT – Fundação para a Ciência e a Tecnologia (Portuguese Foundation for Science and Technology) within project FCOMP-01-0124-FEDER-022701, and under Carnegie Mellon-Portugal Program grant SFRH/BD/52158/2013.

A.P. Moreira et al. (eds.), *CONTROLO'2014 - Proc. of the 11th Port. Conf. on Autom. Control*,
Lecture Notes in Electrical Engineering 321, DOI: 10.1007/978-3-319-10380-8_49

for the mapping task, merging measurements to produce more precise maps. This has great interest, because it allows more robust and fastest mapping [2].

In most literature, however, this is done in a centralized approach. There is a single agent (single point of failure) responsible for merging the individual maps, and sending orders to the robots telling where they should go. This has the disadvantage of not working in environments with limited networking conditions. For that purpose topological maps can be used, because of the limited amount of data that needs to be transmitted. Furthermore, this kind of representation is a high-level description of the world, allowing to use heterogeneous robot teams, and the inclusion of information from humans.

Therefore, the aim of this work is to use a decentralized method, which also allows intermittent and limited communication bandwidth, and scalability in terms of the number of robots that can be used. In this work the objective is only to create a communication strategy using topological maps in order to solve the distributed coordination in the multi-robot mapping tasks.

This kind of solution allows the appearance of global intelligent and efficient behaviors through local decisions and optimization.

2 Related Work

Various map representations can be used for mapping, such as occupancy grids, geometric features, or topological. This last one is used in [4]. Other approach is the concept of manifolds for the map representation, as presented in [8], so as the map is always locally consistent. In [13], an hybrid SLAM solution is proposed, combining occupancy grids and topological maps in a structure inspired in the manifold concept of [8], producing detailed maps without sacrificing scalability.

For unknown environments mapping using a team of robots, SLAM has been the ordinary solution [3]. In multi-robot systems specific problems arise, such as creating a global map from the independent information of each robot, and also the coordination difficulties. In order to coordinate exploration, a good approach makes use of the expected information gain and the movement costs [16,2].

However, these common approaches use a central agent to coordinate the others, which can be a problem. Regarding distributed computation, it facilitates the scalability of the adopted methods (in [9] a team of 80 robots is used for exploration). Besides, this is also advantageous when communication is limited [3], not allowing the transmission of entire individual maps.

Other approaches use information sharing among teammates to improve the robot beliefs, complementing the limited perceptions (due to, for example, uncertainty and hardware limitations) [5]. Nevertheless, the information merging task can be quite challenging, and a promising approach is to use two separate world models, one with the own robot's state (e.g. an individual topological map), and a shared one storing the overall team's state.

Regarding heterogeneous multi-robot systems, in [10] they use not only terrestrial mobile robots of different sizes and characteristic, but also aerial robots, for the purpose of rescuing missions. In this example, they use the different sensing capabilities of each one to attribute and coordinate the tasks given to the robots. In [12] a novel approach is presented for the formation control of heterogeneous robot teams, so as they navigate without colliding with each other.

The use of topological maps is shown to be effective when interacting with humans [15], and it is also important for heterogeneous teams, because their interface can be done using high-level representations of the environment.

A possibility to model humans' availability and costs of interruption to determine when to query them during navigation is presented in [14]. The human knowledge might be difficult to use for mapping, but it is certainly easier when using more abstract features, such as topological maps. So, the human can help robots localize themselves in these high-level maps.

Finally, there are other distributed exploration strategies using topological information. For example, in [6] swarm robots learn topological information about an unknown environment.This method does not consider the possibility of using humans in the systems, so the topological map is not extracted with that in consideration. Besides, this is especially designed for a large number of agents.

In [11], they address the problem of arranging a meeting between two or more robots in a topological environment, creating the possibility of loop-closure for better mapping, under limited communication. In our work, we consider the possibility of circumventing the limited communication problem making use of a multi-hop network of robots, not restricting their behavior so as to move to a predetermined spot.

3   Exploration with Topological Maps

In our solution, robots use a wireless ad-hoc network. In exploration, there is no need to build extremely accurate maps because the global map can be estimated in the end when all robots finish exploring and join themselves again. For example, for coordination a robot only needs to know what others have already explored, not what the map is in those regions. So an algorithm that uses minimum communication to efficiently coordinate the robots is needed.

The solution we propose consists in using topological maps, which will not be used for the global map building and map fusion, but only during exploration. Each individual agent must create its own local map using SLAM (allowing its autonomy and independence), extract topological map from grid map, and then use frontiers to decide where to go. Robots must share only high-level information (topology) among the other robots to generate consensus on what to explore. An hybrid approach (grid and topological maps) allows to create detailed sub-maps until they reach its complexity limit, then becoming a node of the topological map, in a process resembling that of the *manifold* concept.

The topological map is then represented as a graph. The different regions are nodes, that can have information for identification (e.g., area and perimeter), and labels of exploration status (*explored* or *not explored*). Moreover, the regions connectivity is represented by the edges of the graph, as shown in Figure 2.

This message passing protocol contains only the topological information, but for the local exploration problem (inside same region) the frontiers of exploration

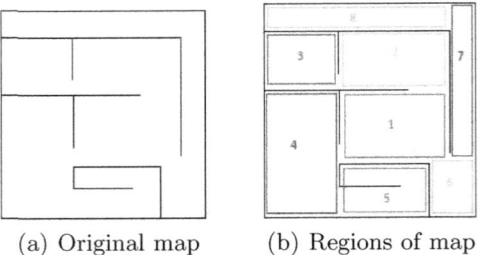

(a) Original map   (b) Regions of map

Fig. 1. Example of a topological map (nodes correspond to the colored regions)

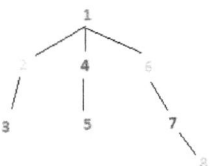

Fig. 2. The graph that would result from world of Figure 1

and local map might also need to be communicated. However, in this work we just rely on obstacle avoidance for local coordination.

4   Coordination Mechanisms

In this section we detail the kind of information that needs to be shared in order to achieve cooperation, and what heuristic rules each robot uses locally when deciding what region needs to be explored.

4.1   Message Structure

So, in this work, each robot stores a graph, with information in each node that is valuable for exploration. Each region, node i of graph G, is characterized as follows on each individual data structure m. This data structure is sent in messages to other robots.

- e^r_{mi}, the local exploration status, boolean variable that is true when region is *explored*, and false for *not explored*;
- e^o_{mi}, the known global exploration status, which is the information received by messages from other robots;
- r^o_{mi}, robot ID that explored region i, when e^o_{mi} is true;
- v_{mi}, true if any known robot has been or is currently in region i;

Furthermore, in each message sent the data structure contains not only the graph information, but also other useful details

- r_m, ID of robot sending message in data structure m;
- t_m, timestamp of message;
- G_m, graph with topological information of exploration status;
- h_m, region which robot is headed to explore;
- p_m^r, the current robot pose;
- p_m^g, the pose of the goal position the robot is trying to reach, in region h_m.

From now on, we consider the graph of own robot as G_A, and the received in messages as graph G_B.

Regarding the graphs received in messages, every robot will use it to update its own graph, which means for each region it will substitute its e_{Ai}^o by the maximum (status is maximum when it is explored) of three possibilities: keep the same, local exploration status, or global exploration status in message.

$$e_{Ai}^o \leftarrow \max\{e_{Ai}^o, e_{Bi}^r, e_{Bi}^o\} \tag{1}$$

Update Graph from Messages. When comparing with e_{Bi}^r, it will always update its own e_{Ai}^o with received information if it is from the same robot ($r_B = r_{Ai}^o$) and timestamp is more recent. This means if the robot is the same, the exploration status can decrease. This is a reasonable behavior because the robot could have said previously some region was explored by mistake, and then find out it was not truly explored.

The same reasoning can be done with e_{Bi}^o, with one exception. In this case it can be information from the robot that is receiving the message ($r_{Bi}^o = r_A$), which is a loop in the information exchange. In this case, it should never keep the information, because the robot itself has the most recent data.

Moreover, it is also easy to find out the importance of e_{mi}^o. With this, it is possible for a robot to receive information from another robot that it could not directly communicate with, through a chain of other agents, like multi-hop routing in mesh networks.

It is interesting to note that timestamps are compared only when belonging to the same robot, to decide when to update information received from others at different times. Thus, there is no need for clock synchronization among robots.

Finally, regarding v_{Ai}, in order to update this variable we only need to apply the OR operation and change its own v_{Ai} to *true* whenever entering a new region.

$$v_{Ai} \leftarrow v_{Ai} \vee v_{Bi} \tag{2}$$

4.2   Coordination Rules

First of all, the frontiers of exploration are sorted according with their information gain, that accounts for the size of frontier, and the cost to get there (distance). After that, the robot checks where each frontier is in the topological map. This could be problematic in the boundary of two regions. Therefore, in order to solve this problem, we introduce the concept of *margin* in the current node of the topological map, which makes frontiers between regions x and y belong to region x when robot is in that one. The implication of this is that robot changes region with confidence, adding hysteresis to the robot localization in the topological domain.

Then, the frontiers are separated in four categories, presented here in decreasing priority order:

- f^{fu}, frontiers in following unexplored regions;
- f^c, frontiers in current region;
- f^{fs}, frontier in following semi-explored regions;
- f^o, the other frontiers that do not fit in any of the previous categories.

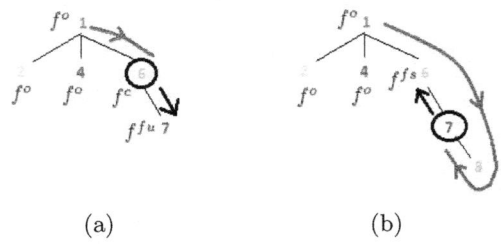

(a)   (b)

Fig. 3. When in a certain region, robot gives preference to the following regions that it still has not explored. In (a) the robot is in regions 6, coming from region 1, and is headed to the following region, 7. In (b) the robot has already reached the leaf of a branch of the graph (region 8), and has come back to region 7, finished exploring that region, and is headed to explore the following region, this time region 6.

So, the robot will choose first frontiers from the first type, f^{fu}, then the second, f^c, etc. Inside each type, they are ordered by information gain, as previously said in the beginning of this section.

The f^{fu} and f^{fs} frontiers belong to regions connected to the current one, but that are not the previous node. For example, as shown in Figure 3, when going from region 1 to 6, the following will be region 7. And when coming back to region 7 from 8, the following is region number 6.

In order to know if regions are unexplored, v_{mi} is used. When this variable is true but region is not completely explored, it is semi-explored, and its frontiers will belong to category f^{fs}.

The idea of using this type of classification is to generate a behavior that makes robots move as far away as possible from the initial region, allowing teams with a large number of robots to explore efficiently. Furthermore, when returning to previously visited regions, the behavior will be to explore those ones completely after moving to another region.

Our aim is to use simple rules in each robot, that result in a complex and organized behavior when considering the whole distributed system.

When there is only the possibility of exploring the same region as other robots, they still do it, with no need for mutual exclusion resolution, because there is no point in stopping when exploration is still going on. Furthermore, it is good to sometimes explore the same region, because that gives common areas in the map to be used later for a more robust map matching between them.

Nevertheless, the individual list of frontiers by type is just a preference list, and a specific goal has still to be chosen from that list.

Finally, a last rule is applied to choose a specific frontier as goal position, which takes in consideration the previously presented parameter h_m. From the list of frontiers ordered by types (individual preferences), it starts checking from the first if there is any robot going there to explore, now giving preference to the ones that are not an exploration goal of other robots.

This strategy might result in choosing one from the bottom of the list, which still is a reasonable choice, even if it is not the preferable region. If someone is already exploring the nearest frontiers, the robot should go to another in order to have efficient exploration. If two or more robots go to the same region, if the distance to the goal of one is less then the other, it will still go there. Therefore, when choosing a frontier to explore in the same region other robot is also headed to explore, each one will use the following rule to decide if it should go there:

$$\|p_A^g - p_A^r\| < \|p_B^g - p_B^r\| \tag{3}$$

One important consideration is that the frontier is not chosen taking into account current regions of other robots, but where they are headed. This means that a robot will probably decide to go to region x even if there is another in there, if that one is going to another region y and the first region is not completely explored.

This exploration strategy could generate temporary incorrect decisions. However, considering the time constant of mechanical systems compared to the communication module that controls them, the incoherent decisions would be negligible, because the time in transitory states of the decision space will still be smaller than the mechanical time constants. The overall decision process is presented in Algorithm 1.

5   Simulation Testbed

For the implementation the Robot Operating System (ROS) and the simulator Stage were used, because it allows the use of other algorithms such as mapping and localization, creating a path to reach a goal position, and follow that path. Another advantage of ROS is the possibility to transparently go from simulation to real scenarios, with little modifications.

The local maps of each robot are occupancy grids, created using an already available SLAM algorithm, Gmapping (a occupancy grid variation of FastSLAM with improved odometry) [7]. In this testbed, frontiers are the central points of clusters of free cells near unexplored cells. The topological map creation and matching are not executed in this application, but simulated using the known ground truth (robot position and known map structure). This, however, could be the result of an "oracle" giving information with no cost, as explained in [1].

Order frontiers by information gain and separate by category;
while *goal undefined* **do**

> Select next frontier from current category;
> **if** *No more frontiers in current category* **then**
>> | Change category and select first frontier;
>
> **end**
> **if** *Current frontier's region being explored by other robot* **then**
>> **if** *Distance to frontier smaller than other* **then**
>>> | Select frontier as goal position;
>>> | Break;
>>
>> **end**
>
> **else**
>> | Select frontier as goal position;
>> | Break;
>
> **end**

end
if *goal undefined* **then**
> | Select first frontier from first category;

end

Algorithm 1. How to choose a goal position from frontiers list

Finally, it is also possible to model network inefficiencies, such as limited communication range and packet losses. The first is done using the relative position between robots in the simulation to limit the range of communication. Regarding packet losses, in our applications we can simply accomplish it using a high period for the exploration package (but still enough to achieve coordination).

It is also important to notice that lack of communication does not result in inconsistencies, but mainly in inefficient exploration by the team.

6 Results

For this project we did not focus on the mapping quality, only the efficiency of exploration. So, we analyze the effect of using N robots on the total time, and how the increasing number of robots affects the exploration time.

We used the environment from Figure 1(a), with simple structured environments, because the graph extraction had to be done manually, which would be impractical for real building maps. So, in this scenario the current position of the robot and frontiers in the topological map is given manually.

With that setup, four different scenarios were tested:

- Exploration with one robot and unlimited communication;
- Exploration with two robots and unlimited communication;
- Exploration with three robots and unlimited communication;
- Exploration with three robots and limited communication with range of 5 meters.

The first test took 450 seconds, the exploration with 2 robots took 75% of the that time, and with 3 robots around 50%.

When using limited communication, two of the three robots started exploring already explored regions, namely regions 6 and 7 of Figure 1(b). They decided to re-explore them because those regions were the last to be explored, so both robots do it until exploration is finished.

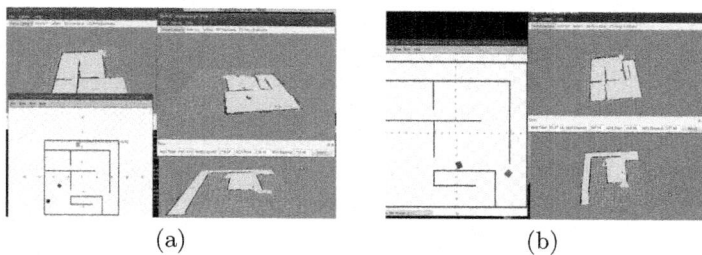

(a)   (b)

Fig. 4. In (a), efficient exploration with 3 robots. Green robot partially explores region 1, and fully explores regions 6, 7 and 8. The red one explores regions 1, 2, 3 and partially region 4. The blue robot explored partially 1, and completely regions 4 and 5. In (b), the final state when exploring with 2 robots. The red robot explored regions 1, 2, 3, 4 and 5. The blue one explored partially region 1, and completely regions 6, 7 and 8.

Finally, it is possible to see the result of exploration in Figure 4, with little overlap of mapping. This demonstrates the efficiency of this coordination strategy, that can later result in a full global map although each robot explored only specific regions of the environment.

7   Conclusions and Future work

The abstraction of world structure through topological maps helps solving the coordination problem, allowing to use it in the future with heterogeneous teams.

With this paper we contributed with a communication framework for coordination, and a heuristic to decide what regions to explore, creating rules that result in a global intelligent behavior of the robots. So, using frontier-based exploration coupled with topological maps and message passing, we achieved a coordination strategy that can be used by robots in exploration and mapping.

As future work, it is important to create a local exploration strategy that can coordinate robots when they are in the same region, and to implement autonomous topological map extraction and graph matching procedures in order to be able to use it with real robots for autonomous exploration.

References

1. Armstrong-Crews, N., Veloso, M.: Oracular partially observable markov decision processes: A very special case. In: 2007 IEEE International Conference on Robotics and Automation, pp. 2477–2482. IEEE (2007)

2. Burgard, W., Moors, M., Stachniss, C., Schneider, F.: Coordinated multi-robot exploration. IEEE Transactions on Robotics 21(3), 376–386 (2005)
3. Carlone, L., Ng, M., Du, J., Bona, B., Indri, M.: Rao-blackwellized particle filters multi robot slam with unknown initial correspondences and limited communication. In: 2010 IEEE International Conference on Robotics and Automation (ICRA), pp. 243–249. IEEE (2010)
4. Chang, H., Lee, C., Hu, Y., Lu, Y.: Multi-robot slam with topological/metric maps. In: IEEE/RSJ International Conference on Intelligent Robots and Systems, IROS 2007, pp. 1467–1472. IEEE (2007)
5. Coltin, B., Liemhetcharat, S., Meriçli, C., Tay, J., Veloso, M.: Multi-humanoid world modeling in standard platform robot soccer. In: 2010 10th IEEE-RAS International Conference on Humanoid Robots (Humanoids), pp. 424–429. IEEE (2010)
6. Dirafzoon, A., Lobaton, E.: Topological mapping of unknown environments using an unlocalized robotic swarm. In: 2013 IEEE/RSJ International Conference on Intelligent Robots and Systems (IROS), pp. 5545–5551 (2013)
7. Grisetti, G., Stachniss, C., Burgard, W.: Improved techniques for grid mapping with rao-blackwellized particle filters. IEEE Transactions on Robotics 23(1), 34–46 (2007)
8. Howard, A.: Multi-robot mapping using manifold representations. In: Proceedings of the 2004 IEEE International Conference on Robotics and Automation, ICRA 2004, vol. 4, pp. 4198–4203. IEEE (2004)
9. Howard, A., Parker, L., Sukhatme, G.: Experiments with a large heterogeneous mobile robot team: Exploration, mapping, deployment and detection. The International Journal of Robotics Research 25(5-6), 431–447 (2006)
10. Luo, C., Espinosa, A., Pranantha, D., De Gloria, A.: Multi-robot search and rescue team. In: 2011 IEEE International Symposium on Safety, Security, and Rescue Robotics (SSRR), pp. 296–301. IEEE (2011)
11. Meghjani, M., Dudek, G.: Multi-robot exploration and rendezvous on graphs. In: 2012 IEEE/RSJ International Conference on Intelligent Robots and Systems (IROS), pp. 5270–5276 (2012)
12. Nascimento, T.P., Conceição, A.G.S., Alves, H.P., Fontes, F.A., Moreira, A.P.: A generic framework for multi-robot formation control. In: Röfer, T., Mayer, N.M., Savage, J., Saranlı, U. (eds.) RoboCup 2011. LNCS, vol. 7416, pp. 294–305. Springer, Heidelberg (2012)
13. Pfingsthorn, M., Slamet, B., Visser, A.: A scalable hybrid multi-robot SLAM method for highly detailed maps. In: Visser, U., Ribeiro, F., Ohashi, T., Dellaert, F. (eds.) RoboCup 2007. LNCS (LNAI), vol. 5001, pp. 457–464. Springer, Heidelberg (2008)
14. Rosenthal, S., Veloso, M.: Modeling humans as observation providers using pomdps. In: 2011 IEEE RO-MAN, pp. 53–58. IEEE (2011)
15. Santos, F.N., Moreira, A.P., Costa, P.C.: Towards extraction of topological maps from 2D and 3D occupancy grids. In: Correia, L., Reis, L.P., Cascalho, J. (eds.) EPIA 2013. LNCS, vol. 8154, pp. 307–318. Springer, Heidelberg (2013)
16. Simmons, R., Apfelbaum, D., Burgard, W., Fox, D., Moors, M., Thrun, S., Younes, H.: Coordination for multi-robot exploration and mapping. In: Proceedings of the National conference on Artificial Intelligence, pp. 852–858. AAAI Press, MIT Press, Menlo Park, Cambridge (1999, 2000)

Implementation and Validation of a Docking System for Nonholonomical Vehicles

João Barbosa[1], Carlos Cardeira[1], and Paulo Oliveira[1,2]

[1] IDMEC/LAETA - Instituto Superior Técnico, Universidade de Lisboa, 1049-001 Lisboa, Portugal
[2] ISR - Instituto Superior Técnico Universidade de Lisboa, 1049-001 Lisboa, Portugal

Abstract. This paper presents the experimental validation of a docking system for a differential drive robot. The docking problem is solved with a smooth, time-invariant, globally asymptotically stable feedback control law which allows for a very human-like closed-loop steering that drives the robot to a certain goal with a desired attitude and a tunable curvature. Simulations of the docking problem are presented that illustrate the performance of the system and it is also validated by performing tests on the aforementioned real robot.

Keywords: Docking, Mobile Robots, Nonlinear Control, Lyapunov Techniques.

1 Introduction

Industrial automation has experienced huge advance in the last decades [1]. Flexible automation manufacturing cells require the use of automatic handling solutions usually resorting to Automatic Guided Vehicles (AGVs). Nowadays the fleets of AGVs must navigate among warehouses, automated workcells, and charging stations. Thus, the automated docking of mobile robotic platforms, such as AGVs, with minimal structuring of the environment is still a active topic of research [2]. The solutions found in the literature to solve this problem vary both in algorithm and sensor payload. One approach, defined as visual servoing, with an early contribution reported in [3], consists of representing a given task directly by an error relative to a goal image to be captured by the vision system. This approach became popular from 1990 onward with works such as [4], with a great contribute of the task function approach [5]. Visual servoing benefits from contributes with out-of-body cameras, i.e., Camera-Space Manipulation (CSM) [6], Mobile Camera-Space Manipulation (MCSM) which extends the latter with body embedded cameras and, more recently, [7] which computes the goal configuration using visual landmarks. Other approaches to the docking problem include the computation of feedback control laws by using Lyapunov and backstepping techniques that lead to an Ultra-Short Baseline (USBL) acoustic positioning system [8] applied on the underwater counterpart of this work, the use of electromagnetic homing systems [9], optical guidance approaches such as [10] and computing the deceleration needed by a robot,

A.P. Moreira et al. (eds.), *CONTROLO'2014 - Proc. of the 11th Port. Conf. on Autom. Control,*
Lecture Notes in Electrical Engineering 321, DOI: 10.1007/978-3-319-10380-8_50

resorting to an estimation of a *time-to-contact* (τ) through optical flow field divergence measurements of an image stream as in [11] and references therein. In [12] a method based on the direction of arrival (DOA) of signals transmitted by RFID transponders is proposed, showing that a robot can dock in a station transmitting through an RFID by using two antennae installed on-board of the vehicle. A method proposing the estimation of the position and orientation of a visual landmark is presented in [13] to help on docking and automatic recharging, thus being similar to the work presented herein.

This work validates a functional docking system using a full state feedback law inspired in [14] which drives a nonholonomic vehicle from any initial position to a certain defined goal with a desired attitude. Mobile robots resorting to vision systems allow for more versatility in an industrial facility regarding docking stations, positioning of cargo pallets, consequently simplifying the overall task of map building and task planning, dropping the need of extreme precision regarding these actions.

The present paper is organised as follows: Details on the feedback control law used in this work are presented in Section 2, which is followed by the results of real-time experiments in Section 3. Finally, some conclusions on the overall performance of the proposed strategies are drawn in Section 4.

2   Docking Problem

In this section, the model of the mobile robot and the operation environment is introduced (see [15] for details) and the implemented control law is described. Suppose the state of the robot is $\mathbf{z} = [e_x \ e_y \ \psi]^T \in \mathcal{R}^3$ composed by quantities depicted in Fig. 1. Representing the docking station position $\mathbf{e}(t) \in \mathcal{R}^2$ and attitude $\psi(t) \in \mathcal{R}$ in the body frame allows for the derivation of linear kinematics and output equations. Furthermore, it is also possible to estimate linear and angular slippages, respectively $b \in \mathcal{R}$ and $s \in \mathcal{R}$, that may occur due to the lack of knowledge of the contact points with the floor as well as the lack of precision in the measurement of each wheel radius or asymmetries in mechanical construction. Both slippages are considered to be slow time-varying or even constant ($\dot{b} = 0$ and $\dot{s} = 0$). The state vectors estimated by the position and attitude estimators are, respectively, $\mathbf{x} = \begin{bmatrix} \mathbf{e}^T & b \end{bmatrix}^T$ and $\boldsymbol{\theta} = [\psi \ s]^T$. Both systems are represented in (1)

$$\begin{cases} \dot{\boldsymbol{\theta}}(t) = \mathbf{A}^{\boldsymbol{\theta}}\boldsymbol{\theta}(t) + \mathbf{B}^{\boldsymbol{\theta}}\omega(t) + \boldsymbol{\nu}(t) \\ y(t) = \psi(t) + \eta(t) \end{cases}, \tag{1}$$

where

$$\mathbf{A}^{\boldsymbol{\theta}} = \begin{bmatrix} 0 & -1 \\ 0 & 0 \end{bmatrix}, \mathbf{B}^{\boldsymbol{\theta}} = \begin{bmatrix} -1 \\ 0 \end{bmatrix},$$

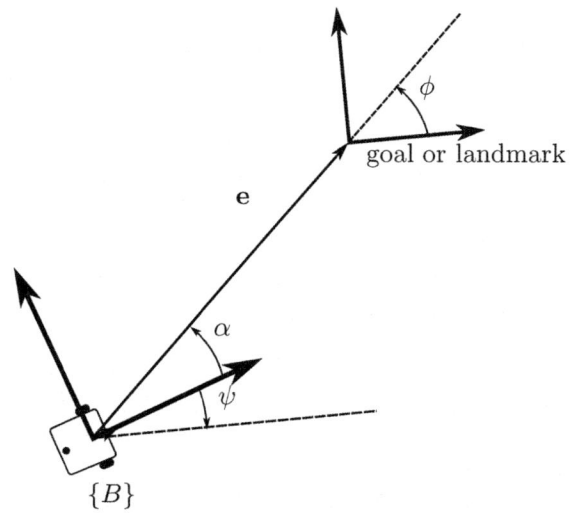

Fig. 1. Depiction of robot and docking station frames

and (2)

$$\begin{cases} \dot{\mathbf{x}}(t) = \mathbf{A}^{\times}(\omega(t))\mathbf{x}(t) + \mathbf{B}^{\times}\bar{v}(t) + \mathbf{v}(t) \\ \mathbf{y}(t) = \mathbf{e}(t) + \mathbf{w}(t) \end{cases} , \qquad (2)$$

where

$$\mathbf{A}^{\times}(\omega(t)) = \begin{bmatrix} 0 & \omega & -1 \\ -\omega & 0 & 0 \\ 0 & 0 & 0 \end{bmatrix}, \mathbf{B}^{\times} = \begin{bmatrix} -1 \\ 0 \\ 0 \end{bmatrix}.$$

Notice that both systems are linear and do not make use of any approximation. Furthermore, (1) originates an optimal observer and (2) a sub-optimal one, as ω is a measured quantity, rather than a known one. Both systems are proven observable and their Kalman Filter application is detailed in [15].

Notice now that $\mathbf{z} = [e_x \; e_y \; \psi]^T$ can be represented in the form

$$\dot{\mathbf{z}} = \mathbf{f}_\omega(\mathbf{z})\omega + \mathbf{f}_v(\mathbf{z})v, \qquad (3)$$

where

$$\mathbf{f}_\omega = \begin{bmatrix} e_y \\ -e_x \\ -1 \end{bmatrix}, \quad \mathbf{f}_v = \begin{bmatrix} -1 \\ 0 \\ 0 \end{bmatrix}.$$

The present work follows Lyapunov's direct method of finding a scalar energy-like function $V(\mathbf{z})$ and devise a control law $\mathbf{u}(\mathbf{z})$ that ensures the resulting closed-loop system is asymptotically stable (the goal is to park the vehicle in a position $\mathbf{z}^*$). This leads to a smooth and time invariant control law. However, a theorem developed by Brockett shown in [16], states that, for systems in the structure

$$\dot{\mathbf{z}} = \sum_{i=1}^{m} \mathbf{f}_i(\mathbf{z}) u_i,$$

with vectors $\mathbf{f}_i(\mathbf{z})$ being linearly independent and continuously differentiable at a point z^*, then there exists a stabilization solution, with a smooth and time invariant feedback law, if and only if $m = n$, where n is the order of the system, meaning there need to be the same number of control parameters as the dimension of the state vector to be controlled. The system in (3) does not respect the last condition and clearly has $\mathbf{f}_\omega$ and $\mathbf{f}_v$ independent at the origin. This would then require the use of time-varying or discontinuous control laws in order to achieve the desired stabilization. Seeing as the need of stabilizing n states at a point $\mathbf{z}^*$ is still the objective, then only a system with singularities is of interest. With this in mind, a new system is proposed in [14]. The said system, represented in (4), is based on a state vector that is isomorphic with the one in (3), characterized by the isomorphism $g : \mathbb{R}^3 \setminus \{\mathbf{0}\} \mapsto \mathbb{R}^3 \setminus \{\mathbf{0}\}$

$$\begin{cases} e = \|\mathbf{e}\| \\ \alpha = atan(e_y/e_x) \\ \phi = atan(e_y/e_x) - \psi \end{cases},$$

variables depicted in Figure 1, leading to the kinematics

$$\begin{cases} \dot{e} = -v \cos \alpha \\ \dot{\alpha} = -\omega + v \frac{\sin \alpha}{e} \\ \dot{\phi} = v \frac{\sin \alpha}{e} \end{cases}. \tag{4}$$

Due to the singularity at the origin, Brockett's theorem no longer applies, since the regularity assumptions do not hold, and so the asymptotic stabilization of (4) is possible. One then cannot formally use the definition of equilibrium point to describe the origin, since it is now located in the frontier of the open set of validity of the system dynamics. The objective of the control law is then to asymptotically drive the system to $\mathbf{z_p}^* = \begin{bmatrix} 0 & 0 & 0 \end{bmatrix}^T$ without attaining $e = 0$ in a finite time, where $\mathbf{z_p} = \begin{bmatrix} e & \alpha & \phi \end{bmatrix}$ is henceforth the notation used for the new state vector. A simple choice for a candidate Lyapunov function is the often used quadratic error form

$$V(\mathbf{z}) = \underbrace{\frac{1}{2}\lambda e^2}_{V_1} + \underbrace{\frac{1}{2}\alpha^2 + \frac{1}{2}h\phi^2}_{V_2}, \quad \lambda, h > 0 \tag{5}$$

where λ and h are positive weighting constants that will help shape the control law. By separating the scalar function in two terms, we have that the first term refers to the error in distance to the target position, and the second term corresponds to the to a "alignment vector" error $\begin{bmatrix} \alpha & \sqrt{h}\phi \end{bmatrix}$. It is clear by now that the a candidate of a scalar function has been chosen and then a function $\mathbf{u}(\mathbf{z_p})$ will be derived in order for the behaviour of V along the trajectory of (4) to drive the state asymptotically to the origin. Taking then the derivative $\dot{V}$, given by

$$\dot{V} = \lambda e \dot{e} + \left(\alpha \dot{\alpha} + h\phi \dot{\phi} \right)$$
$$= \lambda ev \cos\alpha + \alpha \left[-\omega + v \frac{\sin\alpha(\alpha + h\phi)}{\alpha e} \right]. \tag{6}$$

The first term of (6) can be made non-positive by letting

$$v = \gamma e \cos\alpha, \quad \gamma > 0, \tag{7}$$

leading to

$$\dot{V_1} = -\lambda\gamma \cos^2\alpha \, e^2 \le 0. \tag{8}$$

This choice of linear velocity control law ensures that the validity of (4) throughout the parking problem, since V_1 is lower bounded and non-increasing, making it asymptotically converge to a non negative finite limit, thus ensuring e exhibits the same behaviour. The same strategy is applied to the second term, and so expression for the angular velocity control law is

$$\omega = k\alpha + v \frac{\sin\alpha(\alpha + h\phi)}{\alpha e}$$
$$\overset{(7)}{=} k\alpha + \gamma \frac{\cos\alpha \sin\alpha(\alpha + h\phi)}{\alpha}. \tag{9}$$

The derivative of the total Lyapunov function then becomes

$$\dot{V} = -\gamma \left(\cos^2\alpha \right) e^2 - k\alpha^2 \le 0, \tag{10}$$

which is negative semi definite. It is however possible, as described in [14], to prove that the origin is globally asymptotically stable by using LaSalle's theorem and Barbalat's Lemma upon the inspection of the resulting closed-loop system. Note that the objective is to have the vehicle dock in a certain station with positive linear velocity, but it is possible to obtain different trajectories by simply changing the goal objective (to for instance $\mathbf{z_p}^* = [0, \pm\pi, \pm\pi]$). A depiction of the trajectories performed by the system are depicted in Fig. 2, where a simulation was performed with $\gamma = 3$, $h = 1$ and $k = 6$.

As intended, the vehicle always arrives at the target location facing, which goes accordingly with state vector converging to the origin.

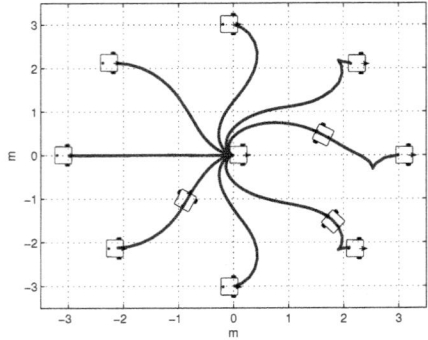

Fig. 2. Trajectories performed with $e(0) = 3$ and $\psi(0) = 0$

3 Experimental Results

The focus of this section is the validation of the docking solution. The robot prototype and landmark setup are shown in Fig. 3. The architecture of the localization system that provides the necessary measurements is represented in Fig. 4. Below is a summary of the parameters and initialization of both Kalman Filters used in the observer.

- Camera noise covariance: $\mathbf{R}^\mathsf{x} = 1 \times 10^{-2}\mathbf{I}_2$ and $\mathbf{R}^\theta = 1 \times 10^{-2}$
- Plant noise covariance: $\mathbf{Q}^\mathsf{x} = diag(4.1 \times 10^{-6}\mathbf{I}_2, 1 \times 10^{-8})$ and $\mathbf{Q}^\theta = diag(2 \times 10^{-5}, 1 \times 10^{-8})$
- Initial covariance matrix: $\mathbf{P}_0^\mathsf{x} = 1\mathbf{I}_3$ and $\mathbf{P}_0^\theta = 0.1\mathbf{I}_3$
- Initial conditions: $\hat{\mathbf{e}}$ and $\hat{\theta}$ were set to the real initial position, and both bias estimates $\hat{b}$ and $\hat{s}$ were set to zero.

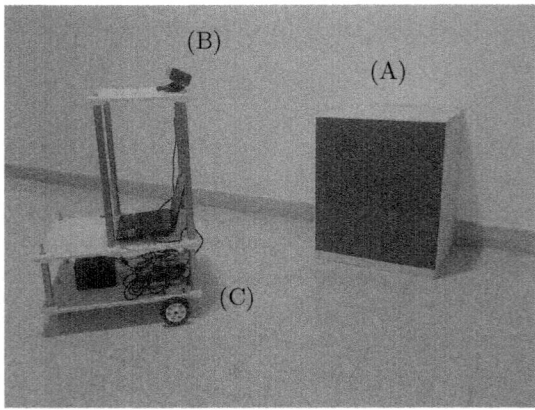

Fig. 3. (A) Docking station, (B) 3D camera and (C) Robot prototype

Sensor Package   **Localization Modules**

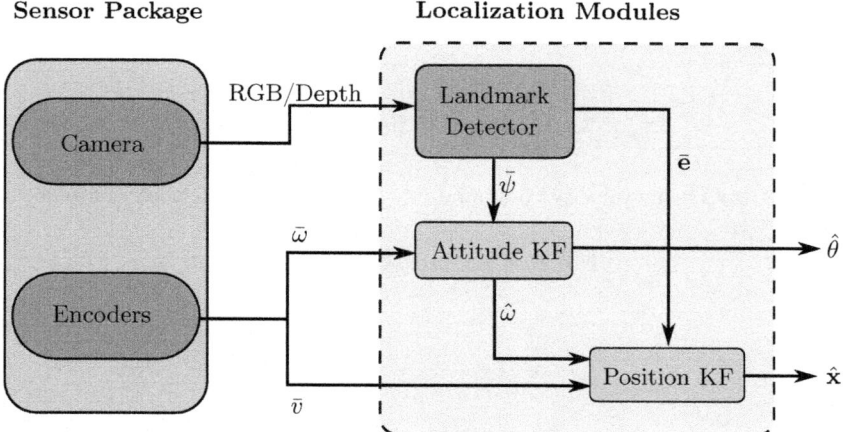

Fig. 4. Estimator Modules

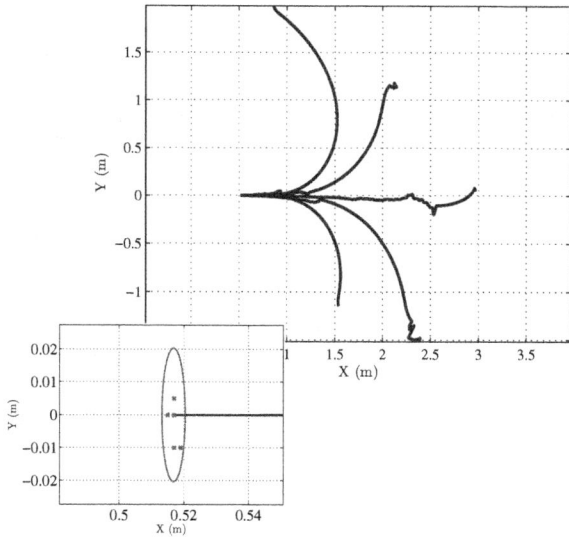

Fig. 5. Depiction of several docking manoeuvres and their 3σ error confidence and ground truth of the final position (zoom)

Table 1. Error in docking manoeuvres

	μ	σ
$e_x\ [cm]$	3.5	0.12
$e_y\ [cm]$	-0.5	0.61
$\psi\ [°]$	0.08	0.285

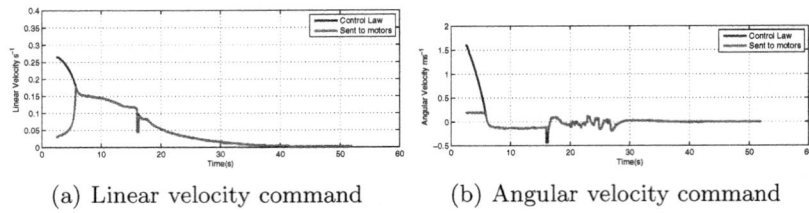

(a) Linear velocity command   (b) Angular velocity command

Fig. 6. Time progression of commands

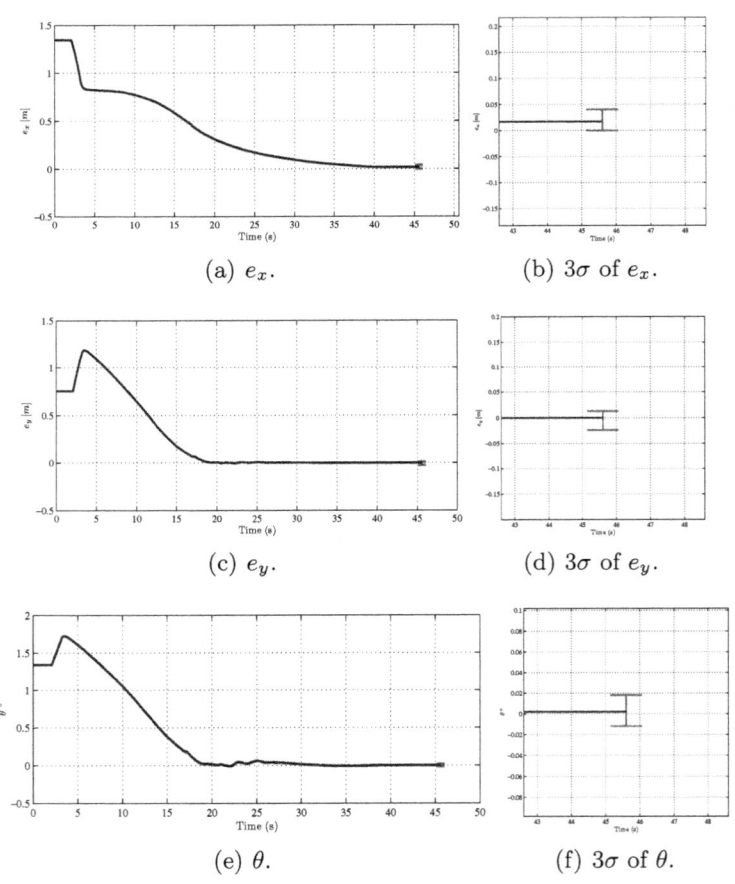

(a) e_x.

(b) 3σ of e_x.

(c) e_y.

(d) 3σ of e_y.

(e) θ.

(f) 3σ of θ.

Fig. 7. Time progression of state variables and the 3σ interval for final position

For the tests presented in this section, the landmark or docking station is considered to be the origin of the inertial frame and the goal of every experiment presented in this section was set $0.5\,m$ in front of the real landmark object being used. Also, in every experiment, unless stated otherwise, the initial estimate of the position was very near the real position of the mobile robot, seeing as the estimation filters were active before the feedback loop was enabled. A saturation of $v_{max} = 0.2\ ms^{-1}$ and $\omega_{max} = 0.2\ ms^{-1}$. Figure 5 depicts the localisation estimate and true final position of several tests carried out in the laboratory environment. In each one of the 6 tests, represented in zoom of Fig. 5, the prototype was able to perform a successful docking manoeuvre, even when the initial estimate had a slight error. The final position error is represented in Table 1.

In Fig. 6 the commands in a particular experiment are shown and the effect that the correction of the estimate in them while Fig. 7 depicts the state variables converging to zero. Notice that, due to the discrete nature of commands, the vehicle never reaches the goal completely, and the error distance to the target will depend on the value of γ and the error in the camera to body transposition and rotation calibrations.

4   Conclusions and Further Work

The implemented feedback control successfully drives the robot to a given goal and is tunable to the needs of the localisation system, seeing as it is possible to require the robot to take the same path under different speeds by just adjusting the parameters that tune the feedback. It is also possible to achieve good results by imposing a saturation in linear and angular velocities as to ensure convergence of the localisation system within the manoeuvre time frame. The robot is able to drive itself to the correct goal even when a wrong initial position and attitude estimate occurs, given that the docking station is, at some point, visible by the sensor package. The error of the docking manoeuvre, within 3σ tolerance, is $3\sigma_x = 0.36\ cm$, $3\sigma_y = 1.83\ cm$ and $3\sigma_\theta = 0.855\ °$, being that most of it was due to uncertainties in the camera transform and due to the quantization of the motor commands. Also note that the 3D camera does not recognize the docking station upon the final approach segment of the manoeuvres, which corresponds to a $1\ m$ distance, resulting in odometry navigation during the last part of each docking. The accuracy may improve in further work which includes other sensors with better accuracy, such as laser range-finders, to avoid the odometry errors during the last segment. This work can go towards 3D manoeuvres (with other vehicles and models) and cooperative manoeuvres among multiple robots.

Acknowledgements. This work was supported by FCT, through IDMEC, under LAETA Pest-OE / EME / LA0022 and partially supported by the project PRODUTECH-PTI (Proj. 3904) under the program COMPETE / QREN / FEDER.

References

1. Groover, M.P.: Automation, Production Systems, and Computer-Integrated Manufacturing, 3rd edn. Prentice Hall Press, Upper Saddle River (2007)

2. Hada, Y., Yuta, S.: A first-stage experiment of long term activity of autonomous mobile robot result of repetitive base-docking over a week. In: Rus, D., Singh, S. (eds.) Experimental Robotics VII. LNCIS, vol. 271, pp. 229–238. Springer, Heidelberg (2001)
3. Agin, G.J.: Real time control of a robot with a mobile camera. SRI International (1979)
4. Espiau, B., Chaumette, F., Rives, P.: A new approach to visual servoing in robotics. IEEE Transactions on Robotics and Automation 8(3), 313–326 (1992)
5. Samson, C., Espiau, B., Borgne, M.L.: Robot control: the task function approach. Oxford University Press (1991)
6. Skaar, S.B., Yalda-Mooshabad, I., Brockman, W.H.: Nonholonomic camera-space manipulation. IEEE Transactions on Robotics and Automation 8(4), 464–479 (1992)
7. Lefebvre, O., Lamiraux, F.: Docking task for nonholonomic mobile robots. In: Proceedings of the 2006 IEEE International Conference on Robotics and Automation, ICRA 2006, pp. 3736–3741. IEEE (2006)
8. Batista, P., Silvestre, C., Oliveira, P.: A two-step control strategy for docking of autonomous underwater vehicles. In: American Control Conference (ACC), pp. 5395–5400. IEEE (2012)
9. Feezor, M.D., Yates Sorrell, F., Blankinship, P.R., Bellingham, J.G.: Autonomous underwater vehicle homing/docking via electromagnetic guidance. IEEE Journal of Oceanic Engineering 26(4), 515–521 (2001)
10. Park, J.Y., Jun, B.H., Lee, P.M., Lee, F.Y., Oh, J.H.: Experiment on underwater docking of an autonomous underwater vehicleisimi'using optical terminal guidance. In: OCEANS 2007-Europe, pp. 1–6. IEEE (2007)
11. McCarthy, C., Barnes, N., Mahony, R.: A robust docking strategy for a mobile robot using flow field divergence. IEEE Transactions on Robotics 24(4), 832–842 (2008)
12. Kim, M., Chong, N.Y.: Direction sensing rfid reader for mobile robot navigation. IEEE Transactions on Automation Science and Engineering 6(1), 44–54 (2009)
13. Luo, R.C., Liao, C.T., Su, K.L., Lin, K.C.: Automatic docking and recharging system for autonomous security robot. In: 2005 IEEE/RSJ International Conference on Intelligent Robots and Systems (IROS 2005), pp. 2953–2958. IEEE (2005)
14. Aicardi, M., Casalino, G., Bicchi, A., Balestrino, A.: Closed loop steering of unicycle like vehicles via lyapunov techniques. IEEE Robotics & Automation Magazine 2(1), 27–35 (1995)
15. Barbosa, J.: Design and Validation of a Docking System for Nonholonomical Vehicles. Master's thesis, Instituto Superior Técnico, Universidade de Lisboa (2014)
16. Brockett, R.W., et al.: Asymptotic stability and feedback stabilization. In: Differential Geometric Control Theory, pp. 181–191. Defense Technical Information Center (1983)

Grígora S: A Low-Cost, High Performance Micromouse Kit

Antonio Valente[1,2,*], Paulo Salgado[2,**], and José Boaventura-Cunha[1,2,***]

[1] INESC TEC
INESC Technology and Science (formerly INESC Porto)
UTAD Pole, Vila Real, Portugal
[2] School of Sciences and Technology
Engineering Department
Universidade de Trás-os-Montes e Alto Douro (UTAD)
Vila Real, Portugal
{avalente,psal,jboavent}@utad.pt

Abstract. There is a need to attract students to science and engineering courses. Robotic contests are one of the most promising ways to attract students to the field of robotics and thereby to science and technology. Micromouse contest is one of promising contests where a small autonomous robot has to navigate its way through an unknown 16×16 cells maze. Since the design and construction of robots is interesting but difficult, this paper presents a high performance, low-cost robot kit for high school and university students participate on micromouse robot contests. This micromouse kit, developed at University of Trás-os-Montes and Alto Douro (UTAD), fits on a 10×8 cm rectangle and uses very small stepper motors allowing a maximum speed of about 4 m/s thus comparing to state-of-the-art micromice. The kit also incorporates a popular microcontroller hardware module (Arduino Leonardo) which facilitates all programming tasks.

Keywords: Micromouse Kit, Micromouse robot, Arduino, Odometry.

1 Introduction

The enchantment of robots for many children (and adults) is noticeable in the surge of robotics animation films, such us *Wall-E* and *Robots*, in the proliferation of affordable robot toys and construction sets, and in the publication of robotics magazines and websites. At the same time, there is a generalized concern about the falling numbers of science and technology students leaving to the need to attract them to these courses as early as possible [1]. Marian Petre [1] described examples of children learning subjects that they previously considered difficult and inaccessible, in order to solve problems in robotics. Furthermore, secondary

---

* ORCID:0000-0002-5798-1298.
** ORCID:0000-0003-0041-0256.
*** ORCID:0000-0002-8406-0064.

© Springer International Publishing Switzerland 2015   535
A.P. Moreira et al. (eds.), *CONTROLO'2014 - Proc. of the 11th Port. Conf. on Autom. Control,*
Lecture Notes in Electrical Engineering 321, DOI: 10.1007/978-3-319-10380-8_51

school students working in teams learned that this programming and engineering knowledge has a social context. Robotic contests are one of the most promising ways to attract students to the field of robotics, since winning a award at a competition not only gives students a sense of accomplishment but also gives pride and visibility to schools. In RoboCup Jnr., Robot World Cup Initiative [2] for primary and secondary school children, there are competitions in specific challenges: *Soccer* - matches between 2-on-2 teams of autonomous robots on a 90 × 150 cm grey-scale pitch; *Rescue* - autonomous robots race to identify 'victims' in a line-following task incorporating obstacles and uneven terrain; *Dance* - one or more autonomous robots perform to music in a competition judged for creativity. In the recent years a 35 years-old contest is become very popular - the Micromouse Contest. These contests are very visual attractive and they be held around the world (UK [3], USA [4], Japan [5], Singapore [6], and Taiwan [7]). In 2013 the University of Trás-os-Montes and Alto Douro starts to organize the Portuguese Micromouse Contest [8] trying to attract students to robotic and engineering courses. In 2013 only high level university students participate but the contest aims the participation of high school students and, in a near future, people from 8 to 80 years old.

Micromouse is a small autonomous microcontrolled robot vehicle that has to navigate its way through an unknown maze. The main challenge for the contestant is to impart to the micromouse an adaptive intelligence to explore different maze configurations and to work out the optimum route for the shortest travel time from start to finish. The maze, depicted in Fig. 1, consists of 16 × 16 squares of 18 cm × 18 cm each. The horizontals and verticals passageways are 16.8 cm wide and the diagonals are 11.03 cm. The walls are 5 cm high, white on the sides and white on the top (in some events the top is red). The floor of the maze is black. The start of the maze is in one corner (S); the goal (G) is the center four squares.

The scoring procedure is quite complicated, rewarding efficient maze exploration algorithms, and penalising inefficient ones. It is not only speed that determines the winner; reliability and intelligence are also taken into account. For example, if a mouse is touched in anyway, then it is heavily penalised. State-of-the-art micromouse can even run at a speed of 3 m/s and in a diagonal path. The combination of easily defined goals, plus a scoring system that rewards efficient and reliable design makes micromouse an ideal student project that can be taken from low to high level students.

Unfortunately, learning the design philosophy of robots is interesting but difficult, because it includes several areas of knowledge, e.g., mechanics and electronics (a robot is a mechatronic device), automatic control theory, software programming of microcontrollers, among others. Building a micromouse, which has dimensions constrains, become even more difficult and time consuming for university level students and almost impossible for high school students. The dimension constrains and necessary good performance also discourages the use of popular modules (e.g. modules from Pololu [9]). But these kinds of microcontroller modules, very popular nowadays with attractive IDE (Integrated

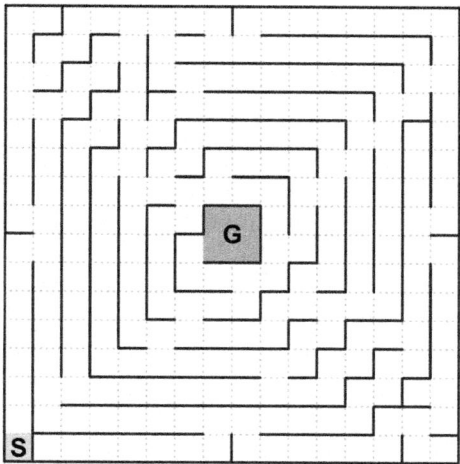

Fig. 1. Typical maze layout

Development Environment) and with significant easy to use amount of software modules on internet, have to be taken in consideration. One of the most popular of these microcontroller module is Arduino [10]. Therefore, the development of a micromouse kit based on Arduino will be a breakthrough for high school and university students.

Some of the problems with commercially available robot kits ([11–13]) or other kits [14] for micromouse contest are that they are usually expensive and have very low performance essentially due to high dimensions and/or motor type. Therefore, a low-cost micromouse kit has been devised in the University of Trás-os-Montes and Alto Douro to help to raise high school students interest to science and technology. The kit is based on Arduino Leonardo, have two wheels differential drive with a stepper motor on each wheel, and infrared hall detection sensors.

2   Micromouse Kit

The micromouse kit, whose block diagram is shown in Fig. 2, is steered with NEMA 8 (very small, only 20mm × 20mm by 30mm long) 1.8 ° steeper motors. The micromouse controls four infrared light emitting diodes (IR LED) in three directions (front, diagonal left and diagonal right), and detects the intensity of the reflected light to determine the maze wall information and to correct robot navigation. The main unit (controller unit) is based on Arduino Leonardo [15] containing a 16 MHz ATmega32U4 microcontroller from ATMEL with 32 kb of flash memory and 2.5 kb of SRAM.

The microcontroller is programmed through a micro USB port who can also be used as serial communication port (useful for debug). The firmware in the microcontroller can interact with the user with buttons, LED, buzzer, and serial

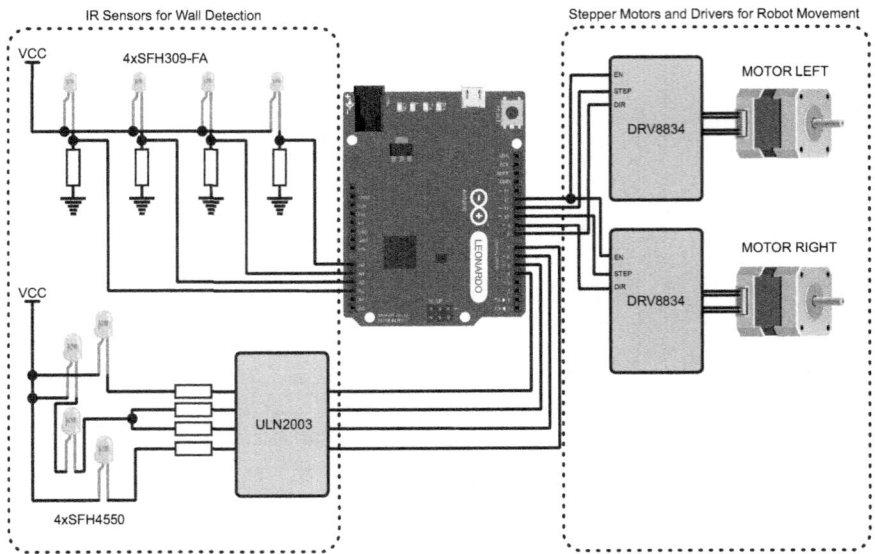

Fig. 2. Micromouse kit block diagram

port connector (could be used to plug a Bluetooth module for debug - not allowed during official runs). By using the functions libraries (C and C++), students should contrive their own maze-solving algorithm to help the micromouse find out the goal and decide and optimal route from the start to de goal.

2.1 Hardware and Mechanics

As state-of-the-art micromouse robots (e.g. Min7 a 2011 winner of the 31st All Japan Micromouse Contest [16]), the micromouse kit uses the printed circuit board (PCB) as chassis, as showed on Fig. 3a. It is possible to see the major blocks of the micromouse kit: the stepper drivers (for left and right motor), the infrared LED driver for the four LED, and the 'Arduino Leonardo'-like block that includes the microcontroller and the remaining of the board (power module, LED, USB port, In-Circuit Serial Programming port, and switches).

The steppers where mounted directly on PCB and screwed on a L-shaped support made of aluminum, as represented on Fig. 3b. The wheels, made of aluminum rod with 24.5 mm diameter and 5 mm thick, are directly attached (internal screw) to the shaft (4 mm). The kit uses standard Mini-Z rubber tires making the total wheel diameter of 27.5 mm.

The motors used are small NEMA 8 size hybrid bipolar stepping motor with 1.8° step angle (200 steps/revolution). Each phase draws 600 mA at 3.6 V, allowing for a holding torque of 180 g · m. These stepper motors only weight 60 g each allowing them to be a good alternative to DC motors. The use of DC motors with encoder, use by state-of-the-art micromouse robots, must be avoided in this kind of kits because they are expensive and very difficult to control. The

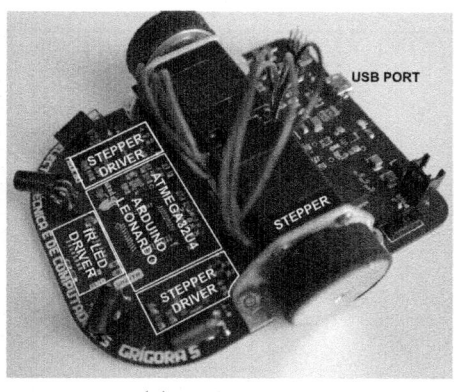

(a) Major blocks   (b) Steeper and wheel mount.

Fig. 3. Grígora S: Micromouse kit photos

students must have some knowledge of control theory, e.g. PID control, in order to control the robot movement. For high school students and even first years university, the use of stepper motors is strongly recommended.

2.2 Maze Wall Detection

For wall detection, the micromouse kit, uses infrared LED emitters (SFH4550) with a narrow emission angle of $\pm 3°$ and a typical peak wavelenght of 850 nm. The infrared LED emitters are driven by ULN8003 Darlington transistor array in order to achieve a short pulse forward current of $I_F = 500$ mA wich led to a typical radiation intensity of 2800 mW/sr. The infrared LED emitters are paired with SFH309-FA phototransistors with 95 % relative spectral sensitivity at 850 nm.

Since the IR phototransistors don't have an ambient light filter it is necessary to do the filtering in the digital domain, i.e., in software. To accomplish this the Alg. 1 is used, which calculates the difference between the 'dark' (IR emitter OFF, ambient light) reads and the 'light' (IR emitter ON) ones. This is a simple and efficient method to filter the ambient light.

The data obtained, showed on Fig. 4, from infrared sensor calibration (using Alg. 1) was used to fill a vector of 1024×3 values (maximum ADC value - 10 bits by three power types of sensor pulse - front, diagonal, and diagonal high power). This vector is used to calculate the micromouse distance from values obtained through ReadIRSensorDistance. With this procedure floating point calculations was avoided which speeds up wall detection. From the data presented on Fig. 4 we can find that wall detection is must efficient between 2 cm to 20 cm which is acceptable as a cell only measure 18 cm and if the robot have is axis in the center of a cell it is possible to detect a front wall placed on the next cell (26.75 cm from the robot axis and 22.5 cm from the IR receiver). Also, as can be showed

---

Algorithm 1: ReadIRSensorDistance

Data: IRsensor = $\{FRONT\_RIGHT, FRONT\_LEFT, DIAG, DIAG\_H\}$
Result: Value
begin

    IR LED off and read dark value;
    pin(IRsensor) ⟵ 0;
    wait to stabilize;
    delay(15 µs);
    darkValue ⟵ **analogRead**(IRsensor);
    IR LED on and read light value;
    pin(IRsensor) ⟵ 1;
    wait to stabilize;
    delay(15 µs);
    lightValue ⟵ **analogRead**(IRsensor);
    Value ⟵ lightValue − darkValue;

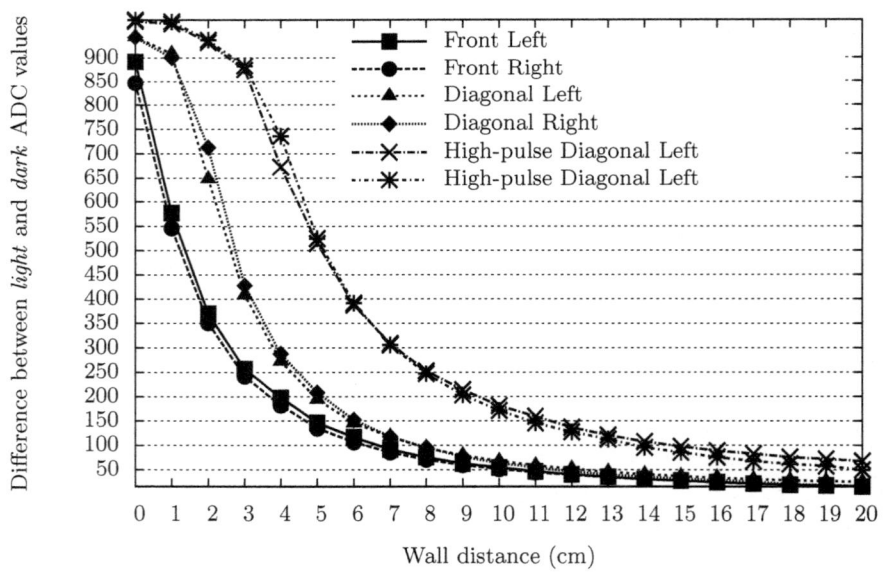

Fig. 4. Infra-red sensor calibration curve

on Fig. 4, the diagonal detection uses a pulse with higher power and have a even higher power pulse to detect lateral walls and long diagonal paths, respectively.

2.3 Motion Control

The micromouse kit uses stepper motors to move itself around in a maze. Therefore, it is very important to control adequately the acceleration and speed of the stepper motors. An integrated circuit DRV8834 from Texas Instruments is used as stepper motor driver (one for each motor). The driver controls the motor current, the direction of rotation, and rotates the motor according to programmed running modes and input pulses. The time interval between these consecutive input pulses determine the speed of step motors, being the acceleration the rate of change of the velocity in the time interval. Consider Fig. 5 as an example showing pulses from microcontroller to stepper driver. Each rising edge of the pulse will rotate the motor one step, according to the direction set by the driver DIR pin. The number of steps required for one full rotation depends on the selected mode (select modes can be from full-step to 32 microsteps/step). Considering K_s a parameter regarding the number of steps required for one full rotation, $K_s = MODE \times 200$ pulse/round, $MODE = \{1, 2, 4, 8, 16, 32\}$, and d_w the wheel diameter, the velocity and acceleration can be calculated by:

$$v_i = \frac{d_w \pi}{K_s T_i} \quad (\text{m/s}), \quad i = 1, 2, 3, \ldots, \tag{1}$$

$$a_i \approx \frac{v_{i+1} - v_i}{T_i} \quad (\text{m/s}^2), \quad i = 1, 2, 3, \ldots, \tag{2}$$

Combining Eq. 1 and Eq. 2 we can determine, Eq 3, the next pulse time interval, T_{i+1}, given the present time interval T_i and the desired acceleration a_i.

$$T_{i+1} = \frac{d_w \pi T_i}{k_s T_i^2 a_i + d_w \pi} \quad (\text{s}), \quad i = 1, 2, 3, \ldots, \tag{3}$$

In pratice, the desired acceleration should be changed gradually and limited by the output torque of the step motors to prevent loss of steps and robot slipping. The velocity and acceleration profiles are stored in the firmware of the micromouse kit to save time for the microcontroller.

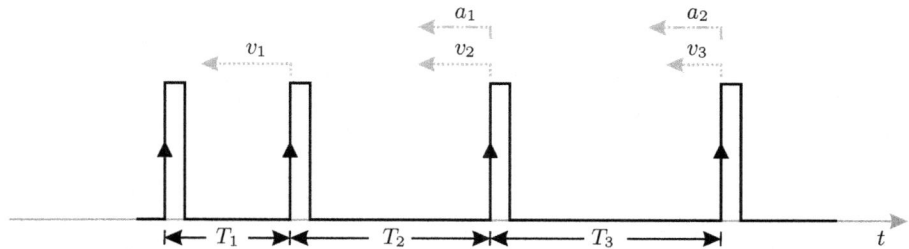

Fig. 5. Microcontroller step pulses sent to the stepper driver

Micromouse stepper tests show that the minimum time interval T_i achieved was 4 µs (maximum allowed step frequency of the DRV8834, 250 kHz) with $MODE = 16$. The diameter of the wheel with tire is 27.5 mm thus giving a maximum velocity of 6.75 m/s with wheels running free (not touching the floor). State-of-the-art micromouse have a straight line speed of approx. 4 m/s.

2.4   Odometry

Signal applied to stepper motors (left and right) will also be used to calculate the position of the micromouse in the maze. Considering Fig. 6, if the robot starts from a position $p(x, y, \theta)$, and the right and left wheels move respectively the linear distances ΔS_R and ΔS_L, the new position $p'(x', y', \theta')$ is given by

$$p' = \begin{bmatrix} x \\ y \\ \theta \end{bmatrix} + \begin{bmatrix} \Delta x \\ \Delta y \\ \Delta \theta \end{bmatrix} \tag{4}$$

where,

$$\Delta x = \Delta S \cos(\theta + \frac{\Delta \theta}{2}) \tag{5}$$

$$\Delta y = \Delta S \sin(\theta + \frac{\Delta \theta}{2}) \tag{6}$$

$$\Delta \theta = \frac{\Delta S_R - \Delta S_L}{L} \tag{7}$$

and

$$\Delta S = \frac{\Delta S_R - \Delta S_L}{2} \tag{8}$$

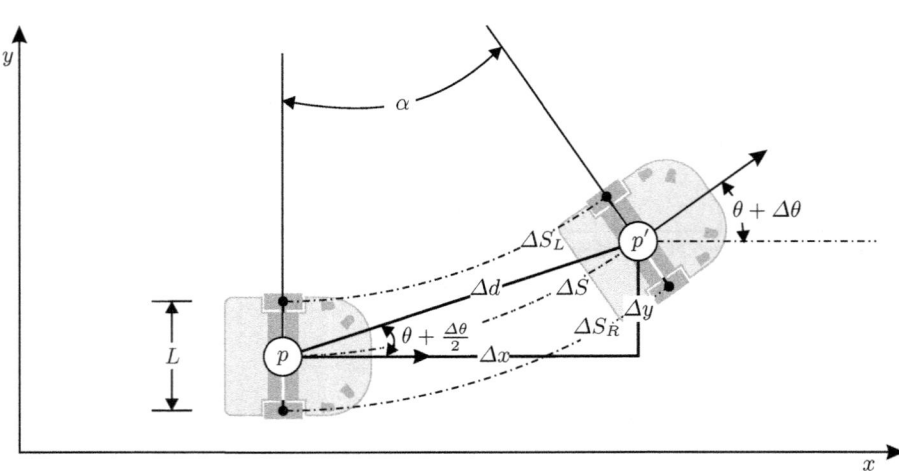

Fig. 6. Position calculation of the micromouse in a maze

Combined equations (4), (5), (6), (7), and (8) will give the odometry equation modelling the micromouse motion,

$$p' = f(x, y, \theta, \Delta S_R, \Delta S_L) = \begin{bmatrix} x \\ y \\ \theta \end{bmatrix} + \begin{bmatrix} \frac{\Delta S_R - \Delta S_L}{2} \cos\left(\theta + \frac{\Delta S_R - \Delta S_L}{2L}\right) \\ \frac{\Delta S_R - \Delta S_L}{2} \sin\left(\theta + \frac{\Delta S_R - \Delta S_L}{2L}\right) \\ \frac{\Delta S_R - \Delta S_L}{L} \end{bmatrix} \quad (9)$$

Considering N_L and N_R the number of pulses used to move the respective left and right stepper motors since last move, ΔS_L and ΔS_R can be calculate from

$$\Delta S_L = N_L \frac{d_w \pi}{K_s} \quad (10)$$

$$\Delta S_R = N_R \frac{d_w \pi}{K_s} \quad (11)$$

3   Conclusions

A high performance, low-cost robot kit for high school and university students participate on micromouse robot contests were developed. The robot fits on a 10×8 cm rectangle and uses very small stepper motors allowing more than 4 m/s maximum speed thus comparing to state-of-the-art micromice. The kit also incorporates a popular microcontroller hardware module (Arduino Leonardo) which facilitates all programming tasks.

The overall cost of the kit is below 100 €. The only drawback is the machining of mechanical parts (wheel and L-shaped stepper mount) that are time consuming since they are not standard or commercial parts easily accessible. In futures developments of the kit it will be considered the use of plastic parts from 3D printing or from custom online CNC machined parts.

Future developments also will consider the use of popular programming with blocks (Alice from Carnegie Mellon [17], Scratch from MIT [18], Blockly from Google [19], among others) removing syntax and therefor facilitating children participation on micromouse contests.

References

1. Petre, M., Price, B.: Using robotics to motivate 'back door' learning. Education and Information Technologies 9, 147–158 (2004)
2. Robocup (Robocup 2014 site), http://www.robocup2014.org/ (visited on January 15, 2014)
3. Micromouse UK (Micromouse UK contest site), http://www.tic.ac.uk/micromouse/ (visited on January 15, 2014)
4. APEC - Applied Power Electronics Conference and Exposition (Micromouse USA contest site), http://www.apec-conf.org/conference/participating-in-micromouse/ (visited on January 15, 2014)

5. NTF - New Technology Foudation (All Japan Micromouse Competition site), `http://www.ntf.or.jp/mouse/` (visited on January 15, 2014)
6. Institute for Technical Education, Robotic Games Society (Singapore), Nanyang Polytechnic, Nanyang Technological University, National University of Singapore, Ngee Ann Polytechnic, Republic Polytechnic, Singapore Polytechnic, Singapore Science Centre and Temasek Polytechnic (The 21st Singapore Robotic Games 2014 site), `http://guppy.mpe.nus.edu.sg/srg/` (visited on January 15, 2014)
7. Ministry of Education - Taiwan (Taiwan Micromouse and Intelligent Robot Contest site), `http://robot.lhu.edu.tw/` (visited on January 15, 2014)
8. UTAD - University of Trás-os-Montes and Alto Douro (Portuguese Micromouse Contest site), `http://www.micromouse.utad.pt/` (visited on January 15, 2014)
9. Pololu (Pololu robotics & electronics), `http://www.pololu.com/` (visited on January 15, 2014)
10. Banzi, M., Cuartielles, D., Igoe, T., Martino, G., Mellis, D.: (Arduino), `http://arduino.cc/` (visited on January 15, 2014)
11. Robot Store (Airat2), `http://www.robotstorehk.com/micromouse/RS-AIRAT2.html` (visited on January 15, 2014)
12. EngageHobby (ASURO Micromouse RobotDIY - Educational Robotics Kits), `http://www.engagerc.com/servlet/the-1234/` `ASURO-micromouse-robotdiy,Educational-Robotics/Detail` (visited on January 15, 2014)
13. Picaxe (Picone: PICAXE Micromouse), `http://www.picaxe.com/Hardware/Robot-Kits/PICAXE-PICone-Micromouse/` (visited on January 15, 2014)
14. Su, J.H., Lee, C.S., Huang, H.H., Huang, J.Y.: A micromouse kit for teaching autonomous mobile robots. International Journal of Electrical Engineering Education 48, 188–201 (2011)
15. Arduino (Arduino Leonardo site), `http://arduino.cc/en/Main/ArduinoBoardLeonardo` (visited on January 15, 2014)
16. (Kiat, N.B.), `https://sites.google.com/site/ngbengkiat/Downhome/Topic1/min7`
17. Carnegie Mellon University HCII research group (Alice - An Educational Software that teaches students computer programming in a 3D environment), `http://www.alice.org/` (visited on January 15, 2014)
18. Lifelong Kindergarten Group from MIT Media Lab (Scratch - Educational Programming Language), `http://scratch.mit.edu/` (visited on January 15, 2014)
19. Google (Blockly - A Visual Programming Editor), `https://code.google.com/p/blockly/` (visited on January 15, 2014)

Introduction to Visual Motion Analysis for Mobile Robots

Andry Maykol Pinto, Paulo Gomes Costa, and António Paulo Moreira

Centre for Robotics and Intelligent Systems - INESCTEC and
Faculty of Engineering, University of Porto, Portugal
{andry.pinto,paco,amoreira}@fe.up.pt

Abstract. Human being has an extraordinary capability for motion perception due to its remarkable visual sensing system that makes it possible to perceive, distinguish and characterize the different moving elements of the environment. Thus, it extracts information through sensory experience and conducts reliable judgments based on intrinsic motion features, namely, location, direction, trajectory, magnitude, colors, boundary and shape.

Unfortunately, the same cannot be said for mobile robots. The critical nature of visual perception for these kinds of systems turns motion detection and analysis as one of the most relevant areas discussed on the literature, existing several models and methods to perform motion analysis in a variety of environments. This paper discusses motion analysis for mobile robots. A brief description about the complexity of motion perception based on moving observations and for surveillance applications is presented. In addition, the most often encountered approaches and future orientations are also discussed.

Keywords: Visual Perception, Mobile Robots, Optical Flow, Egomotion.

1 Introduction

Motion perception plays an important role on human daily interactions, for instance, to communicate with other humans and to drive or walk in a street. Inferring about the direction and speed of moving objects is one of the most critical skills because the brain is constantly analyzing motion for indications of danger. At other hand, mobile robotics applications have certain problems related to visual perception and interpretation of the dynamic scene: unmanned aerial vehicles [4,8], unmanned surface vehicles [2] and unmanned ground vehicles [7]. In these applications, a suitable motion detection is crucial to feed the high level processes with relevant information in order to define external interactions, navigation procedures and to increase their autonomy.

The ability to interpret, understand and interact with dynamic scenes are crucial for a new generation of autonomous mobile robots, especially, in robots designed for surveillance operations. To extend the operation scope of these

© Springer International Publishing Switzerland 2015   545
A.P. Moreira et al. (eds.), *CONTROLO'2014 - Proc. of the 11th Port. Conf. on Autom. Control*,
Lecture Notes in Electrical Engineering 321, DOI: 10.1007/978-3-319-10380-8_52

vehicles, a safe interaction through unknown and dynamic environments must be assured. Therefore, robots require innovative perception algorithms and new behavioral laws in order to achieve this goal. Vision-based technologies are non-invasive and acquire information in a similar manner as the human vision, which make them appealing for mobile robots. In the scientific community, motion perception is one of the most relevant areas under discussion, and there are several models and methods to perform motion analysis in a variety of environments. Visual perception of motion can be divided into three stages [5]: *detection, measurement* and *cognitive* stages. Changes in brightness, texture, color and shapes are used to identify regions of interest that might represent some motion during the detection stage. The measurement phase is responsible for evaluating the intrinsic motion parameters of elements belonging to the scene. Finally, the cognitive phase classifies the type of motion according to features and based on information that is obtained in previous stages.

Motion perception can be performed using two distinct ways which are directly related with the movement of the observer that is capturing the scene: *static observations* or *moving observations*. Motion perception with a static observer is quite different from the moving observer because when a static observer captures the scene, every spatial and temporal variation represents part of the moving object (neglecting illumination changes and noise). The motion pattern obtained from a moving observer exhibits changes in almost every pixels. These changes depend on the presence or absence of external moving objects as well as the relative motion between the visual sensor and scene. Two approaches have specific theoretical assumptions that cause relevant differences in the performance, flexibility and robustness of techniques for motion detection and analysis. These approaches are being studied by the scientific community however, techniques based on moving observations are still in a preliminary stage when compared to static observations. This is a consequence of the egomotion (motion of the observer) since it creates new paradigms that turn the study of motion even more complex and challenging.

The article is organized as follows. Section 2 presents a brief review of the visual motion perception techniques that are commonly used in robotic systems: section 2.1 and 2.2 discusses several approaches for motion perception based on static and moving observations, respectively. In addition, it gives some guidelines for future researching activities. Section 3 presents major conclusions.

2   Visual Motion Analysis

Currently, there is a wide diversity of motion perception methods which is justified by the inherent complexity of the problem. The main goal of visual motion perception is to segment image sequences into *background* and *foreground* regions, which are regions with absence of temporal displacements and regions with moving objects, respectively. The three-dimensional trajectory of a moving object is projected into the image plane which creates two-dimensional trajectories whose derivatives represent two-dimensional velocities. Motion perception techniques can be classified according to viewer's motion relative to the environment, namely: Stationary Observations (SOb) and Moving Observations (MOb). Visual motion detection with static observers has been extensively explored for many surveillance applications. In contrast, detecting moving objects on image sequences based on moving observations is more difficult because of an additive two-dimensional motion field that is created by the egomotion of the vehicle.

2.1   Based on Static Observations

Nowadays, the research based on static observations focuses on how to increase the quality of motion segmentation in scenes with illumination changes, dynamic background objects and temporal occlusions [10]. In the literature it is possible to identify three motion perception approaches for conventional SOb systems [10,17]: temporal differencing, background subtraction and optical flow.

Temporal differencing is the simplest method to detect moving objects. The traditional temporal differencing method is defined by an absolute subtraction of consecutive images. In this way, pixels whose absolute difference is greater than a pre-defined threshold are classified as foreground, otherwise, are classified as background. This method is very sensitive to any kind of movement and adapts quickly to changes in lighting conditions since the reference model is not created, it is not suitable for environments with moving elements belonging to the background because those elements are difficult to distinguish from the foreground. The method is also misleading for foreground objects that stop for a short period of time and if foreground objects have homogeneous colors since the temporal differencing may not detect all the foreground pixels which causes an internal cavity effect. Despite the limitations, there are several research works that resort to temporal differencing as part of a more complex motion detection architecture. Moreover, it is used especially to start other algorithms like background subtraction and optical flow techniques since it can improve the computational efficiency.

A background subtraction technique, as is suggested by the name, separates objects of interest (*foreground*) from the rest of the image (*background*). It is the most typical and traditional approach to segment moving objects based on static cameras because makes it possible to recover shapes and features of the foreground objects. The background subtraction approach is consisted basically by two distinct phases: creation (and update) of the reference model, and the subtraction of the reference model to the current image. The segmentation of moving objects is accomplished by a subtraction operation which defines the classification of each pixel between foreground and background. The literature is rich in background subtraction approaches however, each technique has different strengths and weaknesses in terms of segmentation accuracy and computational requirement. The scientific community recognizes background methods as those that provide the best compromise between performance and reliability [17]. The most common approaches for motion segmentation based on background subtraction are the following: *Running Gaussian Average* [18], *CodeBook* [11], *Mixture of Gaussians* (MOG) [3], *Kernel Density Estimators, Mean-shift based estimation* [14], *Eigenbackgrounds* and *Markov Random Field* [21]. They differ in the way that the reference model is created. Some background subtraction methods have limitations related with two commonly adopted assumptions that are often violated: pixels are independent and the temporal evolution of the background is slow. These assumptions are not truth for real-life situations since they ignore the spatial dependency among neighboring pixels which leads to inconsistent (noisy) predictions, and the background might change too much over time due to, for instance, the wind, the camera jitter, illumination changes, windows and doors. Background subtraction techniques are being implemented on a wide range of hardware and used by a variety of surveillance scenarios however, some of the presented approaches are still not computationally efficient for real-time applications.

Optical flow is one of the most well-known techniques for motion detection. It analyzes the spatial and temporal evolution of pixels and assigns the respective motion vector. Usually, the term "optical flow" are sometimes confused with motion projection of the three-dimensional objects in the two-dimensional image plane however, they are not the same thing. There are factors that affect the estimation of the optical flow (for instance, quantization, reflections and noise), differentiating the estimation from the real value of the motion projection on the image plane. Although, there is a close relationship between both [19]. Optical flow techniques provide a motion vector for each pixel however, they are computationally too expensive to be used in real-time applications and without special hardware [16]. A recent work [16] presented a novel optical flow technique to mitigate this problem. Results show that the proposed technique estimates dense flow fields in short time (100 milliseconds) which makes it suitable for robotic applications equipped with generic computers.

2.2 Based on Moving Observations

Usually, motion perception in surveillance is performed with SOb however, reliable solutions based on MOb make possible to automate new environments by taking advantage of the observer's moving capability. Visual motion perception with MOb is expected to have a large and revolutionary impact in the surveillance of new locations since more conventional methods like the background subtraction and the temporal differencing cannot be applied directly without a method to compensate the egomotion. However, only a few research projects are robust enough to make possible the movement of the viewer. In this case, conventional SOb methods cannot be applied directly because an estimative of the egomotion is required to compensate the movement of the observer.

Interpretation of the Observer's Motion: The three-dimensional movement of the observer influences the process of motion perception because the motion causes two dimensional projections. Hence, the two-dimensional motion that results from the three-dimensional translational and rotational of the observer must be quantified. The mathematical formulation of this section follows closely the fundamentals presented by *H.C Longuet-Higgins and K. Prazdny (1980)* [15].

A monocular observer moving in a static scene has a motion with two components: the translational, $\dot{\Gamma} = (t_X, t_Y, t_Z)^T$, and the rotational velocity, $\Omega = (w_X, w_Y, w_Z)^T$. Both components are described in the camera coordinate system. That coordinate system moves together with the camera and, therefore, the coordinate of a static point $\mathbf{P} = (X, Y, Z)^T$, represented in the camera coordinate system, changes due to this motion.

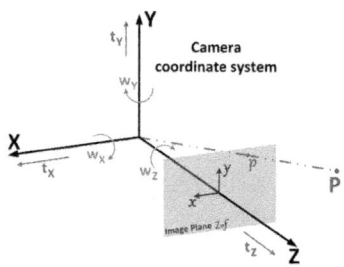

Fig. 1. The geometry of the image formation. The camera coordinate system moves with translational (red) and rotational (green) velocity. A static point is represented by $\mathbf{P}$ in the camera coordinate system and its projection into the image plane is portrayed by $\mathbf{p}$.

The velocity of $\mathbf{P}$ is in opposite direction to the camera movement and can be described by $\dot{\mathbf{P}}$:

$$\dot{\mathbf{P}} = -\dot{\Gamma} - \Omega \times \mathbf{P}. \tag{1}$$

Considering the pin-hole camera model, $x = fX/Z$ and $y = fY/Z$, and the two-dimensional velocity of $\mathbf{p}$, $\mathbf{v} \equiv (u, v)^T$, the equation 1 can be written as:

$$\mathbf{v} = \frac{1}{Z} \begin{bmatrix} -f & 0 & x \\ 0 & -f & y \end{bmatrix} \dot{\Gamma} + \begin{bmatrix} \frac{xy}{f} & -f - \frac{x^2}{f} & y \\ f + \frac{y^2}{f} & -\frac{xy}{f} & -x \end{bmatrix} \Omega, \tag{2}$$

where f is the focal length. Equation 2 does not include the position of the point represented on the three-dimensional camera coordinate system. Therefore, the apparent motion is a vector sum of the translational and the rotational velocity of the camera. The inverse of the depth appear in the translational component of the expression, which means that a scaling factor is assigned to its calculation [5]. This phenomenon is often called as the *parallax effect*: represents the inability to distinguish between a near object that moves slowly from a distant object that moves quickly (and vice versa if the camera moves and the object remains static on the environment). Suppose a visual observation of two static points, P_1 and P_2 at different depths, Z_1 and Z_2. If the observer moves along the environment (with a non-null translational component) the apparent motion vector of each point will be different due to the parallax effect. This is the reason why some researches avoid monocular vision and resort to stereoscopic systems or 3D sensors.

Most approaches that were presented so far assume that images are captured with static observers. Typically, methods for SOb assume the same spatial correspondence over time of each pixel. Therefore, motion is detected by performing a temporal analysis of the brightness. The displacement of the observer increases the complexity of the SOb-based formulation because new aspects need to be considered. The 2D visual information that is captured by a moving observer has two motion components: egomotion (motion of the observer) and movement of the external objects. Therefore, the formulation of motion perception with MOb is more complex because it cannot assume the spatial correspondence of pixels in different frames, for instance, the position of each pixel changes over time and even for scenes without moving objects. In this context, an estimation of the egomotion must be previously obtained to compensate the motion component that does not reflects moving objects. This means, motion analysis is conditioned by the dynamic of the observer that captures the environment.

Thus, the problem of motion perception based on MOb has two steps: the egomotion estimation and disassociation of motion components. The apparent motion of each pixel is a combination of the egomotion, $\mathbf{v}_{ego} = (u_{ego}, v_{ego})^T$, and the objects motion, $\mathbf{v}_{obj} = (u_{obj}, v_{obj})^T$. These motion vectors are two dimensional projections of three-dimensional motions. The motion vector $\mathbf{v}_{motion} = (u_{motion}, v_{motion})^T$ of an image captured by a moving observer can be expressed by $\mathbf{v}_{motion} = \mathbf{v}_{ego}$ for static objects (background) and by $\mathbf{v}_{motion} = \mathbf{v}_{ego} + \mathbf{v}_{obj}$ for moving objects (foreground). These equations depicts the problem using the relative velocities and it shows the importance of knowing the egomotion because the object's motion can be obtained by $\mathbf{v}_{obj} = \mathbf{v}_{motion} - \mathbf{v}_{ego}$.

The foreground and background velocity can be re-written as equation 3 and 4, respectively.

$$\mathbf{v}_{motion} = \frac{1}{Z_{obj}} \begin{bmatrix} -f & 0 & x \\ 0 & -f & y \end{bmatrix} \begin{bmatrix} t_X^{ego} \\ t_Y^{ego} \\ t_Z^{ego} \end{bmatrix} + \begin{bmatrix} \frac{xy}{f} & -f-\frac{x^2}{f} & y \\ f+\frac{y^2}{f} & -\frac{xy}{f} & -x \end{bmatrix} \begin{bmatrix} w_X^{ego} \\ w_Y^{ego} \\ w_Z^{ego} \end{bmatrix}$$

$$+ \frac{1}{Z_{obj}} \begin{bmatrix} f & 0 & -x \\ 0 & f & -y \end{bmatrix} \begin{bmatrix} t_X^{obj} \\ t_Y^{obj} \\ t_Z^{obj} \end{bmatrix} + \begin{bmatrix} -\frac{xy}{f} & f+\frac{x^2}{f} & -y \\ -f-\frac{y^2}{f} & \frac{xy}{f} & x \end{bmatrix} \begin{bmatrix} w_X^{obj} \\ w_Y^{obj} \\ w_Z^{obj} \end{bmatrix} ; \quad (3)$$

$$\mathbf{v}_{motion} = \frac{1}{Z_{scene}} \begin{bmatrix} -f & 0 & x \\ 0 & -f & y \end{bmatrix} \begin{bmatrix} t_X^{ego} \\ t_Y^{ego} \\ t_Z^{ego} \end{bmatrix} + \begin{bmatrix} \frac{xy}{f} & -f-\frac{x^2}{f} & y \\ f+\frac{y^2}{f} & -\frac{xy}{f} & -x \end{bmatrix} \begin{bmatrix} w_X^{ego} \\ w_Y^{ego} \\ w_Z^{ego} \end{bmatrix} , \quad (4)$$

where Z_{scene} is the depth to the scene (static), and Z_{obj} is the depth to the moving object.

The egomotion is normally estimated using a motion model, for instance, translational, Euclidean, similarity, affine and projective. These models can represent several types of movement with different properties. Usually, the affine and projective models enable more complex types of movement (translation, rotation and zooming) because the number of degrees of freedom is higher. The suitable motion model should be selected according to the requirements of each application like, the computational requirement and the reliable representation of motion. Moreover, the estimation of $\mathbf{v}_{motion}$ and $\mathbf{v}_{ego}$ might have numerous errors due to numerical approximations, sensor noise and photometric effects like, reflections, shadows, transparency and changes in lighting.

Techniques for Visual Motion Perception with MOb: Visual motion detection and analysis for moving observers is becoming an active research field and preliminary techniques typically use one of the following approaches:

- Organizing the background into moisacs [8,13]: The background mosaic is created using spatial registration and tonal alignment techniques. Creating a mosaic background presents several disadvantages due to the photometric and spatial misalignments [10];
- Modifying background subtraction methods [10,6]: In some applications, conventional background subtraction methods are extended in order to incorporate the motion of the visual sensor. Usually, this approach adds the spatial information and allows motion perception along some pre-defined movements;
- Optical flow and geometrical models [20,16]: The optical flow approaches resort to the dense optical flow field or to sparse flow using only some features to extract information about egomotion of the visual sensor. The egomotion is computed based on motion models. The cluster techniques can also be applied to segment pixels that correspond to moving objects. This type of approach is commonly used in applications where the observer has several degrees of freedom.

Motion perception and analysis is an extremely important problem for several mobile robotic applications, especially for UAVs. A tracking application that resorts to a pyramidal Lucas-Kanade optical flow is presented in *Jay et al. (2011)* [4]. The researching work identifies and extracts regions where the flow field does not represent the UAV's egomotion, for instance, for tracking a target that moves at different velocity comparatively to the background. *Aryo Ibrahim et al. (2010)* [8] present a mosaic technique for an UAV application. The technique maps the areas and detects moving objects. The authors match invariant features SURF (Speeded Up Robust Feature) or SIFT (Scale Invariant Feature Transform) between frames in order to compute the matrix that describes the geometrical transformation (projective model). The matrix is used by the warping process which aligns the current frame with previous frames. *Jing Li et al. (2011)* [13] focus on monitoring the highway traffic flow using an airbone monocular camera. The goal of this research is to detect moving vehicles. They detect the road by extracting areas with similar intensities. The authors identify the road by assuming that the road is larger than other blobs. A simplified Lucas-Kanade method combined with an image registration technique makes it possible to obtain the motion vector between consecutive frames. This defines the egomotion (affine model) of the vehicle. Finally, the moving objects are retrieved by the temporal differencing approach. *Abhijit Kundu et al. (2010)* [12]

focus on detecting moving objects using a monocular vision system mounted in a robotic platform. The features from accelerated segment test (FAST) corners are extracted at different image pyramidal levels. Only some frames are used to triangulate the three dimensional points and the epipolar geometry gives the initial estimative of the camera's pose. Then, an iterative process redefines the estimation of the cameras' pose by minimizing the first order approximation of the reprojection error, called *Sampson error*, that is calculated from the structure from motion (epipolar geometry). *Ninad Thakoor et al. (2004)* [20] use temporal differencing approach with motion compensation. The egomotion model (affine motion) is computed using the hierarchical Lucas-Kanade optical flow technique over three consecutive frames. Affine motion parameters are computed iteratively using a reweighed least squares. The forward and the backward model is obtained relatively to the middle frame which makes it possible to generate an estimative of the reference model (for background subtraction). The presented approach is interesting since moving objects are detected however, the motion boundary is not extracted completely. *Rita Cucchiara et al. (2004)* [6] use region growing with color information to segment the image in different regions by assuming that each region contains part of one object. Features like the size, the position of the centroid and the bounding-box are also associated in a graph. The translational model for the egomotion is adopted by the authors to enable a real-time estimation of motion vectors. These flow vectors are computed by a region matching procedure that is described in the paper. *Alexiadis and Sergiadis (2009)* [1] use a weighted fuzzy c-means clustering procedure to obtain the velocity estimates for color sequences. The dense optical flow fields are computed using square blocks and the estimated velocity is assigned to the center of the block after the median filter. The authors separate the different types of motion in the two-dimensional hypercomplex Fourier domain and resort to an energy-minimization-based approach. They assume that the velocities of the moving objects (translational motions) are smoothly time-varying. An approach that estimates the egomotion of a monocular camera using feature correspondence and the Lucas-Kanade optical flow with outlier removal can be found in [9]. The Expectation-Maximization (EM) algorithm is used to cluster the particles and, as a result, the moving regions are extracted by thresholding the Gaussian mixture function. The approach was evaluated in different environments and using aerial and ground robots. The technique processes images with a resolution of 320×240 in 5 frames per second.

A robust motion perception algorithm based on MOb enables new market niches for autonomous robotic applications. A better interpretation of the scene makes possible to automate the surveillance processes that nowadays are carried out through remote monitoring. Existing methods do not extract relevant information about motion in a reliable and efficient manner. Moreover, small mobile robots have additional hardware limitations regarding to the computational demands and power consumption that must be also considered. A recent research focused in these concerns was presented in [16]. In this research, a dense optical flow technique designed for robotic applications with generic computers extracts information about motion in a very short time. Overcoming the problem of computational requirements, optical flow techniques will be widely used in robotics. They have a remarkable reliability for perceiving motion without assumptions that restrict the measurement of apparent motion. Therefore, optical flow techniques will revolutionize the way that mobile robotic systems operate.

3   Final Considerations

This research discusses the importance of motion detection and measurement in robotic applications. Visual motion detection based on MOb is inherent difficult because the egomotion of the observer creates a two-dimensional velocity components that depends on the structure of the environment, for instance, the depth of the objects captured by the observer. Robotic applications with MOb include unmanned aerial vehicles (UAVs), unmanned surface vehicles (USV) and unmanned ground vehicles (UGV). In these applications, a suitable motion detection is crucial to feed the high level processes with relevant information in order to define external interactions, navigation procedures and to increase their autonomy. Therefore, motion perception with moving observers (MOb) is a complex and challenging problem with difficult solution because the observer's motion must be quantified and extracted first. Recent methods and algorithms for motion perception based on MOb are presented and discussed in this paper.

Acknowledgements. This work was partially funded by the Portuguese Government through the FCT - Foundation for Science and Technology, SFRH-BD-70752-2010; and by the ERDF European Regional Development Fund through the COMPETE Programme (operational programme for competitiveness) and by National Funds through the FCT within project FCOMP - 01-0124-FEDER-022701.

References

1. Alexiadis, D.S., Sergiadis, G.D.: Motion estimation, segmentation and separation, using hypercomplex phase correlation, clustering techniques and graph-based optimization. Computer Vision and Image Understanding 113(2), 212–234 (2009)
2. Campbell, S., Naeem, W., Irwin, G.W.: A review on improving the autonomy of unmanned surface vehicles through intelligent collision avoidance manoeuvres. Annual Reviews in Control 36(2), 267–283 (2012)
3. Cheng, J., Yang, J., Zhou, Y., Cui, Y.: Flexible background mixture models for foreground segmentation. Image and Vision Computing 24(5), 473–482 (2006)
4. Choi, J.H., Lee, D., Bang, H.: Tracking an unknown moving target from UAV: Extracting and localizing an moving target with vision sensor based on optical ow. In: 5th International Conference on Automation, Robotics and Applications (ICARA), pp. 384–389. IEEE (December 2011)
5. Correia, M.F.P.V.: Técnicas computacionais na percepc ao visual do movimento. PhD in electrical and computer engineering, Faculty of Engineering of the University of Porto, Porto, Portugal (2001)

6. Cucchiara, R., Prati, A., Vezzani, R.: Real-time motion segmentation from moving cameras. Real-Time Imaging 10(3), 127–143 (2004)
7. Di Paola, D., Milella, A., Cicirelli, G., Distante, A.: An autonomous mobile robotic system for surveillance of indoor environments. International Journal of Advanced Robotic Systems 7 (2009)
8. Ibrahim, A.W.N., Ching, P.W., Gerald Seet, G.L., Michael Lau, W.S., Czajewski, W.: Moving Objects Detection and Tracking Framework for UAV-based Surveillance. In: 2010 Fourth Pacific-Rim Symposium on Image and Video Technology, pp. 456–461. IEEE (November 2010)
9. Jung, B., Sukhatme, G.S.: Detecting moving objects using a single camera on a mobile robot in an outdoor environment. In: 8th Conference on Intelligent Autonomous Systems, pp. 980–987 (March 2004)
10. Kim, I.S., Choi, H.S., Yi, K.M., Choi, J.Y., Kong, S.G.: Intelligent visual surveillance: A survey. International Journal of Control, Automation and Systems 8(5), 926–939 (2010)
11. Kim, K., Chalidabhongse, T., Harwood, D., Davis, L.: Real-time foreground background segmentation using codebook model. Real-Time Imaging 11(3), 172–185 (2005)
12. Kundu, A., Jawahar, C.V., Madhava Krishna, K.: Realtime moving object detection from a freely moving monocular camera. In: IEEE International Confer- ence on Robotics and Biomimetics (ROBIO), pp. 1635–1640. IEEE (December 2010)
13. Li, J., Ai, H., Cui, J.: Moving Vehicle Detection in Dynamic Background from Airborne Monocular Camera. Energy Procedia 13, 3955–3961 (2011)
14. Liu, Y., Yao, H., Gao, W., Chen, X., Zhao, D.: Nonparametric background generation. Journal of Visual Communication and Image Representation 18(3), 253–263 (2007)
15. Longuet-Higgins, H.C., Prazdny, K.: The interpretation of a moving retinal image. Proceedings of the Royal Society of London 208(1173), 385–397 (1980)
16. Pinto, A.M., Paulo Moreira, A., Correia, M.V., Costa, P.G.: A ow-based motion perception technique for an autonomous robot system. Journal of Intelligent and Robotic Systems, 1–18 (2013), doi:10.1007/s10846-013-9999-z
17. Spagnolo, P., Orazio, T., Leo, M.M., Distante, A.: Moving object segmentation by background subtraction and temporal analysis. Image and Vision Computing 24(5), 411–423 (2006)
18. Su, S.-T., Chen, Y.-Y.: Moving Object Segmentation Using Improved Running Gaussian Average Background Model. In: Computing: Techniques and Applications (DICTA), pp. 24–31. IEEE (December 2008)
19. Tagliasacchi, M.: A genetic algorithm for optical ow estimation. Image and Vision Computing 25(2), 141–147 (2007)
20. Thakoor, N., Gao, J., Chen, H.: Automatic object detection in video sequences with camera in motion. In: Advanced Concepts for Intelligent Vision Systems, pp. 7–14 (August 2004)
21. Wang, Y., Tan, T., Loe, K.-F., Wu, J.-K.: A probabilistic approach for foreground and shadow segmentation in monocular image sequences. Pattern Recognition 38(11), 1937–1946 (2005)

Robust Outliers Detection and Classification for USBL Underwater Positioning Systems*

Marco Morgado[1], Paulo Oliveira[1,2], and Carlos Silvestre[1,3]

[1] Institute for Systems and Robotics, Instituto Superior Técnico,
Universidade de Lisboa, Av. Rovisco Pais, 1, 1049-001, Lisbon, Portugal
{marcomorgado,pjcro,cjs}@isr.ist.utl.pt
[2] Department of Mechanical Enginnering, Instituto Superior Técnico
[3] Faculty of Science and Technology, University of Macau, Macau

Abstract. This paper presents a data classification algorithm able to detect corrupted measurements as outliers, with application to underwater ultra-short baseline (USBL) acoustic positioning systems. The devised framework is based on causal median filters that are readily implementable, and a set of theoretical analysis tools that allows for the design of the filter parameters is also presented. The design takes into account very specific implementation details of USBL acoustic positioning systems and also inherent non-ideal characteristics that include long period data outages. The outlier classifier is evaluated both in simulation and with experimental data from a prototype USBL acoustic positioning system fully developed in-house.

1 Introduction

Measurement outliers are naturally present in the output of every sensor package available on the market nowadays. The correct identification of these spurious perturbations often stands out as one of the important steps to the correct usage and successful integration of the aforementioned sensor packages into larger systems that are the building blocks of any robotic platform. This paper addresses the design and experimental validation of an outlier detector and classifier for underwater positioning systems.

The fast deployment, less complex hardware of small and compact arrays of receivers and increasing performance of modern factory-calibrated USBL positioning devices makes it suitable for faster intervention missions Napolitano et al. (2005). In generic operating conditions, a conductivity, temperature and depth (CTD) profile is normally required to account for the underwater sound velocity variations. Inverted USBL Vickery (1998) configurations, besides paving the way to future fully autonomous systems without the need to have surface mission support vessels, allows for the sound velocity to be considered constant while operating in the same underwater layer as the transponders (for instance, bottom operation while interrogating bottom placed transponders). The inverted USBL configuration is illustrated in Fig. 1.

* This work was supported by project FCT [PEst-OE/EEI/LA0009/2013], by project FCT PTDC/EEA-CRO/111197/2009 - MAST/AM, and by the EU Project TRIDENT (Contract No. 248497).

Due to several undesired aspects of the underwater sound propagation channel such as acoustic reverberation, layered underwater sound speed profiles, and mostly due to multipath phenomena, these type of acoustic positioning systems are highly susceptible to measurement outliers which need to be correctly identified. Otherwise these position measurement outliers can have a severe impact on systems that use them, degrading their performance downstream, as illustrated in Fig. 2, and worst case leading control and navigation systems to instability. This paper addresses the design of an outlier detector and classifier for a USBL positioning system and validates this classifier using experimental data obtained at sea with a USBL prototype fully developed in-house Morgado et al. (2010).

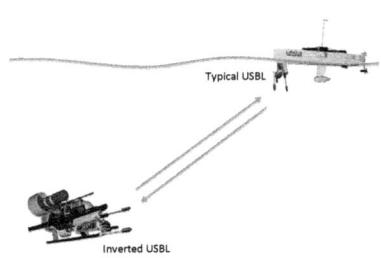

Fig. 1. Inverted USBL configuration as opposed to a typical installation floating on the sea surface - the prototype system developed in-house is also illustrated in the schematic attached to an underwater robotic in the inverted configuration and to the bow of an autonomous surface craft in the typical USBL configuration

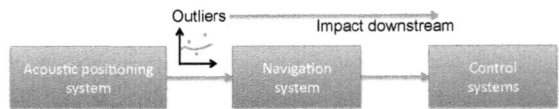

Fig. 2. Outliers from the acoustic positioning systems might have a severe degradation of the navigation system accuracy if not properly identified. In the worst case scenario it might even lead to instability of control systems that use this information downstream.

1.1 Paper Organization

The paper is organized as follows: Section 2.1 provides a review on the concepts of causal median filters that the outlier detector and classifier builds upon. Section 2.2 details the application of the median-based causal outlier detector and classifier to the USBL case. Some simulation results are analysed in Section 3 and Section 4 validates the usage of the devised classifier with real experimental data obtained at sea with the USBL prototype. Finally, Section 5 provides some concluding remarks and comments on future work to be developed within this subject.

2   Online Outlier Detection Algorithms

The detection and identification of possible outliers in the acoustic positioning measurements is of the utmost importance as already pointed out due to the fact that, if not correctly flagged, these spurious outliers might severely degrade the navigation systems performance that use this information, which information can also be critical to vital control systems on-board the underwater robotic vehicle. Albeit other more integrated solutions could be devised, that include designing navigation Kalman filters robust to outliers (see Ting et al. (2007a) and Gandhi and Mili (2010)), the idea of using instead a standalone outlier detector and classifier, that is coupled to the output of the acoustic positioning device, stems from the fact that not all navigation algorithms fit the framework of robust Kalman filters as presented in Ting et al. (2007a) and in Gandhi and Mili (2010). It is often desirable to have an outlier classifier detached from the dynamic filtering framework, thus allowing for several algorithms to be implemented independent of the outlier detection stage. Moreover, this setup allows for the USBL to provide position measurements with outliers correctly classified to a multitude of systems on-board.

The causal median on-line outlier classifier adopted in this work, is presented in this section and is based on the work presented in Menold P.H. and Allgower (1999). Section 2.1 provides an overview of the most important concepts of the causal median filter presented in Menold P.H. and Allgower (1999) and Section 2.2 explains the steps taken to adapt the causal median filter framework to the USBL outlier identification and classification problem. See Ting et al. (2007b) for a recent and alternative approach on the design of outlier detectors and classifiers using a Bayesian approach.

2.1   The Causal Median Filter

Most of the material presented in this section was carefully introduced in Menold P.H. and Allgower (1999) and it is introduced here to give the reader an overview of the theoretical basis for the design of the outlier classifier. Thus this section summarizes the most important concepts for the design of the outlier classifier. Consider the current observation x_k at time instant k and a data window W_k of fixed-width N

$$W_k = \begin{bmatrix} x_{k-N+1} & \cdots & x_{k-1} \; x_k \end{bmatrix} \in \mathcal{R}^N.$$

If the values in W_k are sorted in descending or ascending order to obtain the sorted window R_k

$$R_k = \text{sort}(W_k)$$

the median $x_k^\dagger$ is easily obtained as the mid-point of R_k as

$$x_k^\dagger = \begin{cases} R_k(\frac{N+1}{2}) & \text{if } N \text{ is odd,} \\ \frac{1}{2}\left(R_k(\frac{N}{2}) + R_k(\frac{N}{2}+1)\right) & \text{if } N \text{ is even.} \end{cases}$$

The distance from the current data point x_k to the median value $x_k^\dagger$ of the window W_k is given by

$$d_k = |x_k^\dagger - x_k|. \tag{1}$$

The data cleaning filter first identifies outliers by testing this distance d_k against a specified threshold $T_k \geq 0$ (which might depend on the data inside the window), and if the distance d_k exceeds the threshold T_k, then the current data point x_k is classified as an outlier. If the data point x_k is deemed an outlier, then it may be replaced by a prediction x_k^* to obtain a filtered sequence f_k given by

$$f_k = \begin{cases} x_k & \text{if } d_k \leq T_k, \\ x_k^* & \text{if } d_k > T_k. \end{cases}$$

or simply flagged to be an outlier so that systems that use this data downstream may now that it's not a reliable sample. The authors in Menold P.H. and Allgower (1999) mention several replacement strategies, which include, for instance, replacing the outliers by the current median value $x_k^* := x_k^\dagger$ or by the last valid value inside of the window W_k. If the outlier replacement actually takes place in the filtering framework, such setup is normally called a **data cleaning filter**. On the other hand, if the outliers are simply identified and marked, the setup is called an **outlier classifier**.

In the scope of this work we are not particularly interested in data cleaning filters since these tend to change the input data. We want to be able to provide raw acoustic USBL measurements to a myriad of systems and navigation filters on-board but with some sense of safety by flagging inappropriate data that might lead these navigation and control systems to instability. Navigation systems on-board the considered robotic platforms are typically based on dynamical systems that resemble the kinematics of rigid bodies and are able to provide open-loop numerical integration of other sensors such as accelerometers and rate gyros Morgado et al. (2008) when acoustic positioning systems data is not available or their measurements are suspected to be outliers. Moreover the effect of replacing outlier data points using the aforementioned strategies might introduce delays on the sequence and additional distortions on the noise characteristics of the signals that are difficult to model on the design of control algorithms and navigation filters. Thus, these replacement strategies should be used with appropriate care.

Threshold Selection. The threshold selection strategy adopted in this work and presented in Menold P.H. and Allgower (1999) is actually a combination of two strategies — the median absolute deviation (MAD) scale based threshold and a fixed lower bound for the threshold — and is given by

$$T_k = \max\left(cS_k, T_{min}\right) \tag{2}$$

where T_{min} is the lower bound for the threshold, S_k is an estimate for the MAD, and for some constant $c \in \mathcal{R}^+$, chosen independent of the data in the window W_k. The MAD scale estimate is defined as the median absolute deviation of the data points in the window W_k from the median $x_k^\dagger$, and is simply given by the median of the distances between all the data points in the window W_k and the median $x_k^\dagger$

$$D_k = \begin{bmatrix} d_{k-N+1} & \cdots & d_{k-1} & d_k \end{bmatrix} \in \mathcal{R}^N \tag{3}$$

where d_{k-i} with $i = \{0, 1, \ldots, N-1\}$ is defined similarly to (1)

$$d_{k-i} = |x_k^\dagger - x_{k-i}|, \quad \forall i = \{0, 1, \ldots, N-1\}.$$

Thus, the MAD scale estimate S_k is given by the median of D_k from (3). This un-normalized MAD scale estimate is often normalized Menold P.H. and Allgower (1999) to $\tilde{S}_k = S_k/0.6745 \approx 1.4826S_k$ to make it an unbiased estimate of the standard deviation for Gaussian data Huber and Ronchetti (1981). The choice of the scale parameter c in (2) will be addressed in Section 2.1.

The idea behind the dual strategy combination lies on the practical limitation with the MAD scale estimate being $S_k = 0$ for sequences that have, in a window of width N, at least $(N - 1)/2 + 1$ values (if N is odd, or $N/2 + 1$ if N is even), identical to the current data point x_k. If a lower bound T_{min} was not adopted, the threshold would be $T_k = 0$ in such cases regardless of parameter c. Thus T_{min} should be chosen taking into account the measurement noise level of the input signal and other parameters such as quantization and sensor resolution. Using this threshold selection rule, changes in the input sequence up to T_{min} are invariant, and as such it should not be chosen too large.

MAD Scale Parameter c. The results presented in this section provide a theoretical background on the design choice of the parameter c in (2). Most importantly they also provide lower bounds for c for certain type of sequences to be invariant under the MAD data cleaning filter. The following theorem establishes a lower bound for c under monotonic sequences that satisfy a growth rate restriction.

Theorem 1 ((Menold P.H. and Allgower, 1999, Theorem 5.1)). *Any monotonic sequence $\{x_k\}$ satisfying the growth rate restriction*

$$|x_{i+2} - x_{i+1}| \le m|x_{i+1} - x_i|,$$

for some $m \in [0, 1]$ and $\forall i \in \mathcal{N}$, is invariant under the data cleaning filter of width $N = 4H + 1$ provided $c \ge 1 + m^H$.

Proof: *See the proof of Theorem 5.1 in Menold P.H. and Allgower (1999).*

An illustrative example of a monotonic decreasing sequence and the corresponding lower bound for c under three different window sizes is presented in Fig. 3. From this example it comes that as a rule of thumb, the parameter c should be larger than 2 for this type of sequences, that is $c \ge 2$.

The next set of results provide a basis to the lower bounding of the parameter c for two other distinct types of sequences defined in the following.

Definition 1 ((Menold P.H. and Allgower, 1999, Definition 5.1)). *A sequence of* **Type I** *satisfies the following conditions*

$$x_{k-2H} = x_k^\dagger, \quad 0 < c_1 \le c_2 < \infty,$$
$$c_1(2H - i) \le x_{k-i} - x_k^\dagger \le c_2(2H - i) \quad \forall i \le 2H,$$
$$c_1(2H - i) \ge x_{k-i} - x_k^\dagger \ge c_2(2H - i) \quad \forall i > 2H,$$

for all k and where $x_k^\dagger$ is the median of the window of width $N = 4H + 1$.

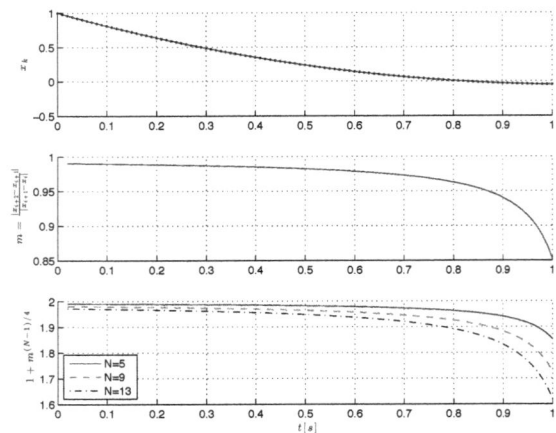

Fig. 3. Growth rate analysis of monotonic sequences and choice of the constant value $c \geq 1+m^H$ with $H = (N-1)/4$

Definition 2 ((Menold P.H. and Allgower, 1999, Definition 5.2)). *A sequence of* *Type II* *satisfies the following conditions*

$$x_{k-2H} = x_k^\dagger, \quad 0 > c_1 \geq c_2 > \infty,$$
$$c_1(2H - i) \geq x_{k-i} - x_k^\dagger \geq c_2(2H - i) \quad \forall i \leq 2H,$$
$$c_1(2H - i) \leq x_{k-i} - x_k^\dagger \leq c_2(2H - i) \quad \forall i > 2H,$$

for all k and where $x_k^\dagger$ is the median of the window of width $N = 4H + 1$.

The following theorem provides a lower bound for the parameter c under this type of sector bounded sequences defined in **Definitions 1** and **2**.

Theorem 2 ((Menold P.H. and Allgower, 1999, Theorem 5.2)). *Any sequence $\{x_k\}$* *of type I or II is invariant under the MAD-based data cleaning filter of width $N =$* *$4H + 1$ with $c \geq 2c_2/c_1$.*

Proof: See the proof of Theorem 5.2 in Menold P.H. and Allgower (1999).

Remark 1. The sequences of **Type II** in **Definition 2** are the decreasing analogous to the increasing **Type I** sequences in **Definition 1**.

In practice, the choice of the parameter c can be accomplished with the aid of the results presented in this section. For this purpose one could analyse sub-sequences of the nominal sequence and apply both Theorems 1 and 2 to compute a set of lower bounds for c, and then choose the largest lower bound that satisfies the invariance for the full sequence. On the other hand, it is recognizable that this procedure might be cumbersome, and probably the simplest way to choose a reasonable value for c is to examine and try out some values on training sets of the contaminated sequences. It is important to emphasize that c should also not be set too large, otherwise the outlier identification function will cease to be effective.

Table 1. USBL Data classification levels and flags

Level	Flag	Description
0	invalid	unrealisable solutions due to physical constraints of the USBL array: exceed the maximal allowed time delay between any two receivers on-board
1	valid	pass the physical limitations validation test but are yet unknown regarding its **good** or **outlier**
2	outlier	valid solution but clearly flagged as an outlier that violates the distance to the median of the window of valid samples
3	good	indicates that at least $2/3$ of the samples on the detection window were classified as valid

Window Size N. The window size is also a very important parameter in the design of the outlier classifier, and it should be chosen to avoid observations from high dynamic range systems to be incorrectly considered outliers. Both the median $x_k^\dagger$ and the MAD scale estimate S_k become less connected to local variations as N becomes too large and the analysis of a new measurement x_k less effective. On the other hand, N should also not be set too small in order to accommodate a reasonable amount of defective data patches. Data patches may occur for instance when a sensor saturates its output or errors in the measurements happen. For a window size of $N = 4H + 1$, both the median $x_k^\dagger$ and the MAD scale estimate S_k are completely set to the value of a patch of $2H + 1$ samples. For instance, for $N = 9$, 5 patched outliers would undermine the effectiveness of the classifier.

2.2 Adaptation to the USBL System

The USBL positioning system provides measurements of the position of a transponder with respect to the reference frame of the robotic vehicle, that is, a $\mathbf{p} \in \mathcal{R}^3$. In order to adapt the outlier detection scheme to the USBL system, the algorithm outlined in Section 2.1 has to be extended to the three-dimensional case. The extension is fairly simple in which the window size is also extended to $W_k \in \mathcal{R}^{3 \times N}$ and the evaluation is performed separately for each of the three Cartesian coordinates.

The next improvement to be incorporated is the introduction of a classifier flag instead of performing outliers replacement, deriving what was named an **outlier classifier** in Section 2.1. A four level classification scheme is adopted that allows to introduce robustness to the classification process and the usage in downstream systems that require the classified data. The four levels can be summarized in Table 1.

Another feature that is needed due to the fact that underwater acoustics are highly susceptible to jamming and periods without actual measurements, is a time-elapsed window reset that removes observations from the window if their time tags are older than R seconds from the current system time. Finally, the global classification flag for each of the N-triplet values is assessed as follows: if any of its three values violates the threshold distance rule then the entire triplet is set as an **outlier**. The final

Table 2. Acoustic outlier classification algorithm

Algorithm *ClassifyData(p, t)*
(∗ detect outliers and classify USBL positioning data ∗)
Input: p - current position measurement
Input: t - current measurement time
Output: class - classification level of current measurement
(∗ persistent W_p positions window size $3 \times N$ init. to **0** ∗)
(∗ persistent t_v last valid measured time tag init. to 0 ∗)
1. **if** $t - t_v > R$
2.   **then** remove from W_p elements older than t_v
3.   **else** insert the new data point in the window
4. $W_v \leftarrow$ select only valid elements from W_p
5. $m_k \leftarrow$ compute the row-wise median of W_v
6. $d_k \leftarrow$ compute the distances $|W_v - m_k|$
(∗ Compute the MAD scale estimate ∗)
7. $S_k \leftarrow$ compute the row-wise median of d_k
(∗ Normalize the MAD scale estimate ∗)
8. $\tilde{S}_k \leftarrow 1.4826 S_k$
(∗ Threshold selection ∗)
9. $T \leftarrow \max(c\tilde{S}_k, T_{min})$
(∗ Test the data point against the threshold ∗)
10. **if** $|p_k - m_k| > T$
11.   **then** class = outlier
12.   remove data point from the window W_p
13.   **else** class = valid
14.   update last valid time tag to current $t_v \leftarrow t$
15. **if** number of valid data points in $W_p > 2/3N$
16.   **then** class = good

algorithm is outlined in Algorithm 2. The adaptation steps of the algorithm can be briefly summarized as follows:

- Extension to three dimensions
- Introduction of a classifier flag instead of performing outliers replacement
- Creation of a four-level classification scheme: **invalid**, **outlier**, **valid**, and **good**
- Introduction of an elapsed-time R window reset.

3   Simulation Analysis

The outlier classifier presented previously was first evaluated in simulation to assess its feasibility and performance. The nominal sequence to be tested was derived from the output of a second order spring-mass-damper system on the presence of small input step changes. Additive white Gaussian Noise was added to the output with a standard deviation of $\sigma = 0.03$. Ten percent of the values on the sequence were disturbed with outliers in random positions, with amplitudes in the interval $\pm[0.25, 0.6]$.

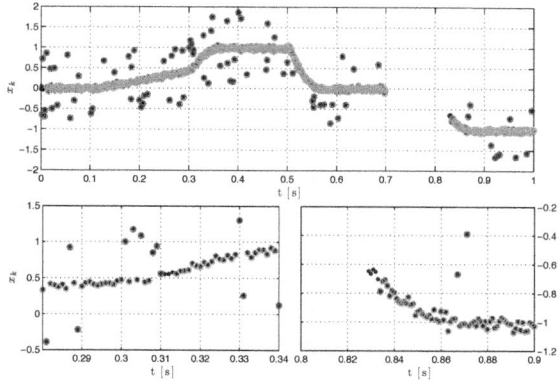

Fig. 4. Classification analysis in simulation results, with $N = 9$ and $c = 5$ — the parameter c was adjusted on this training set so that there are not false positive classifications as **good**

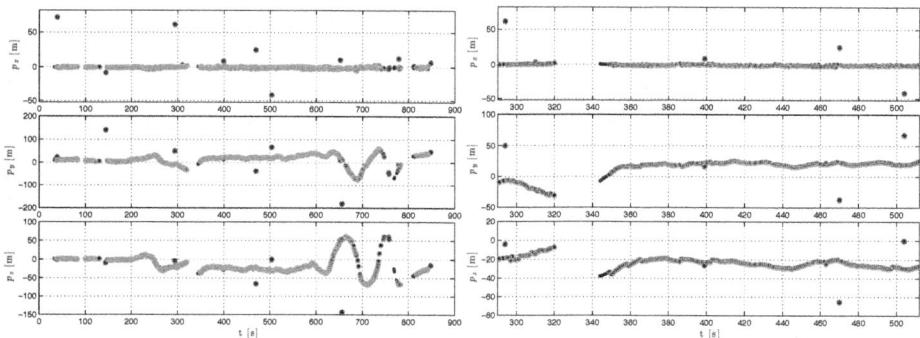

Fig. 5. Experimental results for the USBL positioning system during tests in Roses, Spain in October 2011. On the left: Overview of the total of approx. 900 seconds of data. On the right: Zoom from 290 to 512 seconds of operation.

The lower-bound for the threshold T_{min} was adjusted to accommodate disturbances on nominal, non-dynamically changing sequences up to 3σ of the additive white Gaussian noise perturbation, whereas the MAD scale estimate multiplier was initially set to $c = 6$ with a window size of $N = 9$. This first approach led to the conclusion that some values were being classified as false positive good values. It can be easily argued that is highly preferable to have false identifications of outliers on good values, rather than having outliers being classified as good values. Thus the value for the MAD scale estimate multiplier was adjusted to $c = 5$ and the classifier rerun on the same data. The remastered results are presented in Fig. 4, where it can be seen that there are no more false positives of good values while maintaining the performance on the remainder of the sequence. Lower-bounds for this c value can be found with the aid of Theorems 1 and 2 on sub-sequences of this training set, nonetheless it is always a good practice to adjust this value bearing in mind the overall performance of the classifier on the entire training set.

4   Validation with Real Data

The devised outlier classifier was implemented and applied to a real USBL positioning system, fully developed in-house Morgado et al. (2010) and its outlier detection capabilities and performance are evaluated in this section. The parameters of the classifier were adjusted to: $c = 6$, a window-size of $N = 9$ samples, time reset constant of $R = 20$ seconds, and a threshold lower-bound of $T_{min} = 6$ meters. The outlier detection capability of the system in real world operation scenarios is evidenced in Fig. 5.

5   Conclusions

This paper presented an outlier detection and classifier algorithm with application to underwater acoustic positioning systems. The devised framework is based on causal median filters and a set of theoretical analysis tools that allows for the design of the filter parameters was also presented. Specific details that arise from the implementation of such an algorithm in real-world operation conditions were taken into account and a set of new features, such as a multi-level classification scheme and a time-based moving window reset, was added to cope with periods of acoustic data outage, which are quite typical in underwater scenarios. Interestingly enough, given the necessary window-size and computations, the outlier classifier is easily implementable in low-cost and low-power consumption digital signal processor (DSP) hardware. The outlier classifier was finally evaluated both in simulation and with experimental data from a prototype USBL acoustic positioning system fully developed in-house. Interesting ideas on future directions of research in this subject might include the validation of an adaptive algorithm for the choice of certain parameters in the filter.

References

Gandhi, M.A., Mili, L.: Robust Kalman filter based on a generalized maximum-likelihood-type estimator. IEEE Trans. Signal Processing 58(5), 2509–2520 (2010)

Huber, P.J., Ronchetti, E.: Robust statistics, vol. 1. Wiley Online Library (1981)

Kinsey, J.C., Whitcomb, L.L.: Preliminary field experience with the DVLNAV integrated navigation system for oceanographic submersibles. Control Engineering Practice 12(12), 1541–1548 (2004) (invited paper)

Kinsey, J.C., Eustice, R.M., Whitcomb, L.L.: A survey of underwater vehicle navigation: Recent advances and new challenges. In: Proceedings of the 7th Conference on Manoeuvring and Control of Marine Craft (MCMC 2006), Lisbon, Portugal (2006)

Larsen, M.B.: Synthetic long baseline navigation of underwater vehicles. In: Proc. MTS/IEEE OCEANS 2004, Providence, RI, USA, vol. 3, pp. 2043–2050 (September 2000)

Lurton, X., Millard, N.W.: The feasibility of a very-long baseline acoustic positioning system for AUVs. In: Proceedings of the MTS/IEEE OCEANS 2004 Conference, Brest, France, vol. 3, pp. 403–408 (September 1994)

Menold, R.K., Pearson, P.H., Allgower, F.: Online outlier detection and removal. In: Proceedings of the 7th International Conference on Control and Automation MED 1999, Haifa, Israel, pp. 1110–1134 (1999)

Miller, P.A., Farrell, J.A., Zhao, Y., Djapic, V.: Autonomous underwater vehicle navigation. IEEE Journal of Oceanic Engineering 35(3), 663–678 (2010) ISSN 0364-9059, doi:10.1109/JOE.2010.2052691

Milne, P.H.: Underwater Acoustic Positioning Systems. Gulf Pub. Co. (1983)

Morgado, M., Oliveira, P., Silvestre, C., Vasconcelos, J.F.: Improving Aiding techniques for USBL Tightly-Coupled Inertial Navigation System. In: Proceedings of the IFAC World Congress 2008, Seoul, South Korea, IFAC (July 2008)

Morgado, M., Oliveira, P., Silvestre, C.: Design and experimental evaluation of an integrated USBL/INS system for AUVs. In: Proceedings of the 2010 IEEE International Conference on Robotics and Automation (ICRA), Anchorage, AK, USA, pp. 4264–4269. IEEE (May 2010), doi:10.1109/ROBOT.2010.5509597

Napolitano, F., Cretollier, F., Pelletier, H.: GAPS, combined USBL + INS + GPS tracking system for fast deployable and high accuracy multiple target positioning. In: Proceedings of the OCEANS 2005, Brest, France (June 2005)

Pascoal, A., Oliveira, P., Silvestre, C., et al.: Robotic Ocean Vehicles for Marine Science Applications: the European ASIMOV Project. In: Proceedings of the OCEANS 2000, Rhode Island, USA (September 2000)

Ting, J.A., Theodorou, E., Schaal, S.: A Kalman filter for robust outlier detection. In: IEEE/RSJ International Conference on Intelligent Robots and Systems, IROS 2007, pp. 1514–1519. IEEE (2007a)

Ting, J.-A., D'Souza, A., Schaal, S.: Automatic outlier detection: A bayesian approach. In: Proceedings of the 2007 IEEE International Conference on Robotics and Automation, ICRA 2007, pp. 2489–2494 (April 2007b)

Vaganay, J., Bellingham, J.G., Leonard, J.J.: Comparison of fix computation and filtering for autonomous acoustic navigation. International Journal of Systems Science 29(10), 1111–1122 (1998)

Vickery, K.: Acoustic positioning systems. New concepts. In: Proc. 1998 Workshop Autonomous Underwater Vehicles, AUV 1998, Cambridge, MA, USA (August 1998)

A Centralized Approach to the Coordination of Marine Robots*

Bruno M. Ferreira**, Aníbal C. Matos,
Nuno A. Cruz, and António Paulo Moreira

INESC TEC and Faculty of Engineering, University of Porto
Campus da FEUP, Rua Dr. Roberto Frias, 4200-465 Porto, Portugal
bm.ferreira@fe.up.pt

Abstract. This paper presents a centralized coordination scheme for multiple marine vehicles. The only requirements for proper operation of this method are the presence of bidirectional communication links with a virtual leader and bounded reference tracking errors. By relying on a, lower level, individual position tracking control, coordination is achieved by means of a centralized potential-field that uniquely defines the desired formation geometry as well as its position. The formation can be driven along a path that does not necessarily need to be predefined. Instead, a virtual leader defines the formation position at each instant of time. Furthermore, the possibility of setting stationary points over the path followed by the formation is guaranteed. The approach is illustrated in practice with autonomous surface vehicles in real environment, subjected to disturbances such as wind and waves.

1 Introduction

Integration of multiple robotic systems in several new challenges in marine environments often requires the individual robots to coordinate their actions or motion to achieve coherent behaviors. Such a motion if frequently subjected to disturbances and uncertainties that usually result in tracking errors that must be accommodated to guarantee a proper coordinated operation. This paper deals with the coordination of a team of marine robots under natural disturbances, formalizing the construction of the scheme used in [1], which has also been previously employed to build an acoustic network in [2].

Typical missions where the robots are required to share information with a central entity to achieve a common task have motivated the centralized approach described here. An example of this type of mission can be found in [3] and [1], where vehicles coordinate their motion in order to estimate the position of a

---

* Project "NORTE-07-0124-FEDER-000060" is financed by the North Portugal Regional Operational Programme (ON.2 – O Novo Norte), under the National Strategic Reference Framework (NSRF), through the European Regional Development Fund (ERDF), and by national funds, through the Portuguese funding agency, Fundação para a Ciência e a Tecnologia (FCT).

** B. Ferreira was supported by the Portuguese Foundation for Science and Technology through the Ph.D. grant SFRH/BD/60522/2009.

© Springer International Publishing Switzerland 2015   567
A.P. Moreira et al. (eds.), *CONTROLO'2014 - Proc. of the 11th Port. Conf. on Autom. Control*,
Lecture Notes in Electrical Engineering 321, DOI: 10.1007/978-3-319-10380-8_54

sound source. Other applications in the context of localization and navigation aid have also been proposed by several research groups. The works in [4–7] are just a few number of motivating examples for the development of the present coordination strategy.

Murray [8] classified the approaches to formation control problems into three main categories: potential-field based [9–12], optimization-based [13, 14] and swarm behaviors [15,16]. In addition, consensus-based algorithms have also been applied to formation control (see, for example, [17–20]). In the context of this paper, which concentrates on positioning of coordinated marine robots, potential-field-, consensus- and optimization-based approaches are the most promising choices. Here, the potential-field approach has been chosen essentially because of its simplicity in incorporating new behaviors by using the simple, yet powerful, general idea that desired states originate *attractive forces* while undesired generate *repelling forces*. The applicability of this type of approach is wide and can be employed in different scenarios from simple schooling schemes [9] to highly dynamic environments such as robotic soccer [10]. Aside from their simplicity, artificial potential fields in the context of coordination of marine vehicle has some advantages over optimization-based and consensus-based methods, namely: the nature of the environment where marine vehicles navigate makes it unpredictable with regard to disturbances and hence optimization-based approaches, which are also typically computationally intensive, may fall ill-suited if they use a finite horizon to predict future actions; the theoretical framework behind average-consensus algorithms generally originate specific solutions, such as consensus on an along-path scalar variable [17] or on relative positions while tracking a geometrically changing formation shape [20].

Several researchers have explored and discussed artificial potential fields to achieve coordination of multiple robots. Egerstedt and Xu, in [11], present a solution to keep in formation a team with n elements. The approach conceptually considers $n+1$ virtual robots and is very interesting in the sense that the individual dynamics are decoupled from the formation dynamics using a *formation constraint function*, that can be seen as a potential function. Within the same context, [12] present a method for formation keeping along arbitrary paths defined by a virtual leader, and formation geometries. Ihle *et al.* [21] proposed an elegant and generic formulation resorting to Lagrangian multipliers to reduce constraint functions, which measure general positioning errors, to zero. The approach is also extended to underactuated ships.

Based on the approach in [11], a generalization of the method for inclusion of generic functions as well as their necessary conditions to guarantee stability and convergence of the formation is provided. Furthermore, a particular method that amounts to the definition of functions that guarantee that the formation is kept with a bounded error even in presence of malfunctioning robot (e.g., vehicles whose velocities are below the expected value) is proposed. Finally, this new approach exempts predefined paths for formations of vehicles and makes it possible to handle static formations, that is, formations holding their positions.

In order to validate the theoretical derivation, the coordination scheme is implemented in autonomous surface vehicles (ASVs) and tested in real conditions.

2 Background

This section presents a brief summary of the key concepts in [11], which are subsequently generalized, extended and adapted to the coordination of marine vehicles.

Consider N vehicles moving in formation. The (possibly reduced order) position of the ith vehicle is denoted $\eta_i(t) \in \mathbb{R}^m$, $i = 1, ..., N$, with $m \leq 3$. Similarly, the virtual leader position is given by $\eta_0(t) \in \mathbb{R}^m$, which is continuous and differentiable. Hereafter, the subscript $(\cdot)_0$ is used to denote the vectors or scalars related to the virtual leader. The idea of considering a virtual leader instead of a real one is advantageous because it is unaffected by real disturbances, which could degrade the overall performances of the formation.

Consider a continuous and differentiable path $p_0(s_0(t)) : \mathbb{R} \to \mathbb{R}^m$, parametrized by a scalar function $s_0(t) : \mathbb{R} \to \mathbb{R}$, that defines the virtual leader position at instant t: $\eta_0(t) = p_0(s_0(t))$. Each real vehicle (follower) should assume a specific position in the formation, with regard to the virtual leader. Therefore, a change of coordinates is appropriate. The desired relative position of a vehicle with regard to the virtual leader is given by $\tilde{\eta}_i^*(t) = \eta_i^*(t) - \eta_0(t)$, where $\eta_i^*(t)$ is the desired vehicle position expressed in an absolute inertial frame. Similarly, the (real) position of the vehicle is also expressed with regard to the leader as $\tilde{\eta}_i(t) = \eta_i(t) - \eta_0(t)$.

The desired positions of the vehicles are given by $\tilde{\eta}_i^d(t) \in \mathbb{R}^m$, also defined with regard to the virtual leader position. These vectors, which can be seen as the vectors defining the positions of virtual followers, make it possible for the vehicles to follow a time-varying reference that is not rigidly coupled to the virtual leader position. Instead, its dynamics can be designed so that it converges to $\tilde{\eta}_i^*(t)$, as intended, while still considering the individual tracking errors defined by the difference between this virtual follower position and the corresponding real vehicle position.

To drive the individual references $\tilde{\eta}_i^d(t)$ to their respective desired final positions $\tilde{\eta}_i^*(t)$, a *formation constraint function* is used

$$F : \mathbb{R}^m \times ... \times \mathbb{R}^m \to \mathbb{R}^+. \tag{1}$$

It is assumed that this function is differentiable and strictly convex, and the solution of

$$F(\tilde{\eta}_1^d(t), ..., \tilde{\eta}_n^d(t)) = 0$$

is uniquely determined by $(\tilde{\eta}_1^*(t), ..., \tilde{\eta}_n^*(t))$, or equivalently $F^{-1}(0) = (\tilde{\eta}_1^*(t), ..., \tilde{\eta}_n^*(t))$. This formation constraint function is actually a positive definite potential-field function that creates a measure of the distance between all the reference positions $\tilde{\eta}_i^d(t)$ and their respective desired final positions $\tilde{\eta}_i^*(t)$.

As for the virtual leader, suppose that the evolution of the individual references are defined by paths $p_i(s_i(t)) : \mathbb{R} \to \mathbb{R}^m$ that can be designed so that $\tilde{\eta}_i^d(t)$ converges to $\tilde{\eta}_i^*(t)$. Take

$$\tilde{\eta}_i^d(t) = p_i(s_i(t)), \tag{2}$$

which, differentiating with respect to time, results

$$\dot{\tilde{\eta}}_i^d(t) = \frac{dp_i(s_i(t))}{ds_i(t)} \dot{s}_i(t), \quad i = 1, ..., N. \tag{3}$$

Coherent motion of the vehicles in formation is guaranteed by choosing proper individual path dynamics $\frac{dp_i(s_i(t))}{ds_i(t)}$ and *along-path* evolution dynamics $\dot{s}_i(t)$. Additionally, the *along-path* evolution $\dot{s}_0(t)$ of the virtual leader must be defined at the cost of the tracking errors of the followers. In other words, $\frac{dp_i(s_i(t))}{ds_i(t)}$ defines the instantaneous direction of the reference while $\dot{s}_i(t)$ defines the evolution rate.

3   Coordination Scheme

3.1   Generalization

The method presented in [11] is suitable for most robotic applications with vehicles that are not strongly affected by exogenous disturbances. The only requirement for the algorithm to work properly is ensuring individual bounded tracking errors. This construction is powerful but may be limited when it comes to practical implementation. For example, in a faulty formation, with a vehicle that cannot move or that is unable to communicate its position to the virtual leader, will continue evolving since $\dot{s}_0(t) > 0, \forall \rho_i(t) : \mathbb{R} \to \mathbb{R}^+$ (see [11]). Depending on the requirements, it may be preferable to stop the formation, that is, the virtual leader holds its position and so do all real and virtual followers operating properly, if one of the vehicles is unable to reach its position reference or is unable to communicate. Furthermore, note that the original construction does not allow static formations, that is, formations holding their positions.

This set of reasons leads us to present a more general construction for this centralized coordination problem. Some changes to the control scheme are proposed. Firstly, and following the same sequence as above, the direction of the path does not necessarily need to follow the steepest descent direction as in [11]. Instead, the use of continuous functions of the type $f_i(F) : \mathbb{R} \to \mathbb{R}^m$ that verify

$$(\nabla_{\tilde{\eta}_i^d(t)} F)^T \cdot f_i(F) < 0, \quad \forall\, F \neq 0 \quad \text{and} \quad f_i(0) = \mathbf{0}.$$

is proposed. Then, the path is chosen so that it obeys the following relationship:

$$\frac{dp_i(s_i)}{ds_i} = f_i(F), \quad i = 1, ..., N. \tag{4}$$

Furthermore, the proposed path evolution rates s_i are given by

$$\dot{s}_i = g_i(\rho_i(t)), \quad i = 1, ..., N, \tag{5}$$

where $g_i(\rho_i(t)) : \mathbb{R}^+ \to \mathbb{R}^+$ are continuous non-increasing functions.

Finally, the leader path evolution is dictated by the product

$$\dot{s}_0 = \lambda_0 \prod_{i=1}^{N} \alpha_i g_i(\rho_i(t)), \qquad (6)$$

where λ_0 and α_i are positive constants.

Note that $\dot{s}_i = 0$ and $\dot{s}_0 = 0$ if $g_i(\rho_i(t)) = 0$, for any $i = 1, ..., N$. This means that both the vehicle (virtual) reference and the virtual leader hold their positions, awaiting for the vehicle i to track its reference so that $g_i(\rho_i(t)) > 0$, and so do the remaining references after having tracked their desired positions. Obviously, this is also valid when there are more than one vehicle whose tracking errors imply $g_i(\rho_i(t)) = 0$.

The rest of the proof of convergence follows the same steps as in [11]. Suppose that each vehicle is able to track its reference in finite time so that $g_i(\rho_i(t))$ is non null, that is, there exists a time t', that verifies $g_i(\rho_i(t)) > 0$, $\forall t > t'$. Then, it follows that

$$\dot{F} = \sum_{i=0}^{N} (\nabla_{\tilde{\eta}_i^d(t)} F)^T \dot{\tilde{\eta}}_i^d(t) = \sum_{i=0}^{N} (\nabla_{\tilde{\eta}_i^d(t)} F)^T f_i(F) \dot{s}_i, \qquad \forall t > t'.$$

Since, by definition, $(\nabla_{\tilde{\eta}_i^d(t)} F)^T f_i(F) < 0$, $\forall F > 0$, and $\dot{s}_i = g_i(\rho_i(t)) > 0$ and consequently $\dot{s}_0 > 0$, for all $t > t'$, it yields

$$\dot{F} < 0, \forall t > t', \forall F > 0.$$

This proves that the formation converges to the desired shape and position. In contrast to [11], note that no assumption was made on the leader path. This makes it possible to handle both dynamically changing paths, as well as stationary points and, as a result, station-keeping behaviors can be obtained.

3.2 Particular Method

In several situations contemplating "malfunctioning vehicles", it may be preferable to make the formation hold its position, waiting for the *missing(s) vehicle(s)*[1]. Furthermore, supposing that a given geometry configuration allows robust communication links, it may be desirable that such a configuration is achieved. Nevertheless, the scheme proposed in [11] can not guarantee that the vehicles converge to their desired positions $\tilde{\eta}_i^*(t), i = 1, ..., N$ in the formation if they have bounded velocities, since $\dot{s}_i > 0$, $\forall \rho_i(t)$, $i = 0, ..., N$.

Next, a particular implementation is proposed based on the generalization presented above. Any other choice of function meeting the conditions above would

---

[1] Missing vehicles are seen here as vehicles that are above a predefined threshold distance from their respective references (virtual follower) or that can not communicate with the virtual leader.

be valid as well. Here, a focus on limited tracking errors and rigid formations is given. Empirically, the aim is to guarantee that the vehicles do not deviate too much from their position references in the formation. The objective is to reduce the distance between the desired and the reference positions. Therefore, define the proposed formation constraint function to be given by

$$F = \sum_{i=1}^{N} ||\tilde{\eta}_i^d(t) - \tilde{\eta}_i^*(t)||^2, \tag{7}$$

which is strictly convex and positive definite. As real marine vehicles have bounded velocities, the functions $f_i(F)$ are defined to be a bounded function of the gradient of F:

$$f_i(F) = -\text{sat}\left(\nabla_{\tilde{\eta}_i^d(t)} F, \ \Gamma_i\right), \tag{8}$$

where Γ_i are positive constants and sat $(\cdot, \ \cdot) : \mathbb{R}^n \times \mathbb{R} \to \mathbb{R}^n$ is the saturation function. This definition obviously meets the condition $(\nabla_{\tilde{\eta}_i^d(t)} F)^T f_i(F) < 0$. Note that decreasing the constants Γ_i makes it possible to evolve the position references more slowly.

The next step is to define the functions $g_i(\rho_i(t))$. The aim is to achieve a *rigid* formation where possible failures in both tracking performances and/or communication links cause the formation to stop (virtual leader holding its position and vehicles holding their positions in the formation). This will allow the *missing vehicles* to join the formation before it continues evolving over the desired path. Henceforth, the choice is a simple piecewise linear and nonnegative function as follows:

$$g_i(\rho_i(t)) = \begin{cases} \Lambda_i - \lambda_i \rho_i(t), & \text{if } 0 \le \rho_i(t) \le \frac{\Lambda_i}{\lambda_i} \\ 0, & \text{otherwise} \end{cases}, \tag{9}$$

where $\Lambda_i, \lambda_i > 0$ are positive constants.

This definition implies that the formation evolves only if $\rho_i(t) < \frac{\Lambda_i}{\lambda_i}$ for all $i = 1, ..., N$. Otherwise, the formation will end up stopping, because the virtual leader stops. Hence the *rigidity* of the formation, that is, the error tolerance for the vehicles to track their respective references, is dictated by the ratio $\frac{\Lambda_i}{\lambda_i}$. This can be used to design rigid formations where the positions in the formation are ideally maintained as precise as desired. This, however, may come at the expense of a slower formation evolution.

Additionally, as Λ_i is the maximum value of $g_i(\rho_i(t))$, the evolution rate is directly proportional to this constant: the greater the value of Λ_i, the faster the evolution of the ith position reference along the path that conducts the vehicle to the desired position in the formation.

Remark 1. In the proposed approach, the along-path evolution rates $\dot{s}_i$ are dictated by the functions $g_i(\rho_i(t))$ which only depend on the tracking error, that is, on positional error. For scenarios where the attitude of the followers must be considered, the extension of this scheme is trivial as the attitude error component can easily be added to the function $g_i(\rho_i(t))$ in (9).

Fig. 1. Zarco and Gama ASVs

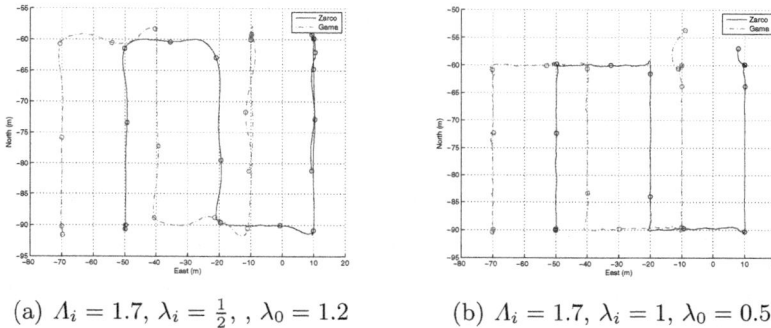

(a) $\Lambda_i = 1.7$, $\lambda_i = \frac{1}{2}$, , $\lambda_0 = 1.2$ (b) $\Lambda_i = 1.7$, $\lambda_i = 1$, $\lambda_0 = 0.5$

Fig. 2. Coordinated mission of Zarco and Gama in La Spezia

4 Experiments

4.1 Setup Description

Zarco and Gama ASVs (see [22] and Fig. 1) have been used here to test the coordination scheme. These are similar vehicles with 1.5 m of length and weighting slightly over 50 kg in their basic configuration. Their shape and payload can be adapted depending on the mission requirements. They are actuated by two thrusters located at the stern granting control on surge (longitudinal axis) and yaw (heading) degrees of freedom (DOFs). Each thruster is capable of generating 50 N. The basic set of navigation sensors includes a global positioning system (GPS) receiver and a inertial measurement unit (IMU). During the experiments presented hereafter, Zarco and Gama were running the same target tracking control scheme locally and communicating with the virtual leader, running on a computer on shore, over radio.

In the sequel, the subscripts $(\cdot)_Z$ and $(\cdot)_G$ are used to denote quantities related to Zarco and Gama, respectively.

4.2 Results

Zarco and Gama were deployed on the La Spezia shore, Italy. Apart from navigation sensors and dry body parts, the vehicles had the same body configuration.

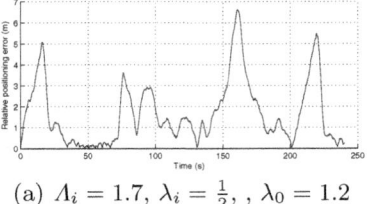
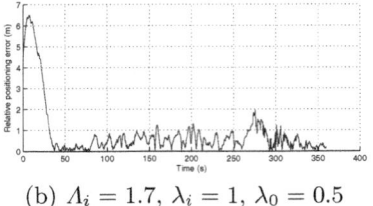

(a) $\Lambda_i = 1.7$, $\lambda_i = \frac{1}{2}$, , $\lambda_0 = 1.2$ (b) $\Lambda_i = 1.7$, $\lambda_i = 1$, $\lambda_0 = 0.5$

Fig. 3. Relative positioning error

In the operation area, natural disturbances impacted on the motion of the vehicles. Namely, the presence of wind, small currents and waves were the main causes for deviations. The position references of the vehicles were set to follow the leader with offsets in the eastward direction. Specifically, Zarco position reference, with regard to the leader, was set to $\tilde{\eta}_Z^*(t) - \tilde{\eta}_0^*(t) = [0, 10]^T$, while Gama's was set $\tilde{\eta}_G^*(t) - \tilde{\eta}_0^*(t) = [0, -10]^T$. Therefore, their relative position should be $\tilde{\eta}_Z^*(t) - \tilde{\eta}_G^*(t) = [0, 20]^T$ along all the path, if perfect tracking was possible. The starting position of the leader was $[-60, 0]^T$.

A short mission composed of five legs, as depicted in Fig. 2, was ran. The parameters were varied to change the *rigidity* of the formation and to assess the impact of the parameters λ_i and Λ_i on the formation. The parameters Λ_i were kept constant over this experiment and set to $\Lambda_i = 1.7$, while λ_i was set first to $\lambda_i = \frac{1}{2}$, $i = Z, G$. The virtual leader parameter was set to $\lambda_0 = 1.2$. Figure 2(a) depicts the trajectories of Zarco and Gama. In this case, the maximum *allowed* deviation of the vehicle with regard to its own reference before the formation "stops" is $\Lambda_i / \lambda_i = 3.4$ meters.

For further rigidity in the formation, the parameters λ_i, $i = Z, G$ were increased to 1 and the virtual leader related parameter was reduced to $\lambda_0 = 0.5$. Hence, the formation ultimately stops if a vehicle deviates more than $\Lambda_i / \lambda_i = 1.7$ meters from its reference. The resulting Fig. 2(b) shows the trajectories of the vehicles. These show accurate motions of the vehicles even in the presence of disturbances induced by wind, small waves and current.

The relative positioning errors, between Zarco and Gama, are shown in Fig. 3. As expected, the second mission with increased λ_i, shows more precise relative positioning since the *rigidity* was augmented.

5   Conclusions

Motivated by several sensor fusion applications that are intrinsically centralized, a centralized coordination scheme has been proposed. Using the approach in [11], the coordination scheme was first generalized and subsequently particularized to accommodate typical performances of marine vehicles. Assuming a lower level target tracking algorithm, no pre-established paths have to be known *a priori*.

Although only two vehicles were used, it has been demonstrated that this coordination scheme is expandable to as many vehicles as desired, as long as the communication means has the capability of handling the transmissions of the states to the virtual leader and of the references to the vehicles. If this is verified, similar results are expected with formations incorporating more vehicles.

The experiments conducted in real conditions have provided very encouraging results, allowing the vehicles to behave robustly under drifts induced by wind and currents. In the future, the convergence and performances of this framework under degraded and low rate communications will be assessed for a natural application to formations of combined underwater and surface vehicles. This analysis will also be complemented with experiments using acoustic communications.

References

1. Ferreira, B.M., Matos, A.C., Campos, H.S., Cruz, N.A.: Localization of a sound source: optimal positioning of sensors carried on autonomous surface vehicles. In: MTS/IEEE (ed.) Proceedings of the MTS/IEEE OCEANS 2013 Conference (September 2013)
2. Cruz, N.A., Ferreira, B.M., Kebkal, O., Matos, A.C., Petrioli, C., Petroccia, R., Spaccini, D.: Investigation of underwater acoustic networking enabling the cooperative operation of multiple heterogeneous vehicles. Marine Technology Society Journal (2013)
3. Ferreira, B.M., Matos, A.C., Cruz, N.A.: Optimal positioning of autonomous marine vehicles for underwater acoustic source localization using toa measurements. In: Proceedings of 2013 IEEE International Underwater Technology Symposium (UT 2013) (2013)
4. Papadopoulos, G., Fallon, M., Leonard, J., Patrikalakis, N.: Cooperative localization of marine vehicles using nonlinear state estimation. In: 2010 IEEE/RSJ International Conference on Intelligent Robots and Systems (IROS), pp. 4874–4879 (October 2010)
5. Pascoal, A., Oliveira, P., Silvestre, C., Sebastiao, L., Rufino, M., Barroso, V., Gomes, J., Ayela, G., Coince, P., Cardew, M., Ryan, A., Braithwaite, H., Cardew, N., Trepte, J., Seube, N., Champeau, J., Dhaussy, P., Sauce, V., Moitie, R., Santos, R., Cardigos, F., Brussieux, M., Dando, P.: Robotic ocean vehicles for marine science applications: the european asimov project. In: OCEANS 2000 MTS/IEEE Conference and Exhibition, vol. 1, pp. 409–415 (2000)
6. Rui, G., Chitre, M.: Cooperative positioning using range-only measurements between two auvs. In: OCEANS 2010 IEEE, Sydney, pp. 1–6 (May 2010)
7. Bahr, A., Leonard, J.J., Fallon, M.F.: Cooperative Localization for Autonomous Underwater Vehicles. International Journal of Robotics Research 28(6), 714–728 (2006), 10th International Symposium on Experimental Robotics (ISER), Rio de Janeiro, Brazil (July 6-12, 2006)
8. Murray, R.M.: Recent research in cooperative control of multivehicle systems. Journal of Dynamic Systems Measurement and Control-Transactions of the ASME 129(5), 571–583 (2007), International Conference on Advances in Control and Optimization of Dynamical Systems, Bangalore, INDIA (February 1-02, 2007)
9. Leonard, N.E., Fiorelli, E.: Virtual leaders, artificial potentials and coordinated control of groups. In: Proceedings of the 40th IEEE Conference on Decision and Control, vol. 3, pp. 2968–2973 (2001)

10. Vail, D., Veloso, M.: Dynamic multi-robot coordination. In: Schultz, A.C., Parker, L.E., Schneider, F.E. (eds.) Multi-Robot Systems: From Swarms To Intelligent Automata, PO BOX 17, 3300 AA DORDRECHT, NETHERLANDS. Proceedings Paper, vol. Ii, pp. 87–98. Springer (2003), 2nd International Workshop on Multi-Robot Systems, NAVAL RES LAB, WASHINGTON, D.C., MAR (2003)

11. Egerstedt, M., Hu, X.: Formation constrained multi-agent control. IEEE Transactions on Robotics and Automation 17(6), 947–951 (2001)

12. Ogren, P., Egerstedt, M., Hu, X.M.: A control Lyapunov function approach to multiagent coordination. IEEE Transactions on Robotics And Automation 18(5), 847–851 (2002)

13. Defoort, M., Palos, J., Kokosy, A., Floquet, T., Perruquetti, W.: Performancebased reactive navigation for non-holonomic mobile robots. Robotica 27, 281–290 (2009)

14. Schwager, M., Rus, D., Slotine, J.-J.: Unifying geometric, probabilistic, and potential field approaches to multi-robot deployment. International Journal of Robotics Research 30, 371–383 (2011)

15. Şahin, E.: Swarm robotics: From sources of inspiration to domains of application. In: Şahin, E., Spears, W.M. (eds.) Swarm Robotics 2004. LNCS, vol. 3342, pp. 10–20. Springer, Heidelberg (2005)

16. Ducatelle, F., Di Caro, G.A., Pinciroli, C., Gambardella, L.M.: Self-organized cooperation between robotic swarms. Swarm Intelligence 5(2), 73–96 (2011)

17. Ghabcheloo, R., Aguiar, A.P., Pascoal, A., Silvestre, C., Kaminer, I., Hespanha, J.: Coordinated path-following in the presence of communication losses and time delays. SIAM Journal on Control and Optimization 48(1), 234–265 (2009)

18. Ren, W., Atkins, E.: Distributed multi-vehicle coordinated control via local information exchange. International Journal of Robust and Nonlinear Control 17(10-11), 1002–1033 (2007)

19. Fax, J.A., Murray, R.M.: Information flow and cooperative control of vehicle formations. IEEE Transactions on Automatic Control 49(9), 1465–1476 (2004)

20. Sun, D., Wang, C., Shang, W., Feng, G.: A Synchronization Approach to Trajectory Tracking of Multiple Mobile Robots While Maintaining Time-Varying Formations. IEEE Transactions on Robotics 25(5), 1074–1086 (2009)

21. Ihle, I.-A., Jouffroy, J., Fossen, T.: Formation control of marine surface craft: A lagrangian approach. IEEE Journal of Oceanic Engineering 31(4), 922–934 (2006)

22. Cruz, N., Matos, A., Cunha, S., Silva, S.: Zarco - an autonomous craft for underwater surveys. In: Proceedings of the 7th Geomatic Week, Barcelona, Spain (February 2007)

Underwater Source Localization Based on USBL Measurements*

Joel Reis[1], Nuno Carvalho[1], Pedro Batista[1],
Paulo Oliveira[1,2], and Carlos Silvestre[1,3]

[1] Institute for Systems and Robotics, Instituto Superior Técnico,
Universidade de Lisboa, Av. Rovisco Pais, 1049-001 Lisboa, Portugal
{joelreis,mcarvalho,pbatista,pjcro}@isr.ist.utl.pt
[2] Department of Mechanical Engineering, Instituto Superior Técnico,
Universidade de Lisboa, Portugal
[3] Department of Electrical and Computer Engineering, Faculty of Science and
Technology of the University of Macau, Macau
cjs@isr.ist.utl.pt

Abstract. This paper presents a new sensor fusion technique for 3-D
tracking of underwater targets based on direction measurements and
a single range measurement. Applications relying on this solution include, above all, marine animal studies, consisting of two Ultra-Short
Baseline (USBL) aided INS navigation systems and an acoustic trans-
ducer of reduced dimensions attached to the body of the target. A Linear
Time-Varying (LTV) system is designed, which presents a globally
asymptotically stable (GAS) error dynamics. The conditions to ensure
observability are derived and the performance of the proposed solution
is assessed through proper simulations.

Keywords: Marine Robotics, Localization Filter, Estimation Theory.

1 Introduction

The problem of underwater source localization has been a challenge for the
scientific community [1],[2]. Where humans can not go, Autonomous Underwater
Vehicles (AUV) go further. Interestingly, the cooperative aspect of navigation
has deserved much of the emphasis, as seen in [3] and references therein. In
addition, the question of local or absolute localization of marine vehicles has
also been a growing field of study: the authors in [4] consider a simultaneous
estimation of both localizations, while Antonelli et. al., in their works in [5],
achieve a solution of relative localization for AUVs and they also study the
observability of the problem. As for simultaneous localization and navigation, in
the presence of an autonomous vehicle, the works in [6] presented a set of filters
with globally exponentially stable error dynamics. However, the total absence of

---

* This work was supported by project FCT PEst-OE/EEI/LA0009/2013 and by
project FCT MAST/AM - PTFC/EEA - CRO/111107/2009.

AUVs, in exchange of elements with random and unpredicted behavior, such as marine animals, reconfigures the mission scenario, especially in the mathematical point of view. Moreover, the inclusion of humans sets aside convenient properties like control inputs to fix a desired velocity. In this paper, the authors propose a novel approach to the problem of relative localization of a tagged underwater target, using exclusively the spatial information obtained from the reception of acoustic signals emitted by the target, complemented by information from an USBL installed at the surface.

This paper is organized as follows. Section 2 describes the framework of the problem and outlines the system dynamics. The filter design and the observability are presented in Section 3. Section 4 includes simulation results along with discussions. A brief set of conclusions and references to future work are reported in Section 5.

1.1   Notation

Throughout the paper, a bold symbol stands for a multi-dimensional variable, the symbol $\mathbf{0}$ denotes a matrix of zeros and $\mathbf{I}$ an identity matrix, both of appropriate dimensions. A block diagonal matrix is represented as $\mathrm{diag}(\mathbf{A}_1,...,\mathbf{A}_n)$ and the set of unit vectors on $\mathbb{R}^3$ is denoted by $S(2)$. $\delta(t)$ represents the Dirac delta function.

2   Problem Statement

2.1   Motivation and Framework

The objective of this work consists in determining the position of a moving target (source) in an underwater environment through the use of acoustic signals. The target is equipped with an acoustic pinger which produces a known signal. In order to determine the position of that source, two receivers are available: i) a portable underwater robotic tool (PURT) carried by a diver (the whole set will henceforth be designated as the agent); and

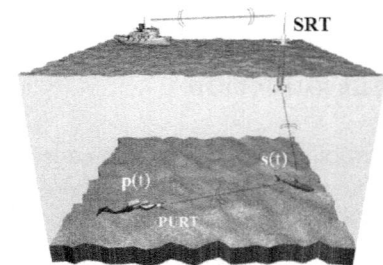

Fig. 1. Graphical representation of the mission scenario

ii) a surface robotic tool (SRT), which may or may not be adrift and can be optionally used for precise target positioning and diver localization in an inertial frame $\mathcal{I}$. Both the PURT and the SRT are equipped with hydrophone arrays in an inverted USBL configuration, which are used to obtain directions of arrival (DOA), and acoustic transponders to interrogate each other. The assumed mission scenario is depicted in Fig. 1.

2.2   System Dynamics

The mission scenario, in its most simplistic approach, dwells in a moving plane defined by three parties: an agent; a randomly moving source with an acoustic pinger attached to its body, commonly a fish under a study routine, periodically emitting a known signal; and lastly, a SRT on the surface, henceforth referred to as a transponder. By interrogating the transponder, the agent learns both the distance and the direction between them, thus determining a relative position. The distance between these last two elements can be resolved resorting to a query-response scheme, with the emission of appropriate signals, by measuring the round-trip time, assuming the response involves a fixed and known delay to resolve ambiguity problems and a constant known speed of sound in the medium. Further, the transponder and the agent determine the direction of the source relative to each one of them.

Consolidating the aforementioned localization paradigm, let $\mathbf{q}(t), \mathbf{s}(t) \in \mathbb{R}^3$ be the positions of the transponder and the source, respectively, w.r.t. the agent body frame $\mathcal{B}$. In accordance with this last assignment, let $\mathbf{v_q}(t) = \dot{\mathbf{q}}(t)$, $\mathbf{v_s}(t) = \dot{\mathbf{s}}(t) \in \mathbb{R}^3$ be the velocities of the transponder and the source, respectively, expressed in $\mathcal{B}$. Given the nature of the problem, one can mildly (and locally) assume $\dot{\mathbf{v}}_\mathbf{q}(t) = \mathbf{0}$ and $\dot{\mathbf{v}}_\mathbf{s}(t) = \mathbf{0}$, i.e. that both velocities are constant. In practice, one can adjust the gains so that it is possible to slowly track time-varying velocities with relatively small error.

From the set of measured directions, two of them are conveniently expressed in $\mathcal{B}$: the directions of the source and of the transponder, hereinafter labeled, respectively, as $\mathbf{d_s}(t), \mathbf{d_q}(t) \in S(2)$. The third direction, which relates the source to the transponder, is firstly expressed in the body of the latter, and it can be written as $\mathbf{d}_{\mathbf{s}|\mathbf{q}}^{\mathcal{Q}}(t) \in S(2)$, where $\mathcal{Q}$ is the transponder body frame. However, to fulfill the problem requirements, the direction needs to go through two rotations in order to be expressed in $\mathcal{B}$. Hence, it follows $\mathbf{d}_{\mathbf{s}|\mathbf{q}}(t) = {}_{\mathcal{I}}^{\mathcal{B}}\mathcal{R}\,{}_{\mathcal{Q}}^{\mathcal{I}}\mathcal{R}\,\mathbf{d}_{\mathbf{s}|\mathbf{q}}^{\mathcal{Q}}(t) \in S(2)$, where ${}_{\mathcal{A}}^{\mathcal{B}}\mathcal{R} \in SO(3)$ is the rotation matrix from frame $\mathcal{A}$ to frame $\mathcal{B}$.

The aim of the problem is thus to estimate the position and the velocity of the source. If the SRT that houses the transponder is equipped with a GPS antenna, the overall scheme of underwater localization can be expressed in absolute terms. In order to devise such paradigm, it suffices that an inertial point of reference be known.

3   Localization Filter Design

This section presents a filter design methodology for the problem stated in Section 2. First, a linear time-varying system (LTV) is introduced in Section 3.1. Afterwards, the observability of this system is studied in Section 3.2. The filter design is then discussed in 3.3.

3.1   System States

In the vast majority of localization problems, the main difficulty is to draw an estimator with global asymptotic stability guarantees. In this paper, the authors propose a strategy for linear filter design, thus guaranteeing a GAS evolution of the estimation errors. In the situation described in Section 2.2, the overall system is devoid of any inputs, whereby the dynamics can be written as a LTV expressed by

$$\begin{cases} \dot{\mathbf{x}}(t) = \mathbf{A}(t)\mathbf{x}(t) \\ \mathbf{y}(t) = \mathbf{C}(t)\mathbf{x}(t) \end{cases}. \tag{1}$$

Contrary to what is usually done, consider first the following measurements vector:

$$\mathbf{y} = \begin{bmatrix} \mathbf{d}_{\mathbf{s}}^T & \mathbf{d}_{\mathbf{q}}^T & \mathbf{d}_{\mathbf{s}|\mathbf{q}}^T & \|\mathbf{q}\| \end{bmatrix}^T \in \mathbb{R}^{10}. \tag{2}$$

Presented therein are three directions and a single range measurement, all w.r.t. $\mathcal{B}$. However, knowing that $\mathbf{d}_{\mathbf{q}}(t)\|\mathbf{q}(t)\| = \mathbf{q}(t)$, one could condense the number of measurements involved by simply considering $\mathbf{q}(t)$ instead of the pair $\{\mathbf{d}_{\mathbf{q}}(t), \|\mathbf{q}(t)\|\}$. But doing so, it could be obliterated the fact that the noise affects each element of the last mentioned pair of measurements, and not $\mathbf{q}(t)$ on its own.

Nevertheless, directions sustain a non-linear relation between the vector numerator and its norm. To circumvent this problem, one can assume that instead of directions, the corresponding measurements are zero, although it remains implicit in the observations matrix that an association between states exists. The new measurements vector, $\mathbf{y} \in \mathbb{R}^{10}$ thus results in

$$\mathbf{y} = \begin{bmatrix} \mathbf{0} & \mathbf{0} & \mathbf{0} & \|\mathbf{q}\| \end{bmatrix}^T$$
$$= \begin{bmatrix} \{\mathbf{d}_{\mathbf{s}}^T\|\mathbf{s}\| - \mathbf{s}^T\} & \{\mathbf{d}_{\mathbf{q}}^T\|\mathbf{q}\| - \mathbf{q}^T\} & \{\mathbf{d}_{\mathbf{s}|\mathbf{q}}^T\|\mathbf{s} - \mathbf{q}\| - (\mathbf{s} - \mathbf{q})^T\} & \|\mathbf{q}\| \end{bmatrix}^T. \tag{3}$$

Therefore, to continue to deal with a linear problem, the states vector $\mathbf{x}(t) \in \mathbb{R}^{15}$ and its time derivative $\dot{\mathbf{x}}(t) \in \mathbb{R}^{15}$ follow as

$$\mathbf{x}(t) = \begin{bmatrix} \mathbf{q}^T(t) & \mathbf{v}_{\mathbf{q}}^T(t) & \|\mathbf{q}(t)\| & \mathbf{s}^T(t) & \mathbf{v}_{\mathbf{s}}^T(t) & \|\mathbf{s}(t)\| & \|\mathbf{s}(t) - \mathbf{q}(t)\| \end{bmatrix}^T \tag{4}$$

and

$$\dot{\mathbf{x}}(t) = \begin{bmatrix} \mathbf{v}_{\mathbf{q}}^T(t) & \mathbf{0} & \mathbf{d}_{\mathbf{q}}^T(t)\mathbf{v}_{\mathbf{q}}(t) & \mathbf{v}_{\mathbf{s}}^T(t) & \mathbf{0} & \mathbf{d}_{\mathbf{s}}^T(t)\mathbf{v}_{\mathbf{s}}(t) & \mathbf{d}_{\mathbf{s}|\mathbf{q}}^T(t)(\mathbf{v}_{\mathbf{s}}^T(t) - \mathbf{v}_{\mathbf{q}}^T(t)) \end{bmatrix}^T. \tag{5}$$

Finally, the dynamics matrix $\mathbf{A}(t) \in \mathbb{R}^{15 \times 15}$ and the observations matrix $\mathbf{C}(t) \in \mathbb{R}^{10 \times 15}$ can be easily derived from (5) and (3), respectively.

3.2   Observability Analysis

Throughout this section, the observability of the problem is analyzed, taking into consideration the directions and range readings. The following proposition [Proposition 4.2, [7]] is useful in the sequel.

Proposition: Let $\mathbf{f}(t) : [t_0, t_f] \subset \mathbb{R} \to \mathbb{R}^n$ be a continuous and i-times continuously differentiable function on $\mathcal{T} := [t_0, t_f], T := t_f - t_0 > 0$, and such that

$$\mathbf{f}(t_0) = \dot{\mathbf{f}}(t_0) = \dots = \mathbf{f}^{i-1}(t_0) = \mathbf{0}. \tag{6}$$

Further assume that there exists a nonnegative constant C such that $\|\mathbf{f}^{(i+1)}(t)\| \le C$ for all $t \in \mathcal{T}$. If there exist $\alpha > 0$ and $t_1 \in \mathcal{T}$ such that $\|\mathbf{f}^{(i)}(t_1)\| \ge \alpha$ then there exist $0 < \delta \le T$ and $\beta > 0$ such that $\|\mathbf{f}(t_0 + \delta)\| \ge \beta$.

From the Peano-Baker series it follows the transition matrix associated to $\mathbf{A}(t)$, denoted by $\phi(t,t_0) \in \mathbb{R}^{15 \times 15}$.

The LTV system (1) is said to be observable only under the condition that the Observability Gramian is invertible, the latter being defined as

$$\boldsymbol{\mathcal{W}}(t_0, t_f) = \int_{t_0}^{t_f} \phi^T(\tau, t_0) \mathbf{C}^T(\tau) \mathbf{C}(\tau) \phi(\tau, t_0) d\tau. \tag{7}$$

Theorem: The LTV system (1) is observable on $\mathcal{T}$ if and only if the transponder, the agent and the source do not have collinear positions in two successive moments. Mathematically speaking,

$$\exists_{t_1 \in \mathcal{T}} : \quad \forall\, i,j = \{\mathbf{s}, \mathbf{q}, \mathbf{s}|\mathbf{q}\}, i \ne j, \ \left| \mathbf{d}_i^T(t_0)\mathbf{d}_j(t_1) \right| < 1. \tag{8}$$

Proof. Let $\mathbf{c} = [\mathbf{c}_1^T \ \mathbf{c}_2^T \ c_3 \ \mathbf{c}_4^T \ \mathbf{c}_5^T \ c_6 \ c_7]^T \in \mathbb{R}^{15}$, with $\mathbf{c}_i \in \mathbb{R}^3$, for $i = 1,2,4,5$, and $c_3, c_6, c_7 \in \mathbb{R}$, be a unit vector. Since it must be $\|\mathbf{c}\| = 1$ for all $\mathbf{c}$,

$$\mathbf{c}^T \boldsymbol{\mathcal{W}}(t_0, t_f)\mathbf{c} \ne 0 \tag{9}$$

in order for the LTV system to be observable. Then,

$$\mathbf{c}^T \boldsymbol{\mathcal{W}}(t_0, t_f)\mathbf{c} = \int_{t_0}^{t_f} \|\mathbf{f}(\tau)\|^2 d\tau, \tag{10}$$

where $\mathbf{f}(\tau) \in \mathbb{R}^{10}, \tau \in \mathcal{T}$ is given by

$$\mathbf{f}(\tau) = \begin{bmatrix} -\mathbf{c}_4 + \left[\mathbf{d}_\mathbf{s}(\tau) \int_{t_0}^{\tau} \mathbf{d}_\mathbf{s}^T(\sigma) d\sigma - (\tau - t_0)\mathbf{I} \right] \mathbf{c}_5 + \mathbf{d}_\mathbf{s}(\tau)c_6 \\ -\mathbf{c}_1 + \left[\mathbf{d}_\mathbf{q}(\tau) \int_{t_0}^{\tau} \mathbf{d}_\mathbf{q}^T(\sigma) d\sigma - (\tau - t_0)\mathbf{I} \right] \mathbf{c}_2 + \mathbf{d}_\mathbf{q}(\tau)c_3 \\ \left\{ \mathbf{c}_1 + \left[(\tau - t_0)\mathbf{I} - \mathbf{d}_{\mathbf{s}|\mathbf{q}}(\tau) \int_{t_0}^{\tau} \mathbf{d}_{\mathbf{s}|\mathbf{q}}^T(\sigma) d\sigma \right] \mathbf{c}_2 + \dots \right. \\ \left. \dots - \mathbf{c}_4 + \left[(\tau - t_0)\mathbf{I} + \mathbf{d}_{\mathbf{s}|\mathbf{q}}(\tau) \int_{t_0}^{\tau} \mathbf{d}_{\mathbf{s}|\mathbf{q}}^T(\sigma) d\sigma \right] \mathbf{c}_5 + \mathbf{d}_{\mathbf{s}|\mathbf{q}}(\tau)c_7 \right\} \\ \int_{t_0}^{\tau} \mathbf{d}_\mathbf{q}(\sigma) d\sigma \mathbf{c}_2 + c_3 \in \mathbb{R} \end{bmatrix}. \tag{11}$$

It can be easily shown that the condition (8) is necessary. For instance, suppose $\mathbf{c} = \begin{bmatrix} \mathbf{0} & \mathbf{0} & 0 & \mathbf{c}_4^T & \mathbf{0} & c_6 & c_7 \end{bmatrix}^T$. It results

$$\mathbf{f}(\tau) = \left[\{-\mathbf{c}_4^T + \mathbf{d}_\mathbf{s}^T(\tau)c_6\} \quad \mathbf{0} \quad \{-\mathbf{c}_4^T + \mathbf{d}_{\mathbf{s}|\mathbf{q}}^T(\tau)c_7\} \quad 0 \right]^T. \tag{12}$$

Taking into account that $\mathbf{c}$ is a unit vector, and defining $c_6 = c_7 = \frac{\sqrt{3}}{3}$ and $\mathbf{c}_4 = k\frac{\sqrt{3}}{3}\mathbf{d}_\mathbf{q}(t_0)$, with $k = \{-1, +1\}$, then under the supposition that the condition in (8) is not verified, it is true that (with implicit collinearity)

$$\mathbf{f}(\tau) = \left[\left\{ k\frac{\sqrt{3}}{3}\mathbf{d}_\mathbf{q}^T(t_0) + \mathbf{d}_\mathbf{s}^T(\tau)\frac{\sqrt{3}}{3} \right\} \quad \mathbf{0} \quad \left\{ k\frac{\sqrt{3}}{3}\mathbf{d}_\mathbf{q}^T(t_0) + \mathbf{d}_{\mathbf{s}|\mathbf{q}}^T(\tau)\frac{\sqrt{3}}{3} \right\} \quad 0 \right]^T = \mathbf{0}$$

if the transponder is in between the source and the agent or if it is the agent who is in the middle, obviously considering the right choice for k. Therefore, the Observability Gramian $\mathcal{W}$ is not invertible and the system is not observable. Notwithstanding, if $c_6/c_7 = -1$, by adjusting the value of k one obtains again $\mathbf{f}(\tau) = \mathbf{0}$, reaching to the same conclusions, proving that if the source is in between the agent and the transponder, and the condition (8) is not verified, the LTV system (1) is not observable.

At this point, (8) was shown to be a necessary condition; another procedure is contemplated to make a statement about its sufficiency: to prove that whatever the value of $\mathbf{c}$ might be, if (8) is verified, then the system is observable. Starting with the evaluation of $\mathbf{f}(\tau)$ at $\tau = t_0$, which is given by

$$\mathbf{f}(t_0) = \left[\left\{ -\mathbf{c}_4 + \mathbf{d_s}(t_0)c_6 \right\}^T \quad \left\{ -\mathbf{c}_1 + \mathbf{d_q}(t_0)c_3 \right\}^T \quad \cdots \right.$$
$$\left. \cdots \quad \left\{ \mathbf{c}_1 - \mathbf{c}_4 + \mathbf{d_{s|q}}(t_0)c_7 \right\}^T \quad c_3 \right], \tag{13}$$

if $c_3 \neq 0$ then $\|\mathbf{f}(t_0)\| > 0$, and it follows from *Proposition* that $\mathbf{c}^T\mathcal{W}(t_0,t_f)\mathbf{c} > 0$. In a similar way, evaluating the derivative of $\mathbf{f}(\tau)$ at $\tau = t_0$ yields for its last entry (the scalar one) the term $\mathbf{d_q}^T(t_0)c_2$. Clearly, if $c_2 \neq 0$, it is implicit that $\| d\mathbf{f}(\tau)/d\tau|_{\tau=t_0} \| > 0$, whereby using *Proposition* twice it follows that $\mathbf{c}^T\mathcal{W}(t_0,t_f)\mathbf{c} > 0$. Proceed now assuming that $c_2 = 0$ and $c_3 = 0$. In case $\mathbf{c}_4 \neq \mathbf{d_s}(t_0)c_6$ or $\mathbf{c}_1 \neq \mathbf{0}$ it follows $\|\mathbf{f}(t_0)\| > 0$, therefore $\mathbf{c}^T\mathcal{W}(t_0,t_f)\mathbf{c} > 0$. Next, by doing $\mathbf{c}_1 = \mathbf{c}_4 = \mathbf{0}$, if $c_7 \neq 0$ $\|\mathbf{f}(t_0)\| > 0$, and again the use of the *Proposition* leads to $\mathbf{c}^T\mathcal{W}(t_0,t_f)\mathbf{c} > 0$. However, to reach the same conclusion when $c_7 = 0$ it implies that $c_6 \neq 0$. On the other hand, $c_6 = 0 \Rightarrow \mathbf{f}(t_0) = \mathbf{0}$; consequently, the time derivative of $\mathbf{f}(t_0)$ becomes

$$\left. \frac{d\mathbf{f}(\tau)}{d\tau} \right|_{\tau=t_0} = \left[\left\{ \left[\mathbf{d_s}(t_0)\mathbf{d_s}^T(t_0) - \mathbf{I} \right] \mathbf{c}_5 \right\}^T \quad \mathbf{0} \quad \left\{ \left[\mathbf{d_s}(t_0)\mathbf{d_s}^T(t_0) + \mathbf{I} \right] \mathbf{c}_5 \right\}^T \quad 0 \right]^T. \tag{14}$$

$\mathbf{c}_5$ cannot be zero since that assumption would violate the fact that $\mathbf{c}$ is a unit vector. The first entry of $d\mathbf{f}(\tau)/d\tau|_{\tau=t_0}$ determines the projection of $\mathbf{c}_5$ over the plan orthogonal to $\mathbf{d_s}(t_0)$. The term is $\mathbf{0}$ only when $\mathbf{c}_5\|\mathbf{d_s}(t_0)$. Nonetheless, if such situation occurs, the third entry of $d\mathbf{f}(\tau)/d\tau|_{\tau=t_0}$ would be different from $\mathbf{0}$. Hence, $\| d\mathbf{f}(\tau)/d\tau|_{\tau=t_0} \| > \mathbf{0}$ and using the *Proposition* twice allows to conclude that $\mathbf{c}^T\mathcal{W}(t_0,t_f)\mathbf{c} > 0$. The proof is thus concluded, evidencing that $\mathbf{c}^T\mathcal{W}(t_0,t_f)\mathbf{c} > 0$ for all $\|\mathbf{c}\| = 1$, which means that the observability Gramian is invertible and as such (1) is observable. ∎

3.3   Kalman Filter

Section 3.1 introduced a LTV system for source localization based on direction and range measurements and its observability was studied in Section 3.2. It was proven that the LTV system (1) is observable if (8) is satisfied. Taking advantage of the fact that (1) is linear, the Kalman filter follows as the natural estimation solution, with GAS error dynamics. As it is widely known, the Kalman filter design is omitted in this paper. The system dynamics, including additive system disturbances and sensor noise, can be written as

$$\begin{cases} \dot{\mathbf{x}}(t) = \mathbf{A}(t)\mathbf{x}(t) + \mathbf{w}(t) \\ \mathbf{y}(t) = \mathbf{C}(t)\mathbf{x}(t) + \mathbf{n}(t) \end{cases}, \tag{15}$$

where $\mathbf{w}(t) \in \mathbb{R}^{15}$ is zero-mean white Gaussian noise, with $E[\mathbf{w}(t)\mathbf{w}^T(t-\tau)] = \boldsymbol{\Xi}\delta(\tau)$, $\boldsymbol{\Xi} \succ \mathbf{0}$, $\mathbf{n}(t) \in \mathbb{R}^{10}$ is zero-mean white Gaussian noise, with $E[\mathbf{n}(t)\mathbf{n}^T(t-\tau)] = \boldsymbol{\Theta}\delta(\tau)$ and $\boldsymbol{\Theta} \succ \mathbf{0}$. The noises are uncorrelated, therefore $E[\mathbf{w}(t)\mathbf{n}^T(t-\tau)] = \mathbf{0}$. Notwithstanding, noises are here considered to be additive, which might not be the reality, and thus the proposed solution is sub-optimal.

4   Simulation Results

To assess the performance of the localization filter, a simulated mission scenario was built that represents a typical spatial arrangement in which most underwater missions fit. Hence, first consider a SRT at the surface moving due to ocean currents, with small variations on its vertical speed. Since the behavior of marine animals is quite unpredictable, the chosen curved and descending source path is an educated guess. The agent (in terms of depth) is placed in between the source and the SRT, featuring a sinusoidal motion along its longitudinal and transversal axes to ensure the condition in (8) is verified. The resulting trajectories are depicted in Fig. 2.

Noise was considered for both the directions and single range measurements [8]. According to (3), the null entries are virtual measurements, therefore no noise is to be added there. Direction readings were assumed perturbed by standard deviations of 0.5° on the elevation and azimuth angles. Zero-mean additive Gaussian noise with unitary standard deviation was considered for range measurements. The Kalman filter parameters were tuned to $\boldsymbol{\Xi} = \mathrm{diag}(10^{-5}\mathbf{I}, 10^{-5}\mathbf{I}, 10^{-1}, 10^{-5}\mathbf{I}, 10^{-5}\mathbf{I}, 10^{-1}, 10^{-1})$ and

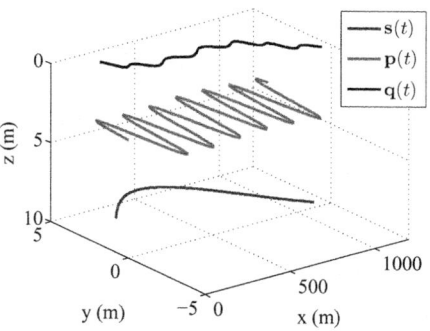

Fig. 2. Simulated Trajectories

$\boldsymbol{\Theta} = \mathbf{I}$. The initial estimates were all set to zero. Results for the source position and velocity estimation errors are presented in Fig. 3 and Fig. 4, respectively. With the chosen filter parameters, the transients in both figures quickly fade out and the estimation error converges to zero with small deviations, proving the efficiency of the proposed solution.

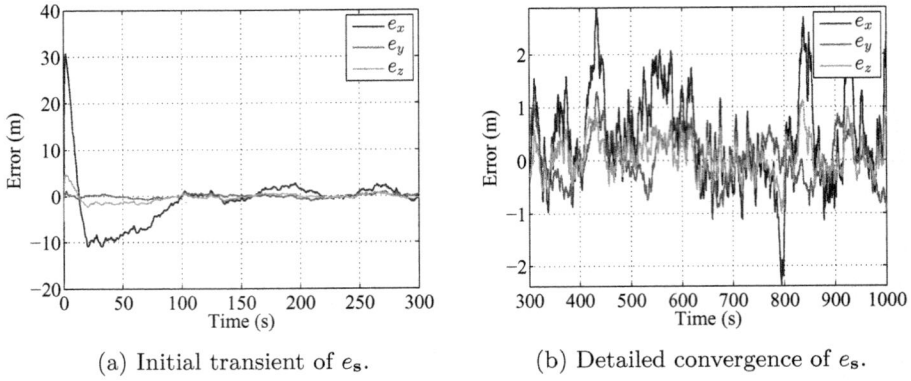

(a) Initial transient of $e_{\mathbf{s}}$.

(b) Detailed convergence of $e_{\mathbf{s}}$.

Fig. 3. Evolution of the source position estimation error

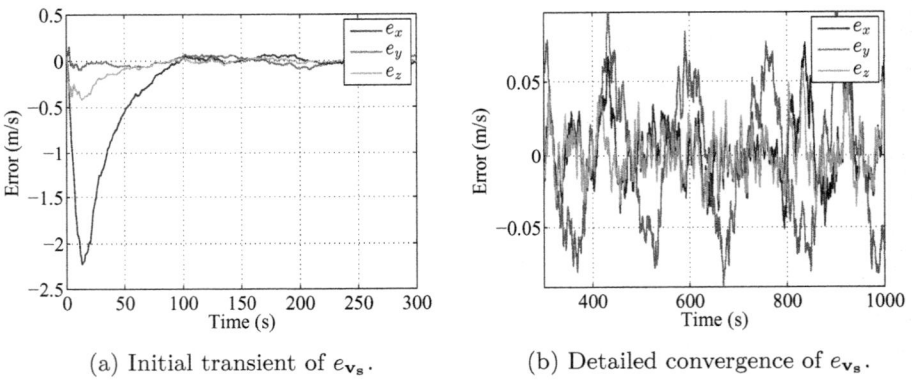

(a) Initial transient of $e_{\mathbf{v_s}}$.

(b) Detailed convergence of $e_{\mathbf{v_s}}$.

Fig. 4. Evolution of the source velocity estimation error

5   Conclusions

This paper presented a novel time-varying Kalman filter with globally asymptotically stable error dynamics for the problem of single source localization based on direction and range measurements. The observability of the system was studied, which allowed to conclude about the asymptotic stability of the Kalman filters. Simulations results were presented that illustrate the good performance achieved by the proposed solution. Future work includes real sea tests to evaluate the overall physical system.

References

1. Hodgkinson, B., Shyu, D., Mohseni, K.: Acoustic source localization system using a linear arrangement of receivers for small unmanned underwater vehicles. In: Oceans, pp. 1–7 (2012)

2. Lohrasbipeydeh, H., Gulliver, T.A., Zielinski, A., Dakin, T.: Single hydrophone passive source range and depth estimation in shallow water. In: 2013 MTS/IEEE OCEANS, Bergen, pp. 1–4 (2013)

3. Papadopoulos, G., Fallon, M.F., Leonard, J.J., Patrikalakis, N.M.: Cooperative localization of marine vehicles using nonlinear state estimation. In: 2010 IEEE/RSJ International Conference on Intelligent Robots and Systems (IROS), pp. 4874–4879 (2010)

4. Moore, D.C., Huang, A.S., Walter, M., Olson, E., Fletcher, L., Leonard, J., Teller, S.: Simultaneous local and global state estimation for robotic navigation. In: IEEE International Conference on Robotics and Automation, ICRA 2009, pp. 3794–3799 (2009)

5. Antonelli, G., Arrichiello, F., Chiaverini, S., Sukhatme, G.: Observability analysis of relative localization for auvs based on ranging and depth measurements. In: 2010 IEEE International Conference on Robotics and Automation (ICRA), pp. 4276–4281 (2010)

6. Batista, P., Silvestre, C., Oliveira, P.: Globally exponentially stable filters for source localization and navigation aided by direction measurements. Systems & Control Letters 62(11), 1065–1072 (2013)

7. Batista, P., Silvestre, C., Oliveira, P.: On the observability of linear motion quantities in navigation systems. Systems & Control Letters 60(2), 101–110 (2011)

8. Morgado, M.: Advanced Ultra-Short Baseline Inertial Navigation Systems. PhD thesis, Instituto Superior Técnico, Universidade Técnica de Lisboa (2011)

Comparative Study of Chattering-Free Sliding Mode Controllers Applied to an Autonomous Underwater Vehicle (AUV)

Mariana Uzeda Cildoz[1,*], Carlos Henrique Farias Dos Santos[1], and Romeu Reginatto[2]

[1] Robotic Research Group of the State University of Western Paraná, Av. Tancredo Neves 6731, Foz do Iguacu, Brazil
mariana.uzeda@gmail.com, chf.santos@uol.com.br
[2] Exact Sciences Center of State University of Western Paraná, Av. Tancredo Neves 6731, Foz do Iguacu, Brazil
romeu.cece@gmail.com

Abstract. This paper carried out a comparative study between three different Sliding Mode Control strategies applied to an Autonomous Underwater Vehicle's (AUV) position control in 6 DOFs, under the influence of wind, waves and ocean currents. The comparative study aims to avoid the Chattering effects looking for an acceptable trade-off between tracking error performance and closed-loop system stability. To this end, stability proofs and computational simulations for the control strategies were performed with satisfactory results.

Keywords: Sliding Mode Control, Robust Control, Autonomous Underwater Vehicles (AUV), Chattering Problem.

1 Introduction

Among the several motion control techniques applied to Autonomous Underwater Vehicles (AUVs), the Sliding Mode Control is a nonlinear and variable structure robust controller[1] that has been successfully implemented in recent years due to advances in computer technology and high-speed switching circuitry [7].

This paper carried out a comparative study between three different Sliding Mode Control strategies applied to an AUV. These comparisons are in terms of stability[2], tracking error performance and Chattering effects. The following sections introduces the dynamical and kinematical models of the vehicle, the sliding surface selection, the stability proofs of the control laws, the simulation results and the conclusions respectively.

---

[*] The authors gratefully the support of Itaipu Technologic Park (PTI) to this contribution.
[1] Sliding Mode Control was developed from relay and bang-bang control theory.
[2] Stable error dynamics in closed-loop.

A.P. Moreira et al. (eds.), *CONTROLO'2014 - Proc. of the 11th Port. Conf. on Autom. Control*,
Lecture Notes in Electrical Engineering 321, DOI: 10.1007/978-3-319-10380-8_56

2   Vehicle Modeling

For the mathematical modeling of the underwater vehicle's movement in six degrees of freedom a Body-attached and Inertial reference frame are used.

2.1   Kinematical Model of the Vehicle

To relate the movement of the vehicle in the two reference frames aforementioned, the Newton-Euler approach is used [1]. Thus, the kinematical model of the vehicle is given by

$$
\begin{bmatrix} \dot{\eta}_1 \\ \dot{\eta}_2 \end{bmatrix} = J(\eta) \begin{bmatrix} \nu_1 \\ \nu_2 \end{bmatrix} = \begin{bmatrix} J_1(\eta) & 0_{3x3} \\ 0_{3x3} & J_2(\eta) \end{bmatrix} \begin{bmatrix} \nu_1 \\ \nu_2 \end{bmatrix} \tag{1}
$$

where $\eta = [\eta_1, \eta_2]^T = [x, y, z, \phi, \theta, \psi]^T$ are the vehicle's positions and orientations in the inertial reference frame, $\nu = [\nu_1, \nu_2]^T = [u, v, w, p, q, r]^T$ are the vehicle's linear and angular velocities in the body-attached reference frame and $J_1(\eta)$ and $J_2(\eta)$ are transformation matrices of linear and angular velocities respectively, given by

$$
J_1(\eta) = \begin{bmatrix} c\psi c\theta & -s\psi c\phi + c\psi s\theta s\phi & s\psi s\phi + c\psi c\phi s\theta \\ s\psi c\theta & c\psi c\phi - s\phi s\theta s\psi & -c\psi s\phi + s\theta s\psi c\phi \\ -s\theta & c\theta s\phi & c\theta c\phi \end{bmatrix}, J_2(\eta) = \begin{bmatrix} 1 & s\phi t\theta & c\phi t\theta \\ 0 & c\theta & s\phi \\ 0 & s\phi/c\theta & c\phi/c\theta \end{bmatrix} \tag{2}
$$

wherein $c = \text{cosine}$, $s = \text{sine}$, $t = \text{tangent}$.

2.2   Dynamical Model of the Vehicle

The modeling of the vehicle use the SNAME[3] pattern and is expressed by

$$
M\dot{\nu} + C(\nu_R)\nu_R + D(\nu_R)\nu_R + g(\eta) = \tau + \tau_d \tag{3}
$$

where M_{6x6} is the Inertial matrix of the vehicle; $C_{6x6}(\nu_R) = C_{RB_{6x6}}(\nu) + C_{A_{6x6}}(\nu_R)$ is the Coriolis and Centripetal matrix; $D_{6x6}(\nu_R)$ is the Damping matrix; $g_{6x1}(\eta)$ is the Gravitational forces and moments vector; τ_{6x1} is the vector of forces and moments of the thrusters; $\tau_{d_{6x1}}$ is a uniformly bounded vector of forces and moments of disturbances[4]; $\nu_R = \nu - \nu_c$ is the vehicle's relative velocity vector[5]; ν_c is the ocean current velocity vector. And considering that the ocean current acceleration vector is $\dot{\nu}_c = 0$, the vehicle's relative acceleration vector is given by $\dot{\nu}_R = \dot{\nu}$, where $\dot{\nu}$ is the vehicle's actual acceleration vector.

---

[3] The SNAME (The Society of Naval Architects) pattern was established in 1950.

[4] In this work, the forces and moments induced by the waves and wind are considered as the main disturbances.

[5] In order to include the ocean currents velocity ν_c, in the dynamical equation of the vehicle, matrices M_{6x6}, $C_{A_{6x6}}(\nu_R)$, $D_{6x6}(\nu_R)$, $g_{6x1}(\eta)$ are parameterized as in [1] and $C_{RB}(\nu)$, is parameterized as in [3].

3   Sliding Surface

This section presents the sliding surface with integral action or desired dynamics to be achieved by the three control strategies in the next section. Initially, define the following scalar measure of tracking error on the inertial frame

$$s = \left(\frac{d}{dt} + \lambda\right)^2 \int \tilde{\eta} d\tau = \dot{\tilde{\eta}} + 2\lambda\tilde{\eta} + \lambda^2 \int \tilde{\eta} d\tau \tag{4}$$

where $\tilde{\eta} = \eta - \eta_d$, is the tracking error, $\eta = [x, y, z, \phi, \theta, \psi]^T$, are the actual positions and orientations of the vehicle, $\eta_d = [x_d, y_d, z_d, \phi_d, \theta_d, \psi_d]^T$ are the desired positions and orientations of the vehicle and λ, is a positive scalar matrix that defines the control's bandwidth and satisfy the Hurwitz[6] condition.

The above expression is defined such that the convergence of s to zero also implies the convergence of $\tilde{\eta}$ to zero. Hence, the goal of the control design reduces to finding a nonlinear control law that ensures,

$$\lim_{t\to\infty} s(t) = 0. \tag{5}$$

By introducing a *virtual reference* trajectory η_r, defined such that

$$s = \dot{\eta} - \dot{\eta}_r \tag{6}$$

where $\dot{\eta}_r = \dot{\eta}_d - 2\lambda\tilde{\eta} - \lambda^2 \int \tilde{\eta} d\tau$. Scalar measure of tracking (4) can be rewritten in the body-attached reference frame by,

$$s_1 = \nu - \nu_r \tag{7}$$

where, $\nu = J^{-1}(\eta)\dot{\eta}$ and $\nu_r = J^{-1}(\eta)\dot{\eta}_r$ are the actual and virtual velocity vectors in the body-attached reference frame. Then, since $\lim_{t\to\infty} s_1 = 0$ implies $\lim_{t\to\infty} s = 0$, it is possible to satisfy the stability conditions on s_1 to achieve the tracking goal. Furthermore, differentiating in (7) with respect to time we have

$$\dot{s}_1 = \dot{\nu} - \dot{\nu}_r, \tag{8}$$

where $\dot{\nu} = J^{-1}(\eta)[\ddot{\eta} - \dot{J}(\eta)J^{-1}(\eta)\dot{\eta}]$ and $\dot{\nu}_r = J^{-1}(\eta)[\ddot{\eta}_r - \dot{J}(\eta)J^{-1}(\eta)\dot{\eta}_r]$ are obtained from the chain rule.

4   Control Strategies

This section presents three different Chattering-free model-based Sliding Mode Control strategies and their respective stability proofs.

---

[6] The eigenvalues of the matrix λ are all negative.

4.1   Conventional Sliding Mode Control Based on Lyapunov Stability Analysis

This subsection introduces a conventional sliding mode control law obtained from the Lyapunov stability conditions for the closed-loop system as in [1].

Stability Proof. Consider the following Lyapunov-like function candidate defined in s_1,

$$V(s_1, t) = \frac{1}{2} s_1^T M s_1 \qquad (9)$$

where $M = M^T > 0$, is the inertia matrix of the vehicle. Then, from (3) and the matrices propriety[7] $s_1^T[\dot{M} - 2C(\nu_R)]s_1 = 0$, the differential of (9) with respect to time can be expressed by,

$$\dot{V}(s_1, t) = s_1^T[M\dot{s_1} + C(\nu_R)s_1] \qquad (10)$$

and, from (3), (7), (8), we can rewrite (10) as

$$\dot{V}(s_1, t) = -s_1^T D(\nu_R)s_1 + s_1^T[\tau + \tau_d - M\dot{\nu}_r - C(\nu_R)\nu_r - \\ -D(\nu_R)\nu_r - g(\eta) + M\dot{\nu}_c + C(\nu_R)\nu_c + D(\nu_R)\nu_c] \qquad (11)$$

since $\nu_R = \nu - \nu_c$. Hence, the control law τ that satisfy $\dot{V}(s_1, t) \leq 0$, was chosen as

$$\tau = \hat{M}\dot{\nu}_r + \hat{C}(\nu_R)\nu_r + \hat{D}(\nu_R)\nu_r + \hat{g}(\eta) - \\ -[\hat{M}\dot{\nu}_c + \hat{C}(\nu_R)\nu_c + \hat{D}(\nu_R)\nu_c] - Ksgn(s_1) \qquad (12)$$

where $\hat{M}$, $\hat{C}(\nu_R)$, $\hat{D}(\nu_R)$, $\hat{g}(\eta)$ are the estimated[8] parameters matrices and $K > 0$ is the switching gain matrix of appropriate dimension. Indeed, by setting

$$\tilde{h}(\nu_R, \nu_r, \dot{\nu}_r, \eta) = \tilde{M}\dot{\nu}_r + \tilde{C}(\nu_R)\nu_r + \tilde{D}(\nu_R)\nu_r + \tilde{g}(\eta) - \\ -[\tilde{M}\dot{\nu}_c + \tilde{C}(\nu R)\nu_c + \tilde{D}(\nu_R)\nu_c] \qquad (13)$$

---

[7] The parametric configurations of the matrices of the vehicle confers them some certain properties, see [1].

[8] It is assumed that the unknown parameters matrices M, $C(\nu_R)$, $D(\nu_R)$ and $g(\eta)$ are limited by a known functions.

where $\tilde{M}$, $\tilde{C}$, $\tilde{D}$, $\tilde{g}$ are the errors between the estimated and unknown parametric matrices, and replacing (12) in (11) we have that

$$\dot{V}(s_1, t) = -s_1^T D(\nu_R)s_1 + s_1^T \tau_d + s_1^T \tilde{h}(\nu_R, \nu_r, \dot{\nu}_r, \eta) - |s_1^T|K, \qquad (14)$$

since $-s_1^T K sgn(s_1) = -|s_1^T|K$. Then, by the triangular inequality

$$\dot{V}(s_1, t) \leq -s_1^T D(\nu_R)s_1 + |s_1^T| \left[|\tau_d| + |\tilde{h}(\nu_R, \nu_r, \dot{\nu}_r, \eta)| - K \right] \qquad (15)$$

where $D(\nu_R)$, is a positive diagonal matrix due to the hydrodynamic characteristics of the vehicle. And, since parametric uncertainties and disturbance are considered bounded by a known functions, selecting K as

$$K \geq \tau_{d_{max}} + F(\nu, \nu_r, \dot{\nu}_r) + \mu, \qquad \mu > 0 \qquad (16)$$

where $|\tau_d| \leq \tau_{d_{max}}$ and $|\tilde{h}(\nu, \nu_r, \dot{\nu}_r, \eta)| \leq F(\nu, \nu_r, \dot{\nu}_r, \eta)$, produces

$$\dot{V}(s_1, t) \leq -s_1^T D(\nu_R)s_1 - |s_1^T|\mu \leq 0. \qquad (17)$$

Thus $V(s_1, t) \leq V(s_1, 0)$, then s_1 is bounded. Hence, $\dot{V}$ is uniformly continuous and by Barbalat's lemma $s_1 \to 0$. Therefore $s \to 0$ and $\tilde{\eta} \to 0$ as $t \to \infty$.

4.2   Adaptive Sliding Mode Control

This subsection introduces an Adaptive Sliding Mode Control strategy, which performs a real-time update of model parameters from the output errors.

Stability Proof. From [1] the dynamical model of the vehicle (3) can be rewritten in the inertial reference frame as,

$$M_\eta(\eta)\ddot{\eta}_R + C_\eta(\nu_R, \eta)\dot{\eta}_R + D_\eta(\nu_R, \eta)\dot{\eta}_R + g_\eta(\eta) = J^{-T}(\eta)\tau. \qquad (18)$$

Then, considering the following Lyapunov-like function candidate

$$V(s, \tilde{\Theta}, t) = \frac{1}{2} \left[s^T M_\eta(\eta)s + \tilde{\Theta}^T \Gamma^{-1} \tilde{\Theta} \right], \qquad (19)$$

where $M_\eta = M_\eta^T > 0$ is the inertial matrix of the vehicle, Γ is a symmetric and positive definite weighting matrix of appropriate dimension and $\tilde{\Theta} = \hat{\Theta} - \Theta$, is a vector of estimation errors of system parameters that will be defined after. Differentiating (19) with respect to time, and by using the matrices properties $s^T M_\eta \dot{s} = \dot{s}^T M_\eta s$ and $s^T[\dot{M}_\eta(\eta) - 2C_\eta(\nu, \eta)]s = 0$, we obtain

$$\dot{V}(s, \tilde{\Theta}, t) = s^T [M_\eta(\eta)\dot{s} + C_\eta(\nu_R, \eta)s] + \dot{\tilde{\Theta}}^T \Gamma^{-1} \tilde{\Theta}. \qquad (20)$$

From (18), (6), (8) and since $\dot{\eta}_R = \dot{\eta} - \dot{\eta}_c$, we can rewrite (20) as

$$
\begin{aligned}
\dot{V}(s,\tilde{\Theta},t) = &-s^T D_\eta(\nu_R,\eta)s + \dot{\tilde{\Theta}}^T \Gamma^{-1}\tilde{\Theta} + s^T[J^{-T}\tau + \\
&+J^{-T}\tau_d - M_\eta(\eta)\ddot{\eta}_r - C_\eta(\nu_R,\eta)\dot{\eta}_r - D_\eta(\nu_R,\eta)\dot{\eta}_r - \\
&-g_\eta(\eta) + M_\eta(\eta)\ddot{\eta}_c + C_\eta(\nu_R,\eta)\dot{\eta}_c + D_\eta(\nu_R,\eta)\dot{\eta}_c]
\end{aligned}
\tag{21}
$$

then, by considering that

$$
\begin{aligned}
&M_\eta(\eta)\ddot{\eta}_r + C_\eta(\nu_R,\eta)\dot{\eta}_r + D_\eta(\nu_R,\eta)\dot{\eta}_r + g_\eta(\eta) \\
&= J^{-T}(\eta)[M\dot{\nu}_r + C(\nu_R)\nu_r + D(\nu_R)\nu_r + g(\eta)],
\end{aligned}
\tag{22}
$$

$$
\begin{aligned}
&M_\eta(\eta)\ddot{\eta}_c + C_\eta(\nu_R,\eta)\dot{\eta}_c + D_\eta(\nu_R,\eta)\dot{\eta}_c \\
&= J^{-T}(\eta)[M\dot{\nu}_c + C(\nu_R)\nu_c + D(\nu_R)\nu_c],
\end{aligned}
\tag{23}
$$

and using the following parameterization,

$$
M\dot{\nu}_r + C(\nu_R)\nu_r + D(\nu_R)\nu_r + g(\eta) \triangleq \Phi(\dot{\nu}_r,\nu_r,\nu_R,\eta)\Theta,
\tag{24}
$$

$$
M\dot{\nu}_c + C(\nu_R)\nu_c + D(\nu_R)\nu_c \triangleq \Phi_1(\dot{\nu}_r,\nu_r,\nu_R)\Theta
\tag{25}
$$

where Θ is the vector of unknown parameters introduced in (19) and Φ, Φ_1 are a known *regression matrices* of appropriate dimensions[9], we rewrite (21) as

$$
\begin{aligned}
\dot{V}(s,\tilde{\Theta},t) = &-s^T D_\eta(\nu_R,\eta)s + \dot{\tilde{\Theta}}^T \Gamma^{-1}\tilde{\Theta} + s^T J^{-T}\tau_d \\
&+[J^{-1}(\eta)s]^T[\tau - (\Phi - \Phi_1)\Theta].
\end{aligned}
\tag{26}
$$

Then, defining the control law as

$$
\tau = (\Phi - \Phi_1)\hat{\Theta} - J^T K sgn(s),
\tag{27}
$$

where $K > 0$, is a symmetric and definite positive regulator gain matrix of appropriate dimension and $\hat{\Theta}$, is the estimated parameter vector, we have that

$$
\begin{aligned}
\dot{V}(s,\tilde{\Theta},t) = &-s^T D_\eta(\nu_R,\eta)s - |s^T|K + s^T J^{-T}\tau_d \\
&+\tilde{\Theta}^T[\Gamma^{-1}\dot{\tilde{\Theta}} + (\Phi - \Phi_1)^T J^{-1}(\eta)s]
\end{aligned}
\tag{28}
$$

---

[9] By the parameterization used in this work, Θ_{33x1} and Φ_{6x33}.

since $-s^T K sgn(s) = -|s^T|K$. Furthermore, assuming that $\dot{\Theta} = 0$, and chosen the parameter update law $\dot{\hat{\Theta}}$, and the control gain K, as

$$\dot{\hat{\Theta}} = -\Gamma(\Phi - \Phi_1)^T J^{-1}(\eta)s \qquad K \geq J^{-T}\tau_{d_{max}} + \mu, \qquad \mu > 0 \qquad (29)$$

where $|\tau_d| \leq \tau_{d_{max}}$ since τ_d is considered limited by a known function, we obtain

$$\dot{V}(s,\tilde{\Theta},t) = -s^T D_\eta(\nu_R,\eta)s - |s^T|\mu \leq 0, \qquad (30)$$

$\forall \nu_R, \eta \in \Re^n$. Hence, from Barbalat's lemma the convergence of s is guaranteed[10]. And, since V is bounded, s and $\tilde{\Theta}$ are bounded also.

4.3   Sliding Mode Control Based on Input-Output Stability

Considering that the system is input-output stable, in this subsection the sliding mode control gains are computed from a regression matrix of the system.

Stability Proof. Considering the Lyapunov-like function candidate (9), given in the stability proof of subsection 4.1, and by using the parameterizations in (24) and (25), we can rewrite (11) as

$$\dot{V}(s_1,t) = -s_1^T D(\nu_R)s_1 + s_1^T[\tau + \tau_d - \Phi(\dot{\nu}_r,\nu_r,\nu_R,\eta)\Theta +$$
$$+\Phi_1(\dot{\nu}_r,\nu_r,\nu_R)\Theta], \qquad (31)$$

where Φ, Φ_1 are a known regression matrices of appropriate dimensions and Θ is a vector of unknown parameters. Then, we define the control law as

$$\tau = [\Phi(\dot{\nu}_r,\nu_r,\nu_R,\eta) - \Phi_1(\dot{\nu}_r,\nu_r,\nu_R)]\hat{\Theta} - \tau_{s_1}, \qquad (32)$$

where $\hat{\Theta}$ are the vector of estimated parameters and the robust term τ_{s_1}, is given by

$$\tau_{s_1} = K_i sgn(s_{1_i}) + s_{1_i} \qquad (33)$$

in which $K_i > 0$, is the respective gain control for each degree of freedom $i = 1,...,6$. Thus, by replacing the control law (32) in (31) we have

$$\dot{V}(s_1,t) = -s_1^T D(\nu_R)s_1 + s_1^T[\tau_d + (\Phi - \Phi_1)\tilde{\Theta} - \tau_{s_1}] \qquad (34)$$

where $D(\nu_R)$ is a positive matrix by the hydrodynamic characteristics of the vehiclea and $\tilde{\Theta} = \hat{\Theta} - \Theta$, is the vector of parametric errors. Furthermore, since

---

[10] Except for the singular point $\pm 90°$.

the system is considered input-output stable, the matrices Φ, Φ_1 and vector Θ are considered limited[11], i.e., $|\tilde{\Theta}_j| \leq \bar{\bar{\Theta}}_j$, $|\Phi_{ij}| \leq \bar{\Phi}_{ij}$ and $|\Phi_{1_{ij}}| \leq \bar{\Phi}_{1ij}$ for each degree of freedom $i = 1, ..., 6$ and each parametric error $j = 1, ..., 33$. And, since the disturbance τ_d is considered limited, i.e., $|\tau_d| \leq \bar{\tau}_d$ where $\bar{\tau}_d$ is a positive known function, we can select K_i as

$$K_i = \sum_{i=1}^{6} \bar{\tau}_{d_i} + \sum_{i=1}^{6}\sum_{j=1}^{33} (\bar{\Phi}_{ij} + \bar{\bar{\Phi}}_{1_{ij}})\bar{\bar{\Theta}}_j \tag{35}$$

to obtain,

$$\dot{V}(s_1, t) = -\sum_{i=1}^{6} s_{1_i} D_{ii} + \sum_{i=1}^{6} s_{1_i}\tau_{d_i} + \sum_{i=1}^{6}\sum_{j=1}^{33} s_{1_i}[\Phi_{ij} + \Phi_{1_{ij}}]\tilde{\Theta}_j -$$
$$-\sum_{i=1}^{6} |s_{1_i}|\bar{\tau}_{d_i} - \sum_{i=1}^{6}\sum_{j=1}^{33} |s_{1_i}|[\bar{\Phi}_{ij} + \bar{\bar{\Phi}}_{1_{ij}}]\bar{\bar{\Theta}}_j - \sum_{i=1}^{6} s_{1_i}^2 \leq 0 \tag{36}$$

that ensure the convergence of s_1 to the sliding surface, i.e., guarantees that $\lim_{t\to\infty} s_1(t) = 0$, thus also guarantees $\lim_{t\to\infty} s(t) = 0$ and therefore $\tilde{\eta} \to 0$ as $t \to \infty$.

5   Simulation Results

The Biointeractive BA-1 vehicle's specifications [8] was used for simulations. The simulations time was 1000 ds. and for the first and third strategies, the system's parametric matrices used to produce the respectives control law had 5% of uncertainty. The control parameters were manually tuned via numerical simulations. The velocities[12] and accelerations considered for the ocean currents were $\nu_c = [0.29, 0.29, 0, 0, 0, 0]^T$ m/s and $\dot{\nu}_c = \mathbf{0}$ m/s^2 respectively. Furthermore, the disturbance relative to the wave and wind forces and moments was set as $\tau_d = [50, 50, 50, 0, 0, 0]^T$ N.

Notice that, all control strategies presents a practically null steady state errors with respect to positions and orientations. And the respective scalar tracking measures converges to a small vicinity of the sliding surface due to an interpolation of the discontinuous control action inside a boundary layer of the sliding surface[13]. Also, note that there is not sign of chattering on the respective control outputs. Furthermore, since the rolling motion of the vehicle is self-adjustable, there is not any control action aging on it as shown in Fig. 1, Fig. 2, Fig. 3 and Fig. 4.

---

[11] Because for a limited input, the system produces a limited output.

[12] Assuming that the tracking performance is held at Itaipu lake. This value was obtained from [5].

[13] The Boundary Layer method was applied in all control strategies in order to smooth out the control law discontinuity within a vicinity of the sliding surface and thus avoid the chattering problem [4].

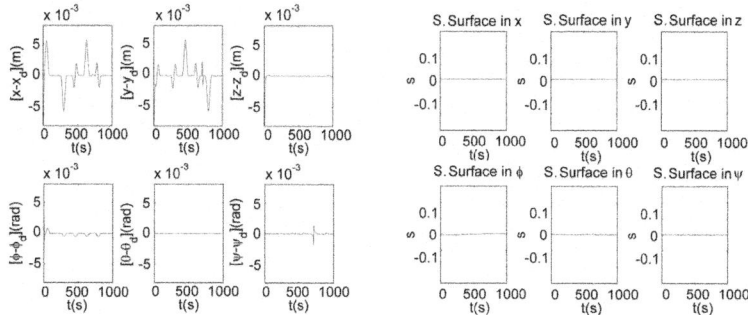

Fig. 1. Tracking error performance and convergence to the sliding surface of the Sliding Mode Control based on Lyapunov Stability Analysis. The control parameters used were $K = 72$, $\lambda = 0.3$ and $\phi_1 = 0.015$.

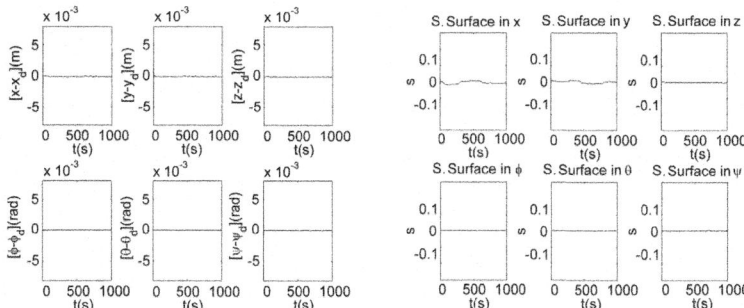

Fig. 2. Tracking error performance and convergence to the sliding surface of the Adaptive Sliding Mode Control. The control parameters used were $K = 112$, $\lambda = 2$ and $\phi_1 = 0.0125$.

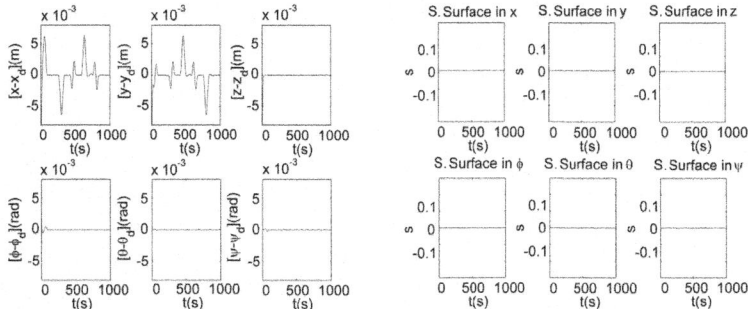

Fig. 3. Tracking error performance and convergence to the sliding surface of the Sliding Mode Control Based on Input-Output Stability. The control parameters used were $\lambda = 0.3$ and $\phi_1 = 0.15$.

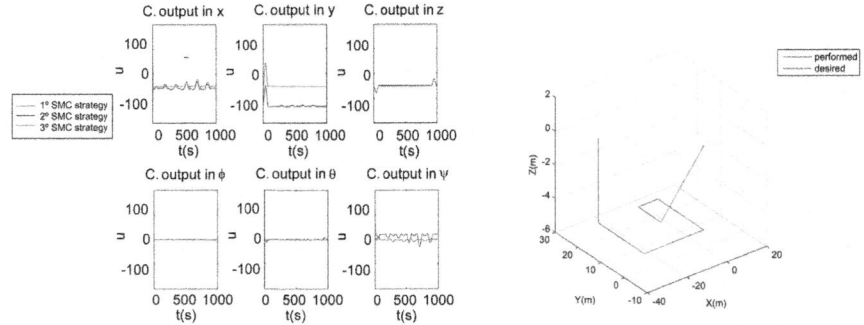

Fig. 4. Control outputs for each degrees of freedom and Tracked trajectory in 3D

6   Conclusions

The control strategies achieve an acceptable free-chattering tracking error performance with low loss of stability and robustness to disturbances. However, the adaptive sliding mode control strategy produces the better tracking error performance.

References

1. Fossen, T.I.: Guindance and Control of Ocean Vehicles. John Wiley and Sons, Chichester (1994)
2. Fossen, T.I., Sagatun, S.I.: Adaptive Control of Nonlinear Underwater Robotic systems. In: Int. Conf. on Robotics and Automation, pp. 1687–1694. IEEE Press, California (1991)
3. Fossen, T.I.: How to incorporate wind, waves and ocean currents in the marine craft equations of motion. In: IFAC Conference on Maneuvering and Control of Marine Craft, Genova (2012)
4. Yoerger, D.R., Slotine, J.-J.E.: Robust trajectory control of Underwater Vehicles. IEEE Journal of Oceanic Engineering 10(4), 462–470 (1985)
5. Chiella, A.B., Dos Santos, C.H.F., Motta, L.H., Rauber, J.G., Driedrich, D.C.: Control strategies applied to autonomous underwater vehicle for inspection of dams. In: Int. Conf. on Methods and Models in Automation and Robotics, Miedzyzdrojie, pp. 319–324 (2012)
6. Liu, J., Wang, X.: Advanced Sliding Mode Control for Mechanical Systems. Press and Springer, Beijing (2012)
7. DeCarlo, R.A., Zak, S.H., Matthews, G.P.: Variable Structure Control of Nonlinear Multivariable Systems: A Tutorial. Procedings of the IEEE 76(3), 212–232 (1988)
8. Choi, J.-K., Kondo, H.: On Fault-Tolerant Control of a Hovering AUV with Four Horizontal and Two Vertical Thrusters. In: Oceans 2010 IEEE, Sydney (2010)

Autonomous Robot Control by Neural Networks

Adriano B. Pinto[1], Ramiro S. Barbosa[1,2], and Manuel F. Silva[1,3]

[1] Institute of Engineering - Polytechnic of Porto (ISEP/IPP), Rua Dr. António
Bernardino de Almeida, 431, 4200-072 Porto, Portugal
adrianobessapinto@gmail.com, {rsb,mss}@isep.ipp.pt
[2] GECAD − Knowledge Engineering and Decision Support Research Center
[3] INESC TEC - CROB

Abstract. This work aims to apply the concepts associated with neural
networks in the control of an autonomous robot system. The robot was
tested in several arbitrary paths in order to verify the effectiveness of
the neural control. The results show that the robot performed the tasks
with success. Moreover, in the case of arbitrary paths the neural control
outperforms other methodologies, such as fuzzy logic control.

Keywords: neural networks, robot, autonomous, control, neural
control.

1   Introduction

The investigation on Artificial Neural Networks (ANN) has been motivated and
developed by the acknowledgment that the human brain processes all the information captured in a very particular way. The brain can be compared to a
very complex, non-linear and parallel computer. It is capable to structure and
organize their process units, known as neurons, with the purpose of performing
a very complex and much faster processing task than any other digital computer
available today [1]. An ANN is inspired in that interpretation of the human
brain, although it is quite different in size, model and on the operating way of
its elements. The ANN can be used to solve a lot of problems encountered in
several application areas, such as classification, diagnosis, analysis of signals and
images, pattern recognition, optimization and control [1, 2]. They are particularly effective in solving problems that do not have an analytical formulation or
explicit knowledge of the problem to be solved [1].

In this work, are applied the concepts associated with neural networks in the
control of an autonomous robot. The robot platform used [3] enabled the test
and validation of the ANN control and its comparison with other widely used
methodologies, such as fuzzy logic control (FLC), already implemented in the
same platform in previous works [3]. For the acquisition of experimental data
from the robot, the platform was endowed with a remote control system developed in an Android Smartphone. Two approaches were taken in implementing
the robot control. In the first approach, the robot was taught to perform similarly to the control by fuzzy logic that was previously developed. The second

© Springer International Publishing Switzerland 2015   597
A.P. Moreira et al. (eds.), *CONTROLO'2014 - Proc. of the 11th Port. Conf. on Autom. Control*,
Lecture Notes in Electrical Engineering 321, DOI: 10.1007/978-3-319-10380-8_57

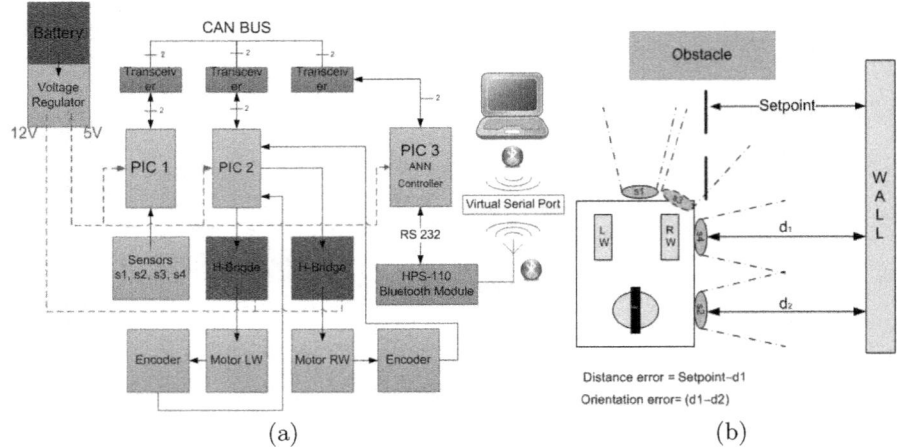

Fig. 1. Robot architecture (a) and schematic of the robot physical structure (b)

approach taught the robot from data taken from the remote control system. Both approaches were initially implemented and simulated in MATLAB, prior to making its implementation in the platform control hardware. Similar works using neural networks for the control of autonomous robots can be found in [4–8].

The rest of the paper is organized as follows. Section 2 describes the architecture of the robot system, while section 3 outlines the neural network training algorithm. Section 4 presents the two approaches adopted for the control of the system and shows the simulation results regarding the training of the ANN for the various paths. In section 5 are presented the experimental results that illustrate the effectiveness of the proposed methodology. Finally, section 6 draws the main conclusions.

2   Robot Platform Architecture

The robot has a modular structure (Figure 1(a)), being constituted by a rigid PVC base structure, two DC motors (with encoder) in the front, four ultrasonic sensors (sonars) and a free-wheel in the rear, to support part of the robot weight [3]. The robot kinematic model is described in [3]. Three PIC 18F4585 microcontrollers are used, with a CAN interface for the communication between the modules of the system. PIC 1 performs the data acquisition from the sonar sensors, PIC 2 is responsible for the velocity control of the left wheel (LW) and right wheel (RW) differential traction (allowing orienting the robot in space), and PIC 3 implements the ANN. The use of the CAN bus aimed to distribute the several tasks across the multiple nodes, being also chosen due to the flexibility that this network offers in the incorporation of future modules into the system. The robot is also equipped with a Bluetooth Module giving it the ability to communicate with remote systems.

Figure 1(b) presents the schematic representation of the robot physical structure, with the location of the sonars. The two right side sonars (S_2 and S_4) are used to determine the orientation and the distance of the robot from the wall. This configuration of the sonars is used for the implementation of the right wall-following behavior. The sonars S_1 and S_3, located on the front of the robot, allow the detection of obstacles and the implementation of the emergency behavior.

3   Neural Network Training Algorithm

The ANN is trained by applying the Levenberg-Marquardt algorithm [9, 10], which is an approximation to the Newton method. Like the quasi-Newton methods, the Levenberg-Marquardt algorithm was designed to approach second-order training speed without having to compute the Hessian matrix. When the performance function has the form of a sum of squares (as is typical in training feedforward networks), the Hessian matrix can be approximated as $\mathbf{H} = \mathbf{J}^T \mathbf{J}$ and the gradient can be computed as $\mathbf{g} = \mathbf{J}^T \mathbf{e}$, where $\mathbf{J}$ is the Jacobian matrix that contains first derivatives of the network errors with respect to the weights and biases, and $\mathbf{e}$ is a vector of network errors.

The Levenberg-Marquardt algorithm uses this approximation to the Hessian matrix in the following Newton-like update:

$$\mathbf{x}_{k+1} = \mathbf{x}_k - \left[\mathbf{J}^T \mathbf{J} + \mu \mathbf{I}\right]^{-1} \mathbf{J}^T \mathbf{e} \qquad (1)$$

where the parameter μ is a scalar controlling the behavior of the algorithm. When μ is zero, this is just Newton's method, using the approximate Hessian matrix. When μ is large, this becomes gradient descent with a small step size. Newton's method is faster and more accurate near a minimum error, so the aim is to shift toward Newton's method as quickly as possible. Thus, μ is decreased after each successful step (reduction in the performance function) and is increased only when a tentative step would increase the performance function. In this way, the performance function is always reduced at each iteration of the algorithm. This algorithm appears to be the fastest method for training moderate-sized feedforward neural networks (up to several hundred weights) [11].

4   ANN Design

This section presents the two approaches for the ANN design. First, the ANN controller creation is based on a FLC previously implemented. Second, the creation of the ANN controller is based on the samples gathered from the remote control.

4.1   ANN Created from the Data Sampled from the FLC

In this case was designed an ANN that aims to have the same behavior as the FLC. For this, it is necessary to supply the input data and the respective desired

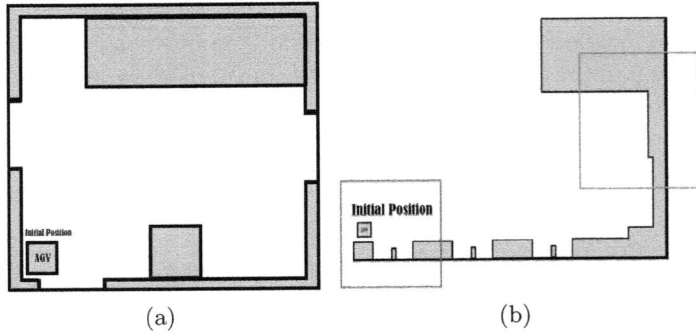

(a)   (b)

Fig. 2. Route sampled through the FLC (a) and random or not trained course (b)

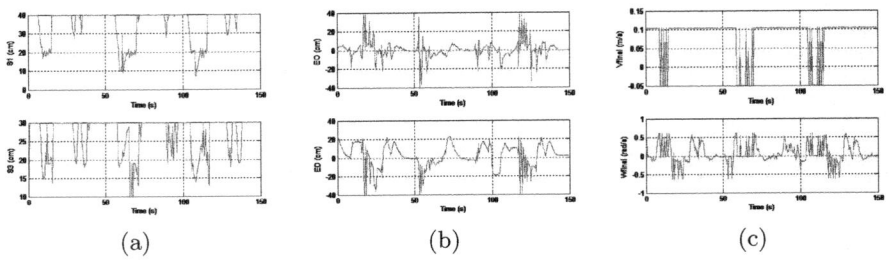

(a)   (b)   (c)

Fig. 3. Input data, S_1 and S_3 (a) and EO and ED (b), and output data (c) of the FLC used to train the ANN

outputs to the ANN, to carry out its training, in order for the ANN to operate as desired.

To design the ANN based on the existing FLC it was selected a route that makes the FLC exhibit all the behaviors implemented on it, such as obstacles avoidance and following walls. The route to be traversed by the robot in order to train the ANN (depicted in Figure 2(a)) is characterized by having sections where the robot will have to reverse its direction in order to follow the desired route. The sections where this occurs are after the square and later after the rectangle (see Figure 2(a)).

As can be seen from Figure 2(a), the robot has to deviate from the square and subsequently, as it tries to approximate the followed wall, detects the front wall; so the robot moves back and then moves to follow the same front wall. Then, after the robot has followed the rectangle, a similar case to the one mentioned above occurs due to the spacing between the wall and the rectangle.

The sampled data of the measurements of the robot front sonars S_1 and S_3, which were collected and received through Bluetooth, are shown in Figure 3(a). In Figure 3(b) are presented the values of the distance and orientation errors, while Figure 3(c) shows the output values of the FLC.

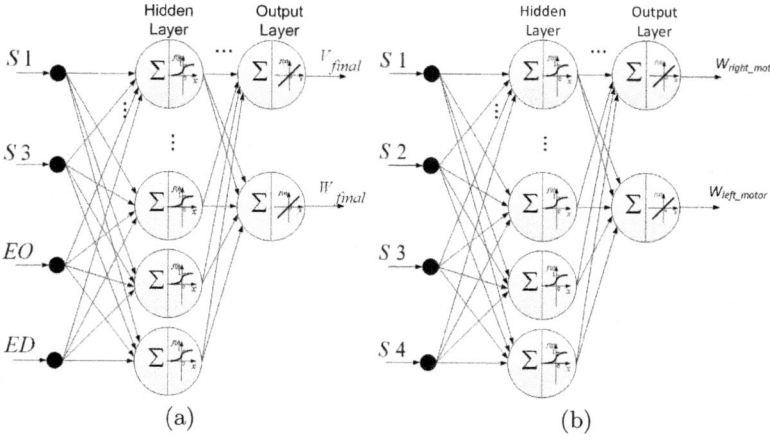

Fig. 4. Diagram of the ANN created from the FLC (a) and ANN implemented to behave as the sampled remote control (b)

These samples were used to perform the training of the ANN. The ANN is a 4-12-2 multilayer feedforward ANN as presented in Figure 4(a). The input layer consists of 4 inputs (corresponding to sonars S_1 and S_3 and errors EO and ED). The hidden layer consists of 12 neurons with a logarithmic sigmoid activation function and the output layer is composed of two neurons (the two output variables), with a linear activation function.

The training of the network yielded the results of the output variables (the system's linear and angular velocities) shown in Figure 5. It is observed that the output values of the ANN are very similar to those of the FLC. The Mean Square Error (MSE) obtained was 3.4146×10^{-4}.

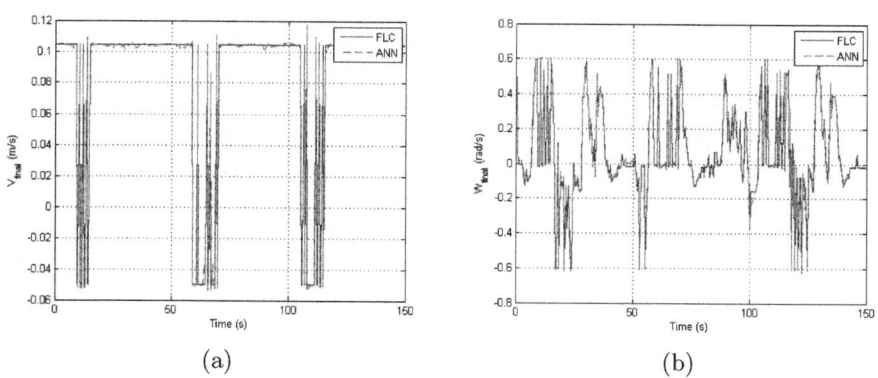

Fig. 5. Simulation results of the ANN in MATLAB: linear (a) and angular (b) velocities

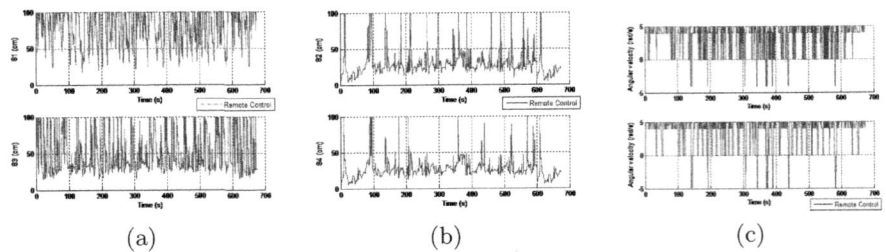

Fig. 6. Input data, S_1 and S_3 (a), and S_2 and S_4 (b), and output data (c) to train the ANN

4.2 ANN Created from the Data Sampled from the Remote Control

Through the remote control available in the robot and implemented on the Android Smartphone application, was designed a system that allows to train the robot with information provided by the sonars and motor encoders, when it travels along a path controlled by the remote control. The objective of this implementation is that the system controlled autonomously by an ANN would behave in a similar way to the case of controlled remotely, thus performing a similar path in both cases.

As in the previous case, the ANN is a proactive multilayer feedforward network with four inputs (values measured by the sonars) and two outputs (values of the reference angular velocities for the motors), as shown in Figure 4(b). The hidden layer is composed of 12 neurons. Thus, to obtain the training data for this ANN was used the robot controlled remotely, performing routes with the same complexity used in previous subsection.

In order for the robot to operate satisfactorily in standalone mode with a ANN trained from the remote control was conducted an experiment composed by three different routes, where one of those routes was the one depicted in Figure 2(a). The sampled data of the robot travelling along the three routes used for training the ANN are shown in Figure 6. Figure 6(a) and Figure 6(b) depict the data referring to the evolution of S_1 and S_3 and the evolution of S_2 and S_4, respectively, during the sampling time. Figure 6(c) displays the velocities of the robot motors, which are the targets of the ANN training, corresponding to the sampled inputs.

By implementing a 4-12-2 multilayer feedforward ANN were obtained the results shown in Figure 7. These figures show the comparison between the training and simulation for the angular speed of the right and left wheels of the robot, respectively. The performance value of the trained ANN is $MSE = 2.6230$, which means that the output values of the ANN are, in general, slightly different of the target values used in the training. This is visible in Figure 7, which depicts the first 200 samples obtained from the simulation of the ANN (magenta curve) and the corresponding first 200 samples used in training as targets (blue curve). It is seen that the values that the ANN presents at its output are not the values used

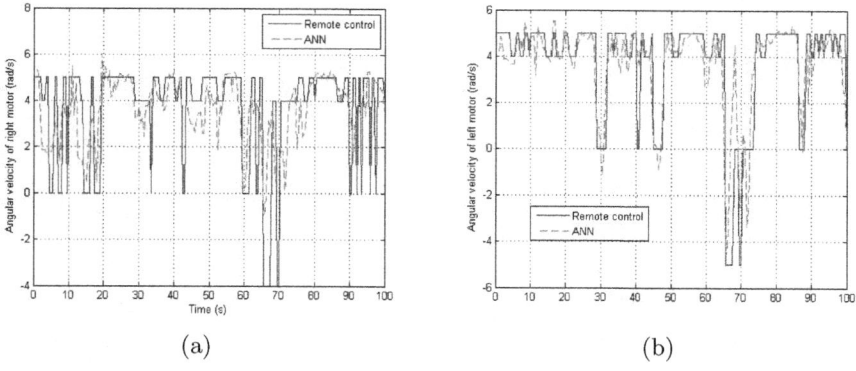

(a)   (b)

Fig. 7. Simulation results in MATLAB of 200 input values used in the ANN training

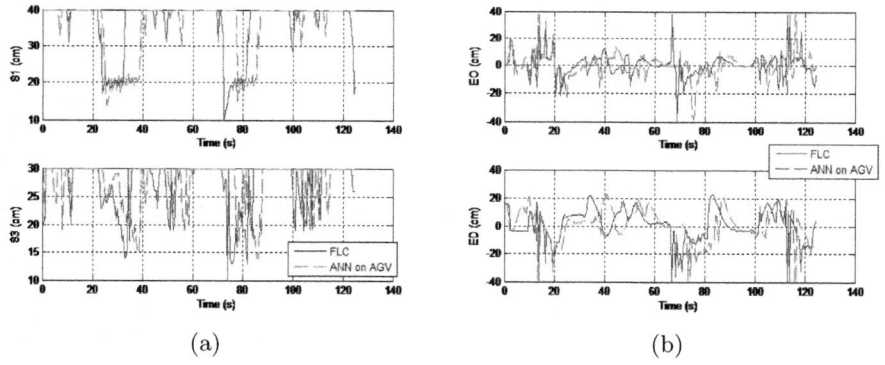

(a)   (b)

Fig. 8. Experimental results obtained with the ANN and the FLC implemented in the robot while performing the training route: obstacle detection sensors (a) and calculated errors (b)

in the training set. This discrepancy was expected and desired because with it the robot will present smoother variation on the velocities than what it would show if the output values of the ANN were exactly those used as targets.

5   Experimental Results

5.1   ANN Created from Data Sampled from the FLC

With the implementation of a 4-12-2 multilayer feedforward ANN on the robot, were obtained the results presented in Figure 8. In Figure 8(a) is displayed the evolution of the input variables S_1 and S_3, while Figure 8(b) presents the evolution of the input variables EO and ED of the ANN. It is observed that the experimental results of the ANN and FLC are similar. This is supported

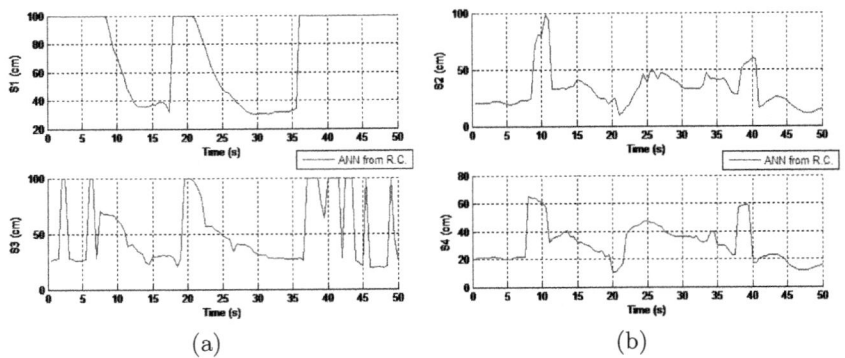

Fig. 9. Experimental results of the robot going through the first section of the random course: inputs S_1 and S_3 (a) and S_2 e S_4 (b)

by similar variations in the measurements of the sensors and errors of both controllers.

5.2   ANN Created from Data Sampled from the Remote Control

The experimental test of the robot when controlled by the ANN designed and trained in subsection 4.2 was performed on the random route shown in Figure 2(b). The test was conducted in two sections of this route. The first section is shown in Figure 2(b), highlighted by the red square and the second section is presented in the same figure but highlighted by the green square. In Figure 9 are shown the results of the first section through the input variables S_1 and S_3, and S_2 and S_4, respectively. As shown in Figure 9(a), the robot never approaches too much an obstacle, keeping a minimum distance of at least 20 cm. In Figure 9(b) it is verified that at the beginning and at the end of the sampled route, the robot keeps a distance of 20 cm when following the wall. In the intermediate zone, where it had to deviate from obstacles in its path, the distance kept to the wall decreased to a maximum of 10 cm.

The results obtained by the robot while performing the second section are shown in Figure 10. The results are similar to those of the first section, with regard to the distance from obstacles - the distance kept to an obstacle was at least 20 cm (Figure 10(a)). In what regards the distance kept to the wall that the robot was following - this maintained a distance of 7 cm, which is lower than in the previous case. Also, the average distance maintained came down from 30 cm to 20 cm. As indicated in Figure 10(b), variations of the distance kept to the wall that the robot is following occur, although the wall was flat. This was caused by the training data used, that is, the sampled data of the robot being operated by the remote control had also different values of the distance kept to the wall, even when the wall was flat.

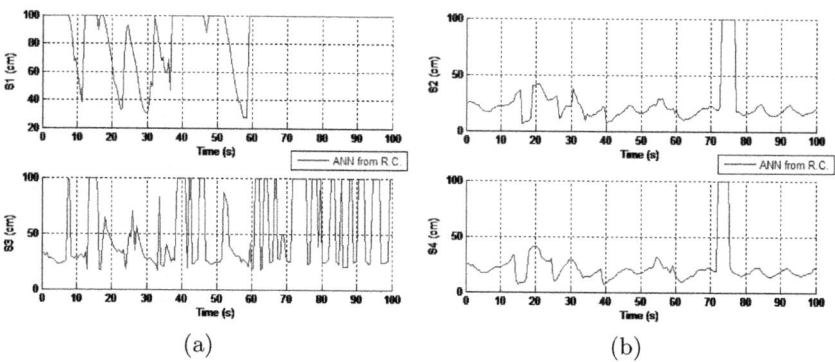

Fig. 10. Experimental results of the robot going through the second section of the random course: inputs S_1 and S_3 (a) and S_2 and S_4 (b)

Table 1. Performance comparison between ANN and FLC

Behaviors	Control			
	FLC		ANN	
	S_{EO}	S_{ED}	S_{EO}	S_{ED}
Wall-following (flat wall)	0.0139	0.1256	0.0123	0.1021
Wall-following with discontinuities	0.1057	0.6554	0.1811	0.4650
Wall-following with discontinuities and obstacles avoidance	1.9345	3.6527	1.8497	3.5684

5.3   Discussion of the Obtained Results

In order to compare the performance between ANN and FLC, it was calculated the sum of squared errors of the distances and orientations, respectively S_{ED} and S_{EO}, given by the following equations:

$$S_{ED} = \sum_{i=1}^{N} ED_i^2, \quad S_{EO} = \sum_{i=1}^{N} EO_i^2 \qquad (2)$$

where N is the number of samples of the performed path.

The results are shown in Table 1 for each of the implemented behaviors. It is observed that the ANN produces lower error values in almost all situations, except for the wall-following behavior with discontinuities, in which the value of S_{EO} for FLC is smaller than ANN. It may be considered that, in general, the ANN performs better than the FLC for the tested paths. This is due to a better adaptation of this type of control over new and unexpected situations than other approaches.

6   Conclusions

This paper presented the control of an autonomous robot through neural networks. A comparison with FLC was also made. It can be concluded that the implementation of the ANN from the FLC that was previously implemented on the robot was successful. Also, the ANN control from the remote system had a satisfactory behaviour.

As future work, the remote control system could be replaced by one with more differentiated levels, this way achieving more distinctive sets of input-output pairs, and thus obtaining a better performance of the ANN training. The use of more sonars, in particular on the left side of the platform will allow a better "view of the environment" from the robot control.

Acknowledgement. This work is supported by FEDER Funds through the "Programa Operacional Factores de Competitividade - COMPETE" program and by National Funds through FCT "Fundação para a Ciência e a Tecnologia" under the project: FCOMP-01-0124-FEDER- PEst-OE/EEI/UI0760/2014.

References

[1] Haykin, S.: Neural Networks: A Comprehensive Foundation, 2nd edn. Prentice-Hall, Singapore (1999)

[2] Rojas, R.: Neural Networks: A Systematic Introduction. Springer, Berlin (1996)

[3] Osório, D., Barbosa, R.S., Silva, M.F.: Fuzzy control architecture for a mobile robot system. In: 31st IASTED International Conference on Modelling, Identification, and Control (MIC 2011), Innsbruck, Austria, pp. 286–293 (2011)

[4] Zhang, Y.: Tracking control for wheeled mobile robots using neural network model algorithm control. Journal of Theoretical and Applied Information Technology 46(2), 794–199 (2012)

[5] Chaitanya, V.: Full-state tracking control of a mobile robot using neural networks. International Journal of Neural Systems 15(5), 403–414 (2005)

[6] Ng, K.: A neuro-fuzzy controller for mobile robot navigation and multirobot convoying. IEEE Transactions on Systems, Man, and Cybernetics-Part B: Cybernetics 28(6), 829–840 (1998)

[7] Fierro, R., Lewis, F.: Control of a nonholonomic mobile robot using neural neural networks. IEEE Transactions on Neural Networks 9(4), 598–600 (1998)

[8] Velagic, J., Osmic, N., Lavecic, B.: Neural network controller for mobile robot motion control. Engineering and Technology 47, 193–198 (2008)

[9] Hagan, M., Menhaj, M.: Training feed-forward networks with the Marquardt algorithm. IEEE Transactions on Neural Networks 5(6), 989–993 (1994)

[10] Marquardt, D.: An algorithm for least-squares estimation of nonlinear parameters. SIAM Journal on Applied Mathematics 11(2), 431–441 (1963)

[11] Beale, M.H., Hagan, M.T., Demuth, H.B.: Neural Toolbox: User's Guide, MATLAB2013b. The MathWorks, Inc. (2013)

Robust Robot Localization Based on the Perfect Match Algorithm

Héber Sobreira[1], Miguel Pinto[1], António Paulo Moreira[1],
Paulo Gomes Costa[1], and José Lima[2]

[1] INESC TEC (formerly INESC Porto) and Faculty of Engineering,
University of Porto, Portugal
{dee09025,dee09013,amoreira,paco}@fe.up.pt
[2] INESC TEC (formerly INESC Porto) and Polytechnic Institute of Bragança, Portugal
jllima@ipb.pt

Abstract. Self-localization of a robot in an indoor plant is one of the most important requirement in mobile robotics. This paper addresses the application and improvement of a well known localization algorithm used in Robocup Midsize league competition in real service and industrial robots. This new robust approach is based on modeling the quality of several measures and minimizing the maching error. The presented innovative work applies the robotic football knowledge to other fields with high accuracy. Real and simulated results allow to validate the proposed methodology.

Keywords: AGV, Mobile Robot, Service Robots, 2D Matching, Laser Range Finder.

1   Introduction

Robust self-localization is one of the main requirements for autonomous mobile robots in industry. It can be defined as the task of estimating the robot's position and orientation in a map of the environment. This has been a research topic over the last years and many different solutions for these localization problems have been presented [1].

There are several localization systems already implemented on industries but usually they use a laser with artificial beacons [17]. It has a disadvantage of requiring a large number of beacons to be spread in the manufacturing plant. Besides the aesthetic, it can be a difficult and expensive task to spread those beacons over the manufacturing plant. Despite not getting aesthetic, it is sometimes a difficult and expensive task to implement those beacons spread over the manufacturing plant. Some problems arise from changes in the environment over time. These changes can be caused by dynamic obstacles like people or other vehicles across the floor. The main goal of this paper is to describe a localization system that does not require artificial beacons but, on the other hand uses natural features of the world. It was based on the robotic football algorithms [2] and implemented on industry and service robots. Moreover,

© Springer International Publishing Switzerland 2015
A.P. Moreira et al. (eds.), *CONTROLO'2014 - Proc. of the 11th Port. Conf. on Autom. Control*,
Lecture Notes in Electrical Engineering 321, DOI: 10.1007/978-3-319-10380-8_58

estimated localization by the developed system is stable even with closer objects. Our example uses a precision and high range specifications laser scanner NAV 350 from SICK. It is desired to pick a palette of material from one side to another in a laboratory with 8 x 25 m. As further presented, precision is good enough to perform that task. The system is developed in ROS [3], a well known Robot Operating System that provides libraries and tools to help software developers create robot applications. The main topic of this work is the robustness of the algorithm to the obstacles (other robots, people and objects) that can affect the localization.

This paper is organized as follows: Section 1 presents a small introduction about laser scanner. Then, section 2, presents some related work in the community. Section 3, describes the algorithm whereas section 4 addresses the results of the developed tests. Finally, section 5 finishes the paper with conclusions and future work direction of the developed system.

2   Related Work

In the matching algorithms the pose estimation is commonly fused with dead reckoning data, using for that purpose, probabilistic methods such as the Kalman Filter family and the Particle Filters [6, 8, 15, 18].

There are matching algorithms that require prior knowledge on the navigation area. This prior knowledge can be an environment map, natural landmarks or artificial beacons.

There are other types of matching algorithms, which compute the overlapping zone between consecutive observations, to obtain the vehicle displacement. One possible matching algorithm to estimate the quantity of angular and linear displacement of a vehicle between two different and consecutive configurations is the Iterative Closest Point (ICP). This type of matching, known as point to point matching, analyses the contribution of each point of the laser scan, in the cost function.

The algorithm is composed of two fundamental steps, which are iterated until the solution convergence: the matching and optimization. The result is the distance between consecutive scans, corresponding to the vehicle configuration that minimizes the cost function.

The problem of this approach is the huge amount of data to be processed. The process of finding the correct correspondence between points (matching) is a difficult and time-consuming task.

Examples of works where the scan alignment is used to perform the registration between consecutive scans are described in the papers [12, 13, 14].

J. Minguez et al. [14], developed the metric-based ICP algorithm (MbICP), improving the standard ICP with a novel distance measure between corresponding points. This measure considers the translation and rotation displacements at the same time, in contrast to the standard ICP. This avoids two different minimization problems and consequently reduces the computational cost. Nevertheless, is still a heavy algorithm when compared with the present work.

The presented work was based on [4], with some different approaches, such as: stopping criteria, sensors, Kalman filter [5] and the application type (from RoboCup environments [7] to a shop floor automation field). Authors address the match algorithm, which is a time saver algorithm [6] and it has a large potential to be implemented in the service robots localization systems. By using a Kalman filter, it is possible to acquire signal from several sensors and implement a sensor fusion strategy [5]. Furthermore, the innovative algorithm approach improves the processing time, robustness and accuracy.

3   Algorithm

The developed localization system uses the result of a Matching algorithm as an observation measurement to be fused with the vehicle's odometry data. There are several sensors and techniques used in mobile robot positioning [11]. The presented work uses laser scanner information instead of image processing data [4].

This algorithm of Matching is based on the light computational Perfect Match algorithm, described by M. Lauer et al in [4]. In this algorithm the vehicle pose is computed using 2D distance points from the surrounding environment. These points are acquired with a Laser Range finder, and are matched with the map of the building previously computed. Therefore, the vehicle pose is calculated by trying to minimize the fitting error between the data acquired and the building's map.

The Perfect Match is based on the steps: 1) matching error and gradient computation; 2) optimisation routine Resilient Back-Propagation (RPROP); and 3) co-variance estimation using the second derivative.

Through the building occupancy grid (based in a SLAM algorithm [10, 19]) it is possible to obtain the distance and gradient matrices. The distance matrix at each cell gives the distance to the closest obstacle. There are two gradient matrices, in order to the x direction and in order to the y direction. The first gives the direction variation of the distance matrix with the variation of the x position. The second shows the direction variation with the y position variation. These three matrices, shown in the following figures (Figure 1, Figure 2 and Figure 3) for a corridor, can be pre computed and so they can speed up the completion of the Perfect Match algorithm. For more details about these matrices see [6] and [8].

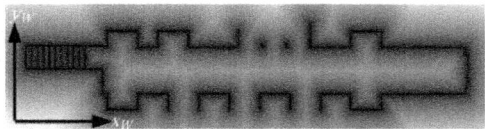

Fig. 1. Maps on a corridor where experiments were conducted: Distance Matrix

Fig. 2. Maps on a corridor where experiments were conducted: Gradient Matrix. Distance variation in direction y

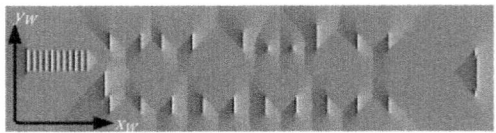

Fig. 3. Maps on a corridor where experiments were conducted: Gradient Matrix. Distance variation in direction x.

Consider now that the list of points of a Laser Range Finder scan $PntList$. The point i of this list in the world frame, is $PntList(i) = (x_i, y_i)$. The cost value is given by:

$$E = \sum_{i=1}^{PntList.Count} E_i, \quad E_i = 1 - \frac{L_c^2}{L_c^2 + d_i^2} \tag{1}$$

where d_i and E_i represent the distance matrix and cost function values of the point i.

The parameter L_c is used to discard points with large error E_i, increasing the robustness of the algorithm to the outliers. The resulting state/pose X_{Match} is given by the RPROP algorithm. The co-variance P_{Match} is computed using the second derivative of the algorithm of the quadratic matching error. For more details about the Perfect Match algorithm see [6].

Another improvement of the algorithm presented in this work, relatively to the authors' previous work [6] is the stop criterion used to stop the RPROP iterations. Instead of a fixed iterations number (ItN), the process can be stopped earlier if RPROP step is small enough meaning that a minimum has been reached.

In non-linear optimization problems, ItN depends on initial solution and cost function, among others. Furthermore, ItN shouldn't be a fixed value: on the one hand, increasing ItN increase processing time and the quality of solution. On the other hand, a few ItN decreases the solution quality. The criterion used is the RPROP step.

The limitation of the RPROP algorithm iterations guarantees that the time spent by the algorithm has ever a limit, that is ever lower than the algorithm cycle imposed by the observation module (Laser Range Finder). The limitation of the convergence guarantees that the minimization algorithm goes to a local minimum with the desired accuracy.

3.1   Kalman Filter Update

The Kalman filter [16] update stage combines the estimated state using the odometry, equal to $X_v(k + 1|k)$ and the Perfect Match resultant state, $X_{Match}(k + 1)$. The Kalman Filter equations can be seen in [5].

The observation model $h_v(X_v, r)$ in the update stage is equal to the vehicle state X_v:

$$h_v(X_v, r) = \begin{bmatrix} x_v + e_{rx} \\ y_v + e_{ry} \\ \theta_v + \varepsilon_{r\theta} \end{bmatrix} \tag{2}$$

where r is modeled as white noise, with a Gaussian distribution with zero mean ($\hat{r} = 0$) and covariance matrix R.

Therefore, in the update stage the observation is equal to the state obtained after the application of the Perfect Match:

$$Z_v(k + 1) = X_{Match}(k + 1) \tag{3}$$

The estimated observation is equal to the present estimative of the vehicle state, propagated during the Kalman Filter Prediction stage:

$$\hat{h}_v(X_v, \hat{r})(k + 1) = X_v(k + 1|k) \tag{4}$$

In that way, the innovation of the Kalman filter ($V(k + 1)$) is equal to:

$$V(k + 1) = Z_v(k + 1) - \hat{h}_v(X_v, \hat{r})(k + 1) \tag{5}$$

In this stage, the propagated covariance matrix, using odometry ($P(k + 1|k)$), and the covariance matrix computed in the Perfect Match procedure ($P_{Match}(k + 1)$), are used to determine the Kalman Filter gain ($W(k + 1)$):

$$W(k + 1) = P(k + 1|k)\frac{\partial h_v}{\partial X_v}^T \left[\frac{\partial h_v}{\partial X_v} P(k + 1|k)\frac{\partial h_v}{\partial X_v}^T + \frac{\partial h_v}{\partial r} P_{Match}(k + 1)\frac{\partial h_v}{\partial r}^T\right]^{-1} \tag{6}$$

The gradient of the observation module, in order to the vehicle state and the observation noise, $\frac{\partial h_v}{\partial X_v}$ and $\frac{\partial h_v}{\partial r}$ respectively, are identity matrices. Therefore, the previous equation can be re-written as the following:

$$W(k + 1) = P(k + 1|k)[P(k + 1|k) + P_{Match}(k + 1)]^{-1} \tag{7}$$

Therefore, after the update stage the new estimated state ($X_v(k + 1|k + 1)$), is given by the expression:

$$X_v(k + 1|k + 1) = X_v(k + 1|k) + W(k + 1) \cdot V(k + 1) \tag{8}$$

The new covariance matrix, decreases with the following equation:

$$P(k + 1|k + 1) = [I^{3x3} - W(k + 1)] \cdot P(k + 1|k) \tag{9}$$

where I^{3x3} is the square identity matrix with the dimension 3x3.

The rejection of data from obstacles that are not present in the map is a key to the localization algorithm robustness. A histogram of measures errors is processed and a threshold allows to select the valid data to the localization algorithm.

4   Results

Firstly, the time requirements are validated: 7 ms to process 1440 beams of laser with an Intel T4300 processor with 1.2 GHz clock speed. This is a compatible time to perform decision tasks and control. Robot could reach a repeatability of position in stop procedures of 5 mm (in the laboratory environment).

The assembled robot was used to perform all measurements and tests. The tests were conducted in the laboratories of the Faculty of Engineering of Porto, as presented in next subsection 4.1, and in the EMAF exposition in Exponor (Porto International Fair), as presented in next subsection 4.2. As it can be seen in next subsection, the developed filter allows rejecting closest objects that could interfere in localization. Moreover, the presented algorithm ensures precision and fast computation time (4 ms).

As we can see in [15], with a number of points acquired by the LRF equal to 682 points, the maximum time spent in the matching localization algorithm is limited. This limit time was measured and is equal to 17 milliseconds. The average execution time is about 12 milliseconds. Both the maximum and mean time spent, allows the algorithm to be used online, in a low computational power computer (these tests were executed in a Mini ITX, EPIA M10000G with a processor of 1.0GHz).

The ROS platform provides a package of mapping called gmapping. Using this package, a 2D grid map can be obtained. Another ROS' package, the adaptive Monte Carlo localization (amcl) algorithm, uses the 2D map grid built by the gmapping package to locate the vehicle, using 2D data. In comparative tests (same conditions than the last referred test), the execution time of the localization provided by the amcl package (with 1000 particles, the accuracy is similar to the proposed algorithm) has a maximum value of 42 ms and an average time of 32ms, almost 3 times more than the proposed algorithm.

4.1   Laboratory

The laboratory tests consists in achieving the localization of the robot in a space composed by 3 rooms (with an average dimension of 9x7 meters), a corridor (with a length of 20 meters) and a larger room with dimension 20x7 meters, as presented in Figure 4. The robot is moving and the obstacle interferes in the localization. The closest the object is, more error will be introduced in localization, as it can be seen in

Figure 5. This Figure is an example of a bad localization. The robot localizes itself in a wrong position of the map. The developed threshold method allows the robot to be correctly localized in the map as further presented in this paper.

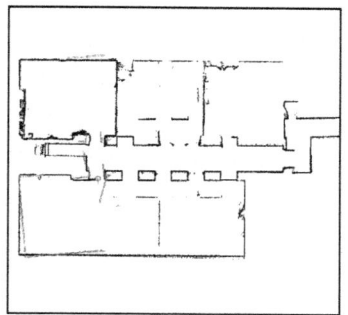

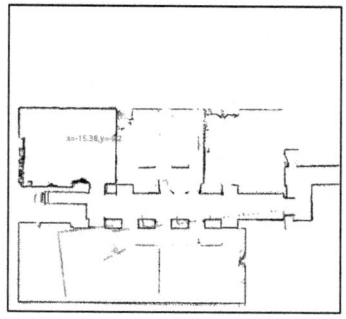

Fig. 4. Algorithm estimated location   **Fig. 5.** Robot localization failure

Red colored dots are the measures used for localization. The graphic presented in Figure 6 shows the laser error in measurements for the map presented in Figure 4. In other words, the distance between laser measure and the closest obstacle that doesn't match with robot posture.

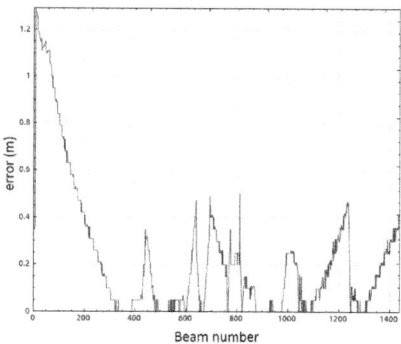

Fig. 6. Error in measure (without thresholding)

With the developed algorithm to reject error measures and as a good demonstration of our algorithm it is shown in Figure 7 green beams that are rejected and makes possible the robot to be localized.

The rejection of data from some obstacles that are not present in the map can be modeled as a function (see Equation (1)) with a threshold (A) operation as presented in Figure 8 (adapted from [4]). By this way, it is possible to drop completely outliers that are present in the measures. Threshold value depends on several map conditions like the number and dimensions of the expected objects that can generate outliers and the maximum dimension of the map.

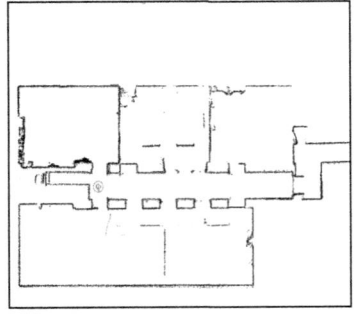

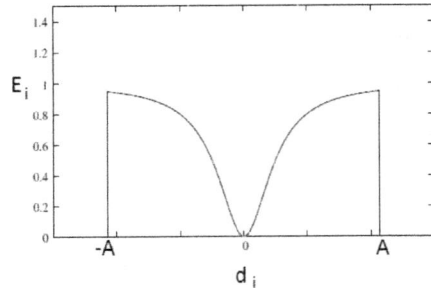

Fig. 7. Algorithm estimated location (with threshold, position 1)

Fig. 8. Error function with threshold (A)

Figure 9 shows the error in measurements, with an error limitation of 0.5 for the map presented in the Figure 7.

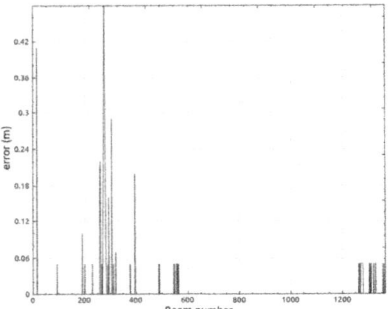

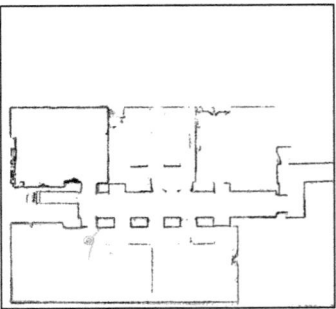

Fig. 9. Error in measure (with 0.5 thresholding filter)

Fig. 10. Algorithm estimated location (with threshold, position 2)

It is possible to notice that our algorithm rejects the fault measurements. Moreover, with the lean object, it is possible to notice that algorithm obtains the localization without error, as presented in Figure 10.

4.2   Professional Robotics Event

The experiments executed in the EMAF event were the final test of our algorithm. The robot navigated during three days, with several people moving around it and the path was planned so that the robot should cross areas with 1 cm of gap. The EMAF environment becomes a validation place different from laboratory with larger areas and far walls.

Figure 11 shows the developed robot moving around the EMAF fair (see map of Figure 12). Of course, there was security plans to avoid collision with humans. The same experiments were done (as subsection 4.1) and the results were compatible. Figure 12 shows the acquired map to plan the path.

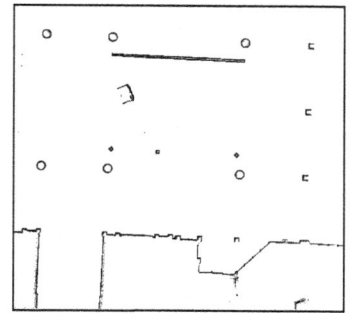

Fig. 11. Robot moving in the EMAF   **Fig. 12.** Acquired map of EMAF

5   Conclusions and Future Work

The localization method is applied in the AGV of test from the robotics laboratory. The localization method used the occupancy grid of indoor environments.

The experiments conducted made it possible to confirm that there are advantages to the proposed localization algorithm: the used observation modules are affordable and the localization algorithm used here works in structured mapped environments, without the need for any kind of artificial landmarks – this is the main advantage of this work.

The Extended Kalman Filter was applied as a multi fusion sensor system in order to combine the odometry information and the result of the Perfect Match. The experiments conducted by M. Lauer *et al* [4], which elects the Perfect Match as the faster algorithm comparatively to the Particle Filter algorithm [9][12], remain valid. As conclusion, the presented algorithm ensures precision and fast computation time (4 ms).

As a future work direction, enhancing the outlier's rejection is a promising research area since robot moves in a map with several obstacles. Moreover, handling more outliers and with bigger dimensions (such as people moving or layout changing) is a challenge to future work.

Acknowledgments. The work presented in this paper, being part of the Project PRODUTECH PTI (num-ber13851) - New Processes and Innovative Technologies for the Production Technologies Industry, has been partly funded by the Incentive System for Technology Re-search and Development in Companies (SI I&DT), under the Competitive Factors Thematic Operational Program,of the Portuguese National Strategic Reference Framework, and EUs European Regional Development Fund.

References

[1] Wulf, O., Lecking, D., Wagner, B.: Robust Self-Localization in Industrial Environmentsbased on 3D Ceiling Structures. In: Proceedings of the 2006 IEEE/RSJ International Conference on Intelligent Robots and Systems (2006)

[2] Liu, J., Yin, B., Liao, X.: Robot Self-localization with Optimized ErrorMinimizing for Soccer Contest. Journal Of Computers 6(7) (2011)

[3] Quigley, M., Gerkey, B., Conley, K., Faust, J., Foote, T., Leibs, J., Berger, E., Wheeler, R., Ng, A.Y.: ROS: an open-source Robot Operating System. In: Proc. Open-Source Software workshop of the InternationalConference on Robotics and Automation, Kobe, Japan (May 2009)

[4] Lauer, M., Lange, S., Riedmiller, M.: Calculating the perfect match: An efficient and accurate approach for robot self-localization. In: Bredenfeld, A., Jacoff, A., Noda, I., Takahashi, Y. (eds.) RoboCup 2005. LNCS (LNAI), vol. 4020, pp. 142–153. Springer, Heidelberg (2006)

[5] Thrun, S., Burgard, W., Fox, D.: Probabilistic Robotics. Massachusetts Institute of Technology (2006)

[6] Pinto, M., Moreira, A., Matos, A., Sobreira, H.: Novel 3D Matching Self-Localisation Algorithm. International Journal of Advances in Engineering & Technology 5(1), 1–12 (2012)

[7] Schulz, H., Behnke, S.: Utilizing the structure of Field Lines for efficient Soccer Robot Localization. Advanced Robotics 26(14), 1603–1621 (2012)

[8] Pinto, M., Paulo Moreira, A., Matos, A., Sobreira, H., Santos, F.: Fast 3D Matching Localisation Algorithm. Journal of Automation and Control Engineering 1(2), 110–115 (2013) ISSN 2301-3702

[9] Grisetti, G., Stachniss, C., Burgard, W.: Improved Techniques for Grid Mapping with Rao-Blackwellized Particle Filters. IEEE Transactions on Robotics 23(1), 34–46 (2007)

[10] Kohlbrecher, S., Meyer, J., von Stryk, O., Klingauf, U.: A Flexible and Scalable SLAM System with Full 3D Motion Estimation. In: Proceedings of IEEE International Symposium on Safety, Security and Rescue Robotics (SSRR), pp. 155–160. IEEE (2011)

[11] Borenstain, J., Everett, H.R., Feng, L., Wehe, D.: Mobile Robot Positioning and Sensors and Techniques. Journal of Robotic Systems, Special Issue on Mobile Robots 14(4), 231–249

[12] Aghamohammadi, A., Jensen, B., Siegwart, R.: Scan Alignment With Probabilistic Distance Metric. In: IEEE/RSJ International Conference on Intelligent Robots and Systems (IROS 2004), Sendai, Japan, September 28-October 2, vol. 3, pp. 2191–2196 (2004)

[13] Minguez, J., Lamiraux, F., Montesano, L.: Metric-Based Scan Matching Algorithms for Mobile Robot Displacement Estimation. In: Proc. of the IEEE International Conference on Robotics and Automation (ICRA), Barcelona, Spain, April 18-22, pp. 18–22 (2005)

[14] Minguez, J., Lamiraux, F., Montesano, L.: Metric-Based Iterative Closest Point Scan Matching for Sensor Displacement Estimation. IEEE Transactions on Robotics (October 2006)

[15] Migueis Pinto, M.: SLAM for 3D Map Building to be used in a Matching Localization Algorithm, PhD Electrical and Computer Engineering, Faculty Engineering of University of Porto (2012)

[16] Gouveia, M., Paulo Moreira, A., Costa, P., Reis, L.P., Ferreira, M.: Robustness and Precision Analysis in Map-Matching based Mobile Robot Self-Localization. In: IROBOT - 14th Portuguese Conference on Artificial Intelligence, EPIA 2009, Aveiro, Portugal, October 12-15, pp. 243–253 (2009)

[17] Ronzoni, D., Olmi, R., Secchi, C., Fantuzzi, C.: AGV global localization using indistinguishable artificial landmarks. In: IEEE International Conference on Robotics and Automation, pp. 287–292. IEEE, Shanghai (2011), doi:10.1109/ICRA.2011.5979759

[18] Jensfelt, P., Kristensen, S.: Active Global Localization for a Mobile Robot Using Multiple Hypothesis Tracking. IEEE Transactions on Robotics and Automation 17(5) (October 2001)

[19] Tipaldi, G.D., Braun, M., Arras, K.O.: FLIRT: Interest Regions for 2D Range Data with Applications to Robot Navigation. In: 12th Int. Symposium on Experimental Robotic (ISER 2010), New Delhi, India (2010)

3D Map and DGPS Validation for a Vineyard Autonomous Navigation System

Olga Contente[1,2], José Aranha[3,4], José Martinho[3,6], José F.M. Morgado[1],
Manuel Reis[2,4], Paulo Jorge Ferreira[2,5], Raul Morais[4,7], and Nuno Lau[2,5]

[1] Escola Superior de Tecnologia e Gestão de Viseu (ESTGV), Viseu, Portugal
ocont@estv.ipv.pt
[2] Instituto de Engenharia Electrónica e Telemática de Aveiro (IEETA), Aveiro, Portugal
[3] Centro de Investigação e de Tecnologias Agro-Ambientais e Biológicas (CITAB), Vila Real,
Portugal
[4] Universidade de Trás-os-Montes e Alto Douro (UTAD), Vila Real, Portugal
[5] Universidade de Aveiro (UA), Aveiro, Portugal
[6] Centro de Geociências da Universidade de Coimbra (CGUC), Coimbra, Portugal
[7] INESC TEC (formerly INESC Porto), Porto, Portugal

Abstract. An autonomous DGPS navigation system must use an accurate three-dimensional (3D) digital map. However, it is crucial to validate it using data collected in the field. One possible way to validate the map is to employ a vehicle driven by an expert to ensure that the trajectory is plotted within the boundaries of navigation paths. It is essential to take this care, especially when the terrain is very highly uneven and small differences in position may correspond to large vertical deviations. A small navigation error can result in a serious fall, which may damage or even destroy the vehicle. In the Douro Demarcated Region, in northern Portugal, the vineyard is planted on narrow terraces built on steep hills along the winding Douro River. This paper presents the results of a dynamic trajectory survey obtained from a real navigation procedure, carried out by an expert driving an instrumented tractor during the spraying of the vineyard. The results were obtained using a DGPS (accuracy = 2 cm) and compared to an existing Digital Elevation Model (DEM) of the vineyard already created by the authors' work group, with an average accuracy of 10 cm. The results are shown in a C# developed interface with OpenGL facilities, which enable the viewing of the 3D vineyard details. The results confirm the validation of the methodology previously adopted for map extraction and respective equipment selection. The trajectory of the tractor, including some maneuvers, is drawn within the inner and outer edges of each terrace or path that exists in the vineyard. The interface can also be used as an important tool in path planning to automatically extract the topology of the vineyard and to select the best path to carry out some vineyard tasks.

Keywords: Instrumentation, Agricultural process, Robotics and automation.

1   Introduction

Technological advances often arise from the need to solve everyday problems. In the case of intensive agricultural production the use of semi-autonomous machinery is

© Springer International Publishing Switzerland 2015
A.P. Moreira et al. (eds.), *CONTROLO'2014 - Proc. of the 11th Port. Conf. on Autom. Control*,
Lecture Notes in Electrical Engineering 321, DOI: 10.1007/978-3-319-10380-8_59

already a reality [1], [2]. Specifically, in the case of viniculture, there are already automatic systems to prepare, maintain (pruning, application of herbicides and pesticides) and harvest the vineyard. Autonomous guidance systems are available for agricultural tractors [3], [4]. To perform these operations, the machines must be equipped with a broad set of sensors [5]. For example, the grape harvesting operation may require the use of a stereo vision system for grape detection among the vine foliage [6]. For autonomous navigation, perception of the elements in the vineyard (such as rows of vines, tools left in the vineyard, among other obstacles) is critical. Monocular charge-coupled device (CCD), 3D stereo vision system or light detection and ranging sensors (LIDAR), with satellite-based global position information have been used to extract 3D maps of vineyard regular environments [7], [8], [9]. Vehicle localization and orientation is another difficulty in autonomous navigation even when the crop machine is equipped with DGPS (Differential Global Positioning System) and inertial sensors. This is emphasized when the vineyard terrain is irregular. In fact, the receiving satellite conditions may become weak, the number signals received from satellites may be insufficient and the measuring equipment might become subject to excessive vibration during its use and, when this happens, the machine positioning may be determined incorrectly. Thus, the terrain characteristics where some vineyards are planted can be one of the major limitations for automated agriculture.

Due to its unique characteristics, the Douro Demarcated Region (DDR) (a UNESCO World Heritage Site and the oldest Wine Demarcated Region in the World) presents very specific challenges, mainly due to its unique topographic profile, pronounced climatic variations and complex soil characteristics. The DDR (as described in [10]) is located in Northeast Portugal and consists mostly of steep hills (slopes reaching 100%) and narrow valleys that flatten out into plateaus above 400 m. The Douro River digs deep into the mountains to form its bed and the vineyards are planted in terraces fashioned from the steep rocky slopes and supported by hundreds of kilometers of dry stone wall. Because of theses specificities, the DDR vineyard has traditionally been harvested by humans. Nowadays, manpower needs are great and it is becoming increasingly difficult to find qualified workers for vineyard management. Autonomous systems might not only reduce harvesting costs and manpower needs, but can also work during the night. There is, however, one important constraint: although it is not essential that the work rate of a harvesting robot surpass that of a human, it is crucial that they satisfy quality control levels at least similar to those achieved by humans. To work in this type of vineyard, an agricultural machine, like a farm tractor, has to be mid-sized (be able to circulate within narrow terraces), sufficiently strong to carry the agricultural apparatus, such as vineyard sprayer, and should have four-wheel drive [11]. The use of an accurate three-dimensional (3D) digital elevation map (DEM) as input data is required in order to support autonomous navigation over rough terrain [12]. In addition to the set of georeferenced points that define the topology of the terrain, more field information can be added to the model to create a more realistic scenario of the vineyard particularities. The preexisting position data can serve with other sensor information to assure that navigation is not compromised. For the region under study, the authors' work group has developed a ground survey methodology that allowed a 10 cm average accuracy Digital Elevation Map (DEM) of the DDR vineyard to be created using a moderately priced DGPS [13].

The aim of this paper is to validate the earlier work through the continuous surveying of the tractor trajectory during a vineyard spraying procedure. In this study, the DGPS system, the same used for the map extraction, is mounted in the middle of the front-end of the tractor. In order to examine the tractor trajectory, the pathways defined in the vineyard and all the vineyard particularities are loaded to a 3D tool viewer. The viewer is part of a broader set of tools that compose a dedicated path planning interface to assist the navigation of an autonomous harvesting robotic system [14]. The map is validated if the acquired sequence of points overlaid on the DEM of the vineyard lie within the defined circulation areas. This work represents part of an effort that is being undertaken by our research team to help with the introduction of Precision Viticulture (PV) in farmers' everyday practices in the DDR [15], [16].

The paper begins with this introduction and section 2 presents the material and methodology used to extract a farm tractor navigation trajectory during a vineyard crop activity. Section 3 presents a description of some features of the simulation interface used to visualize the 3D map. Section 4 presents and discusses the results obtained and conclusions are drawn in section 5 and future work is suggested.

2   Materials and Methods

2.1   Materials

The vineyard is part of the Bateiras property, Ervedosa do Douro, São João da Pesqueira, Viseu, Portugal (Fig. 1), located on a hill (west) on the left bank of the Douro River between the (baselines) 41°10'49,44'' and 41°10'23.52''N parallels and the 7°33'23.04'' and 7°32´57.12'' W Gr meridians.

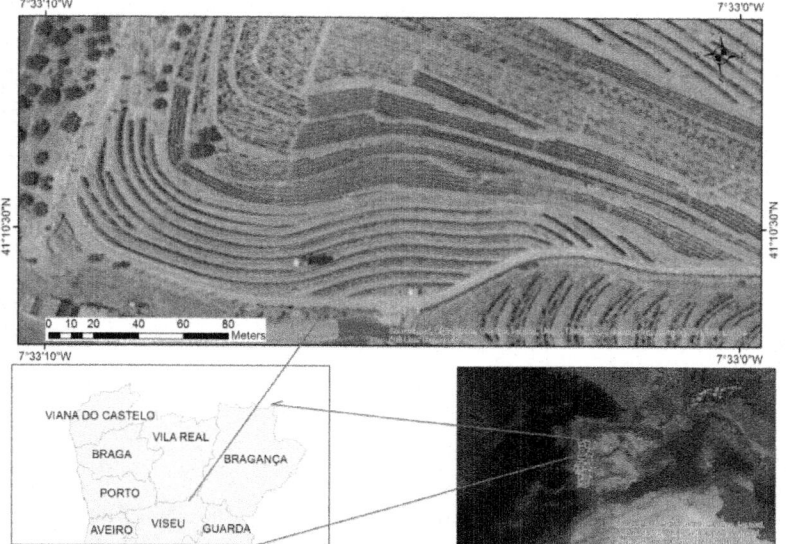

Fig. 1. Localization of the area studied

The area, a 0.623 ha parcel as part of the total 13.895 ha of vineyard, located at the bottom of the hill, as presented in Fig.2, is characterized by:

- long terraces that contour the hill forming wavy curves.
- short terraces known as dead end terraces, i.e., terraces interrupted by natural or artificial discontinuities.
- different vineyard paths which access very uneven terraces and border paths.

Fig. 2. Vineyard photography

The grape vine rows, supported by 3 or 4 wires fixed on 2.10 m wooden stakes approximately 4 m apart, are located on the outer edge of each terrace. The terraces are elevated up to 2 m from each other. The width of the terraces ranges from 1.6 m in tight spaces to 2.1 m at its ends (inbound and outbound access). Some paths that follow the slope of the mountain have a slope angle of almost 20% and a width ranging between 2.5 m and 3.5 m. The tractor trajectory survey was conducted with GNSS Ashtech equipment, Mobile Mapper 100 model, with integrated GPRS/GSM modem, running the Mobile Mapper Field software, connected to an ASH111661 model antenna of the same brand to receive the GPS L1 and L2 frequencies. The assembly equipment was mounted on the front end of a New Holland TCE 50 mid-sized tractor, which is actually used for the crop activities in the vineyard.

2.2   Methodology

The methodology adopted (to validate the previously created 3D digital map of the vineyard) comprises three phases, the preparatory work before field data collection, the field data collection and analysis of the results.

Preparatory Work before Field Data Collection

The preparatory work includes selecting and configuring the DGPS equipment and its software as well as defining a strategy to collect data. The GNSS receiver, the same used for mapping the vineyard, was configured in order to receive GPS L1 and L2 frequencies. The heights were referenced above the ellipsoid (HAE) and used as part of the datum-to-datum transformation. Positions were referenced according to the ETRS1989 – Portugal/TM06 local coordinate system that establishes the projection type, the projection units, spheroid and projected parameters. The tractor trajectory was obtained by dynamic survey, as a line of points. The positions were recorded automatically at two-second intervals.

Among the set of a vineyard's preparatory and maintenance operations, which are simultaneously performed using a tractor and cover the totality of the vineyard area, the spraying operation with a plant protection product was selected. In this operation the tractor had to sweep all sorts of terraces, cross the vineyard transitions (where it is difficult to perform maneuvers) and in some places, run backwards (in the case of dead-end terraces).

Where and how to place the DGPS antenna was also considered. Because the top and the front of the tractor are not flat and a specific support system had to be sized and built, a decision was made at this stage of the work, to find a cheaper solution.

Field Data Collection

The GNSS equipment was mounted on the counterweight, in front of the tractor, as shown in Fig. 3, left.

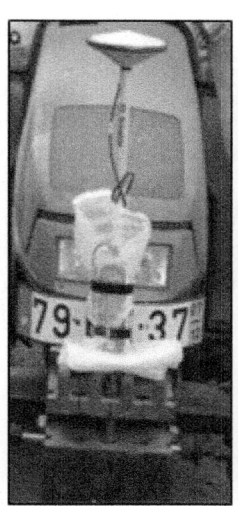 

Fig. 3. GNSS Equipment assembly (left) and tractor maneuvers (right)

For the selected crop activity, the tractor uses a very heavy spray set, with a 300 l tank, sprinkler nozzles, turbine and steel support, which greatly hinders the upward movement of tractor (especially at the beginning of the operation when the tank is full) even when properly balanced. For this reason the initial position for the spraying operation was the northeast corner of the study area, at the steepest slope of the studied vineyard parcel. This position is easily reached by the upper portion of the vineyard through the path which borders the vineyard. A top-down scan was done through the terraces, transitions and paths, covering all the vine row plants. All maneuvers to move from one terrace to another, Fig. 3 (right), as well as backwards movements, were surveyed continuously as a single tractor trajectory.

Analysis of the Results of Field Data Collection

Corrections were made in post-processing mode by Mobile Mapper Office software, using the records of the network stations SERVIR [17]. Under these conditions, the corrected data were filtered using a mask for an estimated positional accuracy better than 20 cm. The field data collected in the previous work, which had allowed us to create an accurate digital elevation map of the vineyard, and the actual tractor trajectory data were overlaid in the some geo referenced coordinate system for validation. As seen in the figure 3 (left), the antenna was placed on a rod, approximately 1.20 m from the ground. During the trajectory survey, due to the vibration and tilting of the tractor, the height position measurement was not considered relevant to validate the map. Thus, to validate the map, it was given more emphasis to the analysis of images from the top. The difference between the elevations maps was analyzed later and used as additional validation criteria of the extracted map.

3   VVPP Interface

The VVPP, still in development, is a Vineyard Viewer and Path Planning tool to assist a vineyard autonomous navigation system for the DDR. The C# developed interface, with OpenGL facilities, enables 3D vineyard details to be viewed, such as terraces, paths and transitions. The viewer, Fig. 4, can display any previously loaded 3D object. It also allows different sets of position points (thereby forming a 3D object) to overlap. More specifically, the map of the vineyard and the spraying tractor trajectory can be loaded. At the top of the interface vertical menu it is possible to select the options which enable an individual view of the terraces or paths or the option to the design the entire vineyard, Fig 4.

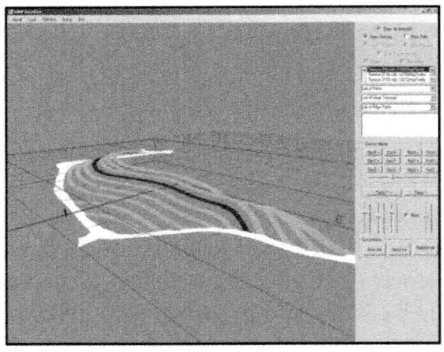

Fig. 4. Vineyard with a selected terrace in the VVPP viewer

The overlay information is detected by using different colors. At the bottom of the vertical interface menu, the displayed object can be rotated and translated, according to the three coordinate axes, as well as have its perspective changed (changing the position of the observer on the screen).

4   Results

As shown in figure 5, the tractor trajectory (dotted line), starts at point A and finishes at point B (near the vineyard entrance). These points belong to the steep path which borders the vineyard (in white) east of the entrance. The DGPS surveyed points regarding tractor maneuvers performed between terraces and through paths, either forwards or backwards, lie, as expected, on the sloping path.

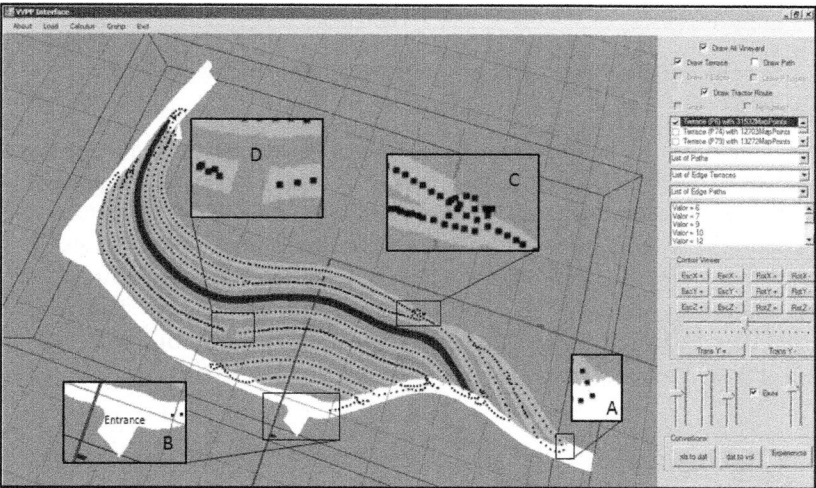

Fig. 5. Vineyard tractor routes overlapped the vineyard map the VVPP tool viewer

As one can observe in the C detail, the set of points outside the background color (cyan) which represents the terrace corresponds to positions where the front of the tractor goes beyond the inner edge of the terrace. This situation occurs when a reversing maneuver is performed near a bottleneck transition (due to the existence of a rainwater drain) and does not invalidate this study because the GPS is mounted on the front of the tractor, in a vertical plane beyond the wheel axis (the wheels of the tractors are never positioned off of the terrace). Another observation is the gap shown in the detail D which is justified by the non-existence of any vine plants in that row. The terrace is too narrow there due to a high tension transformer. So, there is simply not enough room for the tractor to pass. In terms of altimetry analysis the difference between the coordinates measured during the dynamic survey (performed while spraying with the tractor) and the height of each terraces (determined during the vineyard map survey with real time accuracy check) was found to be consistent for all of the vineyard's terraces.

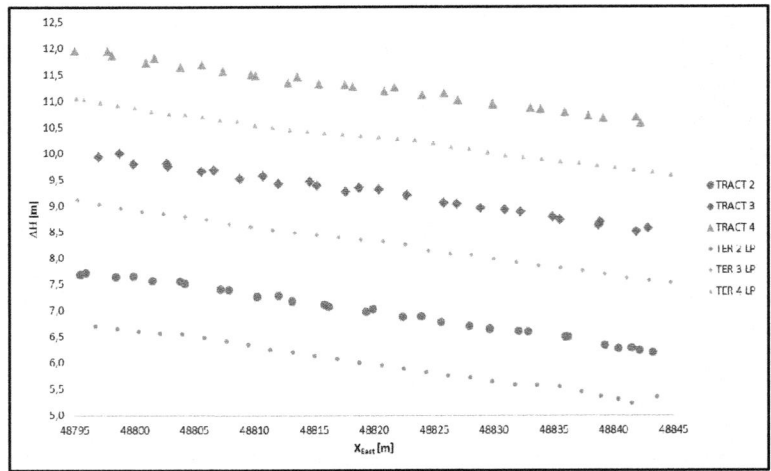

Fig. 6. Tractor route and border relative height points for three terraces

As Figure 6 shows the difference in the slopes of each pair of curves is maintained as well as the spacing between them. The x-coordinates in Figure 6 represent the distance in meters measured in east latitude relative to the ETRS1989 – Portugal/TM06 local coordinate system. The ordinates are relative elevations from the steepest slope vineyard point. The relative height of the borders of each terrace (inner and outer points) and tractor trajectory line of points are plotted as similarly shaped and different-sized dots. A band of the same length (in latitude) with a different longitude (to contain at least 3 terraces) was sampled to be analyzed.

5   Conclusion

The 3D digital map created in a previous work by static survey with an average accuracy of 10 cm has been validated. All post-processing position points using the records of the network stations SERVIR, dynamically surveyed with the same equipment every 2 seconds, lie between the inner and the outer edge of each path or terrace.

The survey trajectory allows the average speed of completing the spraying operation to be estimated as well as the amount of time lost performing maneuvers. It can also allow fuel consumption and CO_2 emissions to be minimized.

The VVPP tool can also be used as an important tool in path planning to identify decision points and to select the best path to carry out a variety of vineyard tasks.

Acknowledgements. The authors wish to thank Abílio Tavares da Silva for allowing us to perform this research in his vineyard.

References

1. Moller, J.: Computer vision - A versatile technology in automation of agriculture machinery. Journal of Agricultural Engeneering 47(4) (2010)
2. Stentz, A., Dima, C., Wellington, C., Herman, H., Stager, D.: A System for Semi-Autonomous Tractor Operations. Autonomous Robots 13, 87–104 (2002)
3. Alonso-Gracia, S., Gomez-Gil, J., Arribas, J.I.: Evaluation of the use of low-cost GPS receivers in the autonomous guidnce of agricultural tractors. Spanish Journal of Agricultural Research 9(2), 377–388 (2011)
4. Reid, J.F., Zhang, Q., Noguchi, N., Dickson, M.: Agricultural automatic guidance research in North America. Computers and Electronics in Agriculture 25(1-2), 155–167 (2000)
5. Lee, W.S., Alchanatis, V., Yang, C., Hirafuji, M., Moshou, D., Li, C.: Review. Sensing technologies for precision specialty crop production. Computers and Electronics in Agriculture 74, 2–33 (2010)
6. Reis, M.J.C.S., et al.: A low-cost system to detect bunches of grapes in natural environment from color images. In: Blanc-Talon, J., Kleihorst, R., Philips, W., Popescu, D., Scheunders, P. (eds.) ACIVS 2011. LNCS, vol. 6915, pp. 92–102. Springer, Heidelberg (2011)
7. Sáiz-Rubio, V., Rovira-Más, F.: Proximal sensing mapping method to generate field maps in vineyards. Agric. Eng. Int.: CIGR Journal 15(2), 47–59 (2013)
8. Rovira-Más, F.: Global-referenced navigation grids for off-road vehicles and environments. Robotics and Autonomous Systems 60, 278–287 (2012)
9. Llorens, J., Gil, E., Llop, J., Queraltó, M.: Georeferenced LiDAR 3D Vine Plantation Map Generation. Sensors (11), 6237–6256 (2011)
10. Andresen, T., Aguiar, F.B., Curado, M.J.: The Alto Douro Wine Region greenway. Landscape and Urban Planning 68(2-3), 289–303 (2004)
11. Santos, F., Azevedo, J.A.: Mecanização das vinhas na Região Demarcada do douro. O passado, o presente e o futuro. Vititecnica 2, 26–33 (2004)
12. Wilson, J.P.: Digital terrain modeling. Geomorphology 137(1), 107–121 (2012)
13. Contente, O., Aranha, J., Martinho, J., Santos, F., Reis, M.J.C.S., Lau, N., Morais, R.: 3D Digital Maps for Vineyard Autonomous Robot Navigation. In: Advances in Artificial Intelligence - Proceedings EPIA 2013 - XVI Portuguese Conference on Artificial Intelligence, pp. 264–275 (September 2013)
14. Hameed, I.A.: Intelligent Coverage Path Planning for Agricultural Robots and Autonomous machines on Three-Dimensional Terrain. Journal of Intelligent Robot Systems (2013)
15. Arnó, J., Martínez-Cassasnovas, J.A., Ribes-Dasi, M., Rosell, J.R.: Review. Precision Viticulture. Research topics, challenges and opportunities in site-specific vineyard management. Spanish Journal of Agricultural Research 7(4), 779–790 (2009)
16. Peres, E., Fernandes, M.A., Morais, R., Cunha, C.R., López, J.A., Matos, S.R., Ferreira, P.J.S.G., Reis, M.J.C.S.: An autonomous intelligent gateway infrastructure for in-field processing in precision viticulture. Computers and Electronics in Agriculture 78(2), 176–187 (2011)
17. Sistema de Estações de Referência GPS VIRtuais (SERVIR), http://www.igeoe.pt/servir/servir.asp (accessed in April 2013)

New Approach to the Open Loop Control for Surgical Robots Navigation

Pedro M.B. Torres[1,2], Paulo J.S. Gonçalves[1,2], and Jorge M.M. Martins[2]

[1] Polytechnic Institute of Castelo Branco, Castelo Branco, Portugal
{pedrotorres,paulo.goncalves}@ipcb.pt
[2] University of Lisbon, Lisboa, Portugal
jorgemartins@tecnico.ulisboa.pt

Abstract. Robotic and computer-assisted technology are increasingly being developed for use in surgery to aid surgeons in providing more precision and accuracy, especially during procedures requiring fine movements. Moreover, reproducibility is another important aspect to be considered to achieve better results. This paper presents a navigation system developed to assist orthopaedic surgery with application in open loop control of the robots used for femur drilling procedures. Although it may have other applications, the system was thought to be used in Hip Resurfacing (HR) prosthesis surgery to implant the initial guide tool. The navigation system does not require any marker implanted in the patient, unlike current systems. The location of the bone in the operating room is obtained through the information from ultrasound (US) images and Computer Tomography (CT) images, obtained respectively in the intra-operative and pre-operative scenarios. During the surgery the bone position and orientation is obtained through a registration process between the US and CT, using an optical measurement system (Polaris) to measure the 3D position of passive markers attached to the US probe and to the drill. The system description, image processing, calibration procedures and results with real experiments are presented and described to illustrate the system in operation.

Keywords: Image-guided surgery, Hip Arthroplasty, Registration, Robotic surgery.

1 Introduction

To increase the performance of surgical procedures, physicians are increasingly looking for equipment to assist them. Especially when surgeons want to perform complex surgeries with high precision, minimizing the invasiveness of surgical procedures and to improve patient outcomes, by reducing complications and improving patient safety. For the past decades, medical robot and computer-assisted surgery (CAS) [1] have gained increasing popularity, to help physicians to overcome these challenges. Surgical robots provide a significant help in surgery, mainly for the improvement of positioning accuracy and particularly for intra-operative image guidance. In orthopaedics, have emerged several

© Springer International Publishing Switzerland 2015
A.P. Moreira et al. (eds.), *CONTROLO'2014 - Proc. of the 11th Port. Conf. on Autom. Control*,
Lecture Notes in Electrical Engineering 321, DOI: 10.1007/978-3-319-10380-8_60

robotic solutions such as ROBODOC [2], for the planning and performance of total hip replacement, Acrobot [3] Sculptor used in knee surgery, or the RIO robotic arm (MAKOplasty) [4], the latest developed system for orthopaedics. In this guideline, surges the HipRob, which is a robotic system for orthopaedic surgery, with application in Hip Resurfacing (HR) [5] prosthesis surgery. Hip resurfacing arthroplasty is a bone-preserving procedure that helps restore comfort and function to patients hips damaged by degenerative joint disease (osteoarthritis rheumatoid arthritis and traumatic arthritis) avascular necrosis or developmental hip dysplasia. It is viewed as an alternative to traditional hip replacements for helping patients return to their active lifestyles. In HR surgery, the femur is prepared by using an alignment guide. A guide wire ensures that the spherical metal cap is positioned correctly on the femoral head. The success of the surgery depends on the correct positioning of the guide wire. Currently this process is made from a very time consuming mechanically procedure. The alignment is made almost by trial and error. The goal of HipRob is automate this task and perform the initial drilling, with the aid of a robot, operating in co-manipulation by the surgeon. Ultrasound images is used for non-invasive real-time bone tracking. For navigation, the system, during surgery, acquires a 3D US bone surface from a sequence of US images. This bone surface will then be registered to the pre-operative bone model, for a precise knowledge of the bone position and orientation. This registration is performed in two steps. The first, a global registration before the surgical procedure, to exactly register the bone. The second is to locally register the femur, which is faster and more suitable for tracking the bone movements. The measured bone movement is used by the robot manipulator to update its drilling position and orientation. Figure 1 shows an overview of the HipRob navigation system.

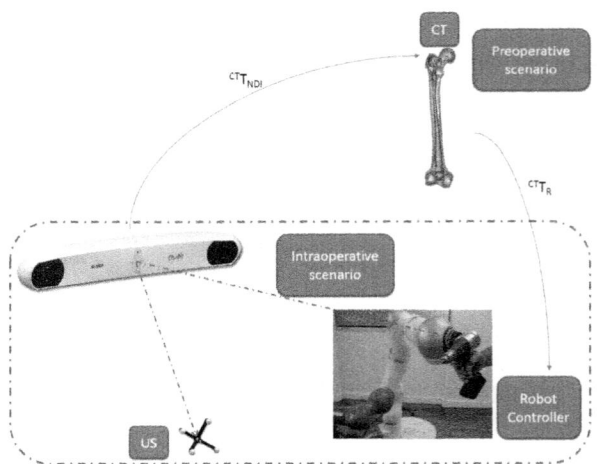

Fig. 1. Overview of the hardware setup and frames transformations of the HipRob navigation system

This paper presents the algorithms and software tools developed for US based orthopaedic surgery navigation, improving the results presented in [6].

2 Bone Tracking Based on Ultrasounds

As there is no fiducial markers on the bone, all movements during the surgery are recorded based on the information extracted from the US images, in other words, the Bone Tracking during the intra-operative scenario is based on contour features extracted slice by slice. The avoidance of fiducial markers implies that the US probe is referenced to the optical measurement system (NDI - Polaris). This referencing is made by placing on marker in the US probe. From this moment, also the images and features are referenced to the tracker. In this section it is proposed a method to track the femur in a sequence of 2D US images. This method allows the 3D real-time measurement enabling computer-assisted surgery for Hip Resurfacing Surgery. After performing the necessary initialization settings for the Optical Tracking System and the image processing algorithms configurations, US images are acquired trough a USB video frame grabber and using methods implemented in OpenCV. All images are acquired in PAL format with resolution of 720×576 pixels at a frame rate of 30 fps. In the acquisition of first image is defined by the surgeon the region of interest (ROI), that includes the bone. This ROI is marked with a square around the bone, as seen in figure 2. This procedure improve accuracy and also speed up the segmentation.

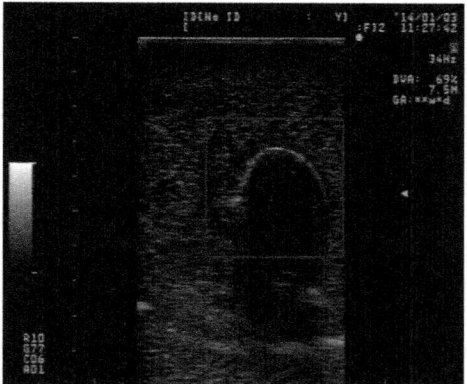

Fig. 2. Region of interest in an US image of the femur

In the next task is ordered to start the NDI Polaris acquisition and synchronize the tracking process. Is read the position and orientation of the US probe and the extremity of the drill. The synchronism between US images and Opto Tracker is established in the acquisition of the first image by marking a point within the bone, the same point is used as the seed of the segmentation algorithm. All images are processed in order to extract the bone surface and to enable bone tracking, based on US images. At this stage, are applied methods to denoise the images and after, to perform bone segmentation, based on the *FastMarching* algorithm [7]. The centroid of the bone contour in each segmentation result is used as input in the image segmentation of the next iteration, i.e., acquired US image. After extracting the 2D contour of each image the upper contour of the bone is identified, extracted the correspondent pixel coordinates and, using the 3D position and orientation of the US probe, obtained the corresponding 3D points. The position and orientation of the drill is sent together with the 3D US point cloud, through different sockets, by UDP to the station that performs the 3D US point cloud registration to the the 3D femur model, obtained pre-operatively, from CT images. This process is continuously repeated until terminated by the surgeon.

3   Registration/Calibration

In a real scenario, inside an operating room, the femur bone must be spatially located, for later identify the point of drilling defined pre-operatively by the surgeon. As the main reference, in the intra-operative scenario, is the referential of the navigation system (POLARIS). This Homogeneous Transformation, relates the intra-operative datasets and the model, CT dataset, needs to be obtained. This transformation ($^{CT}T_{NDI}$) is determined by a initial registration between the CT dataset (Target) and 3D US bone point clouds extracted intra-operatively. With this transformation, all points in the femur are referenced to the CT reference frame. This matrix, obtained using the Iterative Closest Point (ICP) Method [8], is the initial calibration, necessary for the femur location.

3.1   Iterative Closest Point (ICP) Method

The Iterative Closest Point (ICP) method, presented in [8] is the standard method used to perform registration between two set of 3D points. It transforms two sets of points to a common coordinate frame. If the exact correspondences of the two data set could be known, then the exact translation t and rotation R can be found. The main issue of the method is then to find the corresponding points between the two data sets, $Y = (y_1, ..., y_M)^T$ and $X = (x_1, ..., x_M)^T$. The assumption in the ICP method is that the closest points between the data sets correspond to each other, and are used to compute the best transformation, rotation and translation, between them. To obtain the closest point of Y to a point in X, the Euclidean distance is applied:

$$d(Y, X) = \sqrt{(X_x - Y_x)^2 + (X_y - Y_y)^2 + (X_z - Y_z)^2} \qquad (1)$$

When all points of the data set Y are associated to the point in X the transformation is estimated by minimising a mean square cost function:

$$E_{ICP} = \sum_i \|R \cdot x_i + t - y_i\|^2 \qquad (2)$$

From the obtained parameters, the points in the X data set are transformed and the error between them and the ones in Y calculated. If the error is above a pre-defined threshold then the points must be re-associated and the previous steps again performed until the error is below the threshold.

4   Robot Navigation

For a correct and accurate navigation of the robot during the surgical procedure is very important that the navigation system can identify and track the femur movements and transmit them to the robot, to perform a correct update of the target and as soon as possible. Since this procedure takes place in the intra-operative scenario, all processing times have to be minimalists, but ensuring accuracy. To achieve these goals and since in the surgical procedure the femur does not change its position drastically, a local registration is performed on-line to update the target, based on the femur movements. The ideal drilling point, is estimated pre-operatively, but needs to be spatially located within the intra-operative scenario, in order to be updated in case of movement of the femur. The drilling point in the CT referential is calculated at the beginning of the surgery, performed after the intra-operative calibration ($^{CT}T_{NDI}$) and the determination of drilling point by the surgeon in the NDI referential frame ($^{NDI}P_{drill}$), according to equation 3.

$$^{CT}P_{drill} = {}^{CT}T_{NDI} \times {}^{NDI}P_{drill} \tag{3}$$

The same point, but calculated in the robot reference frame, considering the movements that may exist in the femur is given by equation 4.

$$(^{ROB}P_{drill})_k = (^{NDI}T_{ROB})^{-1} \times (^{CT}T_{NDI})^{-1} \times (T_R)_k \times {}^{CT}P_{drill} \tag{4}$$

$(T_R)_k$ is the homogeneous matrix that represent the transformation obtained in the on-line local registration that updates the calibration to compensate the femur movements. The robotic system works on variable impedance control for physical surgeon-robot interaction and the real-time bone tracking is performed in open-loop, as described in figure 3. $^{NDI}T_{ROB}$ represents the drill position measure by the tracker and I, the US image used to track the bone movements.

5   Experimental Setup

The experimental setup for Ultrasound Based Robot Navigation is composed of three workstations, an optical measurement system (NDI Polaris Spectra), a portable ultrasound system with a 7.5 MHz linear probe and a USB video frame grabber, as illustrated in figure 4. The vision-oriented software for bone tracking has been developed in C++ on NetBeans environment running on PC1 under OS

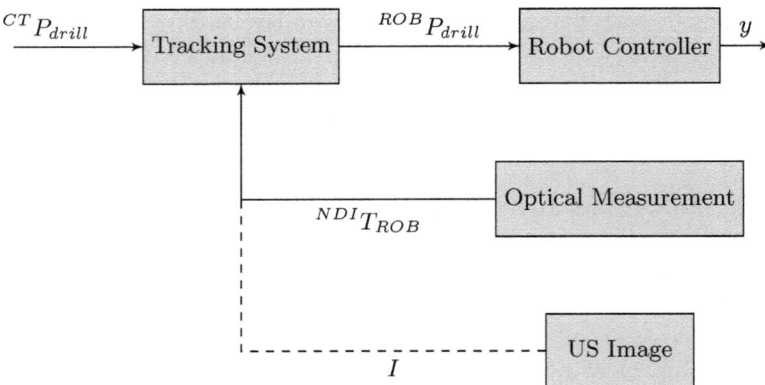

Fig. 3. Block diagram of the open-loop robot control

X operating system. This computer is connected to the NDI Polaris and frame grabber by USB. The calibration, registration and Navigation applications, run on computer PC2 under Ubuntu Linux operating system. Both applications, calibration and Navigation, have been developed in C++ and is used the Point Cloud Library [9] for 3D point cloud processing, registration and visualization. Computer PC3 receives the updates performed in navigation system and implements the trajectory planning for real-time robot control. All computers are connected by Ethernet and communicate via UDP Protocol.

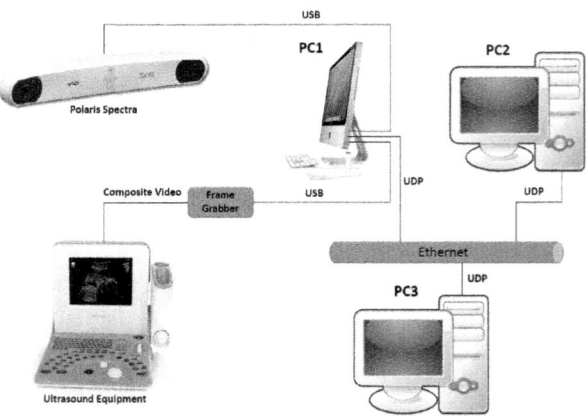

Fig. 4. Experimental Setup

To perform experiments of robot positioning in real scenario (Fig. 5), it was constructed a femur phantom with similar characteristics to a human femur. At the beginning of the experiments it is necessary to define the image scale factors, according to the image plane in which acquisitions will be made. A correct procedure goes through the femur scan with US probe, identify the desired zone and adjust the image plane, probe frequency, brightness and focus. For the acquisitions presented in this section, all images were acquired with an image plane of 10 [cm] (R10), probe frequency 7.5 MHz, as evidenced by the figure 2, which results in scale factors Sx and Sy equal to 0.2858 [mm]. Another important task to perform is to calibrate the system, in order to obtain the $^{CT}T_{NDI}$ transformation matrix. The matrix, automatically enters as input into the navigation software. This procedure takes an average of 5 minutes to be carried out, distributed by Acquisition and registration/calibration. The final task to conclude calibration and initializations, passes for finding the drill tool-tip offset, in order to reference the drill extremity, to the robot wrist. This is referenced to the Polaris system according to the matrix 5, for all experiments. After performing the initializations / calibrations previously described, the surgeon or an assistant makes constant scans with the US probe along the femur. Another assistant starts the image processing station, define the ROI and run the application. Starting from this moment, the system is completely autonomous. It processes sets of 10 slices (US images), to obtain a 3D US point cloud, in order to perform the local registration and the drilling point position and orientation, which is given to the robot controller, as a set-point.

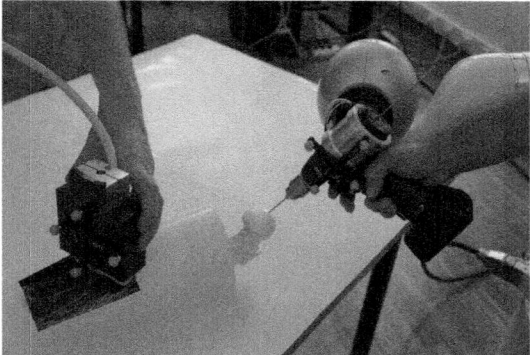

Fig. 5. The Real Scenario of the femur phantom, the Us probe, and the drilling point in the femur head

$$^{NDI}T_{Mk} = \begin{pmatrix} ^{NDI}R_{3\times3} & ^{NDI}t_{3\times1} \\ 0_{3\times3} & 1 \end{pmatrix} \times \begin{pmatrix} 1 & 0 & 0 & -46,82 \\ 0 & 1 & 0 & 0,74 \\ 0 & 0 & 1 & -113,49 \\ 0 & 0 & 0 & 1 \end{pmatrix} \tag{5}$$

where $^{NDI}T_{Mk}$ represents the Homogeneous matrix returned by the tracker with position of markers. These offsets correspond only to the marker placed on the drill.

6   Experimental Results

This section presents some of the results obtained from numerous experiments performed with this setup. Figure 6, illustrate the registration process, in order to determine the calibration matrix $^{CT}T_{NDI}$. The target is the 3D CT femur bone, represented in green, the red point clouds corresponds to the 3D US femur points extracted with Polaris through a scan with the US probe in the central part of the femur. The black is the registered point cloud. Hausdorff Distance, is the metric used to measure the registration errors. This calibration process, is achieved with a registration error of $0.756[mm]$. The bone movements are identified by registration sets of 10 US images. The average update time between registrations is 0.32 seconds. The registration errors obtained in this task are presented in table 1, calculated in 47 experiments.

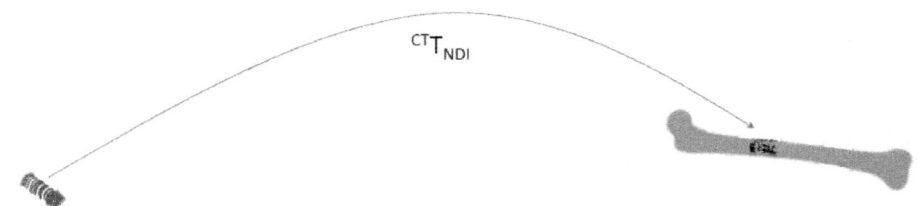

Fig. 6. The registration process - calibration. 3D CT point cloud (green). US point cloud, before registration (red). US point cloud, after registration (black).

Table 1. Results before and after the on-line Registration Process

	Before Registration $H_{dist}[mm]$	After Registration $H_{dist}[mm]$	Time $[s]$
Registration results	1.0667	0.7936	0.32

$$^{NDI}P_{drill} = (^{CT}T_{NDI})^{-1} \times (T_R)_k \times^{CT} P_{drill} \qquad (6)$$

$$\varepsilon = \bigtriangledown^{NDI}P_{drill} - \bigtriangledown^{NDI}T_{ROB} \qquad (7)$$

Figure 7 represents both the drilling point ($^{NDI}P_{drill}$) and the robot associated motions ($^{NDI}T_{ROB}$), measured by the tracker during an experiment, represented in Cartesian coordinates.

$^{NDI}P_{drill}$ is calculated according equation 6 and $^{NDI}T_{ROB}$ is directly measured by the tracking system. In the several experiments performed, the system responded stably, the robot followed the femur movements, randomly performed. The associated motion errors were calculated according equation 7, and presented in table 2.

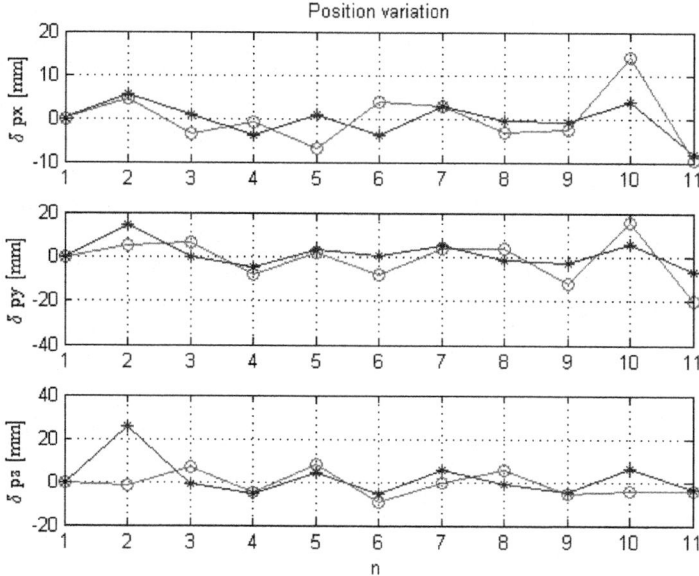

Fig. 7. Bone movements (Red) and robot movements (Blue) during an experiment, represented in XYZ coordinates

Table 2. Errors between the drilling point variation and the corresponding variations of the robot, in the tracker reference frame

	$\overline{norm}$	$\overline{\delta x}$	$\overline{\delta y}$	$\overline{\delta z}$
Position [mm]	10.96	0.22	-2.17	-2.74
Orientation [degree]	18.71	2.81	0.30	-2.39

Table 3. Movements of the drilling point

	$\overline{norm}$	Std	Max	min
Position [mm]	7.64	3.56	14.60	1.46
Orientation [degree]	3.23	2.35	16.41	0.69

Table 3 presents the tracking results obtained in 47 validated experiments. The errors were calculated in the representation of vector / angle, considering the current position and the moved drilling point at each instant, in the robot frame referential ($^{ROB}P_{drill}$).

7   Conclusions and Future Work

This paper described a new approach to the open loop control for surgical robots navigation with application in Hip Resurfacing surgery. The navigation system is based entirely on the information extracted from images obtained from CT pre-operatively and US intra-operatively. Contrary to current surgical systems, it do not use any type of implant in the bone, to track the femur movements. The tracking is performed in real time by registration of 3D US points with the femur model (CT). A KUKA lightweight robot was used to validate the applicability of the tracking system in the experimental tests carried out on a femur phantom. In the robot motion planning was defined a maximum threshold of 10 mm / 5 ° for the admitted error in the drilling point updating. i.e. whether the error was greater than this limit, the robot does not updates its position. In most experiments the update was always validated, with errors of 2 mm / 5°. For future work, it is expected to diminish the image processing and registration errors, and to improve the robustness of the system.

References

1. Adams, L., Krybus, W., Meyer-Ebrecht, D., Rueger, R., Gilsbach, J., Moesges, R., Schloendorff, G.: Computer-assisted surgery. IEEE Computer Graphics and Applications 10(3), 43–51 (1990)
2. Kazanzides, P., Mittelstadt, B., Musits, B., Bargar, W., Zuhars, J., Williamson, B., Cain, P., Carbone, E.: An integrated system for cementless hip replacement. IEEE Engineering in Medicine and Biology Magazine 14(3), 307–313 (1995)
3. Cobb, J., Henckel, J., Gomes, M., Barrett, A., Harris, S., Jakopec, M., Rodriguez, F., Davies, B.: Hands-on robotic unicompartmental knee replacement: a prospective, randomised controlled study of the acrobot system. The Journal of Bone and Joint Surgery B 88(2), 188–197 (2006)
4. Pearle, A.D., Kendoff, D., Stueber, V., Musahl, V., Repicci, J.A.: Perioperative management of unicompartmental knee arthroplasty using the mako robotic arm system (makoplasty). American Journal of Orthopedics 38(2), 16–19 (2009)
5. De Smet, K., Campbell, P., Van Der Straeten, C.: The Hip Resurfacing Handbook A Practical Guide To The Use and Management of Modern Hip Resurfacings. Woodhead Publishing Series in Biomaterials (2013)
6. Gonçalves, P., Torres, P., Santos, F., Antonio, R., Catarino, N., Martins, J.: A vision system for robotic ultrasound guided orthopaedic surgery. Journal of Intelligent and Robotic Systems, 1–13 (2014)
7. Sethian, J.: Level Set Methods and Fast Marching Methods. Cambridge University Press (1999)
8. Besl, P.J., McKay, N.D.: A method for registration of 3-d shapes. Pattern Analysis and Machine Intelligence 14(2), 239–256 (1992)
9. Rusu, R., Cousins, S.: 3D is here: Point cloud library (PCL). In: 2011 IEEE International Conference on Robotics and Automation (ICRA), pp. 1–4 (2011)

Streaming Image Sequences
for Vision-Based Mobile Robots

Andry Maykol Pinto, António Paulo Moreira, and Paulo Gomes Costa

Centre for Robotics and Intelligent Systems of INESCTEC and Faculty of Engineering,
University of Porto, Rua Dr. Roberto Frias, 4200-465 Porto, Portugal
{andry.pinto,amoreira,paco}@fe.up.pt

Abstract. Vision-based mobile robots have severe limitations related to the computational capabilities that are required for processing their algorithms. The vision algorithms processed onboard and without resorting to specialized computing devices do not achieve the real-time constraints that are imposed by that kind of systems.

This paper describes a scheme for streaming image sequences in order to be used by techniques of artificial vision. A mobile robot with this architecture can stream image sequences over the network infrastructure for a device with higher computing power. Therefore, the robot assures the real-time performance with a reduced consumption of energy which increases its autonomy.

The experiments conducted without using specialized computers proved that the proposed architecture can stream sequences of images with a resolution of 640x480 at 25 frames per second.

Keywords: Mobile Robotics, Streaming, Real-Time, Onboard computers, Image processing.

1 Introduction

Vision technology is non-intrusive, does not require the modification of the environment and acquires information in a similar form as the human vision, which makes it appealing for mobile robotics applications. Computer vision is challenging in small mobile robots because of the vehicle itself. The first issue is related to the space that is available onboard for the deployment of computer units and sensors since they are usually placed in a very limited space. Other issue is related to the consumption of power of such robots because it limits their computational capability, which has a direct influence in the performance of the navigation and sensing procedures. However, most of these procedures cannot achieve the real-time constraints that are imposed by mobile systems without resorting to specialized computers.

The digital consumer camera of today is affordable [1] and makes it possible to extract relevant information using computer vision techniques although the image resolution and the complexity of the image-processing algorithms are restricted by the computational power that is available on the embedded computers. One possible approach to deal with this problem is to increase the capability of the embedded

A.P. Moreira et al. (eds.), *CONTROLO'2014 - Proc. of the 11th Port. Conf. on Autom. Control*,
Lecture Notes in Electrical Engineering 321, DOI: 10.1007/978-3-319-10380-8_61

computer at the expense of cost and power consumption. Specialized hardware resources are usually employed for the onboard processing of vision-algorithms since they fulfill the demands for real-time. In some cases, specialized computer devices cannot be used due to the small size of vehicles or because they cause a high consumption of energy that reduces the autonomy of robots. Thus, embedded processing units with real-time capability for complex algorithms of artificial vision are very expense and energy-inefficient for mobile robots.

Another option is to stream image sequences over the network for a device with higher computational power. This external device sends the result back to the robot which decreases the computation power of the embedded computer unit that is required for the robot, while maintaining the real-time demand. This approach can be applied if the network infrastructure is available on the environment where the robot will operate, which is true for most of indoor environments. This last approach is often considered in robotics however, the majority of streaming protocols that are currently available, for instance, RTSP (Real Time Streaming Protocol), RTP (Real-time Transport Protocol) / RTCP (Real-time Transport Control Protocol) and RSVP (Resource reSerVation Protocol), use compression algorithms such as H264. Video compression algorithms reduce the quantity of information that is send over the network making possible the achievement of a high FTR (frame transmission rate); however, they cause severe visual artifacts that corrupt the images. This causes severe problems during the processing of these images, for instance, in three-dimensional reconstructions, registration, classification of objects, motion segmentation and analysis, tracking, identification and recognition of humans [2].

This paper presents an architecture focused in streaming image sequences for mobile robots with real-time constraints. The idea behind this architecture is that, image sequences can be streamed from an embedded computer (in the robot) to an external computer. This external device receives each frame and conducts image processing algorithms. Afterwards, the external computer sends the result back to the robot. Of course, this article assumes that a network infrastructure is available.

The streaming architecture that is proposed in this article is named SIS4RS (Streaming Image Sequences for Robotic Systems) and it has several advantages, for instance, sequences are streamed with high frame rates (~15 to 25 frames per second, fps) in domestic IEEE 802.11n networks where each image has a resolution of 640x480.

In this article, the JPEG (Joint Photographic Expert Group) compression is used for compressing each camera raw image and before sending it over the network. JPEG is a good image compression method since it is possible to control the quality of the compressed image (compression factor, CF) and introduces only a few visual artifacts. Experiments presented in this article show that the JPEG coding/decoding has a low impact in the perceptual quality of the resulting image. Thus, this image compression is satisfactory for robot-vision algorithms. Therefore, the SIS4RS avoids conventional video compressing algorithms since they usually resort to buffering images in order to increase the end-user experience. However, buffering images introduces latency which compromises the time requirements for the whole perception–action cycle. Results show that SIS4RS architecture has virtually no latency since the broadcast delay is less than 60 milliseconds.

The contributions of this paper include:

1. An efficient architecture for streaming image sequences. The proposed SIS4RS is suitable for devices with low computational power;
2. The architecture achieves a FTR of ~25fps for 640x480 images in domestic IEEE 802.11n networks at medium distance to the access point;
3. A study of the visual artifacts in the resulting image that are caused by the compression (JPEG coding/decoding) and transmission over the network;
4. An extensive qualitative and quantitative evaluation under realistic working conditions.

The article is organized as follows. Section 2 presents a brief review of the related works. Section 3 shows the concept and the operation diagram for the SIS4RS. Later, a set of experiments is conducted in order to evaluate the performance of the SIS4RS under realistic surveillance videos (section 3). Finally, major conclusions of this research are presented in section 4.

2   Related Works

Many researches related to video streaming have dealt with wireless video streaming and, more recently, the research is being focused on the development of wireless sensors for streaming video in low-bandwidth networks [4]. However, the majority of the works report several issues related to the frame transmission rate [4], energy consumption [9] and computational power [5].

The work [6] proposed the H.264/MPEG4 compression scheme and the low-latency best-effort transport mechanisms to achieve a low latency video transmission. A client/server topology is presented in [5]; however, the work is focused on sensor nodes and not in streaming images for robotic applications. Some frames are prioritized over others based on their context in [7] and [8], since they defend the concept of context adaptive frames.

Computer vision techniques for robotic applications are computationally more efficient although, this improvement is usually done at the expense of using lower resolution images (320x240). Thus, some authors [9,10] prefer onboard processing for their algorithms however, they use images with low resolution and specialized computer devices.

3   Streaming Image Sequences for Robotic Systems (SIS4RS)

Two operational objects, server and client, form the Streaming Image Sequences for Robotic Systems (SIS4RS). The communication between the server and client is conducted using two distinct port numbers, see Fig. 1, and the messages transmitted in each port number are described in Fig. 4. A mobile robot with the SIS4RS architecture can send sequential images over the network and for an external device that conducts vision-based algorithms and sends the result back to the robot. Therefore, the SIS4RS assures the fulfillment of the time requirements that are imposed by dynamic systems. The User Datagram Protocol (UDP) is considered in this article as the network transport protocol since it is a more time-effectively and simple layer than the TCP (Transmission Control Protocol).

The server/client architecture that is detailed in Fig. 2 and Fig. 3, aims to balance the computational efficiency and the performance of streaming image sequences in generic computer units (for instance, embedded ARM devices). In both schemes, the dark blue represents network procedures where the server/client receives information from the client/server, the light blue represents procedures where the server/client sends packets to client/server and the green means internal procedures with no network interaction.

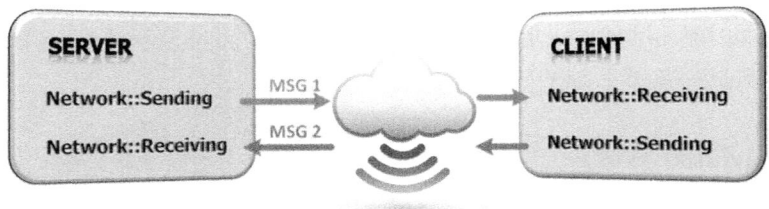

Fig. 1. SIS4RS - Communication interaction between the server and the client

The server's operational scheme is depicted in Fig. 2. Three important features are implemented in the server: decomposition of the image into several sequential packets (packing), the re-sending of all packets that were not successfully received by the client and a timeout mechanism.

These features compensate the limitations of the UDP and assure a real-time communication between server and client. The UDP provides no reliability because it sends the datagrams that the server writes to the IP (Internet Protocol) layer, but there is no guarantee that they ever reach their destination. In addition, the server needs to worry about the size of the resulting IP datagram. If it exceeds the MTU (Maximum Transmission Unit) of the network, the IP datagram is fragmented. Therefore, a packing procedure decomposes the encoded image into several sequential packets (with size equal or lower than the MTU).

As can be seen in Fig. 2, the server grabs an image when all packets of the last frame were successfully received by the client or instead, when a timeout occurs (60 milliseconds). The timeout assures that the server will spent only a pre-defined time for sending (step 5) or re-sending (step 7) the packets associated to a single frame. This means that, if there are packets to be sent at the end of the timeout then they will be disregarded and the entire process restarts. A new raw image is grabbed from the camera device and encoded using the JPEG encoding. The resulting frame is decomposed, afterwards, by taking into consideration the MTU of the network and then, the packets are sequentially broadcasted. The server is listening for messages of type (see Fig. 4) that represent packets that were not received during the broadcasting (step 6). Finally, the server receives a message of type 2 with iFrame + 1 and idxPacket = 0 when the client receives all packets (transition between step 6 and 2). Where iFrame is the index of the current frame that was sent and idxPacket is the index of the packet that should be sent by the server.

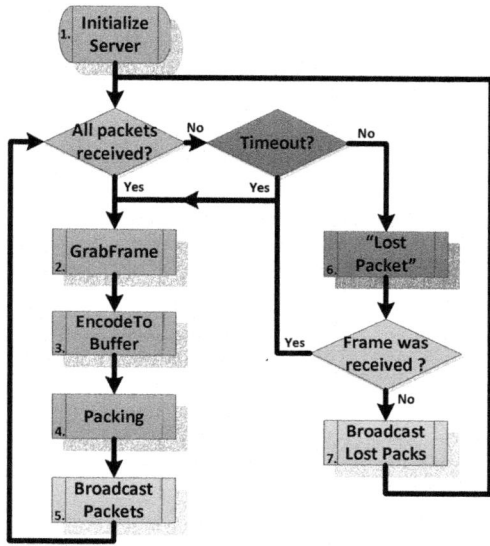

Fig. 2. Operational scheme of the server. The dark blue phase (6) represents a network procedure where the server receives information from the client, light blue phases (5 and 7) represent procedures where the server sends packets to the client and the green phases (1, 2, 3 and 4) are internal procedures.

Fig. 3 represents the logical scheme of the client. The client prepares for the new frame in step 2 and waits for type-1 messages in step 3 with the same iFrame and sequential packet indexes (idxPacket). The client detects the lost packets when indexes (idxPacket) of two consecutive packets that were successfully received are not sequential. Information about lost packets is sent over the network and to the server in step 4. If all packets were received, a JPEG-decoding process is used to obtain the raw image. The decoded-image of step 6 is saved and the client starts a new cycle when it sends a type-2 message with iFrame + 1 (step 5).

Messages transmitted by the server (type-1) and the client (type-2), are depicted in Fig. 4. Both messages have an initiator and terminator character. The type-1 message has an identifier header with 14 bytes where iFrame is the index of the current frame, idxByteFrame is the initial position in bytes of the transmitted packet relatively to the memory buffer that contains the encoded frame, TotalByteFrame is the number of bytes of the encoded frame, idxPacket is the index of the packets, TotalPackets is the number of packets that create the encoded frame, #BytesData is the size of the body Data in bytes and Data is a piece of the encoded frame. A type-2 message is sent from the client to the server and the message has an identifier header with 4 bytes (iFrame and idxPacket) and no Data field. This message is used to announce the packets (idxPacket) that were not received and to alert the server that the entire frame was received (iFrame +1 and idxPacket = 0).

1. Initialize Client
2. New Frame
3. Receive Packet
Sequential Pack?
No
Is New Frame ?
Yes
No
Yes
4. Send Packs Lost Info
No
Packets received?
Yes
5. Send "iFrame +1"
6. DecodeFromBuffer

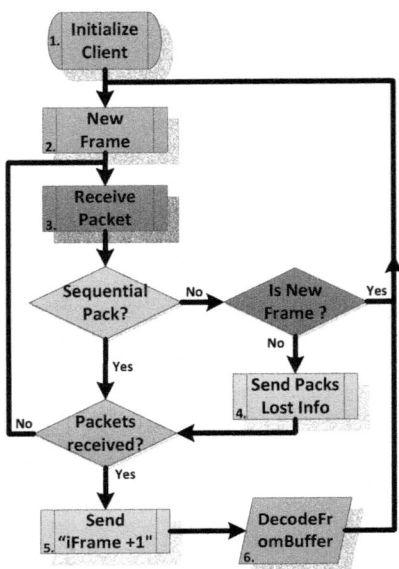

Fig. 3. Operational scheme of client. The dark blue phase (3) represents the network procedure where the client receives sequential packets from the server; the light blue phase (4 and 5) represents a procedure where the client sends messages to the server and the green phases (1, 2 and 6) are internal procedures.

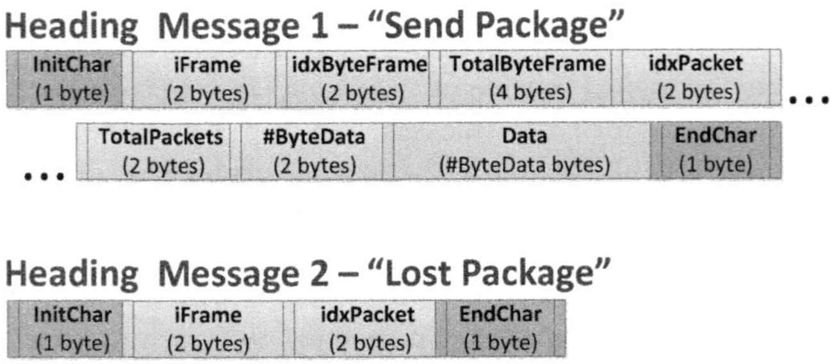

Heading Message 1 – "Send Package"
InitChar (1 byte)
iFrame (2 bytes)
idxByteFrame (2 bytes)
TotalByteFrame (4 bytes)
idxPacket (2 bytes)
...
...
TotalPackets (2 bytes)
#ByteData (2 bytes)
Data (#ByteData bytes)
EndChar (1 byte)
Heading Message 2 – "Lost Package"
InitChar (1 byte)
iFrame (2 bytes)
idxPacket (2 bytes)
EndChar (1 byte)

Fig. 4. SIS4RS - Messages transmitted between the server and the client. Message 1 is sent from the server to the client. Message 2 is sent from the client to server and is used to detect packets regarding to a specific frame that were not received by the client.

4 Results

Extensive experiments were conducted as part of this research. The SIS4RS architecture is implemented in C++ using the OpenCV library (version 2.3.1) and the

Qt library (version 4.8.1). All the results were obtained with an I3-M350 2.2GHz for the client and a Dual-core ARM® Cortex™-A9 MPCore™ at 1.2 GHz for the server.

Experiments to evaluate the perceptual losses on image sequences were conducted and are presented in section 4.1. In these experiments, the original raw image of the server is compared to the decoded image in the client using several evaluation metrics [2] namely, the root mean square error (RMSE), the peak signal-to-noise ratio (PSNR) and the structural similarity (SSIM).

The results presented in this research aim to provide a reliable characterization of the performance and the computational cost involved for streaming sequences using the SIS4RS architecture. The SIS4RS was tested in a realistic surveillance scenario [3] and with a domestic IEEE 802.11 wireless network (Router D-Link DIR-600). These results are presented in section 4.2.

4.1   Visual Quality: Assessment Metrics

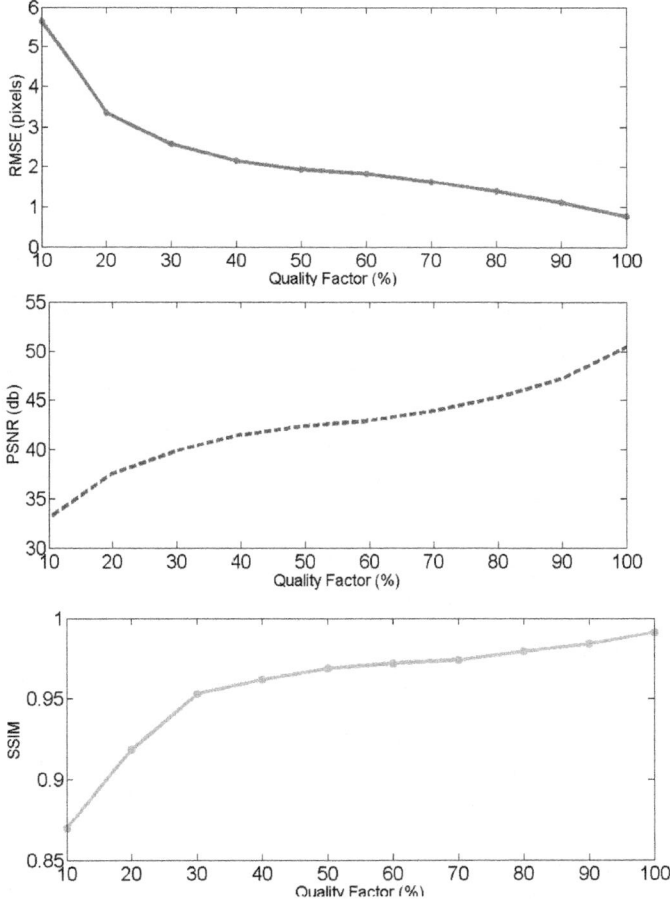

Fig. 5. Quality assessment of the transmission as a function of the quality factor (JPEG encoding). Figures on top, middle and bottom are graphical representations of the performance evolution using RMSE, PSNR and SSIM assessment metrics.

In these graphical representations, the influence of the quality factor (QF) of compression is evaluated taking into consideration its impact in the visual appearance of the image transmitted. As expected by decreasing the quality factor, the image received in the client has more visual artifacts. These artifacts are mainly caused by the compression. The perceptual quality of image sequences transmitted using the SIS4RS with a QF = 60% is about 1.8 pixels (RMSE), 42db (PSNR) and 0.965 (SSIM). For a QF = 90%, the RMSE, PSNR and SSIM values are 1.2 pixels, 47.5db and 0.985. Therefore, Fig. 5 shows an evolution with a linear and small growth rate for values of QF between 60% and 90%, which results in a transmission with a small perceptual loss. This behavior is helpful for defining the frame transmission rate that is desirable for streaming image sequences because the visual degradation of the transmission is not significantly affected by QF values on that interval.

4.2 Transmission Rate

Another relevant feature for streaming sequences is the payload of the transmission, which in this case is also represented as a function of the QF in Fig. 6.

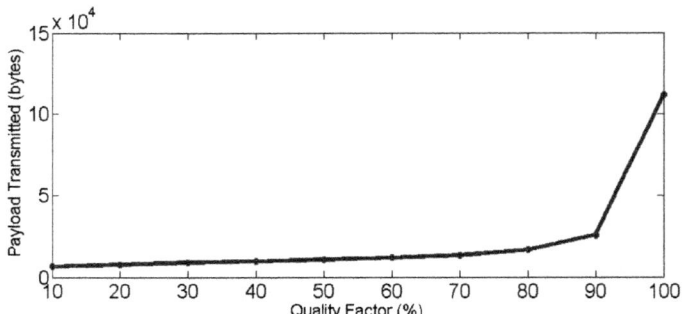

Fig. 6. Payload transmitted as function of the quality factor

As can be noticed, the payload is drastically reduced with the decreasing of QF. The payload for a QF = 60% and QF = 20% is about 1.2×10^4 bytes and 1×10^4 bytes respectively. Most of the reduction occurs from QF=100% to QF =90% since the payload decreases from 12×10^4 bytes to 2×10^4 bytes. This represents a payload reduction of 83% just in 10% of QF difference which results in a perceptual loss of 0.3 pixels (RMSE), 3db (PSNR) and 0.01 (SSIM).

Therefore, the JPEG compression reduces the quantity of information transmitted over the network while preserving the visual quality of each image. This result is demonstrated by selecting a QF value on the interval 60% to 90%, see Fig. 5 and Fig. 6. The visual perceptual quality (SSIM index) in this interval is 0.96 and the payload is practically the minimum case.

Finally, the maximum frame transmission rate (FTR) was analyzed for different QF values and results are presented in Fig. 7. The server is running in a Dual-core ARM® Cortex™-A9 while streaming images in a domestic wireless network (with 5.3 milliseconds of latency in average).

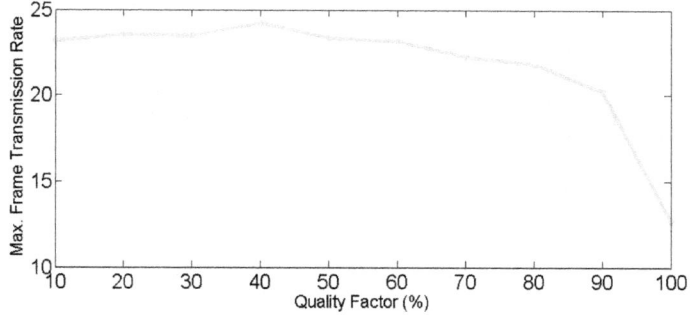

Fig. 7. The maximum frame transmission rate as function of the quality factor

Obviously, increasing the QF leads to lower FTR since the payload is higher (more information to be sent over the network). The graphical representation of the frame transmission rate depicted in Fig. 7 proves that the SISRS can stream sequences of images with 640x480 at ~25fps. Actually, the FTR is close to 25fps for quality factors on the interval 10% to 60%. The FTR decreases less than 4fps to 21fps from QF=60% to QF = 90%; however, it achieves the lower value (12fps) in the highest quality factor.

Better results are expected when using embedded computer devices for running the server scheme of (Fig. 2) with higher computational power because most of the cycle time (~75%) is spent in grabbing and encoding the frame, stages 2 and 3. Other possible solution is to running the grabbing stage in parallel which removes a significant amount of time in the main cycle.

5   Conclusion

This research presents an architecture focused in streaming image sequences from mobile robots and with real-time constraints. Sequential images are streamed from an embedded computer in the robot and to an external computer. This external device receives each frame and conducts image processing algorithms which, afterwards, it sends the result back to the robot.

The streaming architecture that is proposed in this article is named SIS4RS (Streaming Image Sequences for Robotic Systems). It can stream sequential images with a resolution of 640x480 and with high frame transmission rates (~25fps). Moreover, it was designed to be incorporated in embedded computers with a low computational capability and a low consumption of energy.

Experiments conducted with a Dual-core ARM® Cortex™-A9 and in domestic wireless networks have proven that the SIS4RS achieves a frame transmission rate up to 25fps. Moreover, the perceptual loss during the transmission is low. Therefore, the SIS4RS is suitable for small robotic applications equipped with generic computers.

Acknowledgment. This work was partially funded by the Portuguese Government through the FCT - Foundation for Science and Technology, SFRH-BD-70752-2010.

In addition, this work is also financed by the ERDF – European Regional Development Fund through the COMPETE Programme (operational programme for competitiveness) and by National Funds through the FCT within project «FCOMP - 01-0124-FEDER-022701».

References

1. Bräunl, T.: Embedded robotics: mobile robot design and applications with embedded systems. Springer (2008)
2. Pinto, A.M., Costa, P.G., Correia, M.V., Moreira, A.P.: Enhancing dynamic videos for surveillance and robotic applications: The robust bilateral and temporal filter. In: Signal Processing: Image Communication (2013),
   http://dx.doi.org/10.1016/j.image.2013.11.003
3. Pinto, A.M., Moreira, A.P., Correia, M.V., Costa, P.G.: A Flow-based Motion Perception Technique for an Autonomous Robot System. Journal of Intelligent & Robotic Systems (2013), http://dx.doi.org/10.1007/s10846-013-9999-z
4. Wark, T., Corke, P., Karlsson, J., Sikka, P., Valencia, P.: Real-time Image Streaming over a Low-Bandwidth Wireless Camera Network. In: International Conference on Intelligent Sensors, Sensor Networks and Information, vol. 3(6), pp. 113–118 (2007)
5. Chen, P., Ahammad, P., et al.: CITRIC: A low-Bandwidth Wireless Camera Network Platform. In: ACM/IEEE International Conference on Distributed Smart Cameras, pp. 1–10 (2008)
6. Miu, A., Apostolopoulos, J.G., Tan, W.-T., Trott, M.: Low-Latency Wireless Video Over 802.11 Networks Using Path Diversity. In: IEEE International Conference on Multimedia and Expo (ICME) (2003)
7. Ozcelebi, T., Vito, F.D., et al.: An Analysis of Constant Bitrate and Constant PSNR Video Encoding for Wireless Networks. In: IEEE International Conference on Communications, pp. 5301–5306 (2006)
8. Liu, T., Choudary, C.: Content-Aware Streaming of Lecture Videos over Wireless Networks. In: Proceedings of the IEEE Sixth International Symposium on Multimedia Software Engineering, pp. 458–465 (2004)
9. Lee, D.-S.: Effective gaussian mixture learning for video background subtraction. IEEE Transactions on Pattern Analysis and Machine Intelligence 27(5), 827–832 (2005)
10. Jung, B., Sukhatme, G.S.: Detecting moving objects using a single camera on a mobile robot in an outdoor environment. In: 8th Conference on Intelligent Autonomous Systems, pp. 980–987 (2004)

Visual Signature for Place Recognition in Indoor Scenarios

Filipe Neves dos Santos, Paulo Cerqueira Costa, and António Paulo Moreira

INESC TEC - INESC Technology and Science (formerly INESC Porto)
and FEUP - Faculty of Engineering, University of Porto,
Campus da FEUP Rua Dr. Roberto Frias, 378 4200 - 465 Porto, Portugal
{fbnsantos,paco,amoreira}@fe.up.pt

Abstract. Recognizing a place with a visual glance is the first capacity used by humans to understand where they are. Making this capacity available to robots will make it possible to increase the redundancy of the localization systems available in the robots, and improve semantic localization systems. However, to achieve this capacity it is necessary to build a robust visual signature that could be used by a classifier. This paper presents a new approach to extract a global descriptor from an image that can be used as the visual signature for indoor scenarios. This global descriptor was tested using videos acquired from three robots in three different indoor scenarios. This descriptor has shown good accuracy and computational performance when compared to other local and global descriptors.

Keywords: Place recognition, Image processing, Machine Learning, Mobile Robots.

1   Introduction

In the future, robots will work and cooperate with humans. They are moving from high-tech factories to our homes, offices, public spaces and small factories. For safety and interactivity reasons, these robots will require redundancy for localization and mapping processes, as well as a semantic localization and mapping system which includes the human in the mapping and localization loop.

The Hybrid Semantic Localization and Mapping (HySeLAM) is a framework and SLAM extension, proposed in [13], which introduces the human into the localization and mapping loop, and it is an extension that increases/creates the redundancy for the conventional SLAM based approaches. The HySeLAM creates two new layers over the 2D/3D occupancy grid-map of the conventional SLAM approaches, where the first layer is built around an augmented topological map and the second layer is built around an object's map.

The first layer abstracts the 2D/3D occupancy gridmap using an augmented topological map, which is closer to the human description of the place than the gridmap. This topological map represents the world as graph, where each

© Springer International Publishing Switzerland 2015   647
A.P. Moreira et al. (eds.), *CONTROLO'2014 - Proc. of the 11th Port. Conf. on Autom. Control,*
Lecture Notes in Electrical Engineering 321, DOI: 10.1007/978-3-319-10380-8_62

vertex is a distinctive place and the edges are the connections between places. Therefore, the augmented topological map $\mathcal{M}_t$ is defined by an attributed graph:

$$\mathcal{M}_t = (\mathcal{P}, \mathcal{C}) \tag{1}$$

where: $\mathcal{P}$ is the set of vertices (places) and $\mathcal{C}$ corresponds to the edges ($\mathcal{C} \subseteq \mathcal{P} \times \mathcal{P}$). These vertices (places) are augmented with five attributes: semantic words (human words that tag the place), geometric description, visual signatures, area and central position.

This paper shows what kind of visual signatures can be used in this augmented topological map to enable a visual recognition of the place. The answer to this question is constrained to the absence of higher landmarks/references, such as objects or particular references available in the places, as conducted in [9], and it should be possible to mimic the human capacity to recognize a particular place by visual glance, as shown by Torralba in [15].

The process of recognizing/classifying an image can be divided into two stages: information extraction and classification. The information extraction, also known as image digestion, is the stage which extracts primitive features from the image. This stage describes an image using low level features, such as color shape and texture while removing unimportant details. Color histograms, color moments, dominant color, scalable color, shape contour, shape region, homogeneous texture, texture browsing and edge histogram are some of the popular descriptors that are used. These descriptors can be divided into two classes: Local and Global descriptors.

The local descriptors extracts local properties of the image, which are usually attached to a particular feature of the image, such as corners, blobs or lines. These local descriptors are usually associated with a feature detector. SIFT [11], SURF [3], ORB [21], BRIEF [5] and FREAK [1] are some popular approaches for extracting local descriptors. In order to make these descriptors usable by the classifier, the concept of bag-of-words is applied to all descriptors extracted from a single image, as described in [20]. When these descriptors are applied in scenarios where lighting is highly variable, a preprocessing step is required in order to increase the illumination invariance, as performed in [17]. In contrast, the GIST operator [15] is global descriptor extractor, which was developed based on several experiments with humans and based on the concept that *Human can recognize the gist of a novel image in a single glance, independent of it complexity*. The GIST operator was been successfully used in outdoor scenario. [16]. Other global descriptors can be found in the literature, such as: Composed Receptive Field Histograms (CRFH) [10], PCA of Census Transform Histogram (PACT) [22], CENsus TRansform hISTogram (CENTRIST) [23], Pyramid of Histograms of Orientation Gradients (PHOG) [4].

This paper presents the Local Binary Pattern histogram Segmented by Basic colors (LBPbyHSV) approach for the global descriptor extraction and compare the performance of this descriptor against other local and global descriptors for indoor place recognition. The paper is organized as follows:- Section 2 present the theory behind the LBPbyHSV global descriptor. Section 3 presents th performance of the proposed descriptor against other common descriptors using real images from three robots in different places. Finally, section 4 presents conclusions.

2   The LBPbyHSV Global Descriptor

The LBPbyHSV approach is the approach proposed here for the global descriptor extractor that can be used as a visual signature for place recognition. The LBPbyHSV is based on the Local Binary Pattern (LBP) operator and on the uniform patterns.

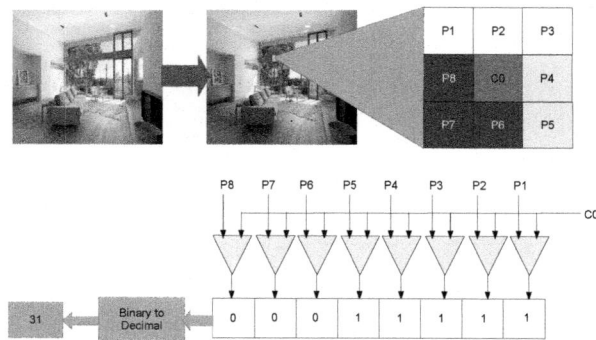

Fig. 1. This image illustrates the $LBP_{8,1}$. Using a code, the LBP operator describes the pattern around a central pixel. To obtain this code, the image is converted to a gray scale image and the central pixel is compared to the neighbor pixels. This comparison results in a binary value of "0" if the central pixel has a higher intensity value, or "1" if the central pixel has a lower intensity value.

The theory behind the LBP operator and other LBP based approaches are described in detail in [18]. The success of the Local binary pattern methods is shown in several computer vision applications, due to the flexibility of the LBP, which makes it easily modifiable and makes this representation more adaptable for certain real world problems. The advantage of using LBP as a descriptor is the fact that it is computationally simple, robust in terms of gray scale variations, and efficient against illumination changes.

The original version of the local binary pattern operator, introduced by [14], works in a 3×3 pixel section of an image. The pixels in this section are thresholded by the center pixel value. Each threshold result is multiplied by powers of two and then summed in order to obtain a label for the center pixel. When the neighborhood consists of 8 pixels, a total of $2^8 = 256$ different labels can be obtained depending on the relative gray values of the center and the pixels in the neighborhood, as shown in figure 1. The label returned by the LBP operator is defined as a natural number which is given by equation 2.

$$LBP_{P,R} = \sum_{p=0}^{P-1} s(g_p - g_c)2^P, \text{with } s(x) = \begin{cases} 1, x \geq 0 \\ 0, x < 0 \end{cases} \tag{2}$$

Where $s(x)$ represents the signal function, g_c is the gray value of the central pixel, g_p is the value of its neighbor pixels and P represents the number of sampling points in the neighborhood.

The sampling process is performed commonly in the clockwise direction, around the central pixel according to a particular radius value R, which determines the spatial resolution of the distributed sampling points. The sampling process is given by:

$$g_p = I\{x_p, y_p\}, p = \{0, 1 \ldots (P-1)\} \tag{3}$$

$$x_p = x + R\cos(2\pi p/P) \tag{4}$$

$$y_p = y - R\sin(2\pi p/P) \tag{5}$$

After the LBP pattern of each pixel in the input image (gray) I_{gray} of size $M \times N$ is identified, it is possible to build a histogram of LBP patterns as follows:

$$H(k) = \sum_{m=1}^{M} \sum_{n=1}^{N} f(LBP_{P,R}(m, n), k), \text{with } k \in [0, K], \tag{6}$$

$$f(x, y) = \begin{cases} 1, x = y \\ 0, otherwise \end{cases} \tag{7}$$

Where K is the maximal LBP pattern values.

In order to increase the robustness of the LBP operator, usually this operator is extended to the approach called *uniform patterns* $(LBP_{P,R}^{u2})$, the superscript "u2" refers to the "*uniform*" patterns with $(U \leq 2)$, which represents the number of spatial transitions allowed in a circular binary form. When measuring uniformity in the U pattern, the uniformity designation merges the bitwise transition from 0 to 1 and vice versa. The LBP pattern is considered uniform if its uniformity measurement is less or equal to 2.

In uniform LBP mapping there is a separate output label for each uniform pattern and all non-uniform patterns are assigned to a single label. Thus, the number of different output labels for mapping patterns of P bits is $P(P-1)+3$. For instance, the uniform mapping produces 59 output labels for neighborhoods of 8 sampling points, and 243 labels for neighborhoods of 16 sampling points.

2.1   LBPbyHSV

LBPbyHSV stands for Local Binary Pattern histogram Segmented by Basic colors which were obtained from clustered HSV spaces. Instead of applying the LBP operator over different color channels, such as OCLBP [12], HSV-LBP and RGB-LBP [2], the LBPbyHSV approach segments the space color representation into n basic colors and this segmentation is used to divide the basic LBP histogram into n histograms. These histograms are then concatenated into a single histogram, as shown in figure 2, and another LBP global descriptor is concatenated to this histogram, forming the LBPbyHSV global descriptor.

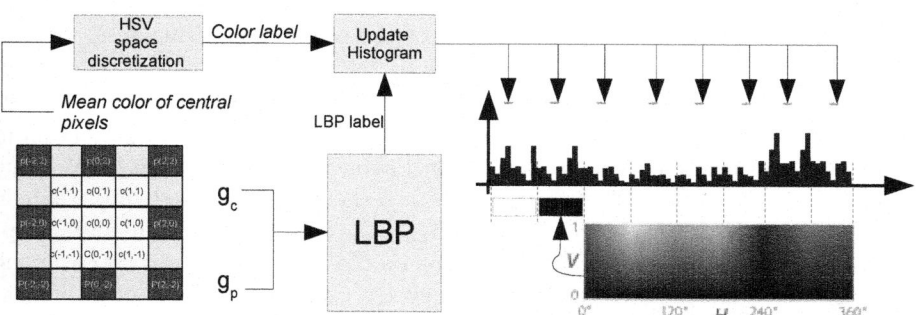

Fig. 2. The LBPbyHSV approach clusters the LBP histogram into n basic colors. Then the LBP label and the basic color of the central pixel are extracted from each image pixel, which then increments the bin related to the LBP label in the color set bin related to the central pixel.

The output descriptor is labeled as LBPbyHSV-XnY, where X refers to the LBP approach used in the first stage (S if the standard LBP operator is used, U f the uniform LBP operator is used), n refers to the number of segmented colors n the HSV space, and Y refers to the LBP approach used in the second stage S if the standard LBP operator is used, U if the uniform LBP operator is used, N if no operator is used). This paper has tested five variants: LBPbyHSV-U8N, LBPbyHSV-U8U, LBPbyHSV-U8S, LBPbyHSV- U16N, and LBPbyHSV-U16S.

In the LBPbyHSV approach, the LBP pattern of each pixel in the input image of size $M \times N$ is identified and the pixel color is associated with the one with he basic colors; thereafter, the histogram for each basic color in the image is uilt as follows:

$$H_c(k) = \sum_{m=1}^{M} \sum_{n=1}^{N} f(LBP_{P,R}(m,n), k, hsvc(m,n), c), \text{ with } k \in [0, K], \quad (8)$$

Where K is the maximal LBP pattern value (which depends on the LBP operator used, uniform or standard, and on the number of sample points P), with:

$$f(x, y, x_c, y_c) = \begin{cases} 1, x = y \wedge x_c = y_c \\ 0, otherwise \end{cases} \tag{9}$$

$$hsvc(m, n) = \begin{cases} 0 & \text{, if } p_V(m, n) < \delta_v \\ 1 & \text{, if } p_S(m, n) < \delta_s \wedge p_V(m, n) \geq \delta_v \\ int(p_H(m, n)/\kappa) + 2 & \text{, otherwise} \end{cases} \tag{10}$$

Where $p_V(m, n)$ $p_S(m, n)$ and $p_H(m, n)$ are functions that return the value of HSV components from the pixel $p_{m,n}$ of the original I_{hsv} image, δ_v defines the limit value for the black color definition, δ_s defines the value of saturation for the white colors, and κ is given by $\frac{360}{(n-2)}$, where n is the number of segmented colors (black and white are always considered) and 360 is the maximum value that the Hue component can obtain (using the opencv library this value should be set as 180). These histograms are then concatenated into a single one, $H = [H_0, H_1, ..., H_n, H_{LBP}]$, where H_{LBP} is the second LBP histogram applied (standard, uniform or none depending on the LBPbyHSV variant) over the $I_{LBPbyHSV}$ image. This intermediate image $I_{LBPbyHSV}$ is a gray scaled image, and the pixel values are obtained as follows:

$$g(x, y) = LBP_{P,R}(m, n) + hsvc(m, n)K \tag{11}$$

In the LBPbyHSV approach, the $LBP_{P,R}$ operator used is also slight modified when $R > 1$, in these cases, the central value g_c is not only obtained from the central pixel value, but also obtained as mean value from the pixels inside radius R, as shown in figure 2.

3 Comparison of Descriptor Performances

Two videos were created with images collected from three different robots and places in order to compare the performance of the LBPbyHSV approach with other descriptors for indoor place recognition.

The Support Vector Machine (SVM) classifier with three kernels (Linear Polynomial and Radial Basis Function) was used for the classification stage However, other techniques can be applied, such as: Neural Networks (NN), Ad aboost, Bayes based classifiers. The theory behind SVM is described in detail in [7]. This work has used an SVM open source implementation, the libsvm [6] which can be found at http://www.csie.ntu.edu.tw/~cjlin/libsvm/.

The descriptors test were conducted by a ROS node application, running in a 2,16 GHz Intel Pentium Dual Core T4300 processor, with 2GB of memory and with a robotic operating system (ROS) (fuerte version) over the Ubuntu O (12.04). The HySeLAM framework implementation, available in hyselam.com was also used to manage the groundtruth of the two videos by publishing

Table 1. Performance comparison of global and local descriptors. The time column presents the average time required by the descriptor approach to obtain the descriptor vector; a variance of this average is presented in brackets. The SVM Accuracy presents the accuracy obtained in the testing video using the descriptor and the SVM kernel, after a training stage with the training video.

Approach	Descriptor size	Time ms	SVM Accuracy (%)		
			Linear	Poly	RBF
LBPbyHSV-U8N	472	13.505(0.001)	71.32	71.91	67.24
LBPbyHSV-U8U	530	27.634 (0.004)	72.49	71.45	71.39
LBPbyHSV-U8S	737	28.376 (0.007)	72.59	72.01	70.27
LBPbyHSV-U16N	944	14.249 (0.002)	72.58	72.91	71.24
LBPbyHSV-U16S	1200	26.645 (0.004)	74.53	74.31	73.24
HSV-LBP	768	29.319 (0.003)	75.27	68.77	66.42
HSV-LBP-u	174	29.670 (0.003)	73.65	74.01	67.15
SLBP(1+2+4) [8]	1218	19.125 (0.004)	75.12	75.22	71.37
HSV-GIST (32x32)	960	8.827 (0.003)	66.08	69.95	19.81
HSV-GIST (64x64)	960	45.187 (0.009)	69.19	70.17	28.79
HSV-GIST (128x128)	960	188.959 (0.101)	69.70	70.05	28.05
BOF-SIFT (500 words)	500	96.859 (0.347)	55.80	59.03	64.70
BOF-SIFT (2000 words)	2000	95.866 (0.334)	61.36	54.57	65.43
BOF-SURF (500 words)	500	140.995(3.697)	57.99	56.20	65.03
BOF-SURF (2000 words)	2000	145.210(3.761)	65.01	58.82	60.01
BOF-ORB (200 words)	200	23.837(0.031)	41.64	45.30	46.16
BOF-ORB (500 words)	500	30.693(0.082)	45.36	39.40	44.81
BOF-SURF+BRIEF (500 words)	500	90.646 (0.513)	49.35	46.43	51.27
BOF-SURF+BRIEF (2000 words)	2000	108.547(1.401)	48.77	34.82	47.62
BOF-SURF+FREAK (500 words)	500	87.055 (0.454)	32.99	35.88	42.19
BOF-SURF+FREAK (2000 words)	2000	108.628(1.503)	31.22	30.99	37.27

Table 2. Performance comparison of global and local descriptors, which encodes in a single descriptor all, half top and half bottom image (these descriptors are identified by (D) extension). And, performance comparison of LBPbyHSV descriptor concatenated with different local descriptors.

Approach	descriptor size	Time ms	SVM Accuracy (%)		
			Linear	Poly	RBF
LBPbyHSV-U8N(D)	1416	14.620 (0.002)	74.93	76.30	76.69
BOF-ORB-500(D)	1500	28.269 (0.121)	44.34	45.64	50.96
BOF-SIFT-500(D)	1500	91.809 (0.388)	55.62	58.27	60.40
BOF-SURF-500(D)	1500	116.750 (2.058)	60.25	62.35	60.69
LBPbyHSV-U8N(D)+BOF-ORB500	1916	43.465 (0.112)	72.69	74.42	75.46
LBPbyHSV-U8N(D)+BOF-SIFT500	1916	105.571 (0.380)	73.84	76.42	75.07
LBPbyHSV-U8N(D)+BOF-SURF500	1916	150.000 (3.384)	75.48	75.21	74.78

semantic localization topic. The localization topic is limited to a set of human words that tags the place, in this work: Freiburg Corridor(ss0), Freiburg Printer office(ss1), Freiburg Room A(ss2), Freiburg Stairs(ss3), Freiburg WC(ss4), Ljubljana Corridor(ss5), Ljubljana Stairs(ss6), Ljubljana Room A(ss7), Ljubljana WC (ss8), FEUP Corridor(ss10), FEUP I-110(ss9), FEUP room I-108(ss11), and, FEUP room I-109(ss12).

The two videos used are characterized as follows:

- Testing video: it contains 7320 frames with a 640x480 resolution. These frames were collected from three robots at three different places: Faculdade de Engenharia da Universidade do Porto (FEUP), University of Freiburg in Germany, and University of Ljubljana in Slovenia. The FEUP video was acquired in a laboratory and it is 337 seconds long at 10fps with a resolution of 640x480. The other two videos are available in the COLD database [19] (http://www.cas.kth.se/COLD/index.php). The Freiburg video is 182 seconds long and the Ljubljana video is 213 seconds long, and both were acquired at 10fps with a resolution of 640x480 on a cloudy day.
- Training video: it contains 6254 frames with the same resolution and frame rate. These frames were collected from the same robots, but with slightly different path, time and illumination conditions. Here the FEUP video is 435 seconds long, the Freiburg video is 171 seconds long, and the Ljubljana video is 194 seconds long.

It is important to highlight that both videos were obtained at different moments, in different paths and weather conditions (cloudy and sunny), and with people present in the scenes. The LBP operator used in LBP based approaches has 8 samples and a radius of 3 ($LBP_{8,3}$). The SVM Poly kernel used is a polynomial function of second order.

4   Conclusions

The proposed approach is more accurate than the local descriptors and more accurate than the GIST based approach, as shown by table 1. However, the LBPbyHSV variant approaches have a slightly worse accuracy than the HSV-LBP and SLBP, but the processing time is faster (more than 2 times) and the descriptor is more compact. This makes the LBPbyHSV approach interesting for real time applications running in embedded systems, because it encodes the color image information on the descriptor faster than the HSV-LBP.

Figure 3 shows that the place transitions between places is a challenge for the classifier, as it is for the humans when they are watching the videos. The descriptor changes smoothly between place transitions, which puts the descriptor in the frontier of two classes where the classifier has less accuracy. To classification accuracy, it is proposed that future work uses the augmented topological map, defined in the HySeLAM framework, in order to create a semantic localization based on graph transitions, where transitions are managed by the classification

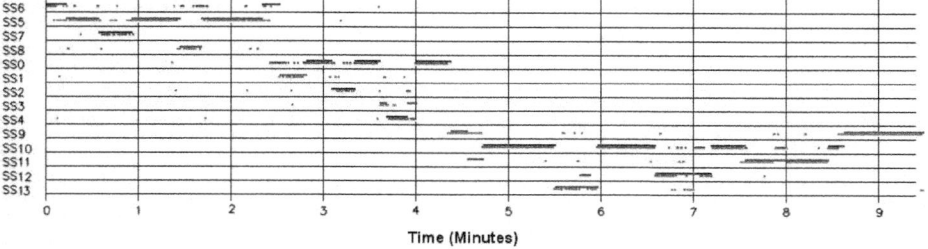

Fig. 3. Semantic localization provided by the LBPbyHSV approach. The semantic localization provided by the HySeLAM (groundtruth) is in blue, the semantic localization provided by the LBPbyHSV-U16S with SVM (Linear kernel) is presented in red.

provided by the visual signature and the by classifier. This approach will make it possible to detect these transitions, which will improve the systems accuracy.

In the figure 3 there are state transitions that are not allowed, most of them are related to a classification of a non descriptive image (for example an image of a white wall) or related to a classification of a similar place. The inclusion of the augmented topological map in classification stage will also allow to remove these outliers.

Acknowledgements. This work is financed by the ERDF European Regional Development Fund through the COMPETE Programme (operational programme for competitiveness) and by National Funds through the FCT Fundação para a Ciência e a Tecnologia (Portuguese Foundation for Science and Technology) within project FCOMP-01-0124-FEDER-022701.

References

1. Alahi, A., Ortiz, R., Vandergheynst, P.: Freak: Fast retina keypoint. In: Computer Vision and Pattern Recognition (CVPR), pp. 510–517. IEEE (2012)
2. Banerji, S., Verma, A., Liu, C.: Novel color LBP descriptors for scene and image texture classification. In: 15th International Conference on Image Processing, Computer Vision, and Pattern Recognition. Las Vegas, Nevada (2011)
3. Bay, H., Tuytelaars, T., Van Gool, L.: Surf: Speeded up robust features. In: Leonardis, A., Bischof, H., Pinz, A. (eds.) ECCV 2006, Part I. LNCS, vol. 3951, pp. 404–417. Springer, Heidelberg (2006)
4. Bosch, A., Zisserman, A., Munoz, X.: Representing shape with a spatial pyramid kernel. In: Proceedings of the 6th ACM International Conference on Image and Video Retrieval, pp. 401–408. ACM Press (2007)
5. Calonder, M., Lepetit, V., Strecha, C., Fua, P.: BRIEF: Binary robust independent elementary features. In: Daniilidis, K., Maragos, P., Paragios, N. (eds.) ECCV 2010, Part IV. LNCS, vol. 6314, pp. 778–792. Springer, Heidelberg (2010)

6. Chang, C.-C., Lin, C.-J.: LIBSVM: a library for support vector machines. ACM Transactions on Intelligent Systems and Technology (TIST) 3(2), 27 (2011)
7. Cristianini, N., Shawe-Taylor, J.: An introduction to support vector machines and other kernel-based learning methods. Cambridge University Press (2000)
8. Hu, J., Guo, P.: Spatial local binary patterns for scene image classification. In: 2012 6th International Conference on Sciences of Electronics, Technologies of Information and Telecommunications (SETIT), pp. 326–330. IEEE (March 2012)
9. Lin, K.-H., Wang, C.-C.: Stereo-based simultaneous localization, mapping and moving object tracking. In: 2010 IEEE/RSJ International Conference on Intelligent Robots and Systems (IROS), pp. 3975–3980. IEEE (2010)
10. Linde, O., Lindeberg, T.: Object recognition using composed receptive field histograms of higher dimensionality. In: ICPR 17th International Conference on Pattern Recognition, vol. 2, IEEE (2004)
11. Lowe, D.G.: Distinctive image features from scale-invariant keypoints. International Journal of Computer Vision 60(2), 91–110 (2004)
12. Mäenpää, T., Pietikäinen, M.: Classification with color and texture: jointly or separately? Pattern Recognition 37(8), 1629–1640 (2004)
13. Santos, F.N., Moreira, A.P., Costa, P.C.: Towards Extraction of Topological Maps from 2D and 3D Occupancy Grids. In: Correia, L., Reis, L.P., Cascalho, J. (eds.) EPIA 2013. LNCS, vol. 8154, pp. 307–318. Springer, Heidelberg (2013)
14. Ojala, T., Pietikäinen, M., Harwood, D.: A comparative study of texture measures with classification based on featured distributions. Pattern Recognition 29(1), 51–59 (1996)
15. Oliva, A., Torralba, A.: Modeling the shape of the scene: A holistic representation of the spatial envelope. International Journal of Computer Vision 42(3), 145–175 (2001)
16. Oliva, A., Torralba, A.: Building the gist of a scene: the role of global image features in recognition. Progress in Brain Research 155 (2006)
17. Petry, M.R., Moreira, A.P., Reisinst, L.P.: Increasing Illumination Invariance of SURF Feature Detector through Color Constancy. In: Correia, L., Reis, L.P., Cascalho, J. (eds.) EPIA 2013. LNCS, vol. 8154, pp. 259–270. Springer, Heidelberg (2013)
18. Pietikäinen, M., Zhao, G., Ahonen, T., Hadid, A.: Computer vision using local binary patterns. Springer (2011)
19. Pronobis, A., Caputo, B.: COLD: COsy Localization Database. The International Journal of Robotics Research (IJRR) 28, 588–594 (2009)
20. Quelhas, P., Monay, F.: Modeling scenes with local descriptors and latent aspects. In: Tenth IEEE International Conference on Computer Vision, ICCV 2005, vol. 1 IEEE (2005)
21. Rublee, E., Rabaud, V.: ORB: an efficient alternative to SIFT or SURF. In: 201 IEEE International Conference on Computer Vision (ICCV). IEEE (2011)
22. Wu, J., Rehg, J.M.: Where am I: Place instance and category recognition using spatial PACT. In: IEEE Conference on Computer Vision and Pattern Recognition CVPR 2008 (2008)
23. Wu, J., Rehg, J.M.: CENTRIST: A Visual Descriptor for Scene Categorization IEEE Transactions on Pattern Analysis and Machine Intelligence 33(8), 1489–150 (2010)

Part XI
Sensing

Online Vision-Based Eye Detection: LBP/SVM vs LBP/LSTM-RNN

Djamel Eddine Benrachou[1], Filipe Neves dos Santos[2],
Brahim Boulebtateche[1], and Salah Bensaoula[1]

[1] University Badji Mokhtar, Department of electronics
BP 12, 23000, Annaba, Algeria
[2] INESC TEC(formerly INESC Porto) and Faculty of Engineering,
University of Porto,
Campus da FEUP, Rua Dr. Roberto Frias, 378, Porto, Portugal
djamelben.univ@gmail.com, fbnsantos@fe.up.pt,
brahim.boulebtateche@univ-annaba.org, bensaoula_salah@yahoo.fr

Abstract. Eye detection is a complex issue and widely explored through several applications, such as human gaze detection, human-robot interaction and driver's drowsiness monitoring. However, most of these applications require an efficient approach for detect the ocular region, which should be able to work in real time. In this paper, it is proposed and compare two approaches for online eye detection. The proposed schemes, work under real variant illumination conditions, using the conventional appearance method that is known for its discriminative power especially in texture analysis.

In the first stage, the salient eye features are automatically extracted by employing *Uniform* Local Binary pattern (LBP) operator. Thereafter, supervised machine learning methods are used to classify the presence of an eye in image path, which is described by an LBP histogram. For this stage, two approaches were tested; *Support Vector Machine* and *Long Short-Term Memory Recurrent Neural Network*, both are trained for discriminative binary classification, between two classes namely *eye / non eye*.

The human eyes were successfully localized in real time videos, which were obtained from a laptop with uncalibrated web camera. In these tests, different people were considered and light illumination. The experimental results are reported.

1   Introduction

Eye detection is a challenging subject, which has shown an expanding interest this last decade and gained large in different fields, it is still widely studied and introduced in several applications such as; driver somnolence prevention and gaze direction estimation for human-machine interfaces. However, that exist numerous commercial eye detection systems, which attest to significant progress achieved in the research field. Despite these achievements, eye detection remains

© Springer International Publishing Switzerland 2015
A.P. Moreira et al. (eds.), *CONTROLO'2014 - Proc. of the 11th Port. Conf. on Autom. Control*,
Lecture Notes in Electrical Engineering 321, DOI: 10.1007/978-3-319-10380-8_63

an active subject in computer vision research, due to the fact that existing systems perform well under relatively controlled environments but tend to fail when variations in different factors such as: (poses, environmental illumination changing and occlusion etc). Therefore, the goal of the ongoing research is to increase the robustness of the systems despite these different factors. The detection of the eyes is interesting for car's driver vigilance monitoring [1], due to the few existing direct measures and tell-tale signs of this phenomenon and the most of them are related to the observed outcome symptoms gathered from human behaviors while driving. Some observations are easily distinguished, mainly in the facial region and especially in eyes, that reflect the individual attention level paid for a particular task. Furthermore, in robotics, the context of human-robot interfaces based eye detection and tracking is also an important topic, it is required to gather the almost information about the human gaze, which considered as a complement to the vocal information supplied by the individual. For example, gaze directions normally tells which object in his/her surrounding a person is interested in. Therefore, eye tracking can be also used as a medium for human-robot interaction, for example instructing a robot's arm to pick a certain object which the user is looking at. Such application for eye tracking is explored through [2], [3], [4]. However, these active eye tracking systems for this kind of applications mentioned above, should be able to work on real-time over embedded systems. This requires that the eye detection approach should be computational efficient. In the context of eye detection, there exist many well know techniques, each one have its own inherent limitations. In this area research could be roughly categorized into four types; *knowledge-based* methods, *template-matching* approaches [5], *Invariant feature-based* methods and *Appearance-based* methods [6].

In the present paper, we compare between two proposed approaches, based on computer vision and machine learning techniques, aiming an online detection of the eye. Both algorithms, are firstly based on appearance method, namely Local Binary Pattern (LBP). Concerning the first approach, LBP feature histograms have been automatically extracted from each video frame, considering both shape and texture informations to represent the eye images, then Support Vector Machine [7]is trained for a binary classification between eye/non-eye classes.

Subsequently for the second scheme, which is slightly different than the first one from viewpoint of classification phase, a combination LBP and particular architecture of *Recurrent Neural Network* namely *Long Short-Term Memory* network *(LSTM-RNN)* has been performed. The network input is presented as a rhythmic sequential forward histogram of LBP features, representing our classes (eye/non-eye). This special neural architecture has been successfully applied for several real world problems, such as protein secondary structure prediction [8],[9], music generation [10], speech recognition [11],[12] and handwriting recognition [13],[14]. This method allows, a temporal classification and the preservation of contextual informational dependencies, thus avoiding to these meaningful informations to be disappear across the temporal learning stage. Finally, an extensive comparison between the schemes is performed, where the execution time of each one is measured, experimental results and statistical valuations are presented.

2   Feature Extraction Using Local Binary Pattern (LBP)

The success of Local binary pattern (LBP) method is shown in several computer vision applications, due to the flexibility of the LBP, which turn it easily modifiable and makes this representation more adaptable for certain real world problems. The advantage of using LBP as a descriptor, for features extraction is computationally simple, robust in terms of gray scale variations, efficient against the illumination changes. In our application, the individual in front of the camera may be subjected to varying illumination conditions, involving a negative impact on the detection results, for those reasons LBPs seem more suitable in this case. The original LBP, introduced by Ojala et al.[15], works in a grid size of 3×3 for a given arbitrary pixel over an input gray-scale image, the LBP code is computed by comparing the gray level value of the central pixel and its neighbors, within the respective grid. Note that the gray value of the neighbor pixels, which are not covered by the grids is estimated by interpolation, the thresholding stage is carried out according to the central pixel, resulting a binary number called LBP code.

$$LBP_{P,R} = \sum_{p=0}^{P-1} s(g_p - g_c)2^p, \begin{cases} 1, x \geq 0 \\ 0, x < 0 \end{cases} \tag{1}$$

The equation above, represents the signs of the computed differences , g_c is the gray value of the central pixel, g_p is the value of its neighbors , P is the total number of sampling points and controls the quantization of the angular space and R is the radius of the neighborhood. The sampling is performed commonly in the clockwise direction, around the central pixel according to a particular radius value, which determines the spatial resolution of the distributed sampling points. As the radius increases, an LBP contains more global information of texture patterns and also needs more neighbors, thus increasing the computational cost. If a large radius R is mixed with small P, it will result in severe artifacts in the re-sampling. In our eye detection application, for the sake of balance between information and computation, the number of neighbors is set to 8, and the radius is fixed to 1.

The LBP pattern of each pixel in the input image I of size $M \times N$ is identified, thereafter the histogram is built for the image textures representation;

$$H(k) = \sum_{m=1}^{M} \sum_{n=1}^{N} f(LBP_{P,R}(m,n), k), k \in [0, K] \tag{2}$$

$$f(x, y) = \begin{cases} 1, x = y \\ 0, Otherwise \end{cases} \tag{3}$$

Where K is the maximal LBP pattern value. In this work, we have employed an extended version of the original LBP, called *uniform patterns* $(LBP_{P,R}^{u2})$ [16], the superscript "$u2$" refers to the "$uniform$" patterns with $(U \leq 2)$, which represents the number of spatial transitions in a circular binary form. The uniformity designation merges, in the uniform measurement of the pattern U, the bitwise transition from 0 to 1 and vice versa. Using the $(LBP_{P,R}^{u2})$, instead another extended representation, because it is more stable, i.e less sensitive to noise [17]. Considering only the uniform patterns allows to deal with the significant LBP labels and the reliable estimation of their distribution requires fewer samples. For an efficient representation of the eye feature map, each input image is empirically and equally divided into a sub-regions, see figure 1. Thereafter, the global spatial histogram is built by concatenate the obtained histogram of each region and used as an eye descriptor. The textures of the eye regions are locally

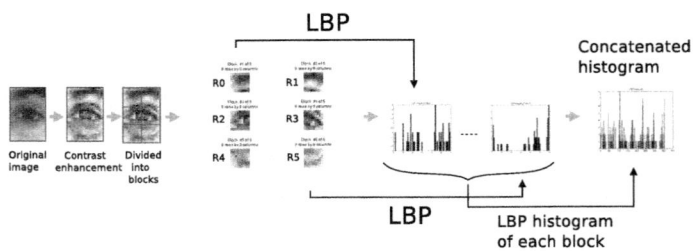

Fig. 1. Feature extraction using local binary pattern

encoded by the LBP patterns, while the whole shape of the eye is recovered by the construction of the eye feature histogram.

3   Long Short-Term Memory

Several varieties of Recurrent Neural Networks (RNNs) have been emerged [18], [19], [20], [21], the most promising of them for sequence classification is *Long Short-Term Memory* (*LSTM*) pioneered by [22], *(RNN)* transducers could be trained using the conventional back-propagation through time algorithm *(BPTT)* [23][24], and the whole system could be optimized with gradient descent. However, train *BPTT* is not a straightforward task and may involves some drawbacks. The error signals in the backward pass tends to either blow up, that may conduct to the oscillation of the weights or vanish, when learning to bridge long time lags, that takes a non acceptable amount of time or does not work at all [22]. Note, the exponential temporal evolution of the back-propagated error is related to the size of the weights [25]. *LSTM* is designed to keep down these error back-flow problems. *LSTM* is built exactly as an *RNN* with an improved architecture, where memory blocks are used instead of hidden layer nonlinear units (sigmod logistic units). The common recurrent neural architecture, imposes a limited access in the contextual information range, the problem arises from the influence of the input conveyed on the hidden layer, therefore on the network output, either vanish or explode exponentially as long as the information cycles throughout the network recurrent connections.

Fortunately, *Long Short-Term Memory* scheme might be able to learn and bridge contextual informations over long time steps even in case of noisy, incomprehensible input sequences, without loss of short time lags capabilities [22]. This is achieved by using an appropriate gradient based algorithm for an architecture enforcing constant error flow through internal state of special units, represented as *Constant Error Carousel (CEC)* and translated through a central self connection linear unit. This architecture exploits the potential of the internal recurrent connection subnets namely Memory blocks. Each block contains one or more self-connected memory cells and three multiplicative units; *Input, Output and Forget gates*. Figure 3 illustrates a single *LSTM* memory cell. These multiplicative gate units allow *LSTM* memory cell to store and access information over long period of time and avoid vanishing of gradient error.

In our experiments, the output activation is normalized using the *Softmax* function [26] which force the output to represent a probability distribution of each output class i.e it ensures all of the output values to be bounded between 0 and 1, and their sum is equal to 1.

$$y_k = \frac{e^{a_k}}{\sum_{k'=1}^{K} e^{a_{k'}}} \tag{4}$$

Here, y_k corresponds to the network output for K classes, a_k is the network input to the output units. Equation (4) shows that the output of each neuron depends on all other neurons adders *Softmax* group.

The network is trained with gradient descent by differentiating the objective function with respect to the output which is referred as a forward pass, then using back-propagation through time to find the partial derivatives with respect to the network weights, this step is referred as a backward pass.

For parametric algorithm such as *Neural Networks*, the appropriate approach for error minimization, is to adjust the algorithm parameters in order to optimize the objective function, for this purpose, Cross-Entropy objective function is employed and represented as follow;

$$CE = - \sum_{x,t' \in S} \sum_{k=1}^{K} t'_k \log y_k \tag{5}$$

Here, (x, t') is a pair of sequence, x is an element of input space X and t' is refereed to the target space T'(the desired output belonging to each element of X). For the optimization phase, we use the conventional momentum operator represented as;

$$v(t) = \alpha v(t-1) - \xi \frac{\partial CE}{\partial w}(t) \tag{6}$$

$$\Delta \partial w = v(t) = \alpha \Delta w(t-1) - \xi \frac{\partial CE}{\partial w}(t) \tag{7}$$

Where, $\frac{\partial CE}{\partial w}$ is the gradient representation, the operator time t is the update of the weights, $\xi > 0$ is the learning rate, $\alpha \in [0, 1]$ or $(1 - \alpha)$ is the momentum constant, which controls the *decay* of the velocity vector v. Recently, Ilya Sutskever

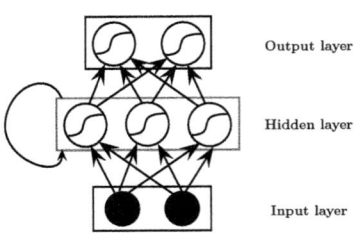

Fig. 2. Recurrent neural network

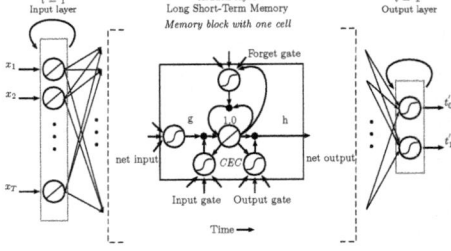

Fig. 3. Long short-term memory representation

et al [27], have suggested a new formulation of the momentum that often works
better and which is inspired by Nesterov method for optimizing convex function
[28].

3.1   Results and Discussion

Through this work, we compare between two algorithms for online eye localization, both are based on LBP feature for coding the details of appearance and
texture of eye patterns with size 27 × 18, these are empirically divided into 6
sub-blocks, only the rectangular regions of the same size 9 × 9 are used, but
other divisions are possible as blocks of different size and shape. Thereafter
we excerpt the histogram of each sub-block, generating a feature vector of 59
distinct elements, the global spatial histogram of the eye image, is built by concatenate each sub-block histogram, involving 354 distinct outputs. This global
histogram takes into account, the micro and the macro structure of the eye. This
is represented in three different standpoints; the labels of the histogram contain
information about the patterns at a pixel-level, the labels are summed over a
sub-blocks level and the sub-block histograms are concatenated to build a global
description of the eye. The first approach is principally based on the combination LBP/SVM, the classifier is trained with two different kernels, in order to
estimate the separability potential of the collected dataset, LBP/SVM (linear
kernel) yielded a best average of 98.1% detection accuracy for $\simeq$ 0.562 seconds
against LBP/SVM(RBF kernel) with 96.8% for $\simeq$ 0.642 seconds. (the mean of
the algorithm execution time calculated over 30 frames). In the second approach
an extended architecture of *RNN-LSTM* has been explored, with *forget gate*
[29] and peephole connections [30] involved. The network is trained with Cross
entropy objective function for error-entropy minimization (EEM), it enables the
convergence of LSTM more frequently than the traditional mean squared error
(MSE) cost function, also dealing with *(EEM)* reduces the number of training
epochs until the convergence [31]. Moreover, We have tested several configuratio
of networks, varying the number of hidden LSTM units. The exposed configuration was found to be a good compromise for our classification task. The networ
has built with an input layer according to the input features dimension, in ou

Table 1. Statistical results

Approach	TP	FP	TN	FN	Precision	Recall	F-score	**Accuracy**	time
LBP+SVM(Linear kernel)	291	10	298	1	0.996	0.967	0.980	98.1 %	0.562 s
LBP+SVM(RBF kernel)	290	17	291	2	0.930	0.993	0.944	96.8 %	0.642 s
LBP+LSTM($\xi = 10^{-3}$)	283	16	292	9	0.970	0.948	0.960	95.8 %	0.697 s

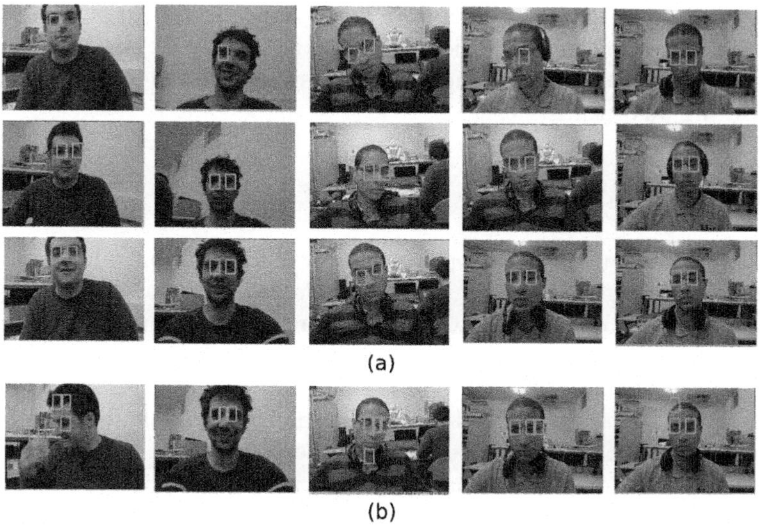

(a)

(b)

Fig. 4. Eye detection with different individuals; (a) Acceptable results, (b) Eye detection results involving false positives

case it is equivalent to 354, one hidden layer of LSTM-blocks, built with 12 *LSTM-blocks "memory cell block of size 1"*, each memory block is composed from one cell unidirectional LSTM fully inter-connected and fully connected to the rest of the network. In the output layer, we used the *Softmax* function, to ensure that obtained outputs are all between 0 and 1, and their sum is equal to 1 at each time step and the number of the output units is equally related to the number of classes. The network is trained with $online - BPTT$ and the weights were randomly initialized to small values. For the optimization phase, regularization terms have been used, improving the convergence velocity, also it is a vital step to get a good performance with *RNNs* and makes them less exposed to over-fitting. The weight-decay is valued to 0.01, momentum is fixed to 0.9 and learning rate equal to 10^{-3}. This approach yielded, a detection accuracy of $\simeq 95.8\%$ for an execution time of $\simeq 0.697$ seconds http://hyselam.com/eye.html.

We notice, that we have a small difference of the detection time, between the algorithms. Nevertheless, this small margin is considerable for an application such as, driver fatigue detection, where the robustness of the monitoring system requires, a precise eye localization in a short time. The whole detection results obtained for each algorithm, are illustrated trough eye localization figure 4, statistical measures are reported in form of true positive (TP), false positives (FP), true negative (TN) and false negatives(FN), involving the calculation of F-score for a final evaluation and the average recognition accuracy for each algorithm, see table 1.

It is important to note, that those algorithms are developed and tested under Linux Ubuntu x86 platform and python environment, in intel Core2Duo^{TM}CPU based laptop system at 2.2 Ghz and 4Gb memory specs. Support Vector Machine implementation, is performed using scikit-learn, which is an open source Python module, integrating a wide range of supervised and unsupervised machine learning algorithms. [32].

4   Conclusion

This paper presents two online approaches for eye localization, which were tested using real face images, captured with a simple web camera, under general uncontrolled, uncalibrated illuminations and complex environmental background. The proposed approaches were designed to work with a high degree of eye's appearance variability caused by the intrinsic dynamic characteristic of the eyes, ambient lighting changes, expression variations, where both shape and photometric appearance of the eye is sensitive to these changes.

Besides the aforementioned challenges, the proposed approaches were tested with persons placed in variable ambient illumination conditions, with myopia glasses, some facial expressions involved, different eye states and in random pose variations (realistic scenario), as shown in figure 4. The proposed algorithms have achieved a significantly high detection rate, with a good localization results http://hyselam.com/eye.html.

The LBP histogram-based descriptor is a powerful technique for capturing the eye features and efficient against illumination changing. Those meaningful informations are preserved by *LSTM-RNN*, during the temporal learning phase. At this stage, it is possible to conclude, that LBP/LSTM-RNN involves a sort of complementarity. Moreover, the localization process is computationally feasible and makes both algorithms, meet the need of real-time application.

As future work, we propose to evaluate a cascade classifier based on SVM/RNN, and test another geometric devision for LBP descriptor, such circular or triangular, which seems to adjust to the eye feature. Another proposal is to test a probabilistic map where the classifier will classify the eye present/ eye absent according to the high probability of the eye existence in the input image. This probabilistic map will be employed for a reliable eye localization and tracking, and could be obtained using the nearest neighbor probabilistic based approach.

References

1. Benrachou, D.E., Boulebtateche, B., Bensaoula, S.: Gabor/pca/svm-based face detection for drivers monitoring. Journal of Automation and Control Engineering 1, 115–118 (2013)
2. André, E., Chai, J.Y.: Introduction to the special issue on eye gaze in intelligent human-machine interaction. ACM Transactions on Interactive Intelligent Systems (TiiS) 1(2), 7 (2012)
3. Atienza, R., Zelinsky, A.: Intuitive human-robot interaction through active 3D gaze tracking. In: Robotics Research, pp. 172–181. Springer (2005)
4. Atienza, R., Zelinsky, A.: Active gaze tracking for human-robot interaction. In: Proceedings of the 4th IEEE International Conference on Multimodal Interfaces, p. 261. IEEE Computer Society (2002)
5. Jian, W., Honglian, Z.: Eye detection based on multi-angle template matching. In: International Conference on Image Analysis and Signal Processing, IASP 2009, pp. 241–244. IEEE (2009)
6. Kith, V., El-Sharkawy, M., Bergeson-Dana, T., El-Ramly, S., Elnoubi, S.: A feature and appearance based method for eye detection on gray intensity face images. In: International Conference on Computer Engineering & Systems, ICCES 2008, pp. 41–47. IEEE (2008)
7. Vapnik, V.: The nature of statistical learning theory. Springer (2000)
8. Hochreiter, S., Schmidhuber, J.: Long short-term memory. Neural Computation 9(8), 1735–1780 (1997)
9. Chen, J., Chaudhari, N.S.: Protein secondary structure prediction with bidirectional lstm networks. In: International Joint Conference on Neural Networks: Post-Conference Workshop on Computational Intelligence Approaches for the Analysis of Bio-data (CI-BIO) (August 2005)
10. Eck, D., Schmidhuber, J.: Finding temporal structure in music: Blues improvisation with lstm recurrent networks. In: Proceedings of the 2002 12th IEEE Workshop on Neural Networks for Signal Processing, pp. 747–756. IEEE (2002)
11. Graves, A., Schmidhuber, J.: Framewise phoneme classification with bidirectional lstm and other neural network architectures. Neural Networks 18(5), 602–610 (2005)
12. Graves, A., Fernández, S., Gomez, F., Schmidhuber, J.: Connectionist temporal classification: labelling unsegmented sequence data with recurrent neural networks. In: Proceedings of the 23rd International Conference on Machine Learning, pp. 369–376. ACM (2006)
13. Liwicki, M., Graves, A., Bunke, H., Schmidhuber, J.: A novel approach to on-line handwriting recognition based on bidirectional long short-term memory networks. In: Proc. 9th Int. Conf. on Document Analysis and Recognition, vol. 1, pp. 367–371 (2007)
14. Graves, A., Liwicki, M., Fernández, S., Bertolami, R., Bunke, H., Schmidhuber, J.: A novel connectionist system for unconstrained handwriting recognition. IEEE Transactions on Pattern Analysis and Machine Intelligence 31(5), 855–868 (2009)
15. Ojala, T., Pietikäinen, M., Harwood, D.: A comparative study of texture measures with classification based on featured distributions. Pattern recognition 29(1), 51–59 (1996)
16. Ojala, T., Pietikainen, M., Maenpaa, T.: Multiresolution gray-scale and rotation invariant texture classification with local binary patterns. IEEE Transactions on Pattern Analysis and Machine Intelligence 24(7), 971–987 (2002)

17. Pietikäinen, M.: Computer vision using local binary patterns, vol. 40. Springer (2011)
18. Elman, J.L.: Finding structure in time. Cognitive science 14(2), 179–211 (1990)
19. Jordan, M.I.: Artificial neural networks, pp. 112–127. IEEE Press, Piscataway (1990)
20. Lang, K.J., Waibel, A.H., Hinton, G.E.: A time-delay neural network architecture for isolated word recognition. Neural Networks 3(1), 23–43 (1990)
21. Jaeger, H.: The "echo state" approach to analysing and training recurrent neural networks-with an erratum note'. Bonn, Germany: German National Research Center for Information Technology GMD Technical Report 148 (2001)
22. Hochreiter, S., Schmidhuber, J.: Long short-term memory. Neural Computation 9(8), 1735–1780 (1997)
23. Williams, R.J., Zipser, D.: Gradient-based learning algorithms for recurrent. Backpropagation: Theory, Architectures, and Applications, 433 (1995)
24. Werbos, P.J.: Generalization of backpropagation with application to a recurrent gas market model. Neural Networks 1(4), 339–356 (1988)
25. Hochreiter, S.: Untersuchungen zu dynamischen neuronalen netzen. Master's thesis, Institut fur Informatik, Technische Universitat, Munchen (1991)
26. Bridle, J.S.: Probabilistic interpretation of feedforward classification network outputs, with relationships to statistical pattern recognition. In: Neurocomputing, pp. 227–236. Springer (1990)
27. Sutskever, I., Martens, J., Dahl, G., Hinton, G.: On the importance of initialization and momentum in deep learning
28. Nesterov, Y.: A method of solving a convex programming problem with convergence rate o $(1/k2)$. Soviet Mathematics Doklady 27, 372–376 (1983)
29. Gers, F.A., Schmidhuber, J., Cummins, F.: Learning to forget: Continual prediction with lstm. Neural Computation 12(10), 2451–2471 (2000)
30. Gers, F.A., Schraudolph, N.N., Schmidhuber, J.: Learning precise timing with lstm recurrent networks. The Journal of Machine Learning Research 3, 115–143 (2003)
31. Alexandre, L.A., de Sá, J.P.M.: Error entropy minimization for LSTM training. In: Kollias, S.D., Stafylopatis, A., Duch, W., Oja, E. (eds.) ICANN 2006. LNCS, vol. 4131, pp. 244–253. Springer, Heidelberg (2006)
32. Pedregosa, F., Varoquaux, G., Gramfort, A., Michel, V., Thirion, B., Grisel, O., Blondel, M., Prettenhofer, P., Weiss, R., Dubourg, V., et al.: Scikit-learn: Machine learning in python. The Journal of Machine Learning Research 12, 2825–2830 (2011)

Fully-Automated "Timed Up and Go" and "30-Second Chair Stand" Tests Assessment: A Low Cost Approach Based on Arduino and LabVIEW

José Gonçalves[1,3], José Batista[1], and André Novo[2,4]

[1] Polytechnic Institute of Bragança, Department of Electrical Engineering, Bragança, Portugal
{goncalves,jbatista}@ipb.pt
[2] Polytechnic Institute of Bragança, Department of Nursing, Bragança, Portugal
andre@ipb.pt
[3] INESC-TEC (formerly INESC Porto), Portugal
[4] Sports Sciences, Health and Human Development Research Center, Portugal

Abstract. In this paper it is described the prototyping of an instrumented chair that allows to fully-automate the "Timed Up and Go" and "30-Second Chair Stand" tests assessment. The presented functional chair prototype is a low cost approach that uses inexpensive sensors and the Arduino platform as the data acquisition board, with its software developed resorting to LabVIEW. The "Timed up and go test" consists in measuring the time spent in the task execution of standing up from a chair, walk three meters with a maximum speed without running, turn a cone and going back to the initial position. The "30-Second Chair Stand" test consists in the count of the number of completed chair stands in 30 seconds. It are agility and strength tests easy to setup and execute although they lack of repeatability, whenever the measures are taken manually, due to the rough errors that are introduced.

Keywords: LabVIEW, Arduino, Sensors, Timed up and go test, 30-Second Chair Stand test.

1 Introduction

The "Timed up and go test" consists in measuring the time spent in the task execution of standing up from a chair, walk three meters with a maximum speed without running, turn a cone and going back to the initial position. It is an agility test of easy execution, needing only equipment of easy access such as a chair, a cone, a tape measure and a stopwatch, making this setup very easy to assemble [1]. The "30-Second Chair Stand" consists in the count of the number of completed chair stands in 30 seconds, requiring as setup a chair with a straight back without arm rests and a stopwatch, making this setup, also, very easy to assemble [2]. Besides the presented advantages some drawbacks arise such as the

© Springer International Publishing Switzerland 2015
A.P. Moreira et al. (eds.), *CONTROLO'2014 - Proc. of the 11th Port. Conf. on Autom. Control*,
Lecture Notes in Electrical Engineering 321, DOI: 10.1007/978-3-319-10380-8_64

lack of repeatability due to rough errors introduced in the process of obtaining data, making difficult to compare different person's tests. The intervention programs results, that many times promote the obtaining of small gains in the person's functional capacity, can be misleading interpreted, leading to erroneous conclusions, due to the rough errors introduced in the manual timing measure. In the next sections it is described an instrumented chair prototyping, presented in Figure 1, that allows to fully-automate the "Timed up and go" and the "30-Second Chair Stand" tests assessment. The presented functional chair allows repeatability, precision and real time assessment to all the interest variables of the referred tests.

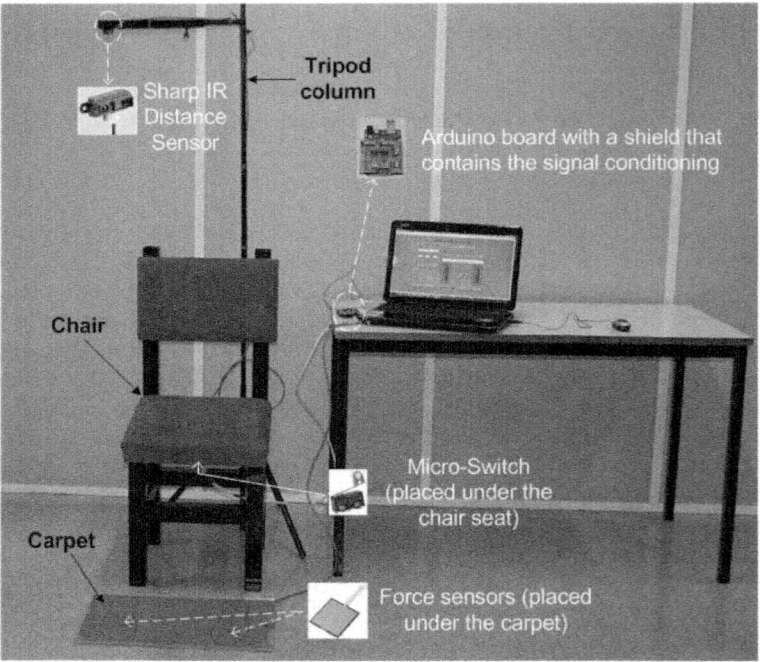

Fig. 1. Functional Chair

The paper is structured as follows, initially the chair instrumentation is explained in detail, then the developed LabView applications are introduced, then some experimental results are presented and finally some conclusions and future work are discussed.

2   Functional Chair Instrumentation

The presented functional chair prototype is a low cost approach, when compared with recent developed prototypes [3], being built with inexpensive sensors and the Arduino platform as the data acquisition board, with its software developed resorting to LabVIEW. The mechanical structure of the functional chair was developed from scratch, in order to better fulfill the project requirements. A patent is pending for the presented approach, being a block diagram of the developed prototype shown in Figure 2.

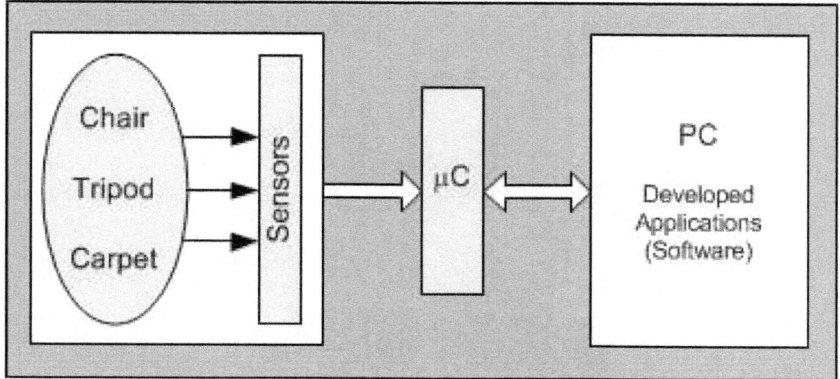

Fig. 2. Functional Chair block diagram

The chair prototype is instrumented with a switch, a distance and a force sensor. Their function and description are detailed in the next subsections. It was developed an Arduino shield that contains the sensors signal conditioning, being presented in Figure 3.

2.1 Switch Sensor

In order to detect if the person is seated the switch sensor D2F-L2 from OMRON was placed underneath the chair seat. The chair seat was provided with springs that allow the sensor to be in off state by default and to turn on when the person is seated. The developed applications prevent debouncing problems associated to this type of switch; this debug is made by software. The information provided by the switch sensor is fundamental for both "Timed up and go" and "30-Second Chair Stand" tests.

2.2 Force Sensor

Force Sensors are applied in both person's feet and the information that they provide is relevant for the "Timed up and go Test" in order to evaluate the time spent to initiate the walk, it also can be used to evaluate the person's equilibrium while seating and standing. The used force sensor was the FSR 06 (Force Sensing Resistor). Although this sensor is not adequate for precision measures, it is widely used in electronic devices that require human touch.

Fig. 3. Developed Arduino Shield

2.3   Sharp Infrared Distance Sensor

The purpose of using a distance sensor was to know if the person is stand up,
being this information relevant for the "30-Second Chair Stand" test. In order
to evaluate this parameter it is necessary to known the person's stature and to
measure the distance to its head, being the sensor placed from above, as shown in
Figure 1. The Sharp family of infrared range finders is very popular for projects
that require cheap and somewhat accurate distance measurements. Some drawback of these sensors is their non-linear response and its mandatory minimum
distance measurement requisites. Their inherently fast response is attractive for
enhancing the systems real-time response. Some IR sensors are based on the
measurement of the phase shift, and offer medium resolution from 5 cm to 10 m
[4], but these are very expensive. This IR sensor is more economical than sonar
rangefinders, yet it provides much better performance than other IR alternatives.
Interfacing to most microcontrollers is straightforward: the single analog output
can be connected to an analog-to-digital converter for taking distance measurements, or the output can be connected to a comparator for threshold detection.

The detection range, suggested by the manufacturer, of the used version is approximately 10 cm to 80 cm, but it offers other models with different measuring ranges. The sensor interface with the microcontroller and its characteristic were previously presented in [5].

3   Developed LabView Applications

The software applications were developed in LabVIEW, their goal is to provide real time assessment and to register the relevant data of the "Timed up and go" and "30-Second Chair Stand" tests. For this purpose it were developed two applications, one for each referred test. After installing the toolkit "LabIEW Interface for Arduino", provided by the National Instruments Company, the Arduino can be used as a data acquisition board [6]. The Arduino Board fulfills the current project requirements, becoming an interesting low cost alternative, even when compared with the cheaper National Instruments Acquisition Boards.

3.1   "Timed Up and Go Test" Application

The "Timed up and go test" application has as goal to register data while a person performs the "Timed up and go" agility test. The person initially is seated in the chair, then receives an order to start walking and initiates the route, finally turns back to the chair, seats and the test is finished. The system registers the time spent to initiate the walk and the total time spent to perform the test. The front panel of the developed application is shown in Figure 4.

On the left side of the panel the data of the monitorized person can be inserted: name, age, weight and stature. If the person is seated, the switch beneath the chair seat is active. In order to initiate the test the person must be seated with its feet pressing the force sensors that are placed under the carpet. The front

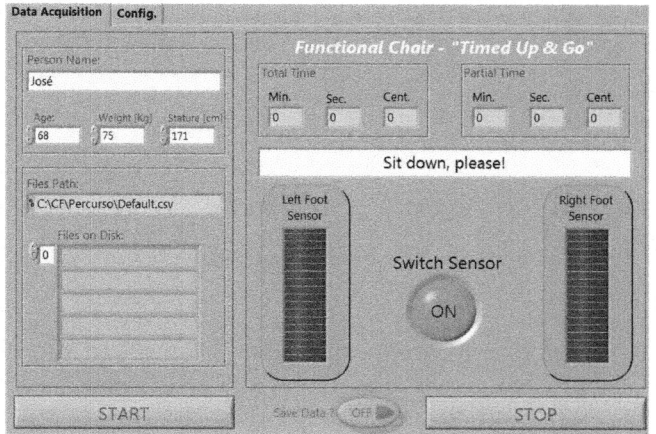

Fig. 4. "Timed up and go test" application front panel

panel shows the pressure that is made by each foot, being visualized the left and the right foot pressure. The information provided by the force sensors and by the switch sensor is used to measure the time spent by the person to initiate the walk. This measure is called "partial time", which corresponds to the time spent from the person's stand (switch turns on) to the walk initiate (pressure release in one of the force sensors). After the walk being initiated the person follows the predefined path. The test is finalized when the person seats in the chair activating the switch sensor, providing information of the total time spent. The test data can be assessed in real-time and stored for posterior analysis. The software can also detect some faults in the test procedure, for example if the person releases pressure from a foot without standing up, the test is considered invalid and must repeated.

3.2 "30-Second Chair Stand Test" Application

The "30-Second Chair Stand" application has as goal to register data while a person performs the "30-Second Chair Stand" strength and endurance leg test. In this test the person stands and seats during 30 seconds, being this value registered. It is also evaluated if the person is standing up correctly, being associated to each stand movement a distance measure that determines if the stand is done correctly. The front panel of the developed application is shown in Figure 5.

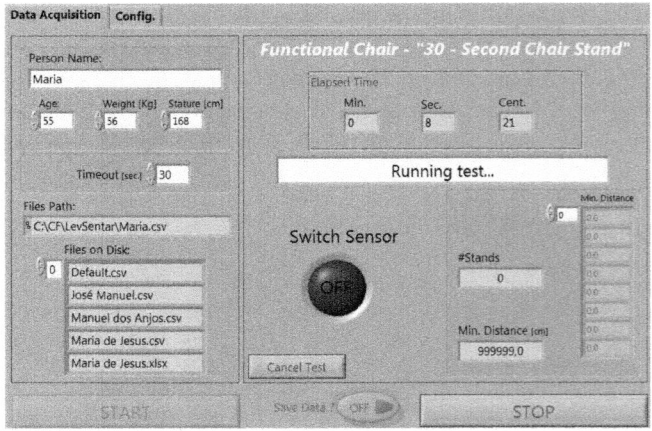

Fig. 5. "30-Second Chair Stand test" application front panel

On the left side of the panel the data of the monitorized person can be inserted: name, age, weight and stature. It can also be inserted the time duration of the test, being 30 seconds the default value. The system counts the number of the person stands during this period, considering the stand valid if it is performed correctly. The distance from the distance sensor to person's head is measured continuously and being stored only its minimum distance for each stand movement.

4   Experimental Results

In order to validate the instrumented chair experimental data was acquired. It was chosen an elder population with its age superior to 65 years, this population was chosen because it is very important to evaluate the functional capacity of elder persons, promoting, as much as possible, the aging with quality of life. The data was acquired at the Betânia Foundation, which is a non profit organization that has as mission to work with elder persons both as home support as well as a nursing home [7]. The acquired data was interpreted only in the perspective of evaluating the instrumented chair as an equipment, rather than the chosen population functional capacity.

In the next subsections it is shown, as an illustrative example, the tests of one person. The selected person is shown in Figure 6, with its personal data shown in Table 1. This person was chosen by the authors from one universe of 4 persons that performed the tests, being this group chosen by the Betânia Foundation physiotherapeut.

Fig. 6. Person seated in the instrumented chair

.1   "Timed Up and Go Test"

he person performs the test as described in subsection 3.1, the registered data
 presented in Table 2.

Table 1. Selected person data

Name	Manuel Paulo Pires
Age	86
Weight(Kg)	65
Stature m	1.68

Table 2. "Timed up and go" test results

Name	Manuel Paulo Pires
Start time	15:02:04
Start walking (s)	10
Total time (s)	39

Table 3. "30-Second Chair Stand" results

Name	Manuel Paulo Pires
Start time	15:10:15
Stands	Minimum distance (cm)
1	34
2	30.1
3	35.4
4	35.4
5	36
6	33.6
7	39.7
8	37.9
9	33.1

In the Table 2 it is shown the "Start time" which is when the person stands up being this information given by the switch sensor placed underneath the chair seat. Then it is presented the "Start walking" which is a partial time between the "Stand up" event and the "Start walking" event, triggered by the information provided by the force sensors placed beneath the person's feet. Finally it is presented the "Total time" that represents the time spent from standing to seating. In a future software version it is also going to be included the "Time to react", which is the time spent to stand, after the order to initiate the walk is given.

4.2   "30-Second Chair Stand"

The person performs the test as described in subsection 3.2. The registered data is presented in Table 3. The distance sensor was placed at a distance of 30 cm above of the person's head while stand up. This distance was chosen having in mind the restriction of the required minimal distance and also to get measure in a range that the sensor has a good sensitivity and less noise. The test is performed during 30 seconds and it is registered for each stand the distance to the sensor in order to evaluate if the person in standing up correctly.

In a future version, the developed application will also provide an alternative to this test, it will count the time that a person needs to seat and stand 5 times. This test is more suitable for an elder population, being easier to perform by persons with reduced functional capacity.

5 Conclusions and Future Work

In this paper it was described the prototyping of an instrumented chair that allows to fully-automate the "Timed Up and Go" and "30-Second Chair Stand" tests assessment. The presented functional chair prototype is a low cost approach that uses inexpensive sensors and the Arduino platform as the data acquisition board, with its software developed resorting to LabVIEW. It are agility and strength tests easy to setup and execute although they lack of repeatability, whenever the measures are taken manually, due to the rough errors that are introduced. The presented prototype was validated for both "Timed Up and Go" and "30-Second Chair Stand" tests, assuring repeatability and accuracy in the presented results.

As future work the authors intend to add more functionalities to the prototype in order to integrate some extra rehabilitation tests assessment. The extra hardware is optional, in order to maintain the cost of the prototype as low as possible, being more attractive for the users that pretend to only use its basics functionalities.

Acknowledgements. The authors would like to thank to the Polytechnic Institute of Bragança for partially funding this project, to the Betânia Foundation for the support in the prototype validation and to Father Paulo for allowing the diffusion of its picture and its tests results in this paper.

The present work is also partially funded by Best Case - Cooperation Project "NORTE-07-0124-FEDER-000060", which is financed by the North Portugal Regional Operational Programme (ON.2 O Novo Norte), under the National Strategic Reference Framework (NSRF), through the European Regional Development Fund (ERDF), and by national funds, through the Portuguese funding agency, Foundation for Science and Technology (FCT).

References

1. Bohannon, R.W.: Reference values for the Timed Up and Go Test: A Descriptive Meta-Analysis Journal of Geriatric Physical Therapy (2006)
2. Millor, N., Lecumberri, P., Gmez, M., Martnez-Ramrez, A., Izquierdo, M.: An evaluation of the 30-s chair stand test in older adults: frailty detection based on kinematic parameters from a single inertial unit. International Journal of NeuroEngineering and Rehabilitation (2013)

3. Frenken, T., Vester, B., Brell, M., Hein, A.: ATUG: Fully-automated timed up and go assessment using ambient sensor technologies. In: 5th International Conference on Pervasive Computing Technologies for Healthcare 2011 (2011)
4. Mohammad, T.: Using Ultrasonic and Infrared Sensors for Distance Measurement World Academy Of Science. Engineering and Technology (2009)
5. Malheiros, P., Gonçalves, J., Costa, P.: Towards a more accurate model for an infrared distance sensor. In: International Symposium on Computational Intelligence for Engineering Systems, 18-19 November 2009. ISEP-Porto, Portugal (2009)
6. National Instruments NI LabVIEW Interface for Arduino Toolkit (October 2013), http://sine.ni.com/nips/cds/view/p/lang/pt/nid/209835
7. Betânia Foundation (2013), http://www.fundacaobetania.com
8. Thompson, M., Medley, A.: Performance of Individuals with Parkinson's Disease on the Timed Up & Go Journal of Neurologic Physical Therapy (2009)

Kalman Filter-Based Yaw Angle Estimation by Fusing Inertial and Magnetic Sensing

Pedro Neto[1], Nuno Mendes[1], and António Paulo Moreira[2]

[1] CEMUC, Department of Mechanical Engineering, University of Coimbra, Coimbra, Portugal
{pedro.neto,nuno.mendes}@dem.uc.pt
[2] INESCTEC, University of Porto, Porto, Portugal
{amoreira}@fe.up.pt

Abstract. Orientation estimation plays a crucial role in robotics. Precise and reliable estimation of orientation, and the yaw angle in particular, is still a challenge and subject of great concern among researchers. This paper presents the development of a platform for yaw angle estimation by fusing inertial and magnetic sensing (a low-cost multi-sensorial system composed by both a digital compass and a gyroscope). A Kalman filter is used to estimate the error produced by the gyroscope. Experimental results indicate that the proposed solution is able to eliminate the drift effect produced by gyroscope data and, at the same time, has the capacity to react to fast orientation changes.

Keywords: Yaw estimation, Kalman, Sensor fusion.

1 Introduction

It is of crucial importance for the robotics field, especially for robot autonomy, to have the capacity to estimate reliable orientations. Different approaches have been employed in estimating roll and pitch angles, trying to create drift-free solutions [1]. In this context, the yaw angle is more difficult to estimate because the gravity measured by accelerometers cannot be used to help to estimate it. Thus, a common solution to estimate yaw angles relies on the fusion of inertial and magnetic sensing. Field *et al.* present a review on motion capture technologies and current challenges associated to their application in robotic systems [2]. In fact, multi-sensor fusion has been applied in many different ways to improve human-robot interaction [3].

In recent years, diverse sensors have become increasingly miniaturized (in size and weight) and cheaper. Inertial sensors (accelerometers and gyroscopes) are no exception. However, only recent advances in micro-electro-mechanical systems (MEMS) have reduced their size and cost considerably, and increased their accuracy. Inertial sensors perform well in motion sensing because they operate regardless of external references, friction, winds, magnetic fields, directions and dimensions. As a drawback, most inertial sensors are temperature dependent and are not very well suited for absolute tracking due to error accumulation. On contrary, they perform better in sensing applications involving relative motion. Inertial measurement units (IMUs) consist

Springer International Publishing Switzerland 2015   679
P. Moreira et al. (eds.), *CONTROLO'2014 - Proc. of the 11th Port. Conf. on Autom. Control*,
Lecture Notes in Electrical Engineering 321, DOI: 10.1007/978-3-319-10380-8_65

of a group of inertial sensors that are combined into a unique system with the aim to measure the orientation and position of the unit throughout space and time.

Magnetic sensors, advantageously, allow obtaining an absolute reference for the system in study and do not suffer from the problem of drift that the inertial sensors suffer. A major drawback of magnetic-based sensors is its sensitivity to magnetic distortions in the Earth's magnetic field. In this way, it is difficult to acquire reliable readings in the proximity of moving magnetic fields, batteries and ferrous materials.

Other sensors such as optical sensors, ultrasonic sensors and GPS-based systems have been used for orientation and position estimation. Many tracking systems are based on the combination of different sensor types, hybrid solutions. There are many possibilities to combine individual sensors into a new multi-sensorial system. The positive aspects of different sensors can be explored and combined, originating a "better" sensor. For example, a digital compass, which provides body orientation, can be composed by accelerometers, magnetometers and a temperature sensor. This combination improves the long-term stability and accuracy of data provided by the compass. The small sensors that combine inertial and magnetic sensing are usually called miniature magnetic and inertial measurement units (MIMUs). Some studies combine accelerometers and gyroscopes to achieve reliable orientation data [4]. The Allan technique can be used in analyzing and modeling the error of inertial sensors [5]. An interesting paper presents a complementary filtering algorithm for estimating orientation based on inertial/magnetic sensor measurements [6]. Vlasic et al. use inertial sensors with ultrasonic detection for a practical outdoor capture technique [7]. Kourogi and Kurata developed a system which estimates poses by integrating data from several sensors attached to a human body using Kalman filter [8]. Miller et al. report the use of a set of inertial sensors (three sensors) to control the robot arm of NASA Robonaut [9]. Also, the combination of inertial and magnetic sensors for motion tracking has been studied [10].

Each different type of sensor has its own advantages and disadvantages. How to effectively integrate/fuse multi-sensor information is the question. Several multi-sensor data fusion methods have been proposed over the years, combining observations from different sensors to achieve "better" descriptions of environments or processes of interest. Error can arise from different factors and situations: misalignment of sensors, calibration, bias, scale factor, thermal drift, unpredictable magnetic fields, etc. Some error sources have a random origin and can only be treated with stochastic processes. In order to achieve the desired performance, great concern must be paid in relation to all the possible sources of error above mentioned. Jurman et al. present several methods to maximize the performance of a MIMU in terms of calibration and alignment of sensors [11].

1.1   Proposed Approach

The method proposed in this paper relies on multi-sensor data fusion (a digital compass and a gyroscope) to make yaw angle estimation more accurate and reliable. The digital compass has embedded a 3-axis magnetic sensor (magnetic sensing) and a 3-axis accelerometer (inertial sensing). It provides roll, pitch and heading angles as well as acceleration data. Heading is the angle between the North direction and the direction of the longitudinal axis of the compass in the horizontal plane. Note that heading data can be easily transformed into yaw data by describing it in relation to a known frame. Digital compasses suffer from some of the weaknesses associated with both magnetic and inertial sensing. It can be reported that, for example, depending on the geographical location and inclination of the compass a tilt measurement can affect heading accuracy. In order to achieve more accuracy in yaw estimation, we are proposing to use Kalman filtering to fuse heading measurements from the compass with integrated angular rates from the gyroscope, Fig.1. A body's angular rate needs to be integrated once to obtain a relative orientation angle, the yaw angle in this scenario. In this context, the gyroscope measures faster changes of orientation (short-term accuracy) compensating the weakness of the compass in this aspect. In the other hand, the compass provides long-term stability of output data. Thus, the individual strength of each sensor is maximized. Roll and pitch angles given by the compass are stable as they are estimated by fusing data from the magnetometers and accelerometers embedded into the compass.

At first sight the proposed solution does not bring any important innovation to the state of the art in yaw angle estimation methods. However, it explores and combines the capacity of different low-cost sensors to achieve a major goal, reliable yaw angle estimation. Experimental results indicate that the proposed solution is able to eliminate the drift effect produced by gyroscope data and, at the same time, has the capacity to react to fast orientation changes.

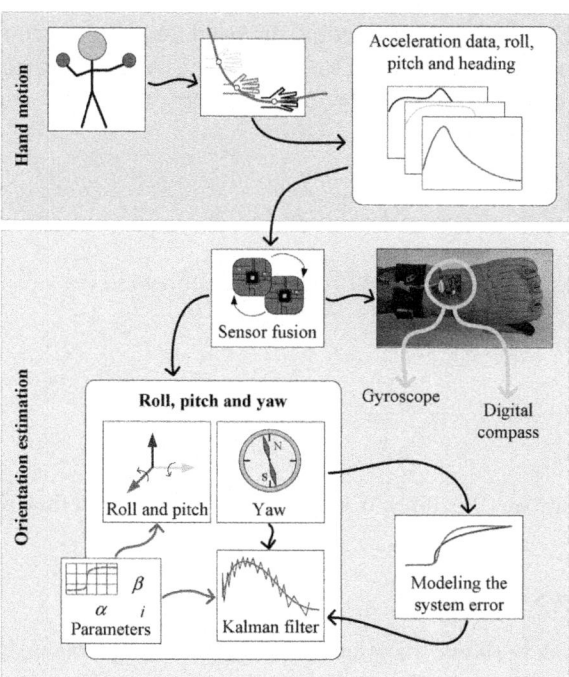

Fig. 1. Layout of the system

2   Yaw Estimation

In order to estimate "more reliable" yaw angles, we are proposing a Kalman filter to fuse heading/yaw data from a digital compass with integrated angular rates from a gyroscope. Thus, the strength of a sensor compensates the weakness of the other, i.e., gyroscope data compensate compass data when it gets disturbed and *vice versa*. In addition, the gyroscope provides data reporting a faster reaction to rotation. On the other hand, heading/yaw angles from the digital compass contribute to determine an absolute angle and to minimize errors (drift) produced by the gyroscope.

Yaw angles estimated by integrating angular rates from the gyroscope have an error that accumulates with time so that after a short period of time the estimated angles can be totally incorrect. In this way, only short-term accuracy can be achieved using gyroscope measurements. In fact, gyroscopes are characterized by its short-term accuracy and long-term drift. Yaw angles from the compass ensure long-term stability and reliability. On contrary, the magnetic sensing characteristics of the compass can lead to distortions in estimated angles.

A body's angular velocity needs to be integrated once to obtain a relative orientation angle (the yaw angle in this case):

$$\psi(t) = \psi_0 + \int \omega(t)\ dt \tag{1}$$

The angular velocity ω sensed by the gyroscope includes components resulting from the Earth's rotation and the sensor motion. However, for the case of the low-cost MEMS with drifts significantly exceeding the magnitudes of the components referred above, these terms can be omitted. In summary, the gyroscope provides discrete angular rates $\dot{\psi}_g$, which are numerically integrated obtain discrete angular increments $\Delta \psi_g$:

$$\Delta \psi_g(k) = \dot{\psi}_g(k)\ \Delta t \tag{2}$$

These angular increments are added to obtain an estimate to the yaw angle $\psi_g(n)$ at a time t_n:

$$\psi_g(n) = \psi_c(0) + \sum_{k=1}^{n} \Delta \psi_g(k) \tag{3}$$

Initial yaw estimate $\psi_c(0)$ comes from the digital compass, so that $\psi_g(0) = \psi_c(0)$.

2.1   Modelling System Error

Kalman filter models should be simple enough to be implemented and at the same time capable to represent the physical scenario in study with accuracy. Modeling gyroscope errors can be a difficult task because there are different error sources and usually, it is necessary to decide what are the most important to take into account. The proposed approach is based on previous studies in the field [12-14]. Fig. 2 shows the proposed solution to estimate yaw angles from gyroscope and compass data. A Kalman filter is used to estimate errors of gyroscope yaw angles, providing such errors to the error updater. The error updater accumulates them and computes the total estimated yaw error $\delta\hat{\psi}_g$, which is subsequently subtracted from the gyroscope yaw angle ψ_g, providing the estimated yaw angle $\hat{\psi}$:

$$\hat{\psi} = \psi_g - \delta\hat{\psi}_g \tag{4}$$

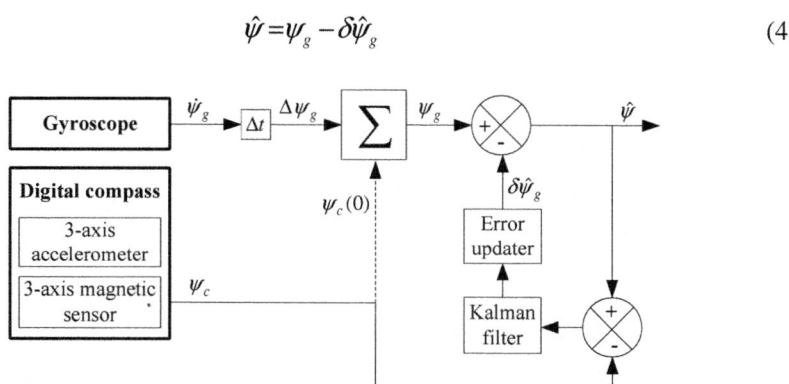

Fig. 2. Estimating yaw angles

2.2   Kalman Filter – Implementation

The system in study can be modeled (discrete model) as follows:

$$\mathbf{x}_{k+1} = \mathbf{\Phi}_k \mathbf{x}_k + \mathbf{w}_k \tag{5}$$

Where $\mathbf{x}_{k+1}$ is the process state vector at step $k+1$, $\mathbf{\Phi}_k$ is the state transition matrix from step k to step $k+1$, $\mathbf{x}_k$ is the process state vector at step k and $\mathbf{w}_k$ is a vector assumed to be a white sequence with known covariance structure (process noise).

The observation/measurement of the process in study is assumed to occur at a discrete time in accordance with:

$$\mathbf{z}_k = \mathbf{H}_k \mathbf{x}_k + \mathbf{v}_k \tag{6}$$

Where $\mathbf{z}_k$ is a measurement vector at step k, $\mathbf{H}_k$ is a measurement matrix and $\mathbf{v}_k$ the measurement error vector.

Error in gyroscope measurements includes scale factor error, bias and Gaussian white noise [12]. Such errors can be modeled by the following:

$$\delta\dot{\psi}_g = \delta k\ \omega + b + u_\psi \tag{7}$$

$$\delta\dot{k} = u_k \tag{8}$$

$$\dot{b} = u_b \tag{9}$$

Where $\delta\psi_g$ is the yaw error from the gyroscope, δk the scale factor error, b the gyroscope bias and u_ψ, u_k, u_b are the forcing functions, i.e., Gaussian white noise with power spectral densities S_ψ, S_k and S_b, respectively [15].

Gyroscope error models (7), (8) and (9) can be grouped in a matrix form:

$$\frac{d}{dt}\begin{bmatrix} \delta\psi_g \\ \delta k \\ b \end{bmatrix} = \underbrace{\begin{bmatrix} 0 & \omega & 1 \\ 0 & 0 & 0 \\ 0 & 0 & 0 \end{bmatrix}}_{F}\begin{bmatrix} \delta\psi_g \\ \delta k \\ b \end{bmatrix} + \begin{bmatrix} u_\psi \\ u_k \\ u_b \end{bmatrix} \tag{10}$$

Being the state vector:

$$\mathbf{x} = \begin{bmatrix} \delta\psi_g & \delta k & b \end{bmatrix}^T \tag{11}$$

In (10) we have a linear continuous dynamic model of the system in study, with the following general form:

$$\dot{\mathbf{x}} = \mathbf{F}\mathbf{x} + \mathbf{u} \tag{12}$$

Where $\mathbf{F}$ is the fundamental matrix of the system and $\mathbf{u}$ a vector of continuous random process disturbances. This continuous model has to be converted into a discrete model of the system (5) [14]. The state transition matrix can be determined by:

$$\Phi = \mathcal{L}^{-1}\left[\left(s\mathbf{I} - \mathbf{F}\right)^{-1}\right]_{t=\Delta t} \tag{13}$$

Where $\mathbf{I}$ is the identity matrix and s the Laplace variable. The Gauss-Jordan method is applied to invert the matrix. Applying the inverse Laplace transform to (13):

$$\Phi = \begin{bmatrix} 1 & \Delta\psi_g & \Delta t \\ 0 & 1 & 0 \\ 0 & 0 & 1 \end{bmatrix} \tag{14}$$

From (5), (10) and (14):

$$\begin{bmatrix} \delta\psi_g \\ \delta k \\ b \end{bmatrix}_{k+1} = \begin{bmatrix} 1 & \Delta\psi_g & \Delta t \\ 0 & 1 & 0 \\ 0 & 0 & 1 \end{bmatrix}\begin{bmatrix} \delta\psi_g \\ \delta k \\ b \end{bmatrix}_k + \mathbf{w}_k \tag{15}$$

Now, we need to achieve the covariance matrix $\mathbf{Q}_k$ associated with $\mathbf{w}_k$:

$$\mathbf{Q}_k = E\left[\mathbf{w}_k \mathbf{w}_k^T\right] \tag{16}$$

Where $E[\ldots]$ represents the expected value. The $\mathbf{Q}_k$ elements are calculated using the transfer function method as shown in [14] and [12]:

$$\mathbf{Q}_k = \begin{bmatrix} \dfrac{S_b \Delta t^3}{3} + \dfrac{S_k \Delta \psi_g^2 \Delta t}{3} + S_\psi \Delta t & \dfrac{\Delta \psi_g S_k \Delta t}{2} & \dfrac{S_b \Delta t^2}{2} \\[3mm] \dfrac{\Delta \psi_g S_k \Delta t}{2} & S_k \Delta t & 0 \\[3mm] \dfrac{S_b \Delta t^2}{2} & 0 & S_b \Delta t \end{bmatrix} \tag{17}$$

As indicated in Fig. 2, the input for the Kalman filter is the difference between the corrected gyroscope yaw $\hat{\psi} = \psi_g - \delta\hat{\psi}_g$ and the yaw angle from the compass ψ_c. This is true assuming that the compass has a zero mean error and the gyroscope presents a considerable drift. Note that the correcting term $\delta\hat{\psi}_g$ is a component of the vector of deterministic inputs $\delta\hat{\mathbf{x}}$. The observation model, in which ψ represents the true value of yaw, is given by the following:

$$z = \psi_g - \psi_c = \left(\psi + \delta\psi_g\right) - \left(\psi + v_c\right) = \delta\psi_g - v_c = \begin{bmatrix} 1 & 0 & 0 \end{bmatrix} \begin{bmatrix} \delta\psi_g \\ \delta k \\ b \end{bmatrix} - v_c \tag{18}$$

By comparing (6) with (18) we can see that the measurement vector $\mathbf{z}_k$ and the vector of measurement noises $\mathbf{v}_k$ become scalars z and v_c respectively. In this way we have that:

$$\mathbf{H}_k = \begin{bmatrix} 1 & 0 & 0 \end{bmatrix} \tag{19}$$

It is also necessary to establish the covariance matrix $\mathbf{R}_k$ of measurement noises $\mathbf{v}_k$, which for this system is a scalar, equal to the variance of the compass measurement errors v_c. $\mathbf{R}_k$ can be computed by taking some off-line sampling measurements.

$$\mathbf{R} = \sigma_c^2 \tag{20}$$

We have now all the parameters necessary to implement the Kalman filter algorithm, Fig. 3, in which:

- $\hat{\mathbf{x}}_{k+1|k}$ is the predicted state vector at step $k+1$ (before the measurement update).
- $\hat{\mathbf{x}}_{k+1|k+1}$ is the filtered state vector at step $k+1$ (after the measurement update).

- $\mathbf{P}_{k+1|k}$ covariance matrix of prediction errors.
- $\mathbf{P}_{k+1|k+1}$ covariance matrix of filtration errors.
- $\mathbf{K}_{k+1}$ Kalman gain.
- $\delta\hat{\mathbf{x}}_k$ vector of corrections from the Kalman filter (error updater).

It is necessary to establish an initial value for $\mathbf{P}$, $\mathbf{P}_{0|0}$. If we are absolutely certain that our initial state estimate $\hat{\mathbf{x}}_{0|0}$ is correct, we would let $\mathbf{P}_{0|0} = 0$. However, given the uncertainty in our initial estimate, by choosing $\mathbf{P}_{0|0} = 0$ would cause the filter to initially and always believe that $\hat{\mathbf{x}}_k = 0$. Thus, we could choose almost any $\mathbf{P}_{0|0} \neq 0$ and the filter would eventually converge. In this case we choose $\mathbf{P}_{0|0} = \mathbf{I}$, being $\mathbf{I}$ the identity matrix. This choice for the identity matrix is due to the fact that this initial covariance is considerably higher than the values for which the filter stabilizes after to converge. Also, it was verified that this initial value is high enough to do not affect the initial convergence velocity.

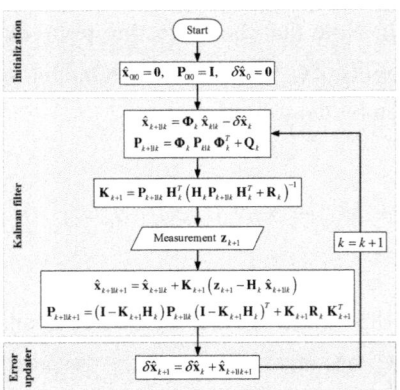

Fig. 3. Kalman filter loop

3   Experiments and Results

The experimental evaluation of the proposed platform to estimate yaw angles from the fusion of inertial and magnetic sensing consisted in a laboratory test in which the MIMU is attached to the user's hand. The MIMU motion tracking system is composed by a digital compass (*OceanServer OS500-US*) and a single-axis gyroscope. The digital compass is tilt-compensated and contains hard- and soft-iron compensation routines.

Initially, the user's hand is oriented to have a yaw angle of about 18°. During this phase the user causes the hand to vibrate slightly in order to analyse the system behaviour in such conditions. Then, the hand is rotated to an angle of about 62°. There follows a period in which the hand is static and after that the hand is re-orientated for an angle of about 21°. The results presented in Fig. 4 contain recordings of yaw angles from the compass, the gyroscope, the Kalman filter and a smoothed Kalman filter. The smoothing function is a very simple one, in which for each point provided by the Kalman filter it is computed the average value among its previous three neighbours.

$$x_{i'} = \frac{2x_i + \sum_{j=1}^{3} x_{i-j}}{5} \qquad (21)$$

Analyzing the first phase of the experiment (until to reach the cycle time 60), the results produced by the Kalman filter and smoothed Kalman filter are stable, in line with the angles provided by the compass. In the end of this initial phase, it is possible to see that the gyroscope reacts faster than the compass to the variation in orientation. In line with the results provided by gyroscope, the Kalman filter also reacts fast to that change in orientation. This is further evidence that the Kalman filter takes advantage of the positive characteristics of the gyroscope (short-term accuracy) and compass (long-term stability). This aspect can be tremendously important for real-time feedback applications. Between cycle time 60 and 100 the user's hand is static, keeping a yaw angle of about 62°. In this situation it is clearly visible the gyroscope drift, Fig. 4. After that, when the user's hand is re-orientated for an angle of about 21° it is possible to see in Fig. 4 that one more time the Kalman filter helps the system to react fast to the change in orientation. As a final remark it can be stated that the Kalman filter helps to preserve the fast response of the gyroscope and the long-term stability of the compass, eliminating the problem of increasing gyroscope errors.

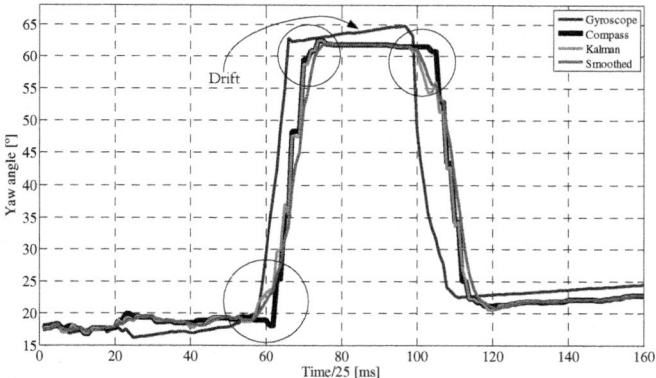

Fig. 4. Estimated yaw angles

Conclusions

A new method for estimating "more reliable" yaw angles has been described. The proposed solution is based on the fusion *via* Kalman filter of magnetic and inertial sensing. The Kalman filter allows to preserve the fast response of the gyroscope and the long-term stability of the compass, eliminating the problem of increasing gyroscope errors (drift). It explores and combines the capacity of different low-cost sensors to achieve a major goal, reliable yaw angle estimation. Experimental results indicate that the proposed solution is able to eliminate the drift effect produced by gyroscope data and, at the same time, has the capacity to react to fast orientation changes.

Acknowledgements. Project "NORTE-07-0124-FEDER-000060" is financed Programme (ON.2 – O Novo Norte), under the National Strategic Reference Framework (NSRF), through the European Regional Development Fund (ERDF), and by national funds, through the Portuguese agency, Fundação para a Ciência e a Tecnologia (FCT).

References

1. Rehbinder, H., Hu, X.: Drift Free Attitude Estimation for Accelerated Rigid Bodies. Automatica 40, 653–659 (2004)
2. Field, M., Pan, Z., Stirling, D., Naghdy, F.: Human Motion Capture Sensors and Analysis in Robotics. Industrial Robot 38, 163–171 (2011)
3. Neto, P., Pereira, D., Pires, J.N., Moreira, A.P.: Real-Time and Continuous Hand Gesture Spotting: an Approach Based on Artificial Neural Networks. In: IEEE Int. Conf. on Robotics and Automation, pp. 178–183 (2013)
4. Luinge, H.J.: Inertial Sensing of Human Movement, Ph.D. dissertation, Twente Univ. Twente, Netherlands (2002)
5. El-Sheimy, H.H., Niu, X.: Analysis and Modeling of Inertial Sensors using Allan Variance. IEEE Trans. on Instrumentation and Measurement 57, 140–149 (2008)
6. Calusdian, J., Yun, X., Bachmann, E.: Adaptive-Gain Complementary Filter of Inertial and Magnetic Data for Orientation Estimation. In: Proc. IEEE Int. Conf. on Robotics and Automation, pp. 1916–1921 (2011)
7. Vlasic, D., Adelsberger, R., Vannucci, G., Barnwell, J., Gross, M., Matusik, W., Popovic, J.: Practical Motion Capture in Everyday Surroundings. ACM Trans. on Graphics 26, 351–359 (2007)
8. Kourogi, M., Kurata, T.: A Method of Personal Positioning Based on Sensor Data Fusion of Wearable Camera and Self-Contained Sensors. In: Proc. IEEE Int. Conf. on Multisensor Fusion and Integration for Intelligent Systems, pp. 287–292 (2003)
9. Miller, N., Jenkins, O.C., Kallmann, M., Mataric, M.J.: Motion Capture from Inertial Sensing for Untethered Humanoid Teleoperation. In: Proc. 4th IEEE/RAS Int. Conf. on Humanoid Robots, pp. 547–565 (2004)
10. Neto, P., Pires, J.N., Moreira, A.P.: 3-D Position Estimation from Inertial Sensing: Minimizing the Error from the Process of Double Integration of Accelerations. In: Annual Conf. of the IEEE Industrial Electronics Society, pp. 4024–4029 (2013)
11. Jurman, D., Jankovec, M., Kamnik, R., Topic, M.: Calibration and Data Fusion Solution for the Miniature Attitude and Heading Reference System. Sens Actuators 138, 411–42 (2007)
12. Kaniewski, P., Kazubek, J.: Integrated System for Heading Determination. Acta Physica Polonica A 116, 325–330 (2009)
13. Kaniewski, P., Kazubek, J.: Operation and Implementation of Heading Reference System Coordinates 7, 14–18 (2011)
14. Brown, R.G., Hwang, P.Y.C.: Introduction to Random Signals and Applied Kalman Filtering. John Wiley & Sons (1997)
15. Petkov, P., Slavov, T.: Stochastic Modeling of MEMS Inertial Sensors. Cybernetics and Information Technologies 10, 31–40 (2010)

Electronic Nose

Edmundo R. Soares, Samuel Cabete, Nuno Miguel Fonseca Ferreira,
and Fernando J.T.E. Ferreira

Instituto Politécnico de Coimbra
Instituto Superior de Engenharia de Coimbra
Departamento de Engenharia Eletrotécnica
Rua Pedro Nunes - Quinta da Nora 3030-199 Coimbra, Portugal
{a21200244,a21200213}@alunos.isec.pt, nunoferreira@ipc.pt,
fernando@isec.pt

Abstract. Currently, we are facing increasingly environmental issues affecting our cities and consequently the health of living beings. A practical example it's the pollution, the excess of different odors and gases.

Thus, our project is to create a sensor integrated in a smart grid that can measure gas pollution. This new sensor has the ability of detecting and classification of different odors that reflect the air quality in a given space.

Keywords: Control, Monitoring, Air pollution, Communication, Wi-Fi.

1 Introduction

This project consists on the development of an electronic nose for the detection and classification of odors that reflect the level of cleanliness or quality of the air in a given space (bathroom, hotel room, etc.), communicating wirelessly with a computer or smartphone.

This system integrates an array of gas sensors, an Arduino board Mega for data processing, as well as a Wi-Fi module, also Arduino, for wireless communication and the implementation of a fuzzy control system.

This project incorporated the following tasks:

- Study of existing gas sensors, and selection of the best sensor array;
- Elaboration of an Electronic Project and construction of an operational prototype;
- Experimental tests;

2 Prototype Architecture

The monitoring and control system consists in distinct parts: the acquisition board, the sensors, the Wi-Fi communication, and the database which is allocated on online server and is used to record all information from the devices.

© Springer International Publishing Switzerland 2015   689
A.P. Moreira et al. (eds.), *CONTROLO'2014 - Proc. of the 11th Port. Conf. on Autom. Control*,
Lecture Notes in Electrical Engineering 321, DOI: 10.1007/978-3-319-10380-8_66

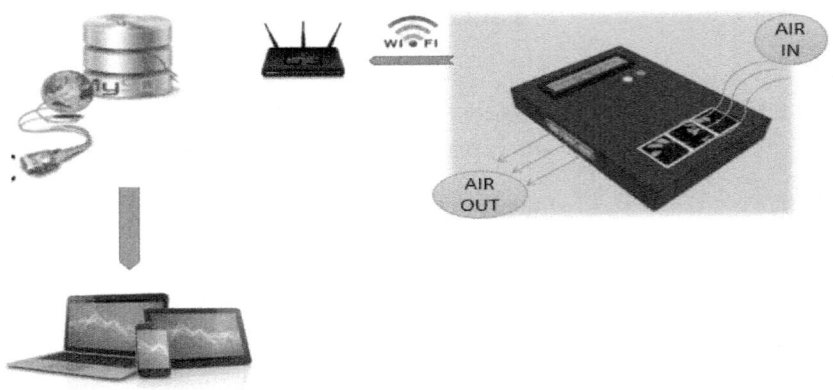

Fig. 1. Prototype Architecture

2.1   Acquisition Board and Data Processing

For this task we used an Arduino Mega 2560 board, since it has 16 analog inputs, 8 more than the Uno, which allowed us to integrate more and use a larger variety of analog sensors.

2.2   Wi-Fi Communication

In order to communicate with the outside, we opted for the wi-fi technology, since most of the buildings already have this network infrastructure, thus being capable of receiving the equipment. Therefore, it is not necessary an additional cost for the creation of the necessary infrastructure for communication.

2.3   Gas Sensors

The sensor array consists on a set of six different gas sensors, which are the ones that we believe to be the most suitable in order to obtain the biggest combination of odors, proper for the project. We chose to use all the sensors of the same brand (Figaro), since it is a brand that works exclusively with gas detection. We used a different brand for the temperature and humidity sensor, opting for the Sensirion.

Name of sensors used:

NGM 2611 – pre calibrated module for Methane
TGS 826 - for the Detection of Ammonia
TGS 2442 - for the detection of Carbon Monoxide
TGS 2600 - for the detection of Air Contaminants
TGS 2602 - for the detection of odorous gases
TGS 2620 - for the detection of Solvent Vapors
SHT15 – for measurement temperature and relative humidity

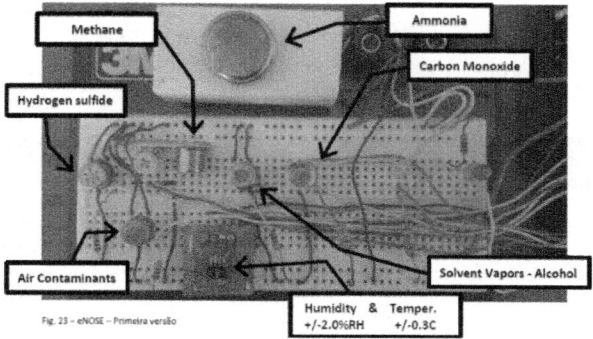

Fig. 2. Gas sensors

2.4   Creation of the Online Platform

Once our project also consisted on the classification of odors, we decided to create an online platform for monitoring the levels of air quality in real time, as well as the possibility to check the historical data over the days in different places, where the various devices are installed.

The platform (http://nariz.host56.com) was developed taking in consideration the point of view of the users, so that these may obtain the necessary information in a simple and effective manner. This also gives them the possibility to access anywhere the level of cleanliness of the place where the devices are located.

The platform developed for this purpose was conducted in PHP, along with HTML, and the MySQL database.

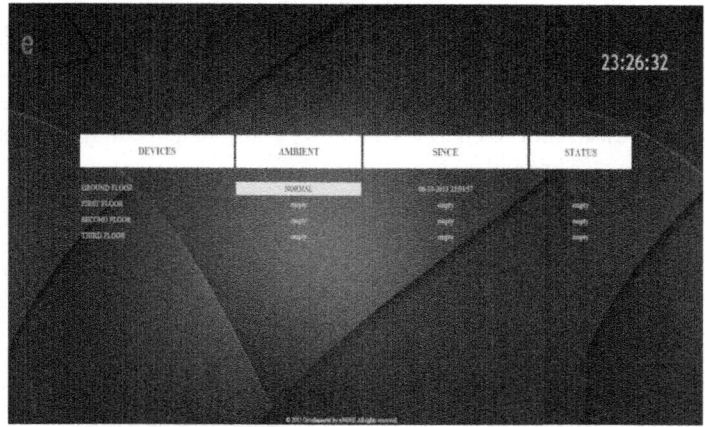

Fig. 3. Monitoring the levels of air quality in real time

In order to view the information that is being obtained over a period of time, we used an existing platform called "xively" where you can find graphs showing the evolution of the information.

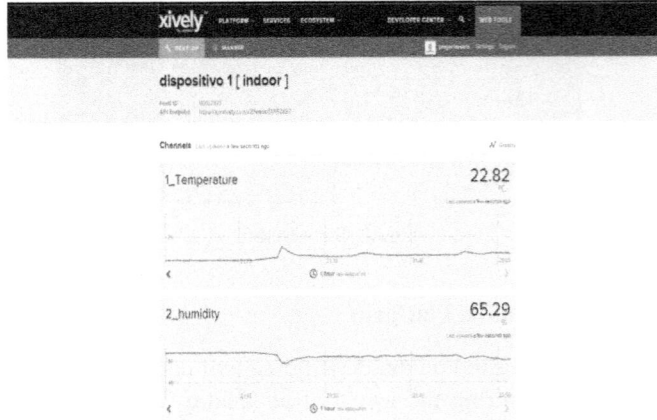

Fig. 4. Historic data

3   Electronic Design and Construction of the Prototype

We developed a small plate, in order to integrate all of the components on a single device. Then, the device was placed in a closed box with two fans, whose main functions are to create an airflow that passes through the different sensors in a controlled way.

We chose to put it in a box with two adjustable speed fans at the top, ensuring a flow of air completely controlled. The adjustment of the fan speed was the solution we managed to find in order to fight the cooling of the sensors, which made it impossible for them to work properly (the sensors have an operating temperature of around 40 ° C).From the viewpoint of energy saving, we chose to activate the fan periodically during the reading (for example, every 5 minutes).

Fig. 5. Prototype

4   Control System

4.1   Fuzzy System

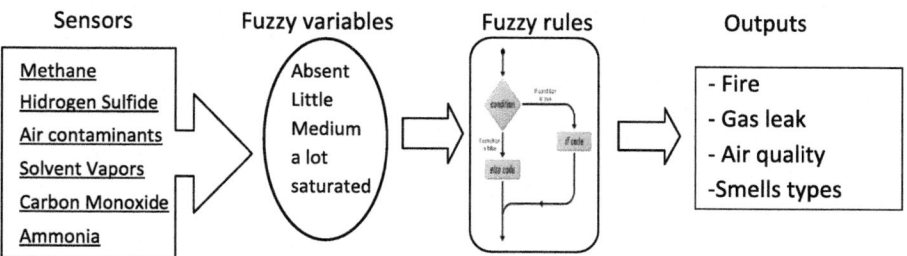

Fig. 6. Implemented fuzzy system

4.2   Fuzzy Rules and Outputs - Different Type of Characters

```
- IF carbon monoxide IS a lot OR saturated THEN Fire;
- IF methane IS medium OR a lot OR saturated THEN gas
leak;
- IF ammonia IS absent OR little AND hydrogen sulphide IS
medium OR a lot OR saturated AND air contaminats IS me-
dium OR a lot OR saturated AND organic vapors IS medium
OR a lot OR saturated THEN bad smell;
- IF ammonia IS medium AND hydrogen sulphide IS absent OR
a little THEN normal smell;
-IF ammonia IS a lot AND hydrogen sulphide IS absent OR a
little THEN good smell;
```

5   Experimental Tests

5.1   Private Bathroom

We can conclude that each time we use the bathroom, there is a rapid and significant increase of the values read by the sensors, such as the methane, air contaminants and the carbon monoxide. The ammonia sensor reaches higher levels when the toilet is flushed. In fact, from these charts, we can't conclude about the state of cleanliness of the facility. This occurs because the bad smell remains only for a short period of time, returning the sensors to baseline levels at the end of this period. We can only conclude that the bathroom needs cleaning if the sensors remain with high values during a longer period of time, therefore the need for testing in a bathroom that has a more intensive use.

Table 1. Private bathroom data

Sensor	clean	After using
Ammonia	90	95
Carbon monoxide	30	50
Air contaminants	165	210
Solvent vapors	35	60
Organic vapors	160	130
Methane	25	50

5.2   Public Bathroom

We concluded that, since it is a bathroom with a more intensive usage, and once its degree of soiling gradually increases throughout the day, there is a bigger variation in the values read by the various sensors and there is still a rapid variation during the time of use. Thus, it is concluded that it is possible to make a proper monitoring of the cleanliness of public bathrooms.

Table 2. Public bathroom data

Sensor	dirty	clean	After using
Ammonia	30	100	45
Carbon monoxide	40	30	50
Air contaminants	190	150	210
Solvent vapors	50	35	60
Organic vapors	140	170	130
Methane	50	25	60

Using these values, we made the code that performs the comparison of the readings obtained by each sensor, in order to update the state of cleanliness in the database.

6   Conclusion

The first conclusion that can be drawn from this work is the versatility of the system. It is found that with an array of sensors suitable for a particular purpose, it is possible to adapt this project for multiple applications. From the control of the cleanliness of toilets, to monitoring air quality in hotel rooms , or a device to control the odour of our breath or BAC, which can be integrated to a smartphone, for example. An application that would be less complex would be a gas leak detector or a fire detector. There are an enormous variety of possibilities.

In this particular situation, we concluded that it is possible to carry out a monitoring of the air quality in the bathroom, in order to reduce the costs of its maintenance. With a device of this kind, the facility doesn't need to be cleaned periodically and shall be only cleaned depending on its actual condition of cleanliness. In some cases this may lead to a reduction in the amount of times the facility is cleaned, resulting in monetary savings and an increase in other cases, preventing dirt and unpleasant odors in the facilities.

In a further phase we could even "teach" the device to detect a certain odor and save it. Every time the sensor detects that certain odor it gives the user an indication that it has been re-detected. This variant would be interesting for an Android application or IOS, helping people monitor the air around them at a personal level.

The project objectives were exceeded and we consider this type of project an asset to us. It made us put into practice various skills acquired throughout our degree, and required us to work autonomously, which also helped us to prepare ourselves for the job market. We intend to continue to develop this device, and perhaps release it in the market.

References

1. https://www.arduino.cc
2. http://www.halito.pt/halitose/biblioteca/09.html
3. Fernandes, D.L.A.: Nariz electrónico para aplicações ambientais, alimentares e clínicas, Universidade de Aveiro (2008)
4. Pimentel, J.G.: Proyeto Final de Carrera Desarrollo Placa Sensores para Monitorización de Gases Tóxicox en Tuneladora en Escudo Abierto, Universidad Carlos III de Madris (2011)
5. Chang, J.B.: Electronic Noses Sniff Success. IEEE (2008)
6. McCormick, Douglas, News for Nose: Machines
7. Use Odor to Diagnose Disease. IEEE (2013)

Modeling of a Low Cost Laser Scanner Sensor

José Lima[1], José Gonçalves[1], and Paulo J. Costa[2]

[1] INESC TEC (formerly INESC Porto) and Polytechnic Institute of Bragança, Portugal
{jllima,goncalves}@ipb.pt
[2] INESC TEC (formerly INESC Porto) and Faculty of Engineering,
University of Porto, Portugal
paco@fe.up.pt

Abstract. A laser scanner is a popular sensor widely used in industry and mobile robots applications that measures the distance to the sensor on a slice of the plan. This sensor can be used in mobile robots localization task.

In this paper, a low cost laser scanner sensor is modelled so that it can be implemented in a simulation environment.

The simulation reflects the laser model properties such as target colour dependences, noise, limits and time constraints. A correction of the laser scanner nonlinearities is proposed. The noise spectrum is also addressed.

1 Introduction

Robot simulators are becoming more and more popular among the researchers and industry. They allow fast development and testing, that is a very important issue nowadays. The more realistic the simulator is, the easiest is to transfer code to real robots. Although the final aim is real robotics, it is often very useful to perform simulations before implementing real robots. This is because simulations are easier to setup, less expensive, faster and more convenient to use. Building up new robot models and setting up experiments is very fast. A simulated robotics setup is less expensive than real robots and real world setups, thus allowing a better design exploration. Simulation often runs faster than real robots while all the parameters are easily debugged. Simulations make it possible to use computer expensive algorithms that run faster than in real robot microcontrollers. Finally, nowadays the simulation results are transferable onto the real robots [1]. Recently, the development of 3D modeling and digitizing technologies has made the model generating process much easier [2]. A laser ranger scanner is a crucial element for a 3D mapping usually used in industry 3D scanning services and robotics. Many indoor robotics systems use laser rangefinders as their primary sensor for mapping, localization, and obstacle avoidance [3]. A remote sensing technology that measures distance by illuminating a target with a laser and analyzing the reflected light is known as *Lidar* sensor.

Usually, *Lidar* are the most expensive component in mobile robots.

There are several well-known Lidar sensors available from Hokuyo (at left) and Sick (at right) manufacturers (Figure 1). The main objective of this paper is to create a model of a low cost distance sensor that can be further used in localization of a mobile robot.

© Springer International Publishing Switzerland 2015
697
A.P. Moreira et al. (eds.), *CONTROLO'2014 - Proc. of the 11th Port. Conf. on Autom. Control*,
Lecture Notes in Electrical Engineering 321, DOI: 10.1007/978-3-319-10380-8_67

Fig. 1. Hokuyo URG-04LX (at left) and Sick TIM551 (at right) Lidar sensors

A very low cost *Lidar* (Piccolo Laser Distance Sensor, presented in Figure 2) is presented and its model is addressed. This laser is only available as a part of an autonomous cleaning robot and there are no paper references available.

Laser model can be introduced in a simulator (SimTwo) allowing that simulation enhances the possibility of test robots with a laser model. Information for the model was based on some experimental setup. This paper also addresses the noise spectrum of this Lidar.

This paper is organized as follows: After introduction of section 1, section 2 presents the communication protocol that interfaces the sensor. Further, section 3 addresses the laser scanner modeling where distance function, noise spectrum and how object color and interfere in measures are presented. Finally, section 4 concludes the paper and points out the direction of future work.

Fig. 2. The Piccolo Lidar sensor

2   Laser Communication

The depth measuring of a scene (based on three-dimension vision hardware) is a topic that is being addressed by several developers and researchers. Examples of that are the stereo vision, the time of flight (TOF) cameras, the well-known Microsoft Kinec and the Asus Xtion based on infrared camera and an IR projector measuring two consecutive IR frames. Structured-light scanner is also a typical method for 3D objec acquisition. Some of these sensors are expensive (TOF cameras) and usually present a low resolution for distant objects. The laser range devices are categorized as an Amplitude Modulated Continuous Wave (AMCW) sensor. The laser emits a laser beam and a rotating mirror changes the beam's direction. Then the laser hits the surface of an object and is reflected. The direction of reflected light is changed again by a rotating mirror, and captured by the photo diode. The phases of the emitted and received light are compared and the distance between the sensor and the object is calculated.

In this paper target colour dependences, noise spectrum, limits and time constraints of this Lidar are addressed. Measures were acquired with a white and non-reflect surface, such as paper in dark conditions at 22° Celsius temperature. Drift effects from the warm-up time, usually presented in lasers ([4] [5]) were neglected for the simulation model. This section summarizes the communication protocol for the presented laser, once there are no papers describing it.

2.1   Communication Protocol

With this setup the Lidar is powered by 5V (from the USB port) and the motor has to be driven at 3.3V (through a voltage regulator). The USB connection was used to acquire digital data from the sensor (through a USB to serial converter).

The laser protocol (data formats) is composed by packets (the length of a packet is 22 bytes) and a full revolution will send 90 packets, containing 4 consecutive readings each. This amounts to a total of 360 readings (1 per degree) on 1980 bytes.

Each packet is organized as follows:

<start byte> <index>   <speed> <D0> <D1> <D2> <D3> <checksum>

where:

start byte is always 0xFA;

index is the index byte in the 90 packets, going from 0xA0 (packet 0, readings 0 to 3) to 0xF9 (packet 89, readings 356 to 359);

speed is a two-byte information. It represents the speed;

D0 to D3 are the readings data. Each one is 4 bytes long, as presented in Table 1.

Table 1. Communication procotol of Piccolo Lidar

byte	data	data	data
0	<distance 7:0>		
1	<"invalid data" flag>	<"strength warning" flag>	<distance 13:8>
2	<signal strength 7:0>		
3	<signal strength 15:8>		

2   Communication Interface

A developed application that communicates (as a serial port) and controls the laser scanner was developed and allows to logger the acquired data. As this application was written in Lazarus IDE, it is possible to compile and run it on Microsoft Windows and Linux operative systems. Figure 3 presents the developed application that allows to read the distances (at left) and computes the radial plan (at right).

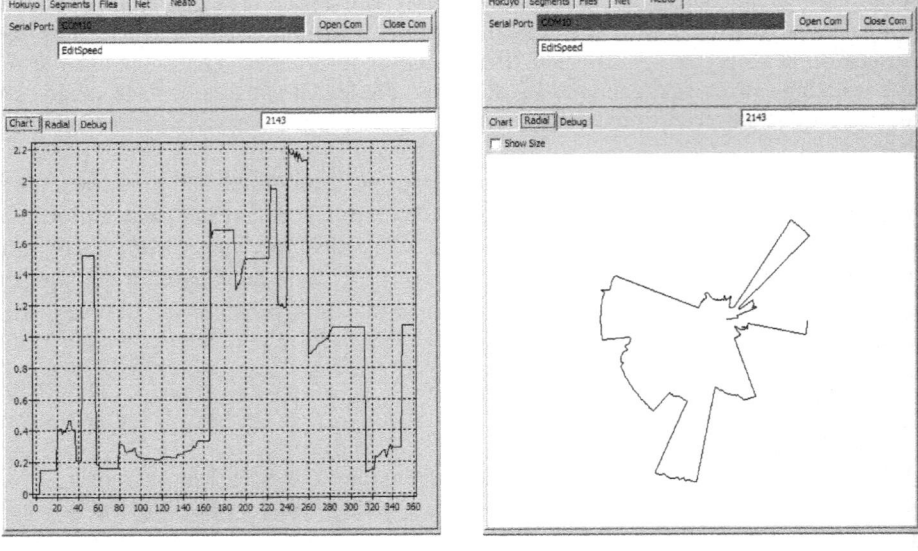

Fig. 3. Developed application that interfaces and acquires data from Piccolo Lidar

3   Low Cost Laser Scanner Modeling

This section describes the model of the Lidar sensor having in mind the distance func
tion and its linearization, the standard deviation (of error), the object color influenc
and finally the noise spectrum. The setup experience was obtained by placing th
targets in front of the laser (5 rotations per second) and several measures were ac
quired and the average was taken.

3.1   Distance Model

In order to measure the linearity and the error introduced as a distance function,
perpendicular object was placed in front of the laser scanner for 30 known distance
from 250 to 3250 mm (see Table 2). For each distance, 32 measures were acquire
and the average was taken. As shown in Figure 4, a non-linear function was obtaine
(presented in red squares) whereas the ideal function is represented by blue rhombus
The distance model of the laser sensor can be approximated by equation (1) based c
least square estimate, where a=-8E-05, b=1.0345, c=- 14.727, x is the real and y th
measured distances.

The corrected function is displayed in Figure 4 through green triangles.

Table 2. Distance and measured values from Lidar (n mm)

Distance	Measure
250	247.7
350	344.8
450	435.
550	529.8
650	623.7
750	717.8
850	811.1
950	883.8
1050	977.6
1150	1072.1
1250	1152.6
1350	1236.2
1450	1306.9
1550	1365.
1650	1442.4
1750	1527.7
1850	1617.3
1950	1697.3
2050	1772.0
2150	1843.4
2250	1918.0
2350	1974.9
2450	2051.1
2550	2115.4
2650	2164.4
2750	2225.0
2850	2293.
2950	2343.5
3050	2372.1
3250	2446.7

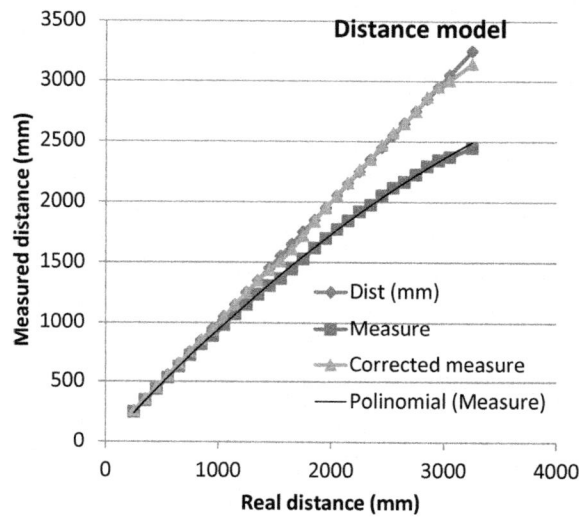

Fig. 4. Measured distance and corrected function

$$y = ax^2 + bx + c \tag{1}$$

The Piccolo Lidar laser sensor increases the error by increasing the object distance. This behaviour can be easily revised with the function $x=f(y)$, as presented in equation (2), where $O_1= 6465.6$, $O_2=41620219.1$ and $O_3=12500$. It allows to linearize the transfer function of this sensor (green triangles of Figure 4).

$$x = O_1 - \sqrt{O_2 - O_3 y} \tag{2}$$

Unlike the Hokuyo laser sensor (where the standard deviation was constant for several distances) the Piccolo Lidar sensor increases the standard deviation of measures for further distances.

The standard deviation is presented in Figure 5.

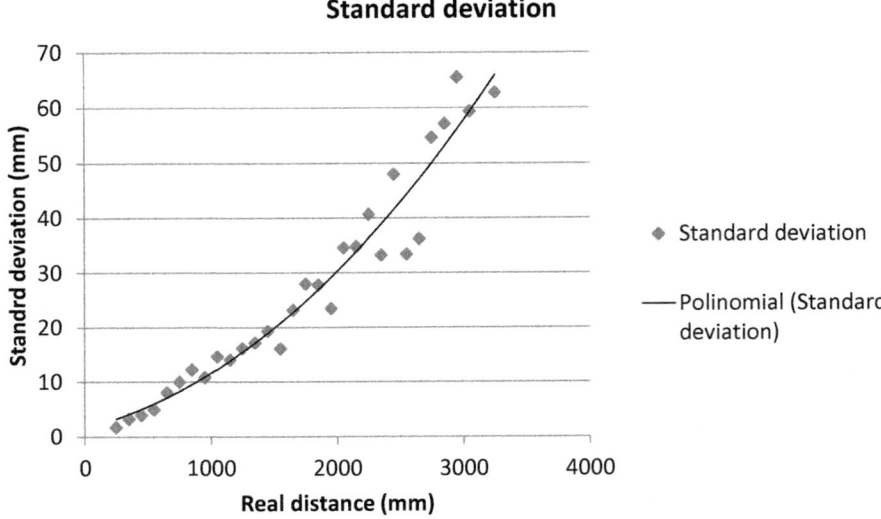

Fig. 5. Standard deviation of measures

The standard deviation model for the measured distances can be approximated by equation (3), where σ is the standard deviation, $k_1 = 4E\text{-}06$, $k_2=0.0057$ and $k_3=1.4999$.

$$\sigma(x) = k_1x^2 + k_2x + k_3 \tag{3}$$

3.2   Object Color Influence

The color of the object does not significantly affect the measured distance [4] other than the black color, as presented in [6] from a Hokuyo laser scanner. To test its influence, three black (8 cm) and white bars (10 cm) printed in a A4 size white planar sheet (Figure 6) was placed in front of the laser 300 mm away from the sensor. Behind, there is a wall 325 mm away from the sensor as presented in the layout of Figure 7.

The Piccolo Lidar laser sensor seems to ignore this color dependence once the measured distance is not affected by the black and white bars pattern (Figure 8).

Fig. 6. Black and white bars pattern

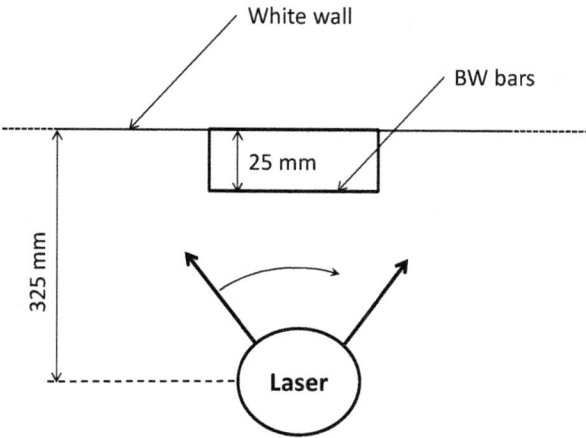

Fig. 7. Layout scheme for black and with bars

The measured distance of the pattern from picture 5 is presented in Figure 8. Object is presented from beam 165 to beam 200.

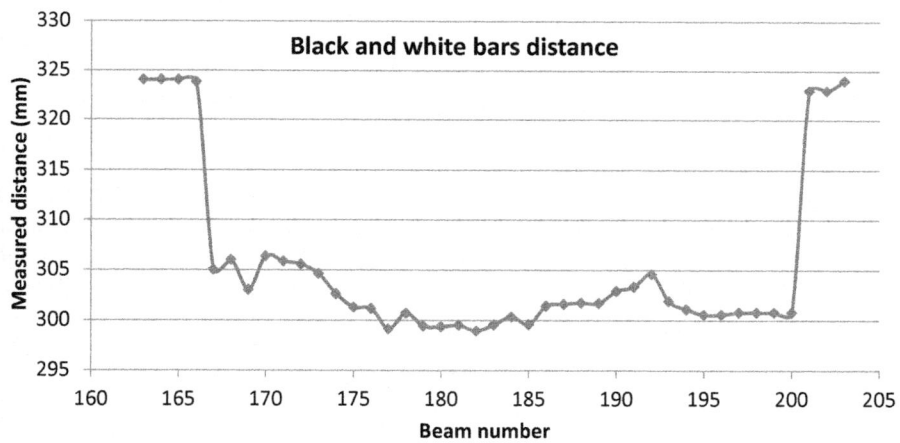

Fig. 8. Measured distance of planar black and white object

It shows that the influence of the black color is null unlike Hokuyo results [6]. This ffect is not visible in the Piccolo Lidar once it uses a linear digital image sensor that 1easures the distance of the reflected object [3].

.3   Noise Spectrum

'he model of a sensor should behave as the real sensor. Sometimes, the frequency ◖pectrum is underestimated. This subsection addresses the noise frequency of the ◖utput.

The noise presented at the output of sensor is a typical white noise. It can be checked through the noise frequency study. The sensor output spectrum was computed based on frequency decomposition using a Fast Fourier Transform (FFT). Figure 9 shows the spectrum analysis of the output measurement. These measures were obtained for 32 samples. For a simple model a white noise generator can be used to model the laser error.

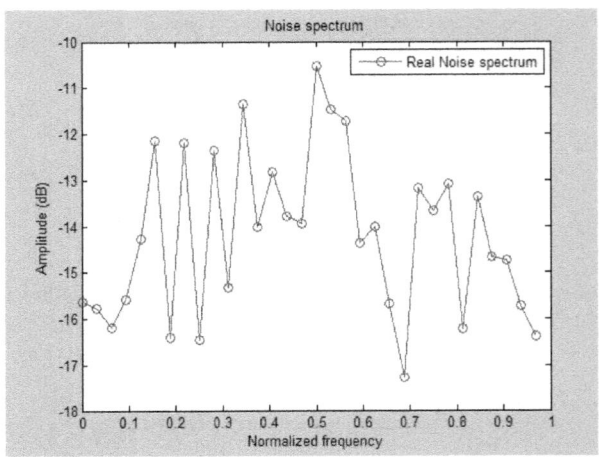

Fig. 9. Noise spectrum

As a final remark, the white noise generator to obtain the same spectral components as the real sensor presented in Figure 10 where H(z)=f(d[z]) is a function that depends on the distance value d[z]. This means, H(z) is tuned by the measured distance according to equation (3), where x is the measured distance.

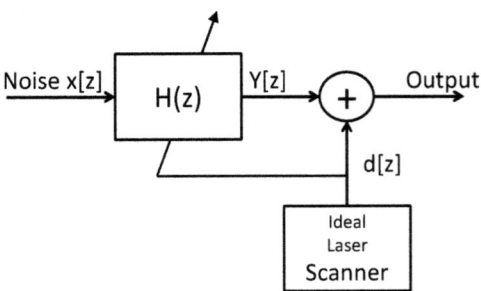

Fig. 10. Diagram block of the laser noise model

4   Conclusions and Future Work

This paper presented the low cost Lidar (Piccolo) modeling in order to simulate and develop applications based on its distance information.

The model was developed having in mind the distance nonlinearities, standard deviation, object colour influence and noise frequency spectrum. An application developed in Lazarus was created to implement the protocol and communicate with the Lidar. This Lidar, unlike well-known others (Hokuyo as example), presents an increasing error for longer distances. Moreover, the linearity of the output is probably related with a calibration issue during the manufacturing process. The linearized function is proposed. As main conclusion of this paper, the Piccolo Lidar is a very low cost laser sensor that can be used to measure distances in a 360 degrees plane.

There are several applications based on the information provided by a laser scanner: 3D scanning, object avoidance and localization, among the others.

As future work, a localization task, based on the perfect match algorithm, can be implemented. The presented model allows to measure the 360 degrees distance and compute the position and orientation of a robot in the known environment. Moreover, the model helps to determine the accuracy of such localization.

Acknowledgments. The work presented in this paper, being part of the Project "NORTE-07-0124-FEDER-000060" is financed by the North Portugal Regional Operational Programme (ON.2 – O Novo Norte), under the National Strategic Reference Framework (NSRF), through the European Regional Development Fund (ERDF), and by national funds, through the Portuguese funding agency, Fundação para a Ciência e a Tecnologia (FCT).

References

1. Michel: WebotsTM: Professional Mobile Robot Simulation. International Journal of Advanced Robotic Systems 1(1), 40–43 (2004)
2. Chen, D.-Y., Tian, X.-P., Shen, Y.-T., Ouhyoung, M.: On Visual Similarity Based 3D Model Retrieval. Computer Graphics Forum 22(3), 223–232 (2003)
3. Konolige, K., Augenbraun, J., Donaldson, N., Fiebig, C., Shah, P.: A Low-Cost Laser Distance Sensor. In: IEEE International Conference on Robotics and Automation, USA (2008)
4. Okubo, Y., Ye, C., Borenstein, J.: Characterization of the Hokuyo URG-04LX laser range finder for mobile robot obstacle negotiation. In: Unmanned Systems Technology XI, Orlando, FL (SPIE 7332) (2009)
5. Kneip, L., Tache, F., Caprari, G., Siegwart, R.: Characterization of the compact Hokuyo URG-04LX 2D laser range scanner. In: IEEE International Conference on Robotics & Automation (ICRA) (2009)
6. Lima, J., Gonçalves, J., Costa, P.J., Paulo Moreira, A.: Modeling and Simulation of Laser Scanner Sensor: an Industrial Application Case Study. In: International Conference Flexible Automation and Intelligent Manufacturing (FAIM), Portugal (2013)

Part XII
Education

A Virtual Workbench Applied to Automation: Student's Response Analysis

José Machado[1], Filomena Soares[2], Celina P. Leão[2], and Carla Barros[2]

[1] R&D CT2M, School of Engineering, University of Minho, Guimarães, Portugal
jmachado@dem.uminho.pt
[2] R&D Centro Algoritmi, School of Engineering, University of Minho, Guimarães, Portugal
fsoares@dei.uminho.pt, {cpl,cbarros}@dps.uminho.pt

Abstract. This paper presents a Virtual Workbench for Automation teaching/learning at undergraduate education level. The methodology defined as well as the formalism and the software tools adopted in the implementation are presented. A case study is detailed following the defined steps. The platform was tested by students attending the Process Control and Automation curricular unit from the 3rd year of the Master in Engineering and Industrial Management of University of Minho. A questionnaire was design to gauge the undergraduate student's acceptance and the platform' effectiveness. The research found that the Virtual Workbench allows new strategies and teaching methodologies related with laboratorial practice providing illustrative real-world examples.

Keywords: Automation, SFC (Grafcet), Ladder Diagrams, Finite Time Automata, Teaching/Learning, Virtual Labs.

1 Introduction

Education programs are changing considering the new teaching/learning methodologies and the new model for a dynamic attitude towards transmitting/absorbing knowledge. In fact, the availability of the Internet, the graphics and animated tools and the paradigm of "to learn where and when you feel up to", calls for an active participation of students. The teacher role changes to a catalyst in the acquisition knowledge process. Nevertheless, motivated and enthusiastic students are not always present. It implies, necessarily, from the teacher an active attitude in order to stimulate students in their learning process. Special important in engineering courses is that the implementation of innovative methodologies should be able to satisfy the industrial needs. Real-world problems should be presented and studied. Cognitive theory states that knowledge learned, and applied in a realistic problem-solving context, is expected to be remembered and properly used when needed later. In fact, these problem-based learning/teaching strategies, case methods and simulations, are useful tools for an effective teaching. Students must make decisions, solve problems and analyze the achieved results [1].

© Springer International Publishing Switzerland 2015
A.P. Moreira et al. (eds.), *CONTROLO'2014 - Proc. of the 11th Port. Conf. on Autom. Control*,
Lecture Notes in Electrical Engineering 321, DOI: 10.1007/978-3-319-10380-8_68

Due to the lack of financial and/or human resources the availability of physical workbenches to test real-world problems is not always possible. To overcome this weakness, emerge the virtual laboratory concept. The case studies are explained and presented through simulators supported by mathematical models and animations, where the User can change parameters, change information and learn from "virtual practice". These virtual labs can run locally or through the Internet. The requisites are only a computer. Regarding automation topic the practical implementation of control programs is particular important. The students need to practice in order to learn. In the Universities and schools there are available physical workbenches and industrial controller to test automation problems. Nevertheless, the availability of a virtual platform anytime and anyplace is really an added value. And there is where this work appears.

The goal of this paper is to present a Virtual Workbench in Automation topic that run in a local computer and with free access. The platform has available several real-world problems programmed in well-known industrial software. The methodology developed divides the problem in two different parts: the modelling of the plant behaviour and the specification of the controller behaviour, considering that they exchange information between them. Each part was separately modelled and simulated following different approaches: the plant (simplification of a real-world process to be study) was modelled by using timed automata [2], and the controller behaviour was modelled by Sequential Function Charts (SFC) or Grafcet [3]. We believe that the benefits of providing such a complementary tool to support the students learning process are an increase in students' engagement, an improvement on students' responsibility, a better communication between students and teachers.

2   Related Work

Industrial automation has emerged in the last years. Commercial suppliers for automation equipment followed this trend regarding new devices and also new software for designing automation industrial solutions. Festo [4] enterprise is one example. Several software for automation solutions are available to the User. Through simulation the User (student, engineer, designer) can plan, test and practice the new solution before implementation. Omron [5] is a well-known industrial automation trade that provides control components and systems for factory automation. In parallel, the User can test his automation problem in animated simulators. Famic Technologies [6] offers products and services in the areas of software engineering and in industrial automation. Automation Studio™ [7] software is a design and simulation solution for education and training purposes. It covers project or machine technologies including hydraulics, pneumatics, electrical, process control, and Human Machine Interface and communications topics.

In Academia there is also an effort to follow the industrial needs. In the context of automation and control teaching several works have been referred in the literature [8,9,10]. Different configurations may be implemented: the plant is a simplification of a real-world process, physically designed in an experimental rig, and the controller is emulated on a personal computer. This approach also allows testing the control in a real industrial controller. Nevertheless, testing a new process requires an investment for a new kit. To overcome this financial restriction, some works implement a virtual platform for testing different real-world plants that can be tested by a virtual controller or by a commercial device.

Summarizing, virtual and real devices, including plant and controller, are designed and implemented in different configurations (depending on financial and human resources) in order to provide educational and training tools.

In [8] the plant is virtually implemented on a personal computer using VBasic Microsoft (R) as a programming language and the control can be implemented by any external control device. The communication between the virtual plant and the real world is performed through the computer parallel port, transparent to the User. The interface registers the inputs and outputs of the virtual plant in a terminal panel. Vargas et al. [9] present an innovative project in the context of virtual and remote experimentation applied to automatic control engineering education. Some examples of remote experiments are detailed as well as the automatic booking system for the access to each didactic setup. In [10], a Web Assisted Laboratory for Control (WALC) engineering on-line education is described; virtual and remote laboratories experiences on automation, process control and numerical methods are available. The platform facilitates the remote access, the monitoring and the control of several laboratories experiments.

The IEC 61131 standard [11] is nowadays widely used in automation and there exist automatic code generators for high-level simulation languages. In [12] an application of model-based tool chain using Matlab/Simulink, a Programmable Logic Controller (PLC) [13] and respective programming languages are presented and applied to a four-tank platform. Experimental results are shown and compared to simulation data. Teaching Industrial Automation includes teaching modeling and control discrete manufacturing processes by using PLC. This process requires a practical approach which includes the availability dedicated equipment and software. In [14] the authors present a solution to the problem based on virtual experimental setups and the use of a remote laboratory with web panels and LabVIEW interfaces [15].

This article presents a different approach for Automation studies. The developed platform implements a virtual real-world environment where it is necessary to model the controller behaviour, the plant behaviour and the interaction between both models. The interaction between both models (done, in real-world by electrical signals) is modelled as Boolean variables that will be the basis for the elaboration of the respective models. The platform runs on the industrial software from Omron Company.

Background on Automation

In order to develop the mentioned platforms it is necessary to create models for the plant and for the controller and connect them using some rules, some formalisms and some software tools. The most complex part is obtaining the "good" model for the intended purposes. For this, SFC (or Grafcet) [3] formalism is used for the specification of the command part of the systems, because it is the one taught during the automation lectures, it is easy to use, easy to learn, by students, and well adapted to the specification behaviour of sequential systems. After the formalisation of the command part of the system (by SFC) this specification can be, latter, translated in algebraic equations that will be the basis for elaboration of the Ladder [11] program, to be inserted into the PLC. [13]. SFC is a specification language for the functional description used to describe the deterministic behavior of the sequential part of an automated system. The SFC fundamental structural elements are: step, transition and link. The SFC fundamental interpreted elements are transition-condition and actions. Here the term interpretation refers to a connection between the sequential system input and output variables, which is established through the transition-conditions and actions.

For modelling the behaviour of the plant, it was used Finite Timed Automata [2]. This formalism is well adapted for modelling the plant behaviour because allows modular modelling approach [16], it is a non-deterministic formalism [17] and has timed aspects formally well-defined and formalised. Finite Timed Automata are finite state machines that can manipulate time, through the application of clocks. These can be used to model and analyze the timing behavior of computer systems, e.g., real-time systems or networks. The Finite Timed Automata fundamental structure elements are location, guard, invariant, transition, message and variable assignments. The location defines the current mechanism where the system is in its evolution. Each transition is labeled by a logical equation called guard, which indicates that the transition can be fired. Variable values can be assigned when the transition is fired by variable assignment conditions. Invariants (conditions that must be always respected) can be used too. Messages are used to link and synchronize different automaton evolutions, when it is intended to have simultaneous evolutions of two or more modular automaton [2].

4   Virtual Workbench/Platform

In order to build a virtual platform for performing simulation, some tasks must be performed in designing a pre-defined methodology, such as study of the system components characteristics, study the operation purpose and the implementation in the simulation environment. Following this approach, the designing time, as well as the likelihood of unforeseen difficulties throughout the construction process, are reduced

The steps of the developed methodology are: (1) Problem Interpretation: definition of the desired behavior for all physical components to be considered on the virtual platform; (2) Construction of the input and output tables of the controller; (3) Conversion of the models to adapted language to be interpreted by adopted software tools for simulation (in this work, OMRON CX-One was used); (4) Construction of the simulation environment using capabilities of CX-One; (5) Simulation of the automatic system in the virtual platform.

In fact, during the development of the virtual workbench, there are developed two PLC programs: one that is the result of the SFC specification describing the system controller behavior; and, other, the program that is the result of translation, to Ladder language, of the plant behavior previously modeled using timed automata. Using CX-one and CX-Simulator, it is possible to establish the connection between both programs and, this way, run the simulation using both software tools together. Moreover it is possible the direct translation, of the obtained programs, to a physical simulator that can be developed for this purpose using two PLCs: one could be the real one (real controller) and the other could be a physical system for simulating the plant behavior.

5   Case-Study and Student's Response Analysis

Fourteen real-world examples were studied. Nevertheless their importance only three examples were selected and presented due to their practical implementation in Automation: Grab station, Apples´ Storage and Drilling Station (Fig 1.).

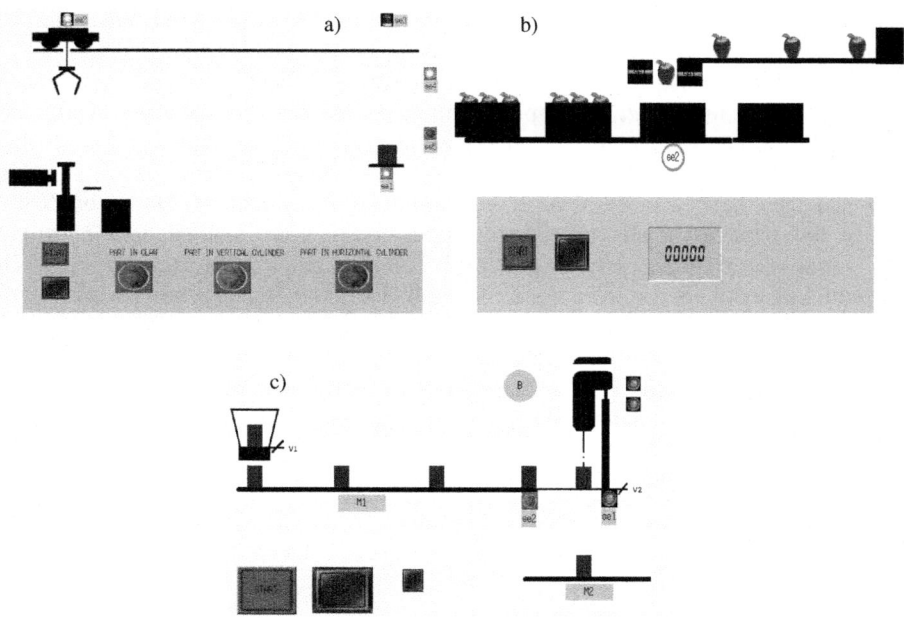

Fig. 1. Examples: a) Grab station in CX-Designer environment, b) Apples´ Storage in CX-Designer environment, c) Drilling Station in CX-Designer environment

To obtain students opinions regarding the automation simulator experimentation, a questionnaire was given to the students before and after the experiment. Due to the limited number of tests performed, the outcomes should be viewed as insights to the effectiveness of the automation simulator and not as hard evidence. The questionnaires were divided into main parts allowing the student characterization, the experiment evaluation, and the simulator assessment. Close, open and multiple choices responses, and the evaluation of several statements were used to analyze students' perspectives, fillings and knowledge before and after carried out the experiment. For the statements evaluation, five-level Likert-scale was used to rank the students' motivation, satisfaction and knowledge about the Grafcet learning. This questionnaire allowed obtaining the students' perception in relation to their own learning style, based on Kolb's theory [18, 19], and the learning style employed during the experiment performance, in accordance with the students' perspective. This factor is important, not only to understand the kind of learning methodology the students used but also, and more importantly, to understand if the platform or the lab procedure is appropriate for the students' learning process on this specific subject. This could help to recognize the necessity of devising and implementing new learning strategies to fulfill students' needs. All in all, the research team verified, through this survey' answers, if the study's objectives set out were achieved. Participants Characterization

The participants in this study (n=14) are students who attend, for the first time, the third year of the Master Course on Engineering and Industrial Management (MCEIM), of School of Engineering, University of Minho, Portugal. The average age is 25 years and 85.7% of the students are male. Before the experiment, 53.8% of them asserted that their expectation and motivation was moderate and the remaining students showed high and very high interest in conduct it.

Analyzing the questionnaire responses concerning students' perception in relation to their own learning style, it transpires that, as per learners' perspectives, the majority of them (about 58.3%) have a converging learning style, as shown in Fig. 2 (third quadrant, green markers). In process field, these students do the things in an active way and they have a more abstract perception of the situations. Based on Kolb's theory, this behavior is characteristic of engineers: they are motivated to investigate how situations are processed; their strength is the practical application of ideas. The instructional methods, such as the interaction, the hands-on, the lab's work, or the assisted computer training, are suitable for the students who employ this kind of learning style [18]. Despite this, 25% of students have a diverging learning style (first quadrant, red markers in Fig. 2). These students enjoy gathering information, to observe everything around them, they are creative and they gain from teamwork [19].

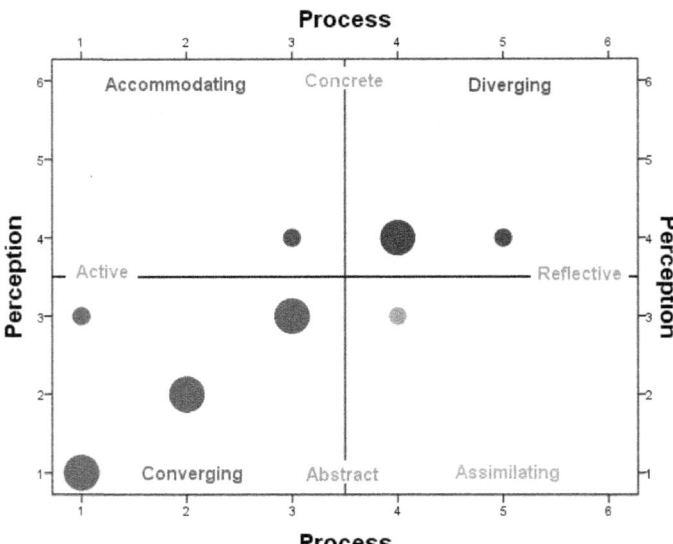

Fig. 2. Learning styles diagram obtained (based on Kolb's theory): 1st quad: diverging style 2nd quad: accommodating style; 3rd quad: converging style; 4th quad: assimilating style.

The MCEIM students' are integrated into a Project Led-Education (PLE) program from the first year of the course [20]. The PLE process involves the solution of a problem that commonly results in an end product, which can last a considerable amount of time. Through the teamwork, the interdisciplinary and the contextualization, the project aims the development of the critical thinking and competencies, the project management, and the autonomous and creative work, promoting an active and deep learning [20]. Since early stage, it is implanted a learning method in students, and the learning style results show that, with a clear predominance of the active processing. The empowerment of thinking, and the necessity of building bridges between different contexts, can explain the significant appearance of diverging style. Thereafter, the analyses of the responses the learning style employed during the experiment performance, in accordance with the students' perspective.

5.1   Participants Insights about Automation Simulator

All students carried out the experience in groups and 92.3% of them asserted that the experiment came up to their expectations. The identification of the learning style employed by each student during the experiment performance shows that 50% of the class presented a converging style (third quadrant, green markers in Fig. 3), while 25% presented a diverging style and 16.7% an accommodating one (respectively, first and second quadrants, red and blue markers in Fig. 3). The remaining 8.3% applied an assimilating learning style (fourth quadrant, yellow markers in Fig. 3).

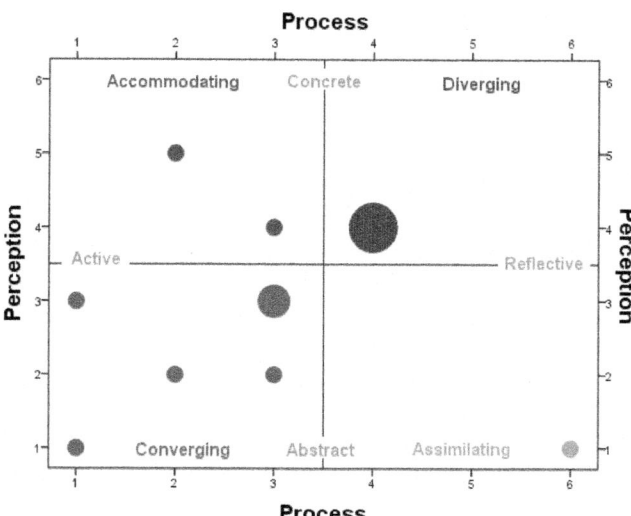

Fig. 3. Learning styles engaged during the experiment by each student: 1st quad: diverging style; 2nd quad: accommodating style; 3rd quad: converging style; 4th quad: assimilating style

During the experiment, some students engaged a style which tends to be slightly more accommodating, which means the experiment was more intuitive and experiential, and they did not have to assimilate a lot of information [18]. Nevertheless, these results do not present statistically significant differences between the students' general learning style and that which was employed during the experiment, both in processing (W=.914, p>.05) and perceiving (W=.705, p>.05) activities. This lack of significance is seen as a positive aspect insofar it shows that the simulator and the experiment designed were meeting with the needs and expectations created, adapting to the students' learning style.

When questioned about some technical features of the simulator, the students attributed a classification to each one, in a 1 to 5 rating scale, where 1 is "very poor", 2 is "poor", 3 is "acceptable", 4 is "good" and 5 is "very good". The students gave their general opinion about the simulator, and more than 75% of them considered it as "acceptable" and "good" (rate ≥ 3). The same result was obtained from the graphical interface assessment. In relation to its utility for automation learning, the majority of the students were consistent, considering the simulator as a useful tool (rate 4), and most of them classified the practical examples as "good".

Other features were evaluated. A group of six questions (Q1, Q2, Q3, Q4, Q5 and Q6) was asked in order to know if the simulator accomplished their function through the features in assessment. The questions, with "Yes" or "No" answers, are:

Q1 – Are the guide texts written in a clear and concise way?
Q2 – Do the animations reflect well the various Grafcet concepts?
Q3 – Does this tool allow a better Grafcet understanding?
Q4 – Do you have difficulty in use the software?
Q5 – Do you consider the software a user-friendly tool?
Q6 – Did you become more motivated to learn the Grafcet?

The results are showed in the Fig. 4, with the percent of positive and negative answers for each question. The majority of the students answered positively to all questions. However, in question Q4, related to the ease in use the software, the students opinions were evenly divided between both answers. They justified the negative verdict, claiming that there were troubles with software installation, and some of them had difficulty using codes and symbols, and in variables definition. It is possible to conclude from these results that this tool enables the easy comprehension and learning of the Grafcet, despite the requiring of some adjustments to make it more user friendly, especially by the improvement some technical features.

One of the experiment's objectives was to give the opportunity to acquire some of the soft skills required in the automation learning, keeping the student motivated by using the automation simulator. For such, the students assessed three important aspects after the experiment: SS1, the motivation for the collaborative work; SS2, the stimulation of the intellectual curiosity; and SS3, the provision of study material. The classification was based on a 1 to 5 rating scale, where 1 is "very poor", 2 is "poor", is "acceptable", 4 is "good" and 5 is "very good". The results show that all assess-

aspects were rated with 4 by most part of the students, which means that both aspects were classified with a positive mark. This result shows the influence of the simulator experimentation on their soft skills acquisition. These results are very encouraging as they demonstrate that the use of simulation tools can have a remarkable effect on these students learning process, especially because they are capable of motivating them, if the experiments are properly designed.

The students were asked for comments about all experience in order to identify some aspects capable of being improved in a near future. The majority of them identify the lack of time to carry out the experiment, and they pointed out the missing of a guide for the software installation. Furthermore, some of them suggested becoming the simulator more intuitive, since, according to them, it is no easy task when they have little or no experience.

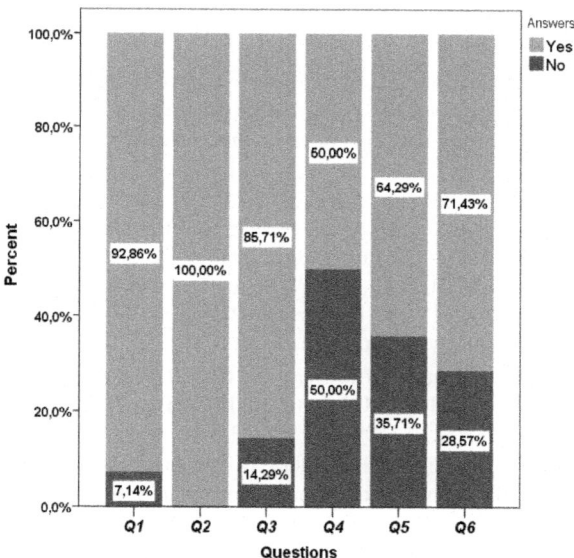

Fig. 4. Visualization of results of the three experiments performed

6   Final Remarks

A Virtual Workbench applied to Automation undergraduate education level is presented. This platform was designed and developed by two Master students for students. While fourteen real-world new applications have been developed, this paper describes the students evaluation and perception' of three examples selected based on their practical implementation in Automation.

The first insights on the Automation simulations were very positive showing a strong interest by students in this way of knowledge transmission. New experiments are under development trying to meet some of the suggestions in this request.

Acknowledgments. The authors would like to express their acknowledgments to all students for their voluntary co-operation and help on the evaluation of the platform, especially to Nuno Canadas and Carlos Barros on their commitment to its development. The authors are also grateful to the Portuguese Foundation for Science and Technology (FCT) for funding through the R&D project PTDC/CPE-PEC/122329/2010.

References

1. McKeachie, W.J.: Teachings tips: Strategies, research, and theory for college and university teachers, 11th edn. Houghton Mifflin Company, Boston (2001)
2. Alur, R., Dill, D.L.: A theory of timed automata. Theoretical Computer Science 126, 183–235 (1994)
3. European Standard EN 60848: GRAFCET specification language for sequential function charts (2002)
4. Festo Company, http://www.festo.com (accessed in January 2014)
5. Omron Company, http://www.omron.com (accessed in December 2013)
6. Famic Technologies Inc., http://www.famictech.com/ (accessed in October 2013)
7. Automation StudioSoftware P6,
http://www.automationstudio.com/pro/en/OnlineDemo/
index.htm#.UvJRba2PP4Y (accessed in January 2014)
8. Seschini, I., Galvez, J.M.: Um Laboratório Virtual para Aplicações em Ensino de Automação e Controle. In: XXXV Congresso Brasileiro de Educação em Engenharia, COBENGE 2007, Brasil, 13 p. (2007)
9. Vargas, H., Moreno, J.S., Jara, C.A., Candelas, F.A., Torres, F., Dormido, S.: A Network of Automatic Control Web-Based Laboratories. IEEE Transactions on Learning Technologies 4, 197–208 (2011), doi:10.1109/TLT.2010.35
10. Leão, C.P., Soares, F., Rodrigues, H., Seabra, E., Machado, J., Farinha, P., Costa, S.: Web-Assisted Laboratory for Control Education: Remote and Virtual Environments. In: Uckelmann, D., Scholz-Reiter, B., Rügge, I., Hong, B., Rizzi, A., et al. (eds.) ImViReLL 2012. CCIS, vol. 282, pp. 62–72. Springer, Heidelberg (2012)
11. Standard: IEC 61131-3: Programming Industrial Automation Systems: Concepts And Programming Languages, Requirements for Programming Systems (1992)
12. Misgeld, B., Pomprapa, A., Leonhardt, S.: Didactic Approach to Multivariable Control Using IEC 61131 Model-Based Design and Programmable Logic Controllers. In: The 10th IFAC Symposium on Advances in Control Education, Part 1, Sheffield, UK, vol. 10, pp. 220–225 (2013)
13. Jones, T.C., Bryan, L.A.: Programmable Controllers: Concepts and Applications. In IPC/ASTEC, Atlanta (1983) ISBN: 0915425009
14. Caldas Pinto, J.R., Sa da Costa, J.: Virtual and Remote Laboratories for Industrial Automation E-Learning. In: The 10th IFAC Symposium on Advances in Control Education, Part 1, Sheffield, UK, vol. 10, pp. 286–290 (2013)
15. LabView Software,
https://www.ni.com/gettingstarted/labviewbasics/pt/
(accessed in January 2014)

16. Kunz, G., Perondi, E., Machado, J.: Modeling and simulating the controller behavior of an Automated People Mover using IEC 61850 communication requirements. In: 9th IEEE International Conference on Industrial Informatics, Lisbon, Portugal, 6 p. (2011), doi:10.1109/INDIN.2011.6034947
17. Machado, J., Seabra, E., Campos, J.C., Soares, F., Leão, C.P.: Safe controllers design for industrial automation systems. Comput. Ind. Eng. 60(4), 635–653 (2011)
18. Kolb, Y., Kolb, D.A.: The Kolb learning style inventory. Version 3.1,Technical specifications (2005)
19. Kolb, D.: Experiential learning: Experience as the source of learning and development. Prentice-Hall, Inc., New Jersey (1984)
20. Lima, R.M., Carvalho, D., Flores, M.A., Hattum-Janssen, N.V.: A case study on project led education in engineering: students' and teachers' perceptions. Eur. J. Eng. Educ. 32(3), 337–347 (2007)

Spanish Control Engineering Challenge: An Educational Experience

Xavier Blasco, Gilberto Reynoso-Meza,
Sergio García-Nieto Rodriguez, and Jesús Velasco

Instituto Universitario de Automática e Informática Industrial - ai2
Universitat Politècnica de València, Spain
{xblasco,gilreyme,sergarro,jevecar}@upv.es

Abstract. The *Concurso de Ingeniería de Control* (CIC) (Engineering Control Challenge) is a student challenge organized by the *Comité Español de Automática* - CEA (Spanish Committee of Automation) from Spain. It was set as an educational and motivational tool for control engineering students. Although originally devised for Spanish pupils, several non-spanish teams ended up participating at the two editions, 2012 and 2013. This article describes details about the challenge and its organization, and it shows, as well, some of the impressions obtained by polls and testimonials from participants.

Keywords: Control teaching, Education, Control Challenge.

1 Introduction and Motivation

During the last two academic years the *Comité Español de Automática* (CEA) has organized a university student challenge in the area of control engineering: the *Concurso de Ingeniería de Control* (CIC). The original idea was to motivate control engineering students all around Spain. As an additional objective the promotion of the control engineering area within the Spanish universities was set. With these objectives in mind, the organization team wondered which could be the key elements of a successful event. The answer was: an interesting process, a complete set of simulation and implementation tools for participants, an appropriate diffusion and, of course, sponsors. Focused on providing participants a control challenge with all the ingredients of a real problem, a platform to test pupils knowledge in control's design and implementation was set by the organizers.

This article describes some details of this challenge, its organization and an overview of the participants' impressions based on their testimonials and on poll results. Besides this introduction, the article is divided in four more sections as follows: section 2 makes a brief description of the challenge itself and the phases that it is constituted by; in section 3 an overview of the organization and the participant teams is performed; section 4 analyses the impact of CIC on team members and tutors and, closing this work, a section 5 is included with some relevant conclusions extracted by the authors.

A.P. Moreira et al. (eds.), *CONTROLO'2014 - Proc. of the 11th Port. Conf. on Autom. Control*,
Lecture Notes in Electrical Engineering 321, DOI: 10.1007/978-3-319-10380-8_69

2   Challenge Description

The trajectory control of an autonomous quadrotor was chosen as the target of the competition (Fig. 1). The work proposed consisted on going through all the steps that must be followed to successfully solve such a control problem. Firstly, students had to obtain a dynamical model of the process (no constraints were imposed regarding the model type, its structure or the methodology to follow). Afterwards, participants should choose a control structure and tune it. This work had to be carried out with the simulation platform supplied by the organization team (based on ©Matlab/©Simulink and available at [1],[2]). To end up with this simulation phase, teams were required to send a short paper describing their work.

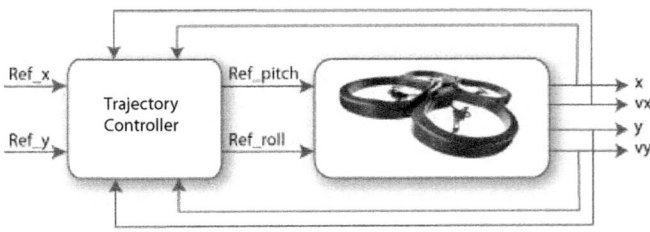

Fig. 1. Control trajectory diagram

With the last two CIC phases in mind, organizers previously created a whole software environment to simulate and actually control the ©Parrot's quadrotor *Ar.Drone*. ©TrackDrone Lite, which is available at [3], can be installed in any windows computer to effectively communicate the drone, acquiring information from on-board sensors and sending back control actions. It also integrates the necessary subroutines to create and modify paths, to calculate the tracking performance, and to graphically represent quadrotor's movements, all this in both simulation and real flight. As a clue characteristic, ©TrackDrone Lite leaves opened the proper space for the teams to include their controlling algorithms as compiled DLL file. Figure 2 shows a screenshot of the application commanding the drone.

Then, the second phase of the challenge included control implementation and performance evaluation. With this tool in their hands, each team was required to implement their solution into the platform (actually programming and compiling their algorithms). It was the organizers task evaluating solutions to decide which participants would go to the final. As a cost function dependent on tracking time, the integrated error and the maximum error, the tracking performance was the real jury.

Once in the final phase, every solution was tested in real conditions and through several paths. The proper conducting of the competition demanded the integration of knowledge in dynamic modeling, systems identification and control, and the development of teamwork skills. Since some of the teams were having their first contact with the real system, different rounds were prepared to allow pupil groups to adjust their controllers. Finally, participants algorithms were confronted to quite difficult trajectories that tested their capabilities. Once again, teams with a smaller value of the cost function, obtained their reward.

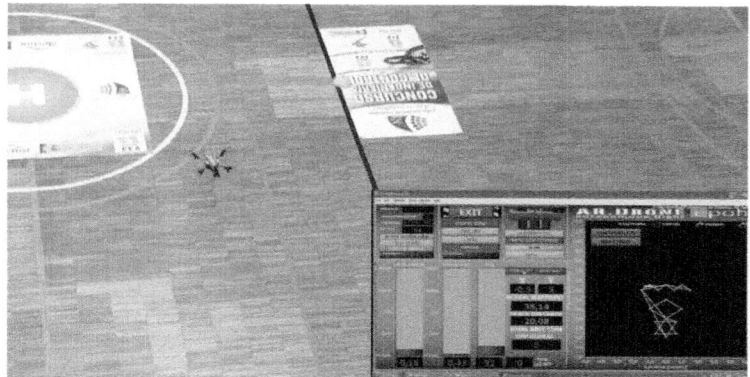

Fig. 2. Screenshot of TrackDrone Lite©(in the bottom right corner) and a quadricopter during the final in Vigo

All details about rules and material supplied in both editions of the challenge an be found in [4],[5],[1],[2].

It is important to remark that the level of difficulty had to be flexible to at-ract both, students with basic control studies and others with more advanced nowledge (master, PhD). In the second edition two categories were set (under-raduates and master/PhD) to level the final classification properly.

Organization and Participants

ince the competition took place during a entire academic year, university teach-rs were required to participate as team tutors. They had the task of enrolling is/her local teams, presenting the challenge as a different teaching tool and eeping the contact with the organization.

The finals (only for the classified teams) were at a specific location during the panish National Control Meeting (Jornadas de Automática). CIC2012 was at e Universidad de Vigo (Vigo, 5th to 7th, September 2012) (Fig. 3) and CIC2013 as at Universitat Politècnica de Catalunya (Terrassa, 4th to 6th, September)13) (Fig. 4).

A total of 22 teams participated in the 2012 edition (CIC2012) whilst 25 did in the 2013 one (CIC2013). Despite they were mainly coming from spanish niversities, there were also some teams from other countries (Table 1).

Table 1. Participants for CIC2012 and CIC2013

Institution	Country	Number of teams CIC2012	CIC2013
Universidad Politécnica de Madrid	Spain	2	1
Universitat de les Illes Balears	Spain	1	
Universidad de Huelva	Spain	1	1
Universidad de Sevilla	Spain	2	
Universitat Jaume I Castelló	Spain	1	1
Universitat Politècnica de València	Spain	3	3
Universitat Politècncia de Catalunya	Spain	2	3
Universitat Autònoma de Barcelona	Spain	1	2
Universidad Pontificia Bolivariana	Colombia	1	
Universidade Federal de Santa Catarina	Brasil	1	
Universidad de Oviedo	Spain	1	
Universidad de Vigo	Spain	2	2
Universidad de La Rioja	Spain	1	2
Universidad Complutense de Madrid	Spain	2	
Universidad Michoacana de San Nicolas de Hidalgo	Mexico	1	
Universidad de Málaga	Spain		1
Universidad de Córdoba	Spain		1
Universidad del País Vasco	Spain		2
Universidad de Girona	Spain		1
Instituto Tecnologico de Veracruz and Universidad del Papaloapan	Mexico		1
Universidad Michoacana de San Nicolas de Hidalgo	Mexico		1
Universidad de Almeria	Spain		1
Eindhoven University of Technology	The Netherlands		1
Universidad Nacional de Educación a Distancia	Spain		1

To attain so many students, a diffusion and outreach phase is crucial in which the challenge must be introduced in the social networks they so frequently use. An important work was driven in this direction to place CIC in social networks:

- Facebook, `https://www.facebook.com/ConcursoIngenieriaDeControl`.
- Twitter, `https://twitter.com/ConcursoIC_CEA`.

In addition to the online broadcast of the finals, two summary videos were produced and uploaded on youtube, and can still be used as motivational material for future students:

- CIC2012 `http://youtu.be/uWYatwnze-A`,
- CIC2013 `http://youtu.be/8eA9yIHG3z8`.

Fig. 3. ARDrone at the location of the final of CIC2012 (University of Vigo)

4   Analysis

Above the scientific level of the proposal, the impact in the motivation and the academic training of our university students has to be the objective in this type of events. To understand the CIC utility, two polls were done. One for the students and another for the tutors.

First of all, it is important to remark that only 2 of the students (over the 26 answering this question) said they had participated because of their own initiative. The rest of them participated by suggestion of their tutors. This means that the role of the tutor is crucial in initiatives as the present one, independently of how motivating is the challenge.

Among other questions, tutors were asked about the impact of CIC in the understanding of automatic control concepts. Three questions were posed:

Fig. 4. ARDrone at the location of the final of CIC2013 (Universitat Politècnica de Catalunya, Terrasa)

T1: I think CIC helped my students to better understand the contents of automatic control subject.
T2: I think the CIC motivated my students to learn new concepts of automatic control subjects.
T3: I believe that CIC allowed my students to have a clearer idea about what is control engineering

Six possible answers were allowed: (SD) Strongly Disagree, (D) Disagree, (N) Neutral, (A) Agree, (CA) Completely agree and (NR) No response. Over 20 answers, they mostly agreed or completely agreed (see Fig. 5).

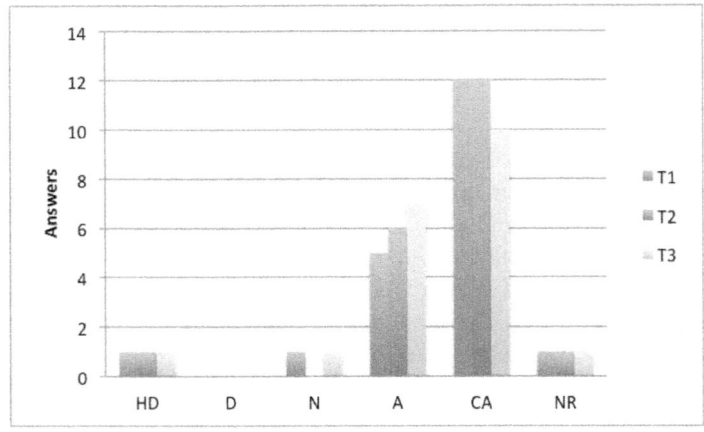

Fig. 5. Tutor's answers about CIC impact

Same questions for students show similar results. Over 33 answers, pupils mostly agreed or completely agreed (see Fig. 6) and the motivation produced is noticeable.

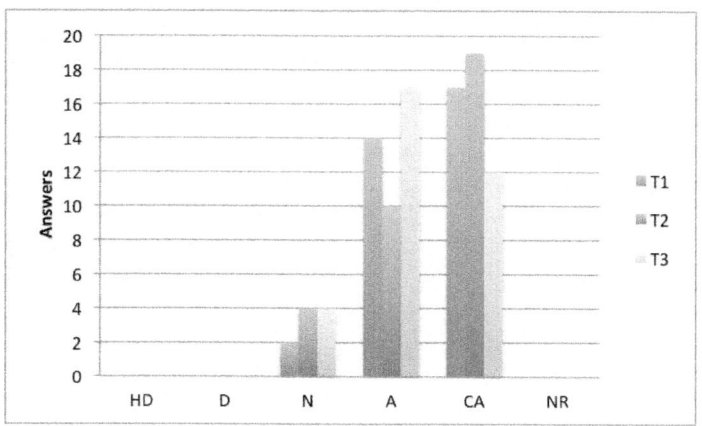

Fig. 6. Student's answers about CIC impact

Two questions were posed to check the impact on teamwork:

S1: I consider teamwork very important to attain goals of CIC.
S2: I think I developed skills for teamwork throughout the development of this CIC.
S3: I got involved in every stage of the developments for CIC.

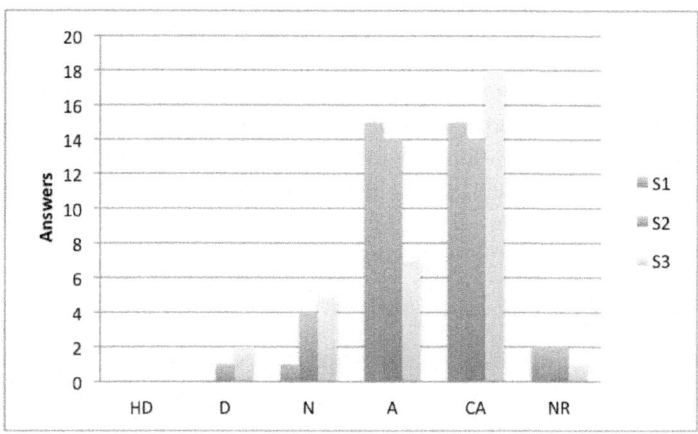

Fig. 7. Student's answers about CIC impact on teamwork

The results show (see Fig. 7) the students considered positive the impact of CIC in his/her teamwork skills, and the majority participated actively in the team.

The students were questioned about the necessity of additional subjects (apart from control processes). Half of the answers showed it is necessary, and the type of additional knowledge they reported as useful was: Identification and modeling, Systems Dynamics, Mechanics, Neural Networks, Artificial Intelligence, Computer programming, Robotics and Optimization.

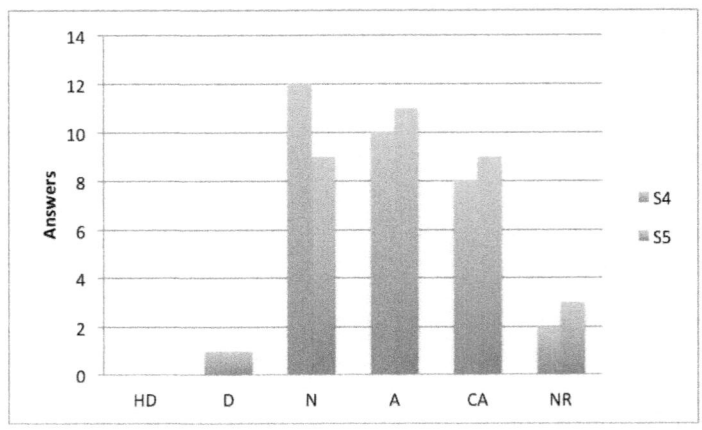

Fig. 8. Student's answers about their results in CIC

Two more questions were performed concerning the students impression about their own results. They were:

S4: I believe that my achievements in CIC were outstanding.
S5: I believe that the achievements of my team in CIC were outstanding.

The answers show they were not completely satisfied with their own achievements, and that they have better valuations of the achievements of their team as a whole. This means that they place high value on teamwork. Even with the feeling they could have done better, in general, they were very happy with the participation. It could be said that they are critical with their own work.

Tutors and students were also questioned if they would recommend other people to participate in future editions (Fig. 9):

S6,T4: I would suggest to other students participating in future CIC editions

In general, their opinions are positive and it seems both students and tutors would encourage other students to participate in this event.

The polls covered other types of question not in the scope of this paper: impressions about organization, delivered material (written documentation, software, etc.), the mechanism of evaluation, and others.

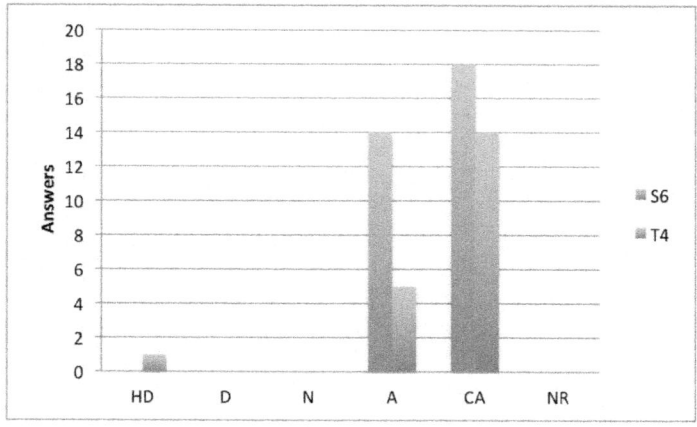

Fig. 9. Tutors (T4) and Students (S6) advise for future students

However, one of those additional questions certainly is worth mentioning. Participants were asked about the current and future works derived from their presence in this challenge. Within the results of this question, it is important to remark that some tutors have encouraged their students to use the material developed during the competition for other purposes. Some teams have published part of their developments as congress contributions and journal papers. Others have used the material for Graduation Thesis and Master Thesis. And finally, several tutors have used the CIC as part of the exercises posed at the subjects they impart. In fact, it seems that the last one is a key point to increase motivation or, at least, to attract students to the CIC.

An anecdote which is also worth mentioning is the start-up of a new research line by one of the tutors. Her team was formed by PhD students and the developments achieved attracted the attention of a company of their locality. They are currently working on a development project with this company.

To finish, sponsors, which supplied cash prizes for the winners, valued very positively this challenge. In their opinion, it is very important for the students to check their knowledge in a realistic environment. With this training, they are more confident with the formation they received and they have more resources to face real problems when they begin to work.

5   Conclusions

To sum up, these two years of experience with this initiative have been very positive in terms of accomplishment of objectives. The challenge achieved to attract a significant number of participants in both editions. The students satisfaction was high and they realized that their participation was very profitable to improve their knowledge on control engineering. Tutors valuated the CIC very positively as well, and they used it to motivate their students.

A huge part of the success is because of tutors. It is noticeable that the motivation supplied by them has been crucial for students participation. Most of the teams were enrolled by suggestion of their teachers. Then, it can be concluded that, in this type of events, it is very important attracting teachers attention. Several formulas exist for this purpose, but the key points observed by the organizers are: giving sufficient material for the teaching aim and setting up an attractive process.

Acknowledgments. This work was partially supported by the Special Action DPI2011-15857-E from the Spanish Ministry of Economy and Competitiveness.

References

1. http://www.ceautomatica.es/og/ingenieria-de-control/benchmark-2011-2012
2. http://www.ceautomatica.es/og/ingenieria-de-control/benchmark-2012-2013
3. Garcá-Nieto, S., Blasco, X., Sanchiz, J., Herrero, J., Reynoso-Meza, G., Martínez, M.: Trackdrone lite. Riunet (2012), http://hdl.handle.net/10251/16427
4. Blasco, X., García-Nieto, S., Reynoso-Meza, G.: Control autónomo del seguimiento de trayectorias de un vehículo cuatrirrotor. Simulación y evaluación de propuestas. RIAI 9(2), 194–199 (2012)
5. Blasco, X., Reynoso-Meza, G., García-Nieto, S.: Resultados del concurso de ingeniería de control 2012 y convocatoria 2013. RIAI 10(2), 240–244 (2013)

Learning Electronics Project Inside a Control Inspired Conceptual Map

Paulo Garrido[1], Aparício Fernandes[1], Paulo Carvalhal[1], João Sepúlveda[1], Nuno Costa[2], and Renato Morgado[3]

[1] Escola de Engenharia, Universidade do Minho, 4800 Guimarães, Portugal
[2] Divisão de Sistemas de Electrónica, Efacec SA, 4471 Moreira da Maia, Portugal
[3] Technical Committee, Efacec SA, 4471 Moreira da Maia, Portugal

Abstract. This paper describes two laboratory and project semester modules in the Electronics Engineering programme of Universidade do Minho. Along time, the learning objectives of these modules evolved from integrating conventional laboratory work across disciplines to include first time experience in project of electronics with collaboration of industry. The design of the modules joins elements of conventional laboratory work, project based learning and cooperation with industry. Cooperation with industry has a crucial role in framing the project work and ultimately the laboratory work. Control is, beyond one of the topics to be learned, a main source of inspiration for the design of the modules at large, as the feedback loop concept is used to unify laboratory work from several disciplines with project into a coherent learning experiment.

Keywords: control education, learning electronics project, learning electronics design, project based learning.

Introduction

The Electronics Engineering programme at Universidade do Minho is an integrated masters degree with five years, ten semesters duration. Historically, this programme adopted a philosophy of joining the laboratory work of different disciplines into so-called Integrated Laboratories modules. After the main reconfiguration of the programme prompted by the Bologna agreement, this philosophy was kept only in the third and fourth years of the programme and a strong bias towards project was given to the modules. This paper describes the design of the two third year modules, currently called Integrated Laboratories and Project 1, for the first semester, and Integrated Laboratories and Project 2, for the second semester. For brevity, we will refer the first module by the acronym LPI1 and the second by the acronym LPI2.

Fundamental concepts of Control must be learned in both LPI1 and LPI2 and control is also the theme by default for the project. In the design of LPI1 and LPI2, we use the concept of the feedback control loop as a unifying frame for the laboratory works of different disciplines. Therefore, this paper pertains to control education in two senses. In one sense, it describes a learning path for control basics in the environment of electronics engineering studies. In the other sense, it shows a use of

© Springer International Publishing Switzerland 2015
P. Moreira et al. (eds.), *CONTROLO'2014 - Proc. of the 11th Port. Conf. on Autom. Control,*
Lecture Notes in Electrical Engineering 321, DOI: 10.1007/978-3-319-10380-8_70

control to establish a mental map for learning electronics along laboratory work culminating in a project.

In the last years, interest in collaboration with industry has grown, both in engineering and control education. There are many potential benefits that can be gathered from such collaboration, both for academia and industry. In the case of LPI1 and LPI2, project based learning ideas inspire the way the project is framed and presented to students. The cooperation with an industrial partner enables the learning of project to develop along guidelines used and practiced in industry. This cooperation is deemed to progress and state practice protocols in laboratory work upon those used in industry.

In section 2, we will describe the overall design of the LPI1 and LPI2 modules. Section 3 presents in detail the laboratory work. Section 4 describes the project and the industry guidelines used in the project. Section 5 describes the assessment methodology. Section 6 concludes and presents the prospects for increasing cooperation with industry.

2   Overall Design of Modules

The LPI1 and LPI2 modules in the Electronics Engineering programme of Universidade do Minho are placed at the third year, first semester, and third year, second semester, respectively. Their aims are twofold. On one hand, they must provide laboratory works for other modules in each semester. On the other hand, they must provide a first time experience of electronics project to students within an adequate procedural framework.

To accommodate both aims, the first fourteen weeks of a semester are dedicated to laboratory works. The remaining six weeks (of the total twenty weeks of a semester) are dedicated to the project – development of an electronic device or a simple electronic system.

To establish synergy between the aims, the laboratory works are designed to b both standalone, in the sense of supporting learning of the modules they refer to, an to provide building blocks for the project, if the need arises.

The design of the Electronics Engineering programme in the third year nears an a tribution of analog subjects to the fist semester and of digital subjects to the secor semester. In fact, the concomitant modules for which LPI1 must provide laboratory works (in the first semester) are:

– Automatic Control
– Signal Processing
– Instrumentation and Sensors
– Electrical Machines

While the concomitant modules for which LPI2 must provide laboratory works (the second semester) are:

– Digital Control
– Digital Signal Processing
– Automation
– Power Electronics

The design for both LPI1 and LPI2 was conceived as a whole with the aim to integrate each module in its semester and to have a coherent sequence from LPI1 to LPI2. To this aim, LPI1 targets mainly concepts of analog electronics and LPI2 targets mainly concepts of digital electronics. The strongest restriction that students have when developing the project in LPI1 is to keep the design of the device or simple system inside the confines of analog electronics. By default, the project in LPI2 (second semester) should be a retake of the project in LPI1 (first semester) with the added freedom to use digital electronics and its capabilities. In this way, the progress along the year from analog to digital is stressed.

In both semesters the concept of feedback control loop is used to create a conceptual map for both laboratory works and project. This is suggested and abstracted from the subjects of the concomitant modules. In each semester, five laboratory works are realized. Each one of four laboratory works is mainly dedicated to a concomitant module, and one laboratory work has an introductory character. When put together, the realization of the laboratory works in each semester can be deemed as "going around" a feedback loop: data sensing and acquisition, signal processing or control computation, and actuation. This "going around" a feedback loop happens in the first semester inside an "analog context" and is repeated in the second semester inside a "digital context".

We turn now to a description of the laboratory works.

3   Detailed Description of Laboratory Work

In both LPI1 and LPI2, students must go through five laboratory works. The first has an introductory character and the other four are dedicated to a concomitant module of theoretical nature.

3.1   Laboratory Works for the First Semester

Laboratory works in the first semester share a common character of dealing with analog devices and related concepts.

Control of a Drying Chamber Model

The theme of the introductory work is control of a laboratory model of a drying chamber. Students must mount an on-off controller, implemented with a difference amplifier and hysteresis comparator. The controller includes a simple circuit to sense temperature through a 100 Ω platinum resistor (PT100) and commands a power transistor that drives the heating element of the model.

With this simple setup, students are able to put at work basic concepts of instrumentation, control and actuation. In the final part of the work, the power transistor is mounted in a variable current source configuration and the hysteresis comparator is disconnected, so that proportional control results. Students can compare the two modes of control. The on-off controller electronic circuit is often reused in project.

Instrumentation Amplifier

In this work, students begin by mounting a PT100 in two different setups: in a Wheatstone bridge and in a current source loop. Following this, they mount an instrumentation amplifier to get a temperature sensor with adjustable gain and offset. This work is in a sense, an elaboration of the simple circuit used in the previous work. The circuit experimented will be used often in project.

Signal Processing

In this work, students begin by obtaining the experimental frequency response of three passive filters: low-pass, high-pass and band pass. This laboratory work is key to the understanding of several basic concepts of signal processing. In the final part of the work students will design an active filter, using available software [1]. Active low-pass filters may be used in the second semester with the aim to avoid aliasing.

Automatic Control

In this work, students practice modeling and simulation of control systems. To begin, they are required to derive the mathematical model of the drying chamber used in the first laboratory work. Next, they will conduct a simple experiment to identify the model parameters. With these, they will simulate the on-off control of the drying chamber temperature, realized in the first laboratory work. We also ask students to mount again the controller, this time with an instrumentation amplifier, and test the simulation results against experimental results.

The simulate-and-test procedure will be repeated for proportional control. This time, instead of a current source, students are required to mount an analog Pulse Width Modulated (PWM) unipolar drive of the heater.

Electrical Machines

In this work, students conduct experiments with several electrical machines with special emphasis on the direct current (DC) motor and the three-phase induction motor Special attention is paid to the model of the DC motor from a perspective of velocity or torque control.

3.2   Laboratory Works for the Second Semester

Laboratory works in the second semester share a common character of dealing with digital or pulsed devices and related concepts. Four of the five laboratory works use a microcontroller board as the core component. The fifth uses an off-the-shelf Programmable Logic Controller (PLC).

User Interface

The introductory work of the second semester has as goals to develop a character oriented user interface for a microcontroller board and to work the concept of a real time system.

Using any available algorithms, students must develop firmware to accept commands that change memory positions or input-output ports of the microcontroller. This firmware must be non-blocking, allowing for the concurrent execution of other processes.

The final part of the work requires students to measure the execution time of a chosen command with the aid of an oscilloscope, by flipping a bit of an output port.

This laboratory work has wide usage in projects, as character-oriented interfaces are almost universal in microcontroller-based projects.

Data Acquisition

The data acquisition laboratory work has several goals. From the standpoint of firmware design, students must ensure periodic sampling of a signal through a microcontroller analog-to-digital channel. Data generated through the analog-to-digital conversion process must subsequently be sent through the serial communication channel. Burst (fixed amount of data) mode or continuous mode of capture and sending must be implemented. Digital low-pass filtering of the input data is required, as well as look-up table linearization.

All the parameters of the data acquisition process should be programmable by the user through the character-oriented interface developed in the previous work.

Bipolar PWM Drive

In this laboratory work, students must mount a bipolar PWM drive composed by a H-bridge commanded by the microcontroller board. The user interface will allow for setting the parameters of the drive and start or stop it.

Bipolar PWM drives are often used to control DC motors in applications that require both directions of rotation. Students are required to develop specific commands envisaged for such applications. The firmware must be tested with a low-power DC motor.

Usage of this work in projects is also quite large.

Proportional-Integral-Derivative (PID) Controller

This laboratory work is dedicated to digital control. Three main goals are set for the development of firmware. First, the firmware must be real-time, allowing for the concurrent execution of the PID control law, a data acquisition channel and a PWM channel. Second, a parallel PID control law with a filter for the derivative action must be implemented. And third, operational commands typical of controllers must be developed: manual to automatic commuting, anti-reset windup and adding a constant value to the command variable.

Similarly to the previous referred works, the parameters of the controller should be set through the character-oriented interface. To this a new requirement is added: students must develop a graphical interface for the PID controller in a personal computer, allowing a much-enhanced usability of the device. The graphical interface and the microcontroller must communicate through the character-oriented interface.

At the end of his laboratory work, students must be in possession of a considerable set of skills and basic components for the project, having developed firmware for data acquisition, PWM generation and PID control, together with a user interface and interfaces to the physical world.

Automation

This laboratory work aims to be an introduction to two off-the-shelf components: a variable speed drive and a PLC. The PLC is used to implement a work cycle in the drying chamber model. Students must develop the solution in the Grafcet language, download the program to the PLC, and test the operation of the PLC connected to the drying chamber as complying with the specifications given.

4   Project and Industry Guidelines

Our design of the project sections of LPI1 and LPI2 is strongly influenced by project based learning concepts (PBL). PBL is a pedagogical methodology that is gaining increased attention and adherence [2]. Three ideas stay at the core of PBL: learning is more effective by doing, by learning in a group rather than in isolation, and by learning through an activity focused in producing something, i.e., a project [3-8]. As one would expect, PBL stands out as a strong candidate to become the methodology of choice in engineering education, in particular in teaching or learning project [9].

We retain all these ideas in the project section of both LPI1 and LPI2. Student must establish groups of four to six people. Each group must produce a prototype of an electronic device or simple system, that is, to realize a project. Projects are evaluated in a public session as detailed in the next section. The group must define its own organization through appointing a leader – and other roles if it is the case – and choose a teacher to work with the group as project adviser.

The group must also define the project theme; eventually, a teacher as a last resort will supply a theme, if the group fails to do it. There are no restrictions put on theme apart from the following three ones. The project must be feasible in the allotted time slice of six weeks and with resources available to students and the school. In the case of the first semester module, the developed solutions must be confined to analog electronics. In the case of the second semester module, the project must be a remaking of the first semester project with the use of digital electronics. The last two restrictions can be lifted in justified cases.

The success of this strategy has been quite satisfactory. Our requirement that the first semester analog project be "digitally re-mastered" in the second semester, so speak, actually makes that along the year there is one only project with a total duration of twelve weeks. While some projects add small, yet sound, value to the control laboratory works, being typically the temperature control of a model of an oven or house, others spring out as quite interesting. As examples of actually functioning devices, one can cite an automatic inductor coil winder, a color mixing machine, motorized skate, a guitar string tuner, a hand washing station, and a coffee machine. As examples of devices or simple systems developed for reduced scale models we can cite several instances of controls for homes, several instances of position and velocity controls for electro-mechanical systems, and small-scale robots. In Figure 1 we try give the reader a flavor of the developed projects character.

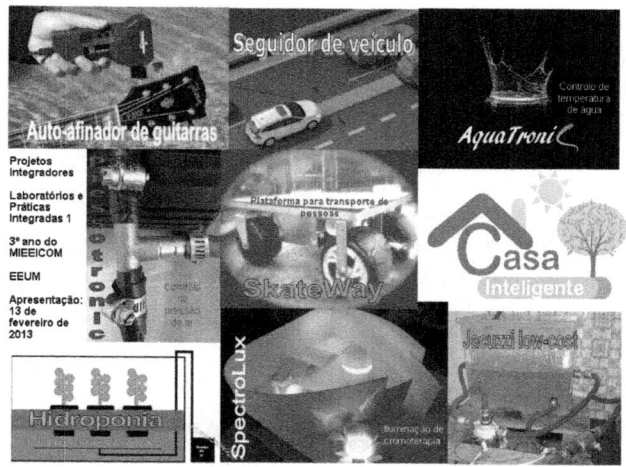

Fig. 1. Poster of projects of LPI1 academic year 2012/13 edition. Left to right and top to bottom one can read: auto guitar tuner, vehicle follower, water temperature control, air pressure control, platform for moving people, intelligent home, hydroponics, illumination for chromotherapy.

While the approach revealed fruitful, procedural guidelines in the project work consistent with industry practice were clearly lacking. The industrial partner Efacec, SA, is actually providing these.

Industry Guidelines for the Project

The guidelines that we are currently implementing in LPI1 and LPI2 refer to project management and product documentation. The implementation of the guidelines must take in account the nature of the school environment and the learning purpose[1].

In project management, project activities are grouped in a four-stage sequence, the names of the stages being self-explanatory:

Project proposal →Viability studies→Project definition and design→Prototype development

The arrows above represent meetings that decide the transition from a stage to the next or abandonment of the project. The events are designated as 'gate reviews'. In an industrial setting, a fifth stage exists corresponding to 'device or system production'. It would be meaningless to try to reflect such stage in the learning practice.

Under the restrictions of the school environment and the learning purpose, the 'project proposal' and 'viability studies' stages must be carried with a low workload before the project section of the modules actually begins. The six weeks section will be then dedicated to the 'project definition and design' and 'prototype development

---

In industry, usually, the ultimate purpose of a project is not to learn but rather to design a device or system that will be produced.

stages'. There will be some overlap of these two stages, as the distinction between them cannot be made completely clear cut.

In spite of the restrictions, the adoption of the above guidelines appears extremely useful, from a learning perspective. Students are exposed to the concepts and language and assimilate it. The "first time project experience" becomes structured in a way that will be useful in the future and gains a distinctive character of increased professionalism.

The product documentation guidelines that we are currently implementing aim to establish a set of documents used in industry as deliverables of the project instead of the traditional "project report". Currently, they are:

– Mechanical drawings (if applicable)
– Electrical schematics
– Component specification (including printed circuit boards, if used)
– Test procedure and test equipment for the prototype
– User manuals

5   Assessment Methodology

In the design of LPI1 and LPI2, assessment is conceived as much as possible as a tool to increase learning.

The final grade of each student is the mean of the grade obtained in laboratory works and the grade obtained in the project, plus a small component of discretionary teacher evaluation.

The grade obtained in the laboratory works is a weighted average of the grades obtained in each work, weights reflecting the time duration of each work. In turn, the grade for each work weights a mark that assesses on-the-fly the attainment of the goals defined for the work and a mark obtained in a written test on the work subject. This way, attainment of goals along lab sessions and long-term learning are balanced in the grade expressing the result of the assessment.

The assessment of projects is made in a public session where each group presents its project. The grade obtained by each group is the mean of the grades given by two panels: the teachers' panel and the students' panel. The teachers assigned to the module edition and guests constitute the teachers' panel. All other project groups constitute the students' panel for each group.

Inside each group, the project grades of members are credited by distributing the project grade earned by the group according to the group assessment of each member and in agreement with the teacher that worked with the group as project adviser.

This set of rules promotes the responsible participation of students in the assessment process of the project both intra and inter-groups. As an interesting observation also reported elsewhere, we remark that, in general, the grades awarded by students' panels are in close agreement with those awarded by teachers' panels.

6   Conclusions and Perspectives

This paper described the design of two modules in the Electronics Engineering programme of Universidade do Minho that integrate laboratory works and first time experience of electronics project.

The basic design of the modules joins conventional laboratory work and project based learning inside a conceptual map supplied by control. Following the restructuring of the programme prompted by the Bologna agreement, the design of the modules has evolved steadily and is now being incrementally adjusted in the frame of cooperation with industry so that students have a learning experience of project that comes closer to industrial reality.

Cooperation with industry has now a crucial role in framing the project work. We want to extend the scope of cooperation to laboratory work. We intend to integrate into laboratory works testing and measuring procedures and protocols currently used by the industrial partner.

As a final note, let we be allowed to say that involvement and collaboration of industry in engineering education are conceptually necessary and, therefore, mandatory for increased efficiency of the education system.

In making this collaboration real, engineering schools will be responsible for promoting and initiating the collaboration process while industrial enterprises will be responsible for assuming an active role in the process.

The collaboration between industry and engineering schools for education aims to attain an increased adequacy of graduates to industry practice. In a scholarly environment, where the majority of teachers do not have professional experience in enterprises, one can expect that there will exist a continuous tendency for education to become astray from industry practice.

Creating spaces of common activities for people of enterprises and schools appears as the best way to counter such tendency and to educate the engineers that industry and society in general need.

Acknowledgements. This work has been supported by FCT – Fundação para a Ciência e Tecnologia within the Project Scope: PEst-OE/EEI/UI0319/2014.

References

Texas Instruments. Active filter design (2011),
http://www.ti.com/tool/filterpro
Graaff, E.D., Kolmos, A. (eds.): Management of Change: Implementation of Problem-Based and Project-Based Learning in Engineering. Sense Publishers, Rotterdam (2007)
Powell, P.C., Weenk, W.: Project-Led Engineering Education. Lemma, Utrecht (2003)
UNESCO, Engineering: Issues, Challenges and Opportunities for Development, United Nations Educational, Scientific and Cultural Organization, Paris, France (2010)
Katajavuori, N., Lindblom-Ylänne, S., Hirvonen, J.: The Significance of Practical Training in Linking Theoretical Studies with Practice. Higher Education 51, 439–464 (2006)

6. Helle, L., Tynjala, P., Olkinuora, E., Lonka, K.: Ain't nothin' like the real thing'. Motivation and study processes on a work-based project course in information systems design. British Journal of Educational Psychology 77, 397–411 (2008)
7. Okudan, G.E., Rzasa, S.E.: A project-based approach to entrepreneurial leadership education. Technovation 26, 195–210 (2006)
8. Helle, L., Tynjälä, P., Olkinuora, E.: Project-Based Learning in Post-Secondary Education - Theory, Practice and Rubber Sling Shots. Higher Education 51, 287–314 (2006)
9. Lima, R.M., Carvalho, D., Flores, M.A., van Hattum-Janssen, N.: A case study on project led education in engineering: students' and teachers' perceptions. European Journal of Engineering Education 32, 337–347 (2007)

Concurrent Projects in Control Laboratorial Classes

Filomena Soares, Gil Lopes, Paulo Garrido, João Sena Esteves, and Estela Bicho

R&D Algoritmi, School of Engineering, University of Minho, Guimarães, Portugal
{fsoares,gil,pgarrido,sena,estela.bicho)@dei.uminho.pt

Abstract. This paper presents a laboratorial curricular unit program running in the 1st semester, 4th year of the Integrated Master in Industrial Electronics at the University of Minho. In Project I curricular unit, students designed, developed and implemented an inverted pendulum platform where different control algorithms were tested. Project specifications were formulated by the teachers and students were allowed to conceive different solutions. Some of the final platforms, as well as the perspectives of the students regarding this learning approach, are presented.

Keywords: Control, Inverted Pendulum, Project-Based Learning.

1   Introduction

The Project-Based Learning (PBL) approach has been applied in high education institutions, with special relevance in engineering courses [1]. This teaching/learning methodology includes developing technical competences as designing, solving and optimizing solutions of real-world problems as well as improving students´ soft skills. All in all, they are a key goal for engineering students [2].

PBL is focused on the integration of several knowledge areas for solving a problem related to real situations, during a curricular semester. This teaching/learning methodology requires the application of practical competences characterized by the "how to do" paradigm based on theoretical concepts previously acquired [3-10].

According to [11], the PBL approach comprises the promotion of students´ working groups as well as the definition of realistic, real world working problems, enabling the self-governing participation of students in their learning process. The teacher serves as a facilitator in this process, since the main role is played by students.

Following this trend, a challenge was proposed to the 4th year students of the Integrated Master in Industrial Electronics at the University of Minho (MIEEIC). In Project I, a laboratorial curricular unit, those students develop a real-world project from the design to the implementation phase. Students are divided in pairs taking into account their specialization area: Control, Automation and Robotics; Embedded Systems; Microelectronics; and Energy.

The goal in this paper is to present the adopted teaching/learning methodology, the problem statement in Control topic, some of the students´ outcomes as well as their perspectives towards this approach.

© Springer International Publishing Switzerland 2015   741
.P. Moreira et al. (eds.), *CONTROLO'2014 - Proc. of the 11th Port. Conf. on Autom. Control*,
Lecture Notes in Electrical Engineering 321, DOI: 10.1007/978-3-319-10380-8_71

The paper is organized in six sections: 1. Introduction, where the outline of the paper is presented; 2. Methodology, which summarizes the guidelines of the course experience; 3. Work-problem, where the inverted pendulum specifications are formulated; 4. Project Outputs, where some of the developed platforms are presented; 5. Final Remarks, which presents the conclusions.

2   Methodology

Project I curricular unit runs in the 1st semester of the 4th year of MIEEIC. It is a compulsory five hours laboratorial class (2+3 hours) a week running in parallel in three laboratories. The curricular unit goal consists of implementing the theoretical aspects presented in the four specialization curricular units available in the 4th year: Control, Automation and Robotics; Embedded Systems; Microelectronics; and Energy. The students work in pairs, assisted by a team of teachers from the four areas.

The semester starts with the presentation of four projects, one from each specialization area. Students must then select and apply for one topic. The overall objectives and the project specification are defined by the teachers. This definition is made in terms of graded topics and it is up to the student to fulfill all or part of the requirements. The student grade is obtained accordingly. This is an open project that allows students to find different but valid solutions to the proposed real-world problem. The projects become concurrent: students must follow the general guidelines and objectives, but they are allowed (and even encouraged) to implement different approaches and solutions.

The main topics include the design and implementation of:

- A DRAM or SRAM memory cell in CMOS 16 (4x4), using Tanner EDA software: S-Edit and T-Spice for the schematic and simulation and L-Edit for the layout of the device;
- Different types of Power Supplies (Linear, Step-Up and Step-Down) controlled by a microcontroller. The project includes computer simulations in PSIM and the comparison of performance and efficiency;
- Analyze the spectrum of an audio signal or song (in MP3) and perform various operations, such as filtering of certain spectra;
- A controlled inverted pendulum platform.

Students are allowed to go to the laboratories outside the official timetable and without supervision.

Table 1 presents the project evaluation. Since class attendance is compulsory, a multiplicative factor (number of attendances/total number of classes) is considered in the final grade calculation.

Table 1. Project Evaluation

Topic	When	Weight
First Practical Evaluation	First week November	15%
Written Report (30 pages maximum)	31st January	30%
Continuous Evaluation in Class	Along the semester	15%
Final Practical Evaluation	3rd week January	40%
		Grade
Multiplicative Factor (MF) = number of attendances/total number of classes		**MF*Grade**

Figure 1 presents the assessment weights regarding practical component, written report and continuous evaluation in class. The evaluation weights and dates were defined in accordance with students. No postponement was admissible.

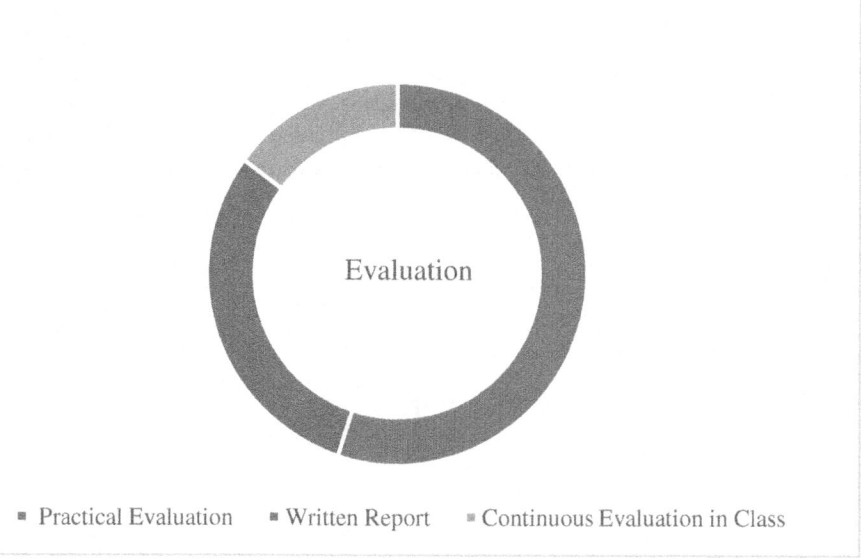

- Practical Evaluation   - Written Report   - Continuous Evaluation in Class

Fig. 1. Evaluation weights

In the following sections, a special attention is paid to the Control, Automation and robotics project of a controlled inverted pendulum platform. The work-problem and me of the students' solutions are presented.

3   Work-Problem

The problem to be solved through the project is presented to the students in a very general setting. This keeps wide open the possibilities for concretely defining the project.

The problem is presented with the statement that the aim of the project is to control a body standing on a mobile platform. Several examples are presented, including the case of an inverted pendulum, able to oscillate in a plane (Figure 2). Furthermore, as there are several groups, each implementing an inverted pendulum, the pendulums being able to move in a coordinated fashion is set as a further goal of the project. Coordinated movements would make feasible a kind of robotic choreographies such as in Figure 3.

Coordinated movement of the inverted pendulums opens interesting possibilities for learning through the project, as it requires servo-control of displacement and reference generation through non-linear dynamic systems, both advanced topics in the concomitant modules of specialization in Control, Automation and Robotics.

Regarding the project learning objectives some tasks are included: integrate electronics and mechanical elements in a device, implement a control loop on an embedded system, and manage microprocessor boards, among other minor functions.

The physical setups to implement an inverted pendulum, readily in reach of ours to supply or of our students to build, use either small autonomous robotic platforms or printers out of use. A schematic representation with the main mechanical and electronic components to assemble into or add to an existing platform (such as the one in Figure 4, which is used as a reference for the following) is provided for a small autonomous robotic platform.

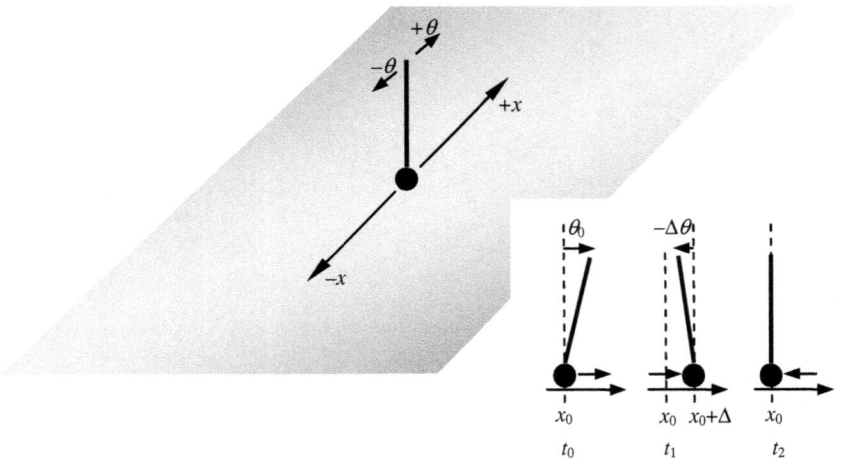

Fig. 2. The inverted pendulum is supposed to move in a plane along a direction x; the angle makes with the vertical is θ. The black circle represents the cart where the fulcrum of the pendulum is fixed. Movements of the cart to equilibrate the pendulum from instants t_0 to t_2 a depicted in the minor figure at the right lower corner.

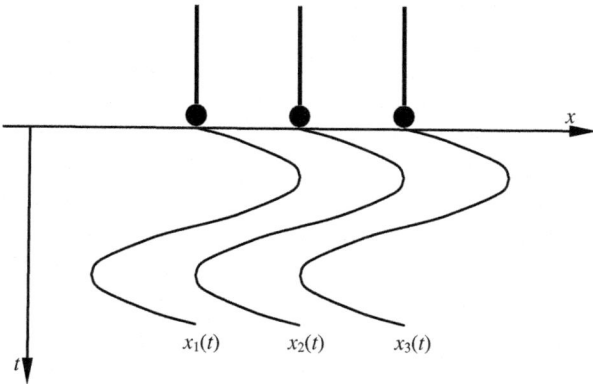

Fig. 3. Three inverted pendulums moving coordinately in the same plane. To achieve coordinate movements, servo-control of the pendulums is needed.

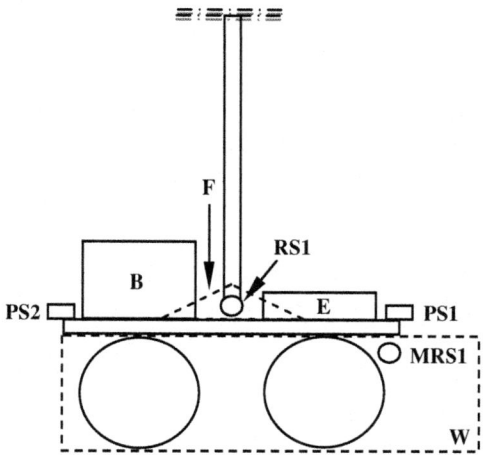

Fig. 4. Schematic representation of an inverted pendulum implemented as a small autonomous robotic platform: W – wheels; F – fulcrum fixture; B – battery; RS1 – encoder for measuring the pendulum's angle; MRS1 – motor and encoder for measuring velocity; PS1 and 2 – proximity sensors; E – electronic boards, microprocessor and pulse width modulation (PWM) bipolar drive

Regarding control issues, five levels of accomplishment are established for projects aiming at implementing an inverted pendulum. These levels are taken as a standard for all cases, although not all levels make sense if students opt by a realization with a printer.

The definition of levels was inspired by [12], the textbook followed in the concomitant module on optimal control. As the basic goal, or level one, students must attain equilibrium of the pendulum, the cart being in a given initial position where the pendulum is manually taken to near equilibrium. In control terms, this is a regulation problem; students are required to design a full-state feedback regulator, either by pole placement or optimal control, and to implement it.

The adjective "basic" in this basic goal, or level one, is somehow misleading. There are many skills and competencies involved in attaining the goal, both in terms of electronics and in terms of control.

In terms of electronics, students must interface two encoders to the microprocessor board, develop firmware to count the pulses and, from the counting, derive estimates of position and velocity, both linear (cart) and angular (pendulum). They must, also, interface the microprocessor board to the PWM bipolar drive, which supplies electric power to the motor, and develop firmware to command it.

In terms of control, students must learn a model for the inverted pendulum. In the literature, there are many models that relate the state variables of the device and the force applied to the cart, neglecting frictions. The parameters of these models are readily identifiable, but students must augment them to include the motor and mechanical transmission, and obtain the related parameters. Having a model of the physical setup, students must choose a sampling period, implement in firmware a full state feedback regulator together with interface functions, design the regulator and test the overall system.

Four further levels are proposed to the students after they have mastered on level one. In level two, the goal is to add an attractive or repulsive behavior of the x initial position. In level one, the initial position is given the value zero; as control is regulatory, the cart will tend always to come to this position in steady state. In level two, this zero reference will be redefined as a function of the distance to an object detected by any of the proximity sensors. If the object is far enough, the cart will approach the object; if the object is too close, the cart will depart from the object. In any case, equilibrium of the pendulum must be maintained and the zero reference position must be redefined. This level allows the students to practice concepts of non-linear system dynamics, also lectured in a concomitant module.

In level three, students must implement servo behavior. The reference position of the cart is allowed to vary and the system must follow step or ramp time varying references. These references can be stored internally to the microprocessor or sent by radio. Together with level two, the implementation of servo behavior will allow coordinated motion of inverted pendulums, as referred above.

In level four, the implementation of a Kalman filter to estimate the four state variables of the model is proposed to the students.

Finally, in level five, a practice in adaptive control is the goal, as one wants the controller to adapt to changes of weight, length, or both, of the pendulum.

While the hardware implementation required is quite small in levels two to five, the full extent of the five levels proposed to students is quite demanding. Although few students, in two academic years, have overtaken level one, teachers are confident that success will increase as their ability to conduct students along this path of control learning will increase.

4   Project Outputs

The student's approach to the proposed problem varied from group to group. In total, three different base configurations were developed: a) Single axis wheeled moving platform; b) Single axis printer-based platform; c) Dual axis using an omnidirectional-wheeled moving platform.

Adding to the challenge, students were asked to create low cost solutions to attain their goal and therefore all the developed solutions used components available in the laboratories or easily accessible moneywise. The single axis configurations are based in the one suggested to students, where the inverted pendulum is fixed to a pivot or fulcrum and can oscillate only in one direction. The platform just has to go back and forward to sustain the pendulum's equilibrium. On the dual axis approach, the inverted pendulum stands on a ball joint and can incline in any direction of the xy plane.

The single axis wheeled moving platform (a three wheel version is shown in Figure 5) is based on a plate over wheels or castors, where platform movement to balance the pivoted inverted pendulum is performed by one or two motorised wheels. This approach was taken by some groups of students. They developed or adapted wheeled platforms using a round or squared hollow aluminium rod as the inverted pendulum. Typically, rod length ranged from 1 m to 1,2 m and rod diameter (round profile) or side (square profile) ranged from 7 mm to 10 mm. The rod is fixed to a pivot in the bottom end having a very low inertia to avoid standing upright on its own if the platform is not moving. In other words, if the platform was halted and the pendulum put in the vertical position by the user it should always fall in one direction when released. This means that the system had to move to create the equilibrium not sustaining the rod vertically when stopped.

Fig. 5. - Diagram of the single axis wheeled platform (left) and picture of the developed platform (right)

The single axis printer-based platform configuration (Figure 6) uses a printer carriage mechanism (shafts, belt, pulley and motor) and the inverted pendulum is pivoted in the carriage. The carriage movement along the shafts equilibrates the pendulum.

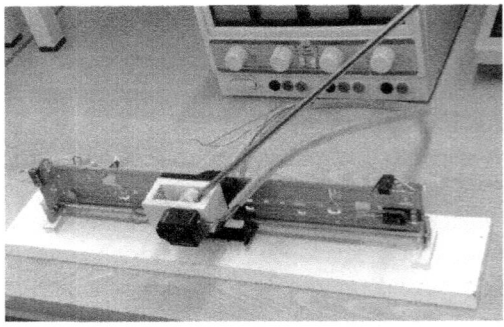

Fig. 6. Single axis printer-based platform developed from an off work printer mechanism

The dual axis configuration uses an omnidirectional-wheeled moving platform (shown in Figure 7) based on a round plate with three Swedish wheels positioned 120° from each other. Applying different speeds to each wheel allows the platform to perform any type of movement (rotation, translation or a combination of both). Thus the equilibrium of the pendulum can be achieved in the xy plane.

Fig. 7. - Dual axis using an omnidirectional-wheeled moving platform (left) and ball joint fixture holding the rod that makes the inverted pendulum (right)

In all the developed platforms, motion is obtained by DC motors with PWM bipolar drive coupled with rotary incremental encoders for feedback (linear encoders on some of the printer-based platforms). For the inverted pendulum position feedback students used rotary incremental encoders mainly in the single axis configurations and inertial sensors attached to the top of the rod in the dual axis configuration. On the wheeled platforms, some of the groups used ultrasonic sensors to get distances from obstacles, whereas the printer-based platforms used end switches to define the limit of the carriage movement. All groups used a microcontroller based on the Arduino development platform. There, sensors signals are acquired, analyzed and the motor drivers signals are generated according to the implemented control methods.

5   Final Remarks

This paper presented a teaching/learning experience based on the PBL approach run in Project I, a 4th year, one semester, compulsory laboratorial curricular unit (5 hours/week) of the MIEEIC.

Students apply for a working project related to their specialization area. Concerning the Control, Automation and Robotics topic, the goal was to test different control approaches based on an inverted pendulum platform designed by students. There were eight concurrent projects, based on three different base configurations: a) Single axis wheeled moving platform; b) Single axis printer-based platform; c) Dual axis using an omnidirectional-wheeled moving platform.

The cart with an inverted pendulum, albeit being a most classical control case study, reveals itself surprisingly rich for control education, since it enables developing a whole one-semester control learning path around the concept. State modeling, full state feedback, pole-placement and optimal regulation, non-linear dynamical systems reference generation, observers and Kalman filtering, servo design and adaptive control are among the many concepts that can be practiced and tested.

At the end of the semester, a questionnaire was given to the students. It was based on a qualitative analysis. Student's feedback was, in fact, very rewarding. They felt that this teaching/learning approach, where they can freely implement in a real-world problem what is taught in class, is very challenging. These are some of the suggestions for improvement: 1) Form groups of three students because the work is quite ambitious. 2) Give students more time to optimize the solution, since the work load is high. 3) Regarding the concurrency between projects, a few students suggest that there should be more than one project under each specialization topic. This means that projects would not compete with each other, since they would have different goals and specifications. However, some teachers defend that the competition approach satisfies the course requirements. 4) A pertinent suggestion includes designing, at the beginning of the semester, a one week course focused on how to define, organize and plan a working project.

All in all, teachers feel that this experience is worth maintaining next year, optimizing the format: include the one week course, reduce the work load and keep the groups of two students. Maintaining or not the concurrency between projects in the future is still under evaluation.

Acknowledgements. This work has been supported by FCT – Fundação para a Ciência e Tecnologia in the scope of the project: PEst-OE/EEI/UI0319/2014. The authors are indebted to their colleague Jorge Martins for the idea of using a printer to make a cart with an inverted pendulum.

References

1. Graaff, E.D., Kolmos, A. (eds.): Management of Change: Implementation of Problem-Based and Project-Based Learning in Engineering. Sense Publishers, Roterdam (2007)

2. Soares, F., Sepúlveda, M.J., Monteiro, S., Lima, R.M., Dinis-Carvalho, J.: An integrated project of entrepreneurship and innovation in engineering education. Mechatronics, Special Issue of Mechatronics on Design-Centric Engineering Education 23(8), 987–996 (2013)

3. Powell, P.C., Weenk, W.: Project-Led Engineering Education. Lemma, Utrecht (2003)

4. Lima, R.M., Carvalho, D., Flores, M.A., van Hattum-Janssen, N.: A case study on project led education in engineering: students' and teachers' perceptions. European Journal of Engineering Education 32, 337–347 (2007)

5. Carvalho, D., Lima, R.M., Fernandes, S.: Learning in Engineering: Projects and Interdisciplinary Teams (title translated from portuguese - Aprendizagem em Engenharia: Projectos e Equipas Interdisciplinares). Presented at the 5° Congresso Luso-Moçambicano de Engenharia (CLME 2008), Maputo - Moçambique (2008)

6. UNESCO, Engineering: Issues, Challenges and Opportunities for Development, United Nations Educational, Scientific and Cultural Organization, Paris, France (2010)

7. Katajavuori, N., Lindblom-Ylänne, S., Hirvonen, J.: The Significance of Practical Training in Linking Theoretical Studies with Practice. Higher Education 51, 439–464 (2006)

8. Helle, L., Tynjala, P., Olkinuora, E., Lonka, K.: 'Ain't nothin' like the real thing'. Motivation and study processes on a work-based project course in information systems design. British Journal of Educational Psychology 77, 397–411 (2008)

9. Okudan, G.E., Rzasa, S.E.: A project-based approach to entrepreneurial leadership education. Technovation 26, 195–210 (2006)

10. Helle, L., Tynjala, P., Olkinuora, E.: Project-Based Learning in Post-Secondary Education - Theory, Practice and Rubber Sling Shots. Higher Education 51, 287–314 (2006)

11. Barron, B., Darling-Hammond, L.: Teaching for Meaningful learning. A Review of Research on Inquiry-Based and Cooperative Learning, Stanford University (2008)

12. Astrom, C., Wittenmark, B.: Computer Controlled Systems: Theory and design. Prentice-Hall, NJ (1997)

Framework Using ROS and SimTwo Simulator for Realistic Test of Mobile Robot Controllers

Tatiana Pinho[1], António Paulo Moreira[2], and José Boaventura-Cunha[1]

[1] INESC TEC (formerly INESC Porto) and Universidade de Trás-os-Montes e Alto Douro, UTAD, Escola de Ciências e Tecnologia, Quinta de Prados, 5000-801 Vila Real, Portugal
[2] INESC TEC (formerly INESC Porto) and Faculty of Engineering, University of Porto, Porto, Portugal
{al30888,jboavent}@utad.pt, amoreira@fe.up.pt

Abstract. In robotics, a reliable simulation tool is an important design and test resource because the performance of algorithms is evaluated before being implemented in real mobile robots. The virtual environment makes it possible to conduct extensive experiments in controlled scenarios, without the dependence of a physical platform, in a faster and inexpensive way. Although, simulators should be able to represent all the relevant characteristics that are present in the real environment, like dynamic (shape, mass, surface friction, etc.), impact simulation, realistic noise, among other factors, in order to guarantee the accuracy and reliability of the results.

This paper presents a ROS (Robot Operating System) framework for the SimTwo simulator. ROS is an open-source library that is commonly used for the development of robotic applications since it provides standard services and promotes large-scale integrative robotic research. SimTwo is a realistic simulation software suitable for test and design of several types of robots. This simulator conducts realistic navigation procedures, since the driving systems, the sensors, the mechanical and physical properties of the bodies are precisely modeled.

The framework presented in this research provides the integration of ROS-based systems with the SimTwo simulator. Therefore, this framework reduces the risk of damage of expensive robotic platforms and it can be used for the development of new mobile robot controllers, as well as for educational purposes.

Keywords: simulation, SimTwo, ROS, framework.

Introduction

obots development is a process that involves several engineering areas such as programming, electronics and mechanics, among others. In this sense technological advanced materials and design methods are needed, which implies that the development of new robotic solutions is often an expensive practice. However, nowadays, the increasing processing capacity allows the development of efficient simulation tools [1].

Considering the fact that a simulator doesn´t take into account hardware aspects the robot, it provides an ideal and efficient way to fulfill preliminary tests of the

Springer International Publishing Switzerland 2015   751
P. Moreira et al. (eds.), *CONTROLO'2014 - Proc. of the 11th Port. Conf. on Autom. Control*,
cture Notes in Electrical Engineering 321, DOI: 10.1007/978-3-319-10380-8_72

developed algorithms, enabling the introduction of different configurations, without materials and/or personal risks [2], leading to better decisions and economic savings. Furthermore, it offers the possibility of test and direct debug in the robots programs [3]. Therefore, a simulator is used to reduce the time and costs involved in the development and validation of a robot model [4].

In addition to the modeling of the robot dynamics, simulators must allow interactions between the robot and the environment by means of sensors and actuators representations [5]. In this sense, the simulation environment can be used in cases in which such real environment cannot be available [2]. The simulation also allows computing variables that are difficult to access in real operation enabling a more exact monitoring of the robot behavior [6].

Another simulation dimension concerns to its educational purpose. Namely, as the robotics teaching requires equipment resources, specialized people for support, laboratorial space, etc., simulation tools provide simple and accessible ways to achieve these requirements [5].

In other words, simulators can be used to research robots technical features, teaching, control manipulation qualities, model a virtual robot with simplified configuration, develop the robot itself, etc. [4]. It should be noted that a simulator must have validated sensor noise models [2].

In conclusion, a robotic simulator can be defined as a software tool which simulates the real world and creates a virtual environment to robots [4]. So, it should incorporate important features of the real world, where the importance of an aspect is different for each case [7]. A simulator will be more reliable, as closer to reality are their results. In the ideal situation, for equal reference inputs, the same results are achieved in real and simulated robots, as represented in Figure 1.

In this way, the control algorithms developed and tested in simulated environments, can be directly transferred to the real operating conditions, with time benefits and avoiding robots damaging [1].

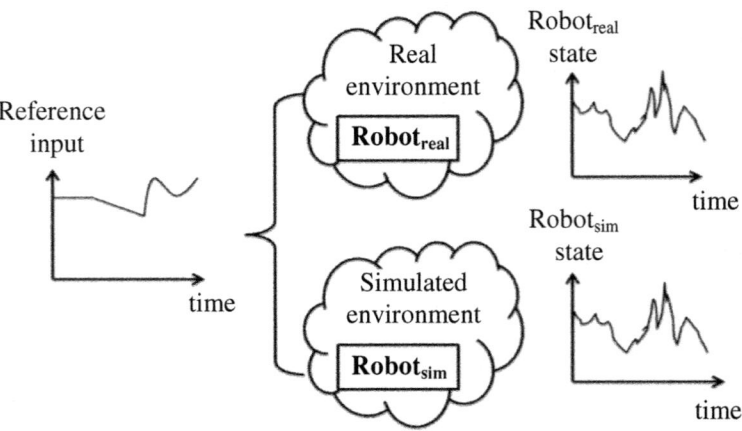

Fig. 1. Real and simulated robots comparison (Source: adapted from [1], p. 30)

Nowadays there are several available simulators for robotic systems, such as Über-Sim [7, 8], SimRobot [3], Webots [9], UCHILSIM [10], USARSim [5], Stage [11], Gazebo [12], ARGoS [13], V-REP [14], Delta3D, X-Plane, Microsoft Flight Simulator, Actin, Microsoft Robotics Developer Studio (MRDS), OpenSimulator, Simbad, FlightGear, Breve, Simspark, MURoSimF e OpenHRP3 [4, 5, 15], among others.

In this work a framework that integrates ROS systems and SimTwo simulation software is proposed, in order to promote the design and test of robot controllers, mainly in the educational context. This simulator choice was based on criteria such as simplicity, installation easiness and cost. This work is organized in 5 sections. In section 1 was made a brief introduction to the theme. Section 2 is dedicated to ROS (Robot Operating System). Section 3 presents the main features of the used simulator, SimTwo. Section 4 describes the proposed framework architecture and the last section (section 5), summarizes the main conclusions.

2   The ROS Framework

ROS is an open-source framework that provides an abstraction level to the complex hardware and software configurations in robotic area [2]. It is composed by a wide range of services and tools, available to users and developers, being included in its concept criteria such as: peer-to-peer communication for data transfer, tools-based, multi-lingual, thin and free and open-source [2, 16, 17]. In this way, ROS emerged as a way to facilitate the development in the robotic area and incorporate common solutions, once that this area involves a high level of complexity and constant update.

One of the ROS advantages is its high capacity of generalization to access to external hardware, either as sensors and actuators. Furthermore, through its modular character, it is possible to incorporate into the framework, new functionalities in a simple way [18]. As ROS is a middleware system, supported by a large community, it possesses shared libraries by the community itself, accessible to all users [16].

ROS uses a proper nomenclature, in which can be defined nodes, messages, topics, services, packages and stacks. Namely, nodes are processes that perform computation, messages are strictly defined structures and a topic consists in a string such as "odometry" or "map". On the other hand, a service is defined by a string name and by two strictly typed messages, one for request and the other for response. A ROS package is a directory that contains a XML file with the package description and stating any dependencies. An organized packages set is defined as stack [2, 17].

In ROS, messages are sent by nodes, as pictured in Figure 2. Publishing nodes send their messages to a specific topic and the subscriber nodes receive that same message. A master node exposes the two types of nodes by a service [19]. It should be noted that publishing/subscribing and client/server are different communication types. The first one is used in general cases, and the second one only in specific cases, particularly when synchronism is needed [17]. It is also important to denote that several nodes can be connected to a robot with different purposes [19]. ROS is a publishing/subscribing system [16], or a client/server in the services case.

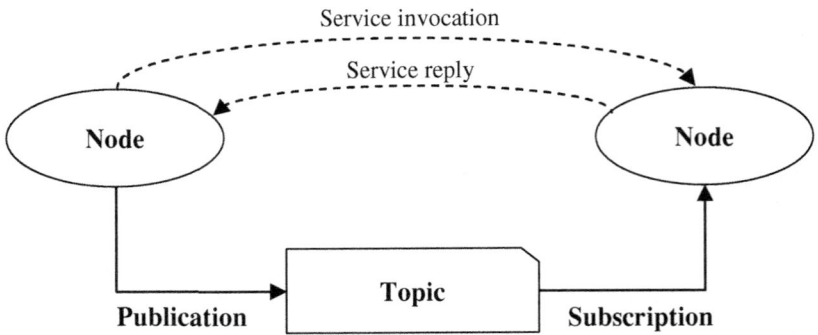

Fig. 2. ROS messaging mechanism (Source: adapted from [19], p. 3087)

3   SimTwo Simulator

SimTwo is a simulation tool developed in Object Pascal that allows the fast test and the construction of several robots types, namely, differential, omnidirectional, industrial, humanoids, among others, defined by the user in a 3D space [15, 20]. In a generic way, it can be said that any terrestrial robot type with rotating joints and/or wheels can be simulated with SimTwo.

For the dynamic simulation of rigid bodies, the Open Dynamics Engine is used, being the robot design and behavior are defined in XML files and the virtual world is represented by GLScene components. The robot control can be made by a script in the simulator itself or by a remote client that communicates by UDP or serial port.

SimTwo simulator has a high level of realism. In terms of dynamic this realism is conferred by robot separation into rigid bodies and electric motors. For each body, the behavior is simulated using physical features as shape, mass, moments of inertia, surface friction and elasticity [20]. Besides that, it possesses non-linear features that are important to the representation of robot real behaviors [21].

The SimTwo graphical interface is a *multiple document interface* (MDI), as represented in Figure 3, where all windows are under the "world view" window control.

In more detail, the "code editor" presents an *integrated development environment* (IDE) for a high-level programming in Pascal; the "configuration" window allows the control of several elements in the virtual scene; in "spreadsheet" can be defined "edit cells" and "button cells" with different purposes; in "chart" window all available variables can be graphically represented for each robot; and in the "scene editor" the scene is defined by several XML files [20].

Concerning to the control, SimTwo possesses two levels. Namely a high-level controller defined by the user, executed with a 40 ms periodicity, and a low-level controller, related to motors control, invoked with a 10 ms period. The model output is updated at a 1 ms rate [15].

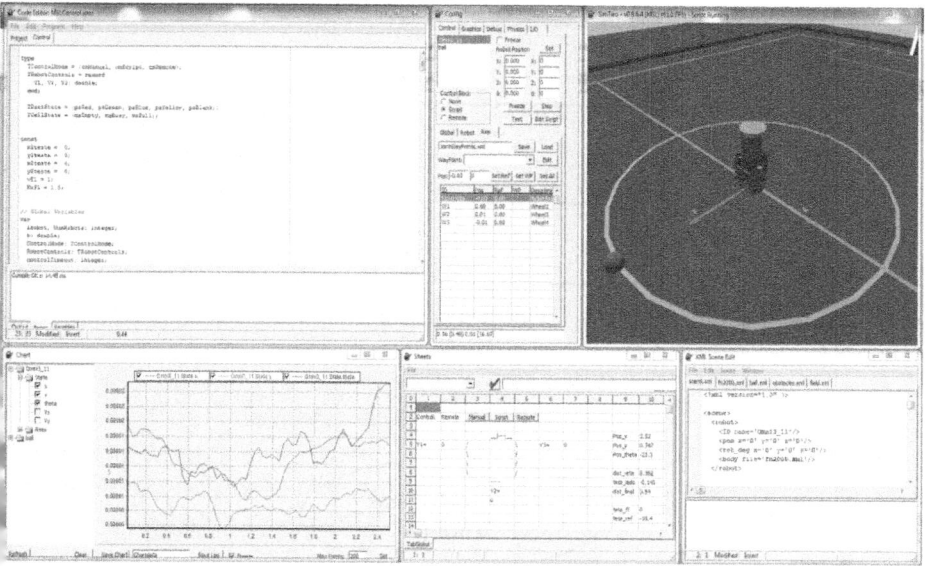

Fig. 3. SimTwo graphical interface

4   ROS/SimTwo Framework

This work proposes a framework development that allows the ROS and SimTwo systems integration. This will facilitate the transition to a real robot, since the developed ROS communicates with the simulator in the same way that communicates with the real robot. Furthermore, this system equally promotes the teaching and test of controllers in academic situation. Figure 4 represents the proposed framework architecture. The mobile robot trajectory control is described in section 4.1.

Considering the reference trajectory to be executed by the robot, together with the characteristic parameters of the model, the optimization is made to determinate which are the controls regarding linear (V, Vn) and angular (ω) velocities of the robot in order to follow the pre-defined trajectory.

Once calculated these controls, they are sent to the real robot by a ROS node and/or to the simulated robot, for which the SimTwo communication is made by UDP protocol, as shown in Figure 5. The synchronism between remote client and SimTwo was guaranteed in order to assure a proper operation of the controller. At the same time the real/simulated robot returns its localization data (x, y and θ), which are feedback into the controller to determine the following controls.

It should be noted that robot velocities V, Vn and ω conversion to each wheel velocity is made only after controls are sent. In other words, this conversion occurs independently of the remote client, giving it a generic character that allows its application to any type of robot.

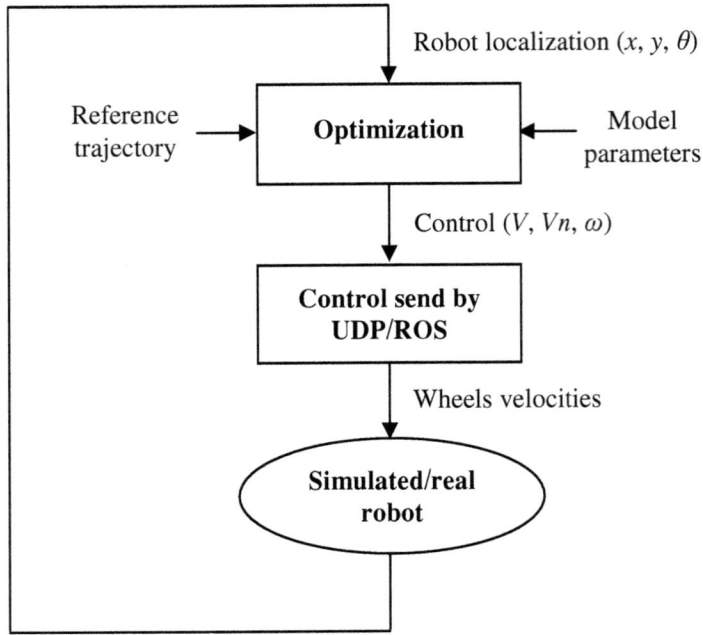

Fig. 4. Generic framework architecture

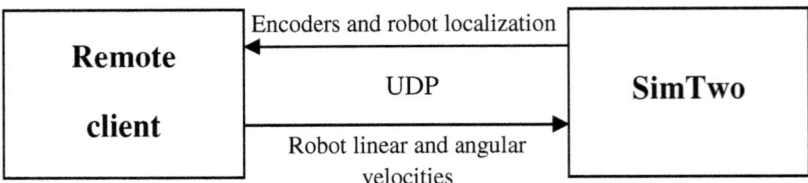

Fig. 5. Remote client – SimTwo communication

4.1   Framework Validation

For test and validation of the developed framework it were used MSL (Middle Si
League) robots from the robotic soccer team of the University of Porto (Figure ?
The characteristic parameters of these robots implemented in SimTwo are describe
in Table 1.

In this work, a proportional controller of the distance and orientation errors relat
to a straight trajectory following was applied. The objective was to track a straig
line coincident with the abscissa axis. To test the controller performance the rot
was initially placed in a pose displaced 3 meters in y, and orientated in parallel to
line. Robot evolution in x and y poses is represented in Figure 6. In the same wა

Table 1. MSL robot constraints

Constraint	Measure	Unity
Robot mass	26	kg
Wheel mass	0.660	kg
Wheel radius	0.051	m
Wheel width	0.042	m
Encoder	12288	PPR
Gear ratio	12	
Resistance of the motor	0.316	Ω
Electric constant of the motor	0.0302	Nm/A
Maximum voltage on motor	24	V
Maximum allowed current	12	A
Moment of inertia (x axis)	0.388	kgm^2
Moment of inertia (y axis)	0.388	kgm^2
Moment of inertia (z axis)	0.705	kgm^2

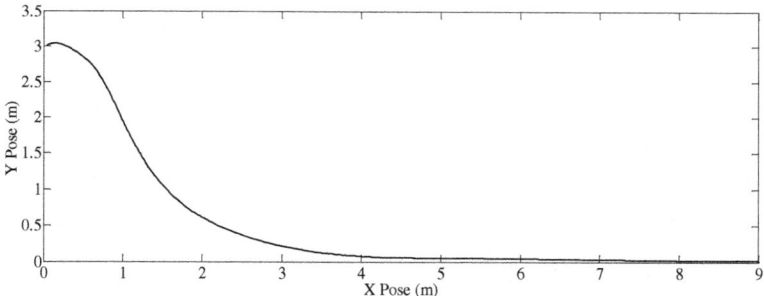

Fig. 6. Robot position, x, y, evolution, in m

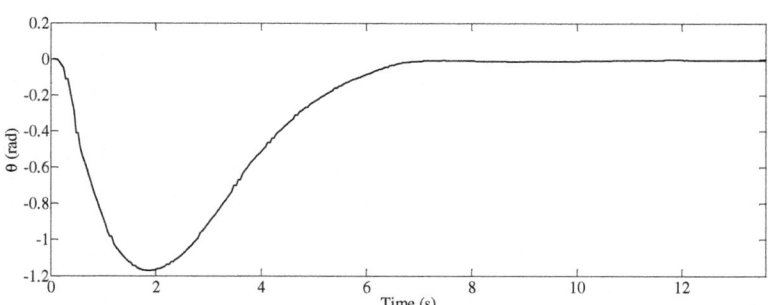

Fig. 7. Robot orientation, θ, evolution, in rad

...s orientation evolution is presented in Figure 7. It should be noted that controller ...eneric character was validated by the implementation of a dimensioned version for a ...ifferential traction robot in an omnidirectional robot.

According to Figures 6 and 7, it is possible to observe that the robot effectively corrects its localization in a relatively fast way, in order to be closer to the pretended trajectory, keeping afterwards coincident to the line. In this way it will be expected that after the transition to the real robot, its behavior will be similar to the simulated robot. This fact occurs because the ROS communicates in an identical way with the robot in the real environment.

5   Conclusions

This work proposes a framework to support the design and validation of robot controllers, mainly in educational context, integrating a ROS system and SimTwo simulation software. With this framework it is pretended to facilitate the passage of simulated to real situations, since ROS system communicates with the simulator in a similar way it communicates with the real robot. Furthermore, this work also aims to support the teaching and test of robotic controllers developed in academic environment.

In order to analyze the proposed framework performance, it was implemented a controller to coordinate a MSL soccer robot in way to follow a pre-defined trajectory. Its efficiency was confirmed by this application.   Since the controls sent to the simulator refer to robot's linear and angular velocities, this work also possesses an abstraction character that provides the possibility of being implemented in any type of robot.

In the future, this work can be extended throughout the incorporation of other phenomena in the models, such as sensors noise, and the implementation of different control strategies aiming the evaluation of distinct control algorithms performances in simulations closer to the real environment robot operation.

Acknowledgements. The author also thanks the FCT for supporting this work through the project PTDC/EEI-AUT/1450/2012 – Optimal Control: Health, Energy and Robotics Applications.

References

1. de Lima, J.L.S.M.: Construção de um Modelo Realista e Controlo de um Robô Humanóide, Tese de Doutoramento em Engenharia Electrotécnica e de Computadores d Faculdade de Engenharia da Universidade do Porto (2008)
2. Balakirski, S., Kootbally, Z.: USARSim/ROS: A combined framework for robotic contro and simulation. In: Proceedings of the ASME 2012 International Symposium on Flexibl Automation, St. Louis, Missouri, USA, pp. 1–8 (June 2012)
3. Laue, T., Spiess, K., Röfer, T.: SimRobot – A General Physical Robot Simulator and I Application in RoboCup. In: Bredenfeld, A., Jacoff, A., Noda, I., Takahashi, Y. (eds. RoboCup 2005. LNCS (LNAI), vol. 4020, pp. 173–183. Springer, Heidelberg (2006)
4. Kumar, K., Reel, P.S.: Analysis of Contemporary Robotics Simulators. In: Proceedings c ICETECT, pp. 661–665 (2011)
5. Carpin, S., Lewis, M., Wang, J., Balakirsky, S., Scrapper, C.: USARSim: a robot simulate for research and education. In: IEEE International Conference on Robotics and Automa tion, Roma, Italy, pp. 1400–1405 (April 2007)

6. Gonçalves, J., Lima, J., Malheiros, P., Costa, P.: Realistic simulation of a Lego Mindstorms NXT based robot. In: 18th IEEE International Conference on Control Applications, Part of IEEE Multi-Conference on Systems and Control, Saint Petersburg, Russia, pp. 1242–1247 (July 2009)
7. Go, J., Browning, B., Veloso, M.: Accurate and Flexible Simulation for Dynamic, Vision-Centric Robots, pp. 1–8. AAMAS, New York (2004)
8. Browning, B., Tryzelaar, E.: ÜberSim: A Multi-Robot Simulator for Robot Soccer, pp. 948–949. AAMAS, Melbourne (2003)
9. Michel, O.: Cyberbotics Ltd. Webots^{TM}: Professional Mobile Robot Simulation. International Journal of Advanced Robotic Systems 1(1), 39–42 (2004)
10. Zagal, J.C., Ruiz-del-Solar, J.: UCHILSIM: A Dynamically and Visually Realistic Simulator for the RoboCup Four Legged League. In: Nardi, D., Riedmiller, M., Sammut, C., Santos-Victor, J. (eds.) RoboCup 2004. LNCS (LNAI), vol. 3276, pp. 34–45. Springer, Heidelberg (2005)
11. Gerkey, B.P., Vaughan, R.T., Howard, A.: The Player/Stage Project: Tools for Multi-Robot and Distributed Sensor Systems. In: Proceedings of the International Conference on Advanced Robotics (ICAR), Coimbra, Portugal, pp. 317–323 (July 2003)
12. Koenig, N., Howard, A.: Design and Use Paradigms for Gazebo, An Open-Source Multi-Robot Simulator. In: Proceedings of 2004 IEEE/RSJ International Conference on Inteligent Robots and Systems, Sendai, Japan, pp. 2149–2154 (October 2004)
13. Pinciroli, C., Trianni, V., O'Grady, R., Pini, G., Brutschy, A., Brambilla, M., Mathews, N., Ferrante, E., Caro, G.D., Ducatelle, F., Birattari, M., Gambardella, L.M., Dorigo, M.: ARGoS: a modular, parallel, multi-engine simulatator for multi-robot systems. Swarm Intell. 6, 271–295 (2012)
14. Freese, M., Singh, S., Ozaki, F., Matsuhira, N.: Virtual Robot Experimentation Platform V-REP: A Versatile 3D Robot Simulator. In: Ando, N., Balakirsky, S., Hemker, T., Reggiani, M., von Stryk, O. (eds.) SIMPAR 2010. LNCS, vol. 6472, pp. 51–62. Springer, Heidelberg (2010)
15. Lima, J.L., Gonçalves, J.A., Costa, P.G., Moreira, A.P.: Humanoid Gait Optimization Resorting to an Improved Simulation Model. International Journal of Advanced Robotic Systems 10(67), 1–7 (2013)
16. De Marco, K., West, M.E., Collins, T.R.: An Implementation of ROS on the Yellowfin Autonomous Underwater Vehicle (AUV), pp. 1–7. IEEE, OCEAN (2011)
17. Quigley, M., Gerkey, B., Conley, K., Faust, J., Foote, T., Leibs, J., Berger, E., Wheeler, R., Ng, A.: ROS: an open-source Robot Operating System. ICRA, 1–6 (2009)
18. Hax, V.A., Filho, N.L.D., Botelho, S.S.D.C., Mendizabal, O.M.: ROS as middleware to Internet of Things. Journal of Applied Computing Research 2(2), 91–97 (2012)
19. Arumugam, R., Enti, V.R., Bingbing, L., Xiaojun, W., Baskaran, K., Kong, F.F., Kumar, A.S., Meng, K.D., Kit, G.W.: DAvinCi: A Cloud Computing Framework for Service Robots. In: IEEE International Conference on Robotics and Automation, Anchorage, Alaska, USA, pp. 3084–3089 (May 2010)
20. Costa, P., Gonçalves, J., Lima, J., Malheiros, P.: SimTwo Realistic Simulator: A Tool for the Development and Validation of Robot Software. Theory and Applications of Mathematics & Computer Science 1, 17–33 (2011)
21. Nascimento, T.P., Moreira, A.P., Costa, P., Costa, P., Conceição, A.G.S.: Modeling omni-directional mobile robots: an approach using SimTwo. In: 10th Portuguese Conference on Automatic Control, Funchal, Portugal, pp. 117–122 (July 2012)

Author Index